Lecture Notes in Civil Engineering

Volume 274

Lecture Notes in Civil Engineering (LNCE) publishes the latest developments in Civil Engineering—quickly, informally and in top quality. Though original research reported in proceedings and post-proceedings represents the core of LNCE, edited volumes of exceptionally high quality and interest may also be considered for publication. Volumes published in LNCE embrace all aspects and subfields of, as well as new challenges in, Civil Engineering. Topics in the series include:

- Construction and Structural Mechanics
- Building Materials
- Concrete, Steel and Timber Structures
- Geotechnical Engineering
- Earthquake Engineering
- Coastal Engineering
- Ocean and Offshore Engineering; Ships and Floating Structures
- Hydraulics, Hydrology and Water Resources Engineering
- Environmental Engineering and Sustainability
- Structural Health and Monitoring
- Surveying and Geographical Information Systems
- Indoor Environments
- Transportation and Traffic
- Risk Analysis
- Safety and Security

To submit a proposal or request further information, please contact the appropriate Springer Editor:

- Pierpaolo Riva at pierpaolo.riva@springer.com (Europe and Americas);
- Swati Meherishi at swati.meherishi@springer.com (Asia—except China, Australia, and New Zealand);
- Wayne Hu at wayne.hu@springer.com (China).

All books in the series now indexed by Scopus and EI Compendex database!

Arjun Sil · Denise-Penelope N. Kontoni ·
Rathish Kumar Pancharathi
Editors

Recent Trends in Civil Engineering

Select Proceedings of ICRACE 2021

Springer

Editors
Arjun Sil
Department of Civil Engineering
National Institute of Technology Silchar
Silchar, India

Denise-Penelope N. Kontoni
Department of Civil Engineering
University of the Peloponnese
Patras, Greece

Rathish Kumar Pancharathi
Department of Civil Engineering
National Institute of Technology Warangal
Warangal, India

ISSN 2366-2557 ISSN 2366-2565 (electronic)
Lecture Notes in Civil Engineering
ISBN 978-981-19-4054-5 ISBN 978-981-19-4055-2 (eBook)
https://doi.org/10.1007/978-981-19-4055-2

This Springer imprint is published by the registered company Springer Nature Singapore Pte Ltd.
The registered company address is: 152 Beach Road, #21-01/04 Gateway East, Singapore 189721, Singapore

Preface

The First International Conference on Recent Advancements in Civil Engineering (ICRACE) 2021 was held at the National Institute of Technology Silchar, Assam, India, during the period 17–19 September 2021. ICRACE aims to bring together experts from academic, scientific, and industrial communities to address new challenges and present their latest research findings, ideas, developments, and perspective of the future directions in the field of civil engineering. The conference serves to foster for presenting novel and fundamental advances in the fields of civil engineering. The main aim of online ICRACE 2021 is to cover a wide range of critical topics concerning the use of recent advances in civil engineering. This includes all fields of civil engineering such as construction, geotechnical, structural, earthquake, environmental, water resources, transportation, and architectural engineering. ICRACE intended to follow up the scientific studies and experiences gained in the developments of the innovative and sustainable construction materials and their applications in the field of concrete infrastructures and transportation engineering during the last decade.

The ICRACE conference 2021 included ten keynote/invited speakers of repute from different backgrounds of the civil engineering community. It involved five parallel sessions with main themes of the conference being primarily focused on (i) Structural Engineering—including Theory and Advanced Technology of Engineering Structure, High-rise Structure and Large-span Structure, Bridge and Tunnel Engineering, New Structure and Special Structure, Advanced Concrete Technology, Civil Engineering Materials, Durable Structures, Sustainable Design in Built Environment, Structural and Construction Engineering, and Sustainable Construction Technologies; (ii) Structural Dynamics and Earthquake Engineering—including Engineering Structure Safety and Disaster Prevention, Building Energy Conservation and Green Architecture, Structural Liability, Durability and Health Monitoring, Coastal Engineering, Performance-based Design, Disaster Management, Structural Model Validation, Fluid–Structure Interaction, and Smart-Structures and Materials; (iii) Geotechnical Engineering—including MEMS/NEMS, Sensors, Actuators, Nanogenerator, Energy Harvester, Micromachining, Microfluidics, Lab-on-Chip, Healthcare Systems, Embedded System Design, Biomedical Systems, IoT and Smart Systems, Robotics and Mechatronics, MEMS Packaging, Reliability,

Semiconductor Memories; (iv) Water Resource Engineering—including Hydraulic and Hydro-Power Engineering, Engineering Water Management, New Technology, Method and Technique in WRE, Hydraulic Engineering, Reservoir Planning and Management, Integrated Water Management, Floods and Droughts Management, Environmental Impact Assessment, Hydraulic Structures, Groundwater Hydraulics, Computational Fluid Dynamics, Ocean Engineering, Numerical Analysis in Water Resources Engineering, and Soft Computing Techniques and Applications in Water Resources; and (v) Transportation Engineering—including Road and Bridge Engineering, Surveying Computing in Civil Engineering, Traffic Guidance and Forecast, Urban Traffic Control and Congestion Pricing, Vehicle Safety and Emissions, Highway Pavements, Railroad and Railway Engineering, Civil and Urban Engineering, Sustainable Transportation Infrastructures, Traffic Operations and Intelligent Transportation Systems, Transportation Safety, Transportation Network Analysis, Pavement Analysis and Design, Use of Recycled and Waste Materials in Pavement Construction, Pavements Mix Design, Pavement Material Characterization, Bituminous and Alternative Materials, Pavement Performance Modelling, and Sustainability in Pavement Engineering.

However, we are thankful to Prof. Sivaji Bandyopadhyay, Director, NIT Silchar, and Dr. Parthajit Roy (former Head, Civil Engineering Department, NIT Silchar), for their encouragement in various steps of the event. We are thankful to the following keynote speakers: Prof. Hemanta Hazarika (Kyushu University, Japan); Prof. B Bhattacharjee Emeritus (IIT Delhi); Prof. Ganeshan N. Emeritus, NIT Calicut, India; Prof. Mohd Rosli bin Hainin, Universiti Malaysia Pahang, Malaysia; Prof. Krishna R. Reddy, University of Illinois at Chicago, USA; Prof. Rajib Kumar Bhattacharjya, IIT Guwahati, India; Prof. Hadi, Muhammad N. S, University of Wollongong Australia; Prof. Elgawady, Mohamed A., Missouri University of Science & Technology, USA; Dr. Denise-Penelope N. Kontoni, University of the Peloponnese, Greece; and Prof. P. Rathish Kumar, NIT Warangal, India. We wish to convey thanks to our colleagues Prof. S Choudhury, Prof. A. I. Laskar, Prof. A K Dey, Prof. P Choudhury, Prof. A K Barbhuiya, Prof. M Ali Ahamed, Prof. D Chakraborty, Prof. U Kumar, Dr. T Rahman, Dr. P Sultana, Dr. N Borthakur, Dr. N Debnath, Dr. B. K. Roy, Dr. P Das, Dr. S Dutta, Dr. M. L. V. Prasad Raju, Dr. B S Sil, Dr. D Bhowmick, Dr. M Hussain, Dr. D K Ghose, Dr. Nitsh A., Dr. Prashanth J, Dr. S Ghosh, Dr. A K Das, Dr. S Jena, Dr. A Kuity, Dr. O Baro, Dr. A Sahu, and Dr. Kulkarni V. V. Such a big event could not have been possible without the help of our students or research scholars' force. We wish to mention the names of research scholars, some of which we especially remember w.r.t the event as Mr. P N Bhowmik, Mr. Naveen Kumar Vanapalli, Ms. Durga Mibang, Mr. Sudipta Debroy, Mr. J Das, Mr. S P Mishra, Mr. S Biswas, and others. We are thankful to Springer Publications for their consent in publishing a SCOPUS indexed proceedings of the selected papers of ICRACE 2021. Springer has not only added extra flavour to the conference, but also encouraged contributors of papers. We are thankful to Prof. Dr. Denise-Penelope N. Kontoni, University of the Peloponnese, Greece, and Prof. P. Rathish Kumar, NIT Warangal, India, for their editorial works and guidance. And last but not least, our special thanks go to all energetic faculty coordinators Dr. M L V Prasad, Dr. B S Sil, Dr. K L Singh, and Dr. M Hussain for their tireless effort throughout this event.

We thank all the authors for their contribution to the ICRACE conference 2021 that has resulted in the proceedings. ICRACE follows its tradition in selecting and reviewing these papers with great rigour, and we hope that the proceedings will provide a glimpse of the state-of-the-art practices followed in different fields related to civil engineering and its allied branches. We would also like to express our sincerest appreciation to the reviewers of the papers and to various technical and financial sponsors for making this event a grand success.

Silchar, India — Arjun Sil
Patras, Greece — Denise-Penelope N. Kontoni
Warangal, India — Rathish Kumar Pancharathi

Contents

Transportation Engineering

About the Editors

Dr. Arjun Sil is currently an Associate Professor and Head of the Department of Civil Engineering, National Institute of Technology Silchar, India. He obtained his B.Tech. from North Eastern Regional Institute of Science and Technology (NERIST), India; M.Tech. from National Institute of Technology (NIT) Silchar; and Ph.D. from Indian Institute of Science (IISc), Bangalore, India. His specific areas of research are interdisciplinary such as earthquake engineering including structural engineering, geotechnical earthquake engineering, seismology, seismic hazard, site response, microzonation, risk evaluation, solid mechanics, condition assessment and health monitoring of structures, reliability and probability, monte carlo simulation, space hazards, Covid-19, GIS and remote sensing, corrosion modelling, chloride, wind, and landslide hazard assessment. He has published so far more than 59 research articles [http://www.nits.ac.in/departments/civil/civil.php] in highly reputed international journals (SCI, Scopus), 12 papers in conferences and one granted patent. He has so far guided 34 M.Tech. scholars, Ph.D. scholars (03 defended, 03 submitted). Currently several Ph.D. and M.Tech. scholars are working under him in various domains. He is a reviewer of various prestigious journals such as *Natural Hazards Review* (ASCE), *Journal of Geomatics, Natural hazards and Risk* (Taylor & Francis), *Journal of Earthquake Engineering* (Taylor & Francis), and *Corrosion Engineering Science and Technology* (Taylor and Francies). He has received several prestigious awards such as Outstanding Structural Engineer of the Year Award-2020, Conferred by the Governing Council of "Indian Association of Structural Engineers (IAStructE)", New Delhi, India, for the overall contribution of Civil Engineering. He is acting as "Research Advisor" of Nanyang Academy of Sciences (Singapore), received IGS-Sardar Resham Singh Memorial Award-2015, conferred by Indian Geotechnical Society (IGS), New Delhi. His Ph.D. thesis was nominated as one of the best thesis for "institute MEDAL" in IISc, Bangalore from Civil Engineering Department during 2014–2015.

Dr. Denise-Penelope N. Kontoni obtained her Diploma in Civil Engineering from the University of Patras, Greece (grade: excellent, ranked 1st), and her Ph.D. (doctor of civil engineering) in structural engineering also from the University of Patras,

Greece. She is currently an Associate Professor in the Department of Civil Engineering of the University of the Peloponnese, Greece. She has also been teaching for 19 years in postgraduate programs of the Hellenic Open University, Greece, and she has supervised many M.Sc. theses in the postgraduate programs such as "Earthquake Engineering and Earthquake Resistant Structures", "Engineering Construction Management", etc. She is the author of 120 scientific articles in refereed international scientific journals and in proceedings of international conferences, which have received many citations from other researchers. She is a reviewer in many international scientific journals (e.g., "Certificate of outstanding contribution in reviewing" from "Engineering Structures"-Elsevier). She is editor and guest editor in refereed international scientific journals and has been a member of the scientific committee in many international conferences. Her research interests are focused on structural dynamics, earthquake engineering, finite element method (FEM), boundary element method (BEM), soil-structure interaction, vibration control, computer-aided structural analysis, wave propagation, elastodynamics, elastoplasticity, computer programming in civil engineering, etc.

Dr. Rathish Kumar Pancharathi is currently a Professor and Head of the Department of Civil Engineering, National Institute of Technology Warangal. He obtained his Ph.D. from the NIT Warangal and Doctor of Engineering and postdoctoral research from Japan. His major areas of interest include new/alternate/supplementary cementitious materials, special concretes, rehabilitation of structures, construction technology and management, structural health monitoring, earthquake engineering and sustainable materials and technologies. He has published 226 technical articles in various international and national journals and conferences of repute. He has contributed 15 book chapters and authored 2 books. He has two patents granted to his credit. He has guided 8 doctoral students and 12 students are presently working with him. He has guided 68 Master's students and 37 Bachelor's student batches. He is a recipient of several awards including Aftab Mufti Medal, Monbusho and JSPS Scholarships of the Japanese Government, Best Engineering Researcher Award, Heritage Scholarship to pursue research at IST Portugal, Danish Government Scholarship, Italy Government Post-Doctoral Scholarship, Heritage Scholarship under Erasmus Mundus, AIT Fellowship, ASEM-Duo India Fellowship, Slovakian Government Scholarship, Jawaharlal Nehru memorial fellowship, Earth leader award and Distinguished Alumnus award. Prof. Kumar is a member of the Research Advisory Council member of the National Council for Cement and Building Materials (NCCBM), Hyderabad, for the past six years. He represents several Bureau of Indian Standard (BIS) committees like CED-4 i.e., Lime and Gypsum Products of the BIS, Preparation of Handbook of lime for the BIS, Convenor for revision of all parts of Indian Standard Code IS 2542-1978 including tests for Gypsum, Plaster, Concrete and Mortars, Member of Revision of Indian Standard Code IS 712:1984 i.e., specification for building limes. He is on the IRC B8 Committee of Inspection, Maintenance and Rehabilitation of Bridges. He is on the editorial board of *Journal of Facta Universtatis, Journal of Cement Wapno Beton* among others and is a reviewer of several International Journals published by Elsevier, Thomas Telford, Taylor and Francis etc.

He has 25 Years of Teaching and Research experience including 17 Years of administrative experience. He does a lot of active consultancy in designs and stability checks and auditing of buildings and bridges.

Structural Engineering

Ductile Detailing Practices Using Confined Masonry—A Case Study of Himachal Pradesh

Ankur Thakur and M. Abdul Akbar

1 Introduction

India has experienced moderate to major earthquakes in the past century. More than 50% of area in India is considered vulnerable to damaging earthquakes. Some of the major earthquakes Indian subcontinent has experienced with magnitude (M) exceeding 8.0 are Assam earthquake of 1897 ($M = 8.7$), Kangra earthquake of 1905 (M = 8.0), Bihar–Nepal earthquake of 1934 ($M = 8.4$), and the Assam–Tibet earthquake of 1950 ($M = 8.7$) (Fig. 1) [1]. It is evident from Fig. 1 that northeastern region and Himalayan belt is more susceptible to earthquakes. These past earthquakes occurring in India and other countries have unveiled many weaknesses associated with both reinforced concrete (RC) frame and unreinforced masonry (URM) construction (Fig. 2a and b). These have revealed the inherently poor performance of non-engineered RC frame and URM buildings, especially in developing nations like India where such construction is more common. In the past earthquakes, non-engineered RC frame and URM buildings were found to be the major contributors to human and economic losses as observed during Bhuj (2001), Kashmir (2005), Sikkim (2011), Nepal (2015), and Turkey (2004) earthquakes [2]. RC buildings not adhering to codal practices are poor in terms of design and construction quality, making them highly vulnerable to earthquake [3–7].

In most parts of India, especially in rural areas, construction works are largely carried out by unskilled laborers. Due to lack of formal training and no or very less technical skills of masons, construction quality has suffered in the country. In the past when these buildings were exposed to a seismic event, it resulted in huge economic

A. Thakur (✉) · M. A. Akbar
Department of Civil Engineering, Dr. B R Ambedkar National Institute of Technology Jalandhar, Jalandhar 144011, India
e-mail: ankurt.ce.20@nitj.ac.in

M. A. Akbar
e-mail: akbarma@nitj.ac.in

A. Sil et al. (eds.), *Recent Trends in Civil Engineering*, Lecture Notes in Civil Engineering 274, https://doi.org/10.1007/978-981-19-4055-2_1

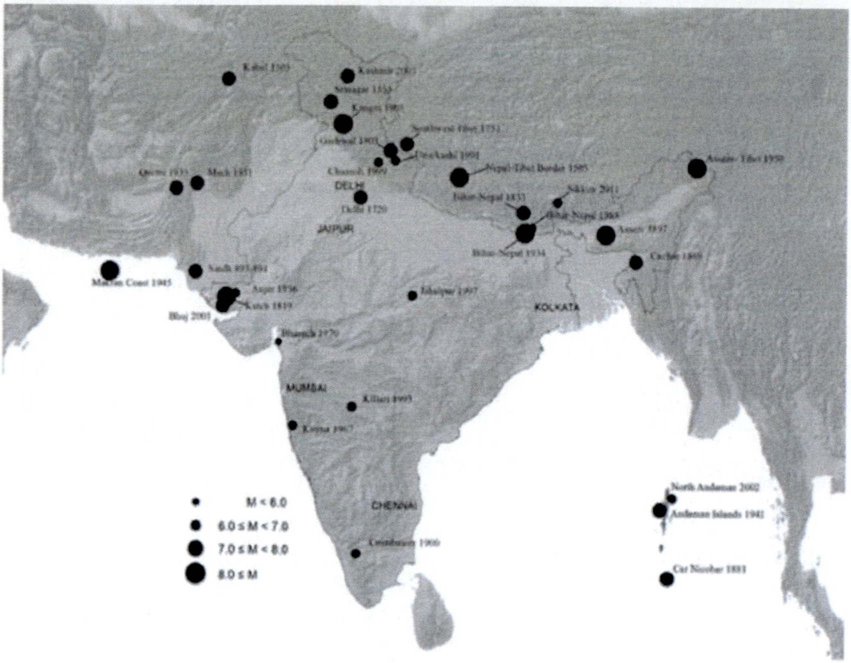

Fig. 1 Location of some past significant earthquakes in Indian subcontinent [10]

Fig. 2 **a** Apartment collapse in 2001 Bhuj earthquake. **b** Destroyed masonry houses

and human losses. Therefore, it is necessary to adopt construction practices which masons can easily follow without much significant change in their conventional style of construction so as to ensure good seismic performance during an earthquake when adequate engineering input is not involved. Confined masonry (CM) construction can be considered as one such viable alternative [8]. Confined masonry has emerged as a promising structural system due to its confining action, owing to the presence of tie-beams and tie-columns at the periphery and its low cost as compared to RC framed structure. Application of CM in buildings does not require advanced construction skills; hence, it can be adopted by masons easily. Study of past earthquakes indicates that good seismic performance can be achieved with CM even without a high level of engineering involved, provided the quality of construction is not compromised [9].

CM buildings are economical when compared with RC framed structure and safer when compared with URM buildings. Low-to-medium rise buildings (generally up to four stories) are the frequent typology for housing, in developing countries all over the world. The key requirement for sustainable building construction demands fulfillment of functional, structural, economical, and environmental criterion. In the past, URM has fulfilled these requirements, but its performance in high seismic regions has always remained a concern. The damages caused by recent earthquakes in India, even in moderate seismicity zones, highlighted the consequences of poorly built masonry structures, comprising around 85% of existing building structures in India. The structural system for seismic resistance needs to be finalized combining sustainability, earthquake resistance, and cost effectiveness, and on these counts confined masonry (CM) appears to be the best alternative for low–medium rise buildings. It can be used as a replacement of non-engineered RC frame and URM construction for low-to-medium rise buildings.

2 Seismic Vulnerability of Himachal Pradesh (HP)

HP is a small northern hill state of India having an area of 55,673 km^2. Nearly 7 million people reside in its twelve districts. The state of HP located in the Western Himalayas has experienced major earthquakes in the past as seismically it is a very active region. The majority of area of the state falls under high seismic zones IV and V (Fig. 3). About 32% of the total geographical area of HP falls under the seismic zone V, while the rest (68%) lies in the seismic zone IV [11]. The entire state is vulnerable to severe seismic events and possibility of occurring of same in the future could not be denied [12]. Some of the major earthquakes experienced by the state in the past are as follows: Kangra earthquake of 1905 ($M = 8.0$), Kinnaur earthquake of 1975 ($M = 6.7$), and Dharamshala earthquake of 1986 ($M = 5.7$). Besides these major earthquakes, the state has experienced more than 250 earthquakes with a magnitude 4.0 and 62 earthquakes with a magnitude more than 5.0. Some of them are given in Table 1 [13]. The state is more vulnerable to seismic events due to active faults and the regional tectonic features of the area [14, 15]. The April 4, 1905, Kangra earthquake in HP killed more than 18,000 people and caused extensive damage to property. This earthquake was felt over a wide area extending about 300 km from Kangra to Dehradun [16]. The Kinnaur earthquake of January 19,1975 ($M = 6.7$) [17] and the Dharamshala earthquake of April 26,1986, ($M = 5.7$) [18] are some other significant earthquakes in respect of damages caused and losses incurred.

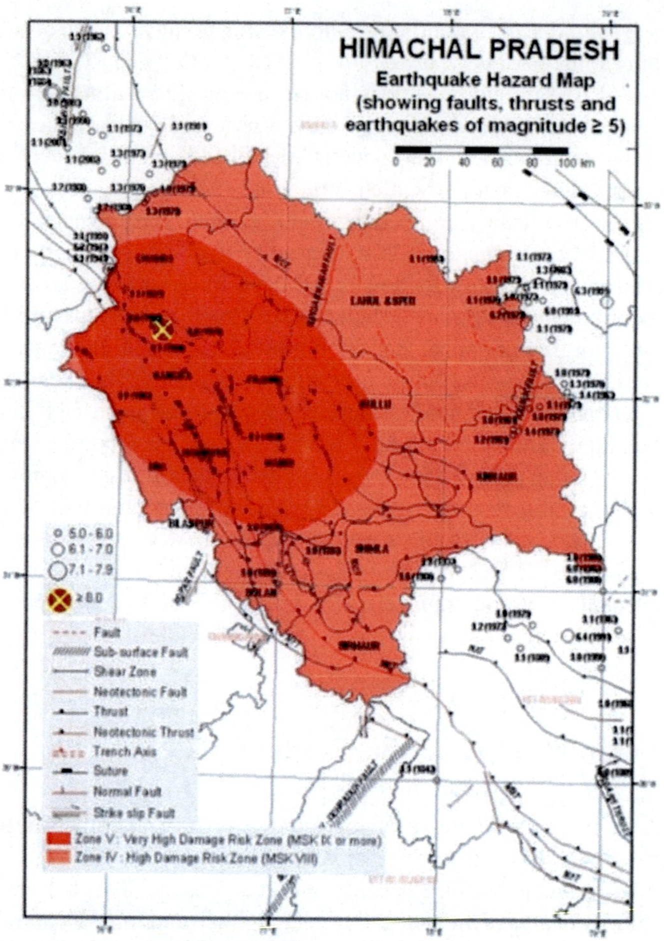

Fig. 3 Earthquake hazard map of HP [20]

In rural areas of HP, which is almost 85% of the total area, most of the construction is non-engineered despite the vulnerability of the state to seismic events. As per vulnerability atlas of India in HP, about 69.5% houses are made of mud, unburnt bricks, and random rubble masonry. The masonry houses do not have earthquake resisting features like seismic bands, etc. Such houses are liable to collapse if earthquake of moderate to major magnitude occurs in the future. This was observed in Dharamshala earthquake of 1986 ($M = 5.7$) [19]. Keeping in view the seismic vulnerability of the state, earthquake-resistant construction practices need to be followed which are discussed in the following section.

Table 1 Past earthquakes in HP from 1900 to 1963 [13]

Sr. No.	Date	Magnitude	District
1	April 04, 1905	8.0	Kangra
2	February 28, 1906	7.0	Kullu
4	October 09, 1914	6.1	Chamba
4	May 11, 1930	5.5	Mandi
5	June 22, 1945	6.5	Chamba
6	July 10, 1947	6.2	Chamba
7	August 12, 1950	5.5	Chamba
8	September 15, 1962	5.5	Kangra
9	April 12, 1963	5.4	Kinnaur
10	November 12, 1963	4.6	Kinnaur

3 Confined Masonry and Detailing Practices

CM is a construction system where the walls are built first, and the columns and beams are poured in afterward to enclose the wall. It consists of masonry walls along with horizontal and vertical RC confining members built on all four sides of a masonry wall panel as shown in Fig. 5 [21]. These vertical members are called as tie-columns and horizontal elements as tie-beams which resemble columns and beams respectively in RC frame construction but generally are of smaller cross-section. These small tie elements improve the ductility and overall seismic performance of building [22]. Though a finished CM building resembles a RC frame building with infill walls, they do not involve sophistication required in RC frame construction. These confining elements strengthen the wall with openings and tie together the floors, walls, and roof of building so that it behaves like a box type enclosed structure during an earthquake [23]. Vertical confining elements, called tie-columns, are cast in situ at the intersection of masonry wall segments. Horizontal elements, called tie-beams or bond-beams, are cast in situ, resting over masonry at different levels (sill, lintel, eves, etc.). Vertical reinforcement bars in tie-columns are anchored into the roof slab. For this reason, contrary to RC structures where infill masonry is built after concrete hardening, in the case of CM, masonry is built first followed by casting of confining elements and hence most of the building weight rests on the masonry panels. In addition, due to concrete shrinkage, the connection between masonry and concrete is very effective, which behaves in unison as a whole up to large deformation levels under lateral loads. Key components of confined masonry buildings are as follows (Fig. 5):

Foundation: It transmits the load from superstructure to underlying soil.

Plinth Beams or RC Plinth Bands: These transfer the vertical and horizontal load from walls to the foundation.

Masonry Walls and Confining Elements: These walls transfer vertical and horizontal loads from the floor and roof slabs above to the foundation. The masonry walls are enclosed by RC elements known as tie-beams and tie-columns at the periphery.

These RC elements provide confinement to the masonry walls and protect them from collapse during major earthquakes. The seismic performance of CM buildings largely depends on the shear resistance of masonry walls. Preferably, toothed edges are left on each side of the wall having projection up to 40 mm to achieve full concrete filling in the teeth space. Good bonding between a masonry wall and adjacent RC tie-columns can be achieved by toothing as shown in Fig. 4a and b. These confining elements provide confinement to masonry walls, so that higher strength and stability can be achieved during in-plane and out-of-plane earthquake forces.

RC Floor and Roof Slabs: These transmit horizontal and vertical loads to walls.

Seismic behavior of a confined masonry wall panel can be explained by composite (monolithic) action of a masonry wall and adjacent RC confining elements. This composite action exists due to toothing between the walls and the tie-columns. This is one of the key features of confined masonry construction. Toothing also helps in enhancing the interaction between masonry walls and confining RC elements [24]. In the absence of toothing, composite action can be achieved by means of horizontal reinforcement (dowels). The seismic response of a confined masonry building and the internal distribution of earthquake forces depend on the type of floor and/or roof system. Floor and roof systems are horizontal elements of the lateral load-resisting system that act as diaphragms. Their primary role is to transfer earthquake-induced lateral forces in the building to vertical elements (shear walls) that resist these forces. A diaphragm can be treated as an I-shaped beam laid in the horizontal plane. The floor or roof functions as the web to resist the shear forces, while the boundary elements (tie-beams in confined masonry buildings) act as the flanges and resist tensile and compressive stresses due to bending moments [25]. Other factors which affect the performance of CM buildings are wall density, wall openings, tie-columns reinforcement, aspect ratio of walls, etc.

The practice of CM construction started in Chile in the 1930s after the 1928 Talca earthquake ($M = 8.0$) that affected a significant number of unreinforced masonry buildings. CM buildings during 2007 Pisco, Peru Earthquake ($M = 8.0$), experienced

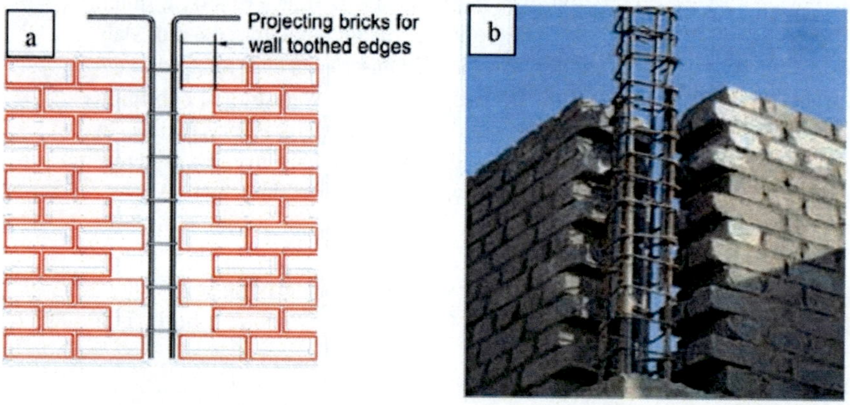

Fig. 4 (**a**, **b**) Toothing in CM walls

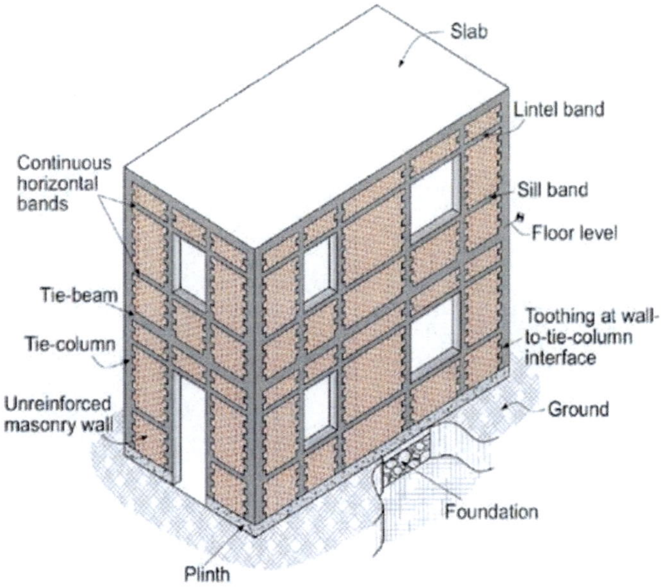

Fig. 5 Typical CM building [21]

very less or negligible damage, whereas complete collapse of nearby adobe house can be observed in Fig. 5. It is pertinent to mention here that CM methodology is followed in high seismic regions around the world and many countries like Chile, Peru, and Mexico have included CM provisions in their building codes [26, 27] (Fig. 6).

CM structure has advantages of both URM and RC structure and provides an economic advantage utilizing the masonry strength in the main load-bearing element. In the major medium rise buildings project of India, it had its first formal application in the construction of 36 CM buildings in the campus of IIT Gandhinagar, Gujarat. It was observed that adoption of CM technology resulted in a cost saving of 10–15% over alternative RC frame construction [28]. The cost savings were mainly due to lesser quantities of concrete and steel on account of smaller member sizes in CM buildings compared to RC frame buildings. As it possesses economic advantage over RC construction and has better seismic performance than URM buildings, it could prove vital for sustainable housing in seismic-prone regions of India [29]. CM is used worldwide for non-engineered low-rise construction and engineered construction such as mid-rise apartment buildings (up to six stories high) [30, 31]. According to Earthquake Engineering Research Institute (EERI), CM construction should be promoted in high seismic regions particularly in emerging economies around the world [32].

Fig. 6 CM building during 2007 Pisco, Peru earthquake [21]

4 Activities of Himachal Pradesh State Disaster Management Authority (HPSDMA)

The Disaster Management Act, 2005, was ratified by Indian government on December 23, 2005, which envisaged creation of the National Disaster Management Authority (NDMA) headed by the Prime Minister, State Disaster Management Authorities (SDMA) headed by the Chief Ministers, and District Disaster Management Authorities (DDMA) headed by the Deputy Commissioners to spearhead and adopt a holistic and integrated approach to disaster management (DM) [33]. HPSDMA was established as per the Disaster Management Act, 2005, to lay down guidelines to be followed by the departments of state government for the purpose of integration of measures for prevention of disasters and provide necessary technical assistance.

Considering the vulnerability of the state to seismic hazards, Himachal Pradesh State Disaster Management Authority (HPSDMA) has started mason training programs with a vision to build a safer and disaster resilient Himachal. The principal aim of these training programs is to create awareness among the construction workers about seismic hazards and provide basic knowledge and skill on earthquake safe construction practices to them. To ensure safe construction of new buildings, training programs on capacity building of masons, carpenters, bar binders, etc., on hazard-resistant construction are being provided by HPSDMA in all the districts of HP in collaboration with Central Building Research Institute (CBRI), Roorkee, and all the Government Engineering and Polytechnic Institutes of the state [34]. DDMA is managing the conduct of training program in its respective district. For example, under this scheme in Kangra district alone, 150 residential training programs of 25 participants in each batch are being organized for the training of 3740 masons,

Fig. 7 Mason training program

carpenters, and bar binders. Photos taken from one of such many training programs held at Kangra district on February 14, 2020, in which first author was one of the Master Trainers are shown in Fig. 7.

The emerging training needs of masons and others construction workers are planned in a simple yet comprehensive way. It is a fair mix of practical hands-on training, audio-visual presentations, models, discussions, etc. A cadre of Master Trainers which includes the faculty of Civil Engineering Departments of Government Engineering and Polytechnic Institutes under the Technical Education Department of HP is anchoring the trainings of all the construction workers. These Master Trainers had undergone training which was conducted by HPSDMA in collaboration with experts from CBRI, Roorkee. Certification to masons is provided after the successful completion of training. HPSDMA is maintaining the database of skilled and trained personnel so that information will be shared with potential employers at a later stage and good construction quality can be maintained in the state [35].

5 New Construction Materials

Conventional clay brick is the most common infill material used for construction in India. The choice of building material has a huge impact on the built environment as well as on the overall cost of the project. Construction using these conventional masonry units is tedious and time consuming. Moreover, burnt solid clay units are brittle in nature and possess low shear strength [36]. So, there is an urgent need to replace this material with one which is more efficient, and this can be achieved by adopting innovative masonry or walling units. We have witnessed some significant research and advancements in construction materials in the past two decades. There are some materials which possess better mechanical properties vis-a-vis conventional clay bricks. Practice of these materials in the construction needs to be intensified in hilly regions like HP to minimize the structural loss during an earthquake. Some of these materials are locally available and can be used in place of clay bricks as an infill material. These materials are briefly discussed here as under:

Autoclave Aerated Concrete (AAC) Blocks: AAC is a lightweight cementitious product of fly ash or sand, water, cement, lime, and aluminum powder mix, used globally for brickwork Although AAC blocks are being used in construction since 1924, its usage in India constitutes only 16–18% of the construction at present. Desirable mechanical properties of AAC blocks such as its low bulk density, better thermal, and acoustic properties along with ease in installation make it an evident choice to replace conventional clay bricks [37]. These are available in different sizes in market [38]. The prominent advantage of AAC is its lightweight, which in high seismic zones is beneficial for the seismic performance of a building [39].

Lightweight Cellular Panels: This cellular lightweight concrete (CLC) is made up by mixing of cement, fly ash, water, and foaming agent. These have desirable properties to be used as an infill material in relation to its lightweight, elasticity, compressibility, resistance to water, and good vibration absorption along with improved thermal and acoustics efficiency. Some other advantages these panels have over the traditional walling units are ease in installation, facilitating rapid construction and reduction of masonry cross-sectional area, etc. [40].

Clay Fly Ash Bricks: These are produced from *fly ash, water, and clay*. These bricks have more strength than conventional burnt clay bricks, consume less energy, provide better thermal insulation, and solve the environmental problem through the utilization of fly ash and industrial waste. These bricks have worldwide application in construction of residential, industrial, and commercial buildings and are easily available throughout the country [41, 42].

Multi-perforated Clay Bricks: These bricks are used in construction due to its characteristics of fire resistance, lightweight, good thermal insulation, and lower construction cost. Masonry made of multi-perforated clay bricks is often used in reinforced and CM constructions, mainly in seismic-prone regions [43].

6 Conclusions

The paper presents the case study of Himachal Pradesh, which on account of its high seismic zone has seen catastrophic failures of poorly designed structures during past earthquakes. It has been observed that one of the major factors contributing to poor construction quality in HP is non-availability of skilled construction workers. The quality of construction is suffering due to lack of formal training of construction workers.

To ensure quality in construction, HPSDMA has come up with mason training programs which are running in all the districts of the state with the help of DDMA and state technical institutes. The purpose of these trainings is to introduce the masons to basic earthquake-resistant construction practices which can lead to more safe buildings. Training and capacity building programs of masons be carried out in other parts of the country also to create awareness about seismic hazards and provide basic knowledge and skill on safer earthquake construction practices to them.

Confined masonry construction technology which does not require advanced construction skills or equipment can be easily adopted by masons. It is also useful in sustainable earthquake-resistant housing since the construction cost of CM buildings is less in comparison with RC buildings and same basic materials are used in CM construction, i.e., concrete, masonry, and steel. Moreover, CM construction does not require a great deal of engineering input, and only with nominal care in design and construction, it can perform very well during earthquakes. Also, when it is used with lightweight infill material like AAC blocks and CLC panels, it could improve the overall earthquake performance of building in high seismic regions like HP.

References

1. Jain SK (1998) Indian earthquakes : An overview. Indian Concr J 72(11):555–561
2. Dougangün A, Ural A, Livaouglu R (2008) Seismic performance of masonry buildings during recent earthquakes in turkey. In: 14th World conference on earthquake engineering, pp 12–17
3. IS 456 (2000) Indian standard specification for plain and reinforced concrete
4. IS 13920 (2016) Ductile design and detailing of reinforced concrete structures subjected to seismic forces—code of practice
5. IS 1893 (PART 1) (2016) Criteria for earthquake resistant design of structures-code of practice
6. IS 4326 (2013) Earthquake resistant design and construction of buildings
7. Meli R, Alcocer SM (2000) Challenges in the implementation of earthquake disaster reduction programs. In: 12th World conference on earthquake engineering, pp 1–8
8. Karantoni F, Pantazopoulou S, Ganas A (2018) Confined masonry as practical seismic construction alternative—the experience from the 2014 Cephalonia earthquake. Front Struct Civ Eng 12(3):270–290. https://doi.org/10.1007/s11709-017-0390-1
9. Tanner JE, Carboni A (2010) Performance of masonry buildings in the 2010 Chile (Maule) earthquake, vol 1997, pp 1–15
10. Jain SK (2016) Earthquake safety in India: achievements, challenges and opportunities. Bull Earthq Eng 14(5):1337–1436. https://doi.org/10.1007/s10518-016-9870-2
11. IIIT Hyderabad (2014) A report on Hazard, vulnerability, risk assessment of Himachal Pradesh, no 02
12. Arya FNA (AS), Damage scenario of a hypothetical 8.0 magnitude earthquake in Kangra region of Himachal Pradesh
13. Chandel VBS, Brar KK (2010) Seismicity and vulnerability in Himalayas: the case of Himachal Pradesh, India. Geom Nat Hazards Risk 1(1):69–84. https://doi.org/10.1080/19475701003643441
14. Earthquake Preparedness (1905) Remembering Kangra earthquake of 4 April 1905, no 311
15. Yeats RS, Lillie RJ (1991) Contemporary tectonics of the Himalayan frontal fault system. Folds, blind thrusts and the 1905 Kangra earthquake. J Struct Geol 13(2):227–233. https://doi.org/10.1016/0191-8141(91)90069-U
16. Thakur VC, Sriram V, Mundepi AK (2000) Seismotectonics of the great 1905 Kangra earthquake meizoseismal region in Kangra-Chamba, NW Himalaya. Tectonophysics 326(3–4):289–298. https://doi.org/10.1016/S0040-1951(00)00126-8
17. Khattri VK, Rai K, Jain AK, Sinvhal H (1967) The Kinnaur earthquake, Himachal Pradesh, India, Of 19 January, 1975. Angew Chemie Int Ed 6, 49(11):951–952
18. Ram VS, Kumar D, Khattri KN (2005) The 1986 Dharamsala earthquake of Himachal Himalaya—estimates of source parameters, average intrinsic attenuation and site amplification functions. J Seismol 9(4):473–485. https://doi.org/10.1007/s10950-005-1418-x

19. A report on Earthquake Hazard Profile of the Himachal Pradesh (1905)
20. SEEDS Technical Services Pvt. Ltd., A report on evolving stratergy for capacity building of Masons, carpenters and wire binders at a Panchhayat level for safe construction
21. Singhal V (2016) Seismic design and behaviour of confined masonry buildings
22. Meli R, Brzev S, Astroza M, Boen T, Crisafulli F, Dai J, Farsi M, Hart T, Mebarki A, Moghadam AS, Quiun D, Tomazevic M (2011) Seismic design guide for low-rise confined masonry buildings
23. Meli R (2015) International guideline for seismic design of low-rise confined masonry buildings in regions of high seismic risk. In: 15th World conference on earthquake engineering, September 2015
24. Singhal V, Rai DC (2014) Role of toothing on in-plane and out-of-plane behavior of confined masonry walls. J Struct Eng 140(9):04014053. https://doi.org/10.1061/(asce)st.1943-541x.0000947
25. Chourasia A (2017) Design guidelines for confined masonry buildings. CSIR-Central Build Res Inst Roorkee 247 667:122
26. Hart TM, Brzev S (2014) The confined masonry network's design and construction guidelines. In: NCEE 2014—10th U.S. National conference on earthquake engineering frontiers earthquake engineer, 2014. https://doi.org/10.4231/D3T43J36Z
27. Rai DC, Review of design codes for masonry buildings by. Earthquake 0–22
28. Jain SK, Basu D, Ghosh I, Rai DC, Brzev S, Bhargava LK (2014) Application of confined masonry in a major project in India. NCEE 2014 - 10th U.S. National conference on earthquake engineering frontiers earthquake engineer, January, 2014. https://doi.org/10.4231/D3B27PS05
29. Borah B, Singhal V, Kaushik HB (2019) Sustainable housing using confined masonry buildings. SN Appl Sci 1(9):1–7. https://doi.org/10.1007/s42452-019-1020-4
30. Bartoli G, Betti M, Galano L, Zini G (2018) Numerical insights on the seismic risk of confined masonry towers. Eng Struct 180:713–727, 2019. https://doi.org/10.1016/j.engstruct.2018.10.001
31. Marques R, Lourenço PB (2019) Structural behaviour and design rules of confined masonry walls: review and proposals. Constr Build Mater 217:137–155. https://doi.org/10.1016/j.conbuildmat.2019.04.266
32. EER Institute (2019) EERI policy white paper, pp 2–6
33. The Disaster Management Act (2005) Government of India
34. Himachal Pradesh State Disaster Management Authority (2017) Himachal Pradesh state disaster management plan. Himachal Pradesh State, p 130, 2017, [Online]. Available: https://hpsdma.nic.in/
35. Guidelines (2018) Scheme for training of masons, carpenters and bar binders on hazard resistant construction
36. Sharma N, Telang D, Rath B (2017) A review on strength of clay brick masonry. Int J Res Appl Sci Eng Technol 5(XII):2620–2626
37. Raj A, Borsaikia AC, Dixit US (2020) Evaluation of mechanical properties of autoclaved aerated concrete (AAC) block and its masonry. J Inst Eng Ser A 101(2):315–325. https://doi.org/10.1007/s40030-020-00437-5
38. BIS:2185 (2005) Indian standard concrete masonry units, part 1: hollow and solid concrete blocks. Bur Indian Stand New Delhi 17
39. Kamal MA (2020) Analysis of autoclaved aerated concrete (AAC) blocks with reference to its potential and sustainability. J Build Mater Struct 7(1):76–86. https://doi.org/10.5281/zenodo.3950489
40. Chourasia A, Singhal S, Parashar J (2019) Seismic performance evaluation of full-scale confined masonry building using light weight cellular panels. J Build Eng 32(November):2020. https://doi.org/10.1016/j.jobe.2020.101473

41. Pawar AS, Garud DB (2014) Engineering properties of clay bricks with use of fly ash. Int J Res Eng Technol 03(21):75–80. https://doi.org/10.15623/ijret.2014.0321016
42. Abbas S, Saleem MA, Kazmi SMS, Munir MJ (2017) Production of sustainable clay bricks using waste fly ash: mechanical and durability properties. J Build Eng 14:7–14. https://doi.org/10.1016/j.jobe.2017.09.008
43. Alcoce SM (1999) Behavior of multi-perforated clay brick walls under earthquake-type loading description of the experimental program. In: 8th North American masonry conference on 1999

Prediction of Twist and Torque of RC Beams of M60 Core Concrete Jacketed with Ferrocement "U" Wraps at Ultimate Stage: Different Methods

G. C. Behera and M. L. V. Prasad

1 Introduction

Many of the structures are subjected to torsional loads. Two types of torsional loads are found in structures. One of them is primary torsion, and the other one is secondary torsion mainly on peripheral beams. Some structures are subjected to torsional loads only, while others are subjected to torsional loading along with axial loads or bending loads. In construction industry, the main material is concrete. Reinforced concrete structures (RC) need to be strengthened when loads are increased or updates in codes or due to manmade disasters or due to natural calamities. Repaired materials like steel fibers, epoxy, FRP or steel jacket can enhance the load sustaining capacity of RC structures to a great extent. The type of repair material to be used depends on its availability in locality, its cost, length of repair time along with and availability of skilled workmanship [1]. In present scenario, FRPs are in driver's seat as a repair material. To enhance capacity of torsion in a deficient RC structure, wrapping in all four sides is a better choice. Practical inaccessibility of wrapping one face compels the designer to opt for "U" wrap. Many research papers covered on full wrap [2], while few studies covered on design and analysis of U-jacket [2–4]. The repair strategy for a distressed structure is to be a cost-effective one and also can serve for a long life period. The repair solution should take into consideration of level of damage in existing structure, shape and size, type of repair material and duration of repair. For enhancing strength of a deficient structure, many repair materials are available with different application techniques. Considering strength, cost and application method, ferrocement is a better choice of repair material over FRP. Enhanced strength in

G. C. Behera (✉)
Department of Civil Engineering, Government College of Engineering, Kalahandi, India
e-mail: beheragb@gmail.com

M. L. V. Prasad
Civil Engineering Department, NIT Silchar, Silchar, India
e-mail: mlvprasad@civil.nits.ac.in

A. Sil et al. (eds.), *Recent Trends in Civil Engineering*, Lecture Notes in Civil Engineering 274, https://doi.org/10.1007/978-981-19-4055-2_2

tension, simple method of application on makes ferrocement a better choice of repair material [5] ACI Committee 549, 1979). Ferrocement also possesses the property of resistance to sulfate attack, frost durability and corrosion protection. All these properties are resulting from due reduced porosity [6, 7].

1.1 Objective of Present Study

Due to torsion, shear is induced on the periphery. To resist torsion induced shear, shear reinforcement is provided on periphery of the structure. Experimental and analytical quantification of torsional parameters on full wrap of FRP are available on few studies [8, 9]. No such literature is available for "U" wraps [10, 11]. The literature provided above proves ferrocement a better choice for developing countries over FRP. Some research works are reported on determination of torsional parameters with jacketing of ferrocement on three sides [12]. The results of torsional parameters are based on experimental and analytical investigation. They do not provide any empirical formulae to predict torque or twist for different cross sections or for any other materials in the section. As mentioned earlier, the experimental method provided result after destruction of structures, and also, it is time consumed and uneconomical one. The analytical method predicts the results based on programming which is also a time consuming one. To overcome these hurdles, this investigation is taken up. The present investigation focuses on application of soft computing methods to find out the torque and twist at ultimate stage with ferrocement as wrapping material on three sides. The soft computing methods employed here are multivariate adaptive regression spline (MARS) and weighted aggregated sum product assessment (WASPAS) method.

2 Some Ways to Determine Torque and Twist

2.1 Based on Experimental Investigation

Beams of width 125 mm, depth 250 mm and length 2000 mm were taken for experimental investigation. Ferrocement wrap of thickness 25 mm with four numbers of mesh layer was provided on three sides leaving the upper part as shown in Fig. 1c. For ferrocement wrap of minimum 25 mm on side and core concrete 75 mm, the width was fixed as 125 mm. Taking aspect ratio 2, over all depth was taken 250 mm. To allow two complete spirals in the central test region of the test beam, a length 1500 mm was kept for observation of crack propagation and pattern. To allow the failure to take place in central length of 1500 mm of the beam, 250 mm length on either side of end zones was reinforced heavily. Thus, the total length of the beam was fixed as 2000 mm with aspect ratio 2.0.

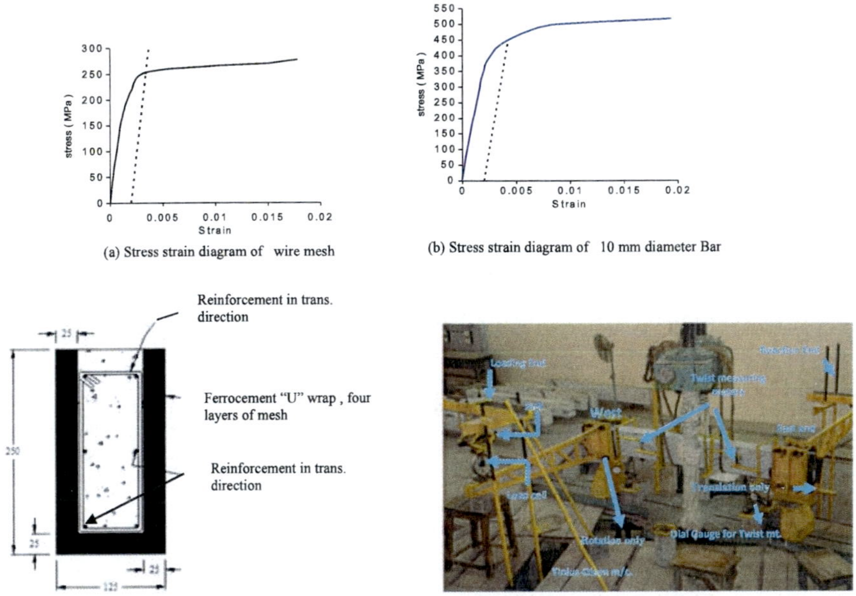

(a) Stress strain diagram of wire mesh

(b) Stress strain diagram of 10 mm diameter Bar

(c) Cross section of beam

(d) Torsion test rig

Fig. 1 **a** Stress–strain diagram of wire mesh. **b** Stress-strain diagram of 10 mm diameter bar. **c** Cross section of beam. **d** Torsion test rig

Stress-strain diagram of wire with 10 mm bar, cross section of the beams and torsion test rig were shown Fig. 1a–d, respectively. Beam dimensions and materials used in the investigation are provided in Table 1. Details of beams and material properties are given in Table 1. All states of torsion were covered in this investigation. Beams without reinforcement, beams with either only longitudinal or only transverse reinforcement and beams with both longitudinal and transverse reinforcement were investigated.

Beam BOH was without any wrap. The grade of concrete is M60, and mortar grade for ferrocement wrap was M55. Ratio of Cement:F.A:C.A:Water is 1.0:0.875:1.625:0.25 with 25 ml of superplasticizer CONPLAST SP-337, procured from FOSROC, Bangalore per kg of binder and 10% replacement of cement with GGBFS for M60 concrete. Ratio of Cement: Sand: Water is 1.0:1.0:0.30 with 30 ml of superplasticizer per kg of binder with 10% addition of silica fume for M55 ferrocement. Ferrocement wrap was provided in "U" shape for all beams except BOH with four numbers of mesh layers. Diameter of mesh was 0.72 mm with square opening 6.35 mm, and yield strength is 250 MPa.

Table 1 Details of the beams tested

Beam designation	Properties of concrete and reinforcement			
	Bars in longitudinal direction		Bars in transverse direction	
	ϕ (diameter) in mm, Numbers	Strength of bar at yield (N/mm^2)	ϕ (diameter), spacing	Strength of bar at yield (N/mm^2)
BOH	"–"	"–"	"–"	"–"
BO4H	"–"	"–"	"–"	"–"
L4H	12, 6	440	"–"	"–"
T4H	"–"	"–"	10 mm dia 70 mm c/c	445
U4H	6, 6	350	6 mm dia 70 mm c/c	350
Lo4H	12, 6	440	6 mm dia. 70 mm c/c	350
To4H	6, 6	350	10 mm dia 70 mm c/c	445
Co4H	12, 6	440	10 mm dia 70 mm c/c	445

2.2 *Based on Model of Analytical Method*

Analytical method of determination of torque and twist at ultimate stage was prepared with modification of softened truss theory incorporating material properties of concrete and ferrocement [10–13].

2.3 *Method Based on Soft Computing*

Experimental method involves in fracturing of proto type structure, long time-consuming method and also involves hectic methods. Analytical model also uses computational works. To overcome these problems for determination of torque and twist of ferrocement "U" wrap, method based on soft computing employed here.

2.3.1 Soft Computing Method—MARS

MARS method is also known as black box or white box method. Approximately 70% of experimental data is taken for testing, and 30% is taken for fitting. It is based on an adaptive procedure. It does not require assumptions to be of any specific type. Very high dimensional problems can be solved here. In analytical method, it is found 22 unknowns to be solved. So, this method is suitable for determining torque and twist at ultimate by providing single type of equations. It was proposed by Friedman [14].

$T_{\text{ultimate}} = 6.752$-maximum [0, 0.32265-spacing of longitudinal reinforcement] * 2.7323-maximum [0, 350-f_y of steel in trans. direction] * 0.002276 + maximum [0, strength of ferrocement mortar-40] * 0.07677.

θ_{ultimate}(rad/m) = 0.03558-maximum [0, Fly-350] * 0.0003376-maximum [0, 350-stress of steel in long. direction] * 0.00008786 + maximum of [0, distance between stirrups in mm] * 0.00102665.

2.3.2 Soft Computing Method WASPAS

For solution of multi-criteria decision-making problems, MCDM method is useful. WASPAS method is based on sum of two MCDM methods. One is weighted sum method (WSM) and another is weighted product method (WPM) and developed by [15–18]. The step by step procedure is presented below.

Setting of matrix for decision at initial stage.
Maximize or minimize the matrix of decision [16].

$$\overline{x}_{ij} = \text{Ratio of } xij \text{ and} \left(\max x_i x_{ij}\right) \tag{1}$$

$$\overline{x}_{ij} = \min_i x_{ij} \tag{2}$$

Use equation of WSM to predict ith alternative based on total relative importance.

$$Q_i^{(1)} = \sum_{j=1}^{n} (\text{Product of } \overline{x}_{ij} \text{ and } w_j) \tag{3}$$

Predict the value of other ith value taking WPM method into consideration.

$$Q_i^{(2)} = \prod_{j}^{n} \overline{x}_{ij}^{w_j} \tag{4}$$

Increase the accuracy ranking in the process of decision making as per [17] and [18] method.

$$\text{determination } Q_i = \lambda * Q_i^{(1)} + (1 - \lambda) * Q_i^2 \tag{5}$$

λ takes the values in between 0 or 0.1 or 1. Usual value of $\lambda = 0.5$. With variation of λ value, relative importance alternatives and ranking alternatives are influenced.

3 Discussion on Results

The torque and twist values obtained by experimental, analytical and soft computing method were compared here along with percentage of error in predicted values in respect of experimental values.

3.1 Behavior of Ferrocement Wrapped High Strength RC Beams

All types' beams pertaining to states of torsion were cast and tested. BOH and BO4H were control specimens without and with wrap, respectively. Control specimen BOH was cast as a plain beam without ferrocement wrap with concrete of M60. Beam BO4H (125 mm * 120 mm * 2000 mm) was provided with a four layer mesh wrap of thickness 25 mm. In core portion M60 grade concrete was there without reinforcement. After cracking of beam due to torsional loading, reinforcement participates in resisting torsional load. When a beam is cast without or with reinforcement in one direction only, the beam is unable to sustain more loads beyond cracking. So, beams with reinforcement in one direction are same as plain beams. The beams with single type of reinforcement were L4H and T4H. Beam U4H was torsionally under-reinforced in longitudinal and transverse direction. Lo4H was under reinforced in transverse direction and over reinforced in longitudinal diction. To4H was opposite to Lo4H. Co4H was provided with over reinforcement in both transverse and longitudinal directions. BOH, BO4H, U4H, Lo4H, To4H and Co4H represent all six states of torsion. As volume fraction of reinforcement was same on sides of a particular beam, crack was found to be 45° to the longitudinal axis while testing under pure torsional loading.

3.2 Ultimate Torque of Ferrocement "U" Wrap Beams with M60 Concrete

BOH was a plain beam of concrete grade M60. Under torsional loading it failed with a crack on longer face as shear induced was more on longer face. Torque at ultimate was also found same as cracking. Beams BO4H, L4H, T4H, U4H, Lo4H, Co4H and To4H were able to sustain more load after cracking due to presence of reinforcement in core or ferrocement wrap or for both. The control specimen BOH was failed at a torque of 4.612 kNm. The same beam with ferrocement "U" wrap BO4H sustains a torsional load much more, i.e., 6.52 kNm. Ultimate torque of beams L4H and T4H was 6.55 kNm and 6.59 kNm, respectively. U4H, the under-reinforced beam was having torque value of 7.68 kNm while completely over-reinforced beam sustained a torque of 12.91 kNm experimentally. Other results were tabulated in Table 2. The ultimate torque of all these beams predicted by different methods such as soft computing, experimental method and analytical method was presented in Fig. 2. From Table 2 and Fig. 2, it was observed enhancement of twist at ultimate is more for states of torsion than ferrocement wrap. Completely over-reinforced section Co4H resists a torque of 12.91 kNm which is approximately (98.01%) double value ultimate torque of control specimen BO4H. After completely reinforced beams, transversely over-reinforced beams exhibit better torque resistance over under-reinforced and longitudinally over-reinforced beams. When a beam was provided with reinforcement in one direction, ultimate torque was same as cracking torque.

The predicted values of ultimate torques were compared with experimental values, and error in percentage was reported in Fig. 3. The value of error in percentage for analytical method was −3.16%, for MARS, it was maximum −10.79%, and for method of WASPAS, error percentage was −12.89% in comparison with experimental values. Maximum error was limited to 12.89%. Predicted values ultimate torque by analytical and soft computing method were found within acceptable limits, so these methods can be employed for prediction of ultimate torque.

Increase of ultimate torque of plain wrapped beam, L4H, T4H, under reinforced, longitudinally over reinforced, transversely over reinforced and completely over

Table 2 Ultimate torque and twist of high strength concrete RC beams with cube strength of concrete and mortar

Ultimate torque (kNm)						
Beams	Conc. cube strength (MPa)	Ferrocement	Expt	Analytical	MARS	WASPAS
BOH	60.27	NA	4.612	4.34	5.074	
BO4H	60.13	55.23	6.52	6.52	6.226	6.341
L4H	60.27	55.22	6.55	NA	7.107	6.395
T4H	60.57	55.37	6.59	NA	7.022	7.373
U4H	59.82	55.37	7.68	7.729	7.904	7.472
Lo4H	60.27	55.07	7.87	7.828	7.904	9.165
To4H	60.13	55.37	8.86	8.58	7.904	9.553
Co4H	60.27	55.37	12.91	12.98	12.91	11.246
Twist at ultimate torque (rad/m)						
BOH	60.27	NA	0.00280	0.00350	0.00483	0.00351
BO4H	60.13	55.23	0.00546	0.00560	0.00483	0.00507
L4H	60.27	55.22	0.00580	NA	0.00520	0.00569
T4H	60.57	55.37	0.00560	NA	0.00560	0.00589
U4H	59.82	55.37	0.13050	0.13736	0.10745	0.11063
Lo4H	60.27	55.07	0.05600	0.05500	0.07706	0.06712
To4H	60.13	55.37	0.09210	0.09158	0.10745	0.09782
Co4H	60.27	55.37	0.07540	0.07361	0.07706	0.05431

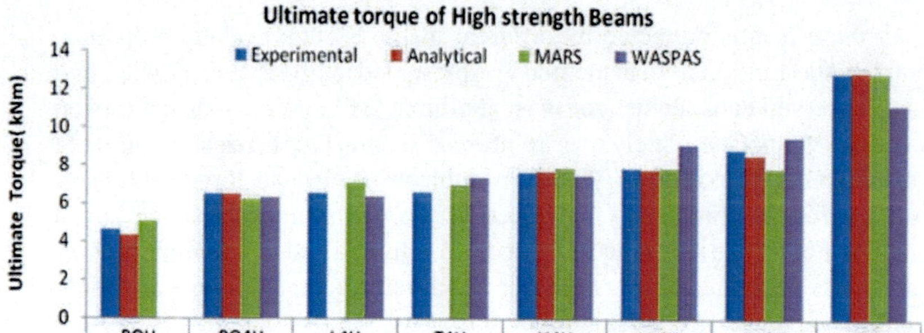

	BOH	BO4H	L4H	T4H	U4H	Lo4H	To4H	Co4H
Experimental	4.612	6.52	6.55	6.59	7.68	7.87	8.86	12.91
Analytical	4.34	6.52	0	0	7.729	7.828	8.58	12.98
MARS	5.074	6.226	7.107	7.022	7.904	7.904	7.904	12.91
WASPAS		6.341	6.395	7.373	7.472	9.165	9.553	11.246

Fig. 2 Ultimate torque of high strength beams by different methods

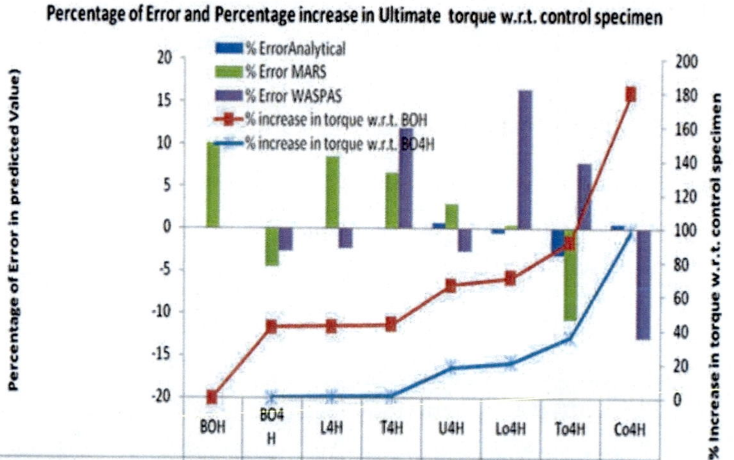

	BOH	BO4H	L4H	T4H	U4H	Lo4H	To4H	Co4H
% ErrorAnalytical	0.00	0.00	0.00	0.00	0.64	-0.53	-3.16	0.54
% Error MARS	10.02	-4.51	8.50	6.56	2.92	0.43	-10.79	0.00
% Error WASPAS	0.00	-2.75	-2.37	11.88	-2.71	16.45	7.82	-12.89
% increase in torque w.r.t. BOH	0.00	41.37	42.02	42.89	66.52	70.64	92.11	179.92
% increase in torque w.r.t. BO4H		0.00	0.46	1.07	17.79	20.71	35.89	98.01

Fig. 3 Error in percentage of predicted values of torque at ultimate and increase in percentage ultimate torque with respect to BOH and BO4H beams

reinforced with respect to control specimen BOH was found to be 41.37%, 42.02%, 42.89%, 66.52%, 70.64%, 92.11% and 179.92%, respectively. This is the effectiveness of wrap of ferrocement in three sides. A wrap of four layer without any reinforcement BO4H was able to enhance ultimate torque 41.37% over unwrapped beam. Increase of ultimate torque of Co4H over control specimen BO4H was found to be 98.01%.

3.3 *Twist Value at Ultimate Torsional Load for RC Beams with "U" Wraps*

The twist value in rad/m found at ultimate torque for the beams L4H, T4H, U4H, Lo4H, To4H and Co4H with control specimens was presented here. The twist of control specimen BOH at ultimate torque was found to be 0.0028 rad/m. BOH when provided with four layers of ferrocement "U" wrap, twist was found to be 0.00546 rad/m. The beams U4H were able to sustain the twist at ultimate torque of 0.13050 rad/m. This beam was subjected to maximum twist in comparison with other beams. Next to under-reinforced beam, transversely over-reinforced beam was showing more twist of value 0.09210 rad/m. The values of twist calculated by different methods were plotted in Fig. 4.

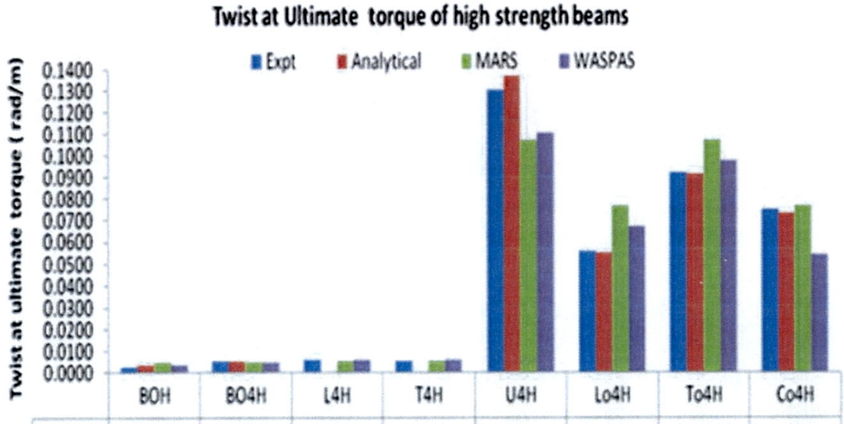

	BOH	BO4H	L4H	T4H	U4H	Lo4H	To4H	Co4H
Expt	0.00280	0.00546	0.00580	0.00560	0.13050	0.05600	0.09210	0.07540
Analytical	0.00350	0.00560	0.00000	0.00000	0.13736	0.05500	0.09158	0.07361
MARS	0.00483	0.00483	0.00520	0.00560	0.10745	0.07706	0.10745	0.07706
WASPAS	0.00351	0.00507	0.00569	0.00589	0.11063	0.06712	0.09782	0.05431

Fig. 4 Estimated values of twist at of RC beams at ultimate torque

The under-reinforced beam U4H has undergone maximum twist as there is less reinforcement. So, the toughness is greater in this under-reinforced beam. The percentage of error in predicted value and ratio of twist with respect to control specimen were presented in Fig. 5. Maximum error in predicted values by analytical method MARS and WASPAS was found to be 5.3, 37.6 and −28%, respectively, for ferrocement "U" wrap beams. The predicted values of twist at ultimate torque are found with errors more than 20% for MARS and WASPAS. Here, analytical model better predicts the twist at ultimate torque. The ratio of twist at ultimate torque of ferrocement "U" wrap beams of BO4H, L4H, T4H, U4H, Lo4H, To4H and Co4H with control specimen BOH was found to be 2, 2.1, 2.0, 46.6, 20.0, 32.9 and 26.9, respectively. This proves efficiency of providing "U" wrap. The same was found to be 1.1, 1.0, 23.9, 10.3, 16.9 and 13.8 times over control specimen BO4H for L4H, T4H, U4H, Lo4H, To4H and Co4H. The twist of U4H was found to be 46.6 times over BOH and 23.9 times over beam BO4H. Next to under-reinforced beam, transversely over-reinforced beams have undergone maximum twist.

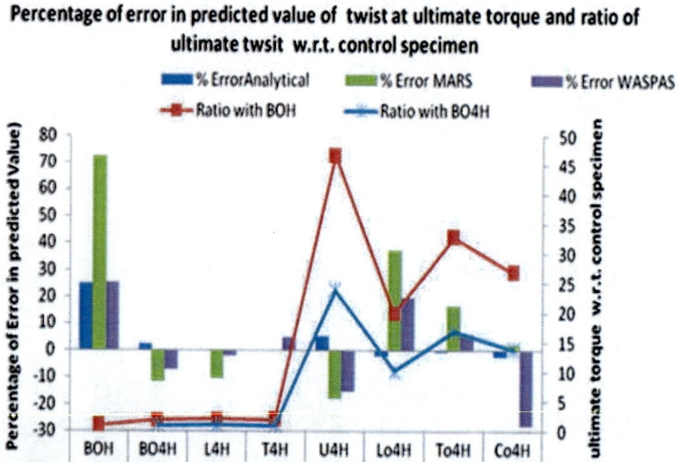

	BOH	BO4H	L4H	T4H	U4H	Lo4H	To4H	Co4H
% ErrorAnalytical	25.0	2.6	0.0	0.0	5.3	-1.8	-0.6	-2.4
% Error MARS	72.6	-11.5	-10.4	0.0	-17.7	37.6	16.7	2.2
% Error WASPAS	25.3	-7.1	-1.9	5.1	-15.2	19.9	6.2	-28.0
Ratio with BOH	1.0	2.0	2.1	2.0	46.6	20.0	32.9	26.9
Ratio with BO4H		1.0	1.1	1.0	23.9	10.3	16.9	13.8

Fig. 5 Predicted percentage of error in twist at ultimate torsional loading and ratio of twist over control specimens

4 Conclusions

Twists and torques at ultimate stage of loading were computed by different methods for control specimens and for all beams pertaining to all six states of torsion with ferrocement "U" wraps using different methods. Following conclusions were drawn from experimental investigation, analytical method and from soft computing methods.

Control Specimen with and without "U" Wrap

The control specimen with ferrocement "U" wrap is having an increase in ultimate torque of 41.37% over its control specimen without any wrap. Wrap enhances the strength to a greater extent.

"U" Wrapped Reinforced Concrete Beams

A. Proving reinforcement in one direction does not enhance the ultimate strength beyond cracking.
B. Beams with over reinforced in both direction have 98.01% more ultimate torque than plain wrapped beams.
C. Transverse reinforcement and longitudinal reinforcement influence the torsional capacity and twist at ultimate stage.
D. Twist at ultimate torque increases with provision of ferrocement "U" wrap. Under-reinforced beams have undergone an ultimate twist 23.9 times than that of the control specimen.
E. The predicted results by soft computing methods are within acceptable limits. All three analytical, MARS and WASPAS, may be employed to predict torque and twist of ferrocement "U" wrap beams.

References

1. Karayannis CG, Chalioris CE, Sirkelis GM (2008) Local retrofit of exterior RC beam-column joints using thin RC jackets—an experimental study. Earthquake Eng Struct Dynam 37(5):727–746
2. Panchacharam S, Belarbi A (2002) Torsional behavior of reinforced concrete beams strengthened with FRP composites. In: Proceedings of First FIB congress, Osaka, Japan, 2002, pp 1–11
3. Chalioris CE (2008) Torsional strengthening of rectangular and flanged beams using carbon fibre-reinforced-polymers—experimental study. Constr Build Mater 22(1):21–29
4. Deifalla A (2015) Torsional behavior of rectangular and flanged concrete beams with FRP reinforcements. J Struct Eng 141(12):04015068
5. ACI Committee 549. Ferrocement-materials and applications. In: ACI symposium proceedings SP-61, Farmington Hills, Michigan
6. Shannag MJ, Mourad SM (2012) Flowable high strength cementitious matrices for ferrocement applications. Constr Build Mater 36:933–939
7. Lalaj O, Yardım Y, Yılmaz S (2015) Recent perspectives for ferrocement. Res Eng Struct Mat 1:11–23

8. Salom PR, Greeley JR, and David YT (2004) Torsional strengthening of spandrel beams with fiber-reinforced polymer laminates. J Comp Constr ASCE 157–162
9. Yasmeen TO, Ahmed MA (2020) Ashteyat, performance of RC beam strengthened with NSM-CFRP strip under pure torsion: experimental and numerical study. Int J Civil Eng 18:585–593
10. Behera GC, Rao TDG, Rao CBK (2008) Torsional capacity of high strength concrete beams jacketed with ferrocement U wraps. Asian J Civil Eng 9(4):411–422
11. Behera GC, Rao TDG, Rao CBK (2014) Analytical model for torsional response of RC beams strengthened with ferrocement U-wraps. SEI, IABSE Struct Eng Int 24(4):509–520. https://doi.org/10.2749/101686614X13854694314847
12. Behera GC, Rao TDG, Rao CBK.A study on post cracking torsional behaviour of high strength reinforced concrete beams with ferrocement "u" wraps. DOI: https://doi.org/10.2478/sjce--0012, Slovak Journal of Civil Engineering,2014;22(3):1 – 12.
13. Behera GC (2018) A model to predict the torsional stiffness of 'U-wrapped' reinforced concrete beams. Struct Build Proc Inst Civil Eng 171(9):676–687
14. Friedman J (1991) Multivariate adaptive regression splines. Ann Stat 19:1–141
15. Alireza A, Javad K (2019) New methods and applications in multiple attribute decision making (MADM), pp 1–233
16. Zavadskas EK, Turskis Z, Antucheviciene J, Zakarevicius A (2012) Optimization of weighted aggregated sum product assessment. Elektron Elektrotech 122:3–6
17. Madic M, Gecevska V, Radovanovic M, Petkovic D (2014) Multicriteria economic analysis of machining processes using the WASPAS method. J Prod Eng 17:1–6
18. Madic M, Antucheviciene J, Radovanovic M, Petkovic D (2016) Determination of manufacturing process conditions by using MCDM methods: application in laser cutting. Eng Econ 27:144–150

Social Network Theory in Construction Industry: A Scientometric Review

Vijayeta Malla, K. V. Prasad, and R. Venkatesan

1 Introduction

Architecture, Engineering, Construction and Operations (AECO) sector is synonymous to characteristics like uniqueness, risk proneness, complex, fragmented work nature (design and production as well as project sites and main office work separately), variation of work, diverse trades and activities, labour inclusive and work in-situ based culture [1, 2]. Over and above, this industry's slackness in adoption of advanced technology and automation is another impediment to its persistent productivity problems [3]. Therefore, immense thrust has been made on the management of construction projects with regards to optimisation of elements of project management 'Iron Triangle', i.e. schedule, cost and quality or the so-called economic aspects, but never questioned or delved into the social or human interaction or behavioural perspective [4, 5]. Also, construction projects observed a high incidence of personnel turnover which posed a dynamic frictional relationship among the construction actors. And this dynamic context is ignored by management who look for improvement in figures and systems. Effective project management should consider the interpersonal relationships from social network point of view rather than mere numbers or establishment of system or documentation [6]. 'A picture conveys a thousand words message, but a network conveys much more than that'. Network theories have been predominantly utilised by research community in social sciences, however, its application in construction industry is quite scant ever since its existence since decades. Construction industry can be viewed as a 'Social Network' as it is a network of different actors or organisations being as 'nodes', the relationship between them as 'ties' and can be utilised in construction management for the analysis in terms of relationship of networks. Construction management's transformation to 'Organic' nature occurs through the usage of Social Network Analysis (SNA) so

V. Malla (✉) · K. V. Prasad · R. Venkatesan
National Institute of Construction Management And Research, (NICMAR), Hyderabad, India
e-mail: vmalla@nicmar.ac.in

A. Sil et al. (eds.), *Recent Trends in Civil Engineering*, Lecture Notes in Civil Engineering 274, https://doi.org/10.1007/978-981-19-4055-2_3

as to identify the key actors and cluster leaders are known through network metrics like degree centralities, cluster coefficients and densities [7].

This study discusses about the paradigm shift in utilising the human dimension through network of relationships with different metrics in construction management, which otherwise would have gone unnoticed. This study has analysed the literature and identified to depict views on intangible aspects like collaboration, coordination, communication [8], project governance [9], diffusion of information [10], knowledge sharing [11], organisational competitiveness [12], relationship network structure analysis [13], Off-Site Construction (OSC) adoption barrier analysis [14], partnership behaviours between contractors, inter-firm collaboration, Job site clusters or networks. In contemporary complex times, there has been a tectonic shift in construction management's transition into 'Network Perspective' from mere economic parameters and systems monitoring. Hence, conducts the scientometric review through computer-assisted text mining tool and categorises the discussion of the literature through content analysis [15] into fives themes are 'individual', 'Inter-organisational', 'Intra-organisational', 'project' and'industry'. The purpose of this study is to understand the extant literature through science mapping for SNA-construction-based research studies. There are two objectives: (1) To examine, understand and map the extant literature for the period (2011–2021) and (2) To identify the themes and future directions of SNA-construction-based research domain.

2 Research Methodology

To achieve the objectives of the study, a two-stage research design is adopted as shown in Fig. 1. In the first stage, the sourcing of relevant research articles from Scopus and google scholar databases is conducted using keyword-based search strategies such as 'social network analysis' AND 'construction management'. Then, the screening, inclusion and exclusion criteria of the articles were adopted using PRISMA [16] protocol. Subsequently, in the second stage, a text mining and visualisation of scientometric review of the final research articles is conducted using VOSviewer. In this stage, the research gaps and prospective research scope is discussed.

3 Findings

3.1 Stage I—Literature Review and Search Strategy

The first objective entails identification of relevant publications related to SNA in construction, wherein the extraction of the documents was from Scopus database with the search query 'social network analysis' and 'construction management' with limitation of the period 2011–2021. This search resulted in 1098 documents. The

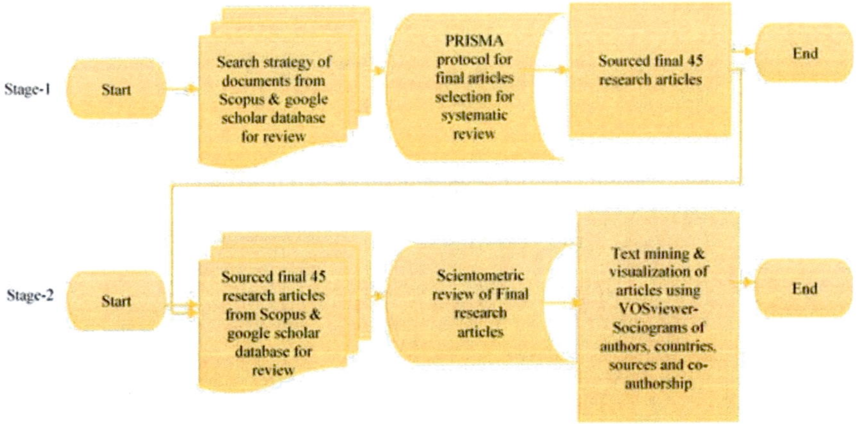

Fig. 1 Overview of research design

Scopus database possesses wider repository of documents hence, was chosen as the source to retrieve the documents. Also, few documents are sourced from google scholar database. On following the PRISMA protocol using the inclusion and exclusion criteria as depicted in Table 1 and Fig. 2. ultimately resulted in total of 45 articles for review.

Table 1 Systematic literature review reporting of inclusion and exclusion criteria

Systematic literature review database reporting	
Search strategy	'***Social network analysis***' and '***construction management***' resulted in **1098** documents. And from google scholar **14** documents Including journal articles, review papers, research reports, conference proceedings
Selection criteria	Based on PRISMA framework. The search is mainly focused on '***construction***' resulted in **644** documents after removing non-relevant and duplicates
Quality assessment	The study is based on original research articles. After the check for duplicate sources, a thorough evaluation of Abstract, Title and keywords was conducted and the search resulted in **137** documents. Subsequently, assessment for the full manuscript was performed and finally, 45 documents were shortlisted for qualitative synthesis. The exclusion criteria at this stage **limiting to English language resulted in 45 documents**
Data extraction	45 articles were selected and the characteristics extracted were: Article must be original paper and journal articles The article must be in English language and from construction domain Extracted articles published between 2011 and 2021

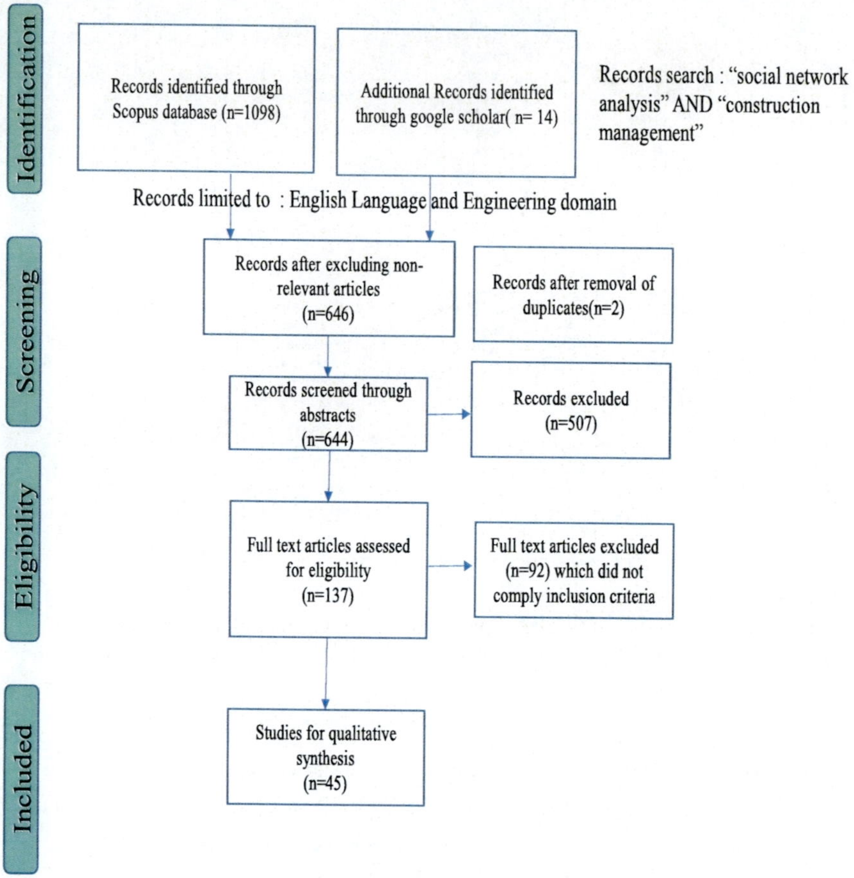

Fig. 2 Stage I sourcing of research articles using PRSIMA protocol

3.2 *Stage II—Scientometric Analysis*

The second objective adopted scientometric review and text mining to analyse the identified articles. Initially, a descriptive statistics are used to understand the publications spread over the last 10 years (spanning from 2011 to 2021) as depicted in Fig. 3. The trend of publications projected a linear increase. Scientometric review comprises visualising of large pool of data with structural associations of a specific domain. This process eliminates the selection of irrelevant articles which might skew the findings. This type of review is used in the research domains such as sustainability, green buildings, BIM and off-site construction, public–private and partnership.

With the use of VOSviewer, a scientometric review is conducted through the database extracted from Scopus. It is a user-friendly software that uses distance-based mode for visualisation [17]. Therefore, the distance between the nodes is indicative of the degree of relatedness it possesses with the nodes. The cluster analysis

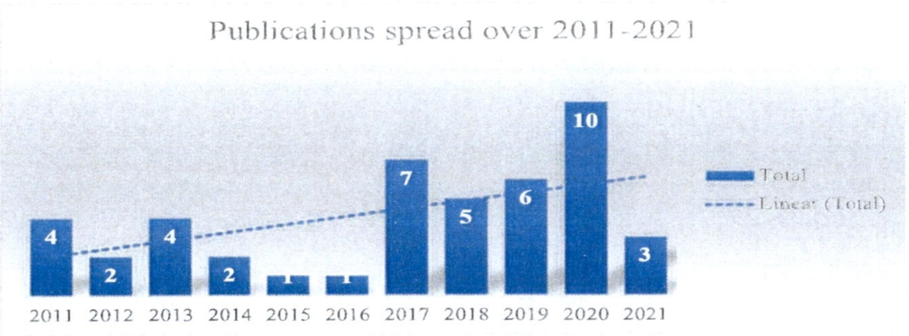

Fig. 3 SNA-construction-based research publications spread over (2011–2021)

of the network or sociogram developed follows an algorithm wherein the items associated with the same cluster are portrayed using the same colour. The current study developed sociogram of co-authorship authors, co-occurrence of keywords, co-authorship countries and citation sources using fractional counting method as recommended by [18].

3.3 Sociogram of Co-occurrence of Keywords

The sociogram in Fig. 4 and the keywords details in Table 2 reflect the popular themes from the extraction of final 45 research articles. The 45 documents comprise of overall 511 keywords wherein the lower limit set is to minimum of three occurrences of keywords and resulted in 35 keywords network map. The sociogram of the keywords co-occurrence as per Fig. 4 is a visualised diagram with nodes (keywords) inter-connected with 293 links and aggregate link strength of 119. This aggregate link strength represents the degree of strength of a parameter to the other parameter.

Each node is portrayed with a colour coding that describes the trend of popularity of the keywords such as purple colour nodes belong to average publication year of 2015, orange colour nodes belong to average publication year of 2017 and yellow colour keywords belong to average publication year of 2015.

3.4 Sociogram of Co-authorship

This sociogram (as per Fig. 5) represents the collaboration network of researchers in the stream of SNA studies in construction. It can be inferred that the collaboration among the researchers associated with their research domain in clusters. There are overall 137 researchers who authored these 45 publications and Fig. 5 depicts the 10 prominent authors and the sociogram possess 64 links and aggregate link strength of

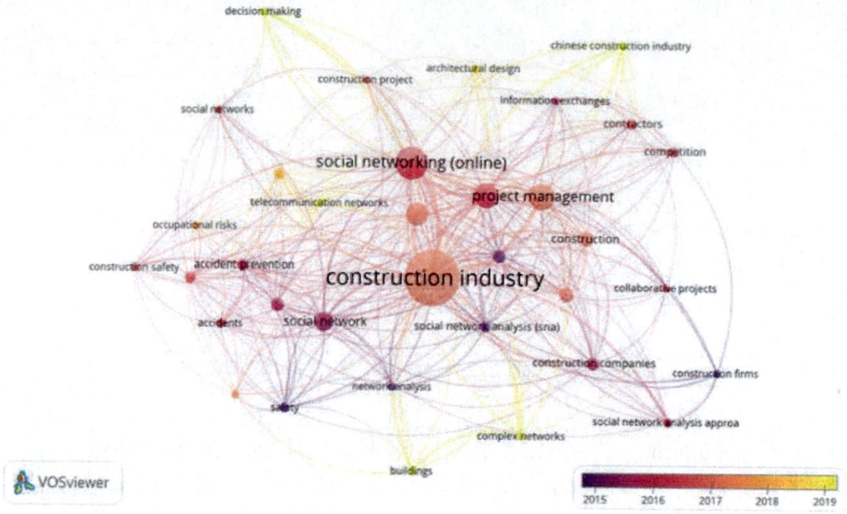

Fig. 4 Keywords co-occurrence sociogram

24.5. The overlay visualisation of the co-authorship sociogram depicts the clusters and the average publication year. The clusters El-adaway, Li. H, Castillo and Alarcon. F are the authors' network whose contemporary (average year of publication is 2018) publications related to social network analysis with emphasis on the construction domain.

3.5 *Sociogram of Countries*

The sociogram in Fig. 6 represents the collaboration network between countries for the research in SNA-based studies pertaining to building sector. The lower limit set for the selection of number of documents from a country is set to 2 and the lower limit for citations is set to 10, which resulted in seven countries meeting the threshold. As per sociogram related to Fig. 6, research affiliations belonging to universities located in United States of America (USA) have contributed substantially towards this research field. This is further followed by China, Hong Kong and Australia.

3.6 *Sociogram of Sources*

The sociogram as depicted in Fig. 7 is a network map that represents the associativity of sources. The lower limit set for the selection of number of documents from a source

Table 2 Frequency of keywords

S. No.	Keyword	Occurrence	Total link strength (TLS)
1	Construction industry	45	43
2	Social networking (online)	22	22
3	Project management	16	16
4	Social network analysis	15	15
5	Construction projects	13	13
6	Social network	10	10
7	construction	7	7
8	Construction management	7	7
9	Communication	6	6
10	construction Companies	6	6
11	Knowledge management	6	6
12	Accident prevention	5	5
13	Competition	5	5
14	Contractors	5	5
15	Human resource management	5	5

is set to 1 and the lower limit for citations is set to 5, which resulted in seven source publications meeting the threshold.

As per sociogram related to Fig. 7, research studies published in Journal of Construction Engineering and Management have contributed substantially towards this research field. This is further followed by Construction Management and Economics and Journal of Management in Engineering.

4 Discussions

The SNA for construction-based research studies gained traction in recent years where more than 67% (30 out 45 articles) of research studies published since 2017. Also, this study highlights the lacuna in diffusion of SNA-based studies in

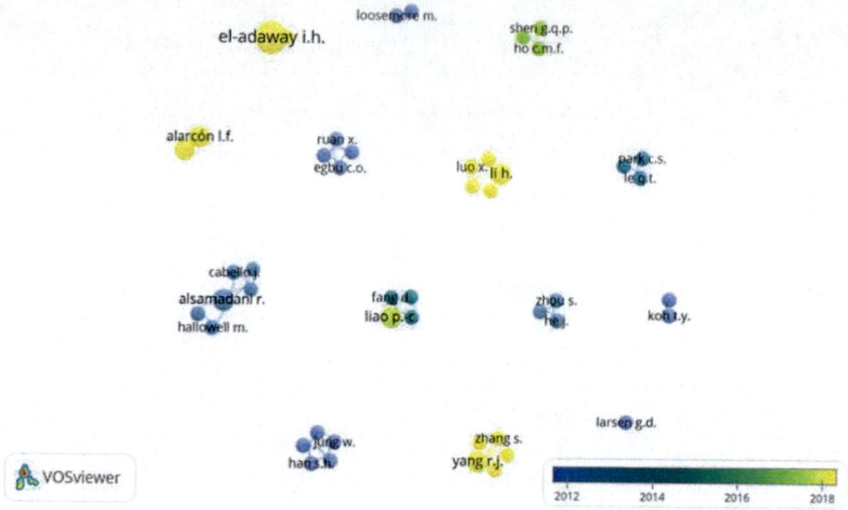

Fig. 5 Co-authorship authors sociogram

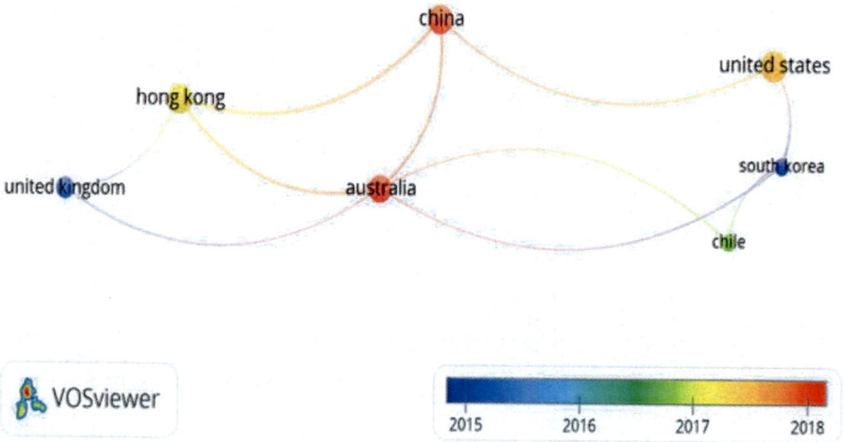

Fig. 6 Co-authorship countries sociogram

the AEC sector. The growing trend of network-based organisations necessitated the use of SNA and this resonates with the findings from Zheng and Xue, X [19, 20]. The current study identified five themes on SNA by performing content analysis of the final 45 documents. The five themes which have been identified were 'individual','Inter-organisational', Intra-organisational', 'project' and'industry'. On examining the timeline trend of the themes as shown in Fig. 8, it is observed that the themes 'individual' and'inter-organisational' category research studies were less concentrated with respect to social network analysis application in construction.

Fig. 7 Citations-sources sociogram

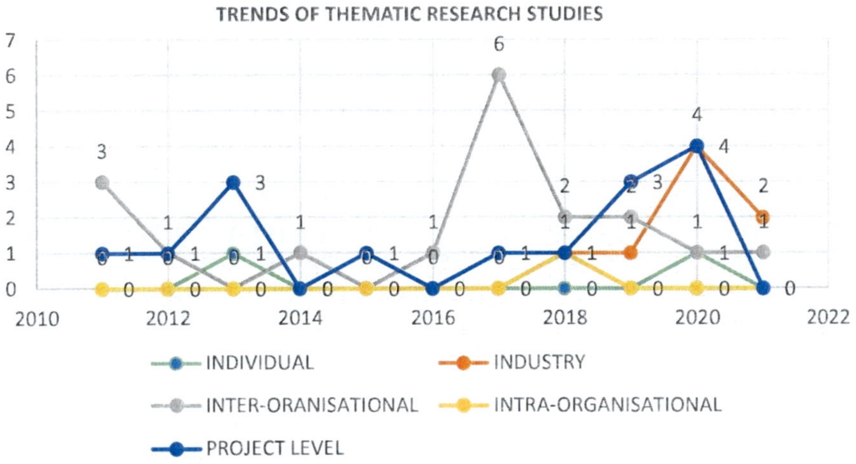

Fig. 8 Trend of thematic research studies

Also, from Fig. 9, out of 45 research articles only two studies were focused on 'mixed methods' and major research studies used 'qualitative' approaches. The author's sociogram from the current study signifies the SNA-based research is in the embryonic stage as there are scattered clusters which indicated the lacuna in collaboration between the researchers.

Also, the co-authorship countries sociogram indicates the maturity level of use of SNA deployment in construction project management investigations is widely adopted in developed nations. Moreover, majority of the construction-grounded

	Year											
Themes	**2011**	**2012**	**2013**	**2014**	**2015**	**2016**	**2017**	**2018**	**2019**	**2020**	**2021**	**Grand Total**
INDIVIDUAL	**0**	**0**	**1**	**0**	**0**	**0**	**0**	**0**	**0**	**1**	**0**	**2**
Qualitative			1							1		2
INDUSTRY	**0**	**0**	**0**	**1**	**0**	**0**	**0**	**1**	**1**	**4**	**2**	**9**
Qualitative				1				1	1	4	1	8
Quantitative											1	1
INTER-ORANISATIONAL	**3**	**1**	**0**	**1**	**0**	**1**	**6**	**2**	**2**	**1**	**1**	**18**
Qualitative	1	1		1		1	5	2	1	1	1	14
Quantitative	2						1		1			4
INTRA-ORGANISATIONAL								**1**				**1**
Qualitative								1				1
PROJECT LEVEL	**1**	**1**	**3**	**0**	**1**	**0**	**1**	**1**	**3**	**4**	**0**	**15**
Mixed Methods							1			1		2
Qualitative	1		3		1				1	1		7
Quantitative		1						1	2	2		6
Grand Total	**4**	**2**	**4**	**2**	**1**	**1**	**7**	**5**	**6**	**10**	**3**	**45**

Fig. 9 Methodology concept matrix for the 45 literature articles

research studies used SNA to the domains such as construction safety (10 articles), knowledge management (5 articles), stakeholder (5 articles) and collaboration management (4 articles). The research in this domain inclines towards adoption of theoretical lenses such as 'graph theory', 'social learning theory' and 'information theory'. Furthermore, the co-authorship countries sociogram demonstrates that the key researchers of top citations hail from developed nations.

Social network analysis is a key tenet is to comprehend the latent complex structural connections among the entities such as individuals, inter-organisational, intra-organisational, industry and project stakeholders through network diagrams for complex construction projects. Research studies in developed countries observed a positive impact through SNA application in project performance assessment on analysing the information exchange during 2007–2011 using network models and network interdependencies at project level [21–23], responsibilities of cultural boundary spanners in project performance for international project networks [24, 25], project internal stakeholders communication [26], cost related interactions in the setting of a network characterised by improved information sharing and coordination [27]. Moreover, numerous investigations are proponent of SNA as it evaluates the latent measures such as relational competence [28], last planner system [29] and structural holes [30]. These measures intangible are softer aspects and possess an indirect effect in the achievement of traditional project management golden triangle attributes such as time, cost and quality.

As SNA applications are considerably apportioned with people skill aspects, this requires special attention in developing nations such as India which is substantially labour intensive. The assimilation of social network theory can be induced through training and followed by mandating the applications of SNA in their regular construction operations processes for effective implementation. Moreover, network theory

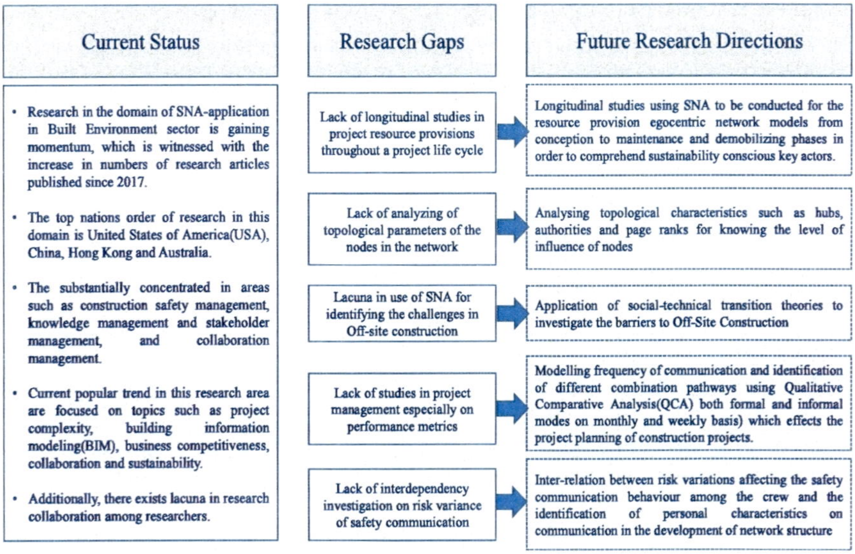

Fig. 10 Gaps and future scope in SNA-construction-based research

can be applied in construction projects of developing nations where human manifestation is indispensable as they enable understanding the complex relations among different workmen in safety management, workforce communication and cultural aspects, information exchange and coordination domains.

4.1 Future Scope of Research

With the review of the literature, there are limited empirical and mixed methods research studies in the SNA-construction-based studies. The summary of research gaps from the literature review and corresponding future directions are displayed in Fig. 10.

5 Conclusions

One of the most influential areas towards the success/ failure of construction projects is the effective, seamless interactions, connections and interrelationships between the individuals, groups and organisations connected with the projects. Social Network Analysis (SNA) approach for construction projects can help in understanding these complex relationships and yield significant benefits, contribute substantially to defragmenting the presently silo structured, isolated interactions in construction projects. The present study has introduced the concept of SNA and has also brought out the various application areas, themes, leading countries and journals coming out with studies on SNA. There are numerous areas and lessons which can be learnt by the developing countries from the developed countries which have already progressed significantly on the research and application of SNA to meaningfully utilise the same for the benefit of the industry. The present study has also identified the areas of present research and gaps such as application in project management, off-site construction, etc. which are potential unexplored areas and application in these areas can bring in improvements to the industry.

References

1. Latham M (1994) Constructing the team: final report of the government/industry review of procurement and contractual arrangements in the UK construction industry. HMSO, London, UK
2. Egan J (1998) Report: department of the environment, transport and the regions (UK): the scope for improving the quality and efficiency of UK construction. DETR Home Pages: http://www.construction.detr.gov.uk/cis/rethink/index.htm
3. Mitropoulos P, Tatum CB (1999) Technology adoption decisions in construction organizations. J Constr Eng Manage 125(5):330–338
4. Braun T, Ferreira AI, Sydow J (2013) Citizenship behavior and effectiveness in temporary organizations. Int J Project Manage 31(6):862–876
5. Limsila K, Ogunlana SO (2008) Performance and leadership outcome correlates of leadership styles and subordinate commitment. Eng Constr Archit Manage 15(2):164
6. Lin SC (2015) An analysis for construction engineering networks. J Constr Eng Manage 141(5):1–13
7. Stephen D (2005) Pryke : towards a social network theory of project governance. Constr Manage Econ 23(9):927–939
8. Hossain L (2009) Communications and coordination in construction projects. Constr Manage Econ 27(1):25–39
9. Pryke S, Pearson S (2006) Project governance: case studies on financial incentives. Build Res Inf 34(6):534–545
10. Al Hattab M, Hamzeh F (2015) Using social network theory and simulation to compare traditional versus BIM–lean practice for design error management. Autom Constr 52:59–69
11. Chinowsky P, Diekmann J, Galotti V (2008) Social network model of construction. J Constr Eng Manage 134(10):804–812
12. Cao D, Li H, Wang G, Luo X, Tan D (2018) Relationship network structure and organizational competitiveness: evidence from BIM implementation practices in the construction industry. J Manag Eng 34(3):1–12
13. Badi S, Wang L, Pryke S (2017) Relationship marketing in Guanxi networks: a social network analysis study of Chinese construction small and medium-sized enterprises. Ind Mark Manage 60:204–218
14. Gan X, Chang R, Wen T (2018) Overcoming barriers to off-site construction through engaging stakeholders: a two-mode social network analysis. J Clean Prod 201:735–747
15. Braun V, Clarke V (2012) Thematic analysis
16. Moher D, Altman DG, Liberati A, Tetzlaff J (2011) PRISMA statement. Epidemiology 22(1):128
17. Van Eck NJ, Waltman L (2011) Text mining and visualization using VOSviewer. ISSI Newsl 7:50–54

18. Van Eck NJ, Waltman L (2010) Software survey: VOSviewer, a computer program for bibliometric mapping. Scientometrics 84:523–538
19. Zheng X, Le Y, Chan AP, Hu Y, Li Y (2016) Review of the application of social network analysis (SNA) in construction project management research. Int J Project Manage 34(7):1214–1225
20. Xue X, Zhang X, Wang L, Skitmore M, Wang Q (2018) Analyzing collaborative relationships among industrialized construction technology innovation organizations: a combined SNA and SEM approach. J Clean Prod 173:265–277
21. Chinowsky PS, Diekmann J, O'Brien J (2010) Project organizations as social networks. J Constr Eng Manag 136(4):452–458
22. Chinowsky P, Taylor JE, Di Marco M (2010) Project network interdependency alignment: new approach to assessing project effectiveness. J Manag Eng 27(3):170–178
23. Chinowsky P, Diekmann J, Galotti V (2008) Social network model of construction. J Constr Eng Manage 134(10):804–812
24. Comu S, Unsal HI, Taylor JE (2011) Dual impact of cultural and linguistic diversity on project network performance. J Manag Eng 27(3):179–187
25. Di Marco MK, Taylor JE, Alin P (2010) Emergence and role of cultural boundary spanners in global engineering project networks. J Manage Eng 26(3):123–132
26. Mead SP (2001) Using social network analysis to visualize project teams. Proj Manag J 32(4):32–38
27. Shields R, West K (2003) Innovation in clean-room construction: a case study of co-operation between firms. Constr Manag Econ 21(4):337–344
28. Pauget B, Wald A (2013) Relational competence in complex temporary organizations: the case of a French hospital construction project network. Int J Project Manage 31(2):200–221
29. Heng HKS, Loosemore M (2013) Structural holes in hospital organisations: facilities managers as intrapreneurial brokers in the tertiary health sector. Eng Constr Archit Manage

Interference Effect Between Tall Buildings with Square Cross-Sections

Varun Aneja and Neelam Rani

1 Introduction

Due to the increase in population of world, there is limited land available to construct new buildings and to provide sufficient accommodation to this large population there is need to construct tall buildings in densely populated cities. But as the height of building is increasing, the wind load consideration becomes more important. To design a building in isolated condition in open terrain, we can take guidance from respective country codes, i.e., AS/NZS: 1170.2 (2002), ASCE: 7-02-2002, EN: 1991-1-4-2005, IS: 875 (Part-3) 2015. Many researchers have carried out their studies to explore the effect of interfering building on the principal building. In the present study, the G+ 10 story building is analyzed for bending moment and deflection in various wind incidence angles 0°, 30°, 60°. Also, wind load on the principal building is calculated by varying the vertical interference distance as ($y = 0$, $0.5b$, b, $3b/2$, $2b$) between interfering building and principal building and a detailed comparison is done. Furthermore, optimum location of shear wall is also found by comparing different locations of shear wall. Shear walls are introduced to limit the deflection because shear walls are a sort of structural system that has lateral resistance to the building or structure. They are vertical components of the structure, i.e., the horizontal force resisting system. Lateral masses that are engaged in the structure are counteracted with the help of shear walls. But there are limited studies available to find out the optimum location of shear wall when interference building is present near reference building and when the wind is incident at angle other than 0°.

V. Aneja (✉) · N. Rani
Department of Civil Engineering, Dr. B.R. Ambedkar National Institute of Technology, G. T. Road, Amritsar Bypass, Jalandhar, Punjab 144011, India
e-mail: varunaneja888@gmail.com

N. Rani
e-mail: ranin@nitj.ac.in

A. Sil et al. (eds.), *Recent Trends in Civil Engineering*, Lecture Notes in Civil Engineering 274, https://doi.org/10.1007/978-981-19-4055-2_4

2 Literature Review

Reference [1] presented the effect of height variation of two interfering buildings on a rectangular-shaped principal building by varying height of one or both of interfering buildings and concluded that interference effect tends to increase as height of interfering building increases. Reference [2] analyzed the previous work on interference effect and concluded that reference building is highly affected if interfering building is closely located also interference effect and depends on arrangement and geometry of buildings. Reference [3] presented the results of wind loading on a '+' plus-shaped tall building. The wind pressure coefficients were studied for wind incidence angle of 0° and 45°. They concluded that wind pressure distribution on irregular plan-shaped building with the change in wind angle negative pressure values changes drastically. Reference [4] studied the effect of variation of height of interfering building and come to conclusion that with increase in height ratio of interfering building value of pressure coefficient also increased, and this increase in pressure is more for oblique configuration. Reference [5] studied the interference effect for various wind incidence angles and for two different configuration of building and concluded that upstream interfering building has significant effect on interference factor than building present at downstream of principal building. Reference [6] analyzed the G+ 4 story building by changing the location of shear wall on different shapes of building and concluded that for square shaped building with shear wall provided on inner core then steel required will be more than in case when shear wall is provided at outer faces. Reference [7] concluded that due to the presence of two interfering buildings can drastically increase the interference effect on reference building. Mean interference effect by two interfering buildings and concluded that interference factor is sensitive to breadth of interfering building. Reference [8] extended their previous work and came to conclusion that when height ratio of interfering building is 0.5 then it acts like an isolated building. Reference [9] studied the wind-induced interference effects on two adjacent high-rise buildings and concluded that mean torsion increased by three times when interference effect is present. Reference [10] presented computational fluid dynamics study for a wind incidence angle of 30 degree and compared the result of pressure coefficients on building walls among the three buildings and an isolated single building. Reference [11] explained the updates in the third revision of the Indian wind loading standard (IS 875 (Part 3), 2015).

3 Modeling and Loading Details

The effect of shear wall with varying wind incidence angles has been analyzed in this study. The building considered is having 4 × 4 bay with 10 storeys. All building frames are made of R.C.C. beams, columns, and slabs. The size of beams and columns is determined through dead and live load analysis. Basic wind speed is taken as 50 m/s and the building is assumed to be in Terrain category 3 (Table 1).

Table 1 Building configuration

S. No.	Particulars	Value
1	Story height	3.5 m
2	No. of story	10
3	Size of beam	400 mm × 400 mm
4	Size of column	450 mm × 400 mm
5	No. of bays along x-axis	4
6	No. of bays along y-axis	4
7	Terrain category	3
8	Thickness of slab	150 mm
9	Thickness of shear wall	250 mm

There are ***8 cases*** on which the present study is mainly focused, for all the cases mentioned below deflection and bending moment at column A which is present at the center of the building are taken.

(a) When wind incidence angle is 0° (Fig. 1).
(b) When wind incidence angle is 30° (Fig. 1).
(c) When wind incidence angle is 60° (Fig. 1).
(d) For interference configuration $y = 0$ (Fig. 2).
(e) For interference configuration $y = b/2$ (Fig. 2).
(f) For interference configuration $y = b$ (Fig. 2).
(g) For interference configuration $y = 3b/2$ (Fig. 2).
(h) For interference configuration $y = 2b$ (Fig. 2).

Different wind incident angle

See Fig. 1.

Different interference configuration

See Fig. 2.

Interference Configuration

See Fig. 3.

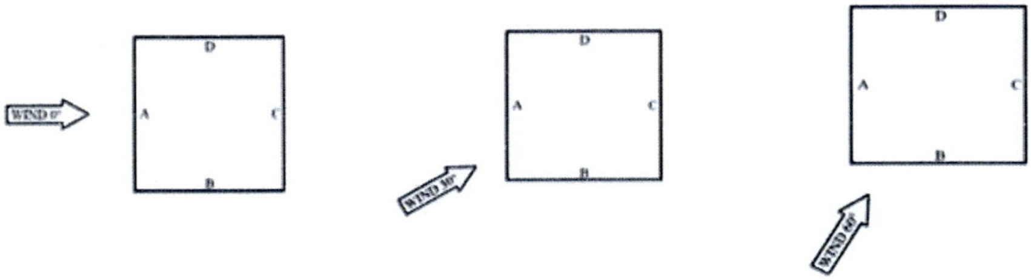

Fig. 1 Showing wind incidence angle of 0°, 30°, 60° (right to left)

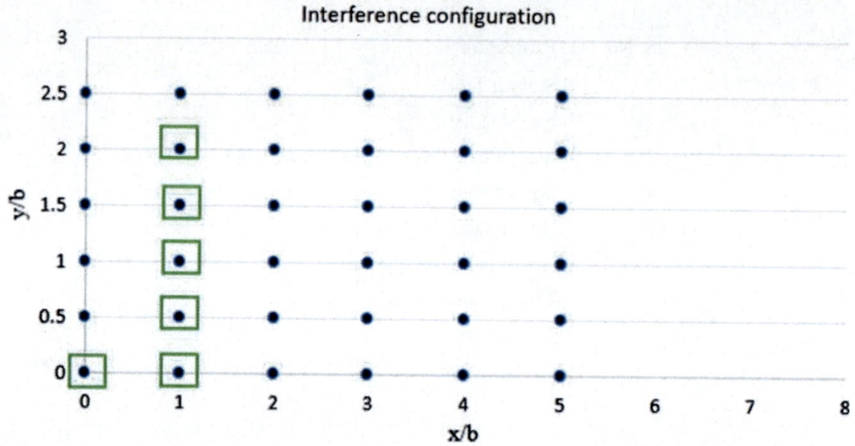

Fig. 2 Interference configuration in graphical form, when principal building is at (0, 0)

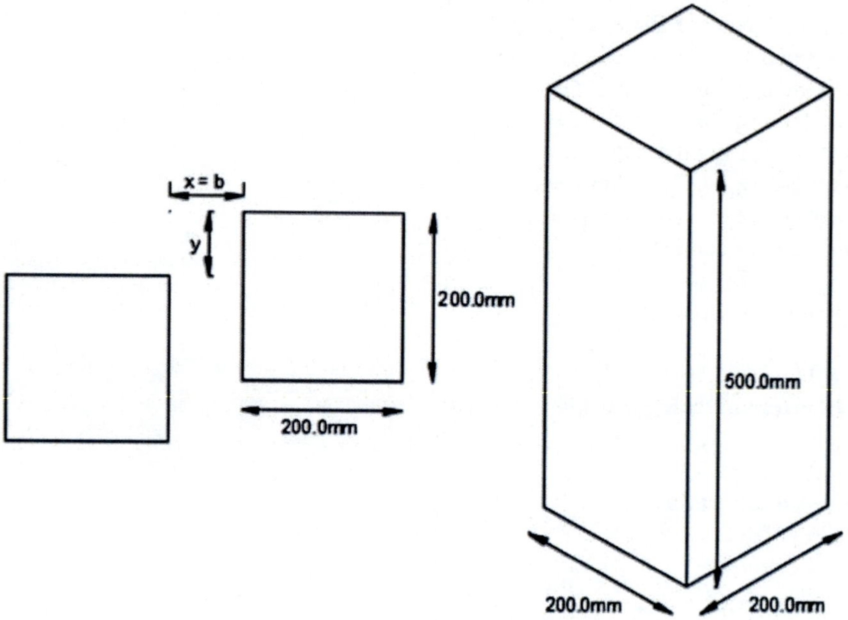

Fig. 3 Interference configuration for model and isometric view of the model to be tested in wind tunnel

Where y is the vertical distance between the top of principal building to top of interfering building and x is the horizontal distance between two buildings and is equal to the width of the building (b) for all cases. Reference [12] effect of wind in presence of an upstream building or several upstream buildings is a very complex scenario and needs careful understanding. Pressure contours for all the cases are obtained with the help of wind tunnel studies and from these pressure contours values of coefficient of pressures (C_p) for different faces of model are obtained and later force at each node is calculated and then using STADD.Pro V8i software deflection and bending moment are calculated at central column A in plan.

In all the above-mentioned cases value of coefficient of pressure is taken from pressure contours that are obtained from wind tunnel test and then with the help of the following formulas.

$$P_z = 0.6 \times (V_d)^2 \tag{1}$$

$$P_d = K_d \times K_a \times K_c \times P_z \tag{2}$$

P_z = wind pressure at height z, in N/m^2; and P_d = design wind pressure

$$F = \left(C_{pe} - C_{pi}\right) A\ P_d \tag{3}$$

where

C_{pe} = External pressure coefficient, C_{pi} = Internal pressure coefficient.

A = Surface area of a structural element or cladding unit, and F = Wind force.

There are **3 cases** on which the present study is mainly focused, for all the cases mentioned below deflection and bending moment at column A which is present at center of the building is taken.

Case 1: Without shear wall

In all the eight cases of wind incident angle and interference configuration, we have three sub-cases and one of them is when no shear wall is provided, this will give the actual results of all parameters (Figs. 4 and 5).

Case 2: Shear wall is provided at all 4 middle spans of the outer faces of the building

As compared to scenario 1 where there is no shear wall, now shear wall is provided and the result difference is observed and different locations of shear wall is also checked for various parameters (Fig. 6).

Case 3: Shear wall is provided at centre core of the building

Shear wall of uniform thickness (250 mm) throughout the height of building is provided at central core of building, and results are compared with case 1 and case 2 (Fig. 7).

Loading details: While calculating wind load risk coefficient factor (***K1***), topography factor (***K3***) and importance factor for cyclonic region (***K4***) are equal to 1.0 according to IS 875 (Part-3) [13]. Factor for variation of wind speed with height (***K2***) is calculated at each story level by doing interpolation and from all these values force generated by wind is calculated. Live load is taken as per IS: 875 (Part-2) [14]. The

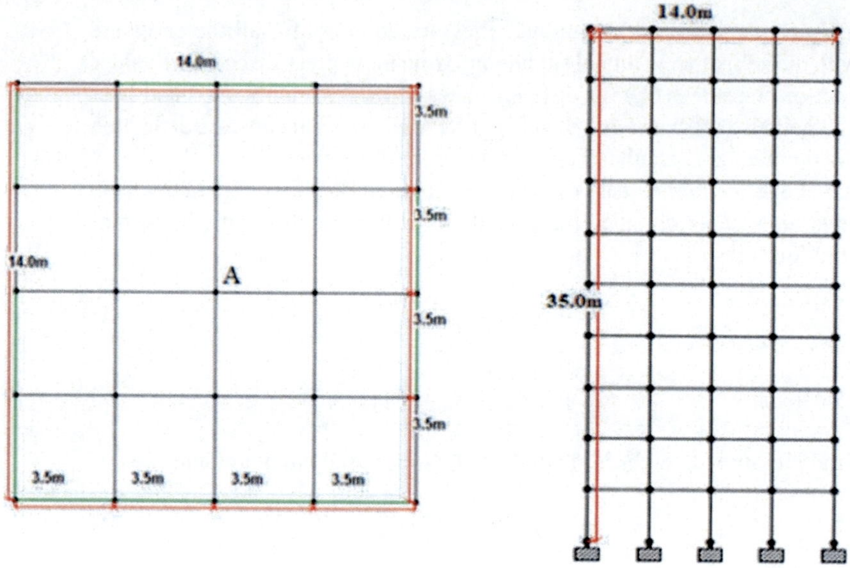

Fig. 4 Plan and elevation view when shear wall is not provided

Fig. 5 Three-dimensional view when shear wall is not provided

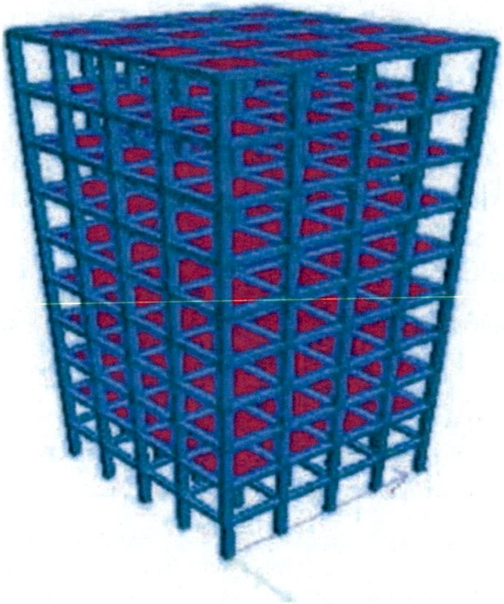

load combination of 1.2 * (DL + LL + WL) is used for analysis purpose. Where DL = Dead Load, LL = Live Load and WL = Wind Load. [15] wind load on the building comes out to be large when it has maximum exposed area.

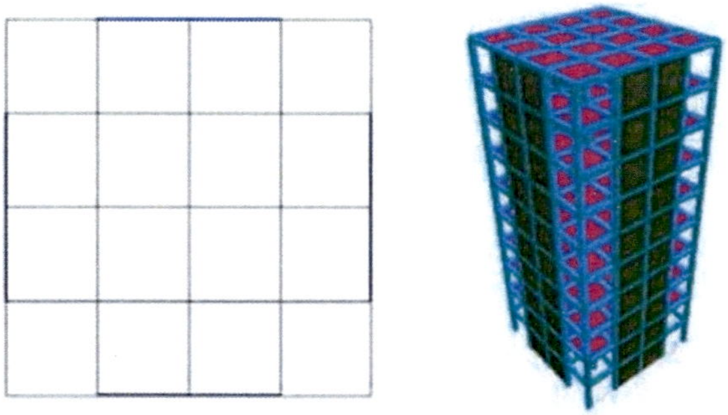

Fig. 6 Plan and 3D view of shear wall

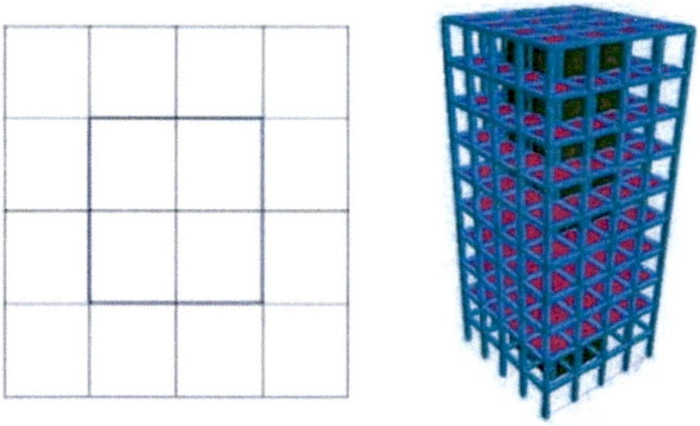

Fig. 7 Plan and 3D view of shear wall at central core of building

Wind pressure varying from 0.869 kN/m^2 at 3.5 m height to 1.21 kN/m^2 at 35 m height with different values of coefficient of pressure is applied on A and C face of building. And the force generated by this wind pressure is shown in Fig. 8.

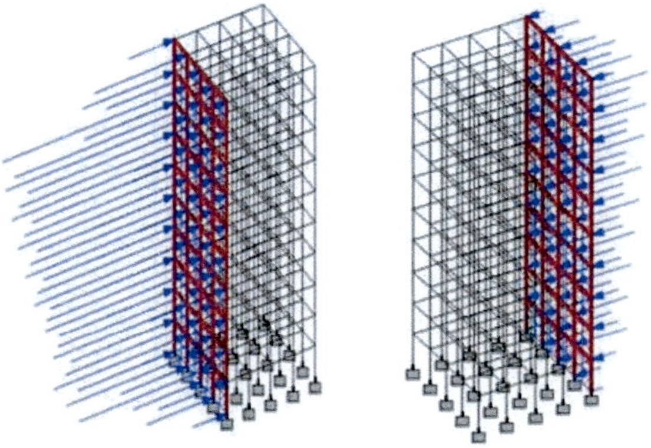

Fig. 8 Wind load in positive x-direction and negative x-direction

4 Observations and Results

In order to study the influence of different wind angles and interference effect on principal building, deflection of central column is observed and compared with deflection when different locations of shear wall is used.

It is clearly observed from Table 2 that when wind hits building at 0°, the deflection produces minimum. As the wind incident angle increase from 0°, the deflection value also increases and the maximum value is for 60° wind (Table 3).

When shear wall is provided at all four middle spans of outer faces of building deflection start to decrease from its normal value whatever be the wind angle. Maximum reduction is seen when wind incident angle is 60°, and its value is 50% (Table 4).

Table 2 Difference in deflection values for different wind angles (0°, 30°, 60°) for case-1

Height (m)	Deflection (0°) (mm)	Deflection (30°) (mm)	Deflection (60°) (mm)
35	13.005	18.269	27.185
31.5	12.891	17.999	26.515
28	12.511	17.353	25.376
24.5	11.867	16.322	23.665
21	10.956	14.915	21.401
17.5	9.778	13.145	18.618
14	8.333	11.031	15.357
10.5	6.622	8.591	11.679
7	4.648	5.846	7.645
3.5	2.409	2.845	3.365

Table 3 Reduction in deflection values for case-2

Height (m)	Deflection(0°) (mm)	Deflection (30°) (mm)	Deflection (60°) (mm)
35	12.184	12.471	13.499
31.5	12.067	12.336	13.301
28	11.715	11.952	12.807
24.5	11.116	11.313	12.028
21	10.274	10.429	11
17.5	9.184	9.302	9.734
14	7.843	7.929	8.241
10.5	6.248	6.306	6.517
7	4.396	4.431	4.555
3.5	2.283	2.296	2.338

Table 4 Reduction in deflection values for case-3

Height (m)	Deflection (0°) (mm)	Deflection (30°) (mm)	Deflection (60°) (mm)
35	6.694	6.831	7.327
31.5	6.703	6.819	7.237
28	6.576	6.669	7.009
24.5	6.318	6.387	6.648
21	5.916	5.967	6.161
17.5	5.367	5.403	5.54
14	4.661	4.685	4.775
10.5	3.783	3.797	3.852
7	2.716	2.724	2.752
3.5	1.435	1.437	1.447

When shear wall is provided at central core deflection decrease from its normal value by large extent whatever be the wind angle. Maximum reduction is seen in case of 60 wind, and its value is 72%.

It is observed from Table 5 that when distance between interference building and principal building is $y = 0$, then the deflection produces maximum and we can say that as the distance between the concerned buildings increases deflection decreases (Table 6).

When shear wall is introduced at all four middle spans of outer faces of building then deflection at central column is decreased by large amount when both the buildings are at same level and maximum deflection is obtained when $y = 0.5b$ (Table 7).

Table 5 Difference in deflection values for different interference configuration (case-1)

Height	Deflection ($Y = 0$)	Deflection ($Y = 0.5b$)	Deflection ($Y = B$)	Deflection ($Y = 3/2b$)	Deflection ($Y = 2b$)
35	15.242	14.527	13.934	13.17	13.041
31.5	15.058	14.369	13.8	13.581	12.917
28	14.583	13.932	13.386	13.162	12.53
24.5	13.828	13.205	12.68	12.46	11.883
21	12.743	12.177	11.673	11.477	10.971
17.5	11.326	10.841	10.373	10.215	9.792
14	9.591	9.2	8.801	8.677	8.345
10.5	7.545	7.259	6.964	6.868	6.633
7	5.203	5.029	4.862	4.792	4.665
3.5	2.601	2.54	2.489	2.459	2.411

Table 6 Reduction in deflection values for different interference configuration (case-2)

Height	Deflection ($Y = 0$)	Deflection ($Y = 0.5b$)	Deflection ($Y = B$)	Deflection ($Y = 3/2b$)	Deflection ($Y = 2b$)
35	12.297	13.499	12.224	12.218	12.183
31.5	12.172	13.301	12.107	12.101	12.065
28	11.808	12.807	11.751	11.744	11.711
24.5	11.199	12.028	11.147	11.14	11.113
21	10.346	11	10.297	10.292	10.27
17.5	9.241	9.734	9.201	9.198	9.182
14	7.887	8.241	7.855	7.853	7.842
10.5	6.279	6.517	6.257	6.255	6.247
7	4.415	4.555	4.403	4.4	4.396
3.5	2.29	2.338	2.286	2.285	2.283

Table 7 Reduction in deflection values for different interference configuration (case-3)

Height	Deflection ($Y = 0$)	Deflection ($Y = 0.5b$)	Deflection ($Y = B$)	Deflection ($Y = 3/2b$)	Deflection ($Y = 2b$)
35	6.755	7.327	6.703	6.72	6.694
31.5	6.753	7.237	6.722	6.719	6.703
28	6.618	7.009	6.592	6.589	6.578
24.5	6.347	6.648	6.328	6.326	6.317
21	5.939	6.161	5.924	5.922	5.915
17.5	5.383	5.54	5.372	5.371	5.366
14	4.672	4.775	4.665	4.664	4.661
10.5	3.79	3.852	3.785	3.785	3.783
7	2.72	2.752	2.718	2.717	2.716
3.5	1.436	1.447	1.435	1.435	1.434

When shear wall is introduced at central core of building, then deflection at central column is decreased by large amount when both the buildings are at same level, i.e., at $y = 0$ and maximum deflection is obtained when $y = 0.5b$.

In Fig. 9, there are two types of comparison shown:

1. With different wind angles and for two different positions of shear wall.
2. With different interference configurations and for two different positions of shear wall.

When the wind is directed at the building at 0°, bending moment value is minimum and is negative. But as the wind incidence angle increases, the value of bending moment also increased and the maximum value occurs for 60° wind. When shear wall is included in building bending moment tends to reduce by large amount as shown in Fig. 9.

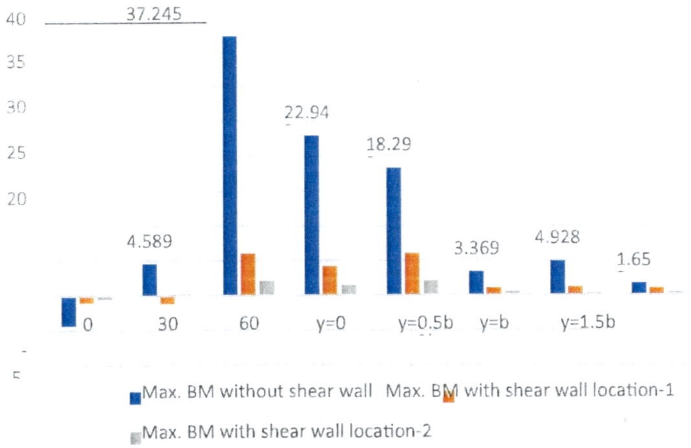

Fig. 9 Maximum bending moment for different cases of wind incident angle and various interference configurations

5 Conclusions

Based on the analysis, the following conclusions have been obtained:

For different wind incidence angles:

(1) Deflection is less when wind incident angle is 0° and is found to increase as we increase the wind angle.
(2) When shear wall is provided at all 4 middle spans of the outer faces of the building then deflection decrease by a significant amount in case of 60°. Small change is observed in values of 0° and 30°.
(3) When shear wall is provided at central core of building deflection is reduced by 50% in all cases of wind incidence.
(4) Maximum reduction in maximum bending moment value is observed when wind incident angle is 60°.
(5) Bending moment is observed to reduce more when shear wall is present at central core of the building.
(6) When shear wall is present at the central core of the building, it increases the amount of area of steel required by the components.

For different interference configurations:

(1) Deflection tends to decrease as the vertical distance between the top of the principal building to the interfering building increased from $y = 0$ to $y = 2b$.
(2) When shear wall is introduced at middle spans of outer faces, then deflection at central column decreases by a small amount.
(3) When shear wall is introduced at central core of the building, then deflection is reduced by a significant amount.

(4) Maximum reduction in maximum bending moment value is observed when interference distance is $y = 0$.
(5) Optimum location of shear wall for square-shaped high-rise building under wind loading is when it is provided at the central core of the building.

References

1. Chauhan BS, Ahuja AK (2017) Height effect of interfering buildings on wind pressure distribution on rectangular plan tall buildings. In: 9th Asia Pacific conference on wind engineering APCWE 2017, December, pp 3–6
2. Khanduri AC, Stathopoulos T, Bédard C (1998) Wind-induced interference effects on buildings—a review of the state-of-the-art. Eng Struct 20(7):617–630. https://doi.org/10.1016/S0141-0296(97)00066-7
3. Chakraborty S, Dalui SK, Ahuja AK (2014) Wind load on irregular plan shaped tall building—a case study. Wind Struct An Int J 19(1):59–73. https://doi.org/10.12989/was.2014.19.1.059
4. Kim W, Tamura Y, Yoshida A (2011) Interference effects on local peak pressures between two buildings. J Wind Eng Ind Aerodyn 99(5):584–600. https://doi.org/10.1016/j.jweia.2011.02.007
5. Hui Y, Yoshida A, Tamura Y (2013) Interference effects between two rectangular-section high-rise buildings on local peak pressure coefficients. J Fluids Struct 37:120–133. https://doi.org/10.1016/j.jfluidstructs.2012.11.007
6. Rajoriya S (2013) Study of the effect of shape of the structure and the position of shear wall in a high rise building under wind load, pp 18534–18546. https://doi.org/10.15680/IJIRSET.2016.0510059
7. Xie ZN, Gu M (2004) Mean interference effects among tall buildings. Eng Struct 26(9):1173–1183. https://doi.org/10.1016/j.engstruct.2004.03.007
8. Kim W, Tamura Y, Yoshida A (2015) Interference effects on aerodynamic wind forces between two buildings. J Wind Eng Ind Aerodyn 147:186–201. https://doi.org/10.1016/j.jweia.2015.10.009
9. Hui Y, Tamura Y, Yang Q (2016) Analysis of interference effects on torsional moment between two high-rise buildings based on pressure and flow field measurement. J Wind Eng Ind Aerodyn 2017(164):54–68. https://doi.org/10.1016/j.jweia.2017.02.008
10. Yu X, Xie Z, Gu M (2018) Interference effects between two tall buildings with different section sizes on wind-induced acceleration. J Wind Eng Ind Aerodyn 182(September):16–26. https://doi.org/10.1016/j.jweia.2018.09.012
11. Kumar KS (2020) J Wind Eng Indust Aerodyn Wind load Tall build Rev Indian Standards Recommen Amendm 204
12. Goliya R (2016) Current status of interference effect studies on tall buildings. https://doi.org/10.3850/978-981-07-8012-8
13. Standards B of I (2015) IS 875: code of practice for design loads (other than earthquake) for building and structures—part 3 : wind loads. BIS New Delhi 875(1):51
14. IS 875 : 1987 (1987) Code of practice for design loads (other than earthquake) for buildings and structures, part 2: imposed loads. Bur Indian Stand New Delhi. Published online 1987:18
15. Roy AK, Singh J (2017) Materials I. Wind Load High Rise build Diff Config A Critic Rev
16. Staad pro V8i, 3-D Structural analysis and design software

Investigation of Crushed Concrete Aggregates as a Replacement for Natural Aggregates in AAB Concrete

Sriman Pankaj Boindala, Kruthi Kiran Ramagiri, and Arkamitra Kar

1 Introduction

The construction industry is one of the most important industries for a developing country like India as it influences national growth and is a major driver of foreign investment in related industries. Since Portland cement (PC) concrete is a popular building material, demand for it rises in tandem, consequently augmenting its constituents' requirements. It is anticipated that India's planned infrastructural development over the next decade will increase PC production to 800 million tons per annum [1]. This huge amount of PC production results in significant carbon emissions. It is established that 1 ton of PC production results in 0.98 tons of equivalent carbon dioxide emission [2]. The emissions associated with PC production can be reduced by using alternative binders that can partially or completely substitute PC. Alkali-activated binder (AAB) is one such potential alternative that has shown comparable or even superior mechanical and durability properties compared to PC [3].

AAB is produced by the reaction between an aluminosilicate-rich precursor such as fly ash (FA) or ground granulated blast furnace slag (SG) and an alkaline activating solution [3]. The chemistry and the associated reaction mechanisms to form hardened AAB vary with the composition of precursor and activating solution [2, 3]. This complexity in chemistry and the lack of information on long-term properties lead to major hindrances in the widespread practical application of AAB concrete [4]. FA-based AAB concrete requires thermal curing in the range of 60–80 °C to achieve

S. P. Boindala
Civil and Environmental Engineering, Technion-Israel Institute of Technology, Haifa, Israel

K. K. Ramagiri · A. Kar (✉)
Department of Civil Engineering, BITS-Pilani Hyderabad Campus, Hyderabad, India
e-mail: arkamitra.kar@hyderabad.bits-pilani.ac.in

K. K. Ramagiri
e-mail: p20170008@hyderabad.bits-pilani.ac.in

A. Sil et al. (eds.), *Recent Trends in Civil Engineering*, Lecture Notes in Civil Engineering 274, https://doi.org/10.1007/978-981-19-4055-2_5

desired mechanical properties, limiting its usage to precast applications [5, 6]. Hence one of the objectives of this study is the development of AAB concrete under ambient-curing conditions.

Another issue associated with the rapid development of global infrastructure is construction and demolition waste (CDW). CDW is dumped into landfills, rendering otherwise usable land unfit for agricultural purposes and impeding sustainable development. Approximately 150 million tons of CDW are generated in India per year [7]. As a result, developing an environmentally sustainable and commercially feasible approach to reuse these CDW is critical. About 90% of the CDW comprises building-derived materials (BDM) such as concrete and bricks. The initiative for recycling CDW began in the early 1980s [8]. The incorporation of recycled coarse aggregates (RCA) in concrete reduced carbon emissions by up to 65% and non-renewable energy consumption by about 58% [9]. According to Serres et al. [10], the energy consumed in RCA concrete and PC concrete productions are 1.39×10^3 MJ and 2.14×10^3 MJ, respectively, and the carbon emissions are 3.35 kg CO_2 /ton and 4.44 kg CO_2 /ton, respectively.

Furthermore, the use of aggregates from CDW has also been investigated with AAB concrete. According to Chindaprasirt and Cao (2015) [11], natural aggregates can be substituted to achieve strengths of more than 70 MPa. The reduction in compressive strength of AAB concrete with recycled aggregates is found to be relatively smaller compared to their conventional concrete counterparts. Analogous to the trend observed in PC concrete, a decline in the strength of FA-based AAB concrete with an increase in percentage substitution of RCA is reported [13]. Nuaklong et al. [14] stated that high calcium FA-based AAB concrete with 100% RCA achieved 78–93% strength compared to limestone-based AAB concrete and suggested optimal molarity of sodium hydroxide (NaOH) as 12. The cracking and susceptibility of recycled aggregate concrete to permeation, diffusion, and absorption of fluids are linked to the cracks and fissures that develop during production. These disadvantages hinder the use of recycled aggregate in reinforced PC concrete at higher percentages (>30%). However, many previous studies reported that AAB concrete has a greater potential for high-volume substitution of RCA than PC concrete without comprising the mechanical properties or durability [11, 15, 16]. The interfacial transition zone in AAB with RCA has higher bond strength compared to AAB with natural aggregates (NA), PC with NA, and PC with RCA, indicating a greater potential to use RCA in AAB [15].

The replacement of NA with RCA in alkali-activated slag mortar significantly reduced the autogenous shrinkage without affecting its strength [12]. The equivalent durability properties with an increase in RCA percentage replacement are attributed to finer binder content in AAB compared to PC. In addition to replacement, researchers focused on improving the properties of BDM with treatments that will improve the properties of concrete made with it. Treatment with supplementary cementitious materials and treatment with carbon dioxide was investigated by Guo et al. [17]. They concluded that treatment with fly ash or other mineral admixtures increases the strength due to reaction between unreacted Ca $(OH)_2$ and fly ash which forms secondary C-S–H matrix. Due to carbonation, additional $CaCO_3$ is formed which can react with the new mortar and increase the strength.

Most of the research with recycled aggregates was done with conventional concrete. The majority of previous studies have indicated that the optimal replacement percentage for PC concrete is about 30%. Untreated BDM is more eco-friendly than PC, and its potential is not fully explored either in PC or AAB concrete. The majority of previous studies on AAB concrete with RCA used class C FA as a precursor or NaOH with high molarity as an activator. Additionally, chemical treatments have also been used to enhance the quality of recycled aggregates, which raises the cost of the concrete. The present study aims to address the existing gaps in the literature. The principal objectives of this study are: (i) to evaluate the mechanical performance of ambient cured AAB concrete with varying precursor proportion and replacement percentages of virgin RCA as coarse aggregate, (ii) to propose an optimal replacement of RCA for ambient cured AAB concrete, and (iii) correlate the mechanical performance with the observed mineralogical and microstructural changes through X-ray diffraction and scanning electron microscopic analyses. The materials and methods are described below.

2 Materials and Methods

2.1 Materials and Mix Proportions

The crushed virgin BDM used as coarse aggregate in this study is obtained by crushing matured M35 grade concrete to 10 mm nominal size. Crushed granite used as natural coarse aggregate (NA) and river sand conform to ASTM C33/C33M [18]. Class F FA complying with ASTM C618 [19] and SG conforming to ASTM C989/C989M [20] are used as precursors. A combination of NaOH and sodium silicate is used as the activator. Industrial grade NaOH pellets of 99% purity are used. The composition of sodium silicate solution is 55.9% H_2O, 29.4% SiO2, and 14.7% Na2O. Polycarboxylate ether-based superplasticizer (SP) conforming to ASTM C494 [21] type F is used. A total of 20 mixes with four different precursor combinations (FA: SG varied as 100:0, 70:30, 60:40, and 50:50) and five different replacement percentages (0%, 25%, 50%, 75%, and 100%) of NA with BDM are studied. The detailed mix proportions are provided in Table 1.

2.2 Methods

X-ray diffraction (XRD) analysis is performed on powdered BDM samples using a RIGAKU Ultima X-ray Diffractometer to obtain the mineralogical characteristics of the BDM aggregates. The 2θ range is varied from 5° to 90°. The CuKα X-rays are

Table 1 AAB mix designs considered for this study

			FA:SG			
		Mix	100:0	70:30	60:40	50:50
		FA	400	280	240	200
		SG	0	120	160	200
Percentage replacement	0%	NA	1290	1290	1290	1290
		BDM	0	0	0	0
	25%	NA	967.5	967.5	967.5	967.5
		BDM	322.5	322.5	322.5	322.5
	50%	NA	645	645	645	645
		BDM	645	645	645	645
	75%	NA	322.5	322.5	322.5	322.5
		BDM	967.5	967.5	967.5	967.5
	100%	NA	0	0	0	0
		BDM	1290	1290	1290	1290

Unit: kg/m^3; other mix constituents are kept constant as follows, sodium silicate: 129.34 kg/m^3, NaOH: 10.57 kg/m^3, water: 67.65 kg/m^3, fine aggregate: 651 kg/m^3, and SP: 3.142 L/m^3

operated at a voltage of 30 kV and current of 30 mA. The scan rate is maintained at 1°/min with a step size of 0.02°. The results are shown in Fig. 1. For the morphological characterizations, SEM analysis is conducted using an FEI Apreo SEM. The micrographs are captured at an operating voltage of 20 kV and magnification of 2500× (Fig. 2).

For this study, the specific gravity and water absorption of the BDM are determined using the guidelines of IS 2386-3[22]. The results are presented in Table 2 and validated against usage requirements as aggregates in concrete applications.

The activator is allowed to rest for 24 h to alleviate additional heat generated by its exothermic nature. The aggregates are then uniformly mixed with the precursors to obtain a uniform dry mix which is blended with the activator and poured into 100 mm × 100 mm × 100 mm cubes after attaining desired workability. For each mix, three samples are made, and the average compressive strength of these three samples is reported. The specimens are demolded after 24–96 h to allow initial hardening and cured underwater until the age of 7 days from casting. After 7 days, the cubes are placed under ambient conditions till the day of testing. All the tests are performed at the age of 28 days. The compressive strength test is performed using an automatic compression testing machine operated in the load control mode. The load is applied at a rate of 2.33 kN/s.

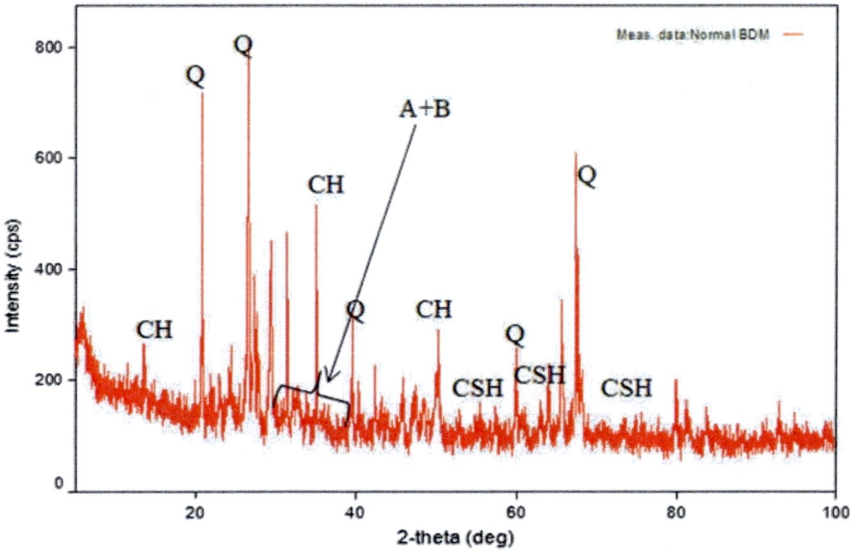

Fig. 1 XRD pattern of the BDM (A = tricalcium silicate, B = di-calcium silicate, CH = calcium hydroxide, CSH = calcium silicate hydrate, Q = Quartz)

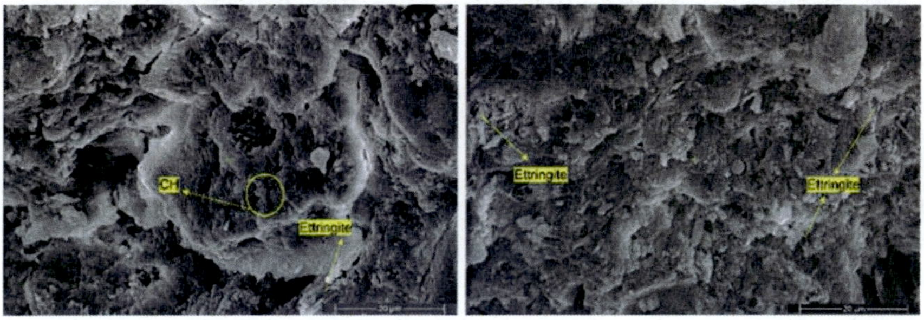

Fig. 2 Scanning electron micrograph of the BDM

Table 2 Water absorption and specific gravity of BDM

Property	Results
Specific gravity	2.54
Water absorption	1%

3 Results and Discussions

Figure 1 presents the XRD pattern of virgin BDM. It reveals the predominance of calcium silicates and quartz in BDM.

The sharp peaks corresponding to 2θ values of 21°, 26°, 39°, 55°, and 66° represent crystalline quartz or silica (Q). The smaller peaks between 2θ values of 30° to 40° represent the tricalcium silicate or alite (A) and dicalcium silicate or belite (B) component of portland cement, whereas the peaks near 2θ values of 16°, 34°, and 49° represent the calcium hydroxide (CH) or portlandite due to hydration of PC. The flatter portions of the diffractogram represented by small humps indicate the presence of an amorphous hydrated cementitious paste component (calcium silicate hydrate, CSH). These results are corroborated with findings from previous research [23].

Figure 2 shows that the morphology of the BDM is composed of a reticular network corresponding to CSH, hexagonal crystals of CH, and slender needle-like crystals of ettringite. These observations are consistent with the findings from hardened pastes of hydrated portland cement [23].

The specific gravity and water absorption results are listed in Table 2. The values obtained for the BDM are within the allowable limits for aggregates used in PC concrete preparation, as specified by IS 2386-3 [22]. The values of specific gravity of aggregates used in PC concrete applications should lie between 2.5 and 3.0. The maximum permissible limit for water absorption of aggregates used in cement concrete applications is 2%.

Figure 3 shows the compressive strength test results of AAB concrete with varying precursor proportions and NA replacement percentages with BDM. With an increase in replacement percentage of NA with BDM from 25 to 50%, the compressive strength of 100:0 and 70:30 decreased by 14% and 11%, respectively. However, with further increase in replacement percentage of NA with BDM to 75% and 100%, the compressive strength of 100:0 increased by 20% and 58%, respectively as the replacement content is increased. Whereas 70:30 exhibited an increase of 10% and 41% in compressive strength, 75% and 100% of NA are replaced with BDM, respectively.

The compressive strength at 100% replacement of NA with BDM is the highest compared to 0% replacement. No significant variation is observed in the compressive

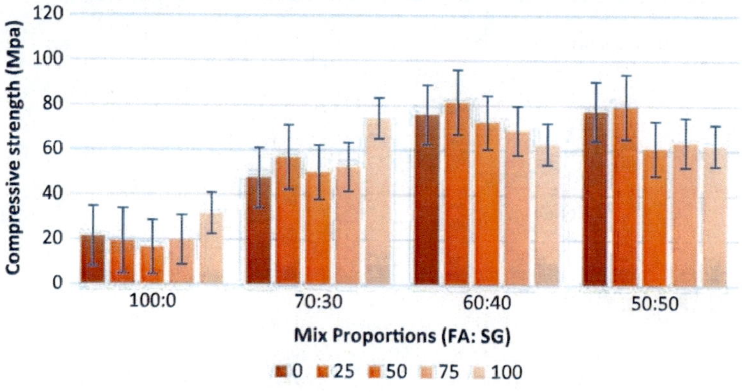

Fig. 3 Compressive strength results for different mix proportions

strength of 60:40 and 50:50 mixes with 25% replacement of NA with BDM. However, a decrease in strength with further replacement of NA with BDM in these mixes is noticed.

According to the compressive strength test data, 100:0 and 70:30 mixes demonstrated an improvement in compressive strength when the NA was entirely replaced with BDM. A comparable pattern is not found in 60:40 and 50:50 mixtures. The authors hypothesize that small spherical FA particles on the surface of crushed aggregates or CSH of virgin BDM reacting with AAB result in a denser matrix. Consequently, the compact matrix leads to an increase in the strength of high fly ash mixes. Further investigation is needed to validate this theory. AAB concrete with FA:SG ratio of 70:30 is proposed as optimum based on the results from the present study. The mix FA:SG of 70:30 exhibits the highest compressive strength at 100% NA replacement with BDM. The selection of AAB concrete with FA:SG of 70:30 is contingent on the primary objective of this study to propose a sustainable concrete mix that can be used for structural applications.

4 Conclusions

In the present study, the mechanical performance of ambient cured AAB concrete with varying precursor proportions, and replacement percentages of NA with BDM is investigated. Based on the findings of this study, the following conclusions are made.

- The primary crystalline phases observed through XRD analysis of BDM are quartz and calcium silicates. SEM analysis of BDM also shows the presence of CSH, CH, and ettringite on the aggregate surface.
- The specific gravity and water absorption show that they are within the limits of IS standard specifications.
- The compressive strength test results indicate that 100:0 and 70:30 blends improve compressive strength by 58% and 41%, respectively, when NA is replaced with BDM at 100%.
- A reduction in compressive strength is observed in 60:40 and 50:50 mixes with increased percentage replacement of NA with BDM.
- Owing to the superior mechanical performance of 70:30 mix at 100% replacement of NA with BDM, it is proposed as the optimum mix.

From the findings of the present study, it is evident that AAB concrete with 100% replacement of NA with BDM can be used for structural applications. Further investigation on the durability of the 70:30 mix and the effect of other BDMs from tiles, bricks, glass, and RAP are recommended.

References

1. BEEINDIA (2021) Cement, by Bureau of Energy Efficiency, Government of India, Ministry of Power. https://beeindia.gov.in/node/166; last accessed on 08 Feb 2021
2. Mindess S, Young FJ, Darwin D (2003) Concrete, 2nd edn. Pearson Education, Upper Saddle River, NJ
3. Provis JL (2017) Alkali-activated materials. Cem Concr Res 114:40–48
4. IS 17452 (2020) Use of Alkali activated concrete for precast products—guidelines, by Cement Matrix Products Sectional Committee, CED 53, Bureau of Indian Standards, New Delhi, India.
5. Pacheco-Torgal F, Castro-Gomes J, Jalali S (2008) Alkali-activated binders: a review. Part 1. Historical background, terminology, reaction mechanisms and hydration products. Constr Build Mater 22:1305–14. https://doi.org/10.1016/j.conbuildmat.2007.10.015
6. Pacheco-Torgal F, Castro-Gomes J, Jalali S (2008) Alkali-activated binders: a review. Part 2. About materials and binders manufacture. Constr Build Mater 22:1315–22. https://doi.org/10.1016/j.conbuildmat.2007.03.019
7. Building materials and technology promotion council (BMTPC). https://www.bmtpc.org/
8. Duran X, Lenihan H, O'Regan B (2006) A model for assessing the economic viability of construction and demolition waste recycling—the case of Ireland. Resour Conserv Recycl 46:302–320. https://doi.org/10.1016/j.resconrec.2005.08.003
9. Hossain MU, Poon CS, Lo IMC, Cheng JCP (2016) Comparative environmental evaluation of aggregate production from recycled waste materials and virgin sources by LCA. Resour Conserv Recycl 109:67–77. https://doi.org/10.1016/j.resconrec.2016.02.009
10. Serres N, Braymand S, Feugeas F (2016) Environmental evaluation of concrete made from recycled concrete aggregate implementing life cycle assessment. J Build Eng 5:24–33. https://doi.org/10.1016/j.jobe.2015.11.004
11. Chindaprasirt P, Cao T (2015) Reuse of recycled aggregate in the production of alkali-activated concrete. Woodhead Publishing Limited. https://doi.org/10.1533/9781782422884.4.519
12. Lee NK, Abate SY, Kim HK (2018) Use of recycled aggregates as internal curing agent for alkali-activated slag system. Constr Build Mater 159:286–296. https://doi.org/10.1016/j.conbuildmat.2017.10.110
13. Shaikh FUA (2016) Mechanical and durability properties of fly ash geopolymer concrete containing recycled coarse aggregates. Int J Sustain Built Environ 5:277–287
14. Nuaklong P, Sata V, Chindaprasirt P (2018) Properties of metakaolin-high calcium fly ash geopolymer concrete containing recycled aggregate from crushed concrete specimens. Constr Build Mater 161:365–373. https://doi.org/10.1016/j.conbuildmat.2017.11.152
15. Ren X, Zhang L (2018) Experimental study of interfacial transition zones between geopolymer binder and recycled aggregate. Constr Build Mater 167:749–756. https://doi.org/10.1016/j.conbuildmat.2018.02.111
16. Pacheco-Torgal F, Ding Y, Miraldo S, Abdollahnejad Z, Labrincha JA (2012) Are geopolymers more suitable than Portland cement to produce high volume recycled aggregates HPC? Constr Build Mater 36:1048–1052. https://doi.org/10.1016/j.conbuildmat.2012.07.004
17. Guo H, Shi C, Guan X, Zhu J, Ding Y, Ling T (2018) Durability of recycled aggregate concrete—a review. Cement Concr Comp Elsevier Ltd 89:251–259
18. American Society for Testing and Materials International, ASTM C33/C33M–16e1 (2016) Standard Specification for concrete aggregates. West Conshohocken, PA
19. American Society for Testing and Materials International, ASTM C618—17a (2017a) Standard specification for coal fly ash and raw or calcined natural Pozzolan for use in concrete. West Conshohocken, PA
20. American Society for Testing and Materials International, ASTM C989/C989M-17 (2017a) Standard specification for slag cement for use in concrete and Mortars, ASTM International. West Conshohocken, PA
21. American Society for Testing and Materials International, ASTM C494/C494M—17 (2017b) Standard specification for chemical admixtures for concrete. West Conshohocken, PA

22. IS 2386:1963 (PART I–VIII), Method of test for aggregate for concrete. Bureau of Indian Standards, New Delhi
23. Suluguru AK, Kar A, GuhaRay A, James N (2019) Experimental studies on physical properties and strength response of construction and demolition wastes. In: Lecture notes in civil engineering. Springer, pp 635–646

Identification and Analysis of Material Management Issues in Construction Industry Using ICT Tools

Vipul Agarwal

1 Introduction

1.1 *Background*

Material management is a combined process consisting of various parties such as individuals, stakeholders, organizations to utilize the technology in identification, inspection, acquisition, recognition, transportation and storage materials, information and equipment effectively throughout the life of the project [3]. Kulkarni et al. [15] defines materials management as a method of scheduling, implementing, regulating and producing the correct source of materials at the correct time with minimum cost, at the correct place and with the required quality. Simply stated, material management is system to manage the ordering, buying, transportation and storage of materials in the supply chain from material producer and supplier to the construction site.

Material management should be taken into account at every stage of construction project as project quality, cost and completion time are greatly affected due to the inefficient management of materials [13]. Some of the common problems such as material unavailability, over-purchasing, under-purchasing, low inventory space, inappropriate delivery and transportation issues are controlled by increasing stock size and ordering extra materials but it leads to increase in wastage of both the material and the capital [8].

While Just in Time (JIT) and implementing logistical planning are some of the methodologies available to be utilized to solve the issues of lack of inventory space, but even in the contemporary times, most of the communication and tracking is done manually which is prone to errors (Fig. 1).

V. Agarwal (✉)
Masters of Construction Management, Deakin Universtiy, Geelong, Australia
e-mail: agl.vipul94@gmail.com

A. Sil et al. (eds.), *Recent Trends in Civil Engineering*, Lecture Notes in Civil Engineering 274, https://doi.org/10.1007/978-981-19-4055-2_6

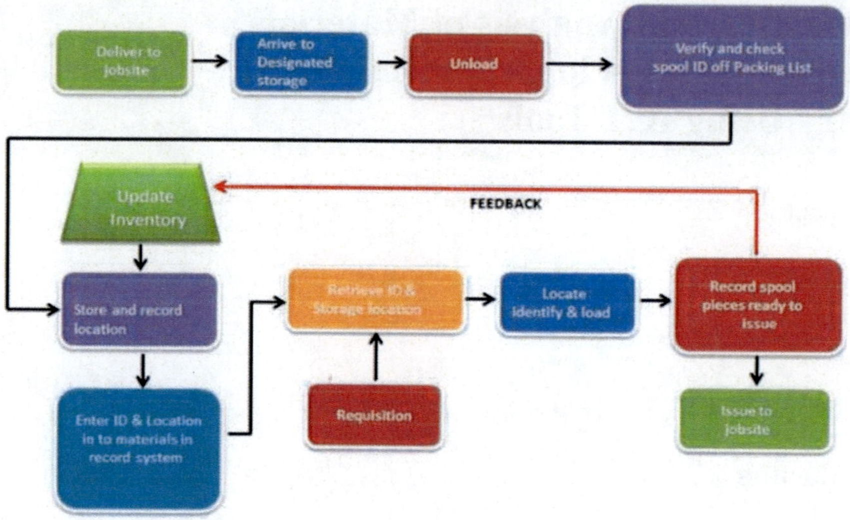

Fig. 1 Flow of materials in management of construction industry

1.2 Importance of Material Management

Construction materials are the major component of a building and construction project. As stated by [7, 14, 19, 21, 24], usually, the cost of the materials in a project ranges from 50 to 80% of the total cost of the project depending on the type of construction. The huge share of the cost makes the monitoring and management of materials a necessity for smooth functioning of the project. Numerous studies have confirmed that inefficient management or problems due to inefficient management such as early material purchase, material degradation and theft results in wastage of materials, quality problems and in turn leads to project delays and cost overruns [11, 12, 21, 25, 26. Therefore, there is a need for efficient materials management.

1.3 Stages of Material Management

The following steps of material management are identified by [3, 12, 13]:

- **Materials planning**: Planning is essential and is done first and foremost. Materials planning comprises quantity assessment and scheduling of orders [9]. Kasim et al. [13] emphasized the importance of planning for record maintenance and evaluation of deviation from target, as it has direct impact on output, profitability and completion time of the construction project.
- **Procurement/Purchasing**: Purchasing is the process of obtaining those goods, services or material that was identified as a requirement in the planning phase. As

the cost of the materials has the direct implication on the project cost [18]. The material procurement from the sources must be such that the product acquired meets the quality standards and within a reasonable price.

- **Transportation**: Through logistics, control and execution of the movement of materials from supplier to construction site is performed [13]. Although, the cost of transportation and logistics are not significant but the effects of logistic delays on the project are consequential.
- **Handling**: It involves proper handling, storage and administration of materials. The materials should not be placed under degrading conditions, precautions should be taken against transportation, loading and unloading damage and maintenance of inventory area are practices of good material handling and reduce the chances of theft and wastage [13].
- **Site management and waste control**: Waste generated at the construction site is a direct indication of inefficiency and contributes to a significant cost [4]. Site stock management ensures that the construction personnel are equipped with the correct materials in the required quantities to carry out the works assigned to them. Implementing systematic stock management helps to enhance efficiency and leads to better waste control at the construction site [5].

2 Research Methodology

2.1 Objective of the Study

The major concern of the research was to identify the most pervasive issues regarding the material management in the construction industry irrespective of the location so to magnify the effects of the research itself. A mixed approach of both qualitative and quantitative is also adopted by this study.

Quantitative research includes data collection often based on surveys directed to a set of individuals, group, or organizations, which validate specific factored parameters to produce implications, inferences which are then applied to a large set of population. The conclusions are often reached through primary data collection and statistical analysis. Qualitative research can be described as research that contains gathering and analyzing subjective data (focus groups, surveys, observations, interviews, conversational analysis, ethnographies) with stress on the situation of the study. Mixed method research involves collecting qualitative and quantitative data by primary or secondary sources, followed by comprehensive analyses of the obtained data in regard to the study [16].

Similar to the previous notion, this research will adopt the methodologies used by various authors such as questionnaire survey and literature review. This research is based on qualitative method and involves collection of secondary data. It mainly involves literature review, collecting data from various books, journals and papers that are related and finally comparative analysis is performed.

2.2 *Data Collection and Analysis*

In order to find the existing problems which are prevalent in material management in construction industry, 28 papers were reviewed for this purpose, out of which 10 papers presented specific problems by location. Although, some of the papers used in this research were issued a long time ago, but they were considered as they provided highly relevant information to this study.

Initially, the research attempted to understand material management and its practices in construction industry. In first stage of the research, exhaustive study of different problems encountered in material management was undertaken. The various problems and factors leading to those problems in effective material management were then identified. Then, the major problems which were common in previous works was compared and identified as they are the most impacting factors in material management. In order to analyze the implementation of ICT tools, a comparative analysis is done on the major problems. Based on the literature review and the analysis, the factors for selection of ICT tool for optimum material management is discussed.

3 Problems in Material Management

For effective material management in a project, identifying the problems and obstacles in the smooth flow of materials is necessary. It is important to understand the impacts and causes of the problems in order to develop approaches to address them. The common approach to identify these problems by various authors is interview and questionnaire survey. Various causes and problems as identified by different authors are presented below:

The following section states the various problems that were identified by five different authors in Malaysia, India, Nigeria, Maldives and Israel.

Hussin et al. [11] identifies some of the common problems in material management as material unavailability, low inventory space, inappropriate delivery, transportation issues, inefficient management of stock, delivery delays which leads to consequences such as wastage of materials, lack of quality and ultimately project delays. Study conducted by [9], suggested that the frequent problems were lack of planning, materials damage during transportation, loss and theft because of improper supervision, lack of material tracking, no information on position and availability of materials on site, wrong material delivered on site, stock update delays, selection of incorrect materials for construction, delays in supplier payment (Table 1).

Navon and Berkovich [20] conducted an interview survey among various construction professionals in Israel and found 13 factors in material management that affect the material, logistics and site operations. The most significant material problems,

Table 1 Material management problems in Israel [20]

Problem category	Problem
Logistics related problems	• Wrong material delivered • Wrong quantity delivered • Material not ordered • Status of orders is not updated • Supplier understood order wrong • Site conditions and unloading equipment availability ignored
Material problems at construction site	• Monitoring materials on site • Follow up of availability of materials around the site and hauling of materials • Wastage due to: double handling, unsuitable storage conditions, theft, loss • Lack of storage space • No information of location of on site stock • Untargeted materials: unknown (extra) materials on site • No inspection of received materials

related to purchase and supply, were wrong material delivered or material delivered in wrong quantity and related to wastage were double handling and loss due to degradation [20] (Table 2).

In Nigerian construction industry, a postal questionnaire survey was conducted by [2], to identify the major problems in the industry. From 132 questionnaire response received from industry professionals, 100 responses were analyzed because of their relevance and ranked according to the mean of their occurrence. 18 problems were identified including theft, loss and poor coordination between contractor and supplier, however, lack of proper work planning and scheduling was found to be the most crucial problem in material management (Table 3).

Based on the questionnaire survey conducted among the experienced professionals in the Maldives construction industry, [1] categorized the problems according to the stage of construction they were incurred on. 29 total problems were allocated in four categories, i.e., material identificaiton, procurement, on site storage and construction. The four stages were similar to the stages of material management. The research findings puts rehandling of materials and late order in storage problems, while it shows that problems such as incorrect item delivered and item not conforming to quality standards are encountered during construction phase.

Another empirical questionnaire survey conducted by [10] among various industry professionals operating at construction sites in India revelead the problems in material management at construciton sites. The research conducted a mean analysis on the occurrence of problems. The identified problems are shown in Tables 4 and 5.

In a case study analysis, [17] conducted interview survey among various construction contractors in Malaysia. Based on the results, the problems were condensed into 11 major problems, for example, the problem of locating material on site and problem of unknown quantity of available material was combined into improper materials

Table 2 Material management problems in Nigeria [2]

Rank	Problem
1	Lack of proper work planning and scheduling
2	Cash flow problems to contractor due to delayed payments
3	Burglary, theft and vandalism
4	Materials shortage during construction and sudden price fluctuations
5	Lack of coordination between contractor and supplier
6	Negligence
7	Insufficient storage space
8	Incompetent material suppliers
9	Site access problems
10	Difficulty in managing materials among sub-contractors due to the limited storage space on site
11	Damage to materials during transportation to the site
12	Lack of security personnel
13	Health and safety procedures implementation on site
14	Hindrance to work progress due to improper stocking of materials
15	Difficulty in distributing materials to sub-contractors on site
16	Rejection of materials due to non-compliance to the specification
17	Conflict among sub-contractors due to acute space available for material
18	Accident/injury occurred at site

Table 3 Material management problems in Maldives [1]

Construction stage	Problems at that stage
Material identification	• Poor planning of material need • Lack of conformance to requirements • Plans and specification differences • Undefined scope • Incomplete drawings • Lack of communication
Procurement (vendor selection phase)	• Uncontrollable bid list • Incomplete proposal • Time wasted on unqualified suppliers
On site storage	• Storage of materials • Theft • Vague stated requirements • Damage of material • Unrealistic delivery date • Late deliveries • Re handling of materials • Poor communication • Late or incorrect order • Lack of conformance to requirements

Table 4 Material management problems in India [10]

Surplus of or missing materials	Lack of storage space for materials on site	Materials whose specifications do not match the ones in the purchase order
Lack of complete and up-to-date information regarding arrival of materials to the site and on site stocks	Materials arriving to the site at the wrong time	Material arriving in the wrong quantity
Burglary, theft and vandalism	Unavailability of information regarding the status of the orders	Waste of man-hours searching for materials and tracking them
Forgetting to order the materials	Non-availability of materials	Lack of conformance to requirement
Lack of communication		

Table 5 Material management problems in Malaysia [17]

Excessive paper-based recording and reporting	Lack of up-to-date information regarding the status of materials	Loss of materials due to theft
Inadequate site storage space	Labor intensive	Waste of labor working hours
Fails to record the movement of materials	Materials at the storage	Labor intensive and tedious work
Fails to order materials due to poor communication	Improper materials record	

record. The research findings states that some of the problems found, such as delay due to labor intensive work, were less significant than others.

Other studies conducted on the issue suggests that the problems are overlapping. Rivas et al. [22] conducted a case study on a Chilean construction company by interview survey. The material related problems were identified as the delivery delays, shortage of materials, unnecessary paperwork and distance of site from the supplier. Other problems at site were identifies as lack of equipment to shift materials, insufficient resources such as tools on the site, availability of equipment and its position on site. In urban regions such as metropolitan Paris, [6] identifies lack of space, improper stock management and lack of preplanning to be major causes of ineffective material management.

The aforementioned section identified the various problems occurring in material management in different parts of the world. The next section discusses the various technologies available to address these problems by means of comparative analysis.

4 Analysis and Findings

4.1 Identification of Common Prevailing Material Management Problems

This section compares the material management problems in 5 countries; Israel, Nigeria, Maldives, India, and Malaysia as identified by [1, 2, 10, 17, 20] to find the common problems that are prevalent in the industry. The problems in Table 6 are listed in order of highest to lowest frequency of occurrence in the works of above authors.

The problems exclusive to only one country is omitted from the above comparison, which includes:

Nigeria—Cash flow problem due to delayed payments, material shortage during construction and sudden price fluctuations, negligence, conflict among sub-contractor due to shortage of material because of low storage space, improper stocking of materials, accidents at site.

Israel—Unknown materials at site and no inspection of materials upon receival of materials.

Malaysia—Paper-based recording and waste of labor working hours.

From the comparative analysis, out of 13 problems stated, it can be noted that there are 6 major problems in material management as these problems occur in all or at least 4 of the 5 countries. These problems include:

Table 6 Common material management problems identified [1, 2, 10, 17, 20]

Problem	Malaysia	India	Nigeria	Israel	Maldives
Wrong material delivered to site	●	●	●	●	●
Inadequate storage space at site	●	●	●	●	●
No update on availability of material on site	●	●	●	●	●
Material not ordered due to improper stock record	●	●	●	●	●
Material missing or loss due to theft	●	●	●	●	●
No update on position of material in storage and double handling	●	●		●	●
Lack of supplier contractor communication		●	●		●
Delivered at wrong time or quantity		●		●	●
Lack of planning and scheduling		●	●	●	
Transportation damages		●	●	●	
No update on status of order		●		●	
Site access problem			●	●	
Incompetent suppliers			●		●

(●) Indicates that the problem is identified in the corresponding country

- Wrong material delivered to site,
- Inadequate storage space at site,
- No update on availability of material on site,
- Material not ordered due to improper stock record,
- Material missing or loss due to theft, and
- No update on position of material in storage.

It is interesting to note that the factor such as no update of material availability on site or storage is a significant cause of material management problems along with wrong material delivered or material not ordered at all. This can be accounted for the fact that if the material is not found in the stock or at site, it becomes unavailable and the percussions are similar to that of wrong material and material missing or lost.

4.2 Method of Selection of ICT Tools for the Professionals

In construction supply chain, materials are purchased and transported from producer to suppliers and then to the site storage area. The whole supply chain is not free from errors. The major problems in material management include: wrong material delivery, update on availability of stock, ordering of materials and missing or lost material.

This real time tracking and information about the material can be obtained by tagging the materials. Data regarding material such as material source and producer, amount and state of material should be accessible to supervisors, project managers and material buyers in the office. Real time inventory information and position can further facilitate the materials management processes. As planners equipped with this real time information can make appropriate decisions and plan according to the situation in terms of ordering, transporting, stocking and use. Through this process, the utilization of storage space can be done in an efficient manner and the problems of inadequate storage space can also be addressed.

All of the process was carried out on paper conventionally, by sending and receiving documents. ICT tools can be implemented to enhance the material management in construction, permitting data sharing to be done in real time by electronic means.

The use of technology can be assessed on following factors [23]:

- *Safety*: People should not be harmed at any place or time while using the technology.
- *Cost*: Simply stated, the cost must be reasonable.
- *Accuracy*: The results from the technology assisted process should be more precise than the manual methods. Due to the complexity and more data carrying capacity, ICT eliminates the human error by a far extent.
- *Easy to learn and use*: The technology must be easy to use and learn, it should be simple and friendly to the operator while reducing human efforts.

- *Ambient environment*: The technology should operate perfectly even in the absence or low presence of natural light, in presence of obstructions and high attenuation to signal could be anticipated.
- *Roughness*: The technology in assessment of use should be able to tolerate the rough environments and must be tough enough to endure the harsh conditions of the construction environments, such as dirt, water. As these adverse conditions are inherent to the construction environment, the technology should be able to operate regardless.
- *Time*: The time taken for initial setup and further operation should be less than the manual methods. The data collection processes must take reduced duration to locate and position the materials compared to traditional methods.

One of the main reasons that construction sector had been slow in adopting ICT tools and harnessing its benefits is the lack of awareness with the technology, therefore, not realizing its fullest potential. Another reason to the hindrance of implementation is the cost barrier.

In response to the stated shortages, it is clear that the industry could benefit from the technology benefits and potential solutions to material management problems in the industry.

5 Conclusion

An effective material management system facilitates the construction projects to be completed within a stipulated time schedule. In the present study common material management problems faced by five developing countries are identified and analyzed from their design to completion stages. The most prominent identified issues at site are: incorrect and inadequate material delivered, inadequate storage space, material missing/loss due to theft, inefficient record keeping of material in store, mismatch of ordering material with available inventory stock and no proper upkeep of material position in the wise spread storage space, respectively. This leads to inconsistencies in time, productivity and overall performance. The effort are made in this study to present a framework using ICT tools for developing some strategic planning and work scheduling which may be helpful to the site decision makers in timely and corrective decision making. Further discussion and research on identification of precise ICT tools and the comparison on their effectiveness in post implementation is needed.

References

1. Ahmed Z (2017) The impact of material management on construction project delivery in Maldives, faculty of engineering and science. Universiti Tunku Abdul Rahman, Universiti Tunku Abdul Rahman
2. Arijeloye B, Akinradewo O (2016) Assessment of materials management on building projects in Ondo State, Nigeria. World Scientific News 55: 168–185. Available at: www.worldscientificnews.com
3. Caldas CH et al (2015) Materials management practices in the construction industry. Pract Period Struct Des Constr 20(3):04014039. https://doi.org/10.1061/(asce)sc.1943-5576.0000238
4. Calkins M (2009) Materials for sustainable sites. John Wiley & Sons, Hoboken, New Jersey
5. Dainty ARJ, Brooke RJ (2004) Towards improved construction waste minimisation: a need for improved supply chain integration? Struct Surv 22(1):20–29. https://doi.org/10.1108/02630800410533285
6. Dakhli Z, Lafhaj Z (2018) Considering materials management in construction: an exploratory study. Logistics 2(1). https://doi.org/10.3390/logistics2010007
7. Dey PK (2001) Re-engineering materials management a case study on an Indian refinery. Bus Process Manag J 7(5):394–408. https://doi.org/10.1108/EUM0000000006002
8. Equere E, Tang LCM (2012) Dearth of automation: the consequences in Nigeria Construction Industry. In: The 2012 3rd international conference on construction and project management, Vol 44, pp 101–107
9. Gulghane AA, Khandve PV (2015) Management for construction materials and control of construction waste in construction industry: a review. J Eng Res Appl 5(4):59–64. Available at: www.ijera.com
10. Harish K (2014) A study of materials management in existing constructions at Coimbatore. Glob J Eng Sci Res 1(3):24–27
11. Hussin JM, Abdul Rahman I, Memon AH (2013) The way forward in sustainable construction: issues and challenges. Int J Adv Appl Sci 2(1):15–24. https://doi.org/10.11591/ijaas.v2i1.1321
12. Kasim N (2008) Improving materials management on construction projects. Loughborough University
13. Kasim NB, Anumba CJ, Dainty ARJ (2005) Improving materials management practices on fast-track construction projects. In: Association of researchers in construction management, ARCOM 2005—proceedings of the 21st annual conference, Vol 2, pp 793–802
14. Kini DU (1999) Materials management: the key to successful project management. J Manag Eng 1(15):30–34
15. Kulkarni V, et al (2017) Factors affecting material management on construction site. Int Res J Eng Technol (IRJET) 4(1):474–478. Available at: https://irjet.net/archives/V4/i1/IRJET-V4I178.pdf
16. Leech NL, Onwuegbuzie AJ (2009) A typology of mixed methods research designs. Qual Quant 43(2):265–275. https://doi.org/10.1007/s11135-007-9105-3
17. Liwan SR, Kasim N, Zainal R (2013) Materials tracking practices for inventory management in construction projects
18. Morris PWG, Pinto JK (2007) The Wiley guide to project technology, supply chain and procurement management. John Wiley & Sons Inc., Hoboken, New Jersey
19. Muehlhausen FB (1991) Construction site utilization: impact of material movement and storage on productivity and cost. AACE Int Trans
20. Navon R, Berkovich O (2006) An automated model for materials management and control. Constr Manag Econ 24(6):635–646. https://doi.org/10.1080/01446190500435671
21. Patel KV, Vyas CM (2011) Construction material management on project sites. In: National conference on recent trends in engineering and technology, pp 1371–1378. https://doi.org/10.22214/ijraset.2018.1207

22. Rivas RA et al (2011) Analysis of factors influencing productivity using craftsmen questionnaires: case study in a Chilean construction company. J Constr Eng Manag 137(4):312–320. https://doi.org/10.1061/(asce)co.1943-7862.0000274
23. Sardroud JM, Limbachiya M, Saremi AA (2010) Ubiquitous tracking and locating of construction resource using GIS and RFID
24. Stukhart G, Kirby D (1995) Construction materials management. M. Dekker, New York
25. Thomas HR, Ellis RD (2005) Fundamental principles of site material management. J Constr Eng Manage 131(July):808–815. https://doi.org/10.1061/(ASCE)0733-9364(2005)131:7(808)
26. Thomas R, Sanvido V, Sanders SR (1989) Impact of material management on productivity—a Case study. J Constr Eng Manag 115(3):370–384

P-M-M Interaction of Bridges Having Different Construction Materials: CC, SCC and FRSCC

Nilanjan Tarafder and M. L. V. Prasad Raju

1 Introduction

Bridge is a delicate structural form. There are different types of bridges, but the T Beam Bridge is one of the most efficient formations concerning the loads that need to be dealt with over time. In general, conventional concrete is used in the construction of bridges. However, after discovering other superior concrete types, it is essential to understand how the new concrete materials will affect the bridge structure. A bridge is most vulnerable under seismic scenarios, and moments generated at the time can be very destructive. These moment's and drift behaviors need to be studied very carefully to ensure the bridge's safety. Drift analysis is an effective process to understand its behavior. Similarly, axial load, moment about the x-axis and moment about the y-axis (PMM) help to understand details about the moments generated in different directions due to an earthquake load.

Accompanied by material science and structural modeling advancements, bridge spans have been progressively growing [1]. It has been an enormous challenge for engineers to find innovative structural forms for long-span bridges with adequate protection, primarily against Seismic hazards [2]. The effect of reinforcing bar buckling was also considered to study the deformation limit states under several seismic ground motion data [3]. Babazadeh et al. simulated a 3D FE model of a bridge structure and thereby predicted the intermediate damage limits for a reinforced concrete (RC) bridge column, which lead to a strain index limit of 0.005 for spalling [4]. Tegos et al. simulated a bridge disabling seismic displacement model in the transverse direction, but the setup proved very useful for precast I-beam type bridges only [5]. A study on the FRC bridge piers showed that the life cycle seismic performance and resilience was improved by around 25% [6]. Another study with FRC showed that fibers' use increased the compressive and flexural strength, and residual

N. Tarafder (✉) · M. L. V. Prasad Raju
National Institute of Technology, Silchar, Assam 788010, India
e-mail: nilanjantf@gmail.com

A. Sil et al. (eds.), *Recent Trends in Civil Engineering*, Lecture Notes in Civil Engineering 274, https://doi.org/10.1007/978-981-19-4055-2_7

strength, energy absorption ability and ductility was also improved [7]. The simple RC column was compared with the FRC column and axial deformation control. The ultimate strength was improved by 199% and 92% with fibers in the regular RC column [8]. Our group has conducted a performance study to quantify the effects of bridge pier type under seismic condition. Base shear capacity and drift results showed single and double pier bridge behavior in longitudinal and transverse directions. It was found that based on the performance level requirements, bridge pier type has to be selected [9]. A new material called self-compacting concrete (SCC), well investigated by the authors in their previous works. SCC can be easily prepared with the proportions prescribed by Meesaraganda et al. The SCC has the property of filling any highly congested reinforcement setup very easily without external vibration [10, 11]. Another study using artificial neural network (ANN) validated the improvement of strength with fiber reinforced self-compacting concrete (FRSCC) [12].

In this study, a t-beam bridge is analyzed with new concrete materials to understand the behavior of the respective bridge formation type under severe seismic conditions concerning the replacement of conventionally used material in the bridge construction with improved performance. Eco-friendly materials: self-compacting concrete and fiber-reinforced self-compacting concrete were chosen as replacement options suitable for constructing any congested reinforced structure.

2 Modeling the Bridge

2.1 3D Model of the Bridge

The present study considers the t-beam bridge over the Lungpuk river in Mizoram, monitored by the PWD, Mizoram [13]. The finite element 3D model of the bridge was developed by using the CSiBridge program [14]. Span length of the 2 span T-beam bridge deck is $L = 24$ m, width $B = 6.4$ m and height $H = 2.1$ m.

2.2 Materials

Conventional concrete (CC), self-compacting concrete (SCC) and fiber reinforced self-compacting concrete (FRSCC) were used in the bridge models. For all the concrete, grade 53 ordinary Portland cement was considered in combination with normal river sand as fine aggregate and 20 mm maximum size of coarse aggregate. The respective concretes' detailed fresh and hardened properties are already available in the previous works [9, 10, 12, 15]. HYSD 500 grade steel was used in structure modeling and kept constant throughout the study along with the amount of steel reinforcement in all the models and shown in Table 1.

Table 1 Steel quantity

Element	Longitudinal reinforcement	Transverse reinforcement	Clear cover (mm)
Abutment	44 Φ20 at top and bottom	Φ12 @ 150 mm c/c at top and bottom	40
Bent cap	16 Φ20 at top and bottom	Φ16 @ 150 mm c/c at top and bottom	40
Pier	32 Φ25 with Φ12 @ 150 mm c/c confinement	10 Φ25 @ with Φ12 @ 150 mm c/c confinement	50

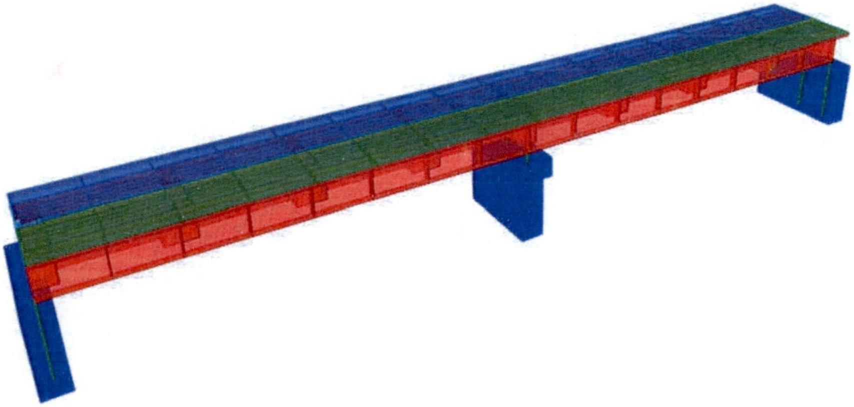

Fig. 1 3D finite element model of the bridge

3 Analysis

The analysis of the bridge models with three different constructional materials was simulated with CSiBridge software, a product of Computer and Structures Inc. (CSi). This software is exclusively used for bridge structure analysis and designing purpose. A 3D finite element model (FEM) of the t-beam bridge was created for each scenario. The basic 3D model is shown in Fig. 1. Same loading conditions were used for all the models and tabulated in Table 2. Further hinge analysis was done on the bridge models to understand its behavior and a comparative result is produced. Also the demand-capacity analysis was done and produced.

4 Results

Analysis results are produced and explained in detail in this section. The provided results clearly shows the differences between the uses of three different materials in the bridges.

Table 2 Loading condition

Load pattern	Load type	Magnitude of load
Dead load	Dead—full structure	5 kN/m
Barrier load	Dead—line load	3 kN/m
Sidewalk load	Dead—area load	1.5 kN/m^2
Pedestrian load	Pedestrian live—area load	1.5 kN/m^2
Asphalt load	Wearing surface—area load	1.5 kN/m^2
IRC class A load	Vehicle live—moving load	As per IRC 7
Pushover load	Seismic response spectrum	As per IS: 1893-2016

4.1 Demand-Capacity Ratio

It is the ratio of the structural member force after the sudden removal of a column to the member strength (capacity), as a benchmark to determine the failure of major structural members by the linear static analysis procedure. The ratio is a very important factor to evaluate the bridges' vulnerability due to the seismic force and its range varies from 0 to 1. If the D/C ratio found to be 0, which means bridge takes no damage and if the ratio is 1, it implies that the bridge is at the verge of failure. Figures 2 and 3 shows the demand-capacity ratio for the bridges in longitudinal and transverse directions. The first figure shows the behavior of the bridges with different concrete materials in the longitudinal direction. From the graph it is visible that using SCC instead of regular CC, performs better and if the CC is replaced by FRSCC, the results are even better. With each replacement, the D/C ratio of the bridge is coming near to zero which indicates less damage was taken by the bridge for the same earthquake condition. The SCC reduced the damage on the bridge by 2.75% which further reduced by 15.84% while adding fibers to the concrete, i.e., by using FRSCC as the bridge material.

Similarly, Fig. 3 shows the behavior of the bridge in transverse direction under same earthquake condition. Here also the replacements reduced the D/C ratio of the bridge which indicates less damage for the bridge with new materials as a constructional built. With the replacement of SCC and FRSCC instead of CC, the reduction of damage against the bridge is 3.14% and 16.36%, respectively.

4.2 P-M-M Interaction Result

It is the axial load-moment interaction that is used to evaluate sufficient design of a reinforced concrete structure subjected to combined axial-moment loading. With the similar setup of loading and reinforcing condition, the models was analyzed and several curves are presented considering the pushover loads from longitudinal and transverse directions. Figures 4, 5 and 6 shows the P-M interaction of the bridge

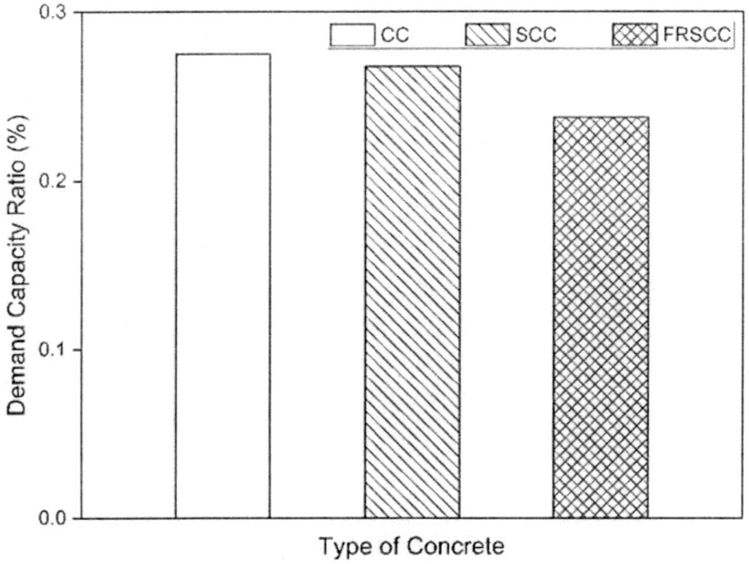

Fig. 2 Demand-capacity ratio in longitudinal direction

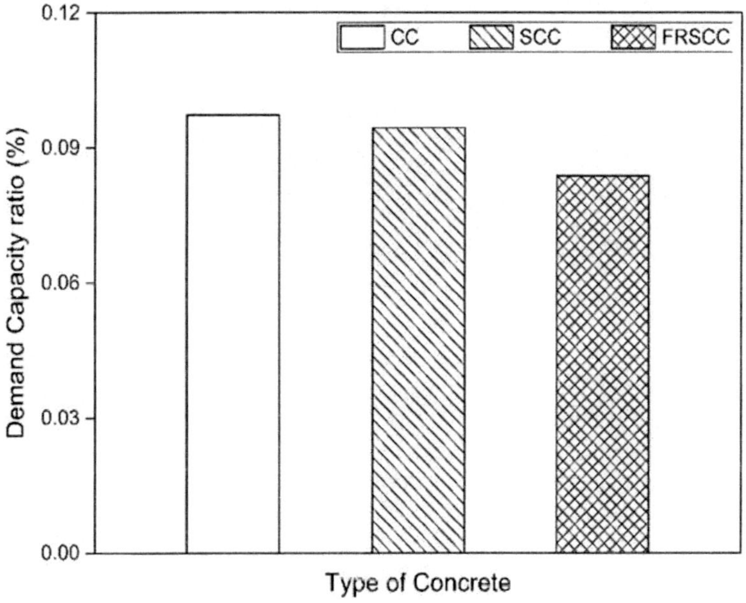

Fig. 3 Demand-capacity ratio in transverse direction

pier in longitudinal direction due to the pushover load in longitudinal direction (X-direction of the bridge). In the figures, the pink dot denotes Beginning of Deflection of the pier due to the seismic force, blue dot indicates the Immediate Occupancy limit, sky dot indicates the Life Safety limit, green dot indicates the Collapse Prevention limit and red dot indicates total Collapse Point of the bridge pier. It can be seen from the figures that the limits for every state is increased after the replacement of CC with SCC and FRSCC. But as the collapse is the last point where anyone can use any structure, comparison of the respective results are presented. After replacing the regular conventional concrete in the bridge structure by SCC and FRSCC, the collapse prevention limit of the piers were increased by 3.08 and 21.48%. SCC replacement improved the limit but the FRSCC provided the huge increment. It is because of the fibers in the SCC which improved the strength of the pier greatly leading to better collapse prevention limit of the piers.

Similarly, Figs. 7, 8 and 9 provides the detailed visual of the behavior of the bridge piers in the transverse direction under the pushover load in that direction (Y-Direction). Here also the pink dot denotes Beginning of Deflection of the pier due to the seismic force, blue dot indicates the Immediate Occupancy limit, sky dot indicates the Life Safety limit, green dot indicates the Collapse Prevention limit and red dot indicates total Collapse Point of the bridge pier. It is seen form the figures that, in transverse direction also the bridge piers' collapse prevention limit was increased by replacing by regular CC with SCC and FRSCC. An increment of 3.21 and 21.45% was observed, respectively, for using SCC and FRSCC instead of regular CC.

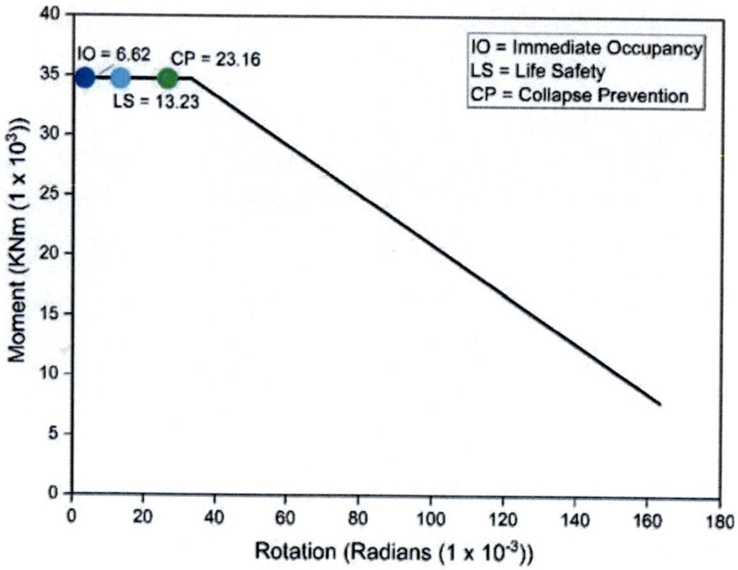

Fig. 4 P-M curve for CC in longitudinal direction

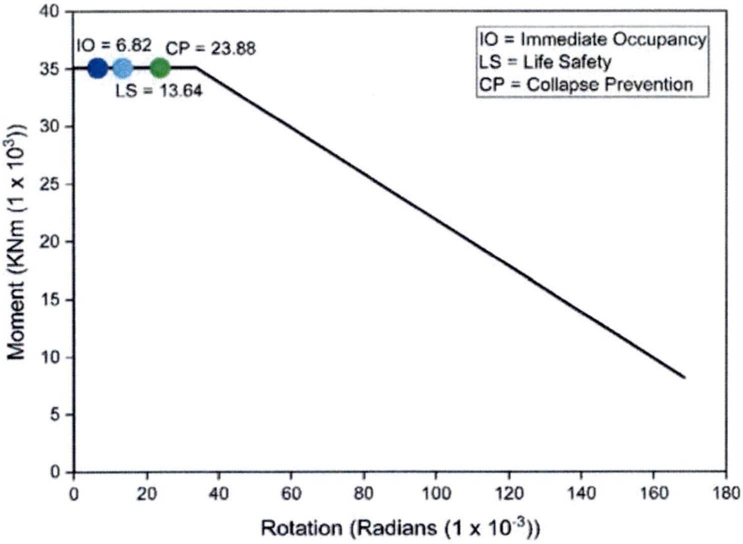

Fig. 5 P-M curve for SCC in longitudinal direction

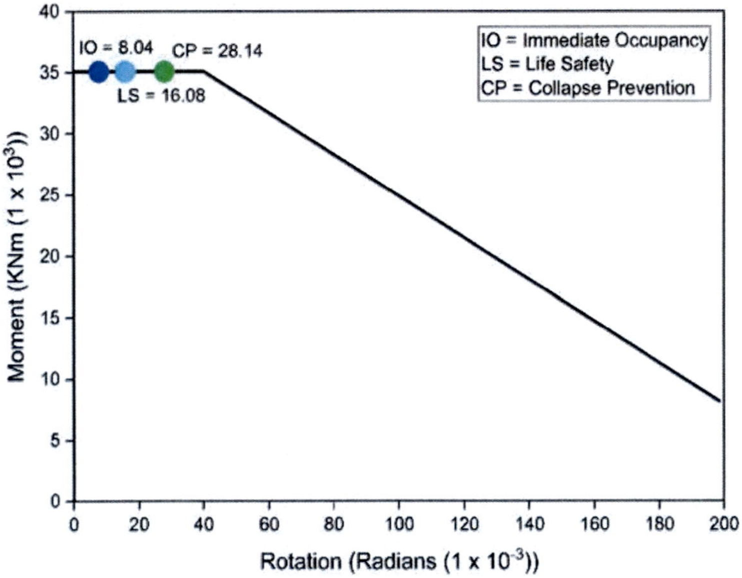

Fig. 6 P-M curve for FRSCC in longitudinal direction

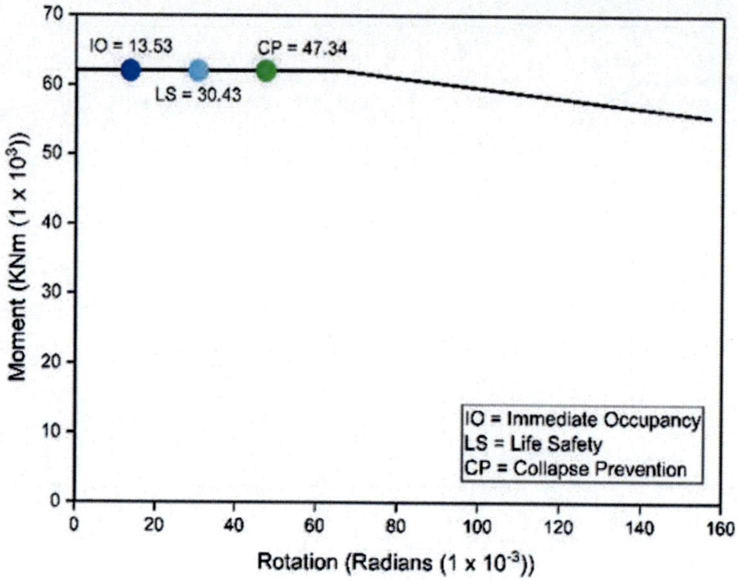

Fig. 7 P-M curve for CC in transverse direction

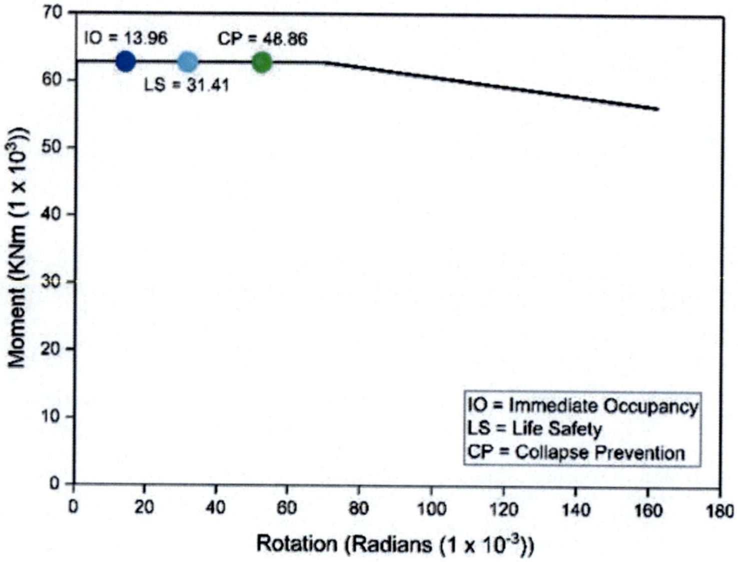

Fig. 8 P-M curve for SCC in transverse direction

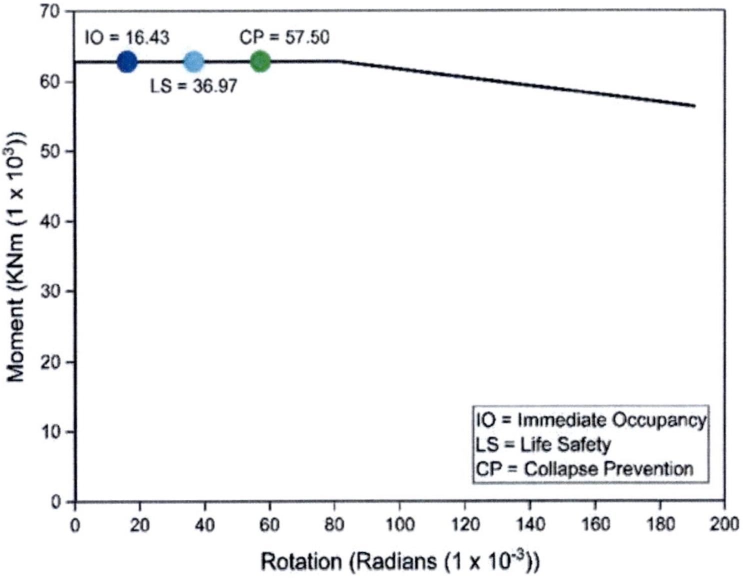

Fig. 9 P-M curve for FRSCC in transverse direction

5 Conclusion

The study investigates new concrete materials: SCC and FRSCC as a replacement of the traditional conventional concrete in the bridge structure to deal with the seismic forces. Results showed promising data and hence the following points are concluded from the study:

- SCC and FRSCC both is more suitable for construction purpose of bridge structure and they are also eco-friendly which reduces sound pollution, saves traditional raw materials of construction.
- Demand-Capacity Ratio was reduced by using the new replacement concrete types which improved the strength of the bridge structure against seismic loads leading to withstand in the situation much longer.
- P-M-M interaction study proved that by using SCC and FRSCC instead of CC, increased the moment bearing capacity of the piers and also the collapse prevention limit of the structure, providing more time under a seismic event to move to safety.

References

1. El-Sheikh A (1996) Development of a new space truss system. J Constr Steel Res 37:205–227. https://doi.org/10.1016/0143-974X(96)00004-1
2. Wang H, Fan F, Qian H, Zhi X, Zhu E (2011) Full-process analysis of pretensioning construction of Dalian gym. Adv Mater Res 163–167:200–204. https://doi.org/10.4028/www.scientific.net/AMR.163-167.200
3. Feng Y, Kowalsky MJ, Nau JM (2015) Effect of seismic load history on deformation limit states for longitudinal bar buckling in RC circular columns. J Struct Eng 141:1–13
4. Babazadeh A, Burgueño R, Silva PF (2015) Use of 3D finite-element models for predicting intermediate damage limit states in RC bridge columns. J Struct Eng (United States) 141. https://doi.org/10.1061/(ASCE)ST.1943-541X.0001253.
5. Prof IT, Tegou S, Eng C, Markogiannaki O, Tegos I, Tegou S et al (2013) Seismic Design Precast I-Beam Bridges Based Ductility 23:176–186. https://doi.org/10.2749/101686613X13439149157434
6. Pang Y, Wei K, Yuan W (2020) Life-cycle seismic resilience assessment of highway bridges with fiber- reinforced concrete piers in the corrosive environment. Eng Struct 222:111120. https://doi.org/10.1016/j.engstruct.2020.111120
7. Hasani M, Moghadas F, Sobhani J, Chini M (2021) Mechanical and durability properties of fiber reinforced concrete overlay: experimental results and numerical simulation. Constr Build Mater 268:121083. https://doi.org/10.1016/j.conbuildmat.2020.121083
8. Vasumathi AM, Rajkumar K, Prabhu GG (2014) Compressive behaviour of RC column with fibre reinforced concrete confined by CFRP strips. Adv Mater Sci Eng 2014:10
9. Tarafder N, Prasad LVM (2020) Seismic performance of box girder bridge with non-linear static pushover analysis. Int J Emerg Technol 11:897–904
10. Meesaraganda LVP, Tarafder N (2019) Durability studies of environmental friendly self compacting concrete with and without fiber. KEM 803. https://doi.org/10.4028/www.scientific.net/KEM.803.207
11. Kassimi F, Khayat KH (2021) Mechanical properties of fiber-reinforced concrete with adapted rheology.pdf. Cem Concr Compos 118:103958
12. Meesaraganda LVP, Saha P, Tarafder N (2019) Artificial neural network for strength prediction of fibers' self-compacting concrete, Vol 816. https://doi.org/10.1007/978-981-13-1592-3_2
13. Public Works Department, Mizoram (n.d.)
14. Computers & Structures Inc. CSiBridge (2021)
15. Nilanjan Tarafder RS (2016) Durability and case study of fiber reinforced polymer (FRP). Int Organ Sci Res Mech Civ Eng 13:53–62. https://doi.org/10.9790/1684-1306035362

An Experimental Study on Flexural Strength of Corroded-Reinforced Concrete Beam with Fly Ash

Sandeep Sathe, Sudhir Patil, and Vikram Shete

1 Introduction

Reinforcement (rebar) corrosion in concrete seriously affects the serviceability and durability structures [1]. The experimental study done by Bazant [2] shows that the volume of the reinforcement bar increases 2–4 times after the corrosion starts which causes cracking of concrete. The corrosion alters the mechanical properties of steel bars and concrete resulting in reduction in ductility ratios. The flexural strength, bond strength, and shear strength of structures are also affected due to corrosion of reinforcement. Many investigators have asserted to calculate corrosion levels from steel coupons obtained by cutting steel bars. This process was also demonstrated by Malumbela et al. [3] on steel coupons. Few research work is being done by calculating the corrosion levels by extracting the corroded reinforcements from the concrete [4, 5] and had lesser samples under observation than the current ones. As in the corrosion of reinforcements goes on increasing, its load carrying capacity decreases. In the investigation, it is seen that per 1% increase in percentage of corrosion level, there is decline in the load carrying capacity for all beam model specimen [6]. The corrosion behavior of high-performance concrete having 25% each of slag and fly ash shows high protection to steel in comparison with ordinary concrete [7]. If the water cement ratio is lesser, the rebar corrosion is less, there is decline in the loss of weight of rebar, and it is as a result of increasing impermeability and lesser electrical conductivity [8]. Wu et al. [9] examined and studied the rates of corrosion from corrosive crack widths and demonstrated the linear relation between width of crack and corrosion rate. As higher water cement ratio, concrete shows high permeability, and high-strength concrete shows more crack width as compared to low-strength concrete for same corrosion level [10]. In earlier works it was found that samples with steel fibers had lesser corrosion [11]. Jin and Zhao [12] studied

S. Sathe (✉) · S. Patil · V. Shete
MIT World Peace University, Pune, India
e-mail: sandeepsatheresearch@gmail.com

A. Sil et al. (eds.), *Recent Trends in Civil Engineering*, Lecture Notes in Civil Engineering 274, https://doi.org/10.1007/978-981-19-4055-2_8

corrosion effects on bond behavior and bending strength of RCC beams and found that increasing reinforcement corrosion reduces the bending strength of corroded RCC beams. Kishore Kumar et al. [13] showed that the initial starting time of corrosion was delayed by using certain amount of fly ash, and initial starting time for normal concrete is less in comparison with fly ash mixed concrete and increases up to 30% replacement of cement with fly ash. Currently, study is pertaining on totally corroded RC beams to learn load-deflection relationship with fly ash and have undertaken considerable efforts to get the precise corrosion levels after the reinforced bar was extracted from concrete.

2 Materials and Methodology

2.1 Material Specifications

2.1.1 Cement

53 Grade O.P.C. Tested Conforming IS Codes: IS: 12,269-1987. Density of Cement: 2400 kg/m^3. Specific Gravity of Cement = 3.15 (IS-2720-Part-3).

2.1.2 Fine Aggregate

Natural river sand passing through 4.75 mm was used as fine aggregate and was tested as per Indian standard code (IS-383-1970 and IS-2386-Part-I-1963). The sand conformed to zone II.

2.1.3 Coarse Aggregate

The coarse aggregate of the maximum size 20 mm is used for experimentation. The aggregates were tested as per Indian Standard Specifications (IS-283-1970 and IS-2386-Part-4).

2.1.4 Fly Ash

The specifications of fly ash class-C conformed to Indian standard code (IS-3812 Part-I-2003). Test methodology adopted was as per IS-1727:2004.

Table 1 Concrete mix proportion for 0, 10, and 15% fly ash

Mix proportion with 0% fly ash	Mix proportion with 10% fly ash	Mix proportion with 15% fly ash
Cement: 474 kg/m^3	Cement: 426.6 kg/m^3	Cement: 402.9 kg/m^3
Mixing of concrete as per IS-4634:1968	Mixing of concrete as per IS-4634:1968	Mixing of concrete as per IS-4634:1968
Coarse aggregate: 1085 kg/m^3	Coarse aggregate: 1085 kg/m^3	Coarse aggregate: 1085 kg/m^3
Fine aggregate (river sand): 748 kg/m^3	Fine aggregate (river sand): 748 kg/m^3	Fine aggregate (river sand): 748 kg/m^3
Water: 110 kg/m^3	Water: 110 kg/m^3	Water: 110 kg/m^3
Fly ash: nil	Fly ash: 47.4 kg/m^3	Fly ash: 71.1 kg/m^3

2.2 Reinforced Concrete Beams

A steel bar of 12 mm diameter (Fe500) is used as main reinforcements for M25 grade of concrete. An 8 mm diameter bar is used as stirrups. All the reinforced bars were weighted properly, and all the properties were noted down properly. Steel bars were embedded in the concrete beams of dimensions 150 mm × 150 mm × 700 mm. M 25 grade of concrete was used for the casting of beams with proper calculations. In the present work, Indian Standard method (IS-10262-2009 and IS-456-2000) is used for concrete mix design with fly ash (Table 1).

2.3 Accelerated Corrosion Process

It is an electrochemical process to accelerate corrosion in beams. The corrosion of beams is a time consuming process, and thus, this process accelerates the corrosion of beams to the required value in stipulated time. All the specimens are immersed in 5% NaCl solution in full saturation condition. The stainless-steel plate is used as a cathode, and the reinforced bar is used as an anode. A plastic tank is used to mount full apparatus in solution. External voltage is used direct current power supply. Calculation of amount of current required to obtain different corrosion levels is done by

$$M_{\text{th}} = W.I_{\text{app}}.T/F.$$

where

M_{th} mass of rust (theoretical) per unit surface area of the bar in g/cm^2.
W equivalent weight of steel (27.925 g).
I_{app} applied current density in amp/cm^2.

Table 2 Time required to achieve the desired percentage of corrosion

Corrosion percentage	Current in amps	Flow of current in hours
10	4	112
15	4	167

Fig. 1 Accelerated corrosion connections

Fig. 2 Corroded beams sample

T time of induced corrosion (sec).
F Faraday's constant (96,487 amp-sec) (Table 2; Figs. 1 and 2).

2.4 Electrical Resistance of Reinforced Steel Bars

The electrical resistance of steel bars gives the exact and continuous health monitoring of the bar. It gives the value of the cross-sectional area loss, and also the difference in the resistance loss gives the actual value of weight loss.

2.4.1 Cross-Sectional Area Reduction (RCS)

$$\mathrm{RCS} = \mathrm{CS_o}R/R_\mathrm{o}\left(\mathrm{mm}^2\right) \tag{1}$$

where

$\mathrm{CS_o}$ is cross-sectional area.
R is the difference in the esistance.
R_o is the initial resistance.

2.4.2 Mass Loss (MLc)

$$\mathrm{MLc} = \mathrm{RCS}.L_\mathrm{A}.S_\mathrm{IR} \tag{2}$$

where

L_A is the length of the bar.
S_IR is specific weight of the bar.

3 Experimental Work

3.1 Compression Test

Specimens of size 150 × 150 × 150 mm from the mix were casted and tested for the compression test. The results of compression test with different proportions of fly ash and nominal mix after 28 days of curing of samples as follows (Table 3):

Table 3 Cube compressive strength with fly ash

Cement replaced by fly ash in percentage (%)	Failure load in kN	Compressive strength in N/mm^2	Average compressive strength in N/mm^2
0	695.25	30.90	30.51
	700.87	31.15	
	663.75	29.50	
10	875.25	38.90	39.03
	893.25	39.70	
	866.25	38.50	
15	810.00	36.00	35.56
	801.00	35.60	
	789.75	35.10	

3.2 Flexure Strength (Three Point Bending Test)

Three point bending test is used to obtain the flexural strength results in terms of load applied and the corresponding resultant strain on the beam. The bending test is conducted on beam with varying percentage of fly ash as cement replacement materials in 0, 10, and 15%. The results obtained are as follows (Tables 4, 5 and 6; Figs. 3, 4 and 5):

Table 4 Flexural strength of beams with 0% corrosion

Fly ash (%)	Ultimate load (kN)	Average ultimate load (kN)	Flexural strength (N/mm^2)	Deflection (mm)	Average deflection (mm)
0	226.30	220.00	58.66	12.90	12.40
	224.50			12.40	
	211.20			12.00	
10	230.20	228.60	60.96	11.20	10.90
	227.40			10.70	
	228.30			10.80	
15	231.30	237.00	63.20	9.90	10.03
	248.30			10.10	
	233.40			10.10	

Table 5 Flexural strength of beams with 10% corrosion

Fly ash (%)	Ultimate load (kN)	Average ultimate load (kN)	Flexural strength (N/mm^2)	Deflection (mm)	Average deflection (mm)
0	186.30	186.90	49.84	9.10	8.83
	186.20			9.00	
	188.30			8.40	
10	191.20	195.50	52.13	8.70	8.60
	197.30			8.80	
	198.00			8.40	
15	201.20	201.30	53.13	8.20	8.10
	202.00			8.10	
	200.80			8.00	

Table 6 Flexural strength of beams with 15% corrosion

Fly ash (%)	Ultimate load (kN)	Average ultimate load (kN)	Flexural strength (N/mm^2)	Deflection (mm)	Average deflection (mm)
0	150.00	145.40	38.77	8.60	8.40
	144.20			8.40	
	144.10			8.30	
10	155.00	156.80	41.81	8.20	8.30
	157.20			8.30	
	158.30			8.40	
15	162.00	161.40	43.04	7.90	8.00
	159.30			8.00	
	163.00			7.90	

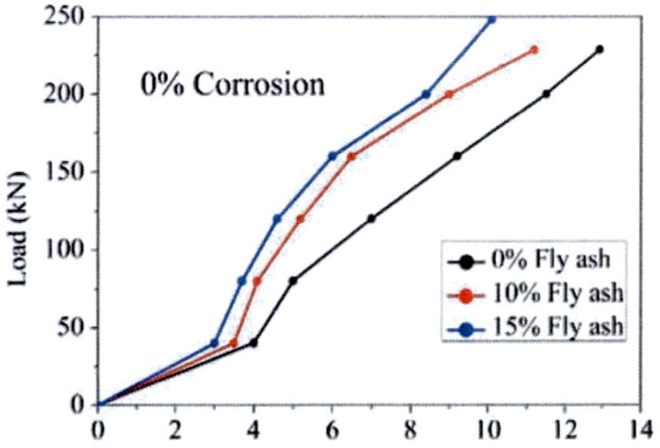

Fig. 3 Flexural strength of beams with 0% corrosion

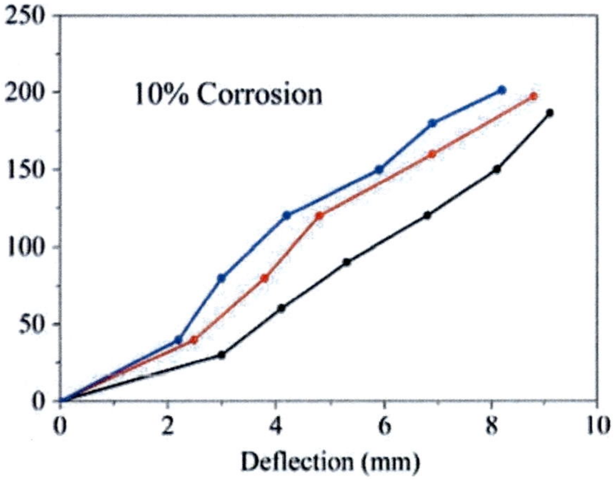

Fig. 4 Flexural strength of beams with 10% corrosion

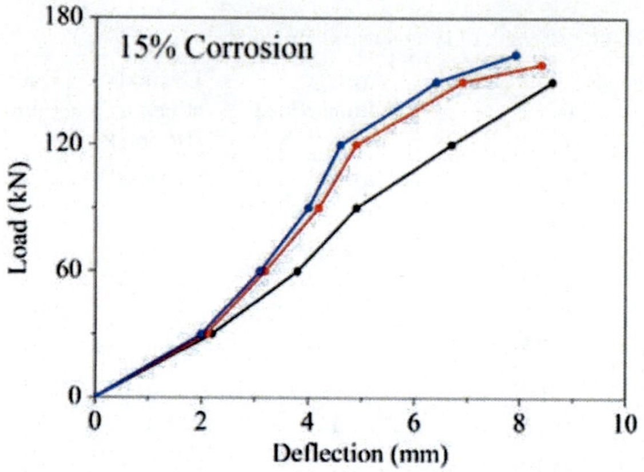

Fig. 5 Flexural strength of beams with 15% corrosion

Table 7 Calculated weight loss (MER method)

Corrosion (%)	Fly ash (%)	Calculated weight loss MLc in gram	Actual weight loss in gram
10	0	103.90	99.63
	10	88.20	86.01
	15	87.10	85.73
15	0	120.30	119.30
	10	103.20	98.12
	15	100.00	96.30

3.3 *Weight Loss in Reinforced bar*

Weight loss in reinforced concrete bar was determined by method of electric resistivity (MER method). Also actual weight loss was measured by extracting the reinforcement from beam after corrosion testing (Table 7; Fig. 6).

4 Conclusion

From the results, the 15% fly ash mix gives enhanced flexural strength growth by 7.75% on an average, while 10% fly ash replacement gives a growth of 3.67% as compared to the normal mix without fly ash for M25 grade of concrete. Also 15% fly ash mix gives lesser amount of weight loss of reinforcement bars as compared to 10% fly ash mix. Thus for overall mixes, it can be concluded that 15% fly ash mix limited till 15% maximum corrosion may give enhanced results than normal mix. This gives a future scope for more percentage of fly ash replacement and to study the enhancement of various strength parameters. Also the 10% fly ash shows higher

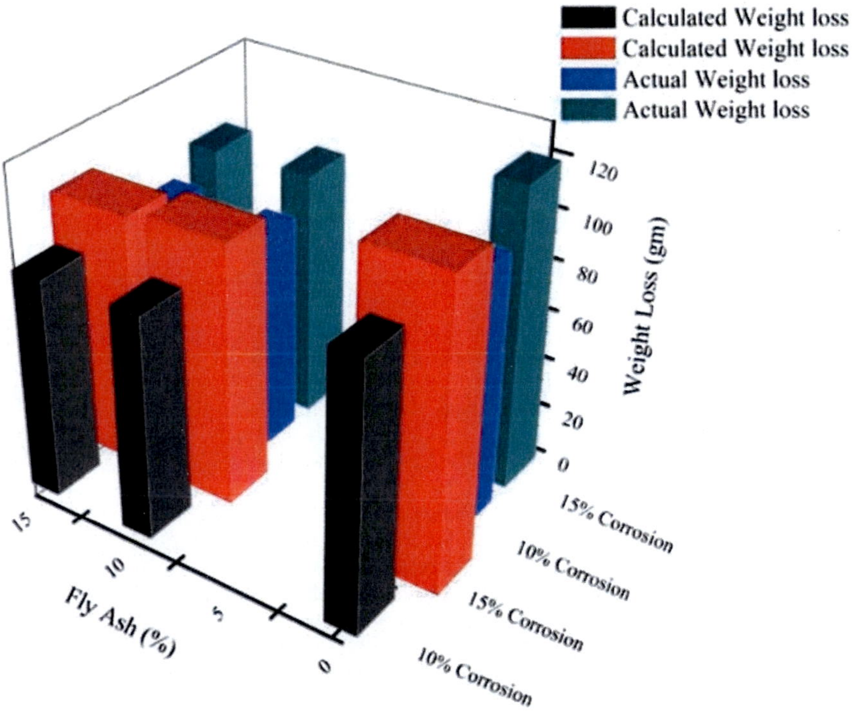

Fig. 6 Weight loss in reinforced bar

compressive value than 15% fly ash mix. Calculated weight loss MLc and actual weight loss in steel reinforcement due to corrosion are nearly same.

References

1. Zhu W, François R (2014) Corrosion of the reinforcement and its influence on the residual structural performance of a 26-year-old corroded RC beam. Constr Build Mater 51:461–472
2. Bazant ZP (1979) Physical model for steel corrosion in concrete sea structures theory. J Struct Div 105(6):1137–1153
3. Malumbela G, Alexander MG, Moyo P (2010) Variation of steel loss and its effect on flexural capacity of RC beams corroded and repaired under load. Constr Build Mater 24(6):1051–1059
4. Azad A, Ahmad S, Al-Gohi B (2010) Flexural strength of corroded reinforced concrete beams. Mag Concr Res 62(6):405–414
5. Ahmad S (2017) Prediction of residual flexural strength of corroded reinforced concrete beams. Anti-Corros Meth Mater 64(1):69–74
6. Shetty A, Venkataramana K, Babu Narayan KS (2015) Experimental and numerical investigation on flexural bond strength behaviour of corroded NBS RC beam. Int J Adv Struct Eng 7:223–231

7. Poursaee A, Hansson CM (2008) The influence of longitudinal cracks on the corrosion protection afforded reinforcing steel in high performance concrete. Cement Concr Res 38:1098–1105
8. Sangoju B, Gettu R, Bharatkumar BH, Neelamegam M (2011) Chloride—induced corrosion of steel in cracked OPC and PPC concretes: experimental study. J Mater Civil Eng 23(7)
9. Wu F, Gong J, Zhang Z (2014) Calculation of corrosion rate for reinforced concrete beams based on corrosive crack width. J Zhejiang Univ Sci A (Appl Phys Eng) 15:197–207
10. Yalciner H, Eren O, Sensoy S (2012) An experimental study on the bond strength between reinforcement bars and concrete as a function of concrete cover, strength and corrosion level. Cem Concr Res 42:643–655
11. Sadeghi-Pouya H, Ganjian E, Claisse P, Muthuramalingam K, Corrosion durability of high performance steel fibre reinforced concrete. In: 3rd International conference on sustainable construction materials and technologies
12. Jin W, Zhao Y (2001) Effect of corrosion on bond behavior and bending strength of reinforced concrete beams. J Zhejiang Univ (Sci) 2:298–308
13. Kishore Kumar M, Rao PS, Swamy BLP, Chandra Mouli C (2012) Corrosion resistance performance of fly ash blended cement concretes. Int J Res Eng Technol 01(03)
14. Song PS, Hwang S, Sheu BC (2005) Strength properties of nylon-and polypropylene-fiber-reinforced concretes. Cem Conc Res 35(8):1546–1550
15. Roesler JR, Altoubat SA, Lange DA, Rieder KA, Ulreich GR (2006) Effect of synthetic fibers on structural behavior of concrete slabs-on-ground. Mater J 103(1):3–10
16. Michels J, Waldmann D, Maas S, Zürbes A (2012) Steel fibers as only reinforcement for flat slab construction–experimental investigation and design. Constr Build Mater 26(1):145–155
17. Michels J, Christen R, Waldmann D (2013) Experimental and numerical investigation on postcracking behavior of steel fiber reinforced concrete. Eng Fract Mech 98:326–349
18. Pujadas P, Blanco A, Cavalaro S, Aguado A (2014) Plastic fibres as the only reinforcement for flat suspended slabs: experimental investigation and numerical simulation. Constr Build Mater 57:92–104
19. Rosidawani II, Sugiri S, Pane I (2015) Behaviour of macro synthetic fiber reinforced concrete columns under concentric axial compression. Proc Eng 125:987–994
20. Sofi A, Phanikumar BR (2015) An experimental investigation on flexural behaviour of fibre-reinforced pond ash-modified concrete. Ain Shams Eng J 6(4):1133–1142
21. Park Y, Abolmaali A, Beakley J, Attiogbe E (2015) Thin-walled flexible concrete pipes with synthetic fibers and reduced traditional steel cage. Eng Struct 100(1):731–741
22. Spadea S, Farina I, Carrafiello A, Fraternali F (2015) Recycled nylon fibers as cement mortar reinforcement. Constr Build Mater 80:200–2009
23. Fraternali F, Spadea S, Berardi VP (2014) Effects of recycled PET fibres on the mechanical properties and seawater curing of Portland cement-based concretes. Constr Build Mater 61:293–302
24. Won JP, Lim DH, Park CG (2006) Bond behaviour and flexural performance of structural synthetic fibre-reinforced concrete. Mag Concr Res 58(6):401–410

UHPC Steel Composite Girder: Numerical Studies on Flexural Behaviour in Negative Moment Region

Soorya M. Nair, P. Parthiban, and M. J. Anju

1 Introduction

The characteristics of good mechanical performance, low weight and expedited design make up steel-concrete composite beams common in engineering projects. Under stress, normal strength concrete (NSC) flange plates break quickly in the hogging moment zones of continuous steel-concrete composite beams, resulting in stiffness and durability losses [1, 2]. Consequently, the vulnerability of NSC flanges in hogging areas is one of the major issues that need to be addressed when it comes to the construction of continuous steel-concrete composite beams. Various strategies for improving crack resistance include prestressing of concrete slab, increasing the percentage of steel reinforcement in the hogging moment region and use of retrofitting techniques like CFRP sheets.

New kinds of cement materials, such as reinforced steel fibrous concrete (SFC), cemented composites (CCG) and UHPC, have been developed and extensively used in civil engineering applications in the last few years. Ultra-high-performance concrete seems to have a lot of untapped potentials to solve the above problem faced due to the reduced activity of normal strength concrete (NSC) in the hogging moment region. Hence, this area is having a lot of unexplored research significance. Shao & Cao et al. [3] proposed a new type of steel UHPC composite bridge deck structure as an initiative. Wang et al. [4] numerically and experimentally studied the behaviour of a continuous girder composed of steel box-girder and ultra-high-performance concrete waffle slab. The test findings demonstrated that the bending capacity for

S. M. Nair · P. Parthiban (✉) · M. J. Anju
Department of Civil Engineering, Jyothi Engineering College, Thrissur, India
e-mail: parthichevaux@gmail.com

S. M. Nair
e-mail: msooryanair@gmail.com

M. J. Anju
e-mail: anjumj@jecc.ac.in

A. Sil et al. (eds.), *Recent Trends in Civil Engineering*, Lecture Notes in Civil Engineering 274, https://doi.org/10.1007/978-981-19-4055-2_9

steel-high-performance continuous composite beams is 1.2 times that of the continuous composite beam of steel-concrete [4]. Flexural testing on composite steel UHPC beams with SU-S and bolt connections (SU-B) connections at the interface was performed by Zhang et al. [1]. Cracking resistance, ultimate flexural capacity, mechanisms of failure and hogging deformation properties were examined. The test results demonstrated outstanding cracking and bending characteristics of steel UHPC composites in the hogging region [1].

The purpose of this research work is to carry out the numerical analysis of the bending behaviour of continuous stainless steel-concrete composite girders by employing ultra-high-performance concrete (UHPC) in place of normal strength concrete (NSC) in the hogging zone. The study's main objective was to analyse the increase in strength and deflection capacity, owing to UHPC presence, of a continuous composite girder made up of an I section steel girder and a UHPC slab [5, 6]. The influence of numerous factors on a UHPC composite girder, including flange depth, breadth and steel grades, have been explored in this work.

Based on the literature review, a simplified numerical model was developed using ANSYS 18.0 software to predict the flexural behaviour of the negative moment region by isolating it from the rest of the structure [7]. To ensure the reliability of the proposed model, the obtained results were also compared with the experimental studies on similar structures conducted by Zhang et al. [1].

2 Finite Element Analysis (FEM)

The numerical model was simulated using ANSYS 18.0 software, to study the nonlinear flexural performance of UHPC steel composite girder. It was used to stud the impact of different aspects affecting girder capacity in the negative moment region [8]. Using the Newton-Raphson method, the applied load was iterated step by step. A bonded type contact was used to simulate the connection at the interface of slab and girder. All components were meshed in cubes of maximum 50 mm × 50 mm × 50 mm. The boundary conditions were established using simply supported ends.

2.1 Element Selection

The concrete was modelled with the element of Solid 65. This element contains eight nodes in each node x-, y-, and z-axes as illustrated in Fig. 1, with three-degree freedom. This element enables plastic deformation, crushing, and cracking in 3 orthogonal dimensions [9]. LINK 180 elements were used to model the reinforcements as they can be incorporated into the SOLID 65 element with ease. It is truss elements that can handle the 3 degrees of freedom along the x-, y-, and z-axes of the node, as illustrated in Fig. 2 [9].

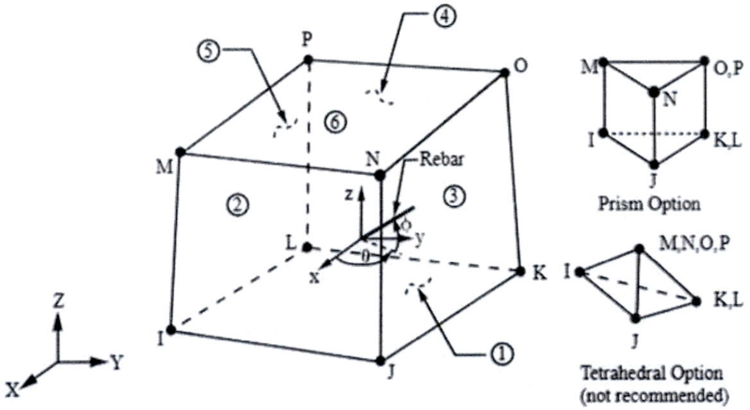

Fig. 1 Solid 65 element

Fig. 2 Link 180

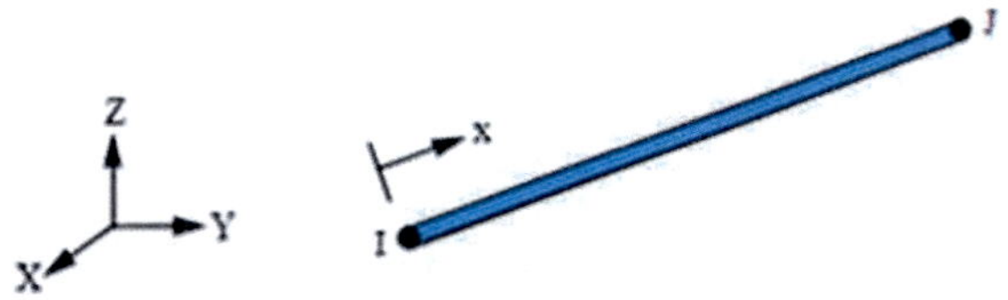

SOLID186 element with 20 nodes containing 3 degrees of freedom per node was used to define steel girders and stiffeners Fig. 3 [9]. The element was deemed appropriate as it supports great deflection and extensive strain capacity. The interaction at the interface of the UHPC slab-steel girder was simulated using TARGE170 and CONTA174 elements [10]. The two material contacts model the friction and cohesion behaviour of the connected elements.

2.2 *Material Modelling*

Concrete was modelled as a homogenous isotropic material. The properties of the concrete are given in Table 1. Normal strength concrete (NSC) of grade M50 was used for the validation and comparative studies. Ultra-high-performance concrete of compressive strength 144 MPa was utilized to carry out the parametric studies of hogging moment region. Material properties of NSC were adopted based on studies conducted by Zhang et al. [1] and UHPC from Ibraheem [11] as cited by Al-Azzawi et al. [12] Kent-park model was adopted to simulate the behaviour of concrete in ANSYS.

Bilinear isotropic stress–strain relationship was used to characterize the behaviour of steel materials, including rebars, stiffeners and girders. Table.2 gives the properties

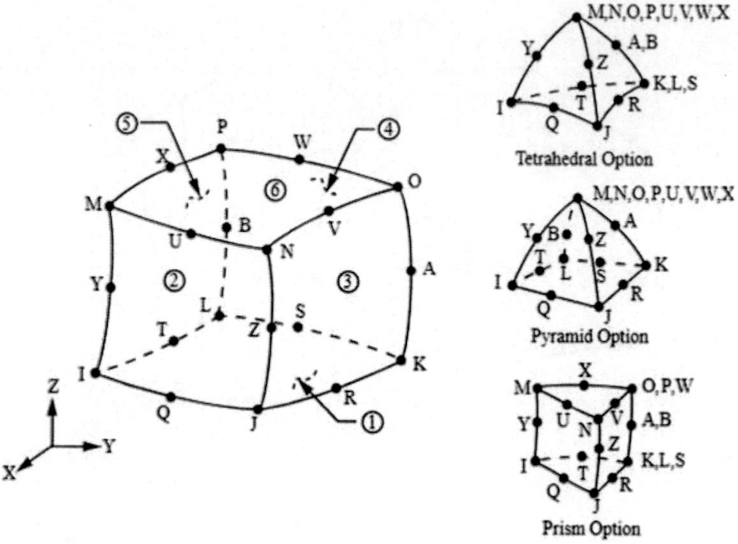

Fig. 3 SOLID 186 element used for steel girder

Table 1 Mechanical properties of normal strength concrete and ultra-high-performance concrete

Material	Compressive strength (f_c) (MPa)	Poisson's ratio (μ)	Modulus of elasticity (E_c) (GPa)
NSC	50	0.15	36.57
UHPC	144	0.20	40.50*

*Ibraheem [11] as cited by Al-Azzawi et al. [12]

summary of the steel elements used in the study. Q35 alloy steel was used to model the I section girders for the studies.

The steel model is based on the assumption that it is an elastic-linear strain hardening material with similar tension and compression characteristics. The elastic modulus was divided by 200 to get the strain hardening modulus.

Table 2 Mechanical properties of steel elements

Material	Yield strength (F_y) (MPa)	Tensile strength (F_y) (MPa)	Modulus of elasticity (E_c) (GPa)
Steel Girder (Q35)	418.21	540.35	206
Reinforcement	466.72	572.38	200
Carbon steel	345	450	204
Stainless steel	205	515	193

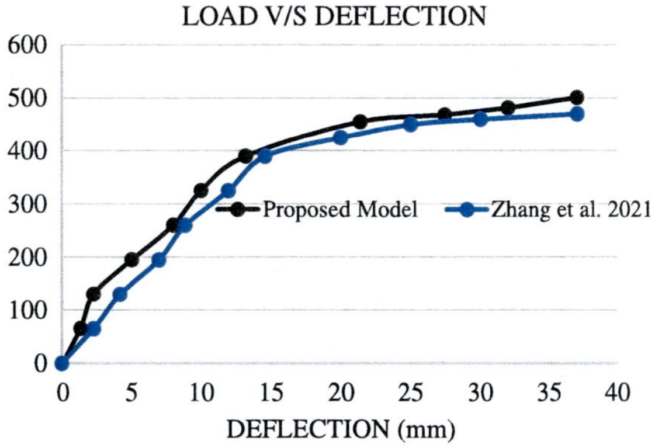

Fig. 4 Validation curve

2.3 *Model Validation*

Zhang et al. [4], the study was chosen for validation because it focuses on the capacity of ultra-high-performance concrete (UHPC) to improve flexural strength in the hogging moment region of composite beams. In this research, the girder indicated as SC-S in the base paper was simulated, and the load–deflection curves obtained are as given in Fig. 4.

Ultimate load obtained from Zhang et al. was 470 kN and the proposed model showed an ultimate value of 500 kN. On comparing the results for ultimate load corresponding to a deflection of 37 mm, an error of only 6% was obtained. Therefore, a numerical study on the behaviour of the hogging part of the UHPC steel-concrete composite girder may be performed using the suggested model.

3 Numerical Studies

Several studies have shown that the use of a high-performance concrete slab with the steel girder in the hogging zone improves the ultimate moment capacity of the beam. A parametric investigation based on the future scopes of Zhang et al. is implemented using the numerical model developed.

3.1 *Specimen Details*

Three steel-concrete composite girder specimens were modelled to determine the impact of using UHPC concrete on the hogging moment region of a continuous girder. Specimen SC-1 used normal strength concrete (NSC) having a compressive

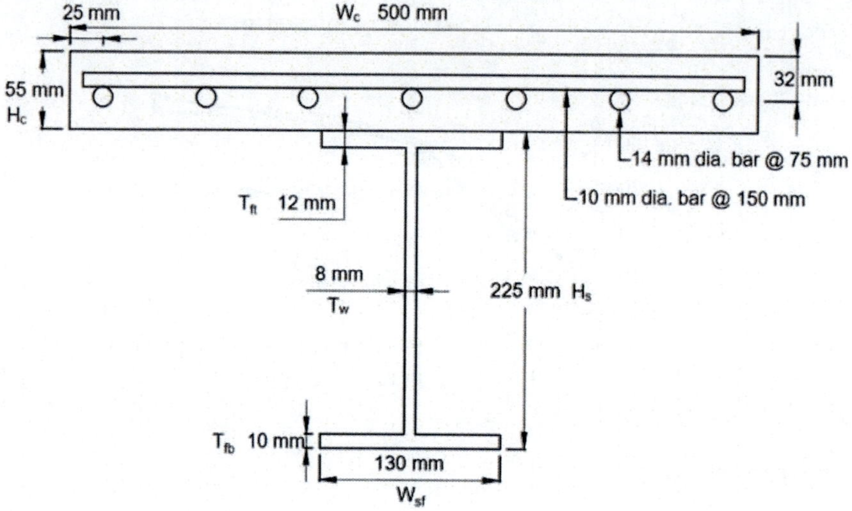

Fig. 5 Cross-sectional details of SC-2

Table 3 Cross-sectional details of the specimen modelled

Specimen	Material	W_c (mm)	H_c (mm)	W_{sf} (mm)	H_s (mm)	T_{ft} (mm)	T_{fb} (mm)	T_w (mm)
SC-1	NSC	500	55	130	225	12	10	8
SC-2	UHPC	500	55	130	225	12	10	8
SC-3	UHPC	500	55	140	250	12	10	8

strength of 50 MPa and SC-2 and SC-3 were modelled using UHPC of 144 MPa. Figure 5 and Table 3 provide the cross-sectional features of the modelled specimen.

The mechanical properties of steel and concrete were adopted as same as those given in the reference paper. The percentage of longitudinal reinforcement was kept as 3.91%, and the length of the hogging moment region was taken as 2000 mm. Web stiffeners have been provided at the loading points and at the supports using the same grade of steel used for the girder and a thickness of 8 mm. Simply supported condition was used to simulate the behaviour of the negative moment region.

To create the hogging effect in the part isolated from a continuous girder, incremental loads were applied from bottom side of the girder in the upward direction as shown in Fig. 6. Three-point loading method was used to obtain the flexure behaviour at the mid-span of the specimen and load–deflection curves were obtained up to a deflection of 40 mm.

Fig. 6 Loading arrangement

3.2 Effect of Concrete Strength

Load deflection behaviour of specimens was analysed by providing incremental loads. The ultimate loads obtained corresponding to a deflection of 40 mm were used to compare the effect of concrete in the hogging moment region (Fig. 7). The flange stress, strain at the interface and the concrete strain obtained for the corresponding loads were also plotted to investigate the varying behaviour of normal and high-strength concrete in the region [13].

The deflected shapes of the girder components are represented using Fig. 8. The ultimate load obtained for the NSC girder was 495kN, and UHPC girder was 600.5 kN and 635 kN for SC-1 and SC-2, respectively. Nearly 25–30% increase in the load carrying capacity was shown by the UHPC girder. This suggests that using ultra-high-performance concrete instead of NSC improves the load–deflection behaviour of the hogging moment zone.

Not only that it reduces the concrete strain as well as the tensile stress and strain on the girder flange at the interface as shown in Fig. 9a, b and c. The turning points in the curve indicate the cracks, yielding and the impending failure. The studies

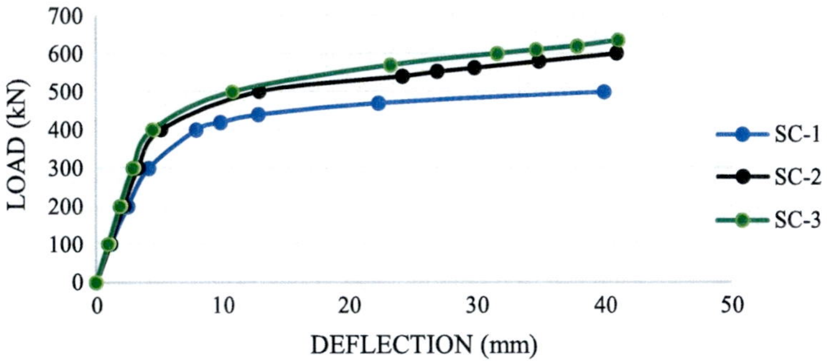

Fig. 7 Load-deflection curve

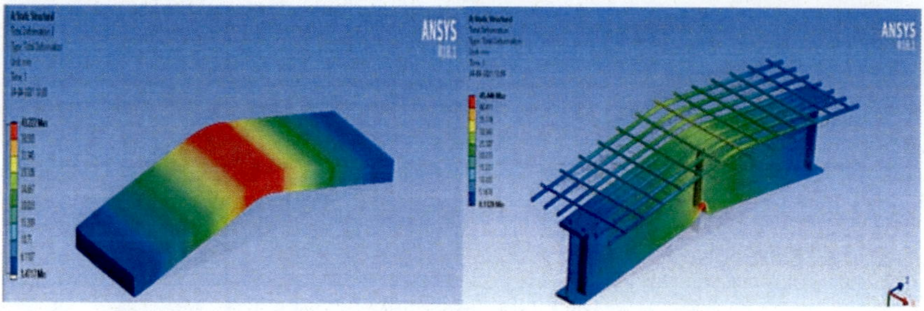

Fig. 8 Deformed shape of UHPC slab and steel girder and reinforcements

point towards the fact that the use of UHPC improves the contribution of concrete in handling the tensile stresses developed at the hogging region.

3.3 *Effect of Slab Depth*

Under this study, the depth of the slab was varied from 45 to 100 mm to identify the impact of the slab depth on the behaviour of the hogging moment region in the girder. The other dimensions of the slab and girders were not varied. Also, the reinforcement percentage was kept constant as 3.91% by varying the number and spacing between the rebars.

The load-bearing capacity of the hogging moment zone improved as the slab depth increased, as shown in Fig. 10. However, we can see that as the depth of the slab increases the difference between the capacities of the two specimens is getting negligible. It can be inferred from the deformation shape that this behaviour is due to the shift in failure mode from concrete slab deformation to the yielding of steel girder yielding. Further experimental studies and detailed modelling of the properties of the steel girder used are necessary to substantiate the findings.

3.4 *Effect of Slab Width*

The steel girder interacts with the concrete flange width at the interface of the composite structure. Width of the concrete girder was increased from 200 to 700 mm, and the load corresponding to standard deformation was selected. The dimensions of the steel girder and percentage of reinforcements were kept constant throughout the study.

Increase in the width of the concrete slab resulted in enhanced capacities of UHPC steel–concrete girders as shown in Fig. 11. The load-bearing capacity of the specimen increased by 4–5% at first, but as the slab's breadth increased, the capacity decreased. We can see that the enhancement in capacities becomes negligible.

Fig. 9 **a** Flange stress in the steel at the interface; **b** steel flange strain at the interface; **c** concrete strain

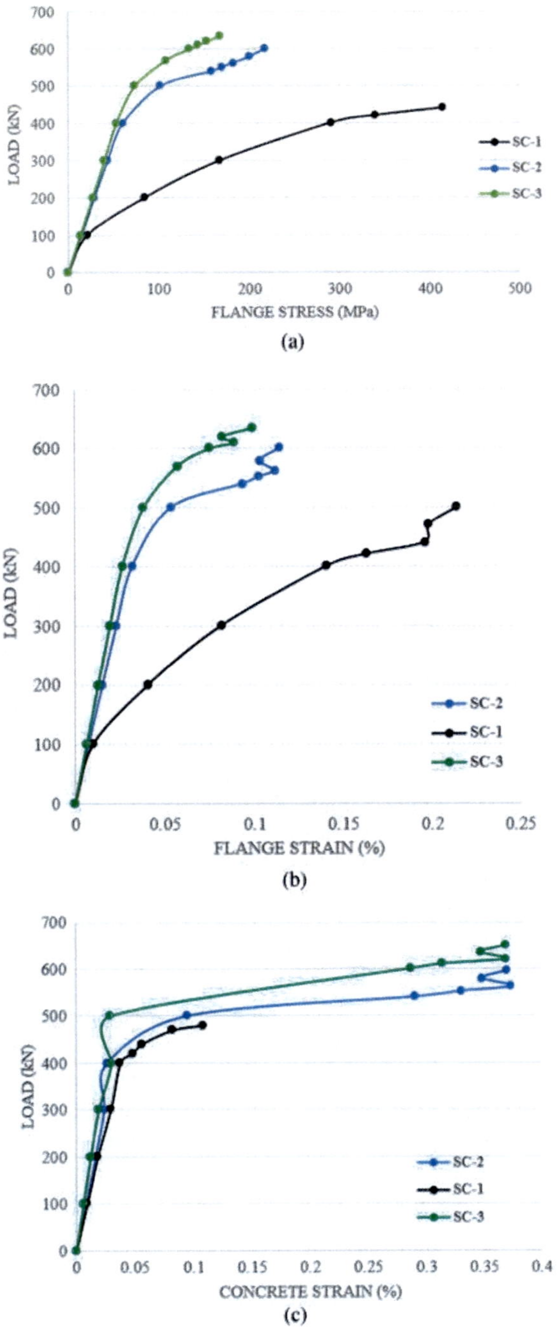

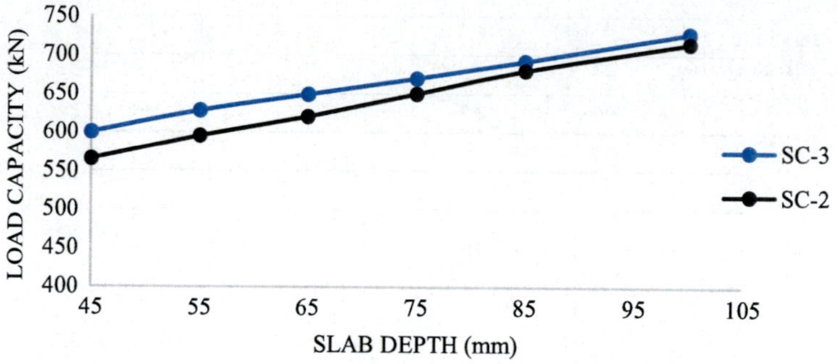

Fig. 10 Effect of the concrete slab depth on the girder strength

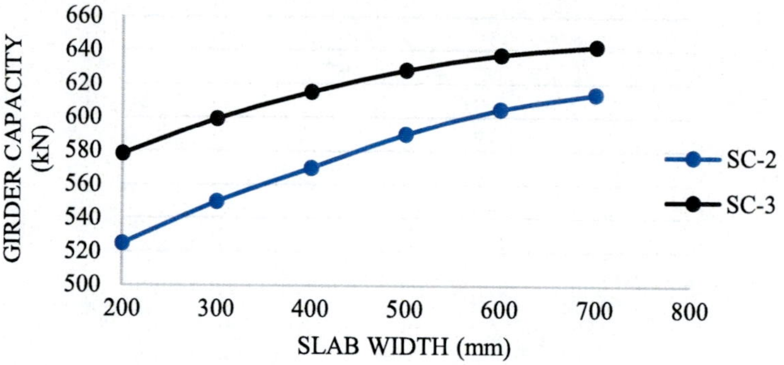

Fig. 11 Effect of the concrete slab width on the girder strength

3.5 *Effect of Steel Grade*

In this parametric study, ultra-high-performance concrete was combined with an I section girder made up of three different steel grades. Carbon steel is a material that is most commonly used in I section modelling. Stainless steel was chosen owing to its superior corrosion resistance and ductility properties [14]. SC-2 specimen was selected and the load–deformation behaviour when the steel girder used was changed from alloy steel to carbon steel and stainless steel was studied and compared (Fig. 12).

It was found that the optimum results were obtained when UHP concrete was combined with alloy steel. The results obtained for carbon steel were found to be comparable to those obtained for alloy steel. The alloy steel reference specimen was more resistant to deformation than the carbon steel specimen by 16%. When girder capacity was compared to stainless steel, alloy steel offered nearly 50% higher capacity. It was possible to conclude that the performance of UHPC slabs combined with stainless steel girders was inferior to that of NSC slabs combined with alloy steel sections.

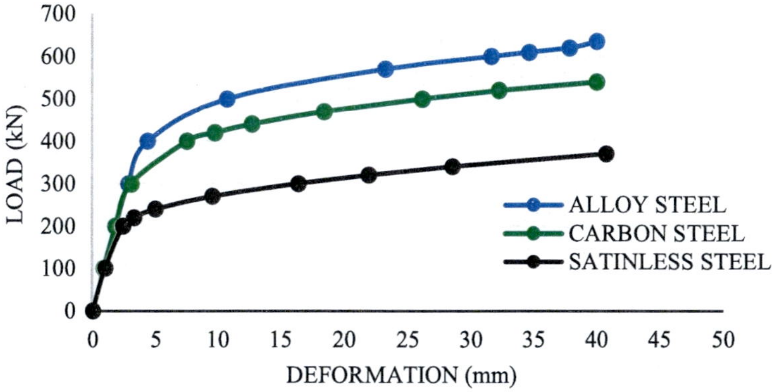

Fig. 12 Load–Deformation curves

4 Conclusions

In this work, a numerical model was made to examine the behaviour of the hogging moment area of a continuous steel–concrete composite girder made up of steel girder and UHPC slab. Static tests were carried out by isolating the hogging moment region and the results obtained were analysed to calibrate the proposed model. A detailed parametric study was carried out to examine the impact of some factors that affect the load-bearing capability of the region under examination. We came to the following conclusions:

1. The girders modelled using UHPC exhibited 25–30% increase in the ultimate capacity than the conventional NSC steel girders. The studies indicate towards the fact that the use of UHPC improves the contribution of concrete in handling the tensile stresses developed at the hogging region.
2. Up to a certain point, increasing the breadth of the concrete slab improves the capabilities of composite girders. As we go on increasing the width of the slab, it was observed that the enhancement in capacities becomes negligible.
3. The hogging moment region's load-bearing capability increased as the slab's depth increased. However, we can see that as the depth of the slab increases the difference between the capacities of the two specimens was found to be negligible. This behaviour might be explained by the shift in failure mode from concrete deformation to steel girder yielding as the slab becomes deeper.
4. The UHPC slab combined with the alloy steel girder was found to contribute more towards the enhancement of the load-bearing capacities of the hogging moment region than the one combined with carbon steel and stainless steel girders. For stainless steel specimens, it was observed that the deformation was reached predominately due to the yielding of steel than due to the deflection of concrete in the hogging moment region.

The model developed using ANSYS successfully represents the flexural behaviour in the hogging moment zone of a continuous UHPC steel composite girder in terms of deflection and capacity. Many assumptions are employed in the suggested model. Additional experimental research is needed to validate the findings and confirm the influence of various factors in real-world scenarios.

References

1. Zhang Y, Cai S, Zhu Y, Fan L, Shao X (2020) Flexural responses of steel-UHPC composite beams under hogging moment. Eng Struct 206:110134
2. El-Zohairy A, Salim H (2017) Parametric study for post-tensioned composite beams with external tendons. Adv Struct Eng 20(10):1433–1450
3. Shao X, Cao J, et al (2018) Static and fatigue properties of the steel-UHPC lightweight composite bridge deck with large U ribs. https://doi.org/10.1016/j.jcsr.2018.05.011
4. Wang H, Sun T, Tang C, Wang J (2020) Experimental and numerical investigation of steel-ultra-high-performance concrete continuous composite beam behavior. Adv Struct Eng 23(10):2220–2236
5. Ryu HK, Chang SP, Kim YJ, Kim BS (2005) Crack control of a steel and concrete composite plate girder with prefabricated slabs under hogging moments. Eng Struct 27(11):1613–1624
6. Nair SM, Mohan N (2019) Numerical studies on strengthening of continuous steel-concrete composite girders using CFRP. In: National conference on structural engineering and construction management. Springer, pp 435–444
7. El-Zohairy A, Salim H, Shaaban H, Mustafa S, El-Shihy A (2015) Finite-element modeling of externally posttensioned composite beams. J Bridg Eng 20(12):04015018
8. Tavakkolizadeh M, Saadatmanesh H (2003) Strengthening of steel concrete composite girders using carbon fiber reinforced polymers sheets. J Struct Eng 129(1):30–40
9. ANSYS (2008) Element manual, 10th edn., Release 4.1. Swanson Analysis Systems, Canonsburg, PA
10. Sharif AM, Samaaneh MA, Azad AK, Baluch MH (2016) Use of CFRP to maintain composite action for continuous steel–concrete composite girders. J Compos Constr 20(4):04015088
11. Ibraheem SKH (2008) Stress-strain relationships of reactive powder concrete. Ph.D. Thesis, University of Technology, Baghdad, Iraq, p 166
12. Al-Azzawi A, Sultan Ali A, Risan KH (2011) Behavior of ultra high performance concrete structures. ARPN J Eng Appl Sci 6(5):95–109
13. Alagunatrayan T, Mohanraj P, Manishankar S, et al (2021a) Engineered cement composite enhanced reinforced concrete beams using flexural response: an experimental study. Mater Today Proc. https://doi.org/10.1016/j.matpr.2021.07.130
14. Parthiban P, Kar S, Kumar Mondal A, et al (2021b) Stress-strain behaviour of (HPFRC) high-performance fibre reinforced concrete: an experimental study. Mater Today Proc. https://doi.org/10.1016/j.matpr.2021.07.175

Parametric Studies and Optimization of Grid Shell Structures Using Genetic Algorithm

S. Gokul Santhosh, A. P. Singh, and M. Abdul Akbar

1 Introduction

Grid shell structures have grown in popularity in recent years as a result of their flexibility and aesthetics. They are basically reticulated shell structures which are divided into smaller grids made of beam elements. Even though they are made up of beam elements, they do have a structural load transfer action similar to continuous shell structures due to their shape and spatial arrangement. Hence, the spacing and arrangement of the grids become a matter of structural importance. They are lightweight structures like regular shell structures and are able to cover large spans without the need for columns and are extremely material and resource efficient.

Timber is one of the commonly used materials for construction of grid shell structure (Fig. 1). However, advances in construction techniques have recently opened up the possibility of using other polymer-based composite materials (Fig. 2). Grid shells are made by mapping the grid onto a shell structure. As per IS 2210–1988 [3], shell structures can be broadly classified based on their Gaussian curvature into 4 categories:

1. Monoclastic: Surfaces having zero Gaussian curvature.
2. Synclastic: Surfaces having positive Gaussian curvature.
3. Anticlastic: Surfaces having negative Gaussian curvature.
4. Other special types.

S. Gokul Santhosh (✉) · A. P. Singh · M. Abdul Akbar
Department of Civil Engineering, Dr. B R Ambedkar National Institute of Technology Jalandhar (NITJ), G. T. Road, Amritsar Bypass, Jalandhar, Punjab 144011, India
e-mail: gokulss.sc.19@nitj.ac.in

A. P. Singh
e-mail: singhap@nitj.ac.in

M. Abdul Akbar
e-mail: akbarma@nitj.ac.in

A. Sil et al. (eds.), *Recent Trends in Civil Engineering*, Lecture Notes in Civil Engineering 274, https://doi.org/10.1007/978-981-19-4055-2_10

Fig. 1 Mannheim Multihalle [1]

Fig. 2 Grid shell in composite materials for Solidays festival in Paris [2]

In the work reported through this paper, grid shells are mapped to shell structures belonging to the three major classification based on Gaussian curvature (Fig. 3). They are:

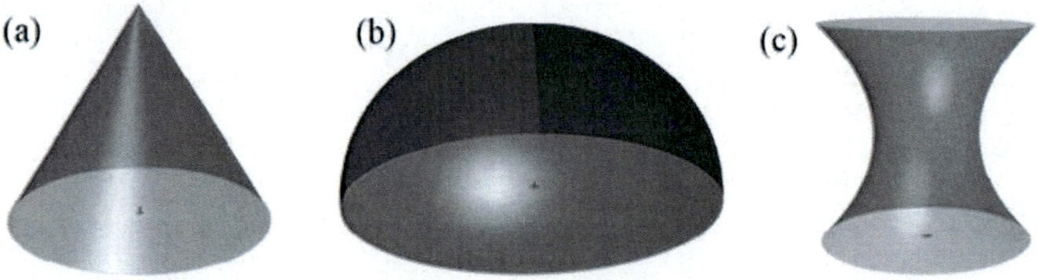

Fig. 3 Types of shells structures: **a** conical shell; **b** ellipsoids of revolution and **c** hyperboloids of revolution of one sheet

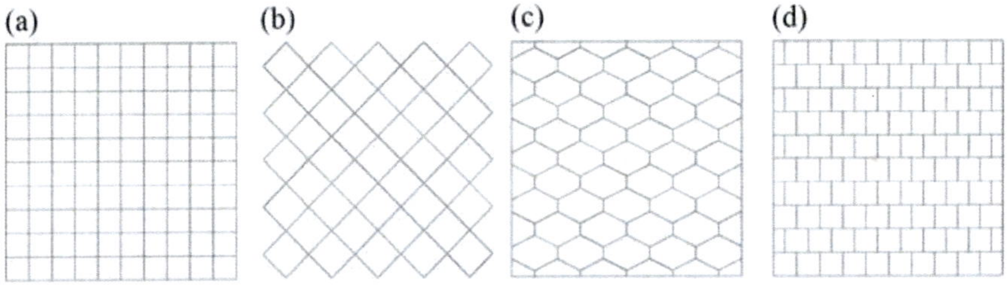

Fig. 4 Types of grid patterns: **a** quad pattern; **b** diamond pattern; **c** hexagonal pattern and **d** staggered pattern

- Conical shell—Monoclastic structure.
- Ellipsoids of revolution—Synclastic structure.
- Hyperboloids of revolution of one sheet—Anticlastic structure.

Furthermore, each structure is further studied in detail considering the maximum displacement, tension, compression, kicking force and material usage using four different patterns of grid arrangements namely (Fig. 4).

1. Quad pattern.
2. Diamond pattern.
3. Hexagonal pattern.
4. Staggered pattern.

The paper has taken into account rise to span ratio, different materials and different types of grid patterns and the variations in their distribution within each type of structure. While this opened up a plethora of design variants and an opportunity for exploration of a highly efficient and economic design, it has also projected a challenge in the workflow to be adopted to generate the various design variants and the necessity to evaluate the resultant design effectively and gain a good qualitative insight among the different variants. One of main challenges in the initial stage of the work was modeling of the different variants. To tackle this problem, a parametric approach was adopted. It was found that parametric approach implementation in a similar problem helped in optimizing the structure better for its surrounding and for reducing the overall material consumption [4].

Integrating the principles of structural analysis and optimization at the conceptual stage and into the architectural design process will aid in arriving at more energy and resource efficient design solution for the built environment [5]. Parametric design tools build using Rhinoceros 3D [6] and Grasshopper [7] for structural analysis of shallow shells have shown that real-time insights into the structural behavior provides an intuitive way to experiment with various design alternatives. The work also studied how the structural behavior of shell structures relates to its geometry [8]. For structural problems, genetic algorithms have proven to be an effective optimization technique [2, 9, 10]. Along with a good optimization method, having a proper visual display of data and parameters involved in design can help structural engineers and architects make better assessments [11]. In this paper, parallel coordinates plot was used to compare the performance and the feasibility of various design variants.

Upon review of literature, it is seen that analysis of different types of grid patterns in grid shells based on regular shell structures as well as their optimization has not been taken up in detail. This paper aims to fill that void while also encouraging other researchers to implement similar parametric and design optimization workflows for more efficient resource use.

2 Methodology

A parametric model generation system was built which made the process of generation of a wide spectrum of models from within a specified range of parameters quite easy. This saved a significant amount of time and effort that would have otherwise been spent in manually constructing each model variant needed for analysis. There are three main grid shell systems, each with four different grid patterns, amounting to a total of 12 different geometries (Figs. 5, 6 and 7).

The parametric system was set up using Rhinoceros 3D and Grasshopper. Rhinoceros 3D (also known as Rhino) is a commercial CAD and 3D computer graphics software that facilitates the creation of 3D geometry. Grasshopper on the other hand is a visual programming interface plugin for Rhino which utilize the features of Rhino in a procedural and parametric fashion. In a way, it facilitates a non-destructive workflow through the usage of "components" which are essentially small snippets of actions or code. These are then connected to one another by "wires" to create a meaningful sequence of actions (Fig. 8).

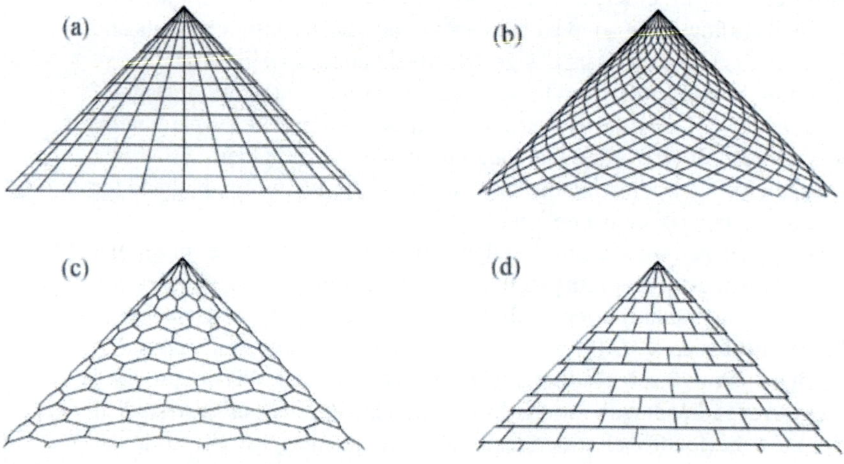

Fig. 5 Conical grid shell with: **a** quad pattern; **b** diamond pattern; **c** hexagonal pattern and **d** staggered pattern

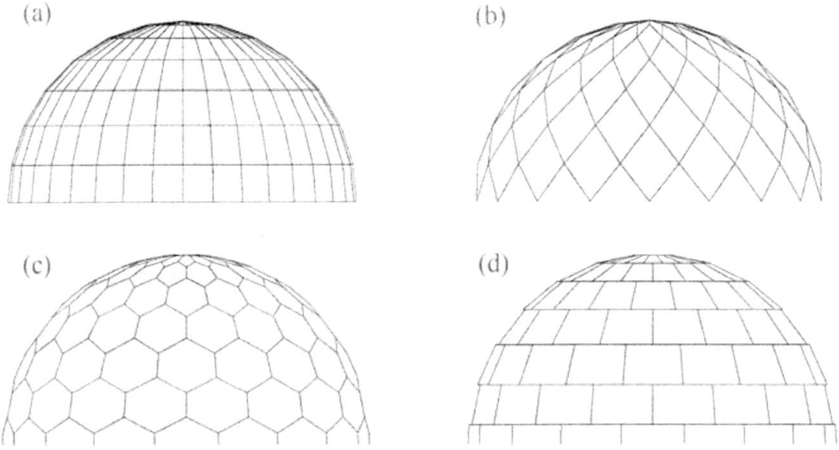

Fig. 6 Ellipsoids of revolution grid shell with: **a** quad pattern; **b** diamond pattern; **c** hexagonal pattern and **d** staggered pattern

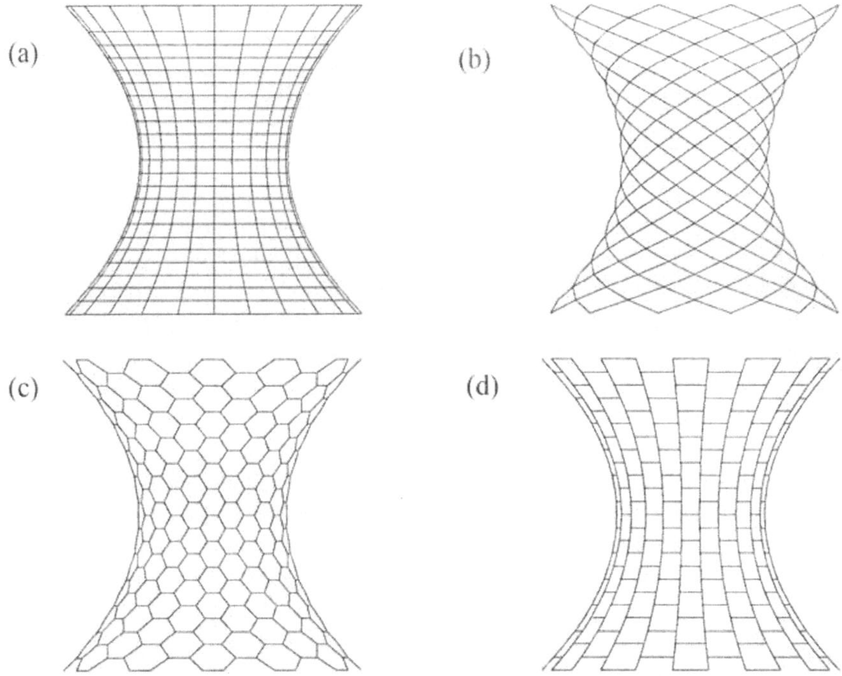

Fig. 7 Hyperboloids of revolution of one sheet grid shell with: **a** quad pattern; **b** diamond pattern; **c** hexagonal pattern and **d** staggered pattern

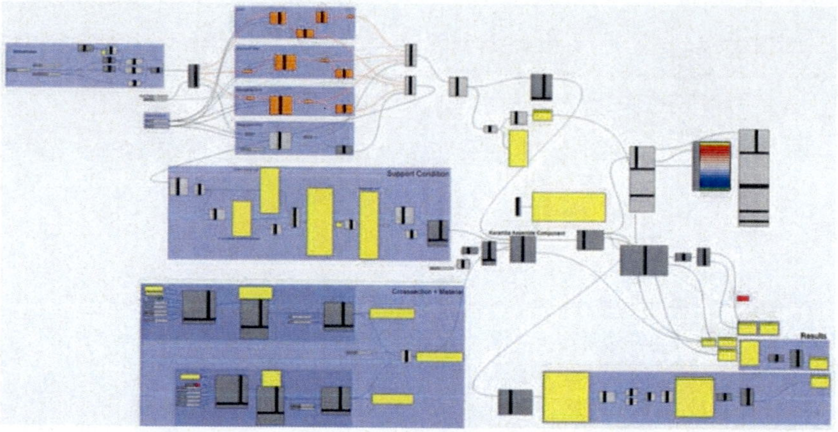

Fig. 8 Grasshopper definition used for defining the hyperboloids of revolution of one sheet grid shell

For grid shells based on conical shell and ellipsoids of revolution, the parameters adopted for the study were rise to span ratio, number of panels required in both directions and the type of material. The span was fixed as 40 m, and the rise was increased from 5 m up to 40 m in increments of 5 m, and the rise to span ratio was noted accordingly. The materials used in the analysis were glued laminated timber (GL32h) [4] and glass fiber-reinforced polymer (GFRP) [12], both of which had previously published material properties and are as given in Table 1.

In order to make a proper comparison on how the structure performs with these two materials, the cross-section was kept a constant. A solid cross-section of 10 cm × 10 cm was adopted for both the materials. For the hyperboloids of revolution of one sheet instead of rise to span ratio, *r*/*R* was adopted where *r* represents the middle radius and *R* represents the top and bottom radius. The top and bottom radius were fixed at 10 m and the middle radius varied from 1 to 10 m in increments of 1 m. The total height considered for the structure was 20 m. The rest of the parameters adopted were similar to the previous two cases.

The grids were created with the aid of a Grasshopper plugin called LunchBox [13], which assisted in mapping the grid patterns onto the various shell structures. The number of grids required along the horizontal and vertical direction (*U* and *V*)

Table 1 Material properties adopted for study

Properties	GL32h	GFRP
Young's modulus (kN/cm^2)	1180	3000
Shear modulus (kN/cm^2)	400	1171.87
Yield strength (kN/cm^2)	3.2	43.3
Specific weight (kN/m^3)	4.3	17.65
Poisson's ratio	0.475	0.28

(Fig. 9) was specified by means of a number slider which ranges from 0 to 100 along both directions. A slider was also set up to switch between the various grid patterns within a particular type of grid shell structure. The implementation of sliders within Grasshopper helps to quickly swap between the various designs variants. The support conditions adopted were pinned condition, and the number of supports was related to the number of grids along the U direction and has been assigned automatically with the help of a custom Python script. In terms of loading, only self-weight has been considered. The structural analysis was performed using Karamba3D [14]. Karamba3D is a commercial structural engineering plugin for Grasshopper. It runs within the same parametric design environment of Grasshopper and has a native access to all of Grasshopper geometry and components. Like Grasshopper, Karamba3D also comes with a set of its own "components" for performing various types of structural analysis. In this paper, Karamba3D was used for calculation of maximum displacement, tension, compression, kicking force and material usage. The line elements forming the geometry of the grid shells were fed into Karamba3D's "line to beam" component to convert them to beams. All other structural parameters and attributes are given to the model through the various other Karamba3D components.

Since analyzing a total of 1,60,000 possible combinations per grid shell takes a long time, genetic algorithm (GA) was used to fine-tune and obtain the best design variants in a much lesser number of iterations, saving a lot of computational resources and time. Galapagos [15] was used to carry out the study's optimization (Fig. 10). Galapagos is an evolutionary solver plugin for Grasshopper which also runs in the same parametric design framework and uses genetic algorithm for optimization of the problem at hand. It was used to fine-tune the grid shell system's parameters to achieve the least displacement. It is based on Darwin's theory of evolution and is based on basic biological growth laws. It consists of a fitness function, and the solver's goal is to satisfy the function by adjusting the design variables ("genome"). The genomes were the rise to span ratio, number of grids along the U and V directions and type of material, and the fitness function was to reduce displacement. The

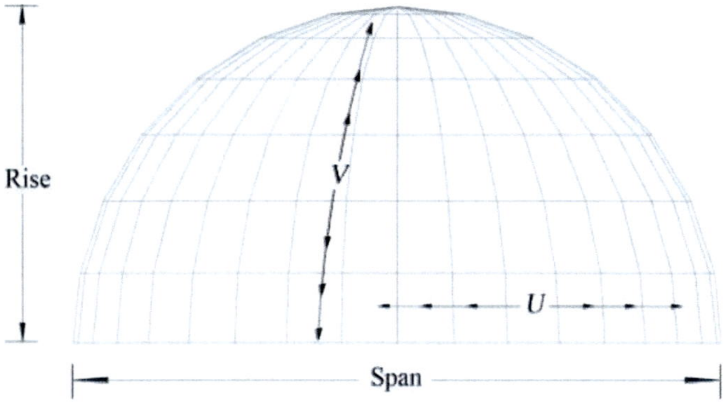

Fig. 9 *U*, *V*, rise and span in ellipsoids of revolution grid shell

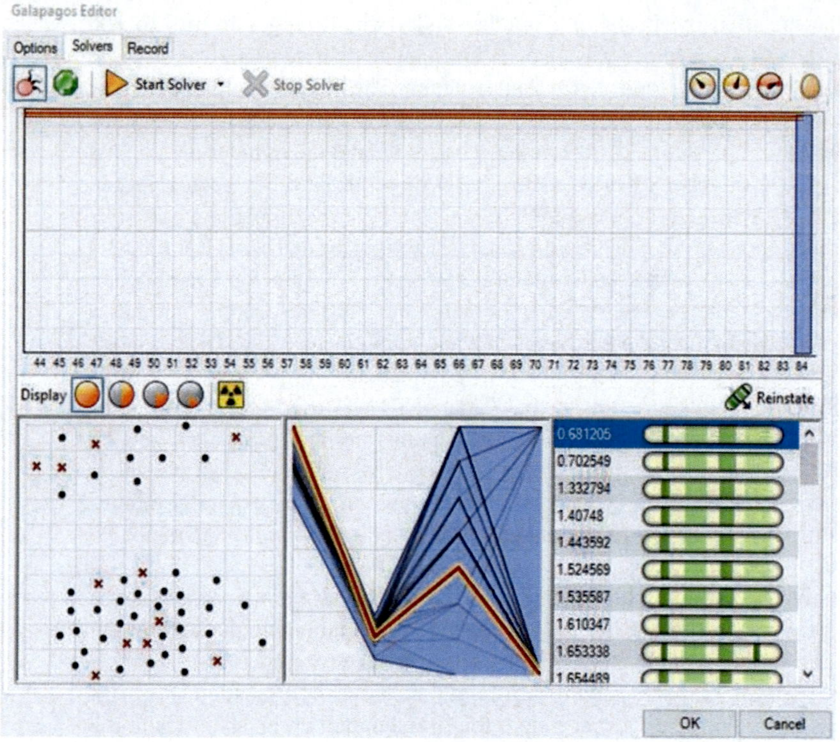

Fig. 10 Galapagos editor iterating over the various genomes

method significantly reduced the number of iterations needed to achieve a satisfactory displacement.

Even though Galapagos provides a suitable solution with least displacement for each case, oftentimes from a practical standpoint, it might not be the most appropriate one. In order to overcome this problem, an open source tool named Design Explorer [16] is used for exploring multi-dimensional parametric studies. Using the plugin TT Toolbox [17] for Grasshopper, the various iterations performed by Galapagos was stored in as a comma separated values (.csv) file and was used as the input in Design Explorer to generate an interactive web-based parallel coordinates plot of all the cases. Using this plot (Fig. 11), results were fine-tuned and filtered to the required variants as per requirement.

3 Results and Discussion

Validation studies were performed for both circular dome and conical shell where the horizontal and vertical reactions of both obtained from Karamba3D were compared

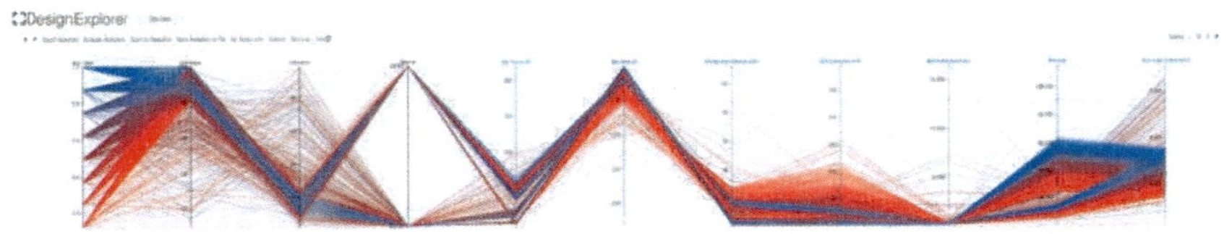

Fig. 11 Parallel coordinates plot generated using design explorer

with values calculated as per standard formulas. The horizontal and vertical reactions are calculated by equating it to the horizontal and vertical components of the meridional thrust, respectively. The results from both were found to be comparable.

For validation of the circular dome (Table 2), a shell of 10 cm thick made of concrete having a unit weight of 25kN/m3 was adopted. The rise and span of the dome adopted were 1.34 m and 10 m, respectively. A total of 74 supports have been taken at the base of the dome in the pinned condition. The meridional thrust for circular dome is as given in Eq. (1)

$$\text{Meridional thrust for circular dome} = wR/(1 + \cos\theta). \tag{1}$$

where w, R and θ represents the weight per surface area, span and semi-central angle of the dome, respectively.

For validation of conical shell (Table 3), a shell of 10 cm thick made of concrete having a unit weight of 25 kN/m^3 was adopted. The height and radius of the cone were 5 m and 2 m, respectively. A total of 25 supports have been taken at the base of the cone in the pinned condition. The meridional thrust for conical shell is as given in Eq. (2)

$$\text{Meridional thrust for conical shell} = 0.5\, wh \sec^2\theta. \tag{2}$$

where w, h and θ represents the weight per surface area, height and semi-vertical angle of the cone, respectively.

Table 2 Validation study for circular dome

Circular dome	Karamba3D (kN)	Manual calculation (kN)
Vertical reaction	2.841	2.843
Horizontal reaction	4.878	4.925

Table 3 Validation study for conical shell

Conical shell	Karamba3D (kN)	Manual calculation (kN)
Vertical reaction	3.371	3.383
Horizontal reaction	1.300	1.353

Table 4 Summary of optimization study done for conical grid shell

Parameters	Quad pattern	Diamond pattern	Hexagonal pattern	Staggered pattern
Rise/span	0.625	0.5	0.5	0.625
U	95	100	100	100
V	18	29	17	6
Material	GL32h	GL32h	GL32h	GL32h
Displacement (cm)	0.0408	1.6880	0.6812	0.3713
Max. tension (kN)	0	0	0	0
Max. compression (kN)	−2.3615	−3.3347	−3.1201	−2.1614
Max. resultant moment (kNm)	0.0125	0.3574	0.2401	0.5661
Max. kicking force (kN)	1.4552	3.3039	2.2851	1.3333
Total length (m)	4235.0716	3432.8361	2473.7812	1977.7101
Mass (kg)	18,210.808	14,761.1954	10,637.2594	8504.1537

The results of the optimization study performed for achieving the minimum displacement for each type of grid shell is summarized as given in Tables 4, 5 and 6. A tolerance factor of 0.001 was adopted as running further iterations will only result in very minimal reduction of displacement. Tolerance factor represents the permissible difference between the displacements of two consecutive iterations. For the tolerance factor adopted, approximately about 3000–5000 iterations were required per pattern per grid shell in order to arrive at the suitable displacement result.

Analyzing the results, it is clear that the diamond pattern performed the poorest in all of the grid shell structures, making it the least suitable pattern of the four. It is also noted that displacements in ellipsoids of revolution grid shell are on the slightly higher side in comparison with the other two grid shells. This is primarily due to the constraint imposed to limit the subdivisions of the grid along both directions to 100 for a consistent comparative analysis between models. Also, based on the properties of the materials used, it seen that GL32h outperformed GFRP in the majority of cases for a similar cross-section. Bending moment diagrams, displacement plots and stress contours were also generated for each case (Figs. 12, 13 and 14).

It is also worth noting that the best configuration is not necessarily the one that results in the least amount of displacement. Other factors that must be considered before making a final decision include material use, economy, axial forces, outward kicking forces, to name a few. To make the process of selecting the optimal design solution from the many iteration results easier, tools like Design Explorer can be used to produce parallel coordinates plots, which can be used to make an informed decision, as discussed in the methodology adopted in this work.

Table 5 Summary of optimization study done for ellipsoids of revolution grid shell

Parameters	Quad pattern	Diamond pattern	Hexagonal pattern	Staggered pattern
Rise/span	0.75	0.625	0.625	0.625
U	89	100	100	96
V	34	82	50	24
Material	GL32h	GL32h	GL32h	GFRP
Displacement (cm)	0.0673	11.1074	4.7170	1.6563
Max. tension (kN)	1.5716	0	0.3989	5.5361
Max. compression (kN)	−2.329	−3.1264	−3.0776	−9.7723
Max. resultant moment (kNm)	0.0130	1.4767	1.1303	3.0615
Max. kicking force (kN)	0.0676	0.7197	0.4021	0.2942
Total length (m)	9888.2683	4911.3132	3575.5904	2656.472
Mass (kg)	42,519.554	21,118.647	15,375.038	46,886.745

Table 6 Summary of optimization study done for hyperboloids of revolution of one sheet grid shell

Parameters	Quad pattern	Diamond pattern	Hexagonal pattern	Staggered pattern
Middle radius/bottom radius	1	0.8	0.9	1
U	95	100	100	98
V	10	32	37	10
Material	GL32h	GL32h	GL32h	GL32h
Displacement (cm)	0.0101	0.2666	0.0734	0.0087
Max. tension (kN)	0.0004	0.1729	0.2032	0.0001
Max. compression (kN)	−1.1595	−1.7318	−1.4070	−1.0052
Max. resultant moment (kNm)	0.0017	0.1283	0.0722	0.0019
Max. kicking force (kN)	0.00009	1.0594	0.6095	0.0001
Total length (m)	2591.0235	2693.4601	2862.5584	2305.515
Mass (kg)	11,141.4014	11,581.8786	12,309.0011	9913.718

Fig. 12 Shape of bending moment diagram of an ellipsoids of revolution grid shell

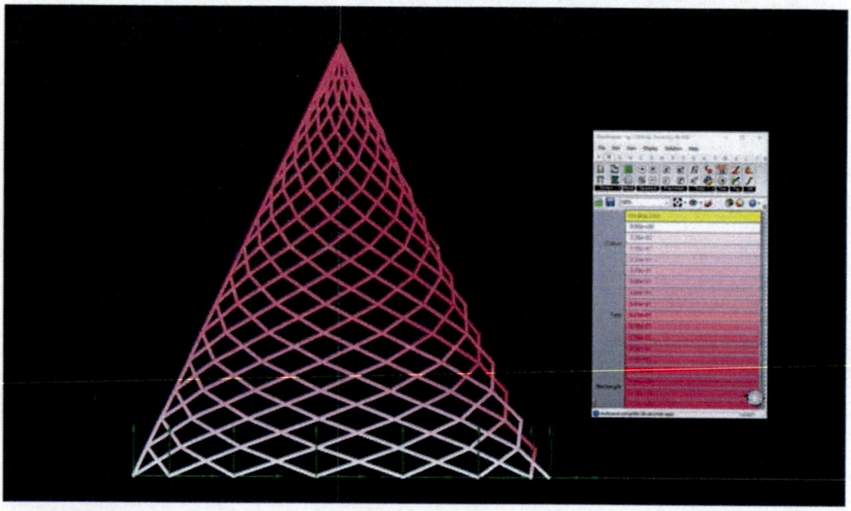

Fig. 13 Displacement plot of a conical grid shell

4 Conclusion

The paper has presented how various geometric patterns, and their arrangements directly influence the structural behavior and stability of grid shells. Out of the four different patterns analyzed, it was found that diamond patterns have a poor structural performance, while the performance of the other pattern were found to be comparable. In case of ellipsoids of revolution grid shells, the most optimum

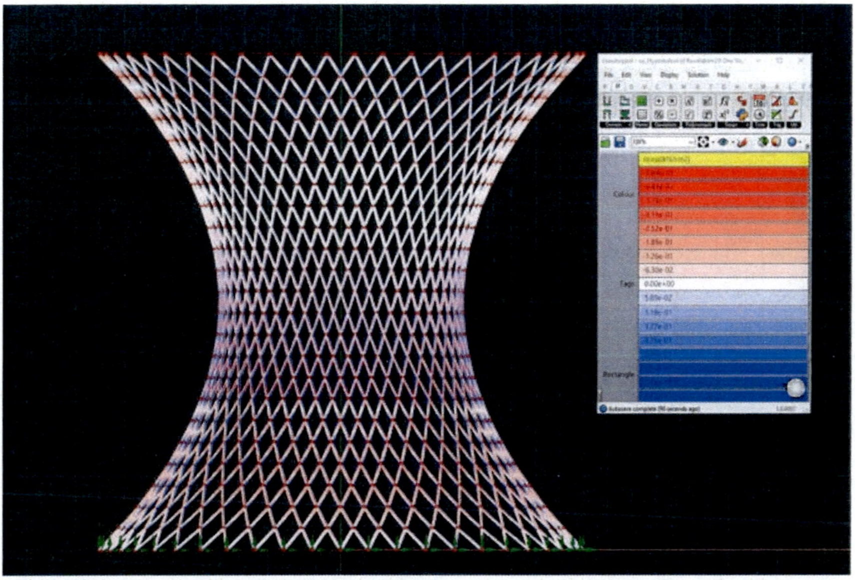

Fig. 14 Stress contour of a hyperboloids of revolution of one sheet grid shell

diamond grid shell produced a displacement of about 11cms, whereas other patterns produced displacements just below 5cms. Further, studies are required to ascertain whether addition of cross braces to diamond patterns will help in improving the structure's performance. It was also found that GL32h had a superior performance in comparison with GFRP in the majority of cases. The optimal r/R ratio for ellipsoids of revolution grid shells was found to be closer to 1, bringing the geometry closer to that of a cylinder. In the design optimization phase, the work used a parametric workflow, which significantly streamlined the process. Using genetic algorithm, optimization studies based on various grid patterns were carried out on three major types of grid shell structures, and it has been demonstrated how proper data visualization would aid in gaining fast insight into structural results. Further, studies may be done by adopting unique rise to span ratios, such as the golden ratio and analyzing them.

References

1. Mannheim Multihalle. [Online]. Available: https://www.archilovers.com/projects/151389/roof-for-the-multihalle-in-mannheim.html
2. Bouhaya L, Baverel O, Caron JF (2014) Optimization of gridshell bar orientation using a simplified genetic approach. Struct Multidiscip Optim 50(5):839–848. https://doi.org/10.1007/s00158-014-1088-9
3. IS 2210-1988 (1992) Criteria for reinforced concrete shell structures and folded plates. Bureau of Indian Standards, New Delhi, India

4. Eliassen M, Huseby A (2018) The digital workflow of parametric structural design developing grid shells in a Nordic climate, p 201
5. Kovaleva D, Gericke O, Kappes J, Tomovic I, Sobek W (2019) Rosenstein Pavilion: design and structural analysis of a functionally graded concrete shell. Structures 18:91–101. https://doi.org/10.1016/j.istruc.2018.11.007
6. https://www.rhino3d.com. [Online]. Available: https://www.rhino3d.com
7. https://www.grasshopper3d.com. [Online]. Available: https://www.grasshopper3d.com
8. Riemens K (2015) A parametric structural design tool for shell structures
9. Winslow P, Pellegrino S, Sharma SB (2010) Multi-objective optimization of free-form grid structures. Struct Multidiscip Optim 40(1–6):257–269. https://doi.org/10.1007/s00158-009-0358-4
10. Vincenti A, Ahmadian MR, Vannucci P (2010) BIANCA: a genetic algorithm to solve hard combinatorial optimisation problems in engineering. J Glob Optim 48(3):399–421. https://doi.org/10.1007/s10898-009-9503-2
11. Loos L, Verbeeck K, De Laet L (2019) Data visualisation as a tool for informed structural design. Comput Des 115:267–276. https://doi.org/10.1016/j.cad
12. Veenendaal D, Augustynowicz E, Tang G (2017) Magnolia: a glass-fibre reinforced polymer gridshell with a novel pattern and deployment concept. In: Proceedings of the IASS annual symposium on 2017—interfaces architecture engineering science, pp 1–9
13. https://www.food4rhino.com/app/lunchbox. [Online]. Available: https://www.food4rhino.com/app/lunchbox
14. Karamba3D. [Online]. Available: https://manual.karamba3d.com
15. Galapagos. [Online]. Available: http://grasshopperdocs.com/addons/galapagos.html
16. Tomasetti T (2015) Design explorer, CORE studio, p 2019. [Online]. Available: http://tt-acm.github.io/DesignExplorer/
17. Studio C (2017) Tt toolbox, p 5. [Online]. Available: https://www.food4rhino.com/app/tt-toolbox

Study of Moment Curvature Response of Beam as Per IS456 Code and Hognestad's Stress–Strain Curve

Pavan Pohekar and Archana Tanawade

1 Introduction

For understanding the failure of flexural members, moment–curvature response is the fundamental property of RC beam as it reflects the whole behavior of beam from the instance when loading starts to the failure [1]. This relationship also plays an important role in earthquake design [2, 3]. It gives the clear idea about yielding and ultimate values of moment and curvature of a beam. Computer analysis indicates that the curvature ductility increases with the increase of longitudinal reinforcement [4]. The complete path-dependent moment–curvature relationship is similar to the material stress–strain relationship with stress-path dependence [5]. The important structural parameters governing the flexural ductility of singly reinforced concrete beams are the tension steel ratio and, the tension steel to balanced steel ratio and the concrete grade [6]. As the flexural reinforcement ratio in beams increased the ductility of beams was observed to increase [7], whereas deflection ductility index decreases as the longitudinal reinforcement increases. Deflection ductility increases with increase in compression reinforcement. Ductility index decreases with increase in l/d ratio [8, 9, 1]. The reinforcement properties largely affect the element behavior, increasing the applied deformation up to the final failure of the structure [10]. Before the section curvature reached maximum value, both the ultimate moment and curvature increase with an increase in reinforcement ratio [11]. Present study mainly focuses on understanding moment–curvature response of RC beam (singly and doubly) as per IS456-2000 and Hognestad's concrete stress–strain model. Effect of addition of both the tension and compression reinforcement on ductility of the beam and the curvature difference by IS456 model and that by Hognestad's model is studied in this project.

P. Pohekar (✉) · A. Tanawade
Department of Civil Engineering, Vishwakarma Institute of Information Technology, Pune, Maharashtra, India
e-mail: pavan.219m0038@viit.ac.in

A. Sil et al. (eds.), *Recent Trends in Civil Engineering*, Lecture Notes in Civil Engineering 274, https://doi.org/10.1007/978-981-19-4055-2_11

1.1 Stress–Strain Models of Concrete

Hognestad's curve shows the actual behavior of the concrete under compression. It shows the values of stress for corresponding strain when the cylindrical specimen of the concrete is tested under compression [12]. Study shows that the strength of concrete cube is greater than strength of concrete cylinder [13].

$$fc = fc' \times \left[\frac{2 \in}{\in_0} - \left(\frac{\in}{\in_0} \right)^2 \right] \tag{1}$$

where,

f_c stress in concrete at strain, $\in$.
f_c' Characteristic strength of cylindrical specimen $= 0.8 \times fck$.[13]
f_{ck} Characteristic strength of cube.
$\in_0$ Strain at peak stress, 0.002.
$\in$ Strain in concrete at considered point.

The relationship between stress and strain is the basis for analysis of any RC component. It governs the behavior of an individual material. Concrete is the important material governing the failure of beam. According to IS456-2000, concrete's stress–strain relationship governing equation is given as[14];

$$fc = 0.446 \times fck \times \left[\frac{2 \in}{\in_0} - \left(\frac{\in}{\in_0} \right)^2 \right] \quad \text{for} \quad (\in \leq 0.002) \tag{2}$$

$$fc = 0.446 \times fck \quad \text{for} \quad (\in > 0.002) \tag{3}$$

The difference in two concrete models can be illustrated from Fig. 1. It is observed, curve has been made straight after peak stress according to IS456-2000 and the value of peak stress has been reduced by applying shape, size and safety factors, effectively taken as 0.445.

2 Problem Statement

The moment–curvature response of beam varies with the concrete stress–strain model considered. As the gap between the values of curvature tells the safety margin, comparative analysis of moment–curvature according to both the models is necessary. This section highlights the configurations of beam analyzed in this study. Following parameters are considered for the analysis.

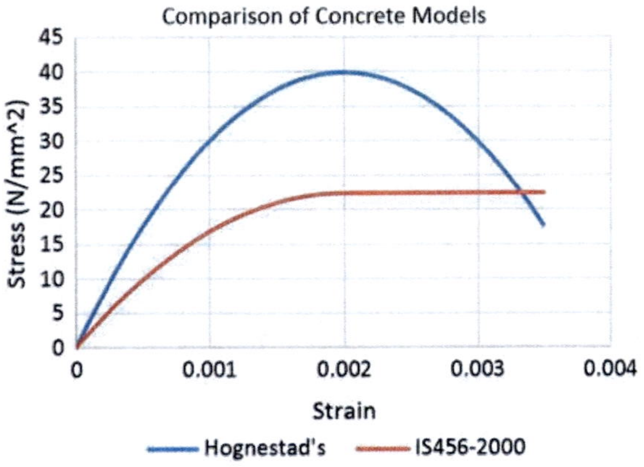

Fig. 1 Comparison of concrete models

2.1 *Singly Reinforced Beam*

Width of beam = 300 mm

Overall depth of beam, $D = 450$ mm
Effective depth of beam, $d = 400$ mm
Effective cover to the tension reinforcement = 50 mm
Concrete characteristic strength, $fck = 25$ MPa
Steel yield strength, $fy = 415$ Mpa
Values of percentage of tension reinforcement [14], pt
Under-Reinforced = 0.8%
Balanced section = 1.201%
Over-Reinforced = 1.5%
Concrete models = IS456-2000 & Hognestad's model.

2.2 *Doubly Reinforced Beam*

Width of beam, $b = 300$ mm

Effective depth of beam, $d = 400$ mm
Overall depth of beam, $D = 450$ mm
Cover to compression reinforcement, $d' = 50$ mm
Yield strength of steel, $fy = 415$ Mpa
Characteristic strength of concrete, $fck = 25$ Mpa
Percentage of tension steel, pt = 1.5%
Values of percentage of compression steel [14], pc
Over-Reinforced = 0.15%
Balanced = 0.321%

Under-Reinforced = 0.5%
Concrete models = IS456-2000 & Hognestad's model
Note- All the data considered for the analysis were assumed by the author.

2.3 Parametric Study

The need for the parametric analysis is studied for evaluating the influence of various key parameters on the moment–curvature response of beam. Increase in the characteristic strength of concrete increases the neutral axis depth, hence increases the moment capacity of the section [15].

- Singly reinforced beam;

Width of beam = 300 mm
Overall depth of beam, $D = 450$ mm
Effective depth of beam, $d = 400$ mm
Effective cover to the tension reinforcement = 50 mm
Concrete models = IS456-2000 & Hognestad's model

Table 1 shows the values for the parameters like percentage of tension reinforcement, characteristic strength of concrete and yield strength of steel for doubly reinforcement beam.

- Doubly reinforced beam;

Width of beam, $b = 300$ mm
Effective depth of beam, $d = 400$ mm
Overall depth of beam, $D = 450$ mm
Cover to compression steel, $d' = 50$ mm
Concrete models = IS456-2000 & Hognestad's model

Table 2 shows the values for the parameters like percentage of tension reinforcement, characteristic strength of concrete, percentage of compression reinforcement, and yield strength of steel for doubly reinforcement beam.

Table 1 Values for singly reinforced beam

pt (%)	*Fck* (MPa)	*Fy* (MPa)
0.5	20	250
1	25	415
1.5	30	500
2	35	600*
2.5	40	700*

Table 2 Values for doubly reinforced beam

pt (%)	*Fck* (MPa)	pc (%)	*Fy* (MPa)
0.5	20	0	250
1	25	0.2	415
1.5	30	0.4	500
2	35	0.6	600[a]
2.5	40	0.8	700[a]

[a] = bar diameters are not available in the market but they are considered for the analysis purpose only

3 Analytical Program

This explains specifications of beam models and analytical model which has been created and utilized in this project for performing actual work.

3.1 *Specimen Description*

For moment–curvature analysis, by considering all parameters, 106 configurations of beam are analyzed using the analytical model built in MATLAB R2020a. It includes singly reinforced, doubly reinforced, under-reinforced, balanced, over-reinforced with varying parameters such as percentage of tension steel and compression steel, and grades of concrete and steel with both the IS456-2000 and Hognestad's concrete model.

3.2 *Materials*

Two main materials considered are steel (Fe250, Fe415, Fe500) and concrete (M20, M25, M30, M35, M40). Two stress–strain models of concrete are used for the analysis on the basis IS456-2000 and Hognestad's curve.

3.3 *Analytical Model*

A dynamic model is built using MATLAB R2020a. This particular model is intended to produce values of moment and curvature for corresponding values of strain in concrete till the failure. This model incorporates two concrete models, viz., IS456-2000 and Hognestad's. These two models are embedded in the same analytical model and can be switched according to the requirement by just specifying the input for corresponding model. The output from this model comprises of the table with 2 columns and 36 rows. Each of these rows contains value of moment and curvature for particular strain value in concrete from 0.0001 to 0.0035.

Figure 2 and 3 show the flowchart for algorithm of model.

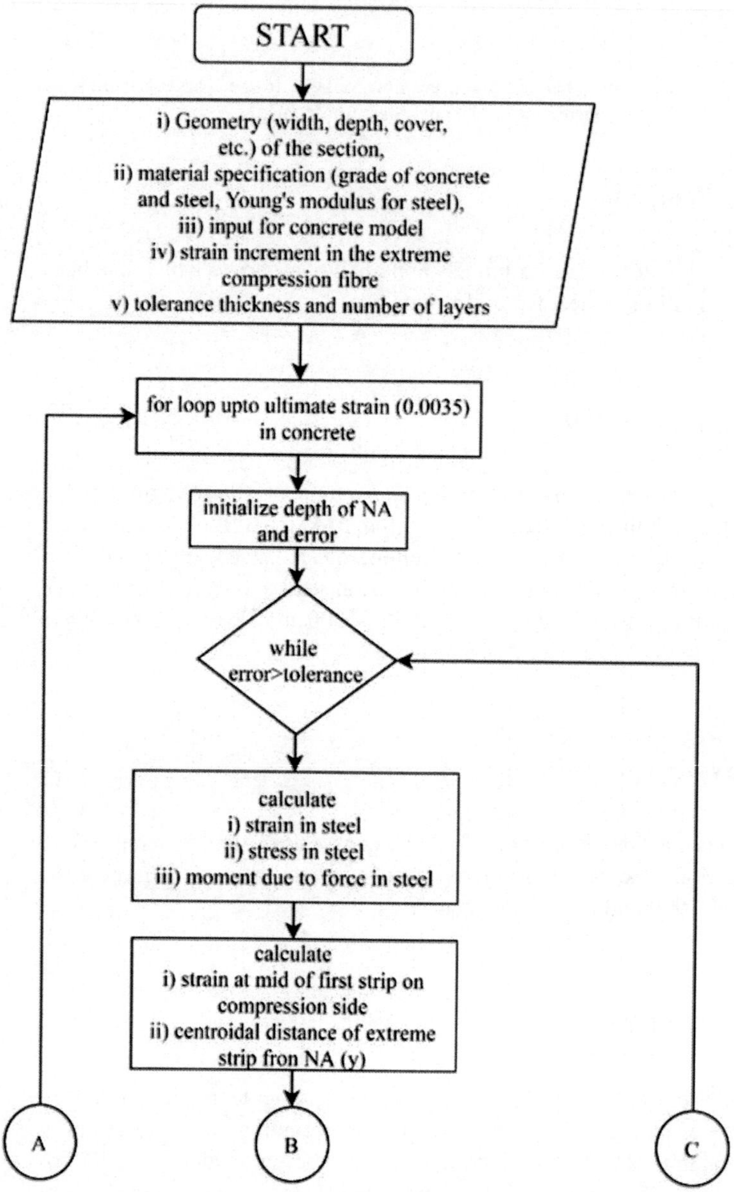

Fig. 2 Flowchart showing the algorithm (part I)

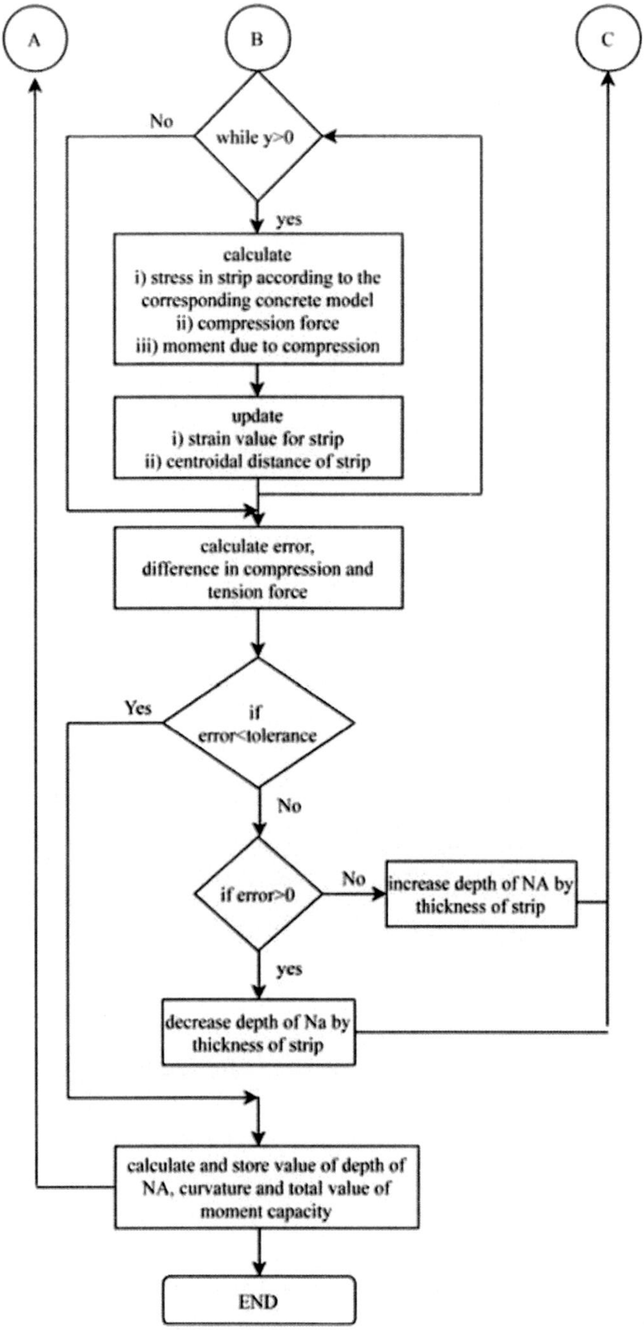

Fig. 3 Flowchart showing the algorithm (part II)

```
%calculation for stress in Fe415 & Fe500 grade compression steel
function steelstr=fs(eps,fy)
if eps<=0
    steelstr=eps*2*10^5;
else
    if fy==250
        steelstr=2*1e5*eps;
    elseif fy==415
        if (eps<=0.00144)
            steelstr=288.7*eps/0.00144;
        elseif eps>0.00144&&eps<=0.00163
            steelstr=288.7+(306.7-288.7)*(eps-0.00144)/(0.00163-0.00144);
        elseif eps>0.00163&&eps<=0.00192
            steelstr=306.7+(324.8-306.7)*(eps-0.00163)/(0.00192-0.00163);
        elseif eps>0.00192&&eps<=0.00241
            steelstr=324.8+(342.8-324.8)*(eps-0.00192)/(0.00241-0.00192);
        elseif eps>0.00241&&eps<=0.00276
            steelstr=342.8+(351.8-342.8)*(eps-0.00241)/(0.00276-0.00241);
        elseif eps>0.00276&&eps<=0.0038
            steelstr=351.8+(360.9-351.8)*(eps-0.00276)/(0.0038-0.00276);
        elseif eps>0.0038
            steelstr=360.9;
        end

    else
        if (eps<=0.00174)
            steelstr=347.8*eps/0.00174;
        elseif eps>0.00174&&eps<=0.00195
            steelstr=347.8+(369.6-347.8)*(eps-0.00174)/(0.00195-0.00174);
        elseif eps>0.00195&&eps<=0.00226
            steelstr=369.6+(391.3-369.6)*(eps-0.00195)/(0.00226-0.00195);
        elseif eps>0.00225&&eps<=0.00277
            steelstr=391.3+(413-391.3)*(eps-0.00226)/(0.00277-0.00226);
        elseif eps>0.00277&&eps<=0.00312
            steelstr=413+(423.9-413)*(eps-0.00277)/(0.00312-0.00277);
        elseif eps>0.00312&&eps<=0.00417
```

Fig. 4 Stress in steel

Figure 4 and 5 show the code for user defined functions for calculating stress in steel and concrete.

4 Results

MATLAB analysis showed the clear difference between the moment–curvature response by IS456-2000 model and Hognestad's model. The analysis also benefited with the strong reason to adopt under-reinforced section in the design. Analysis of doubly reinforced beam showed that incorporation of compression reinforcement

```
%function definition for various concrete models
function concstr=concrete(fck,epc_temp,concrete_model)
if (concrete_model==1)
    if (epc_temp<=0.002)
                concstr=(0.446*fck*((2*epc_temp/0.002)-(epc_temp/0.002)^2));
            else
                concstr=0.446*fck;
    end
else
    concstr=(0.8*fck*((2*epc_temp/0.002)-(epc_temp/0.002)^2));
end
end
```

Fig. 5 Stress in concrete according to the stress–strain model

also returns many advantages in terms of curvature, moment capacity, and life of beam.

4.1 Analysis

Firstly, singly reinforced beam was analyzed and moment–curvature response was obtained. Analysis clearly showed that curvature ductility reduces with the increase in tension reinforcement as shown in Fig. 6. MATLAB analysis performed on singly reinforced beam showed that the difference in curvature for under-reinforced beam is greater than balanced and over-reinforced section. The difference in curvature for under-reinforced, balanced, and over-reinforced section is 16.80, 11.20, and 8.93 per meter and in moment capacity it is 6.08, 13.72, and 21.48 KN-m, respectively.

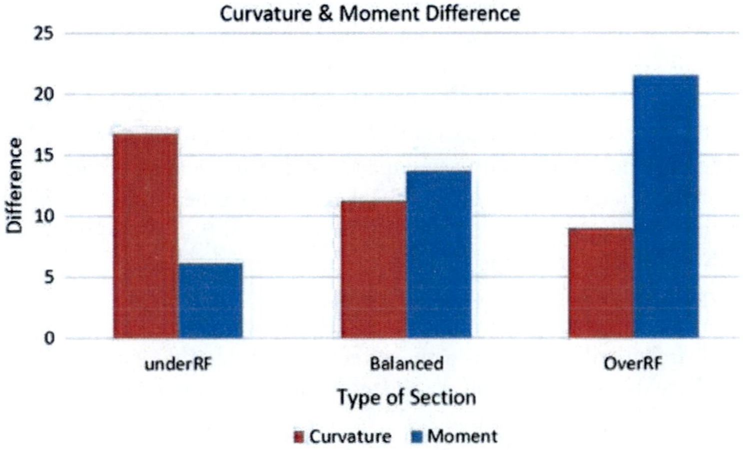

Fig. 6 Difference in curvature and moment of singly reinforced beam by two concrete models

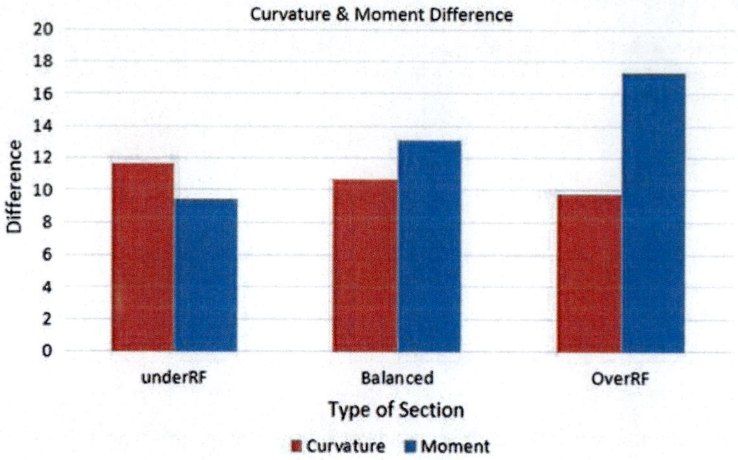

Fig. 7 Difference in curvature and moment of doubly reinforced beam by two concrete models

In doubly reinforced beam was configured and analyzed and compared for two concrete models. Here also, three configurations were fixed as over-reinforced, balanced, and under-reinforced section. As compression steel goes on increasing, beam goes toward becoming under-reinforced section. As can be seen from Fig. 7, curvature difference for doubly reinforced section in under-reinforced, balanced, and over-reinforced condition is 11.61, 10.64, 9.72 per m and moment capacity difference is 9.44, 13.13, 17.27 KN-m, respectively.

4.2 Parametric Analysis

As shown in Fig. 8, percentage of tension reinforcement was varied and moment–curvature variation was studied. Yield curvature for singly reinforced beam is found to be higher according to IS456 model than Hognestad's concrete model. But, it was observed that as the tension reinforcement is being increased, ultimate curvature value and difference decreases, whereas moment capacity value and difference increases. For the same increment in percentage of tension reinforcement, the average percentage difference in ultimate curvature for singly reinforced beam is 54.02%, whereas for doubly reinforced beam it is 43.42%.

As shown in Fig. 9, increment in concrete characteristic strength increases both the curvature value and difference in the values according to two models.

Increasing yield strength of steel has negative effect on curvature of beam. With the increase in steel yield strength, curvature reduces significantly and reduction in its difference shows that increase in yield strength of steel brings the system closer to actual behavior which is harmful. In singly reinforced condition, for FE250 the difference in ultimate curvature is 27.84 per meter and that for FE700 is 9.10 per

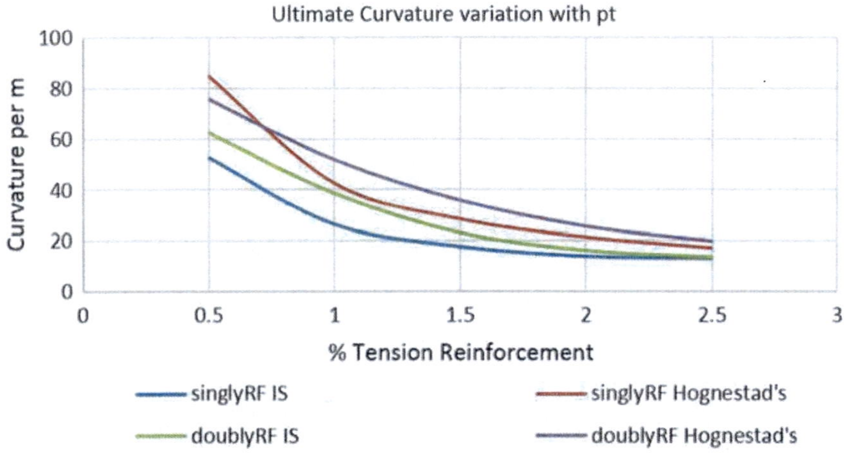

Fig. 8 Variation of curvature with percentage of tension reinforcement

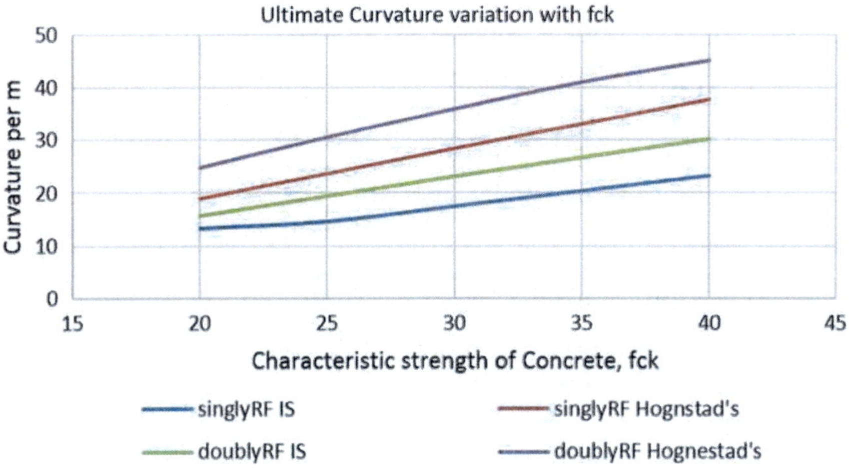

Fig. 9 Variation of curvature with characteristic strength of concrete

meter. In doubly reinforced condition, the differences are 6.13 per meter and 6.89 per meter, respectively. See Fig. 10 for reference.

Presence of compression reinforcement in beam also reduces the adverse effect of increment of steel yield strength greatly and this can be visualized from Fig. 10. Average difference in ultimate curvature without compression reinforcement is 59.96%, whereas with compression reinforcement is 13.29%.

Increment in amount of compression reinforcement shows positive effect on both curvature and moment capacity. This also enhances the difference in the values of

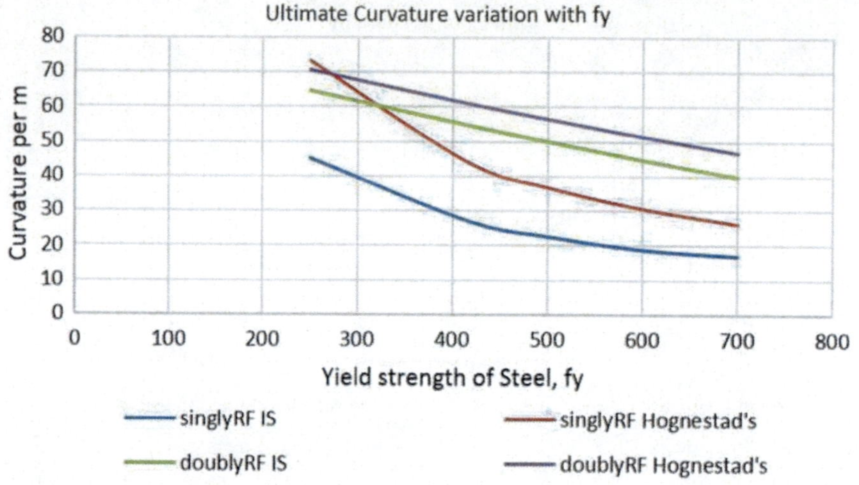

Fig. 10 Variation of curvature with yield strength of reinforcement

ultimate curvature by IS456 concrete model from the Hognestad's model. Mathematically, the difference in ultimate curvature for 0.2% compression reinforcement is 5.01 per meter, whereas for 1.3% compression steel it is 10.02 per meter, Fig. 11.

The difference in moment capacities with the increase in compression steel is seen to be reduced from over-reinforced section to under-reinforced section from Fig. 12. For over-reinforced section, it is from 80.13 to 11.30 KN-m, for balanced section,

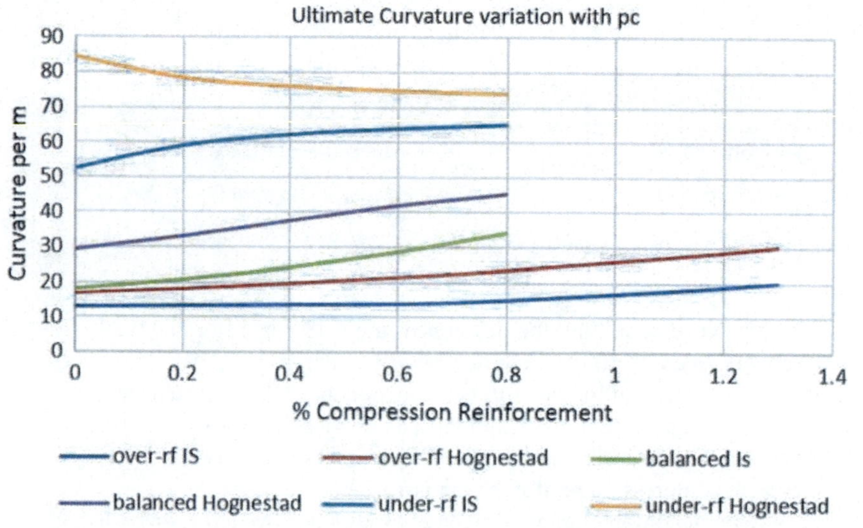

Fig. 11 Variation of curvature with percentage of compression reinforcement

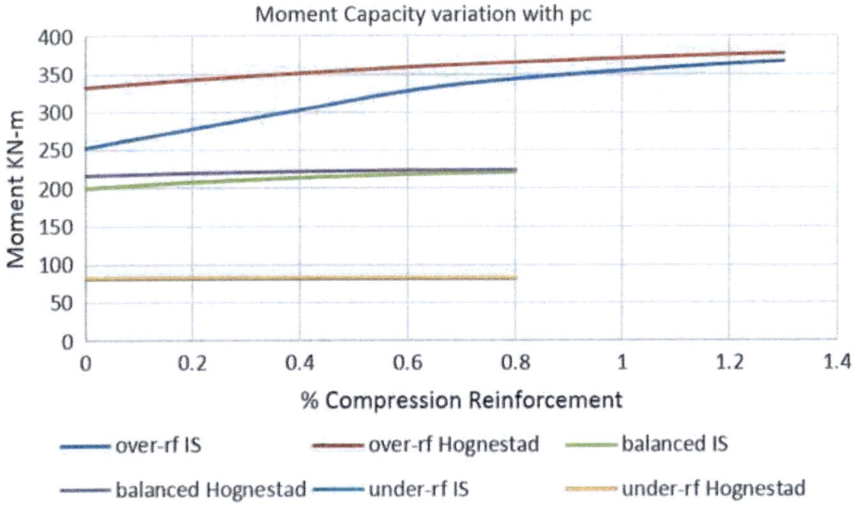

Fig. 12 Variation of moment in doubly reinforcement beam with percentage of compression reinforcement

it is from 16.47 to 3.71 KN-m and for under-reinforced section, it is from 1.97 to 2.32 KN-m.

5 Conclusion

From the analysis of singly reinforced and doubly reinforced beam, under-reinforced section, according to IS456-2000 model, keeps margin of 16.80 per m in curvature from that according to Hognestad's model, which is greatest among that for balanced (11.20 per m) and over-reinforced (8.93 per m) section giving nearly equal resistive moment as that according to Hognestad's concrete model. This implies that adopting under-reinforced section, IS456-2000 provisions can be utilized at their best.

With the increase in concrete compressive strength in singly reinforced section, the average difference in ultimate curvature is 57.53%, and average difference in moment capacity is 11.42%. For doubly reinforced beam, the average difference in ultimate curvature is 54.84%, and average difference in moment capacity 4.68%. This implies that doubly reinforced section can be effectively used to maintain the moment capacity at the value by Hognestad's model.

Increment in steel yield strength decreases the ductility of the beam, but the rate of this reduction can be lowered by the introduction of compression reinforcement.

It can be conclude that when tension steel goes on increasing, the beam tends toward over-reinforced condition decreasing ultimate curvature. A point arrives after which there is no significant increase in moment capacity and decrease in ultimate curvature with increase in tension steel and that can be referred to as the maximum percentage and un-economical usage of tension reinforcement for that particular section.

It can also be concluded that when compression steel goes on increasing, a beam tends toward under-reinforced condition increasing moment capacity and ultimate curvature. A point arrives after which there is no significant increment in moment capacity and curvature and that can be referred to as maximum percentage and economical usage of compression steel for corresponding configuration. Addition of compression steel prevents the brittle failure of over-reinforced section. Hence, for enhancing the capacity of beam with restricted dimensions, addition of reinforcement on compression side rather than on tension side must be preferred.

With the presence of compression reinforcement, behavior of beam according to IS456-2000 concrete model resembles with that according to Hognestad's concrete model.

The analytical model produced in this study can also be used for other materials than concrete or steel just by incorporating their stress-strain model into the existing model.

Acknowledgements Author would like to thank industry mentor Er. Sanket M. Saharkar for his continuous guidance and involvement in the project work. He helped throughout the project work and provided with all the necessary knowledge. Author would also like to extend thanks to Vishwakarma Institute of Information Technology, Pune for continuous support and making essential resources available on time.

References

1. Nogueira CG, Rodrigues ID. Ductility analysis of RC beams considering the concrete confinement effect produced by the shear reinforcement: a numerical approach
2. Naik P, Annigeri S (July 2016) Moment curvature analysis of RC column as per IS456:2000
3. Tipnis OP (Aug 2015) Numerical and analytical modeling of concrete confined with FRP wraps
4. Olivia M, Mandal P (2015) Curvature ductility of reinforced concrete beam
5. Au FTK, Bai BZZ, Kwan AKH (2005) Complete moment-curvature relationship of reinforced normal- and high-strength concrete beams experiencing complex load history
6. Pam HJ, Kwan AKH, Islam MS (Nov 2001) Flexural strength and ductility of reinforced normal- and high-strength concrete beams
7. Rao GA, Vijayanand I, Eligehausen R (Jan 2007) Studies on ductility of RC beams in flexure and size effect
8. Design of tall buildings: trends and achievements for structural performance (Nov 7–11, 2016)
9. Manjunath HR, Prabhakara R (2016) Analysis of flexural behavior of doubly reinforced high-strength concrete beams using ANSYS
10. Grimaldi A, Rinaldi Z. Influence of the steel properties on the ductility of RC structures
11. Sun Z, Yang Y, Yan W, Wu G, He X (2017) Moment-curvature behaviors of concrete beams singly reinforced by steel-FRP composite bars

12. Wight JK, Richart FE, Macgregor JG. Reinforced concrete mechanics and design
13. Kumari R, Malviya MM. Review paper based on the relation between the strength of concrete cubes and cylinders.
14. SP16:1980—design aids for reinforced concrete to IS: 456-1978
15. Mandal P (1993) Curvature ductility of reinforced concrete sections with and without confinement

Effect of Silica Sand on the Mechanical and Durability Characteristics of Concrete

Kushal Ghosh, Subhashis Bose, Bikram Paul, and Partha Ghosh

1 Introduction

As the world is witnessing heightened urbanization, there is an increase in the rate of construction and with that a huge surge in the demand for building materials. According to a study, material extraction across the world stands at a staggering 48.5 billion tonnes, in which around 16.2 billion tonnes/year is contributed by construction materials. This leads to a huge depletion in natural resources. Amongst construction materials concrete is the most popular one whose demand is increasing with each passing day [1, 2]. Sand is a major constituent of concrete, and increased consumption of concrete leads to the depletion of natural sand sources, e.g. river beds [3–5]. Governments of various countries have also come out with strict laws preventing the dredging of river beds. In this context, the construction industry is looking at alternative sources of fine aggregates with sustainable characteristics [6–20]. Copper slag, stone dust, and brick dust have been explored extensively as an alternative to natural sand but data on silica sand as a replacement for natural sand is still not available. The current study aims to fill the abovementioned void in literature.

K. Ghosh (✉) · B. Paul
National Institute of Technology, Sikkim, India
e-mail: kushal@nitsikkim.ac.in

S. Bose · P. Ghosh
Jadavpur University, Kolkata, India

A. Sil et al. (eds.), *Recent Trends in Civil Engineering*, Lecture Notes in Civil Engineering 274, https://doi.org/10.1007/978-981-19-4055-2_12

2 Experimental Programme

2.1 Materials Used

PPC cement under the brand name ultratech conforming to IS 1489-1991 Part-I was used. Coarse aggregate with a maximum size of 20 mm was used. The fine aggregate used in the work conformed to grading zone II as per IS 383. Tap water was used for mixing. A commercially available superplasticizer Sikament 2004 NS was used to ensure that the mix had reasonable workability. For durability study, H_2SO_4 solution and $MgSO_4$ powder with 98% purity were used.

2.2 Mix Design Procedure

The mix proportion was done for a M35 concrete mix according to IS 10262-2019. Five separate mix design proportions were prepared accounting for 0%, 15%, 30%, 45%, and 100% replacement of natural sand by silica sand, respectively.

3 Results and Discussion

3.1 Compressive Strength

From the above observations, it can be seen that the compressive strength of the specimens in unexposed condition is meeting the criteria of the target compressive strength of M35 grade of concrete. When the specimens are subjected to $MgSO_4$ solution for a period of 28 and 56 days, it can be seen that there is a slight increase in the compressive strength. Though exposure to $MgSO_4$ for a long period of time is detrimental for concrete, in this case, the period of 28 and 56 days acts as an extended curing period, thus leading to a slight increase in the strength. But this trend does not hold true when the specimens are subjected to exposure in H_2SO_4, where there is a marked decrease in the compressive strength as the acid has an erosive effect which leads to a decrease in weight as well. The compressive strength of the samples also falls when the specimens are subjected to elevated temperature irrespective of the percentage of sand replacement (Figs. 1, 2, 3 and 4).

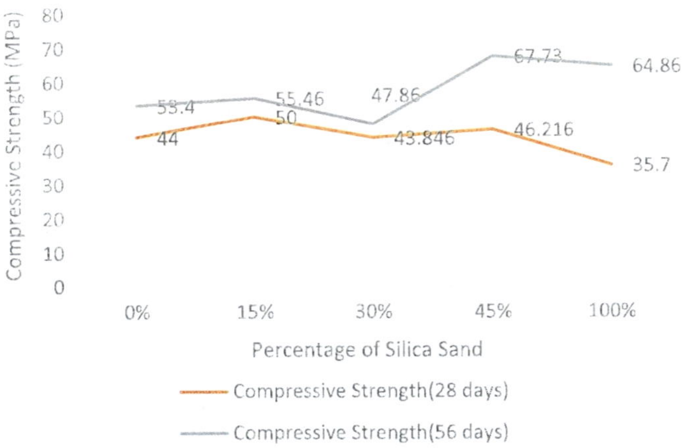

Fig. 1 Compressive strength of unexposed specimens

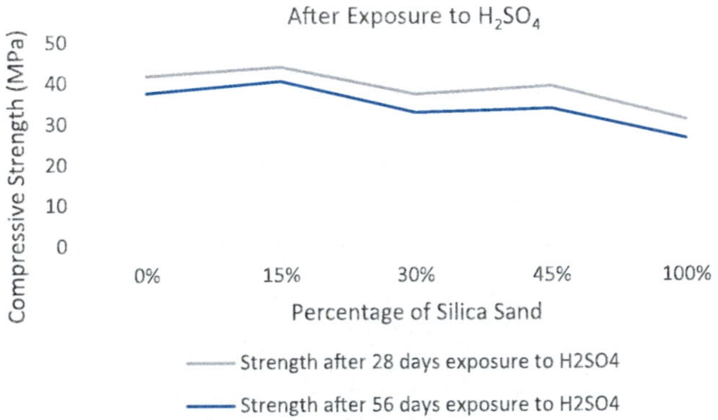

Fig. 2 Compressive strength of specimens exposed to H_2SO_4

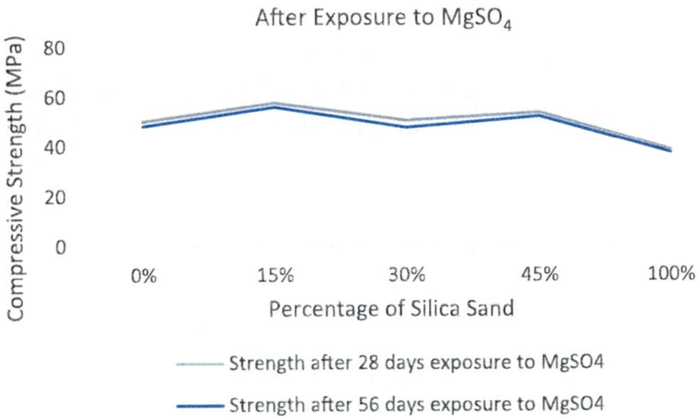

Fig. 3 Compressive strength of specimens exposed to $MgSO_4$

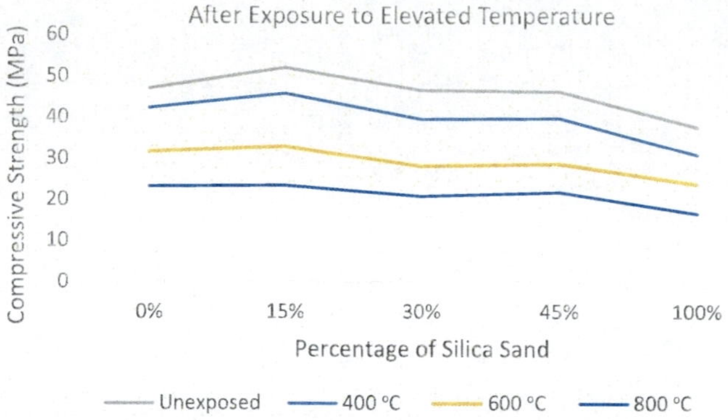

Fig. 4 Compressive strength of specimens exposed to elevated temperature

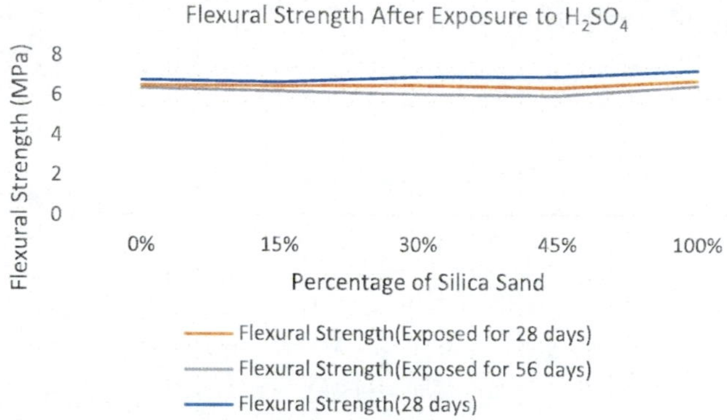

Fig. 5 Flexural strength of specimens exposed to H_2SO_4

3.2 *Flexural Strength*

In keeping with the trend observed in the compressive strength of the specimens and due to the reasons mentioned above, the flexural strength also increases slightly when the specimens are subjected to exposure in $MgSO_4$ and decreases when exposed to H_2SO_4 (Figs. 5 and 6).

3.3 *Ultrasonic Pulse Velocity*

The ultrasonic pulse velocity values of specimens exposed to $MgSO_4$ do not show any clear trend, but they are all in the range of 4.5 km/sec to 5.5 km/sec signifying good to excellent quality of concrete. However, the ultrasonic pulse velocity (upv) values of specimens subjected to exposure in H_2SO_4 show a clear decrease in upv

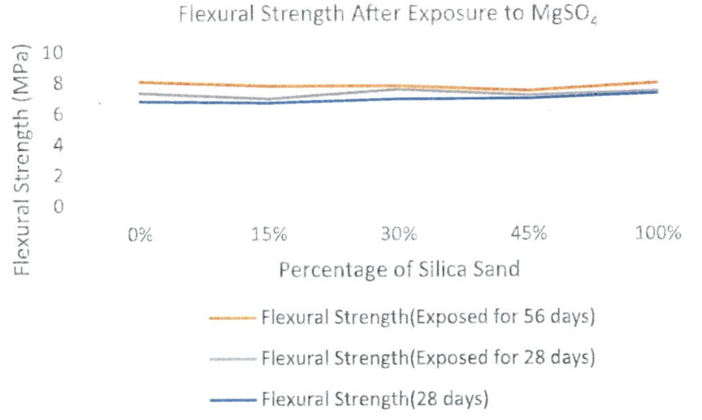

Fig. 6 Flexural strength of specimens exposed to $MgSO_4$

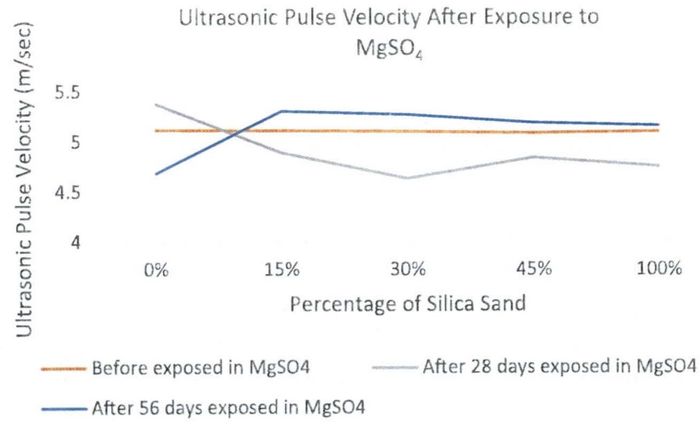

Fig. 7 Ultrasonic pulse velocity of specimens exposed to $MgSO_4$

values which conforms the erosive action of the acid solution. But it can be said that the performance of the silica sand based concrete is at par with normal concrete when subjected to exposure in H_2SO_4 for 28 days (Figs. 7 and 8).

3.4 *Half-Cell Potential Values*

The half-cell potentiometer readings pointed to the fact that all the samples with silica sand had values less than 200 mV, which denotes a lower probability of corrosion taking place. But when it was submerged in H_2SO_4 solution for a period of 56 days, the readings were more than 200 mV suggesting that the samples were more susceptible to corrosion (Fig. 9).

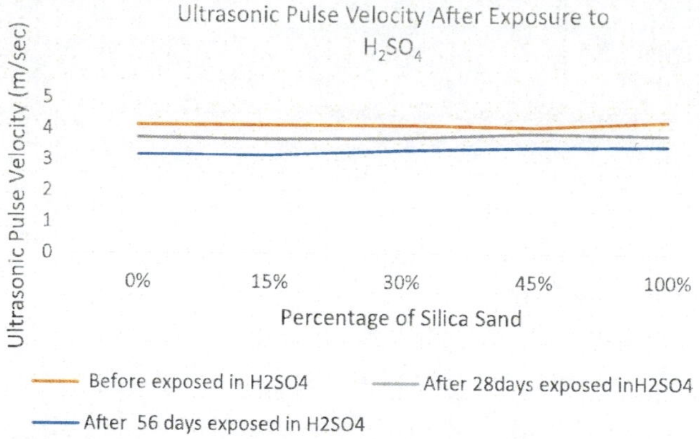

Fig. 8 Ultrasonic pulse velocity of specimens exposed to H_2SO_4

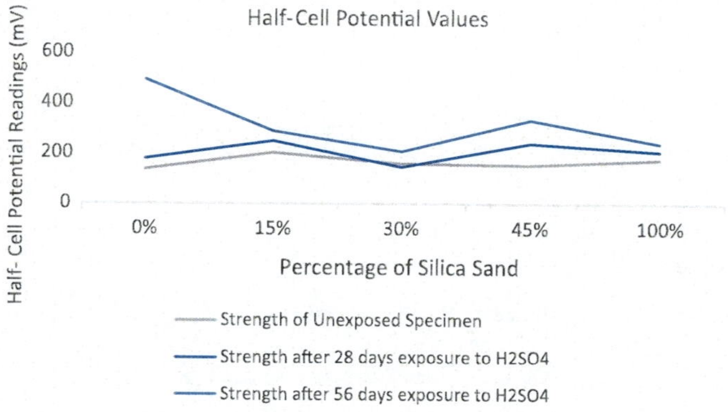

Fig. 9 Half-cell potential values of specimens exposed to H_2SO_4

4 Conclusion

No clear trend can be established between the percentage of silica sand and the compressive strength of specimens but it may be said that sand may be replaced by 45% of silica sand without compromising on the mechanical and durability characteristics of the concerned concrete.

References

1. Singh M, Siddique R (2016) Effect of coal bottom ash as partial replacement of sand on workability and strength properties of concrete. J Clean Prod 112(Part 1):620–630
2. Silva RV, de Brito J, Dhir RK (2016) Establishing a relationship between modulus of elasticity

and compressive strength of recycled aggregate concrete. J Clean Prod 112(Part 4):2171–2186
3. Farahani JN, Shafigh P, Alsubari B, Shahnazar S, Mahmud HB (2017) Engineering properties of lightweight aggregate concrete containing binary and ternary blended cement. J Clean Prod 149:976–988
4. Mithun BM, Narasimhan MC (2016) Performance of alkali activated slag concrete mixes incorporating copper slag as fine aggregate. J Clean Prod 112(Part 1):837–844
5. Tiwari A, Singh S, Nagar R (2016) Feasibility assessment for partial replacement of fine aggregate to attain cleaner production perspective in concrete: a review. J Clean Prod 135:490–507
6. Hanif A, Kim Y, Lu Z, Park C (2017) Early-age behavior of recycled aggregate concrete under steam curing regime. J Clean Prod 152:103–114
7. Shi C, Zheng K (2007) A review on the use of waste glasses in the production of cement and concrete. Resour Conserv Recycl 52(2):234–247
8. Thomas BS, Chandra Gupta R (2016) Properties of high strength concrete containing scrap tire rubber. J Clean Prod 113:86–92
9. Alwaeli M (2016) The implementation of scale and steel chips waste as a replacement for raw sand in concrete manufacturing. J Clean Prod 137:1038–1044
10. Alwaeli M (2017) Investigation of gamma radiation shielding and compressive strength properties of concrete containing scale and granulated lead-zinc slag wastes. J Clean Prod 166:157–162
11. Alwaeli M (2013) Application of granulated lead-zinc slag in concrete as an opportunity to save natural resources. Radiat Phys Chem 83:54–60
12. Alwaeli M, Nadziakiewicz J (2012) Recycling of scale and steel chips waste as a partial replacement of sand in concrete. Construct Build Mater 28(1):157–163
13. Patil AY, Banapurmath NR, Shivangi US (2020) Feasibility study of epoxy coated poly lactic acid as a sustainable replacement for river sand. J Clean Prod 267:121750
14. Bhardwaj B, Kumar P (2019) Comparative study of geopolymer and alkali activated slag concrete comprising waste foundry sand. Constr Build Mater 209:555–565
15. Nithya M, Priya AK, Muthukumaran R, Arunvivek GK (2017) Properties of concrete containing waste foundry sand for partial replacement of fine aggregate in concrete. Indian J Eng Mater Sci 24:162–166
16. Lee YH, Lim MH, Lee YL, Lee YY, Tan CS, Mohammad S, Ma CK (2018) Compressive strength of lightweight foamed concrete with charcoal as a sand replacement. Indian J Eng Mater Sci 25:98–108
17. Thiruvenkitam M, Pandian S, Santra M, Subramanian D (2020) Use of waste foundry sand as a partial replacement to produce green concrete: Mechanical properties, durability attributes and its economical assessment. Environ Technol Innov 19:101022
18. Ling JH, Lim YT, Leong WK, Jusli E, Sia HT (2019) Properties of cement brick with partial replacement of sand and cement with oil palm empty fruit bunches and silica fume. J Civil Eng Forum 5(3):289–300
19. Mustafa MAT, Hanafi I, Mahmoud R, Tayeh BA (Aug 2019) Effect of partial replacement of sand by plastic waste on impact resistance of concrete: experiment and simulation. Structures 20:519–526
20. Steinberger JK, Krausmann F, Eisenmenger N (2010) Global patterns of materials use: a socioeconomic and geophysical analysis. Ecol Econ 69(5):1148–1158

Impact of Optimum Waste Content and NaOH Molarity Dosage on Self-Compacting Geopolymer Concrete Efficiency

Endow Ayar Mazumder and L. V. Prasad M.

1 Introduction

Every day, new types of structures are emerging in this fast growing world. Concrete is one of the prime requirements of almost all the structures. The word 'concrete' originates from the Latin word 'concretus' meaning to grow together. Concrete's advantages being good compressive strength and durability, but it also has a drawback due to low tensile strength and brittle failure. Coming to the ingredients of concrete, it mainly includes pozzolanic material like cement, water, fine aggregate or sand, coarse aggregate, or stone. In 1983, Japan has faced a major challenge in the durability of those structures which are made up of concrete. Adequate compaction done by unskilled labors became immensely problematic particularly in structures having highly congested reinforcement. After vast research work, it was Dr. Hajime Okamura who developed 'Self Compacting Concrete' as the solution to improper compaction and hence today he is known as the father of SCC technology. By virtue of elimination of external compaction and having self flowing ability, SCC became quite popular and fulfilled the need of the hour successfully. Cement is the most important constituent of concrete which is responsible for the process of hydration. However, 1 ton production of cement leads to the emission of more than half a ton of carbon dioxide (CO_2) which is a major greenhouse gas and significantly contributes to global warming. This led to the development of geopolymer concrete which significantly reduces the demand of OPC since it is utilizes the waste products like rice husk ash, fly ash, GGBS, etc., as cementitious material. Compressive strength of concrete increases with the increase in the content of slag [1]. Geopolymer concrete significantly shows better mechanical strength as well as early strength development

E. A. Mazumder (✉) · L. V. Prasad M.
NIT Silchar, Silchar, Assam, India
e-mail: endow_rs@civil.nits.ac.in

L. V. Prasad M.
e-mail: mlvprasad@civil.nits.ac.in

A. Sil et al. (eds.), *Recent Trends in Civil Engineering*, Lecture Notes in Civil Engineering 274, https://doi.org/10.1007/978-981-19-4055-2_13

with very low energy consumptions as compared with OPC concrete, preparation of geopolymer concrete also emits very small amount of harmful gases as compared to OPC concrete [2–4]. The conventional geopolymer concrete needs an external elevated temperature curing to achieve the expected results, in this process, there is an additional requirement of heat energy and is not suitable for cast-in-situ concreting work, hence in the current study the waste materials are arranged in such a manner that it abolished the requirement of external heat curing for gaining its strength, which makes the self compacting geopolymer concrete with optimum dosage of waste materials, an feasible alternative to the conventional geopolymer concrete. Researchers investigated the outcome of temperature and curing under ambient environment on the mechanical properties of SCGPC using GGBS and rice husk ash concluding that GGBS is effective in enhancing the strength properties of concrete while RHA is extremely cost effective and also helps in the safe disposal of industrial as well as agricultural wastes [5]. Self compacting geopolymer concrete mainly used for high early strength requirements and for high rise buildings which has congested reinforced elements, the SCGPC can enter the congested elements and fill the formwork properly [6–8]. Workability of concrete mixes tends to improve upon decrement of alkali activator to binder ratio while strength properties holds a direct relationship with alkali activator to binder ratio [9]. Researchers also observed that workability decreases with the increase of slag content and decrease of alkali activator ratio while strength properties increased with the increase of slag content and decrease of alkali activator ratio [10]. Molar concentration of sodium hydroxide plays a crucial role in the rheological and mechanical properties of SCGPC. Research works have suggested that strength properties increase with the increase of molar concentration up to a particular limit after which it decreases with the further increment in molarity. On the other hand, fresh properties are reduced very slightly with the increment in molar concentration of NaOH [11]. Studies have also been done to know the impact of Si to Al ratio on the process of geopolymerization [12].

2 Materials Used and Methods

Low calcium fly ash (Class F) and GGBS obtained from astra chemicals, Chennai is used as the binder content. The elemental composition of the fly ash and GGBS are exhibited in Tables 1 and 2, respectively. Fine aggregate used is prepared according to Zone II specification as per IS383. Locally available coarse aggregate is used with 50% aggregates passing through 20 mm sieve and retained on 10 mm sieve and remaining 50% aggregates passing through 10 mm and retained on 4.75 mm. Table 3 exhibits the physical properties of materials such as fine and coarse aggregate. Alkaline activator solution comprised of NaOH and Na_2SiO_3. Chryso Optima K9314 is used as superplasticizer (SP) and Chryso Quad 60 is used as the viscosity modifying agent (VMA). The dosage of SP is kept at 0.8% and that of VMA is 0.2%.

Table 1 Chemical composition of Fly Ash

Oxide	Composition (%)
K_2O	0.9
MgO	0.5
Al_2O_3	27.5
CaO	1.97
Fe_2O_3	3.2
SO_3	0.06
Tio_3	1.06
Na_2O	0.3
SiO_2	62.19

Table 2 Chemical composition of GGBS

Oxide	(%) Content
Sulphide/sulphite	0.9
Magnesia	9
Alumina	12
Manganese	3.3
Silica	35
Calcium Oxide	39.8

Table 3 Physical properties of fine & coarse aggregate

Property	Specific gravity	Water absorption (%)	Free moisture	Bulk density
Fine Aggregate	2.65	1.22	Nil	1520 kg/m^3
Coarse Aggregate	2.67	0.80	Nil	1548 kg/m^3

In the present experimental work, molar concentration of NaOH is varied between 6 and 12 M at an interval of 2 M. Alkaline activator ratio (AAR) and alkaline activator-to-binder ratio (AAB) is fixed at 2.5 and 0.45 respectively for the experimental purpose. The mix proportion of SCGPC mixes are calculated from Nan Su Mix design process [13] shown in Table 4. For each molar concentration, eleven SCGPC samples are casted by varying the ratio of fly ash and GGBS between 100:0 and 0:100 at an interval of 10%. For example, SCGPC 1 indicates 100% fly ash and 0% GGBS, while SCGPC 2 indicates 10% fly ash and 90% GGBS.

Table 4 Mix design mix proportion of SCGPC mix

Mix designation	Fly Ash (Kg/m3)	GGBS (Kg/m3)	Fine aggregate (Kg/m3)	Coarse aggregate (Kg/m3)	NaOH (Kg/m3)	Na_2SiO_3 (Kg/m3)	Extra water (%)
SCGPC 1	550	0	880	770	70.71	176.78	5
SCGPC 2	495	55	880	770	70.71	176.78	5
SCGPC 3	440	110	880	770	70.71	176.78	5
SCGPC 4	385	165	880	770	70.71	176.78	5
SCGPC 5	330	220	880	770	70.71	176.78	5
SCGPC 6	275	275	880	770	70.71	176.78	5
SCGPC 7	220	330	880	770	70.71	176.78	5
SCGPC 8	165	385	880	770	70.71	176.78	5
SCGPC 9	110	440	880	770	70.71	176.78	5
SCGPC 10	55	495	880	770	70.71	176.78	5
SCGPC 11	0	550	880	770	70.71	176.78	5

3 Preparation and Placement of Concrete

As per the mix proportion shown in Table 4, the materials are measured using the weight batching technique. Sodium hydroxide solution of a particular molarity is prepared one day prior to casting. Before the commencement of preparation of concrete, the NaOH solution is combined with Na_2SiO_3 solution to get the activator in requisite quantity. Then materials, namely, fly ash, GGBS, fine aggregate, and coarse aggregate is poured in the electrically operated concrete mixer and is allowed to mix properly approximately for 1.5 min. After initial mixing is over, the liquid content such as activating solution, superplasticizer, and viscosity modifying admixture are added one by one in about two to three intervals. The concrete mixer runs continuously for about 6 to 8 min allowing the dry and the liquid content mix thoroughly for the better production of concrete. Finally, extra amount of water is mixed thoroughly for one minute to obtain the uniform color, consistency, and desired workability. After the completion of mixing, the concrete is taken in a tray and various workability tests such as slump flow, J-Ring, U-Box, and V-Funnel are performed, the acceptance criteria for SCC as per EFNARC [14] is shown in Table 5. Then the fresh concrete is placed in the molds of 150 mm side cube, 150 mm × 300 mm cylinder, and 100 mm × 100 mm × 500 mm prism for the determination of compressive strength, split tensile strength and flexural strength, respectively. The preparation and placement of the SCGPC can be seen in the Fig. 1. The specimens are kept in the molds for 24 h and after opening the molds, concrete specimens are kept at ambient curing conditions till the testing age.

Table 5 Acceptance criteria for SCC as per EFNARC specification

Test methods		Range of values
Slump flow test (mm)		650–800
J-ring test (mm)		0–10
U-box test (mm)		0–30
V-funnel test	T_f (sec)	6 to 12
	$T_{5\ min}$ (s)	$T_f + 3$

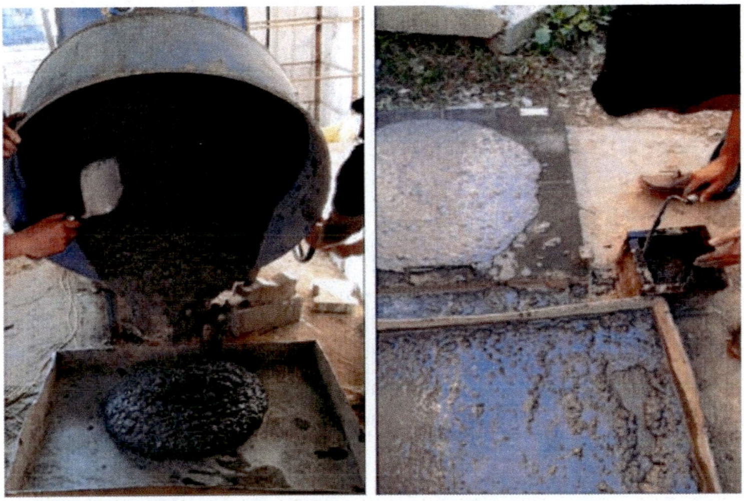

Fig. 1 Preparation and placement of SCGPC

4 Results and Discussion

After testing the samples on completion of desired time, it is found that as the molarity of the mix is increased, gradually the strength of the specimen increases. Strength and workability properties of SCGPC mixes for 10 M sodium hydroxide solution are shown in Table 6. In current experiment, the strength increase is from 6 to 10 M, but there is downward drop is observed when the molar concentration of sodium hydroxide exceeds 10 M. The graphical representation for strength of specimens having 6, 8, 10, and 12 M of sodium hydroxide is shown in Fig. 2.

During the testing practice, it is noted that the specimens having high GGBS content tends to show better strength results as compared to specimens having low GGBS content. It is also noted that the strength for specimen SCGPC 6, having about 50% of fly ash and 50% of GGBS shows a 7 day strength of 32.22 MPa which is about 82% of its 28 days strength in case of compression. Same kind of early strength is noticed in case of flexure as well as split tensile strength testing. It is also noticed that with small quantity of GGBS in any specimen, it shows very good early strength under ambient curing condition.

Table 6 Mechanical properties and workability properties

Mix designation	Compressive strength (MPa)		Split tensile strength (MPa)	Flexural strength (MPa)	Workability properties				
	7 Days	28 Days			Slump flow (mm)	J-ring (mm)	U-box (mm)	V-funnel	
								T_f (sec)	$T_{5\,min}$ (sec)
SCGPC 1	2.44	3.55	0.27	1.39	782	4	8	5	6
SCGPC 2	9.56	14.44	1.11	2.8	768	5	12	5	7
SCGPC 3	14	20	1.6	3.28	736	6	14	6	8
SCGPC 4	18	24.44	1.94	3.62	705	6	16	8	11
SCGPC 5	24	30.67	2.45	4	694	8	21	9	12
SCGPC 6	32.22	39.11	3.09	4.51	676	9	25	10	13
SCGPC 7	41.56	48.9	3.96	5.1	658	10	28	12	15
SCGPC 8	48	54.67	4.48	5.5	609	12	31	14	18
SCGPC 9	53.33	60	5.04	5.76	590	13	34	14	18
SCGPC 10	60	65.33	5.36	5.86	573	14	36	15	19
SCGPC 11	65.33	70.22	5.9	6.24	548	16	38	16	21

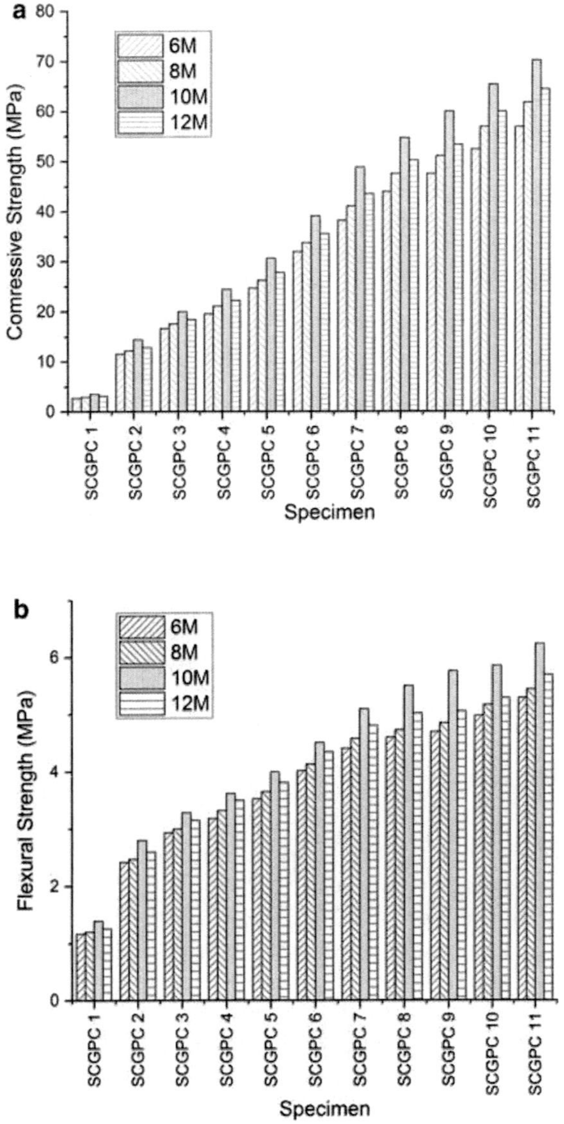

Fig. 2 **a** Compressive strength of specimens having 6, 8, 10, 12 M NaOH solution. **b** Flexural strength of specimens having 6, 8, 10, 12 M NaOH solution. **c** Split tensile strength of specimens having 6, 8, 10, 12 M NaOH solution

As the strength enhancement is noted with increase of GGBS content in the specimens, but there is a prominent decrease of workability is also observed. Additionally, it is noticed that with the increase of GGBS amount, the workability goes on decreasing and beyond 60% dosage of GGBS in any specimen, European Federation of National Associations Representing for Concrete (EFNARC) specifications are not satisfied for achieving the self compactibility of concrete. Therefore, for the making self compacting geopolymer concrete from both strength and workability point of

Fig. 2 (continued)

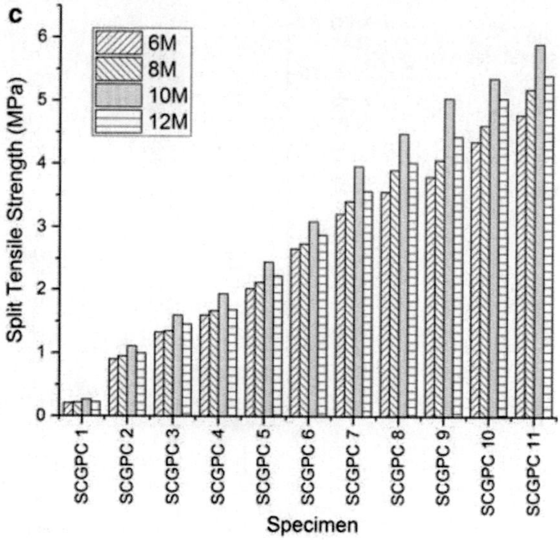

view it is recommended to use 10 M sodium hydroxide solution and FA:GGBS as 40:60, i.e., 40% of fly ash and 60% of GGBS, which will help in disposing both the fly ash and GGBS properly.

5 Conclusion

i. It has been noticed that the increment in the amount of GGBS increases the strength behaviors of SGCPC.
ii. Early strength gain of SCGPC mixes increases with the additional increment of GGBS amount, it occurs because with an increase of GGBS contents, the calcium and silicate both make an effort to react for the formation of C-S-G gel which im parts an elevated early strength.
iii. For 100% fly ash based mix, 7 day strength is 65% of the 28 day strength while for 100% GGBS based concrete, 7 day strength rises to 92% of the 28 day strength.
iv. GGBS increases the strength properties while fly ash gives good workability.
v. With the increase of molarity from 6 M to 10 M, the strength properties increase but tends to decrease on further increment of molar concentration of NaOH.
vi. With the increase of molarity, workability of the concrete mixes slightly decrease.
vii. Although with the increase of GGBS content, strength increases constantly but workability reduces and beyond 60% dose of GGBS, EFNARC specifications are not satisfied for achieving the self-compactibility of concrete.

viii. Hence for the preparation of self compacting geopolymer concrete from both strength and workability point of view it is recommended to use 10 M solution and FA:GGBS as 40:60.
ix. Therefore, SCGPC 7 with 10 M NaOH solution has been finally optimized.

References

1. Dr SE, Reddy CJ (2018) Geopolymer concrete with self compacting: a review. Int J Civ Eng Technol 8:163–172
2. Collins F, Sanjayan JG (2000) Effect of pore size distribution on drying shrinking of alkali-activated slag concrete. Cem Concr Res 30(9):1401–1406
3. Bakharev T, Sanjayan JG, Cheng YB (2003) Resistance of alkali-activated slag concrete to acid attack. Cem Concr Res 33(10):1607–1611
4. Cheng TW, Chiu JP (2003) Fire-resistant geo-polymer produced by granulated blast furnace slag. Miner Eng 16(3):205–210
5. Yamini J, Patel NS (2018) Development of self compacting geopolymer concrete as a sustainable construction material. Sustain Environ Res 28:412–421
6. Memon FA, Nuruddin MF, Demie S, Shafiq N (2011) Effect of curing conditions on strength of fly ash-based self-compacting geopolymer concrete. Int J Civ Environ Eng 3:183–186
7. Memon FA, Nuruddin F, Shafiq N (2011) Compressive strength and workability characteristics of low-calcium fly ash-based self-compacting geopolymer concrete. Int J Civ Environ Eng 3:72–78
8. Noushini A, Castel A (2016) The effect of heat-curing on transport properties of low-calcium fly ash-based geopolymer concrete. Constr Build Mater 112:464–477
9. Shivaranjan NS, Shiva Kumar KS, Nagarak VK (2017) A study on self-compacting geopolymer concrete with various water to geopolymer solids ratio. Int Res J Eng Technol 3:2064–2069
10. Chaterjee P, Adil M (2016) Sodium hydroxide activated ground flyash geopolymer cured at ambient temperature. J Build Eng 2118–2124
11. Menon FA, Nuruddin MF, Khan S (2015) Effect of sodium hydroxide concentration on fresh properties and compressive strength of self-compacting geopolymer concrete. J Eng Sci Technol 8:44–56
12. Duxon P, Mallicoat SW, Lukey GC (2007) The effect of Alkali and Si/Al ratio on the development of mechanical properties of metakaolin based geopolymers. Colloids Surf, A Physiochem Eng Aspects 292:8–20
13. Nan S, Hsu K-C, Chai HW (2001) A simple mix design method for self-compacting concrete. Cem Concr Res 31:1799–1807
14. EFNARC (2002) Specification and guidelines for self compacting concrete

Study and Analysis of Irregularities in RC Buildings

Praful Patel, Tejas Pathak, and Shardul Joshi

1 Introduction

1.1 General

Currently many buildings are designed and constructed with irregular configurations. Due to lack of availability of lands in cities and also to impart lights and ventilation in the structure, architects are preferring irregular structures and this irregularity is may be either vertical irregularity or plan irregularity. The plan irregularities consist of asymmetrical plan shapes, re-entrant corners and irregular distribution of mass, strength, stiffness along plan, etc. And the vertical irregularity consists of mass, strength, stiffness and setback.

These irregular structures are highly vulnerable to earthquakes under the action of dynamic loads. Regular buildings having no irregularity are not subjected to critical effects like overturning moment, excessive displacement and storey drift, but irregular buildings are subjected to all this, above the optimal values. So it is necessary to study and analyse the buildings considering all the effects occurring on the buildings. Therefore, to fulfil the need and demand of the current situations, an extensive study is required to analyse the effects occurring on the structure due to irregularity.

In this study, the effects due to irregularity is analyse using pushover analysis method. As to forecast the effects on the structure due to earthquake, it is not possible to consider the effects due to earthquake up to linear elastic limits. The linear approach of analysis is based on the response reduction factor having a value five, and it means that capacity of the structure is reduced by five times than actual capacity of the structure to assure the economy in the structure. Further, the deflection in the structure is take by the ductile capacity of the structure. In reinforced concrete structures, the beams and columns in the structure are designed such that it a take

P. Patel (✉) · T. Pathak · S. Joshi
Department of Civil Engineering, Vishwakarma Institute of Information Technology, Pune, India
e-mail: praful.219m0032@viit.ac.in

A. Sil et al. (eds.), *Recent Trends in Civil Engineering*, Lecture Notes in Civil Engineering 274, https://doi.org/10.1007/978-981-19-4055-2_14

full impact beyond elastic limit and up to the ductile limit of the structure. Therefore, the drawback in the linear approach is that the response of the structure beyond limit state is neither simple extrapolation or nor it is pure ductile, though it gives the indication of the yielding point of the structure, whereas the nonlinear analysis methods give the appropriate results because it considers the inelastic behaviour of the structure. Nonlinear methods of analysis give the exact results of the progression of failures occurs in the structures. Push over analysis is done by assigning the hinges to the structure. Hinges are the points in the structure which have high flexural displacement where cracking and yielding occur. Basically, hinges represent localized force–displacement and moment-rotation relationship in the members at its elastic and non-elastic states.

There are three types of hinges in the structure flexural hinges, shear hinges and axial hinges. These hinges have nonlinear states defined as 'immediate occupancy' (IO), 'life safety' (LS) and 'collapse prevention' (CP) within its ductile range. Therefore, after push over analysis of the structure, we gate the information about the details of the hinges and what stage between the above three condition they are in. So, from this analysis, we clearly get the idea that where our structure is weak and from where the failure of the structure will occur.

1.2 Literature Review

Habib et al. [1], in this study, seismic zone-I of Bangladesh National Building Code-2006 is taken for a study, author performed a three-dimensional finite element analysis based on the design-built layout plan. The rectangular, square, inverted L, T, U and L shapes buildings have been modelled and analysed in ETABS. In these study, seismic load is imposed on the structure, therefore, lateral maximum displacement has observed in the X-direction and in Y-direction so based on the results author found that the lateral displacement is almost equal in all the buildings, whereas it is 74.5% more in a rectangular building in the Y-direction. Base shear is varying by large value as there are changes in the geometry of the structure. U-shaped building is subjected to maximum base shear also the author found that the time period recommendation given in Bangladesh code is 40% higher than the results obtained in ETABS and the torsional irregularity in the rectangular.

Building is beyond the permissible limits, whereas the other shapes buildings are in permissible limits and the maximum overturning moment has occurred in T shape building followed by U shape, L shape, R shape and S shape.

Sayyed et al. [2], in this study, author carried out the analysis of eight regular configuration and irregular configuration buildings considering two types of the vertical irregularities specifically stiffness irregularity and setback irregularity. In stiffness irregularity, author modelled G + 10 storey building having three bays in both the directions. Stiffness irregularity in the model is made by increasing the height of floor, whereas for setback irregularity, offsets have been made on eight, fifth and second floor in the X-direction. From the results, it is obtained that the storey displacement of a ground floor of the irregular building is 1.5 times more than the ground displacement of a regular building. In stiffness, irregular buildings storey drift is maximum at irregular storeys. In setback irregular buildings, overturning moment and base shear are less than the regular buildings. Due to the setback irregularities, it is observed that as the setback in the building increases the stiffness of the building decreases.

Ahmed et al. [3], in the present study, the author studied the effects of irregularity occurring on the structure having a shape like rectangular, T, E, F and S for both vertical and horizontal irregularities. A 32mX40m, 15 storey structure is modelled having 4mX4m bays is modelled and analysed in ETABS. To obtain the storey drift and displacement at each floor, the models have been analysed using the equivalent static method and response spectrum method. From the results, it is obtained that the fundamental time period for the regular buildings is less than the time period of irregular buildings also it is found that the value of base shear in response spectrum analysis is less than the base shear occurring in the response spectrum method. Also, it is found that the displacement and drift coming in T shape model are very large in response spectrum method and linear equivalent static method.

Chopra et al. [4], in this paper, modal pushover analysis method for estimating the seismic demands is used for the analysis of unsymmetrical plan buildings. The modal pushover analysis method estimates the seismic demand due to the intense ground motions is almost have equal precision for unsymmetrical plan as symmetrical plans. Author concluded this based on the comparison of demand results of modal pushover analysis and response spectrum analysis for four structural systems. Author has considered three buildings for the analysis, first is torsional stiff building. Second is torsion ally flexible building, and the third is torsion ally similar stiff building; from the analysis, they obtained that the results for first and second building are similarly accurate as they are for symmetrical plan buildings; however, the results are varying due to stronger coupling of elastic modes and underestimation of roof displacement; from this analysis, they found that the CQC modal combination rule will not give an accurate results for second building for the highest response, even though the system is linearly elastic.

1.3 Objective

[1] To prepare a model having irregularities in the structure as per IS 1893:2016
[2] To prepare a model with irregularities and analyse it using SAP2000
[3] To find the critical condition in the structure due to which there is chances of the collapse of the structure
[4] Also comparing analysis results of building with different irregularities with regular building and providing suggestion to avoid critical effects in structure.

2 Methodology

[1] Study of literature review for understanding the effects of irregularities due to the earthquake.
[2] Study of the Fema440 and the definitions and concepts given regarding the pushover analysis.
[3] Learning of response spectrum analysis.
[4] Learning of pushover analysis.
[5] Study of the norms given in IS 1893:2016 regarding plan irregularities and vertical irregularities.
[6] Modelling of the structure in SAP2000 and analysis using pushover method of analysis.
[7] Comparison of the results of pushover analysis of building with irregularities to normal building with no irregularity.

2.1 Modelling of the Structure

Type I: There are 15 structures have been modelled having G + 15 storey building with floating or stub column irregularity and stiffness irregularity.

Type II: There are nine structures have been modelled having G + 15 storey building with mass irregularity and vertical geometric irregularity.

Type III: G + 15 storey building with no irregularity in the building.

Model data is as follows:

Number of floors: G + 15.

Number of bays: 7 bays of 5 m each in X-direction and 10 bays of 5 m each in Y-direction.

Floor to floor height: 3.6 m.

Beam size: 300 mm × 600 mm.

Column size: 600 mm × 800 mm.

Earthquake zone: zone III.

Importance factor: 1.2

Response reduction factor: 5

Load on the beam: Wall load + Plaster load.

Wall load = 7.81 Kn/m.

Plaster load = 3 Kn/m.

Total load on the beam = 10.81 Kn/m.

Load on the floors: Live load + Floor finish load.

Live load = 2 Kn/m.

Dead load = 1.16 Kn/m^2

Tile weight = 0.35 Kn/m^2

Total dead load on slab: Mortar load + Tile weight = 1.51 Kn/m^2 (Figs. 1, 2 and 3).

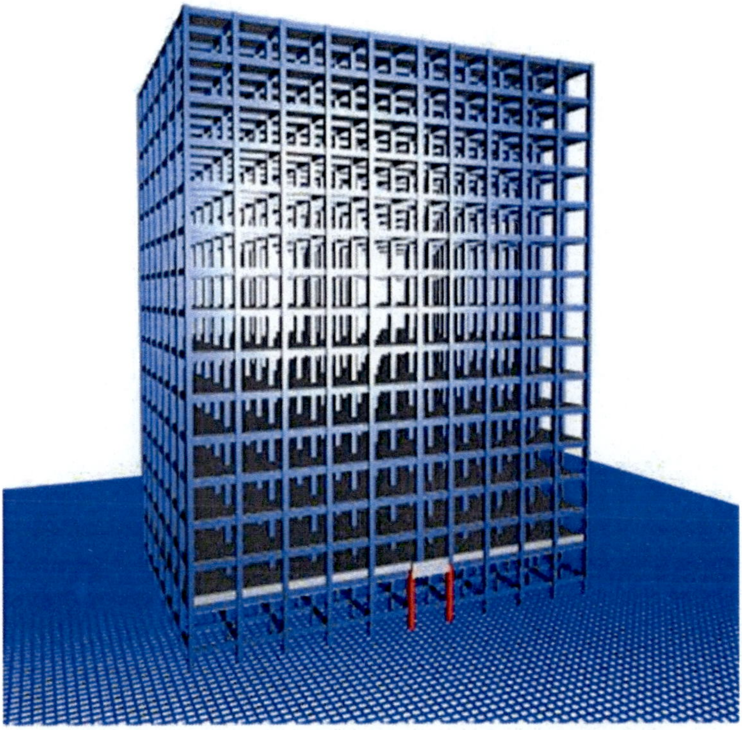

Fig. 1 Type 1 model having floating column till 1st floor. Floating column is modelled till each floor, therefore, total 15 models have been modelled

2.2 *Application of Hinges in the Structure*

Hinges are the localized points in the structure, which represents the force–displacement and moment and rotation relationships. To study the behaviour of the structure, hinges should be assigned to each element of the structure as shown below (Fig. 4).

3 Results and Descussion

3.1 *Pushover Method of Analysis*

The pushover analysis method is a nonlinear static method of analysis of structures that is used to determine yield displacement and succeeding inelastic behaviour of buildings. In the pushover analysis method, the hinges are defined to each element of the structure, hinges are the points in the structure where cracking and yielding are expected to occur in the structures. There are various types of hinges such as

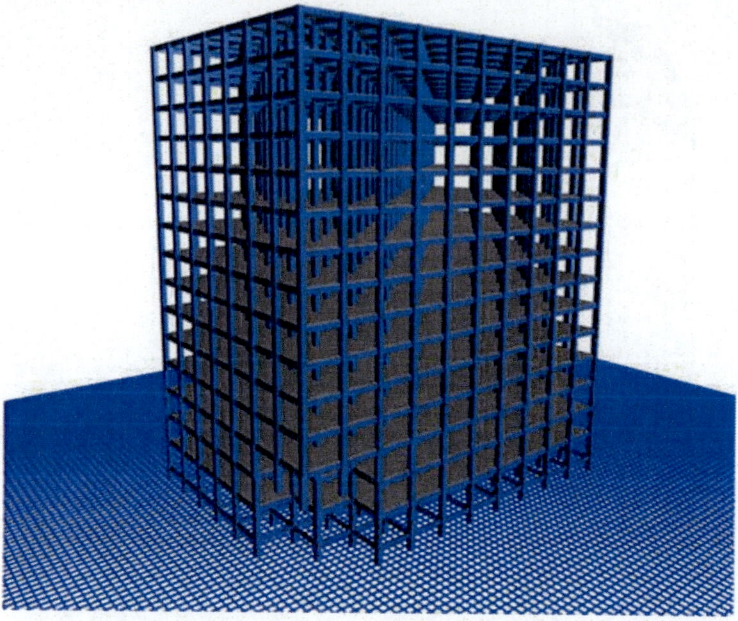

Fig. 2 Type 2 model having mass irregularity and vertical geometric irregularity. Total nine models have been modelled

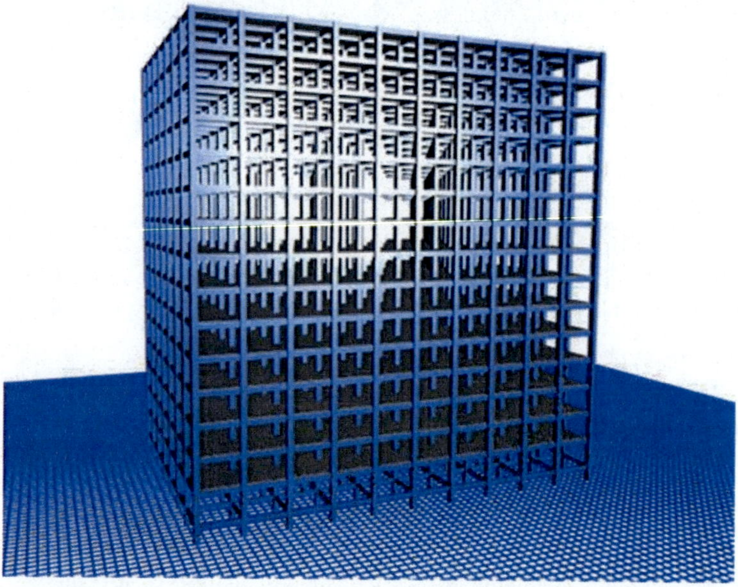

Fig. 3 Type 3 building with no irregularities

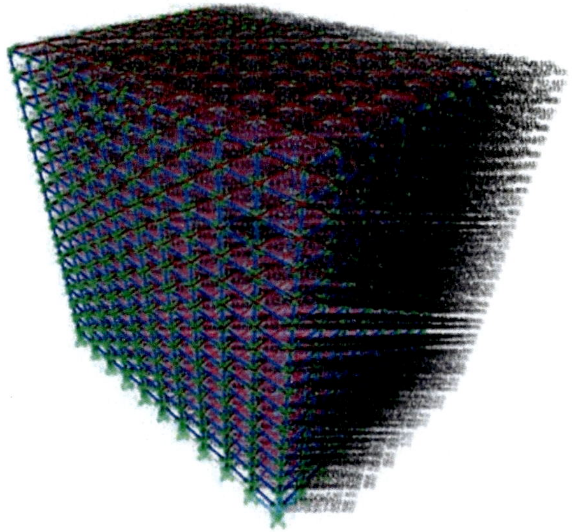

Fig. 4 Hinges applied to model at all point

flexural axial and shear, and this hinges represent the relation between force and displacement in an elastic and inelastic zone of the structures. The seismic performance of a building can be evaluated in terms of pushover curve, performance point, displacement ductility, plastic hinge formation, etc.

The capacity curve obtained by performing pushover analysis is transformed into a capacity spectrum, and the response spectrum is also determined of the structure for the required building performance level. The performance point of the structure analysed is obtained from the intersection of demand and capacity spectrum given (Tables 1 and 2, Figs. 5, 6, 7, 8, 9, 10, 11, 12, 13, 14, 15 and 16).

4 Conclusion

[1] From the results for type I, it has been observed that the base shear of model 9 is minimum in X-Direction as well as in Y-Direction, and hence, model 9 is more critical considering the base shear.

[2] The time period of model 12 is minimum in X-direction and of model 4 is minimum in Y-direction, therefore, the frequency of model 12 and model 4 maximum, hence, there is a chance of failure of the structure.

[3] From the above results, it has been observed that the ductility of model 13 is minimum in X-direction and the ductility of model 3 is minimum in Y-direction, and model 15 has maximum ductility in X-direction and Y-direction.

[4] From the results of type II model, it has been observed that the base shear of model 3.3 is minimum in X-direction, whereas the base shear of model 2.2 is minimum in Y-direction.

Table 1 Results at performance point of type I model

	Push-X	Push-Y	T Sec in X-direction	T Sec in Y-direction	Ductility X-direction	Ductility Y-direction
Model 1	15,151.064	13,849.624	4.415	4.659	4.526	4.546
Model 2	15,193.849	14,122.445	4.428	4.573	4.599	4.449
Model 3	15,180.487	14,147.743	4.436	4.582	4.589	**4.444**
Model 4	15,222.824	14,562.591	4.412	**4.561**	4.504	4.564
Model 5	13,167.396	12,947.493	4.799	5.013	5.397	5.728
Model 6	13,172.387	13,130.433	4.794	5.12	5.387	6.139
Model 7	15,196.236	15,015.817	4.394	4.608	4.457	4.862
Model 8	15,172.744	14,863.269	4.395	4.597	4.453	4.857
Model 9	**11,144.441**	**11,275.309**	5.347	5.549	6.778	7.153
Model 10	15,153.132	14,791.301	4.394	4.629	4.442	4.823
Model 11	15,149.13	14,785.477	4.395	4.633	4.439	4.823
Model 12	15,153.645	14,793.853	**4.392**	4.635	4.43	4.830
Model 13	13,454.963	12,950.291	4.779	5.058	**4.367**	4.79
Model 14	15,142.371	14,785.85	4.42	4.636	4.532	4.828
Model 15	11,193.109	11,310.693	5.284	5.524	6.732	7.406
Regular	11,676.853	11,560.54	5.162	7.243	6.191	5.669

Table 2 Results at performance point of type II model

	Push-X	Push-Y	T Sec in X-direction	T Sec in Y-direction	Ductility X-direction	Ductility Y-direction
Model 1.1	9778.334	8769.612	5.794	6.030	6.168	6.074
Model 1.2	10,500.495	9265.961	5.350	5.965	5.614	6.186
Model 1.3	11,309.09	10,002.062	5.007	5.212	5.425	**5.113**
Model 2.2	10,314.18	**8596.015**	5.498	6.067	5.881	6.042
Model 2.3	9827.223	11,875.716	5.427	**4.793**	**5.186**	5.163
Model 2.4	12,038.55	9725.150	4.810	5.537	5.506	5.822
Model 3.3	**8877.553**	10,724.037	5.198	4.885	5.315	5.603
Model 3.4	11,430.771	9833.266	4.892	5.670	5.214	5.946
Model 3.5	12,636	10,129.843	**4.741**	5.097	5.743	5.365
Regular	11,676.853	11,560.540	5.162	7.243	6.191	5.669

[5] The time period of model 3.5 is minimum in X-direction, and the time period of model 2.3 is minimum in Y-direction, therefore, the frequency of model 3.5 and 2.3 is maximum, and hence, the chances of failure of structure are high.

[6] From the results obtained of type II model, it has been observed that the ductility of model 2.3 is minimum in X-direction and that of model 1.3 in Y-direction, therefore, model 2.3 and 1.3 have less capacity to bear the load.

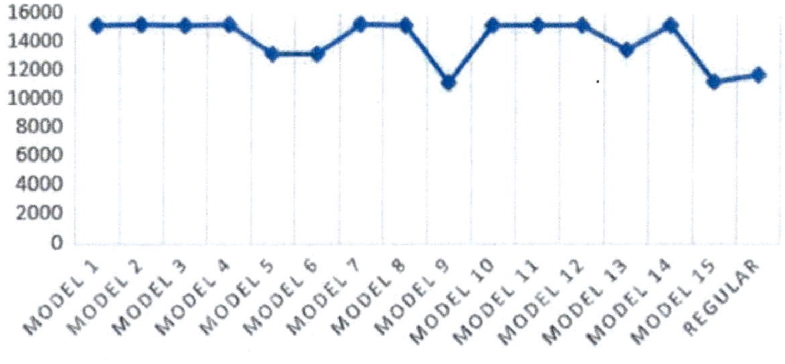

Fig. 5 Push-X type 1 structure graph

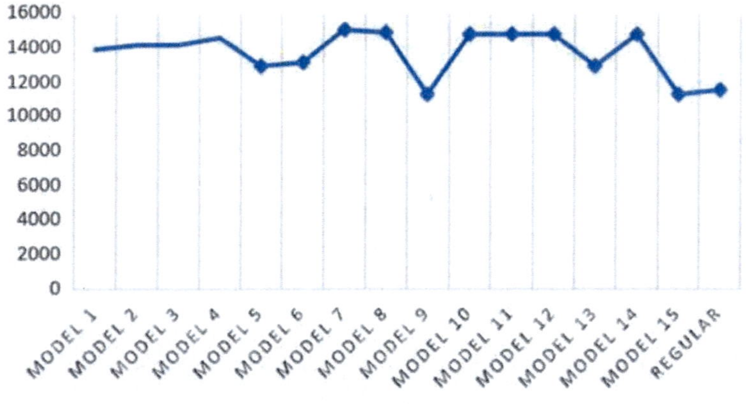

Fig. 6 Push-Y type 1 structure graph

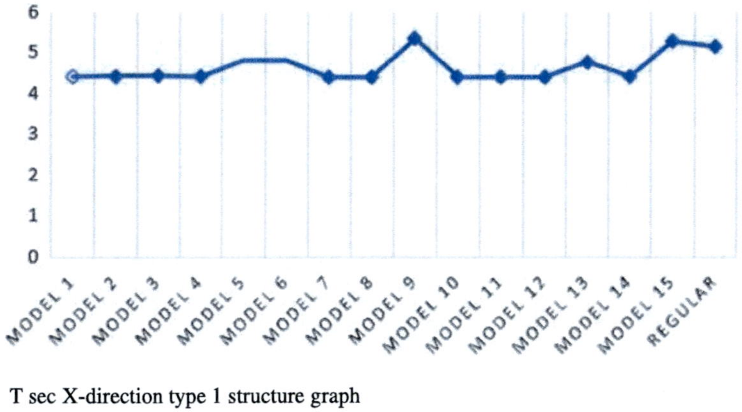

Fig. 7 T sec X-direction type 1 structure graph

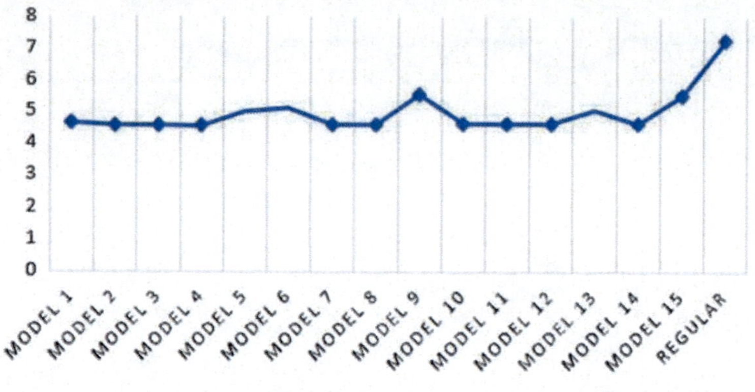

Fig. 8 T sec Y direction type 1 structure graph

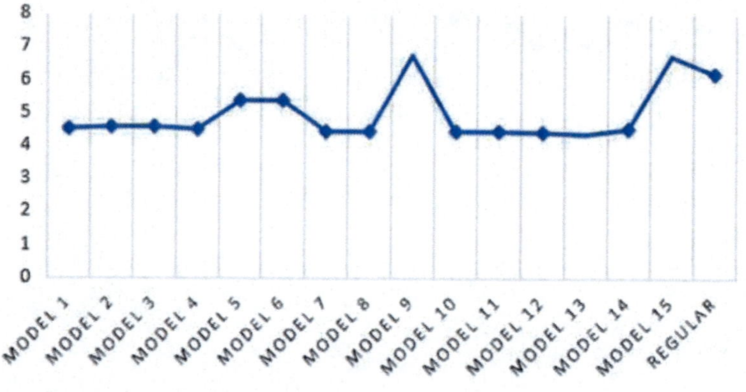

Fig. 9 Ductility X-direction type 1 structure graph

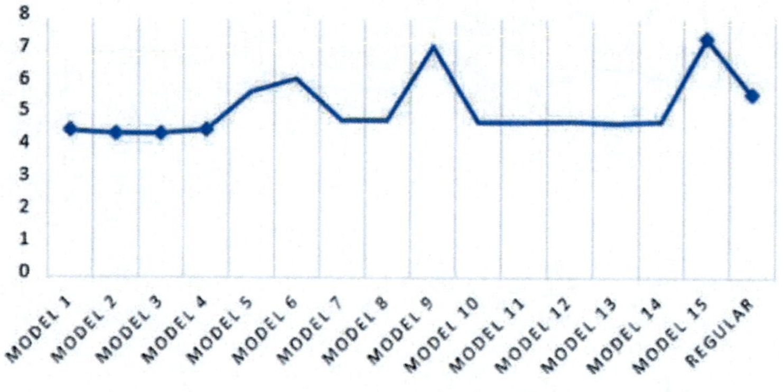

Fig. 10 Ductility Y-direction type 1 structure graph

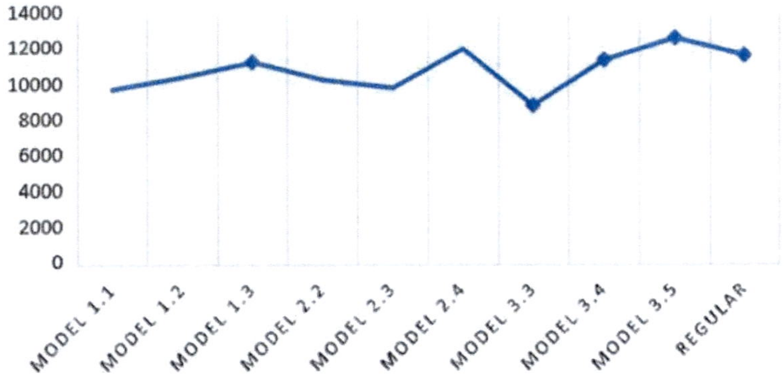

Fig. 11 Push-X type II structure graph

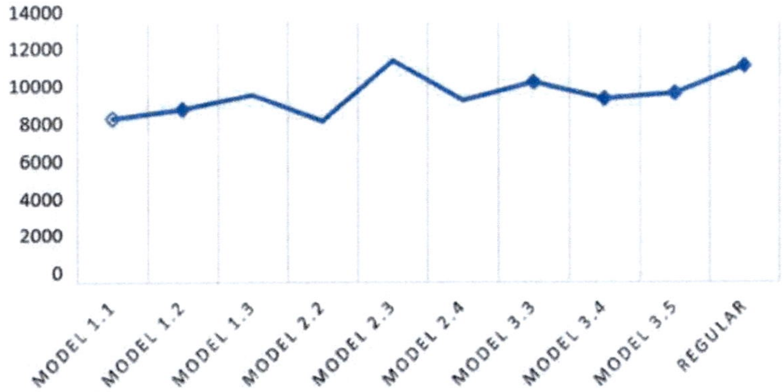

Fig. 12 Push-Y type II structure graph

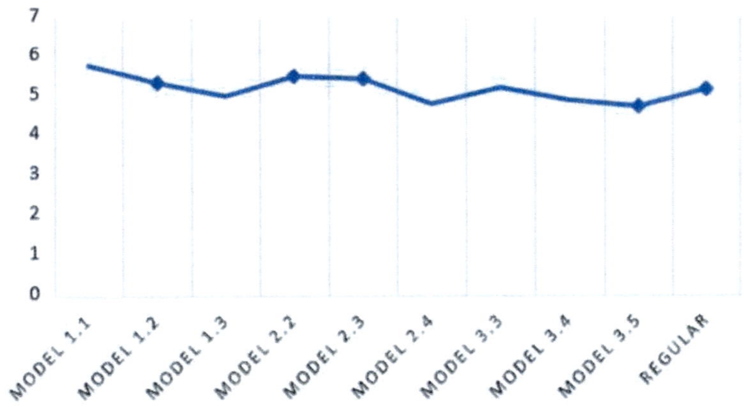

Fig. 13 T sec X-direction type II structure graph

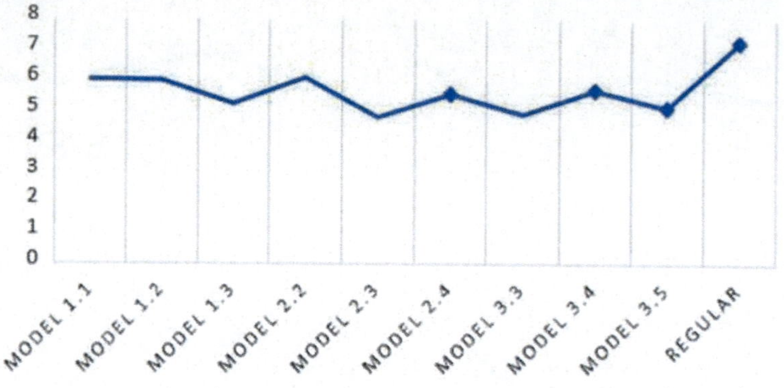

Fig. 14 T sec Y-direction type II structure graph

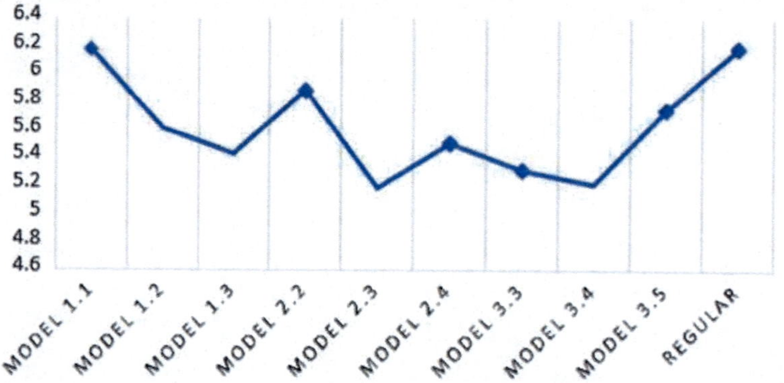

Fig. 15 Ductility X-direction type II structure graph

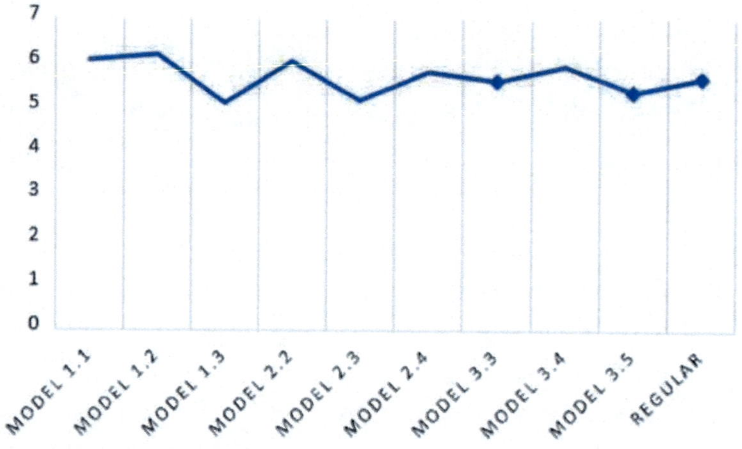

Fig. 16 Ductility Y-direction type II structure graph

References

1. Habib MZ, Alam MA, Barua S, Islam MM (Jan 2016) Effect of plan irregularity on RC buildings due to BNBC-2006 earthquake load. Int J Sci Eng Res 7(1):761. ISSN 2229-5518
2. Sayyed O, Kushwah SS, Rawat A (Jan–Feb 2017) Seismic analysis of vertical irregular RC building with stiffness and setback irregularities. IOSR J Mech Civil Eng (IOSR–JMCE) 14(1):Ver. VI. ISSN: 2320-334X
3. Ahmed A, Abdulla S, Arfat S (Sept 2017) Seismic analysis on buildings with horizontal and vertical irregularities. Int J Adv Eng Res Dev 4(9)
4. Chopra AK, Goel RK. A modal pushover analysis procedure to estimate seismic demands for unsymmetrical plan buildings: theory and preliminary evaluation. A report on research conducted under grant no. CMS-0336085 from the national science foundation
5. Yang P, Wang Y. A study on improvement of pushover analysis
6. IS 1893 (part-1): 2016 criteria for earthquake resistant design of structures
7. IS 875: (part-3): 2015 design loads for buildings and structures code of practice
8. IS 456: plain and reinforced concrete code of practice
9. Fema-357: global topics report on the pre-standard and commentary for the seismic rehabilitation of buildings

Enhancement of Mechanical Properties of Cement Mortar Using Ground Granulated Blast Furnace Slag as a Partial Replacement

Geethu Elsa Thomas, P. V. Indira, and A. S. Sajith

1 Introduction

GGBS is a by-product procured in the creation of pig iron. The basic constituents include silicates and alumino-silicates of calcium [1]. The produced slag particles are made finer than the cement particles, so as to obtain a good cementitious activity. As is notable that response rate increments with enhancing the fineness of pozzolans [2]. Literature reveals that GGBS shows cementitious properties and can be utilized efficiently with Portland concrete when they are finely grounded [3]. This paper is an experimental study to understand the utilization and efficiency of GGBS on the properties of mortar. In this study, initially, the fresh properties are discussed in term of workability, setting time and consistency. Then, effect of GGBS usage on mechanical, properties such as compressive strength, split tensile strength and flexural strength are discussed. Cement enterprises are one of the fundamental drivers of global warming as they are the principal donors of carbon dioxide. Henceforth, there is an earnest need to lessen the utilization of cement. It is the need of the hour to adopt a material that reduces the carbon emission and consumes less energy during their production, like GGBS. It is anticipated that India will deliver about 128.6 mega tons per annum of steel by 2021. Approximately, 0.45–0.50 tons of GGBS will be created per ton of steel. Disposal of such a large quantity of waste will be a matter of concern and a danger to the environment. Using GGBS in conventional construction practices helps to reduce carbon dioxide emission, and it is also an effective utilization of industrial by-products [4]. This investigation mainly is on confirming the available results and creating awareness about the utilization of GGBS for sustainable construction and environmental protection.

G. E. Thomas (✉) · P. V. Indira · A. S. Sajith
Department of Civil Engineering, National Institute of Technology, Calicut, India
e-mail: geethu_p180132ce@nitc.ac.in

A. Sil et al. (eds.), *Recent Trends in Civil Engineering*, Lecture Notes in Civil Engineering 274, https://doi.org/10.1007/978-981-19-4055-2_15

2 Experimental Investigation

2.1 Constituent Materials

Ordinary Portland cement (OPC 53) conforming to IS 12269:2013 [5] is used. GGBS used in this study is purchased from Quality PolyTech, Mangalore, containing 34.68% CaO, 33.11% SiO_2, 21.95% Al_2O_3, 9.50% MgO, 1.28% Fe_2O_3 and 0.10% SO_3 (composition is provided by the supplier). The specific gravity of GGBS is 2.91 with a surface area of 400 m^2/kg and particle size of 97.1 microns. M-sand supplied by POABS India (Ltd.), Calicut is used as fine aggregate, which conforms to guidelines provided by IS 383–1970 (reaffirmed 2002) [6]. It is found that M-sand used comes under grading zone II with a specific gravity of 2.68 and fineness modulus of 2.81. Conplast SP430 is used as superplasticizer to impart workability to the mix. Potable water is utilized in this experimental study.

2.2 Mix Proportions

Guidelines for the preparation of mortar is according to IS: 2250-1981 (reaffirmed 2000) [7]. Proportioning is done using trial and error method depending on the parameters namely amount and properties of binding materials, concentration of superplasticizer and flow value. A mix proportion of 1:1.5 is taken for the study, which is approximately equal to M20 grade, that can be conventionally used as per the recommendation of IS 456-2000 [8] and IS 10262-2009 [9]. A water cement ratio of 0.35 and superplasticizer content as 0.5% of binder are taken. In this investigation, the cement was replaced with 20%, 30%, 40%, 50%, 60% and 70% GGBS on a mass for mass basis. For this entire experimental investigation, the mixes are denoted in Table 1. The amount of fine aggregates, water cement ratio and superplasticizer were kept constant in all the mixes.

Table 1 Notations for various proportions of mix

Mix	Notation
100% OPC	OPC
100% GGBS	GGBS
80% OPC and 20% GGBS	G20
70% OPC and 30% GGBS	G30
60%OPC and 40% GGBS	G40
50% OPC and 50% GGBS	G50
40% OPC and 60% GGBS	G60
30% OPC and 70% GGBS	G70

2.3 Preparation of Test Specimens

Cement, GGBS and M-sand are taken in the required proportions and mixed dry in a mechanical mixer for 3 min. After this, the required quantity of water and superplasticizer is added at regular intervals and the mortar is mixed for 3 min to produce a workable consistency. After mixing, workability of the fresh mortar is determined by conducting flow table test. In order to determine the hardened properties standard cubes of size 70.6 mm, cylinder of diameter 100 mm and height 200 mm and prisms of size 40 × 40 × 160 mm were prepared. The moulds were removed after 24 h, and the specimens were kept for curing by immersing in water at a temperature of 27 °C for a period of 3, 7, 28 and 56 days.

3 Study on Fresh Properties

Experiments were conducted to understand the influence of replacement of cement with GGBS at varying percentage on the fresh properties like consistency, initial and final setting time and workability.

3.1 Consistency and Setting Time

Consistency and setting time were found as per IS 4031 Part IV (1998) [10] using a Vicat apparatus. Results are tabulated in Table 2. It was observed that addition of GGBS does not significantly affect the setting mechanism. The initial setting time for G50 mix is 17% more than that of OPC mix. The reason for this is increasing the period of hydration process due to increased pozzolanic content. Consistency increases as percentage of GGBS content increases.

Table 2 Consistency and setting time of different mixes

Mix	Consistency (%)	Initial setting time (min)	Final setting time (min)
OPC53	30.0	200	425
GGBS	35.0	243	445
G20	30.5	210	400
G30	31.5	222	405
G40	33.0	228	410
G50	**34.5**	**234**	**415**
G60	35.0	238	420
G70	35.5	240	425

Table 3 Flow value of different mixes

Mix	Flow (mm)
OPC	102
G20	103
G30	104
G40	105
G50	106
G60	115
G70	117

3.2 Workability

Flow table test was conducted as per IS4031 (part-7-1988) [11] to determine the workability of fresh mortar mix. The flow values are tabulated in Table 3. GGBS included mortar mix demonstrates enhanced workability compared to mortar mix without GGBS. Increase in replacement level increases workability. According to ACI-233 R03 [2], little water is only absorbed during mixing giving enhanced workability compared to cement, this is due to the better cementitious molecule scattering and surface attributes of the GGBS.

4 Study on Hardened Properties

The mechanical behaviour of mortar mix with 20%, 30%, 40%, 50%, 60% and 70% GGBS as replacement to cement was investigated. Tests were conducted to determine the compressive strength, flexural strength and split tensile strength.

4.1 Compressive Strength

Cubes of size 70.6 mm of different mixes were taken at 3, 7, 28 and 56 days of curing and tested for compression test as per the procedure specified in IS4031 part 6–1988 (reaffirmed 2005) [12]. Table 4 compares the compressive strength attained by OPC mix and its varying GGBS replacement percentages at different ages of curing. The results of 28th day compressive strength test show that addition of GGBS up to 50% gives strength almost equal to OPC mix. Also, later age strength of GGBS mixes increases than that of OPC mix. G50 mix has 18.32% more strength than OPC mix at 56th day compressive strength test. GGBS has higher diffusion rates than OPC at early ages, which reduces as the age increase. The reduction in diffusion rate at ages more than 56 days leads to an increase in the strength of mortar [13]. The

Table 4 Compressive strength of OPC and varying GGBS replacement percentages

Compressive strength (N/mm^2)				
Mixture	3-day	7-day	28-day	56-day
OPC	37.59	42.31	50.87	51.89
G20	35.82	45.04	51.23	55.34
G30	39.75	46.71	50.25	53.21
G40	31.28	39.88	49.31	50.56
G50	**25.91**	**38.18**	**50.91**	**61.40**
G60	22.83	34.11	44.87	45.31
G70	22.63	31.03	41.03	42.63

proportion of the strength enhancing calcium silicate hydrates (C-S-H) is higher in GGBS concrete than in OPC concrete at later stages [14].

4.2 *Flexural Strength*

For determining the flexural strength, mortar prisms of size 40 × 40 × 160 mm are tested under symmetrical three-point bending in a deformation-controlled setup according to ASTMC-348-08 [15] using universal testing machine. The rate of loading adopted is 1 mm/minute. The load is increased till the prism fails by cracking, and the maximum load is recorded as P. The specimens were tested after curing for 28 and 56 days. Table 5 compares the flexural strength attained by the OPC mix and its varying GGBS replacement percentages.

The variation of flexural strength for different mixes shows that addition of GGBS up to 50% increases the flexural strength and decreases it thereafter. Later age strength is also increased by the addition of GGBS. G50 mix has 28.41% more strength than OPC mix at 28th day. Also, it has 29.07% more strength than OPC mix at 56th day. As it is realized that flexural strength is surprisingly delicate to micro-cracks, the

Table 5 Flexural strength of OPC and varying GGBS replacement percentages

Flexural strength (N/mm^2)		
Mixture	28-day	56-day
OPC	5.63	6.19
G20	6.73	7.50
G30	6.89	7.71
G40	7.00	7.82
G50	**7.23**	**7.99**
G60	4.82	4.99
G70	3.80	4.81

Table 6 Split tensile strength of OPC and varying GGBS replacement percentages

Split tensile strength (N/mm^2)		
Mixture	28-day	56-day
OPC	2.562	3.374
G20	2.500	3.320
G30	2.540	3.450
G40	2.249	3.480
G50	**2.600**	**3.560**
G60	2.562	3.405
G70	2.062	2.499

expectation is that the inclusion of GGBS to mortar alters the reaction products and pore structure and thus enhances the flexural strength especially at the later ages.

4.3 Split Tensile Strength

Split tensile strength test on mortar was performed on 200 mm long and 100 mm diameter cylinder specimens at 28 and 56 days of curing as per the procedure specified in ASTMC-348-08 [15]. Universal testing machine (UTM) is used for testing with a loading rate of 2.4 MPa/min. The load was increased till the specimen failed by splitting along the diameter. From Table 6, it is observed that OPC and G50 mix has a comparable split tensile strength property. Later age split tensile strength of mortar mix improves with the addition of GGBS. The splitting tensile strength of G40 mix is lesser than that of OPC mix by 12.21%; however, the strength of G40 mix increases by 5.51% than OPC mix on 56th day test.

5 Microstructural Analysis

Morphology helps to identify the inter-bonding of molecules, shape, size and structure of materials. The form of OPC mortar mix and G50 mortar mix is studied using microstructural analysis techniques like scanning electron microscopy (SEM analysis) as per ASTM C1723 [16] and X-ray diffraction (XRD).

5.1 Scanning Electron Microscopy (SEM)

SEM images of OPC mortar mix and G50 mortar mix is shown in Fig. 1. Table 7 explains the various labels used in the SEM image. Smooth surfaces were observed

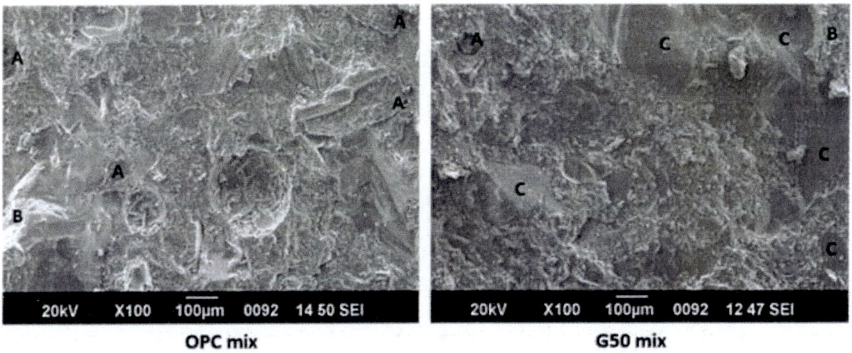

Fig. 1 SEM image for OPC mix and G50 mix

Table 7 Index notation and its description

Label	Description
A	Pores
B	Unreacted or partially reacted slag
C	Smooth surfaces

in G50 mix, and this is expected to be due to the fineness of GGBS. Less number of pores was observed in G50 mix than the OPC mix. The hydration reaction of GGBS is poorer than cement, and also particle size for GGBS grains is less than cement and gets occupied in the spaces between the cement particles, lessening frictional forces and resulting in a less porous matrix. Some unreacted/ partially reacted particles were observed in both the mixes.

5.2 *X-Ray Diffraction (XRD)*

XRD pattern of OPC mix and G50 mix after 28 days of curing is shown in Figs. 2 and 3, respectively. Calcium silicate hydrate (C-S-H) and calcium aluminate silicate hydrate (C-A-S-H) are dominant in G50 mix. Coexistence of these bonds gives higher strength to G50 mix. C2S, C3S, C3A and C4AF are identified in the OPC mix. In addition, there are some peaks that correspond to CaO, MgO, gypsum and basanite.

6 Conclusions

As pointed out in the introduction, the present study aims at confirming the results and to raise an awareness on the utilization of GGBS as a potential material for environmental protection and for sustainable construction. The incorporation of GGBS

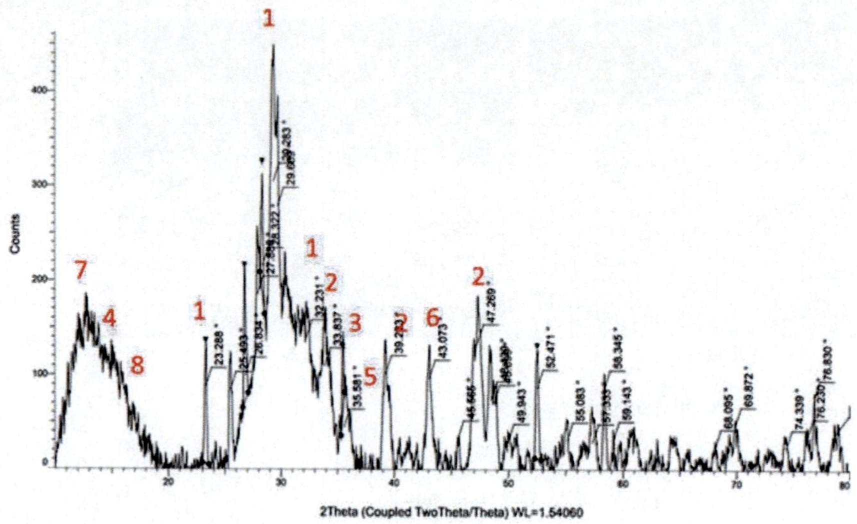

Fig. 2 XRD analysis of OPC mortar mix (1—C_2S, 2—C_3S, 3—C_3A, 4—C_4AF, 5—CaO, 6—MgO, 7—gypsum and 8—basanite)

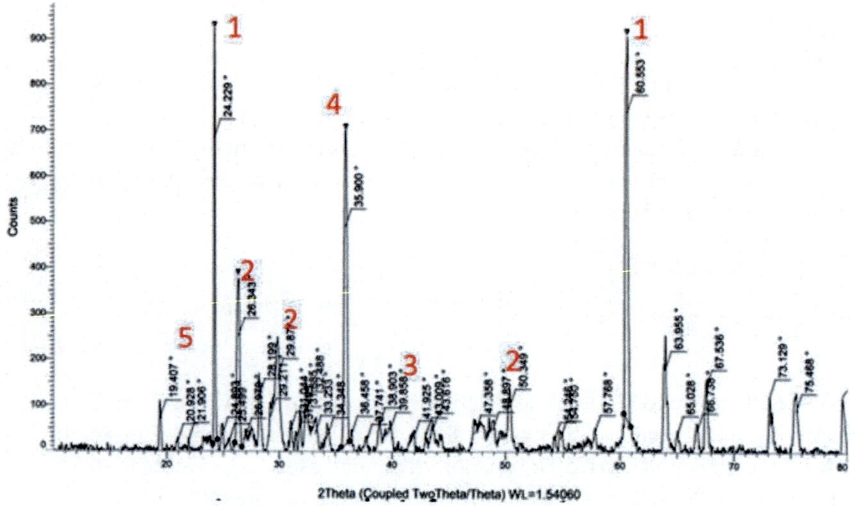

Fig. 3 XRD analysis of G50 mortar mix (1—SiO2, 2—C-S–H, 3—calcite, 4—C-A-S–H and 5—calcium hydroxide)

diminishes the unit cost of cement production altogether. Fresh and hardened properties of cement mortar partially replaced by GGBS are experimentally studied. Additionally, the microstructure of OPC mix and G50 mix is investigated by conducting

SEM analysis and XRD analysis. From test results, the following conclusions can be made:

- The inclusion of GGBS in mortar improves workability and consistency.
- Replacement of cement with GGBS by 50% in mortar gives almost similar compressive strength on 28th day, and this strength is found to increase in the later age which is not the case with cement mortar.
- Flexural strength of mortar improves by adding GGBS up to 50% both on 28th day 56th day.
- Split tensile strength for OPC mix and the one with 50% GGBS is almost similar. Later age strength for mortar incorporated with GGBS increases when compared with the mix without GGBS.
- SEM analysis shows that G50 mix has a less porous structure and has more smooth surfaces, indicating a more hardened structure.
- XRD analysis revealed C-S-H and C-A-S-H bonds in G50 mix, which helps in enhancing the mechanical properties.
- The addition of GGBS up to 50% improves the hardened properties significantly, and the strength decreases when the replacement percentage is more than 50%, this is due to the augmentation of slag.
- It is observed that G50 mix enhances the mechanical properties of mortar significantly, and hence, it can be considered the optimum mix.

Acknowledgements The authors would like to thank DST-SAIF, CUSAT, Kerala for providing the testing facilities and National Institute of Technology, Calicut for supporting the work.

References

1. ACI 116R00 (2005) Cement and concrete terminology, Detroit, Michigan
2. ACI 233 R03 (2000) Slag cement in concrete and mortar, Detroit, Michigan
3. Hwang CL, Lin CY (1986) Strength development of blended blast furnace slag cement mortars. In: Malhotra VM (ed) Proceedings of the 2nd international conference on Fly Ash, silica fume, slag and natural pozzolans in concrete, Madrid, Spain, SP 91–65, Detroit, MI, ACI, pp 1323–1340
4. Elsa Thomas G, Indira PV, Sajith AS (2022) Experimental investigations on durability characteristics of ground granulated blast furnace slag mortar. E3S Web of conferences, vol 347
5. IS 12269-2013 (2009) Ordinary Portland cement, 53 grade-specification, 1st revision, Bureau of Indian Standards, New Delhi
6. IS 383-1970 (reaffirmed in 2002) (2009) Specification for coarse and fine aggregates from natural sources for concrete, 9th reprint, Bureau of Indian Standards, New Delhi
7. IS 2250-1981 (reaffirmed 2000) (2009) Code of practice for preparation and use of masonry mortars, 5th revision, Bureau of Indian Standards, New Delhi
8. IS 456:2000 (2009) Plain and reinforced concrete—code of practice, 4th revision, Bureau of Indian Standards, New Delhi
9. IS 10262-2009 (2009) Concrete mix proportioning—guidelines, 1st revision, Bureau of Indian Standards, New Delhi

10. IS 4031-4 (1998), (reaffirmed 2005) (2009) Methods of physical tests for hydraulic cement, 2nd revision, Bureau of Indian Standards, New Delhi
11. IS 4031-7 (1998), (reaffirmed 2005) (2009) Methods of physical tests for hydraulic cement, 2nd revision, Bureau of Indian Standards, New Delhi
12. IS 4031-6 (1998), (reaffirmed 2005) (2009) Methods of physical tests for hydraulic cement, 2nd revision, Bureau of Indian Standards, New Delhi
13. Basheer PA, Gilleece PR, Long AE, Mc Carter WJ (2002 Oct 1) Monitoring electrical resistance of concretes containing alternative cementitious materials to assess their resistance to chloride penetration. Cement Concr Compos 24(5):437–49
14. Regourd M (1980) Microanalytical studies (X-ray photo electron spectrometry) of surface hydration reactions of cement compounds. Philos Trans R Soc London Ser A 310(1511):85–91
15. ASTM C348-08. Standard test method for flexural strength of hydraulic-cement mortars
16. ASTM C1723-16. Standard guide for examination of hardened concrete using scanning electron microscopy

Structural Dynamics and Earthquake Engineering

Seismic Hazard Assessment of Chennai City Using Deterministic Approach

R. Kiruthika, V. Vasugi, and V. Jagen

1 Introduction

A seismic hazard is the hazard caused by tremors. Seismic hazard assessment is used to evaluate the seismic hazard and to layout seismic hazard values for risk assessment. Seismic hazard assessment is needed for determining the earthquake estimates for projects, areas, case studies or earthquake risk management purposes. Nearly 60% of area is affected by earthquakes in India. Chennai is the capital of Tamil Nadu, which is located in the southeast coast of India which has been considered for the seismic hazard assessment in this study. Chennai is one of the metropolitan city of 1177 km^2 area coverage. Chennai experienced moderate earthquakes; however, the evaluation of seismic hazard is needed to reduce the destruction caused by earthquakes. In IS: 1893, the Chennai was enhanced to zone III., which increases the ground motion design. There are two types of assessment, namely probabilistic seismic assessment and deterministic seismic hazard assessment. Evaluation of seismic hazard using probabilistic method is useful only when there is plentiful accessibility of seismic data while, if there are more active faults but less seismic data, deterministic method is useful to make the assessment. Therefore, this study performs the deterministic seismic hazard assessment for the distance of 100 km around Chennai city. This will help design engineers in designing the earthquake resistant buildings.

R. Kiruthika (✉) · V. Vasugi
School of Civil Engineering, VIT University, Chennai, Tamil Nadu 600127, India
e-mail: r.kiruthika2020@vitstudent.ac.in

V. Vasugi
e-mail: vasugi.v@vit.ac.in

V. Jagen
Structural Consultant, Ramaniyam Real Estates Pvt., Ltd, Chennai, India

A. Sil et al. (eds.), *Recent Trends in Civil Engineering*, Lecture Notes in Civil Engineering 274, https://doi.org/10.1007/978-981-19-4055-2_16

1.1 Probabilistic Seismic Hazard Assessment

Probabilistic seismic hazard assessment identifies unpredictability in place, frequency, results of earthquake and mingle all together to calculate probabilities of dissimilar levels of earth shaking. Its goal is to assess transcending several earth movement intensity.

1.2 Deterministic Seismic Hazard Assessment

Deterministic seismic hazard assessment (DSHA) is the uncomplicated and clear outlook to perform seismic hazard assessment. This seismic structure can be used if an earthquake occurs at particular place, all feasible faults of seismic should be recognized and their strong earth movements must be assessed. This method is widely used for power plants, dams and bridges. The pictorial representation of the DSHA is shown in Fig. 1. The following four-step are the process of DSHA [1]:

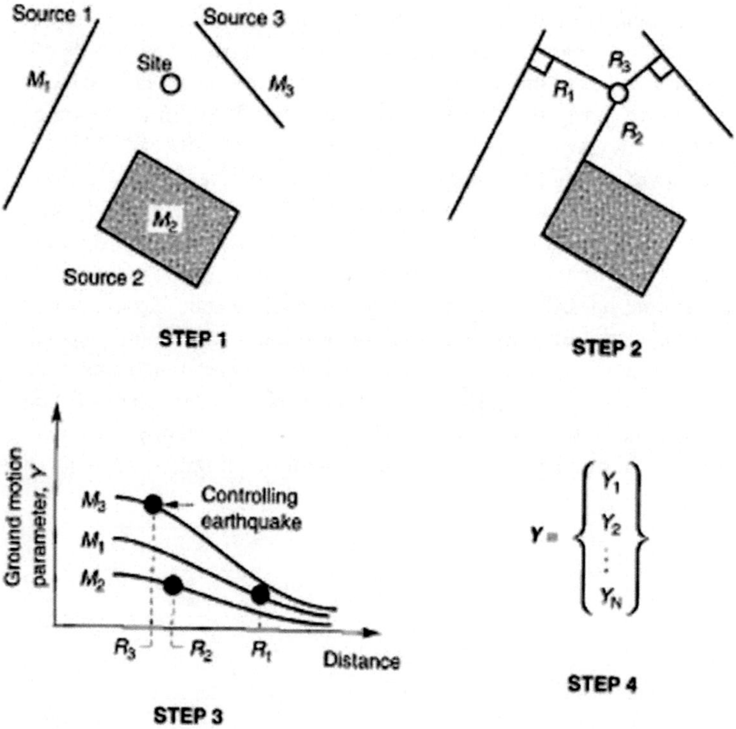

Fig. 1 Steps of a deterministic seismic hazard assessment [1]

1. Recognition of earth tremor sources

The sources of tremor which can bring about definite ground motions are recognized and characterized.

2. Selection of source to site distance:

Smallest distance obtained between each fault and site is selected.

3. Deciding of controlling earthquake:

In this step, the earthquake is selected on the basis of strongest ground motion The earthquake which has the highest magnitude i.e., strongest earthquake is considered.

4. Definition of hazard using strongest earthquake:

Using M and R, the parameters such as peak ground acceleration and spectral acceleration are determined.

2 Seismicity and Seismotectonics of the Region

From the earthquake which occured in the past, the seismotectonic datas expresses the features of seismic tectonics and susceptibility should be analyzed. Chandra [2] dispensed the seismicity of South India. Seismicity details and properties of the region were gathered. From the National Earthquake Information Center of the US Geological Service, the earthquake statistics was collected within 100 km radial distance within the city. The ground motion parameters can be intiated by gathering the seismologicalcharacteristics. These datas are also used by Ghosh (1994) [3]. To find out the faults all over Chennai, a complete exploration and investigation has been studied. From the Seismotectonic Atlas of India (2000) [4], the source information are collected. All the faults are considered to be active in the respective region. The information of familiar faults for study area around 100 km distance with tectonic is given in the Table 1 (Fig. 2).

3 Methodology

1. Covering radial distance of 100 km around Chennai (12.75°–13.25°N and 80.0°–80.5°E) was taken as course of study.
2. The seismicity and seismotectonic features likely to generate significant sources or fault lines have been recognized for the respective area.
3. All faults are identified within 100 km radial distance of Chennai and they all are considered as active faults.

Table 1 Fault datas in the proximity of Chennai (Banerjee and **Boominathan**) [5]

Name of fault	Distance (Km)	Moment magnitude (Mw)
Fault 15	10	4.0
Fault 24	10	4.4
Fault 53	32	4.1
Kilcheri Fault	33	4.0
Fault 15	46	4.5
Neotectonic Fault	48	3.8
Palar Fault	59	4.0
Tambaram Fault	61	4.4
Fault 15 a	67	3.7
Fault 52	68	3.6
Fault 15 e	68	4.5
Fault 54	70	3.8
Mahabalipuram Fault	75	4.0
Kalkulam Fault	82	3.6
Muttukadu Fault	95	3.5
Fault 26 d	96	4.5
Fault 56 e	97	4.5
Fault 26	98	4.9

4. Shortest distance between the source and site has been identified. The hypocentral distance and the maximum moment magnitude was tabulated and used.
5. Maximum and strongest earthquake has been selected. This step is used in the estimation of peak horizontal acceleration from the maximum magnitude.
6. Ground motion hazard parameters such as peak acceleration and spectral acceleration of this region has been evaluated.

4 Earthquake Source Identification and Characterization

Many active faults are situated around Chennai and basically the faults are considered as sources. For designing, either lines/sources are used. Fault details presented by Indian Institute of Technology, Banerjee et al. (2017) [5]. The Fault **26** yielded maximum magnitude potential Mw = 5. Table 2 represents the fault lenth and maximum moment magnitude for the respective tectonic features.

5 Source to Site

In this step, distance between seismic source and site is determined by circle method for triangulation of hypocenter. From the studies regarding the definition of faults, hypocentral distance can be used for the calculation of horizontal acceleration. The distances for the respective faults of Chennai was given in Table 3.

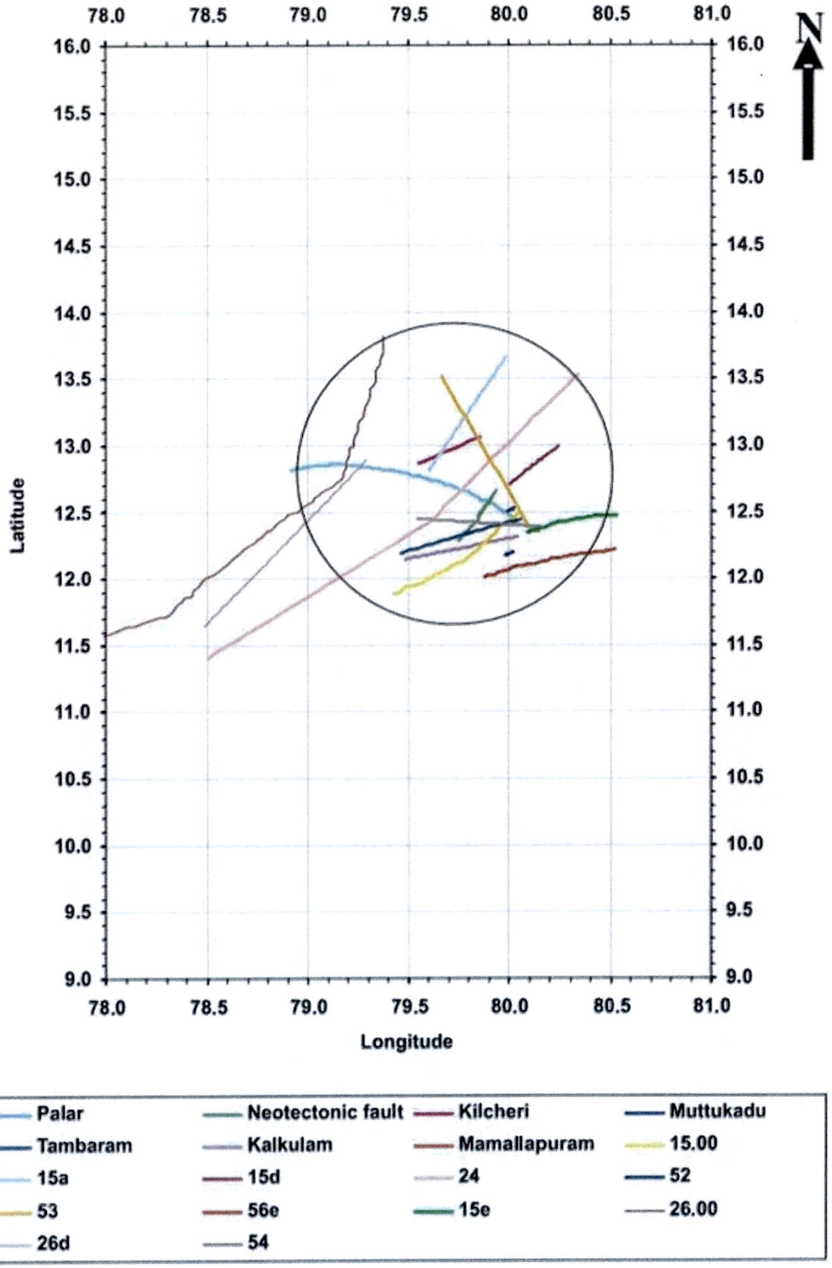

Fig. 2 Fault map of Chennai city, Indian Institute of Technology Madras (2008)

Table 2 Fault length and moment magnitude (Mw)

Tectonic feature	Fault length (Km)	Moment magnitude (Mw)
Fault 26	80	5
Fault 56 e	100	4.5
Fault 15 a	105	4.5
Fault 15 e	50	4.5
Tambaram fault	10	4.5

Table 3 Hypocentral distance

Sources (Faults)	Hypocentral distance
Fault 26	99
Fault 56 e	98
Fault 15 a	47
Fault 15 e	69
Tambaram Fault	60

6 Selection of Controlling Earthquake

In this step, the strongest earthquake is decided and this , step involves the estimation of peak horizontal acceleration using definite attenuation relationship. The hypocentral distance and the highest magnitude are required for the calculation of horizontal acceleration using definite attenuations. The attenuation laws developed from an analysis from Bulletin of the Seismological Society of America[6] have been adopted to identify the peak horizontal acceleration of this region alone. From the analysis, the resultant Eqs. (1) and (2) are obtained (Table 4).

$$\log \mathrm{A} = -1.02 + 0.249\mathrm{M} - \log r - 0.00255r + 0.26\mathrm{P} \tag{1}$$

where,

$$r = \left(\mathrm{d}^2 + 7.3^2\right)^{1/2}$$

Table 4 Critical tectonic faults and peak horizontal acceleration

Tectonic feature	Moment magnitude (Mw)	Hypocentral distance	Peak horizontal acceleration
Fault 26	5	99	0.20
Fault 56 e	4.5	98	0.16
Fault 15 a	4.5	47	0.23
Fault 15 e	4.5	69	0.19
Tambaram Fault	4.5	60	0.21

$$\log V = -0.67 + 0.489M - \log r - 0.00256\, r + 0.17S + 0.22P \quad (2)$$

where, $r = \left(d^2 + 4.0^2\right)$

where

A peak horizontal acceleration (g)
V peak horizontal velocity
M moment magnitude
D closest distance to the surface of fault rupture (Km)
S 0 (at rock sites) and S = 1 (at soil site)
P 0 (50 percentile values) P = 1 (84 percentile values).

From the assessment, it was deducted that the tectonic feature for Chennai is Fault 26 that produces the controlling earthquake of magnitude 5 at 99 km.

7 Evaluation of Ground Motion Hazard Parameters

With the help of controlling earthquake, the hazard can be described clearly. From the empirical relationships, one or two ground motion parameters are acquired, and its features are basically expressed by these parameters. This step involves the estimation of ground motion paramaters such as peak ground acceleration and spectral acceleration. From the earlier history of earthquake, the moment magnitude is obtained using strong ground motion equations proposed by Lee et al. [7], and the peak acceleration (PGA) is assessed. The site classification used in this study was divided into two sites, i.e., rock sites (B types and C types) and soil sites (D types and E types).

The PGA attenuation equation for rock sites is

$$\ln(\text{PGA}) = -2.5 + 1.205M - 1.905 \quad \ln R + 0.516e^{0.6325M}(+0.0075H + 0.275Z_t) \quad (3)$$

The PGA attenuation equation for soil sites is

$$\ln(\text{PGA}) = -0.9 + 1.00M - 1.90 \quad \ln R + 0.9918e^{0.5263M} + (0.004H + 0.31Z_t) \quad (4)$$

where

PGA peak ground acceleration
M moment magnitude
R hypo central distance
H focal depth (kilometers)
Z_t indicates the subduction zone earthquake type; $Z_t = 0$ (interface earthquake) and $Z_t = 1$ (intra- slab earthquakes).

Attenuation relation by Sharma et al. (2009) [8]:

$$\log\ A = b_1 + b_2 M + b_3 \log \sqrt{R^2 + b4^2} + b_5 S + b_6 H \tag{5}$$

where, $b1$, $b2$, $b3$, $b4$, $b5$ and $b6$ = (regression coefficients);

A spectral acceleration, ms^2
S 1 rock site and 0 otherwise
H 1, strike-slip mechanism and S = 0, reverse mechanism.

For spectral acceleration calculation, site coefficients has been taken from Iyengar and Raghunath [9]: b1 = 1.7816, b2 = 0.9205, b3 = − 0.0673 and b4 = 0.0035.

From the calculation, Fault 26 of Chennai recorded the maximum magnitude potentialusing deterministic method, that is the g value of 0.011 g at bedrock level, Spectral acceleration of 0.8 g (assumed the spectral acceleration at 0.04 s for the GMPE derived here) with those estimated by GMPEs using Mw: Ambraseys et al. (2005), Akkar and Bommer (2007) and Boore and Atkinson (2008).

8 Comparative Analysis

Seismic hazard analysis is of mainly two methods, they are probabilistic and deterministic. For the Chennai city, Kumar et al. [10] has reported the seismicity of the area by covering the radial distance of 300 km by probabilistic method. The standard Cornell-McGuire method has been used for hazard computation. The horizontal PGA assessed in Chennai is 0.125 g, whereas in this paper, a study has been carried out to evaluate the hazard using deterministic approach for Chennai city for radial distance of 100 km. The peak ground acceleration of 0.011 g was evaluated from the strongest earthquake at the bedrock level. Probabilistic method involves the levels of ground shaking over a certain return period (in years), whereas peak ground acceleration using attenuation relationship are evaluated in deterministic method. By comparing these two methods, it is known that the value changes slightly according to its area coverage, and Chennai comes under low seismic prone area.

9 Conclusion

Deterministic seismic hazard assessment has been performed for Chennai city. This study shows that the peak ground acceleration of 0.011 g and spectral acceleration of 0.8 g have been evaluated using the attenuation relation by considering the 18 seismic sources within 100 km from the city. Low values of ground shaking for Chennai has been predicted from the evaluation of PGA (horizontal component) by DSHA.

References

1. Reiter L (1990) Earthquake Hazard analysis: issues and insights. Surv Geophys. Springer Link 13:297–298
2. Chandra U (1977) Earthquake of peninsular India—a seismotectonic study. Bull Seismological Soc America 67(5):i387-1413
3. Ghosh SK, Munshi JA (1998) Analysis of seismic performance of a code designed reinforced concrete building. Elsivier 20(608–616)
4. Dasgupta S, Narula PL, Acharyya SK, Banerjee J. Seismotectonic Atlas of India and its environs, geological survey of India
5. Banerjee S, Boominathan A (2008) Seismic Hazard assessment of the city of Chennai, Indian institute of technlogy, Madras, ASME17
6. Joyner WB, Boore DM (1981) Peak horizontal acceleration and velocity from strong motion records. Bull Seismological Soc America 71(6):2011–2038
7. Lin P-S, Lee C-T (2008) Ground-motion attenuation relationships for subduction-zone earthquakes in Northeastern Taiwan. Bull Seismological Soc America 98(1):220–240
8. Sharma ML, Douglas J, Bungum H, Kotadia J (2009) Ground motion prediction equations based on data from Himalayan and Zagrous region. J Earthquake Eng 13
9. Iyenger and Raghukanth (2009) Attenuation of strong motion in peninsular India. Seismological Res Lett 75(4):530–540
10. Kumar GK, Sreedhar U, Dodagoudar G (2009) Probabilistic seismic hazard assessment. IGC Guntur

Stability Analysis of Lift Shaft Tower with Guys for Wind Load

Shreekiran and K. S. Babu Narayan

1 Introduction

Coke-drum is a type of pressure vessel that uses heat and pressure to refine complex hydrocarbons into lighter, more useful products such as gasoline, diesel, and jet fuel. Coke-drum unit is connected to lift shaft/stair tower at various locations, and these towers are provided for operational inspection of the coke-drum unit. Working operations of the coke-drum unit transfers vibrations to the steel tower which in turn affects the safety, stability, and serviceability of the lift shaft/stair tower. An earlier study by Kapat [1] leads to the conclusion that the stand-alone structure is vulnerable to lateral loads, i.e., wind load at the site. In this study to make the lift structure safe, it is separated from vibrating coke-drum unit and analyzed for other loads, dead load, live load, wind load at the site.

This leads to the investigation of the safety, stability, and serviceability of the structure with the help of lateral load-supporting guy units. The guy is a cable or a rope that is tensioned to add lateral stability to the free-standing structure. For towers that are not self-supporting, guy cable is used as external support to make them stand.

2 Methodology

The problem of concern here is a tower with a rectangular plan, having dimensions of 4.53×3.94 m. The height of the tower is 72.0 m. The model with a different set of guys at different levels is made with SAP 2000 v20. The story displacements at 72.00,

Shreekiran (✉) · K. S. B. Narayan
Department of Civil Engineering, NITK, Surathkal, Mangalore, Karnataka 575025, India
e-mail: shreekiran95@gmail.com

K. S. B. Narayan
e-mail: shrilalisuta@nitk.edu.in

A. Sil et al. (eds.), *Recent Trends in Civil Engineering*, Lecture Notes in Civil Engineering 274, https://doi.org/10.1007/978-981-19-4055-2_17

48.37, 26.32 m of height of the tower are tabulated for different sets of guys. The number of over-stressed members for various cases is compared. The best suitable case for the problem is identified, and strengthening is done for the over-stressed members.

To make the lift shaft safe, it is separated from the staircase unit and the coke-drum unit as well. The final separated lift shaft model is shown in Fig. 1. For this model, guys are provided at various heights with initial tension in the cable as 60kN and diameter of guy cable as 50 mm. A coordinate system is defined for the anchorage location of the guy cable. Anchorage location of the guy is located at an offset of 42.42 m from all four corners of the model along the diagonal direction. Four anchorage locations are selected for which guys are attached as shown in Figs. 2, 3, and 4. Before drawing the guy, a cable profile is assigned to them. For guys, pinned support is assigned.

Guy's configuration has been selected with functional requirements as constraints with minimum modifications of tower elements for satisfaction of stability, strength, and serviceability requirements.

The original design concept of connection of lift shaft to the coke-drum unit is the root cause of the problem of vibration transmission. As the very objective is to avoid transmission of vibrations from coke-drum structure to lift shaft, isolation has been

Fig. 1 Lift shaft model

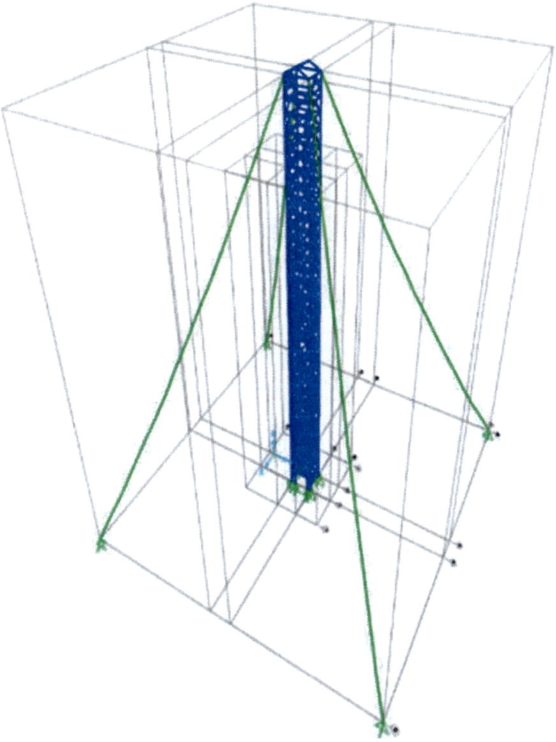

Fig. 2 Model with a single set of guys at top

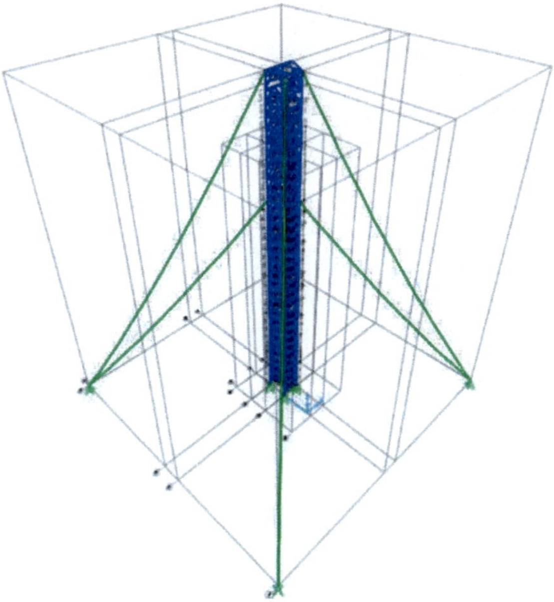

Fig. 3 Model with two sets of guys at 72.00 m and 48.37 m

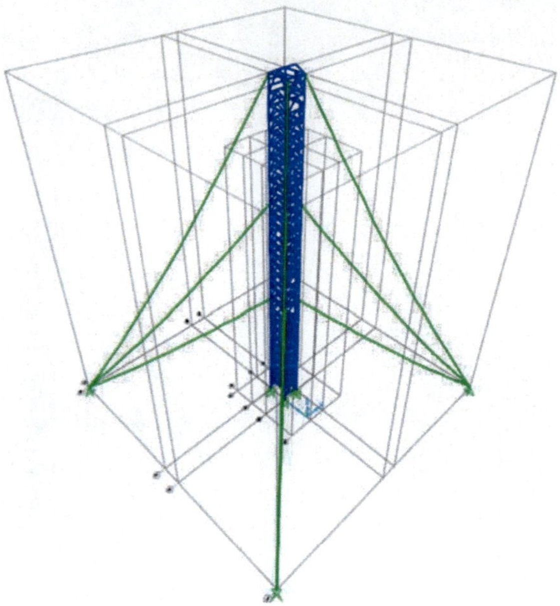

Fig. 4 Model with three set of guys at 72.00 m, 48.37 m, and 26.2 m

done with the provision of guys. Without guys, stand-alone lift shaft is structurally in adequate.

2.1 Loading

The structure is 72 m in height, and wind load plays an important role in the analysis of the structure. So the structure is analyzed for dead load, super imposed and wind load in both x- and y-directions as per IS codes.

Wind pressure coefficients.

Pressure coefficients for lift shaft towers are 1.12 for the positive x-axis and negative y-axis. Wind load along positive y-axis direction is neglected due to the connection of lift shaft structure with that of staircase unit, and wind load along negative x-axis direction is neglected due to the presence of coke-drum unit in that direction.

2.1.1 Load Combinations

Following load combination is used for the analysis as per IS codes.

$$\text{COMB1} = 1.2\ \text{DL} + 1.2\ \text{LL} + 1.2\ \text{WL}_X$$
$$\text{COMB2} = 1.2\ \text{DL} + 1.2\ \text{LL} + 1.2\ \text{WL}_Y$$

$$\text{COMB3} = 1.2\ \text{DL} + 1.2\ \text{LL} - 1.2\ \text{WL}_\text{X}$$
$$\text{COMB4} = 1.2\ \text{DL} + 1.2\ \text{LL} - 1.2\ \text{WL}_\text{Y}$$

where

COMB	Combination
DL	Dead load
LL	Live load
WL_X	Wind load in X-direction
WL_Y	Wind load in Y-direction

2.2 *Analysis*

The capacity of lift, 20 kN is assigned at the top floor of the structure as point load on the column as dead load. Live load is based on IS 875 (Part 2) as occupancy class G (Industrial Building). The location of the building suggests a basic wind speed of 39 m/s and belongs to terrain category IV. The building belongs to category C as per classification by IS 875 (part 3): 2015. Hence, wind load is considered accordingly. Nonlinear static analysis is carried out using the software SAP 2000.

2.3 *Structural Design*

Design checks have been done for the members for the above-mentioned load cases. The results and list of members that need to be strengthened are tabulated in the results and discussion.

3 Results and Discussion

At first, analysis is carried out for three models, they are Case 1: One set of guys at the top without moment release (Fig. 2), Case 2: Two sets of guys (one set at top and one set at mid-height) without moment release (Fig. 3), Case 3: Three sets of guys (one set at top, one at mid-height, and one at one-third height) without moment release (Fig. 4), and it fetched the results where most of the column members were failing in minor axial bending. As the column is failing in minor axial bending, moment releasing is done for the above three cases which added three more cases for the analysis. Moment release case is attained by releasing moment of the beam members in the major axis direction. For a better comparison of the above six cases, results are presented such that the moment release case immediately follows the without moment release case. For six cases, which includes moment release (three)

Table 1 Story displacement at various levels

Cases	Displacements in mm					
	Wind X at heights			Wind Y at heights		
	72 m	48.37 m	26.2 m	72 m	48.37 m	26.2 m
Case 1	365	433	322	332	343	194
Case 2	406	632	493	394	402	216
Case 3	264	185	174	272	185	117
Case 4	271	185	236	265	181	120
Case 5	260	162	82	267	174	80
Case 6	268	161	80	261	170	78

and without moment release (three) nonlinear static analysis is done and the result includes story displacement of the model at the top, two thirds and one-third height, the tension in guy cable for the six cases, number of members failed due to overstress in the six cases. The best possible solution for the problem is selected based on the minimum number of stressed members as well as story displacement well within IS code limits. Strengthening of the over-stressed member is done for that best possible solution (Table 1).

where

Case 1: One set of guys at the top without moment release.

Case 2: One set of guys at top with moment release.

Case 3: Two sets of guys (one set at top and one set at mid-height) without moment release.

Case 4: Two sets of guys (one set at top and one set at mid-height) with moment release.

Case 5: Three sets of guys (one set at top, one at mid-height, and one at one-third height) without moment release.

Case 6: Three sets of guys (one set at top, one at mid-height, and one at one-third height) with moment release.

The lateral displacement limit from the IS800 code is H/200, where H is the height, (H = 72 m) which implies that the lateral displacement limit is 360 mm. In cases 3, 4, 5, 6, values are well within the permissible limits (Table 2).

The criteria for the number of failed members are P-M interactions. If the P-M ratio is less than 1 member is safe for that particular load combination. Here, P is axial force and M is bending moment. Based on P-M interactions case 6 has the least number of failed members.

Guy tension in different cables for different load cases are listed in Tables 3, 4, 5 and 6. Cables from 1 to 4 corresponds to Cases 1 and 2, cables 5 to 8 corresponds to Cases 3 and 4 and cables 9 to 12 corresponds to Cases 5 and 6, respectively. The plan for guy cables at various heights is shown in Fig. 5. The direction of axes is shown in the same figure.

Table 2 Number of members failed

Cases	Number of members failed				Total failed members
	Beam parallel to the x-axis	Beam parallel to the y-axis	Column	Angle	
Case 1	21	16	15	21	73
Case 2	34	11	12	18	75
Case 3	8	3	12	5	28
Case 4	11	3	9	7	30
Case 5	0	2	6	7	15
Case 6	4	2	7	0	13

Table 3 Guy tension in cables for case 1

Cable/Comb	Guy tension in kN			
	Comb1	Comb2	Comb3	Comb4
Cable 1	274.010		−274.010	
Cable 2	293.947	268.306	−293.947	−268.306
Cable 3		268.889		−268.889
Cable 4				

Table 4 Guy tension in cables for case 2

Cable/Comb	Guy tension in kN			
	Comb1	Comb2	Comb3	Comb4
Cable 1	328.807		−328.807	
Cable 2	353.427	273.485	−353.427	−273.485
Cable 3		273.720		−273.720
Cable 4				

A similar list can be made for Cases 3 and 4, due to page constraints, they are not listed here. Guy tension in cable1 for Case1 for combination 1 is 274.010 kN, guy tension in the same guy for Case 5 is 69.219 kN, and hence, there is a reduction in cable tension by a factor of 3.95. When Case 1 and Case 5 are compared for all combinations at least there is three times reduction in tension for the top cables. Guy tension in cable 2 for combination 2 in Case 1 is 268.306 kN, for the same cable, and for the same combination in Case2, the value is 273.485 kN which implies that there is an increase of guy tension of **2%** from Case 1 to Case 2. Comparing all the six Cases, Case 6 fetches the least number of over-stressed members so it is the best possible solution with the help of guy cable. It is observed that for Case 6, 13 members failed and require strengthening. Strengthening of the member is done by upgrading the sections, for example, ISMC 150 is upgraded to 2ISMC 150. Angle

Table 5 Guy tension in cables for case 5

Cable/Comb	Guy tension in kN			
	Comb1	Comb2	Comb3	Comb4
Cable 1	69.219		−69.219	
Cable 2	76.890	76.150	−76.890	
Cable 3		77.100		
Cable 4				
Cable 5	211.200		−211.200	
Cable 6	221.130	237.090	−221.130	−237.090
Cable 7		236.870		−236.870
Cable 8				
Cable 9	150.760		−150.760	
Cable 10	159.740	80.410	−159.740	−80.410
Cable 11		84.030		−84.030
Cable 12				

Table 6 Guy tension in cables for case 6

Cable/Comb	Guy tension in kN			
	Comb1	Comb2	Comb3	Comb4
Cable 1	89.400		−89.400	
Cable 2	98.610	78.940	−98.610	−78.940
Cable 3		78.890		−78.890
Cable 4				
Cable 5	219.870		−219.870	
Cable 6	230.780	236.800	−230.780	−236.800
Cable 7		236.640		−236.640
Cable 8				
Cable 9	152.110		−152.110	
Cable 10	160.410	88.350	−160.410	−88.350
Cable 11		90.145		−90.145
Cable 12				

sections are upgraded to the next higher strength sections of the steel table. After upgrading sections analysis is carried out, and a design check is done (Table 7).

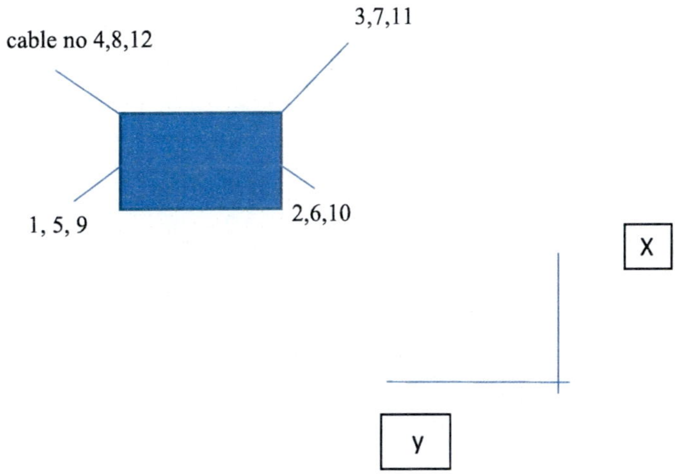

Fig. 5 Plan showing the location of guy cable concerning global x and y

Table 7 List of strengthened members

SI No.	Member designation	Type of member	Original section	Upgraded section	Length (m)
1	1185	Column	ISMC 150	2 ISMC 150	2.90
2	1182	Column	ISMC 150	2 ISMC 150	2.90
3	1223	Column	ISMC 150	2 ISMC 150	0.99
4	1121	Column	ISMC 150	2 ISMC 150	2.90
5	1124	Column	ISMC 150	2 ISMC 150	2.90
6	1258	Column	ISMC 150	2 ISMC 150	0.99
7	1255	Column	ISMC 150	2 ISMC 150	0.99
8	1266	Column	ISMC 150	2 ISMC 150	0.99
9	221	Beam	ISMC 300	2 ISMC 300	3.94
10	219	Beam	ISMC 300	2 ISMC 300	3.94
11	894	Beam	ISMC 300	2 ISMC 300	3.94
12	892	Beam	ISMC 300	2 ISMC 300	3.94
13	674	Beam	ISMC 300	2 ISMC 300	4.47
14	759	Beam	ISMC 300	2 ISMC 300	4.47
15	718	Angle	65*65*6	75*75*8	2.99
16	1038	Angle	65*65*6	75*75*8	2.99

4 Conclusions

Based on the present study, the following conclusions are drawn:

- Three sets of guy cables are required to make the lift shaft structure stable against wind load.
- Failed members are strengthened to meet the required strength.
- Story displacement decreases by 35% when the set of guy cables are increased from one set to three-set.
- Guys carried at least 2% more tension in moment release cases than that of without moment release cases.
- Story displacement is 0.62% less in moment release cases than that of without moment release cases.

Reference

1. Kapat R (2019) Analysis of coke-drum supporting structure master thesis submitted to NITK Surathkal, Karnataka

Comparative Analysis of Diagrid Structural System with Conventional Systems for Tall Buildings

Bellamkonda Sai Teja, V. Vasugi, and V. Jagen

1 Introduction

India has become the second-largest populated country in the world. Every year the population increases gradually mainly in cities and towns because people would like infrastructure and facilities, and many of them moving from villages to towns for employment [1, 2]. So we need to develop constructions of tall buildings and skyscrapers in major cities [3].

The first tall structure was constructed in the nineteenth century in the United States of America with 138 feet in height [4]. In tall building structures, additional strength for safety was enhanced by using outrigger structures, tube structures, exterior braced frame structures, mega-frames, diagrid structures, etc. [5–7].

The diagrid structures are more effective than conventional structural systems [8]. Due to the majority of the external vertical columns are provide diagonally, the diagonal grid element alone itself carries all laterals and vertical loads, but conventional structure carries only vertical loads [9, 10].

The triangular configuration of diagrid on the perimeter contributes increase instability and shear forces in a systematic way [11, 12]. Members of the diagrid provide rigidity of shear bending, efficiently because of their unique diamond model arrangement [13].

The structural ability of the system (diagrid) is also used in preventing exterior columns and accordingly allowing significant flexibility with the floor plan [14]. Since it eliminates the necessity of the use of columns, the diagrid can be useful

B. S. Teja · V. Vasugi (✉)
School of Civil Engineering, VIT Chennai, Chennai, India
e-mail: vasugi.v@vit.ac.in

B. S. Teja
e-mail: bellamkondasai.teja2019@vitstudent.ac.in

V. Jagen
Structural Consultant, Ramaniyam Constructions Pvt. Ltd., Chennai, India

A. Sil et al. (eds.), *Recent Trends in Civil Engineering*, Lecture Notes in Civil Engineering 274, https://doi.org/10.1007/978-981-19-4055-2_18

in constructing large roofs [15, 16]. Additionally, it can be quicker and stronger for developing buildings of elaborate geometry and undulating shapes.

In the present work, the focus is on a diagrid structure for 40 storied tall buildings system is compared with a conventional structural system. Modelling and analysis were made using Etabs software. The structure considered in the study was analysed adopting wind load and response spectrum analysis.

2 Building Modal Data

In this study, we modelled and analysed 40 storied steel structure buildings with conventional and diagrid structures, assuming that the building is located in Srinagar at Zone-5, building preliminary data and seismic analysis data are shown in Tables 1, 2 and 3.

Table 1 Data for 40 storied steel diagrid and conventional structures for tall building [17–19]

S. No.	Description	Data
1	Type of structure	Steel
2	Number of storey	40
3	Building location	Srinagar, J&K
4	Number of bays in X-direction	8
5	Number of bays in Y-direction	8
6	Length along x-direction	5 m
7	Length along y-direction	5 m
8	Height of each floor	3 m
9	Size of beam, column and diagrid	B1 = ISMB 450, B2 = ISMB 400 B3 = ISLB 500, C = D = ISHB 450
10	Thickness of slab	150 mm
11	Angle of diagrid	50.194 degree
12	Grade of concrete for slab	M30
13	Grade of steel	Fe 345
14	Thickness of wall	230 mm
15	Live load and dead load	4 KN/m^2 and 5.25 KN/m^2

Table 2 Seismic analysis data for 40 storied steel diagrid and conventional structures for tall building [20]

S. No.	Description	Data
1	Building located in zone	5
2	Seismic zone factor	0.36
3	Type of soil	2 (medium soil)
4	Response reduction factor (R)	5 (steel building with special moment resisting frame)
5	Importance factor	1.2 (occupancy more than 200 persons)
6	Damping ratio	5%

Table 3 Wind load analysis data for 40 storied steel diagrid and conventional structures for tall building [21]

S. No.	Description	Data
1	Wind speed (Srinagar)	39 m/s
2	Risk coefficient (K1)	1
3	Terrain category (K2)	1.22
4	Topography (K3)	1
5	Importance factor for the cyclonic region (K4)	1

3 Modelling and Analysis

In this study for modelling and analysis of building structures, we used Etabs software and it is used to model, create and analyse a multi-storey diagrid building with various angles, using a diagrid angle of 50.194. For the analysis of the building structure, we used Etabs software. It is more user-friendly software when compared with other software, we can easily analyse building structure by using this software.

The response spectrum analysis and wind load analysis of 40 storied steel structure buildings with conventional and diagrid structures are done by using Etabs software. The total height of the building structure in the Z-direction is 120 m, and each floor height is 3 m. The length of the building in X-direction and Y-direction is 35 m.

Two-dimensional plan and elevation views as well as three-dimensional elevation views of two different building structures are shown below (Figs. 1, 2 and 3).

3.1 Diaphragm

Mostly diaphragm (Fig. 4) is used to transfer the lateral load to the structure. In some buildings, floor diaphragms cannot give definite horizontal action of the diaphragm in its plane. So, that storey shear of the design is distributed to different elements of the resisting system which consider in-plane flexibility of the diaphragm [22].

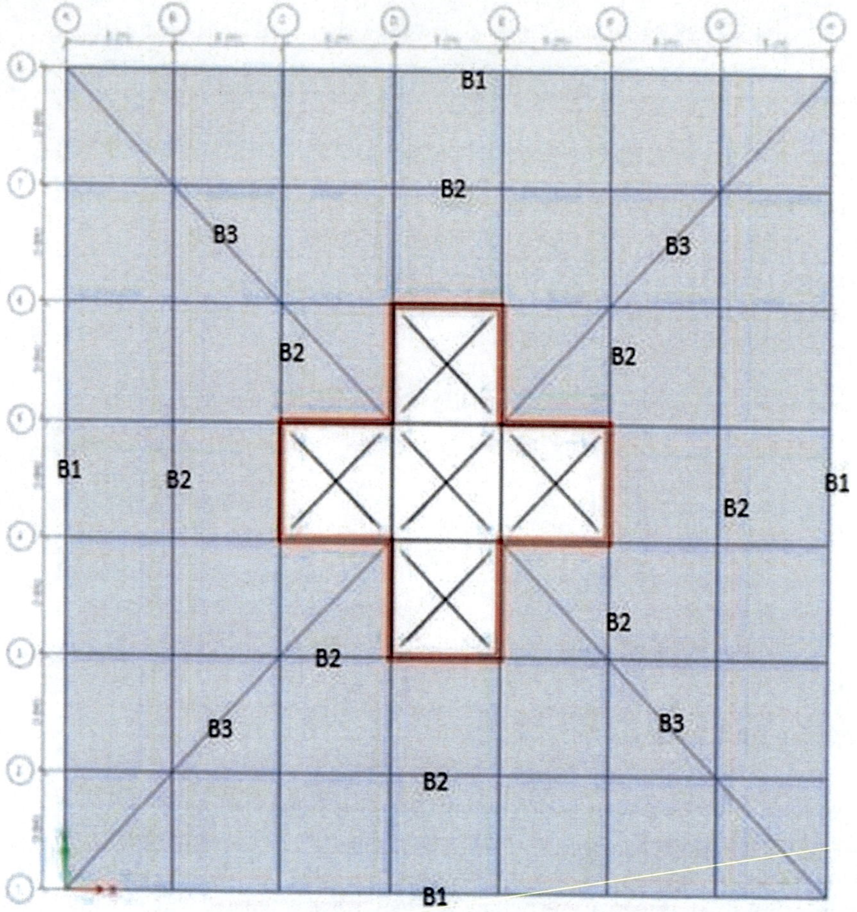

Fig. 1 Plan of conventional and diagrid structures

The floor diaphragm must be flexible, and if undergoes deformation, then maximum displacement in a lateral direction.

3.2 *Load Combinations*

Load combinations considered in this analysis as shown in Table 4 which include wind load, earthquake load and response spectrum parameters [23].

Whereas

DL dead load, LL=live load.

EQL-X earthquake load in X-direction,

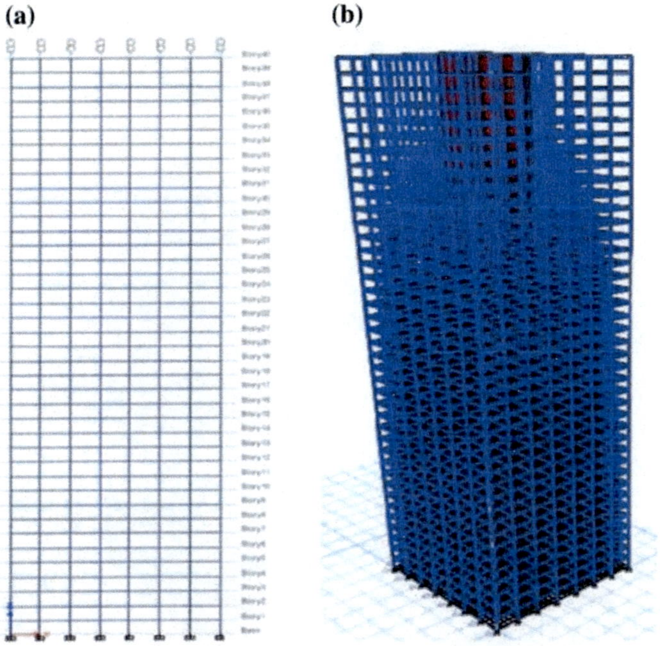

Fig. 2 **a** 2D elevation view **b** 3D elevation view of conventional structure of conventional structure

EQL-Y earthquake load in Y-direction.
WL-X wind load in X-direction,
WL-Y wind load in Y-direction.
RSP response spectrum.

4 Results

After doing the analysis, we got maximum storey displacement, drift, storey shear and stiffness for both conventional and diagrid structures. We compared results for structures; it is clearly shown in Figs. 5, 6, 7, 8, 9, 10, 11 and 12.

4.1 Storey Displacement

The maximum storey displacement of both structures for response spectrum and wind load is as shown in Figs. 5 and 6.

From Figs. 5 and 6, storey displacement is taken as in millimetre (mm). In response spectrum analysis, maximum storey displacement of conventional structure and diagrid structure is 137.2 mm and 70.7 mm, respectively. Maximum displacement

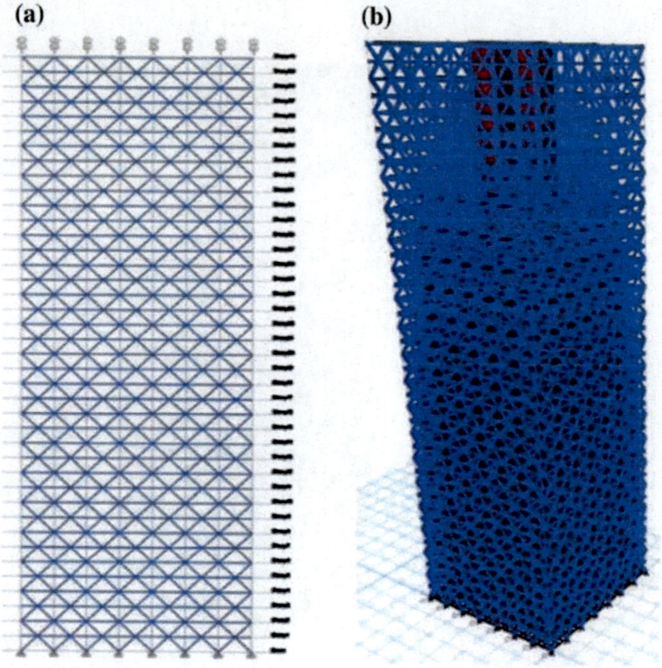

Fig. 3 **a** 2D elevation view of diagrid structure **b** 3D elevation view of diagrid structure

of wind load (WL) analysis on both conventional and diagrid structures is 109.8 mm and 67.1 mm, respectively. Storey displacement values (Table 5) are significantly low for diagrid structures when compared to the conventional structure.

According to IS 456-2000, the maximum limited storey displacement is H/500, where H is the total height of the building structure. In this work, our building height is H = 120 m, so 120/500 = 240 mm. we got less than 250 mm displacement in both conventional and diagrid structures. Hence, it is safe.

4.2 Storey Drift

The maximum storey drift of both structures for response spectrum and wind load is as shown in Figs. 7 and 8.

Storey drift is a lateral displacement of storey with respect to the bottom floor of the structure, in comparison with the current floor. Maximum storey drifts on both structures in response spectrum analysis and wind load analysis are between 20 and 30 storey in Figs. 7 and 8. Storey drift values are very low (Table 6) for the diagrid structure when compared to the conventional structure.

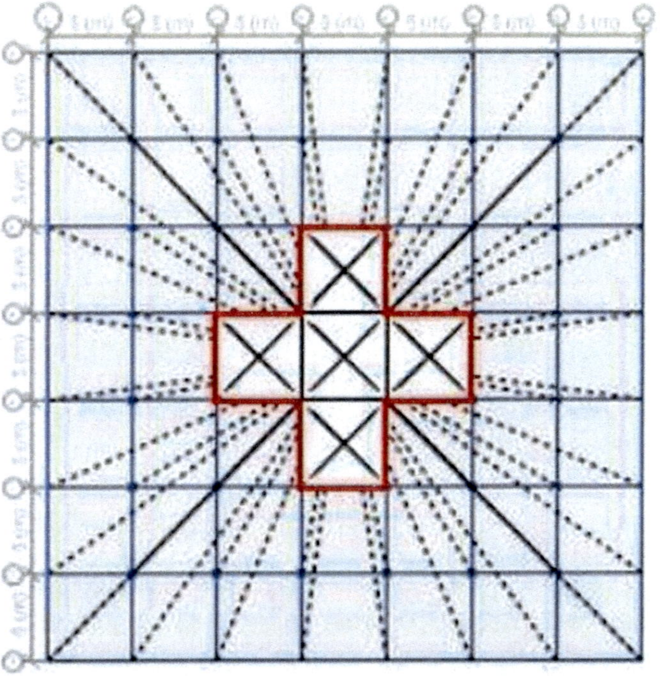

Fig. 4 Diaphragm

According to IS 1893-2016, the maximum limited storey drift is 0.004 × h, h is the storey height. Here, each storey height is h = 3 m. So, the maximum limited storey drift of each storey is 0.004 × 3 = 0.012. We got less than 0.012 of building drift in both conventional and diagrid structures. Hence, it is safe.

4.3 Storey Shear

Storey shear of conventional and diagrid structures is as shown in Figs. 9 and 10.

Figures 9 and 10 show the storey shear at the base of both structures conventional type and diagrid structures. In response spectrum analysis, at the base of the conventional structure is 7146.641 KN and the diagrid structure is 6990.108 KN. Storey shear of wind load (WL) analysis on both conventional and diagrid structures is 5332.642 KN and 5282.831 KN, respectively. Storey shear is low for diagrid structures when compared to the conventional structure. Hence, the diagrid structure resists more lateral loads (Table 7).

Table 4 Load combinations

Load combination on (EQL)	Load combination on (WL)	Load combination on (RSP)
1.5 (DL+LL)	1.5 (DL+WL-X)	1.5 (DL+RSP)
1.5 (DL+EQL-X)	1.5 (DL-WL-X)	1.2 (DL+LL+RSP)
1.5 (DL-EQL-X)	1.5 (DL+WL-Y)	0.9 DL+1.5 RSP
1.5 (DL+EQL-Y)	1.5 (DL-WL-Y)	
1.5 (DL-EQL-Y)	0.9DL+1.5WL-X	
0.9DL+1.5EQL-X	0.9DL-1.5WL-X	
0.9DL-1.5EQL-X	0.9DL+1.5WL-Y	
0.9DL+1.5EQL-Y	0.9DL-1.5WL-Y	
0.9DL-1.5EQL-Y	1.2 (DL+LL+WL-X)	
1.2 (DL+LL+EQL-X)	1.2 (DL+LL-WL-X)	
1.2 (DL+LL-EQL-X)	1.2 (DL+LL+WL-Y)	
1.2 (DL+LL+EQL-Y)	1.2 (DL+LL-WL-Y)	
1.2 (DL + LL-EQL-Y)		

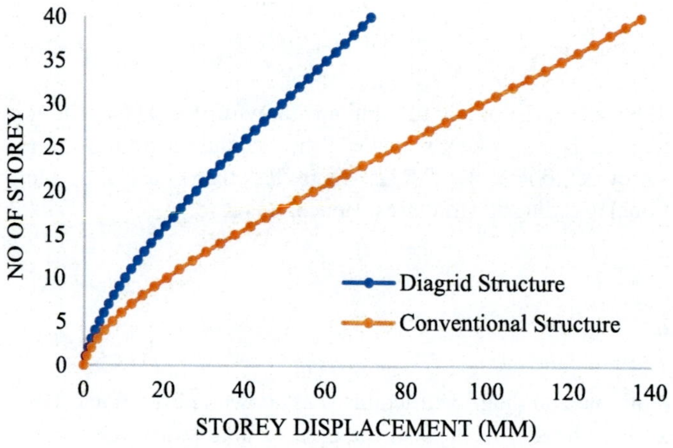

Fig. 5 Storey displacement in response spectrum analysis

4.4 *Storey Stiffness*

Storey stiffness of conventional and diagrid structures is as shown in Figs. 11 and 12.

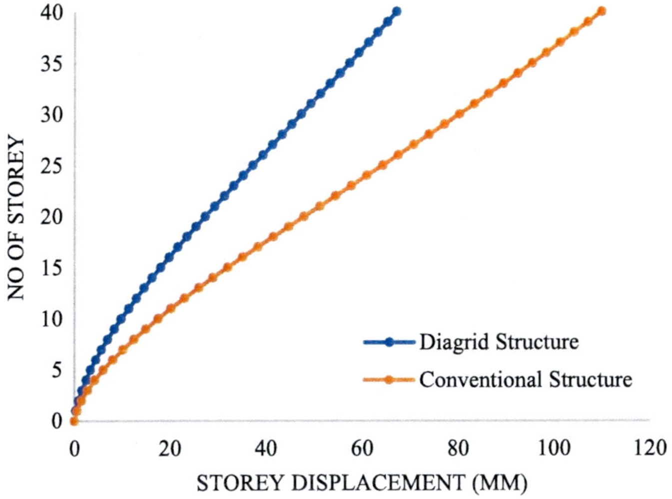

Fig. 6 Storey displacement in wind load analysis

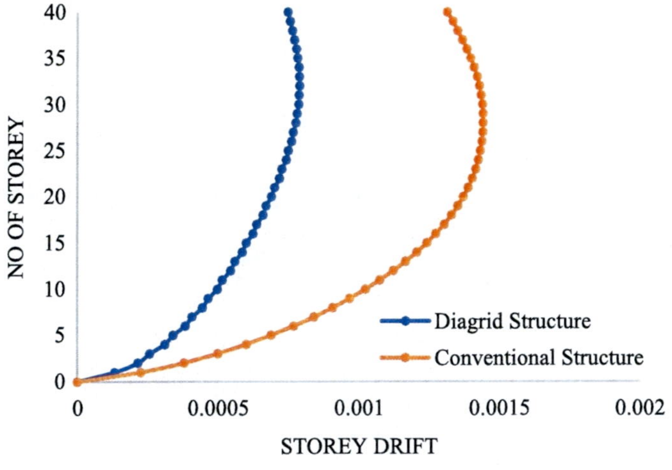

Fig. 7 Storey drift in response spectrum analysis

Figures 11 and 12 show storey stiffness versus a number of the storey for both conventional and diagrid structures. It is observed that storey stiffness is maximum in response spectrum analysis at storey 1 of both structures conventional and diagrid structures, 10.30×10^6 KN/m and 17.74×10^6 KN/m, respectively. Storey stiffness of wind load (WL) analysis on both conventional and diagrid structures is 96.26×10^5 KN/m and 16.46×10^6 KN/m, respectively. Compared to diagrid structure, conventional structure storey stiffness is less, in both wind and response spectrum analysis as the storey level increases (Table 8).

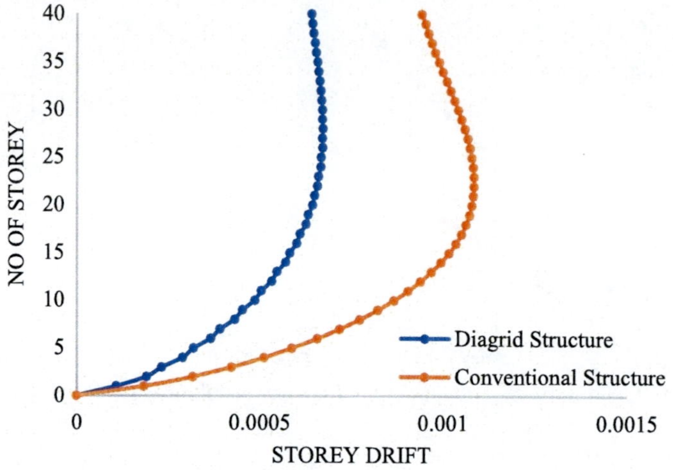

Fig. 8 Storey drift in wind load analysis

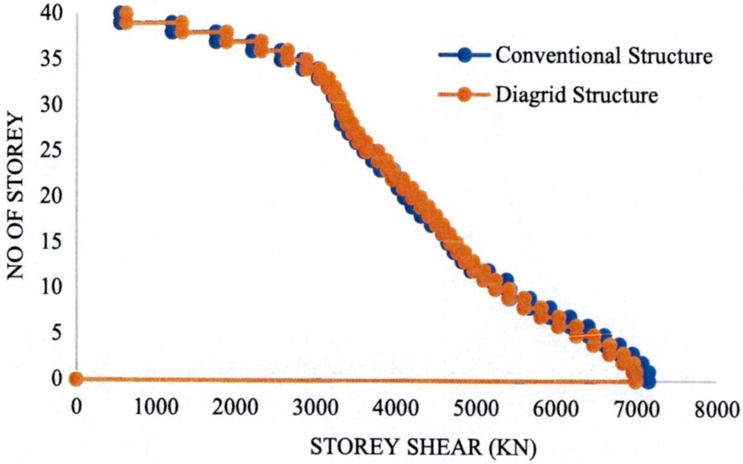

Fig. 9 Storey shear in response spectrum analysis

5 Conclusion

After completing the analysis of conventional and diagrid structure buildings, we understood few points.

That is:

- Maximum storey displacement in response spectrum method got 48.46% higher in conventional structure, and in wind load analysis, it shows 38.8% much as compared with diagrid structural building.

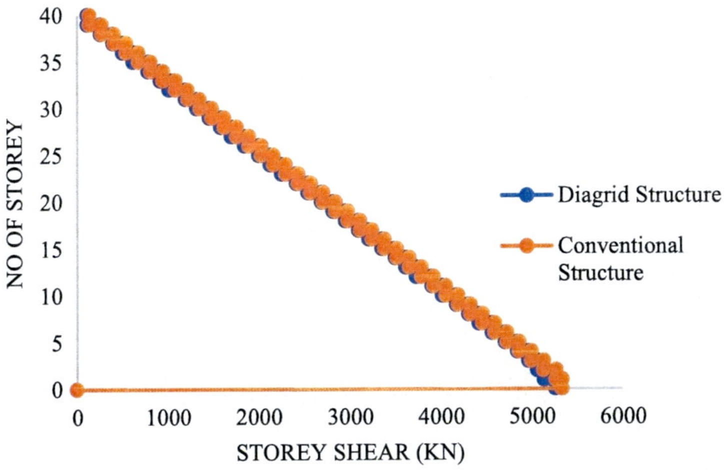

Fig. 10 Storey shear in wind load analysis

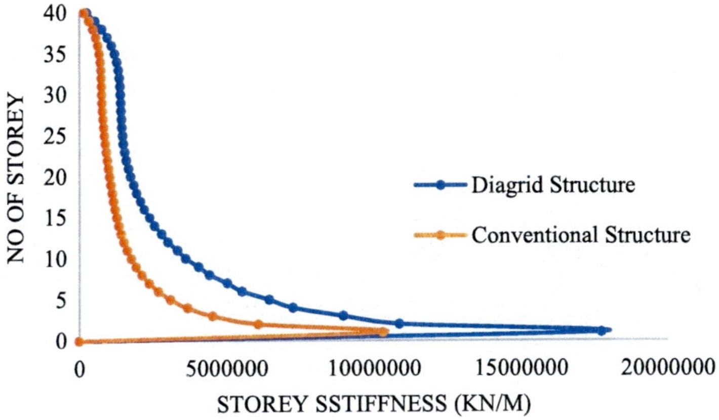

Fig. 11 Storey stiffness in response spectrum analysis

- Maximum storey drift values occur in intermediate storeys for both buildings.
- The stiffness of the first storey of diagrid structure is shown 40% much in both response spectrum and wind load analysis compared with conventional structure.
- Storey shear values got higher in conventional structure compare with the diagrid structure.

Finally, we conclude that the diagrid structural building is more effective to resist lateral loads on building structure as compared with conventional system building.

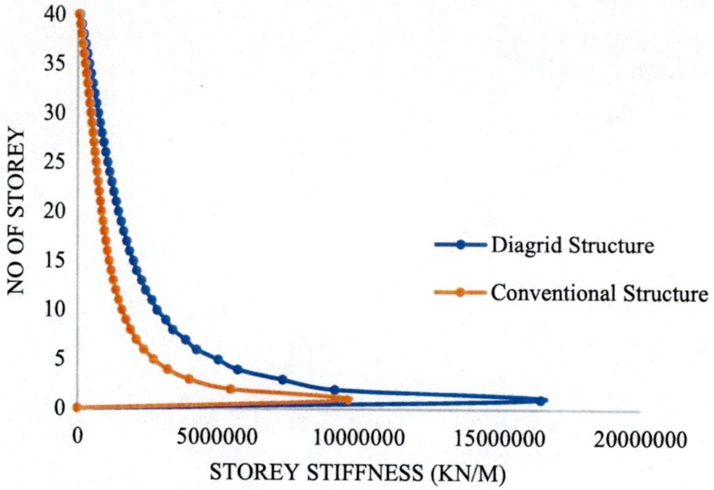

Fig. 12 Storey stiffness in wind load analysis

Table 5 Storey displacement

Storey level	Conventional structure	Storey diagrid structure	Displacement (MM) conventional structure	Diagrid structure
	(RSP)	(RSP)	(WL)	(WL)
Storey 0	0	0	0	0
Storey 10	20.5	10.2	17.6	9.8
Storey 20	56.7	27.7	48.0	27.4
Storey 30	97.5	48.7	80.2	47.4
Storey 40	137.2	70.7	109.8	67.1

Table 6 Storey drift

Storey level	Conventional structure	Store diagrid structure	y drift Conventional structure	Diagrid structure
	(RSP)	(RSP)	(WL)	(WL)
Storey 0	0	0	0	0
Storey 10	0.001023	0.000499	0.000867	0.000488
Storey 20	0.001371	0.00069	0.001080	0.000645
Storey 30	0.001439	0.000786	0.001043	0.000671
Storey 40	0.001312	0.000748	0.000943	0.000641

Table 7 Storey shear

Storey level	Conventional structure	Store diagrid structure	y shear (KN) Conventional structure	Diagrid structure
	(RSP)	(RSP)	(WL)	(WL)
Storey 0	7146.641	6990.108	5332.642	5278.831
Storey 10	5387.164	5244.729	4050.214	4010.692
Storey 20	4204.385	4107.985	2700.00	2680.547
Storey 30	3293.528	3280.193	1350.10	1310.120
Storey 40	628.142	551.537	135.410	112.632

Table 8 Storey stiffness

Storey level	Conventional structure (RSP)	Storey stiffness diagrid structure (RSP)	(KN/M) conventional structure (WL)	Diagrid structure (WL)
Storey 1	10,301,282	17,745,068	47,710.87	70,237.18
Storey 10	1,764,104	3,624,756.5	474,640.3	737,757.2
Storey 20	1,048,280	1,740,671	874,882.9	1,464,920
Storey 30	774,390.9	1,401,267.8	1,609,525	2,858,949
Storey 40	10,301,282	17,745,068	9,626,492	16,467,308

References

1. Khadka GB, Karim HF (2019) Analysis of multi-storey building with and without diagrid system using ETABS. Int J Eng Res Technol IJIRT 148174 5(12):564–570
2. Kambale MS (2018) Evaluate the performance of diagrid building for bracing angle and shape. Int J Innovative Res Sci, Eng Technol 7(12):12211–12217
3. Heshmati M, Shakib H (2020) Seismic performance assessment of tubular diagrid structures with varying angles in tall steel buildings. Structures 113–126
4. Dethe G, Banagar M, Kenjale P (2018 March) Analysis of diagrid structure. Int Res J Eng Technol 5(3):3072–3075
5. Nishith B, Vinubhai R (2019) Diagrid structural system: strategies to reduce lateral forces on high-rise buildings. Int J Eng Res Technol 03(04):374–378
6. Snehal S (2017) Response of high rise building with different diagrid structural system. Int J Sci Technol Eng 4(04):144–150
7. Harsha S, Raghu J (2015) Analysis of tall buildings for desired angle of diagrids. Int J Res Eng Technol 4(4):635–639
8. Somvanshi TR, Kolase K (2019) Comparative analysis and design of diagrid structure and orthogonal structure. Int J Eng Dev Res 7(03):136–141
9. Asadi E (2018 April 25) Seismic performance factors for low- to mid-rise steel diagrid structural systems. Struct Des Tall Special Buil 27(15):1–18. https://doi.org/10.1002/tal.1505
10. Venkatesh P (2018) A comparative study on the structural analysis of diagrid structural systems with conventional structural systems for different plan configurations. Int J Res Appl Sci Eng Technol 6(6):433–441
11. Chetan S (2018) Analysis of diagrid structures with plan irregularity. Int Res J Eng Technol 05(08):435–438

12. Moon K-S, Jerome J (2007) Diagrid structural systems for tall buildings: characteristics and methodology for preliminary design. Struct Design Tall Special Build 16:205–230
13. Shubhangi V, Kakamare MS (2017) Earthquake and wind analysis of diagrid structure. Int J Res Appl Sci Eng Technol 5(7):1729–1739
14. Shroti AM (2017) Constructability of diagrid structures. Int J Eng Res Technol 6(5):135–139, IJERTV6IS050041
15. Akhand J (2019) Comparative study of different shapes of diagrid structure system with conventional system using response spectrum analysis. Int Res J Eng Technol 6(4):2126–2133
16. Nawale UA, Kakade DN (2017) Analysis of diagrid structural system by E-Tab. Int Adv Res J Sci Eng Technol 4(6):193–196. https://doi.org/10.17148/IARJSET.2017.4634
17. IS 800:2007 Indian standard code of practice for general construction in steel. Bureau of Indian standards New Delhi
18. IS 456:2000 (2000) Indian standard code of practice for plan and reinforced concrete, Bureau of Indian standards, New Delhi
19. IS 1893 (part 1):2016 (2016) Indian Standard code of practice criteria for earthquake resistant design of structures, Bureau of Indian standards, New Delhi
20. IS 875(part 2):1987 (1987) Indian standard code of practice for design loads for buildings and structures: part 2 imposed loads, Bureau of Indian standards, New Delhi
21. IS 875(part 3):1987 (1987) Indian standard code of practice for design loads for buildings and structures: part 3 wind loads, Bureau of Indian standards, New Delhi
22. Moolimane NB (2018) Study on rigid and semi rigid diaphragm in multistoried structure using E-tabs. Int Res J Eng Technol 5(8):1216–1222
23. Shereef MI (April 2013) Analysis of tall building subjected to wind and seismic loads. National conference on emerging technology in civil engineering (ETCE'13)

Performance Evaluation of RC Framed Hill Buildings with and Without Shear Walls

Zaid Mohammad, Abdul Baqi, and Mohammed Arif

1 Introduction

Buildings built on hill slopes demonstrate distinct seismic behaviour than conventional buildings resting on a plain ground. Due to asymmetric elevation and plan, irregular buildings exhibit variation in mass and stiffness at different floor levels, leading to the development of additional shear forces and moments in the frame members apart from lateral loads [1–10]. Further, short column effect in hill buildings due to columns of varying lengths at same storey level makes the structure more vulnerable to earthquake. The shorter columns on the uphill side have higher stiffness and attract much more forces as compared to that of the columns on downhill side [11].

Previous research studies conducted on irregular structure buildings presented different techniques regarding mathematical modelling formulation and earthquake analysis of hill buildings. Cheung and Tso [1] investigated setback buildings under lateral loads. Shahrooz and Moehle described a mathematical model for the analysis of buildings with setbacks when subjected to seismic loads [2]. Paul [3] proposed a simple one-dimensional approach for dynamic analysis of asymmetric buildings. Kumar and Paul developed a method of analysis in which each storey of the building was modelled with multi-degree of freedom (D.O.F.) per floor; then, the obtained results were compared with the Codal provisions [4, 5]. Further, Kumar [6] and Kumar and Paul proposed a three-dimensional modelling approach for the dynamic analysis of irregular and asymmetrical hill structures [7, 8]. Birajdar and Nalawade [9] analysed various configurations of hill buildings and parametrically compared dynamic properties of the buildings and suggested the suitability aspect. Singh et al. [10] applied time history analysis and evaluated seismic properties in along and across slope of a stepback buildings. The obtained parameters in the study were

Z. Mohammad (✉) · A. Baqi · M. Arif
Department of Civil Engineering, Z.H.C.E.T., Aligarh Muslim University, Aligarh 202002, India
e-mail: zmohammad_co@myamu.ac.in

A. Sil et al. (eds.), *Recent Trends in Civil Engineering*, Lecture Notes in Civil Engineering 274, https://doi.org/10.1007/978-981-19-4055-2_19

compared with the damage pattern in case study of damaged hill building in Sikkim earthquake (2011). Mohammad et al. [11–13] investigated the seismic performance of hill buildings and observed that stepback-setback configurations showed better resistance than the stepback building configuration under earthquake loads.

It has been observed that the shear walls impart restraint to the extensive damage in buildings in earthquakes. A well-designed system of shear walls in the building enhances its seismic performance. Past studies simplified the modelling and analysis of the structural system with shear walls and evaluated seismic response of the buildings. Medhekar and Jain [14] discussed seismic behaviour, modes of failure and factors influenced by shear walls on structural response. Wallace [15] presented an analytical approach to determine the requirement of transverse reinforcement in RC structural walls with various cross sections and found that the strain distribution in shear wall is significantly affected by its aspect ratio, configuration and reinforcement ratios. Patel et al. [16] studied and discussed the effect of varying columns and seismic response of RC frame building on sloping ground with shear walls at different positions in the structure.

The position and form of shear walls play significant role in structural performance of the building [17–19]. Generally, shear walls are provided in the symmetrical fashion, hence do not generate any torsional forces and moments due to any kind of asymmetry in the structure. Therefore, it is more important to analyse buildings with vertical irregularities like hill buildings. There is a need to explore the seismic response of hill buildings with shear walls present at different locations in the structure, to opt what or which position is best for structures constructed on hill slopes.

Thus, in present study, the effect of shear walls on the seismic performance of two hill building configurations was studied. Both stepback and stepback-setback hill buildings have been modelled with shear walls at three different locations, viz. at corners, at mid-edge and at the centre of building. The finite element models were analysed using both equivalent static as well as response spectrum methods, and the dynamic parameters have been presented and compared in terms of fundamental time period, storey drift, storey shear and shear forces induced in the columns at foundation level in buildings. Further, for better comparison and suitability of position of shear walls, the total concrete volume and number of shear walls were kept equivalent in the respective geometric variations as well as in different direction of hill slopes.

2 Materials and Methods

The present study investigates the effect of reinforced concrete shear walls on two hill building configurations, viz. stepback and stepback-setback. All the configurations are modelled in three dimensions without (bare frame) and with the inclusion of shear walls. Seismic analyses of the structures have been performed using equivalent static and response spectrum method with SRSS combination in a finite element software. Important dynamic parameters, viz. fundamental time period, top storey displacement, story shear, story drift and base shear at ground level in along as well as across hill slope directions, are obtained and compared with the respective bare frame configurations.

As per IS 1893 (Part 1): 2002, the zone factor V was taken as for hilly regions. The importance factor (I) and response reduction factor (R) were taken as 1.5 (for important buildings) and 5 (for SMRF system), respectively. The gravity and imposed loads were taken as per IS 875 (Part 1 and 2): 1987, respectively, for a typical residential building. All the models have been analysed, designed and checked for any failure in the frame structure. The reinforcement in the columns was varied from 1% to 3.5%, whereas in beam and slab, nominal designed percentage of rebar was provided in both the directions (see Table 1).

The bare frame configurations without shear walls have been modelled for stepback (BSTEPALS) and stepback-setback (BSETALS) configurations. All buildings have similar geometrical and material properties (see Fig. 1 and Table 1). The modulus of elasticity of concrete is taken as 25,000 N/mm^2 and value of Poisson's ratio to be 0.2. The foundation is assumed as a fixed support system. Both building configurations have six bays in length along the hill slope; however, width of all the models was kept constant to four bays (Fig. 4) [11, 13]. The thickness of the slab at all floors in all the models was considered as 125 mm.

Shear walls resist large portion of horizontal forces originating from wind or earthquake. The position of shear walls in the structure, influence its overall response of the structure. The hill buildings are modelled with shear walls at three different types of positions. In the numerical study, shear wall is modelled as reinforced concrete bar-bell-shaped shear wall of M 25 grade of concrete using four node shell elements (Figs. 2 and 3). The shear wall is 150 mm thick with the minimum percentage of reinforcement as 0.25%. The shear walls are positioned in three ways, namely at corners, at middle of edges and at the centre of the building (Figs. 5 and 6). To achieve better comparison and keeping the economical view, the surface area and the concrete volume of shear wall in each category were kept equal.

3 Results and Discussion

The present study investigates the influence of shear walls on the seismic performance of two hill building configurations. In all, eight models were geometrically modelled in a finite element software and analysed for gravitational as well as seismic loads. Further, for better assessment of suitability of position of the shear walls, the concrete volume and number of shear walls were designed to be equal in the respective geometric variations as well as in different direction of hill slopes.

Table 1 Geometrical properties of different configurations of hill building

Building configuration	Designation of models				Column size (mm)	Beam size (mm)
	Bare frame	Frame with shear wall				
		In corners	In middle	In centre		
Stepback	BSTEPALS	WCRSTEPALS	WMDSTEPALS	WCESTEP-ALS	450 × 450	Along slope: 300 × 500 Across slope: 300 × 450
Stepback-setback	BSETALS	WCRSETALS	WMDSETALS	WCESET-ALS	400 × 400	

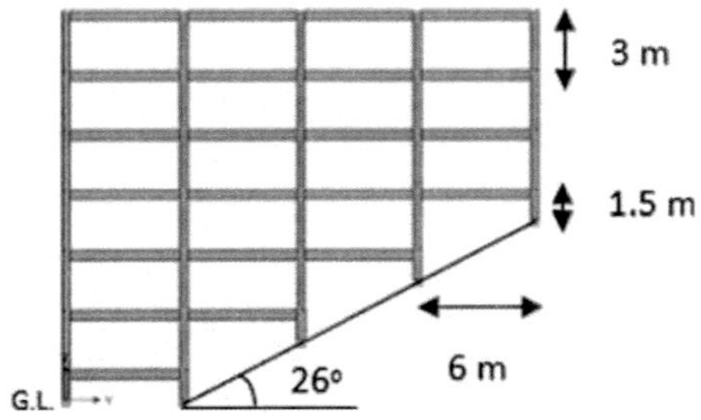

Fig. 1 Terrain properties of hill slope

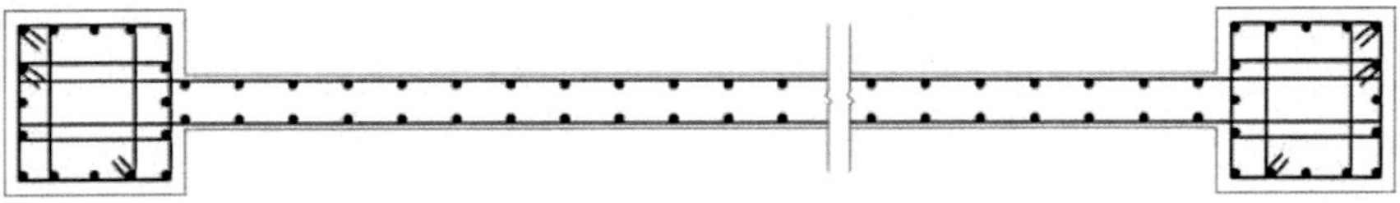

Fig. 2 Typical section of a bar-bell-shaped shear wall

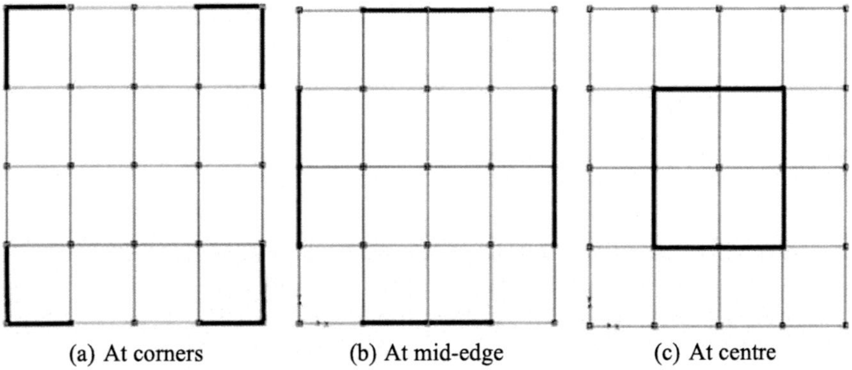

Fig. 3 Different positions of shear walls

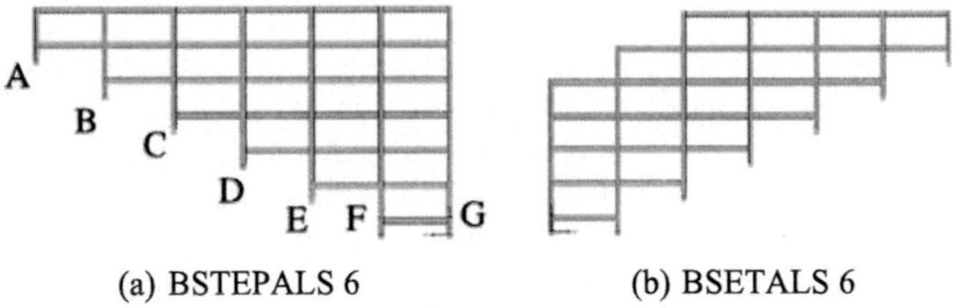

Fig. 4 Bare frame models; **a** stepback configuration and **b** stepback-setback configuration

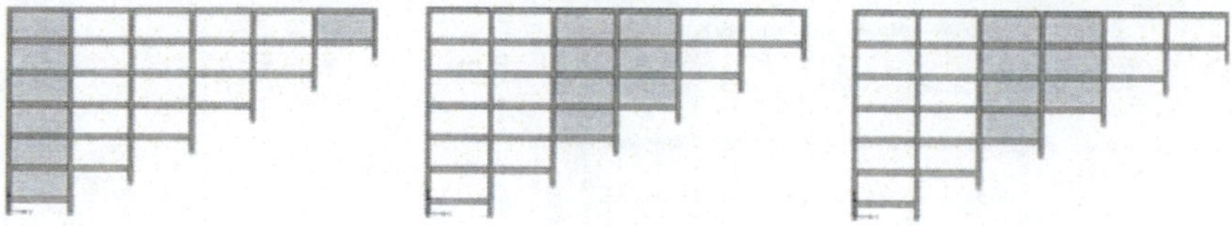

WCRSTEPALS 6 (a) At corners (outer frame)

WMDSTEPALS 6 (b) At mid-edge (outer frame)

WCESTEPALS 6 (c) At centre (inner frame)

Fig. 5 Stepback buildings with shear walls

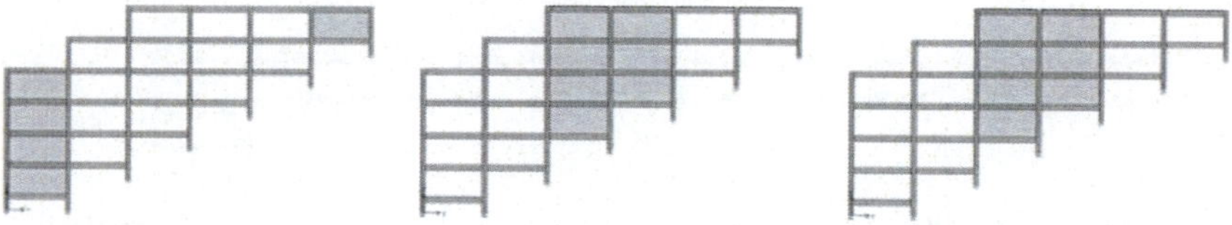

WCRSETALS 6 (a) At corners (outer frame)

WMDSETALS 6 (b) At mid-edge (outer frame)

WCESETALS 6 (c) At centre (inner frame)

Fig. 6 Stepback-setback buildings with shear walls

3.1 Seismic Response of Stepback Configuration

The dynamic properties of bare frame models and models with shear walls at three considered locations, subjected to seismic forces along and across hill slope direction, have been mentioned in Tables 2 and 3. It was observed that shear walls significantly reduced the time period as well as top storey displacement of the building as compared to the values obtained in bare frame. In case of buildings subjected to seismic load in across slope direction, both fundamental time period and top storey displacement values were found to be increased.

Table 2 Seismic response of stepback building along hill slope

Designation	FTP* by RSA (*sec*)	FTP as per IS 1893 (sec)	Max. Top storey displacement (mm)	Base Shear ratio (λ)
BSTEPALS 6	0.313	0.293	6.37	1.345
WCRSTEPALS 6	0.225	0.293	2.60	1.154
WMDSTEPALS 6	0.22	0.293	2.25	1.111
WCESTEPALS 6	0.222	0.293	2.10	1.080

* FTP: Fundamental Time Period

Table 3 Seismic response of stepback building across hill slope

Designation	FTP by RSA (sec)	FTP as per IS 1893 (sec)	Max. Top storey displacement (mm)	Base Shear ratio (λ)
BSTEPALS 6	0.574	0.392	31.61	1.654
WCRSTEPALS 6	0.329	0.392	8.78	1.368
WMDSTEPALS 6	0.311	0.392	7.65	1.351
WCESTEPALS 6	0.326	0.392	6.41	1.148

Figure 7 shows the storey drift variation of stepback models in along hill slope direction. In case of hill buildings with shear walls at corners, the maximum reduction in the value of storey drift was found to be 77.7% and 67.81% in along and across slope direction, respectively, when compared with bare frame model. Whereas, in models with shear wall at mid-edge, this reduction was found to be 79.16 and 73.21% in along and across slope direction, respectively. Following the same pattern in this geometric variation, the maximum reduction in the values of storey drift was observed as 86.76% in along slope direction and 87.94% in across slope direction, when the shear walls were located in the centre. The peaks in the storey drift variation show the presence of soft storey at that level, which was due to the absence of shear walls at foundation level.

The maximum storey shear was obtained at the highest foundation level at the second last storey, in along slope direction, whereas in across slope direction, the maximum increase in the storey shear was observed at mid-storeys of the structure. The maximum increase in the storey shear was seen in hill buildings with shear walls at centre, at mid-storeys in both the directions. While at lower foundation levels, the maximum storey shear was observed in case of models with shear walls at corners, along hill slope direction, whereas in across slope direction, the maximum value of storey shear was observed in hill building with shear walls at mid-edge (see Fig. 8).

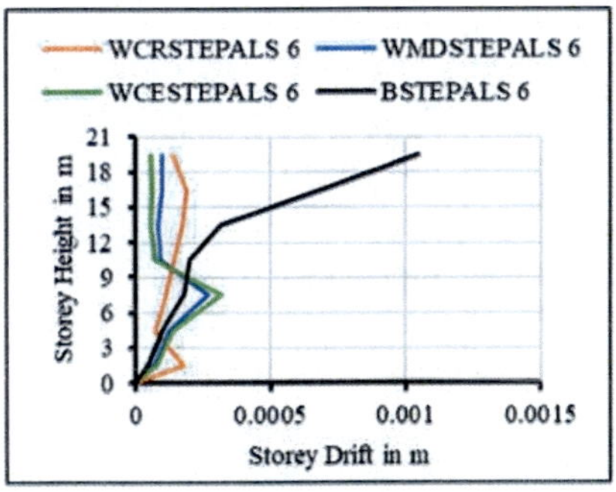

(a) Along Slope

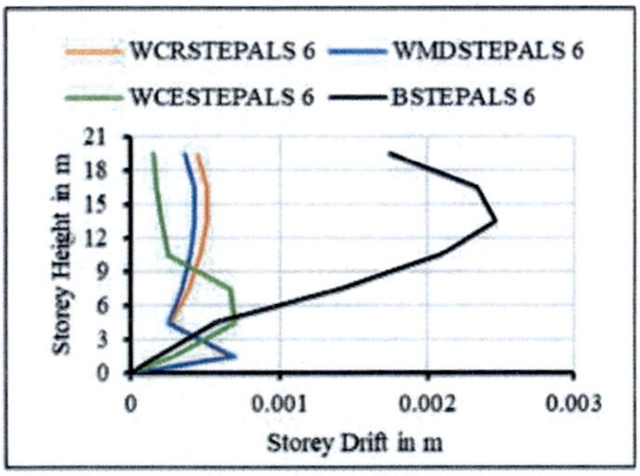

(b) Across slope

Fig. 7 Comparison of storey drift variation in stepback configuration

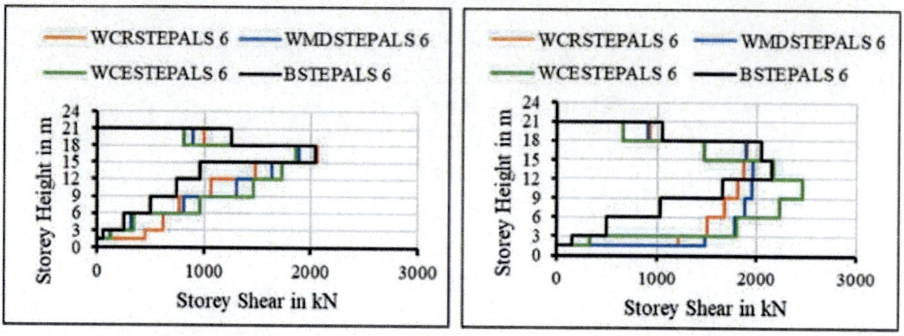

(a) Along Slope (b) Across slope

Fig. 8 Comparison of storey shear distribution in stepback configuration

In Fig. 9, the bar graphs show the comparison of shear force at foundation level in stepback configuration models with and without shear walls, in along slope direction. Following the similar pattern as in previous geometric variations. The maximum reduction in column shear force was observed at frame 'A', at which the highest reduction was shown by models with shear wall at mid-edge (WMDSTEPALS), but in other frames at the middle of building, an increase in the values of shear force was observed in the models with shear walls at mid-edge and centre.

However, in across the slope direction (Fig. 9), the shear force in columns at middle frames experienced a substantial decrease, as shear walls were incorporated in the buildings. Also, an abrupt increase was observed in the case of WCESTEPALS at frame 'F', which could be due to increased axial forces in the shear wall and hence, lead to high shear demand in column at that frame. Also, in larger frames 'G', 'H' and 'I', a significant increase in the values of shear force was observed in case of models with shear wall at corners and at mid-edge.

3.2 *Seismic Response of Stepback-Setback Configuration*

The seismic parameters obtained in analysis of stepback-setback models were found to be reduced in comparison with stepback configuration with same geometric variation. The dynamic properties for stepback-setback configuration with and without shear walls subjected to seismic forces in along as well as across slope direction have been mentioned in Tables 4 and 5. In case of model with shear walls at corners, a small reduction was observed in time period and top storey displacement, as compared with the values of bare frame model. On the other hand, model with shear walls at mid-edge showed highest reduction and found to be 25.96% in time period and 57.19% in storey displacement, when compared with respective bare frame model. However, in case of hill building with shear walls at centre, this reduction was observed to be 24.91% in time period and 61.51% in top storey displacement. Following the similar

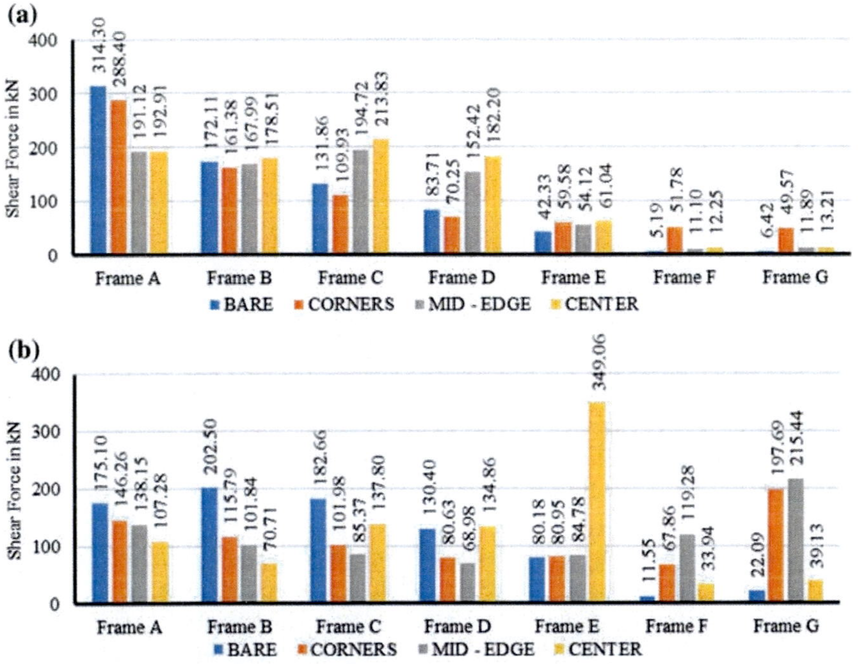

Fig. 9 Base shear distribution at foundation in stepback configuration in **a** along and **b** across hill slope

pattern as in the case of stepback models, the values of seismic parameters obtained in the analysis of stepback-setback models in across hill slope direction were found to be substantially increased as compared with the seismic response in along slope direction.

The storey drift variation in stepback-setback configuration was found to be entirely different from the previous geometric variations of stepback configurations (Fig. 10). This extreme change may be due to the asymmetric geometry of stepback-setback hill buildings. However, the values of storey drift were found to be minimum

Table 4 Seismic response of stepback-setback building along hill slope

Designation	FTP* by RSA (sec)	FTP*as per IS 1893 (sec)	Max. Top storey displacement (mm)	Base Shear ratio (λ)
BSETALS 6	0.285	0.293	5.69	1.328
WCRSETALS 6	0.214	0.293	2.71	1.238
WMDSETALS 6	0.203	0.293	1.88	1.084
WCESETALS 6	0.207	0.293	1.79	1.066

* FTP: Fundamental Time Period

Table 5 Seismic response of stepback-setback building across hill slope

Designation	FTP* by RSA (sec)	FTP* as per IS 1893 (sec)	Max. Top storey displacement (mm)	Base Shear ratio (λ)
BSETALS 6	0.455	0.392	17.40	1.573
WCRSETALS 6	0.311	0.392	8.30	1.458
WMDSETALS 6	0.28	0.392	5.52	1.306
WCESETALS 6	0.297	0.392	3.65	1.170

* FTP: Fundamental Time Period

in all previously considered geometric variations. In model with shear walls at mid-edge (WMDSETALS 6), the reduction in the values was observed to be 90.92% and 57.69% in along and across slope direction, respectively. On the hand, model with shear walls at centre showed significant reduction in storey drift and the values get reduced to 8.0% along hill slope direction and 15.6% across hill slope direction, respectively.

The storey shear distribution in stepback-setback configuration showed entirely different behaviour than previously considered configurations (Fig. 11). The maximum storey shear obtained at upper and lowest foundation levels, in along slope direction, was observed in models with shear wall at corners. However, in mid-storeys, the maximum storey shear is observed in hill building with shear walls at centre. On the other hand, in across slope direction, the maximum storey shear at mid-storeys was observed in WCESETALS. Whereas, at lower foundation levels, this increase was seen in the models with shear wall at mid-edge.

Figure 12 shows the base shear distribution of stepback-setback models. In along hill slope direction, the shear force distribution shows the similar pattern of variation as seen in previous cases. However, in frame 'A', an increase in the shear demand was observed in case of models with shear wall at corners. Whereas, at middle frames,

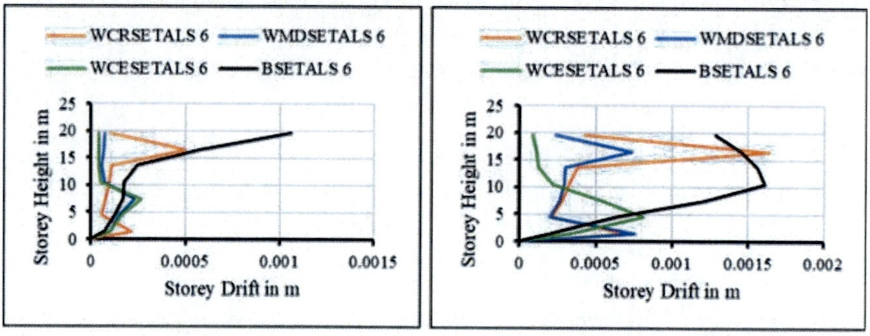

(a) Along Slope (b) Across slope

Fig. 10 Comparison of storey drift variation in stepback-setback configuration

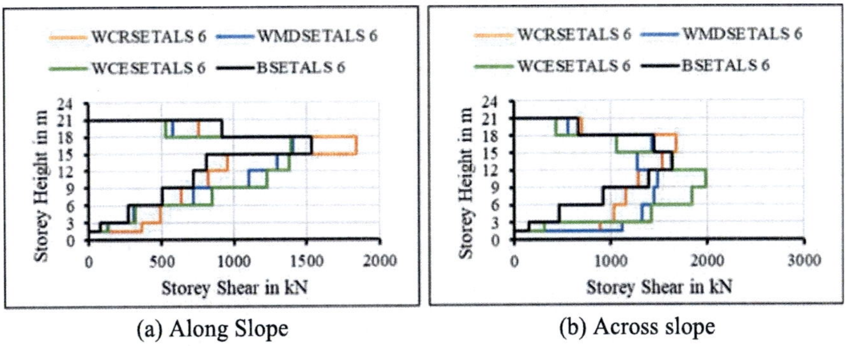

Fig. 11 Comparison of storey shear distribution in stepback-setback configuration

an increase in the shear force was observed in case of models with shear walls at mid-edge and at centre.

However, when models were subjected to seismic forces in across slope direction, the base shear distribution show different variation than in along slope direction. At frame 'A', an abrupt increase was observed in case of WMDSETALS, whereas, at

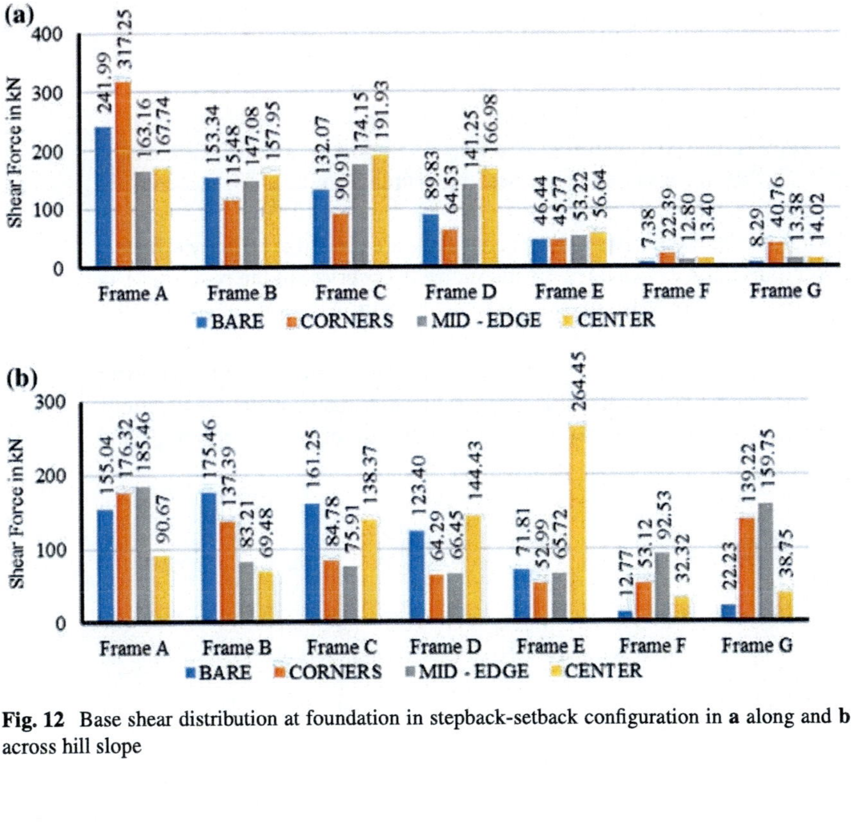

Fig. 12 Base shear distribution at foundation in stepback-setback configuration in **a** along and **b** across hill slope

middle frames, a significant reduction was observed in case of models with shear walls. Also, following the same pattern in previous geometric variation, the maximum shear force was developed at frame 'F' in WCESETALS. It was also found that stepback-setback buildings experience less base shear forces than that in stepback configuration of hill buildings.

4 Conclusion

This study investigates the influence of shear walls on the seismic performance of two hill building configurations. The buildings were modelled with shear walls at three different locations, viz. at corners, at mid-edge and at centre of building. Further, for better comparison of suitability of position, the total concrete volume and number of shear walls were designed to be equal in the respective geometric variations as well as in different direction of hill slopes.

In all the considered configurations, models with shear walls at centre showed maximum reduction in time period and top storey displacement in both hill slope directions. Following the same pattern, maximum reduction in storey drift was observed in buildings with shear wall present at centre and was found to be reduced by 85% as compared with bare frame models. The maximum base shear was observed in stepback models, at upper most foundation level in along slope direction, whereas in transverse direction, the maximum value was found in middle storeys, when the shear walls were kept at mid-edge position. The base shear at foundation level in all the cases was found to be reduced at frame 'A' (shortest frame) in along the slope direction, whereas a marginal increase is observed in other frames. In across the slope direction, a mixed response of increase and decrease in the shear force values was observed.

Thus, the models with shear walls present at centre are proven to be more effective than other configurations. However, a drastic increase observed in the base shear, in across slope direction, should be considered in the structural design, whereas buildings with shear walls at corners and mid-edge show reduced values of base shear at the foundation level and also provide extra stiffness to the structure therefore, reducing the torsional moments. Thus, a combination of shear walls at different positions should be provided for achieving better dynamic response of the hill buildings and to counter their asymmetric nature.

References

1. Cheung VWT, Tso WK (1987) Lateral load analysis for buildings with setbacks. J ASCE Struct Div 113(2):209–227
2. Shahrooz BM, Moehle JP (1990) Seismic response and design of setback buildings. J Struct Eng ASCE 116(5):1423–1439

3. Paul DK (1993) Simplified seismic analysis of buildings on hill slopes. Bull Indian Soc Earthq Technol 30(4):113–124
4. Kumar S, Paul DK (1994) 3-D analysis of irregular buildings with rigid floor diaphragm. Bull Indian Soc Earthq Technol 31(3):141–154
5. Kumar S, Paul DK (1994) Dynamic analysis of stepback and setback buildings. Proc Tenth Symp Earthq Eng 1:341–350
6. Kumar S (1996) Seismic analysis of stepback and setback buildings, thesis, earthquake engineering. University of Roorkee, Roorkee
7. Kumar S, Paul DK (1998) A simplified method for elastic seismic analysis of hill buildings. J Earthquake Eng 2(2):241–266
8. Kumar S, Paul DK (1999) Hill buildings configuration from seismic consideration. J Struct Eng 26(3):179–185
9. Birajdar BG, Nalawade SS (2004) Seismic analysis of buildings resting on sloping ground. Processes in thirteenth world conference on earthquake engineering (13WCEE). Vancouver, Canada
10. Singh Y, Gade P, Lang DH, Erduran E (2012) Seismic behaviour of buildings located on slopes: an analytical study and some observations from Sikkim earthquake of September 18, 2011. Processes in fifteenth world conference on earthquake engineering, 15 WCEE, Lisbon, Portugal
11. Mohammad Z, Baqi A, Arif M (2017) Seismic response of RC framed buildings resting on hill slopes. 11th International symposium on plasticity and impact mechanics (IMPLAST 2016). Procedia engineering, vol 173. Elsevier, New Delhi, pp 1792–1799. https://doi.org/10.1016/j.proeng.2016.12.221
12. Mohammad Z (2019) Effect of unreinforced masonry infills on seismic performance of hill buildings. VW Appl Sci 1(1):37–47
13. Mohammad Z, Razi MA, Baqi A (2021) Influence of masonry infill panels on the seismic performance of irregular buildings. In: Advances in geotechnics and structural engineering. Lecture notes in civil engineering, vol 143. Springer, Singapore, pp 1–11
14. Medhekar MS, Jain SK (1993) Seismic behaviour, design and detailing of RC shear walls, part I: behaviour and strength. Indian Concr J 311–318
15. Wallace JW (1994) New methodology for seismic design of shear walls. J Struct Eng ASCE 120:863–884
16. Patel MUF, Kulkarni AV, Inamdar N (2014) A performance study and seismic evaluation of RC frame buildings on sloping ground. IOSR, J Mech Civil Eng 51–58
17. Pawar SP, Pise CP, Pawar YP, Kadam SS, Mohite DD, Deshmukh CM, Shelar NK (2016) Effect of positioning of RC shear walls of different shapes on seismic performance of building resting on sloping ground. IJCIET 7(3):373–384
18. Anjali BU (2017) Gopisiddappa: effect of positioning and configuration of shear walls on seismic performance of RC building resting on hilly and plain terrain. IRJET 4(6):2501–2506
19. Dane A, Pendharkar U (2019) Effective positioning of shear wall in G+5 storey building on sloping ground. IJEAT 9(2):3602–3606

Structural Behaviour of RC Silo under Gravity and Infill Loads

Zaid Mohammad, Saad Shamim Ansari, Mohd. Anwar, and Abdul Baqi

1 Introduction

Silo structures have been used for industrial as well as domestic storage purposes. The silos are often subjected to static and dynamic loading conditions such as gravity load, granular infill pressure, thermal load due to storage material, wind as well as earthquake loads. As the density, flow and frictional properties of grains, cement, coal, carbon black, etc. vary widely, failure stresses may vary in different parts of silo.

Silo structures are also prone to buckling due to external pressures caused by wind force. Empty silos have been found to be more vulnerable to wind forces. Further, the occupancy levels inside the silos play major role in defining the variable stress development inside the concrete walls. The values of wall stresses and the critical locations have been observed to vary with the amount of infill material inside the silo structure.

The state-of-the-art studies [1–11] have provided various empirical and analytical models to ascertain the internal pressure inside the silo and analysed the vulnerability of the steel as well as concrete silo structure against earthquake and wind loads. Jansen [1] in an experimental study related the saturation pressure with the height of a silo and calculated the internal pressure carried by the walls. An economical design approach was suggested. The horizontal pressure distributions at various depths in cement silos predicted by Jansen's [1] and Airy's [2] theories are compared with the experimental values reported by Faber and Mead [3]. It was observed that non-cohesive material behaves in accordance with Janson's theory. Whereas, the pressure curve for storage material like cement varied widely with experimental results, due to the inconsistent cohesive nature of the material. Further, full-scale experiments of cement silos indicated that the lateral pressures are very much underestimated

Z. Mohammad (✉) · S. S. Ansari · Mohd. Anwar · A. Baqi
Department of Civil Engineering, Z.H.C.E.T, Aligarh Muslim University, Aligarh 202002, India
e-mail: zmohammad_co@myamu.ac.in; zaidzhcet@gmail.com

A. Sil et al. (eds.), *Recent Trends in Civil Engineering*, Lecture Notes in Civil Engineering 274, https://doi.org/10.1007/978-981-19-4055-2_20

towards the upper half of the silos by the Jansen's theory. On the contrary, the lateral pressures are very much over-estimated towards the lower half of the silo by the Airy's theory. Ooi and Rotter [4] proposed simple finite element technique which includes appropriate wall friction characteristics that can accurately model the pressures exerted by stored solids on cylindrical silo walls. Also, it was observed that an increase in the coefficient of wall friction decreases the wall pressure but increases the frictional drag on the wall. Nielsen [5] discussed the effect of internal pressure from flowing granular solids on the silo structure. It was found that silo pressures were quite unsymmetrical even in symmetrical silos and thus affect the safety of the silo structure. Khouri [6, 7] compared the mathematical formulation for calculating the lateral pressure in a silo proposed by Jansen et al. [8]. The former three methods were then compared with the finite element method employing water behaviour approach and found comparative results. Dogangun et al. [9] presented damage and failures that occurred in various types of silos across the world. Most common silo failures were occurred due to explosion and bursting, asymmetrical load, soil pressure, corrosion of metal silo, deterioration of concrete silos due to silage acids, internal structural collapse and thermal ratcheting. Nateghi [10] numerically investigated the earthquake effect on a reinforced concrete silo. The behaviour of reinforced concrete silo under earthquake excitation was compared with a model without considering granular material-structure interaction. The obtained results showed that considering the effective mass of granular material equal to 80% of total granular material mass, results in more severe tension damage in silo wall and flexural crack developed in the lowest part of silo wall. Raeesi et al. [11] discussed the behaviour of steel silo under wind load. ABAQUS/CAE was used for the modelling of wind load effect on different structural parameters, viz. geometric dimensions, stiffeners patterns and wind girders of the steel silo. The study found that the wind girders significantly increased the buckling strength of the silo and critical buckling wind speeds of the field silo, was increased about 44%.

Previous research works have explored the behaviour of infill material on the lateral pressure and proposed various analytical closed-form solutions as well as empirical relationships to calculate the infill pressure. However, the influence of occupancy level of the infill material on the development of critical stresses has not been studied and a thorough investigation is very much needed to effectively design the RC cement silo structures. Thus, in the present study, in addition to the gravity loads, the influence of lateral pressure was studied due to infill material at various occupancy levels, viz. 33%, 50%, 66% and 100%. The silo structure has been modelled in ABAQUS/CAE module and a numerical investigation has been carried out. The concrete structural components were modelled using solid brick elements (C3D8R), whereas, truss elements (T3D2) were used to model the internal reinforcement. The Concrete Damage Plasticity material model was used to simulate the material behaviour of the concrete. The results were obtained and discussed in terms of critical stresses and lateral deflection in the silo wall along the height, with different occupancy levels of infill material. At last appropriate solutions that improve the design of reinforced concrete silo have been suggested.

2 Modelling and Analysis

The present study investigates the effect of lateral pressure due to infill material at various occupancy levels, viz. 33%, 50%, 66% and 100%, on the structural performance of an RC silo. A reinforced concrete silo with 18 m external diameter, 0.4 m of wall thickness and 36.34 m height was undertaken from a practical case study (see Fig. 1). The RC silo was designed from conventional method of analysis using Jansen's theory for calculating the internal pressure. The finite element model of the silo was modelled in ABAQUS/CAE software and material properties and boundary conditions were assigned in the provided modules. In addition to the self-weight of the structure, the primary loads, viz. horizontal pressure (p_h) and vertical shear stress (p_v) were applied inside the silo. These values were determined from Eqs. (1) and (2) proposed by Janssen [1] as follows:

$$p_h = \frac{\gamma D}{4\mu}\left(1 - e^{\frac{-4\mu K j z}{D}}\right) \tag{1}$$

$$p_v = \mu ph \tag{2}$$

where,

p_h Normal pressure acting perpendicularly to silo wall surface.
p_v Shear stress acting along wall surface;
γ Specific weight of grain bulk = $\rho \times g$.

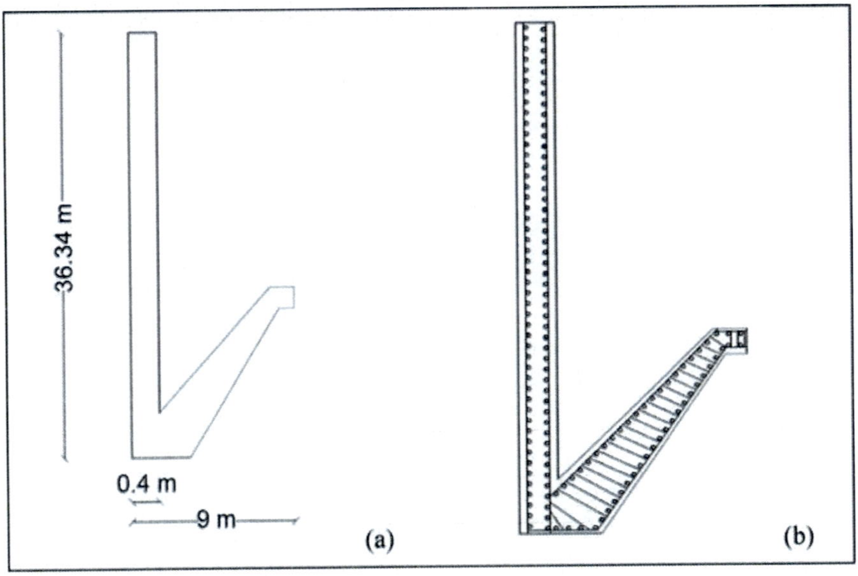

Fig. 1 Geometry of RC silo: **a** sectional elevation and **b** reinforcement details

ρ Bulk density of grains.
g Acceleration due to gravity.
D Inside diameter of silo bin.
μ Coefficient of sliding friction between bulk solid and wall surface; $\mu = \tan\text{Ø}'$
K_j Janssen ratio of horizontal pressure to vertical pressure; $K_j = 1 - \sin(\text{Ø})$.
z Vertical coordinate $=$ the depth of bulk inside of silo.
Ø′ Angle of wall friction between bulk solid and wall surface.
Ø Angle of internal friction between bulk solids.

The Concrete Damage Plasticity material model was used to simulate the material behaviour of the concrete. The material parameters of the silo were taken from previously published study [10]. The geometrical and material properties of the silo have been tabulated in Table 1. The concrete structural components were modelled using solid brick elements (C3D8R), whereas, truss elements (T3D2) were used to model the internal reinforcement. The rebar elements were embedded in the concrete material and required contact properties between the structure and filler material have been assigned. The boundary condition at the foundation of silo was assigned as fixed or *ENCASTRE*. Since the RC silo has been modelled as a quarter model, the walls have been constraint to move in the circumferential direction of the silo structure, however outward or inward translation or rotation has been allowed (see Fig. 2). A mesh analysis study was also conducted for obtaining the optimum element size of the mesh. Reduced integration was used to reduce the computational time. Meshing was applied separately to every single component because of the difference in geometry and size of components (Table 2).

Table 1 Geometrical and material properties of RC silo

Property	Details/quantity
Geometry:	*Height: 36.34 m* *Inner diameter: 17.2* *Outer diameter: 18 m*
Material:	*Concrete grade: M30* *Density: 2400 kg* m^{-3}
Rebar detail:	*Grade: Fe500* *Meridional reinforcement: 2 layer of 16ϕ @ 180 mm c/c* *Circumferential reinforcement: 2 layer of 16ϕ @ 150 mm c/c*
Plasticity parameters of concrete [10]:	*Dilation angle: 25* *Flow potential eccentricity: 0.1* σ_{bo}/σ_{co}*: 1.16* *k: 0.667*

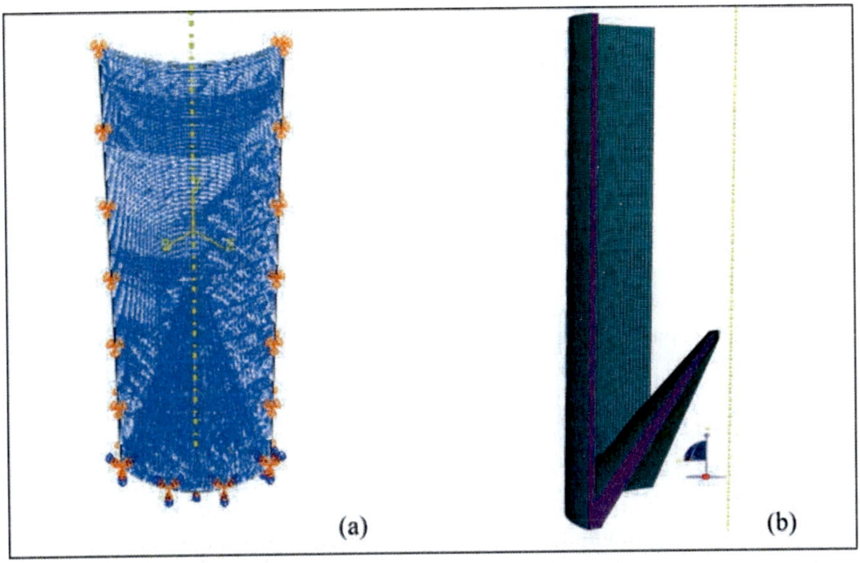

Fig. 2 Finite element model of RC silo: **a** boundary condition and **b** mesh details

Table 2 Mesh properties of RC silo

Part/instance	Mesh/element size (mm)	Element type
Concrete wall	200	C3D8R
Reinforcement	8	T3D2

3 Results and Discussion

Three-dimensional modelling and analysis have been carried out of RC cement silo. In addition to the gravity loads, the effect of internal pressure load due to different occupancy levels was observed on structural performance of the silo by using Jansen's lateral pressure equations. The silo models were analysed in a finite element software for different load conditions, viz. Case 1; for self-weight of silo and Case 2; for the silo filled with infill material at 33%, 50%, 66% and 100% of interior volume of the silo. The Concrete Damage Plasticity material model was used to simulate the material behaviour of the concrete. The results were obtained and discussed in terms of critical stresses and lateral deflection in the silo wall along the height.

3.1 Behaviour of an empty RC Silo due to Gravity Load

The variation of maximum principal stress developed due to self-weight of the structure at the mid-centre of silo wall with the height has been shown in Fig. 3. It was

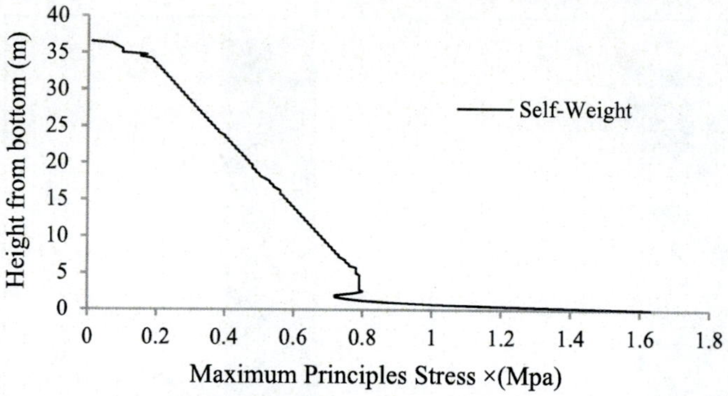

Fig. 3 Maximum principal stress variation in empty silo

found that the maximum principal stress increased from top to bottom of the silo due to increase in the weight of the structure. The observed stress pattern corroborated the lateral pressure variation given by Janssen's theory. However, at the most bottom portion of silo wall, stress was increased suddenly due to the stress concentration zone created with the conical dome.

3.2 *Behaviour of Concrete Silo under Gravity and Internal Pressure*

The structural performance of concrete silo was observed under gravity as well as lateral internal pressure due to infill material. The pressure distribution along the height was calculated from Jansen's theory of lateral pressure and applied accordingly. The effect of internal pressure on the structural parameters was analysed and compared with the results of Case 1, in which the concrete silo was analysed for gravity load only.

Figure 4a shows the variation of maximum principal stress due to gravity and internal pressure load of infill at different levels of filling of silo. It was observed that stress pattern was almost same for different levels of filled conditions, however, the peak value of stress increases with height from top. The stress value at 66% of filled conditions was found to be less as compared with that at 100%, 50% and 33% occupancy. This may be possible because the variation of pressure first increases exponentially and after some height, it remains constant. The pressure pattern was found approximately similar to Janssen's theory for different filled conditions. However, at the bottom of silo, stress increases suddenly approximately double of critical condition as seen in Fig. 4b. The critical locations at which the maximum value of stress to be developed have been shown in Fig. 5. The critical stress values were calculated at various occupancy levels and have been compared in Table 3.

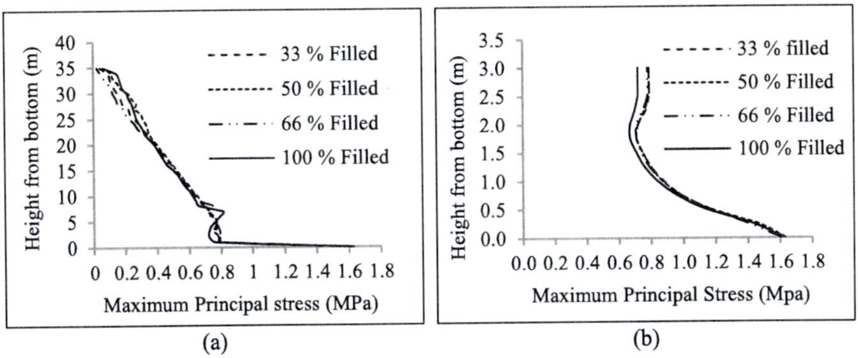

Fig. 4 Maximum principal stress variation in filled silo

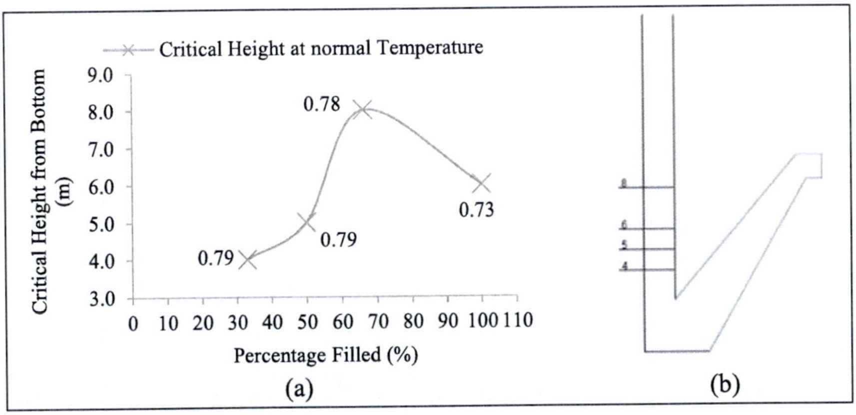

Fig. 5 Critical locations for maximum principal stress in filled silo

Table 3 Maximum principal stress (MPa) at different filled conditions

Critical location from bottom (m)	Filled conditions			
	33%	50%	66%	100%
4	**0.79**	0.75	0.74	0.72
5	0.71	**0.79**	0.76	0.66
6	0.71	0.71	0.75	**0.73**
8	0.69	0.70	**0.78**	0.71
At cone 12.36	0.62	0.62	0.62	0.62

The variation of lateral deflection in silo walls along the height at different occupancy levels has been shown in Fig. 6. It was noted that lateral deflection increased by approximately 1 mm at the height of 10 m, then the deflection was decreased with increase in the height of silo, however at the top most, silo even deflected inside of

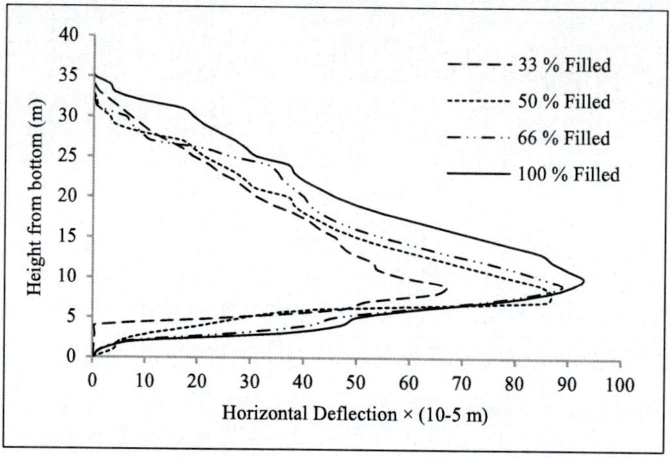

Fig. 6 Lateral deflection variation in filled silo

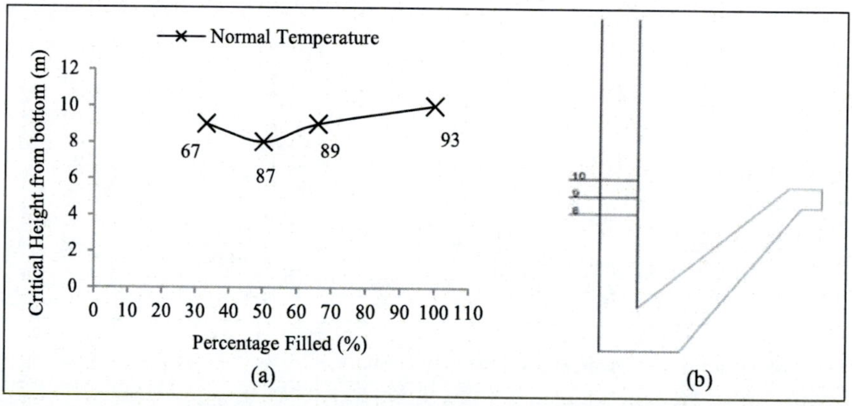

Fig. 7 Critical locations for lateral deflection in filled silo

the wall. Further, when the silo was filled at 100%, the deflection was found to be approximately 28%, 6% and 5% more as compared with filled 33%, 50% and 66%, respectively at critical locations (see Fig. 7 and Table 4).

4 Conclusion

The present study investigates the influence of lateral pressure at different occupancy levels of infill material on the structural performance of RC silo. The silo structure was modelled in a finite element software and a numerical study has been carried out on empty as well as filled silo. The Concrete Damage Plasticity material model

Table 4 Maximum deflection (10^{-5} m) at different filled conditions

Critical Location From bottom (m)	Filled conditions			
	33%	50%	66%	100%
8	65	**87**	84	86
9	**67**	85	89	91
9	67	85	**89**	91
10	59	79	85	**93**

was employed to simulate the material behaviour of the concrete. The following conclusions have been drawn from the study:

(a) The effect of lateral pressure due to infill material was observed at various occupancy levels, viz. 33%, 50%, 66% and 100%. It was observed that the variation of maximum stress values obtained from the numerical study corroborated with Jansen's theoretical model.
(b) The stress variation along the height of the silo wall was found to be approximately similar in various occupancy levels.
(c) The lateral deflection was found to be decreased linearly with increase in height.
(d) The variation in the lateral deflection values at critical locations was found to be non-linear and the deflection was found to be increased with the occupancy of infill material.

References

1. Janssen HA (1885) Versuche uber getreidedruck in silozellen. VDI Zeitschrift, Dusseldorf 39:1045–1049
2. Airy W (1897) The pressure of grain. Minutes Proc, Inst Civil Eng, London 131:458–465
3. Faber J, Mead F (1961) Reinforced concrete. E. & F. N. Spon, London
4. Ooi JY, Rotter JM (1990) Wall pressures in squat steel silos from simple finite element analysis. Comput Struct 37(4):361–374
5. Nielsen J (1998) Pressures from flowing granular solids in silos. Phil Trans R Soc A 356:2667–2684
6. Khouri MF (2005) Comparison of various methods used in the analysis of silos without wall friction. WIT Trans Model Simul, WIT Press 41:425–441
7. Khouri MF (2010) Analysis of silo walls without friction. AES Tech Rev Int J, Part B, IJAMAIM 1(1):69–79
8. Reimbert M, Reimbert A (1976) Silos-theory and practice, 1st edn. Trans Tech Publication, Claustal, Germany
9. Dogangun A, Karaca Z, Durmus A, Sezen H (2009) Cause of damage and failures in silo structures. J Perform Constr Facil 23(2):65–71
10. Nateghi F, Yakhchalian M (2011) Seismic behaviour of reinforced concrete silos considering granular material-structure interaction. Procedia Eng 14:3050–3058
11. Raeesi A, Ghaednia H, Zohrehheydariha J, Das S (2017) Failure analysis of steel silos subject to wind load. Eng Fail Anal 79:749–761

Experimental and Numerical Study on the Behaviour of Grouted Sleeve Coupler Joint Used in Precast Member Connection

S. Paramathmeka and M. Rajeswari

1 Introduction

The rate of growth of population and technology is exponential in the fast-moving world. To meet the demand of the growing population, all the fields of science and technology are striving hard. This thirst of the people led to the development of many new alternate solutions, which are far better in performance than the conventional ones. One such extraordinary alternate solution for rapid construction of building, minimization in the cost of construction without a lack in the strength and performance of the building is the emergence of the prefabricated structures. Precast concrete structures are being widely accepted as a better choice than conventional monolithic concrete structures in terms of construction efficiency, product quality and labour savings. Many residential and industrial buildings have been constructed using precast concrete techniques in many countries including the US, Japan, New Zealand, India and China.

Joints and connections should be designed with adequate strength and stability to transfer the loads from one precast member to the supporting structure or the adjacent elements in the framed system. As specified in the codal provisions, the strength and the serviceability criteria such as strength, ductility, stability and deflection should be ensured. The biggest challenge in the precast concrete frame structure is the lack of structural integrity. Ensuring the effectiveness and integrity amongst the connection system in a precast structural member is the solution to this problem. To achieve the desired strength of a framed structure, usually the strong column-weak beam principle is adopted during the design. Various stresses such as lifting stress, erection stress and the handling stress should be considered, while designing the precast members.

S. Paramathmeka (✉) · M. Rajeswari
Department of Civil Engineering, PSG College of Technology, Coimbatore, India
e-mail: aathmivino.1223@gmail.com

M. Rajeswari
e-mail: mra.civil@psgtech.ac.in

A. Sil et al. (eds.), *Recent Trends in Civil Engineering*, Lecture Notes in Civil Engineering 274, https://doi.org/10.1007/978-981-19-4055-2_21

Grouted sleeve coupler is one of the most frequently used connection techniques in precast building construction. Many researches are done on the behaviour and the failure mechanism of the grouted sleeve coupler connection.

Chen et al. [3] conducted a study to find out the mechanical behaviour of a steel half-grouted sleeve connections with deficiencies in the water/binder grout ratio. Steel half-grouted sleeve specimens were manufactured and tested ascertain the failure mode, ultimate bearing capacity and strain distribution. All the specimens were similar in the load–displacement response. Higher the water/binder ratios higher was the bond failure and the damage depth at the grouting end.

Liu et al. [10] studied a group of grouted sleeve splices that were tested under tension loading. The failure mode was found out. Using finite element study, the influence of the clearance between sleeve and rebar, as well as the anchorage diameter ratio was investigated. From the results, an anchorage-to-diameter ratio greater than 7 is sufficient for preventing the slippage failure.

Ling et al. [11] conducted tests on the grouted splices connected on two types of sleeves, namely welded bar sleeve and tapered head sleeve. The sleeve was subjected to tension with various bar embedded lengths and sleeve diameter. Lower the diameter of the sleeve higher is the degree of confinement generated in the sleeve.

Wang et al. [13] studied the behaviour of a grout-filled pipe sleeve with bolts for large-diameter deformed bars grouted with a novel ultra-high-performance concrete (UHPC). Using UHPC's high material properties the embedded length of the rebars were shortened, this led to a delay in crushing. From the results it is evident that the proposed pipe sleeve is feasible to connect large-diameter deformed bars with an embedded length of 5 dB.

Xu et al. [14] conducted experiments on the grouted sleeve couplers with four types of insufficient grouting configuration. The bond failure mode and bond stress-slip curves were analysed based on the defect level in insufficient grouting. With the increase of defect level for all types of grouting defects the specimens failed in a shifting failure mode from tensile fracture to pull out failure of the rebar.

Yuan et al. [16] studied the tensile properties of grouted sleeve connection. Bar diameter, sleeve dimensions and rebar offset were considered as the main parameters of the study. The specimens exhibit three categories of failure, namely rebar tension fracture, rebar pull out due to bond failure and thread shear failure.

Zheng et al. [18] conducted experiments to ascertain the mechanical performance due to insufficiency in grouting the sleeve which had a predesigned vertical grouting defects by considering three loadings: uniaxial tensile loading, cyclic loading at high stress and cyclic loading at large strain. From the results, the repaired sleeve connections by refilling grout material are considered as well.

The main objective of this study is to ascertain the behaviour of the half-grouted sleeve coupler, grouted with a non-shrink high-strength free-flow cementitious grout, when subjected to a uniaxial tension test. Usage of grouted sleeve coupler connections in a precast structure, improves the structural integrity of the structural member. Important structural component connections such as beam-column connection, column-footing connection and column-column connection can be connected using grouted sleeve couplers. Experimental and numerical studies are carried out to on the specimens. Half-grouted sleeve couplers are used in the study. A half-grouted sleeve, has one end to be threaded end and the other is used as a connecting end in which the rebar is embedded and the grouting agent is grouted. Four grouted sleeve specimens are tested experimentally, out of which, two specimens are used 20 mm diameter rebar on both sides of the grouted sleeve and the other two specimens are used 25 mm diameter rebar on both the side. Thus, these specimens are classified into two types (1) Type A-20 mm rebar and (2) Type B-25 mm rebar.

Based on the position of the rebar in the connecting end, the two major types are further classified as (1) Type A-20 mm concentric, (2) Type A-20 mm eccentric, (3) Type B-25 mm concentric and (4) Type B-25 mm eccentric. Totally four specimens are testes. In the concentric type specimens, the rebar in the connecting end is concentrically placed whereas in the eccentric specimens the rebar in the connecting end is placed eccentrically with a particular offset from the centre.

2 Experimental Study

2.1 *Materials*

In the experimental work, only three materials are used grout, rebar and grouted sleeve coupler. The details of the material are mention below.

2.1.1 Grout

Different types of grouting agents are available in the market. The type of grout used for the study is a high-strength non-shrink free-flow cementitious grout. The water-binder ratio used to mix the grout is 0.14. The grout is mixed using a grout mixer. Compression test is carried out on the grout cube as per IS 4031. The mixed grout is poured in a mortar testing cube of size 70 * 70 * 70 mm and is put under curing for 28 days. On the 28th-day, compression test is carried out on the specimen. As per the 28th-day result, the strength of the grout cube on the 28th day is 40 MPa.

2.1.2 Rebar

The rebars are the major element for connecting the members. 20 and 25 mm diameter rebars are used in the study. Rebars of grade Fe500 is used. The rebar is fixed in the threaded end by fastening it tightly into the thread, whereas in the connecting end the rebar is embedded into the grouted sleeve. Later this end is grouted with the grouting agent. The anchorage length of the rebar for 20 mm diameter rebar is maintained as 145 mm and for 25 mm rebar is 165 mm.

2.1.3 Grouted Sleeve Coupler

The grouted sleeve coupler used for the study is made of cast iron. Two different sized grouted sleeve couplers are used. One is used to connect 20 mm diameter rebar and the other is used to connect 25 mm diameter rebar. One side of the rebar has a threaded end and the other side has a connecting end in which the rebar is embedded. Totally, four different specimens, namely (1) Type A-20 mm concentric, (2) Type A-20 mm eccentric, (3) Type B-25 mm concentric and (4) Type B-25 mm eccentric are studied. The rebars are fixed in their position and grouting is done. The mixed grout is poured into the grouted sleeve and is fully compacted. After the compaction, the assemblage is let under curing for 28 days. After 28 days a uniaxial tension load is applied to the specimen. The details of the four different types of specimens are tabulated in Table 1. Figure 1 represents a grouted sleeve coupler which has a threaded end in one end and a connecting end at the other end, before the assemblage.

2.2 *Experimental Setup*

Each sleeve coupler is assembled completely and is tested under a tension loading. The test is conducted until the specimen failed. The specimens mentioned in Table 1 are loaded in tension by a 100 Ton universal testing machine. The two ends of the specimen are fixed in the clamps of the testing machine and the tension load is applied by moving the upper clamp upward. The ultimate load of the specimen under

Table 1 Details of the specimen

Specimen id	Dia of the rebar (mm)	Length of the sleeve (mm)	Anchorage length (mm)	Position of the rebar
D20-con	20	180	145	Concentric
D20-ECC	20	180	145	Eccentric
D25-con	25	205	165	Concentric
D25-ECC	25	205	165	Eccentric

Fig. 1 Grout sleeve couplers

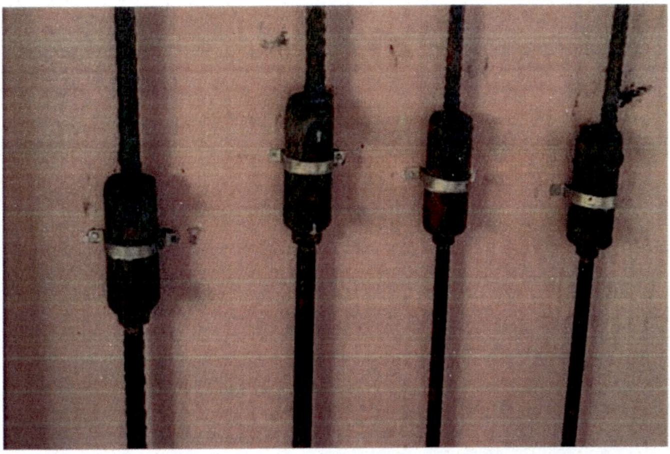

Fig. 2 Specimen before testing

uniaxial tensile loading is measured using the universal testing machine. The test is done in displacement control (Fig. 2).

2.3 *Test Set Up*

All four specimens observed a similar type of failure. The mode of failure is bond failure between the rebar and the grout within the sleeve. The tensile capacity of the rebar and the bond strength between the rebar and the grout are the major factors that contribute to the tensile capacity of the grouted sleeve connection. Even though in two of the specimens the rebar is placed eccentrically, those specimens are also subjected to the same failure mode. This shows that the position of the rebar in the connecting end does not greatly affect the behaviour of the grouted sleeve connection. The anchorage length of the rebar is one of the most important parameters that determines the bond strength between the rebar and the grout. The number of ribs on the rebar and the contact area of the rebar with the grout determines the bond strength. The ultimate load of failure for all the specimens is found out. The results are tabulated in Table 2. From the results, it is evident that the load carried by eccentrically placed rebar in the grouted sleeve coupler is less than that of the concentrically placed. This is because there is a slight deviation in the path the load is distributed. The stress distribution for the two different types of specimens varies tremendously (Figs. 3 and 4).

Table 2 Results of the numerical study

Equation number	Predicted peak load kN	*R*-value	Type of correlation
1	87.68	0.29	Positive weak
2	167.88	0.533	Positive moderate
3	97.645	0.4	Positive moderate
4	143.54	0.517	Positive moderate
5	127.59	0.87	Positive strong
6	132.05	0.91	Positive strong
7	135.77	0.93	Positive strong

Fig. 3 The specimen under tension loading

3 Numerical Study

A numerical study is also carried out on the test specimens using the statistical software MINITAB 19. MINITAB 19 software collects, interprets, and analyse data. The study incorporates the concepts of regression analysis to validate the numerical results with the experimental results. Amongst the following types of regression

Fig. 4 Bond failure in the specimens

analysis, namely linear regression, exponential regression, curvilinear regression, logistic regression, polynomial regression and rigid regression, the concepts of linear regression analysis are suitable for the study. Due to its simplicity in understanding the concepts and interpretation and accuracy in the results, this is more appropriate. In the exponential regression, the dependent and the independent variables should be in the exponential form. In such case exponential regression can be performed. In the binary logistic regression, only variables with two distinct data can be used.

The data for the numerical study is collected from the above-mentioned journals. The data from all the journals are tabulated in the worksheet of the MINITAB 19 software. The data are to be arranged column-wise. The data is categorized as author's name, specimen ID, type of grouted sleeve, type of loading, type of grout, compressive strength of the grout, flexural strength of the grout, water/binder ratio, a material used for the grouted sleeve length of the grouted sleeve, the inner diameter of the grouted sleeve, the outer diameter of the grouted sleeve, the thickness of the sleeve, anchorage length, diameter of the rebar, grade of the rebar and peak load—totally 18 columns. About 235 specimen details are collected from the journals. Two types of linear regression are available

i. Simple linear regression
ii. Multiple linear regression.

Simple linear regression: In this type of regression analysis, the relationship between the two variables is modelled by fitting a linear equation to the data. Amongst the two variables, one variable is considered to be an independent variable, and the other is considered to be a dependent variable. The simple regression equation is written as $\mathbf{Y = a + bX}$. Y is the value of the dependent variable which is to be predicted. 'a' or Alpha, a constant equals the value of Y when the value of $X = 0$. 'b' or Beta, the coefficient of X; the slope of the regression line.

Multiple linear regression: In this type of regression analysis, the relationship is modelled between more than two variables by fitting a linear equation to the data. Amongst the list of variables, one variable is considered to be an independent variable, and the other is considered to be a dependent variable. The multiple regression equation is written as **Y = a + bX1 + cX2 + Dx3**. Where *Y* is the dependent variable, *X*1, *X*2, *X*3 are independent variables, '*a*' is the intercept, '*b, c, d*' are slopes.

"Coefficient of Correlation" is considered as the most important parameter in interpreting the accuracy in the results. It is denoted as '*R*'. The correlation between the dependent and the independent variable is calculated based on the *R*-value. The *R*-value always ranges from − 1 to + 1. When the value of '*R*' falls more than 1 or less than − 1, This interprets there is an error in the correlation measurement. When the value of *R* is greater than 0.7, then the correlation between the independent and the dependent variable is considered to be very strong. When the value of *R* ranges from 0.4 to 0.7, then it indicates, that a moderate correlation exists between the variables. When the *R*-value is less than 0.4, it shows a very poor correlation between the variables. The slope of the regression line directly determines the classification of the regression type. The type of regression can be either a positive or a negative correlation. When the regression line slopes from the bottom left to the top right side in the plot, it is a positive correlation. Whereas if, the regression line slopes from top left to bottom right in the plot, it is a negative correlation (Fig. 5).

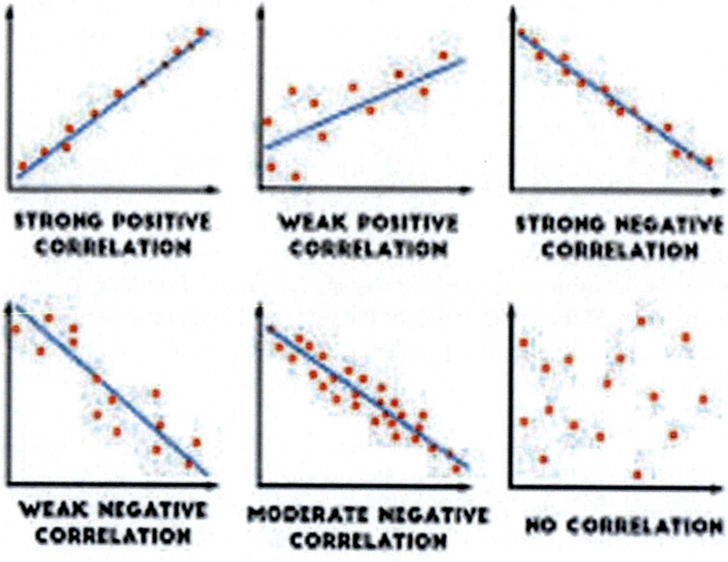

Fig. 5 Coefficient of correlation

3.1 Results of the Regression Analysis—Simple Linear

In this study, the simple linear regression analysis is carried out by considering various variables as an independent variable. In all the analyses, peak load is considered as the dependent variable. The specimen D-20-con is taken for verifying the results. The experimental value of the peak load for the D-20-con specimen is determined to be 122.01 kN. The results of the regression analysis are nearer to that of the experimental results in most of the cases.

3.1.1 Peak Load Versus Compression Strength of the Grout

This is a simple linear regression analysis. Peak load is considered as the dependent variable and compression strength of the grout is considered as the independent variable. The regression equation is

$$\text{peak load} = 11{,}847 + 1896\,\text{compression strength} \tag{1}$$

The value obtained for the peak load from this equation is 87.68 kN. The confidence interval is maintained as 95%. The *R*-value is 0.29. This interprets that the correlation between the compression strength of the grout and peak load is very poor. Similarly, the predicted value is also far away from that of the experimental value. The large variation is contributed due to the data available in the journal. The type of grout material used is high-strength non-shrink free flow cementitious grout. In the journal, various other grout materials are used with different water/binder ratios. This wide range of data affects the numerical result. A scatter plot is plot Fig. 6 by having compression strength in the '*x*' axis and peak load in the '*y*' axis. A fitted line plot Fig. 11 is a plot by having compression strength in the '*x*' axis and peak load in the '*y*' axis.

3.1.2 Peak Load Versus Inner Diameter of the Grout Sleeve

This is a simple linear regression analysis. Peak load is considered as the dependent variable and the inner diameter of the grout sleeve is considered as the independent variable. The regression equation is

$$\text{peak load} = 11,685 + 4620\,\text{inner dia} \tag{2}$$

The value obtained for peak load from this equation is 167.88 kN. The confidence interval is maintained as 95%. The *R*-value is 0.533. This interprets that the correlation between the inner diameter of the grout sleeve and peak load is moderate. The difference between the predicted and the experimental value is nearly 45 kN. Since this has only moderate correlation only this much accuracy can be expected from

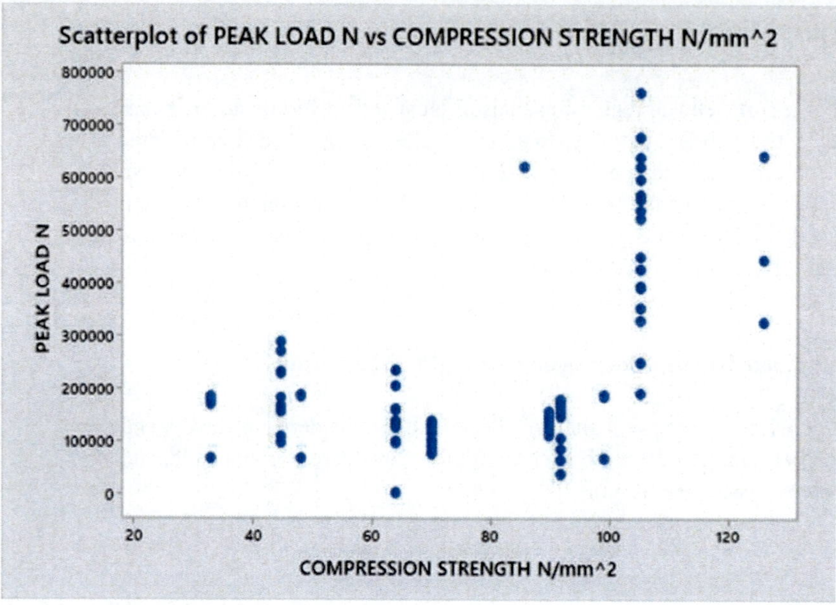

Fig. 6 Scatter plot of peak load versus compression strength

this equation. A scatter plot Fig. 7 is plot by having the inner diameter of the grout sleeve on the '*x*' axis and peak load on the '*y*' axis. A fitted line plot Fig. 12 is a plot by having an inner diameter of the grout sleeve on the '*x*' axis and peak load on the '*y*' axis.

3.1.3 Peak Load Versus Length of the Sleeve

This is a simple linear regression analysis. Peak load is considered as the dependent variable and length of the grout sleeve is considered as the independent variable. The regression equation is

$$\text{peak load} = -26{,}823 + 857.9\ \text{length of sleeve} \tag{3}$$

The value obtained for the peak load from this equation is 97.645 kN. The confidence interval is maintained as 95%. The *R*-value is 0.4. This interprets there exists a moderate correlation between the length of the sleeve and the peak load. The type of sleeves mentioned in the journals is not confined. The variations are high. Nearly 20 different dimensions are available for the length of the sleeve. This affects the predicted value. Because as the values are confined more is the accuracy. A scatter plot Fig. 8 is plot by having the length of the sleeve on the '*x*' axis and peak load on

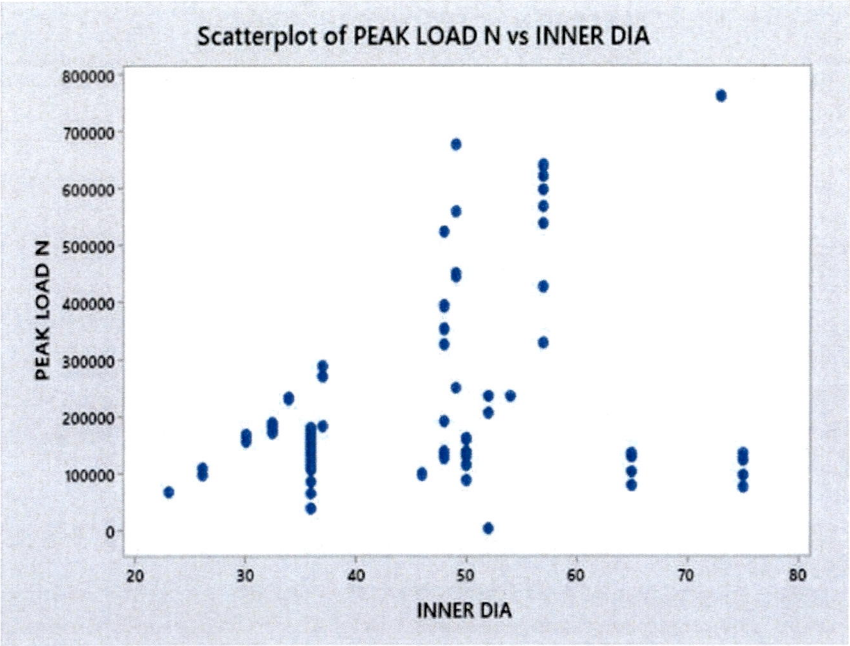

Fig. 7 Scatter plot of peak load versus inner diameter of the sleeve

the '*y*' axis. A fitted line plot Fig. 13 is a plot by having the length of the sleeve on the '*x*' axis and peak load on the '*y*' axis.

3.1.4 Peak Load Versus Anchorage Length

This is a simple linear regression analysis. Peak load is considered as the dependent variable and the anchorage length of the rebar in the grout sleeve is considered as the independent variable. The anchorage length is the length of the rebar that is embedded into the grout sleeve. The regression equation is

$$\text{peak load} = -199,545 + 2573\,\text{anchorage length} \tag{4}$$

The value obtained for the peak load from this equation is 143.54 kN. The confidence interval is maintained as 95%. The *R*-value is 0.517. This interprets there exists a moderate correlation between the anchorage length of the rebar in the grout sleeve and the peak load. The type of sleeves mentioned in the journals is not confined. The variations are quite acceptable. From the data collected and tabulated from the journal's anchorage length, the anchorage length of the rebar in the grout sleeve is fixed as 7.25 diameter of rebar. Because as the values are confined more is the accuracy. A scatter plot is plot Fig. 9 by having the anchorage length of the rebar in the

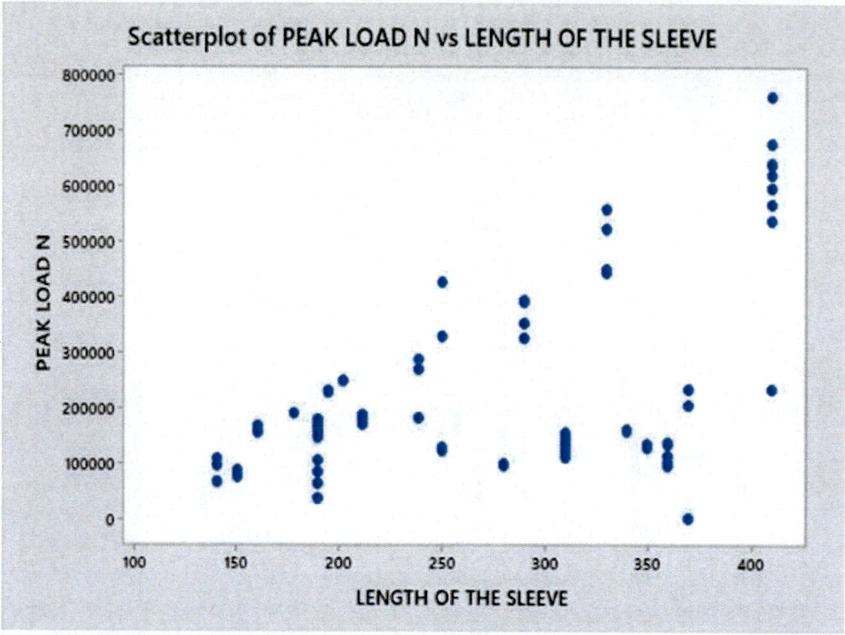

Fig. 8 Scatter plot of peak load versus length of the sleeve

grout sleeve on the 'x' axis and peak load on the 'y' axis. A fitted line plot Fig. 14 is a plot by having anchorage length of the rebar in grout sleeve in 'x' axis and peak load in 'y' axis.

3.1.5 Peak Load Versus Diameter of the Rebar

This is a simple linear regression analysis. Peak load is considered as the dependent variable and the diameter of the rebar in the grout sleeve is considered as the independent variable. The diameter of the bar is 20 mm. diameter of the rebar is the most important parameter in determining the peak load. The regression equation is

$$\text{peak load} = -209{,}180 + 18828 \text{ dia of the rebar} \tag{5}$$

The value obtained for the peak load from this equation is 127.59 kN. The confidence interval is maintained as 95%. The R-value is 0.87. This interprets there exists a strong correlation between the diameter of the rebar in the grout sleeve and the peak load. Amongst all the analyses this equation gives the value very closer to that of the experimental value. This equation is more accurate because from the data collected most of the specimens have the same 20 mm diameter rebar in the study. Since the values are nearly the same that of the values in the data more is the accuracy. A

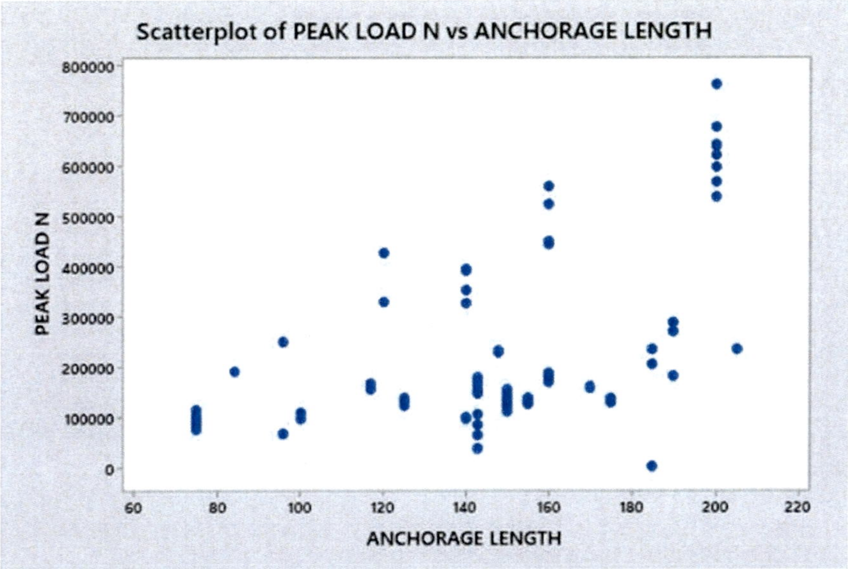

Fig. 9 Scatter plot versus anchorage length of the rebar

scatter plot is plot Fig. 10 by having the diameter of the rebar in grout sleeve in the 'x' axis and peak load in the 'y' axis. A fitted line plot Fig. 15 is a plot by having a diameter of the rebar in grout sleeve in 'x' axis and peak load in 'y' axis.

In these scatter plots, the data are plotted with high accuracy. In Fig. 6, the compression strength of the grout in the collected data ranges from 30 to 125 N/mm^2. The graph clearly shows that most of the value lies nearer to 100 N/mm^2. The data lies under a very wide range, which is the reason for less accuracy in the final value. In Fig. 7, the inner diameter of the sleeve ranges from 20 to 75 mm. The thickness of most of the sleeves are 4, 6 and 8 mm. From the graph, it is known that most of the inner diameter falls nearer to 35–40 mm and nearer to 60 mm. In Fig. 8, the length of the sleeve ranges from 135 to 400 mm. in the plot, the dispersion in the data is clearly shown. The maximum number of data points lies near 400 mm. In Fig. 9, the anchorage length of the rebar from the connecting end of the sleeve ranges from 70 to 210 mm. The anchorage length of most of the specimens is 6 diameter of the rebar, 8 diameter of the rebar. In Fig. 10, the diameter of the rebar ranges from 12 to 40 mm. Most of the specimens fall in 20 mm dia. The diameter used the also 20 mm. This is one of the reasons for the accuracy of the equation.

All these graphs are the same as that of the scatter plot, which has a regression line fitted in it. All the plots have a positive correlation. This is clearly shown as the line slopes from bottom left to top right. In Fig. 11, the data points are very far away from that of the regression line, this shows a very weak correlation. Hence, the correlation between peak load and compression strength of the grout is a positive weak correlation. In Fig. 12, few data points lie on the regression line and a few

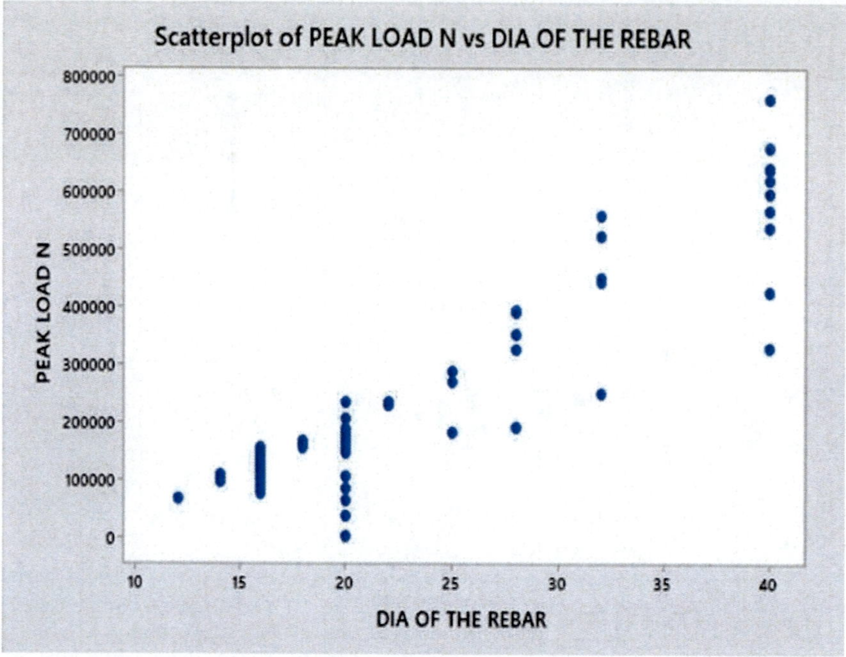

Fig. 10 Scatter plot peak load versus diameter of the rebar

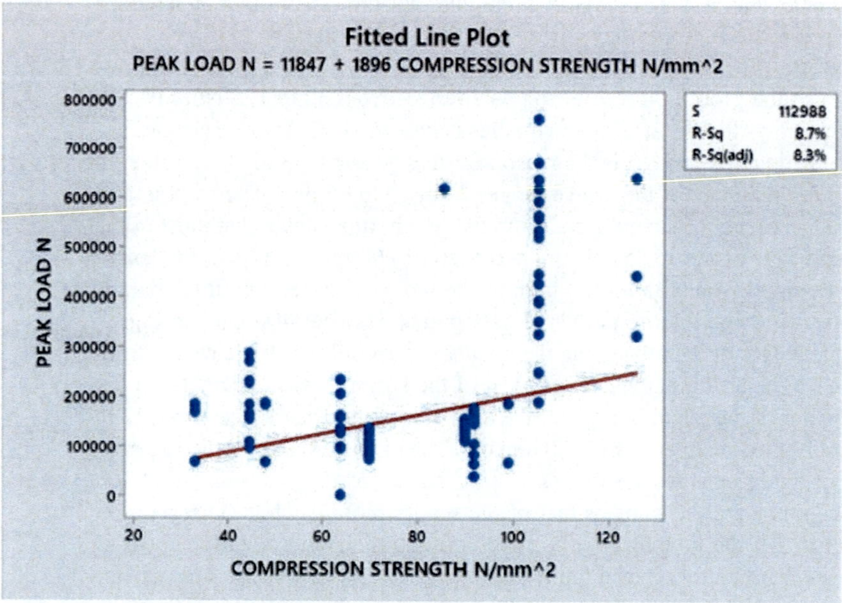

Fig. 11 Fitted line plot for peak load versus compression strength of the grout

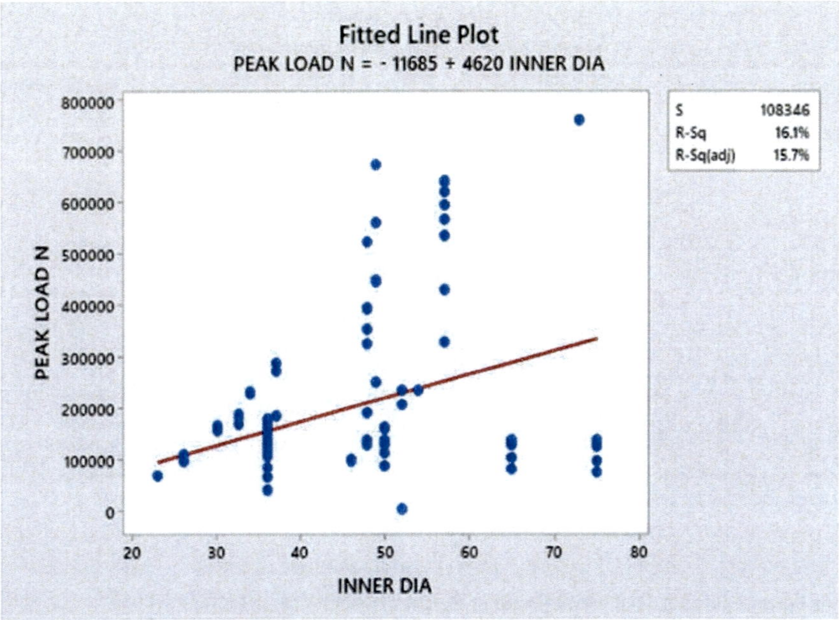

Fig. 12 Fitted line plot for peak load versus inner diameter of the grout sleeve

nearer to the regression line and a few away from the line. Hence this shows a positive moderate correlation. In Fig. 13, the point is similar to that of the points in Fig. 12. Hence, this also falls under a positive moderate correlation. In Fig. 14, the points are dispersed both above and below the regression line. This represents a positive moderate correlation. In Fig. 15 most of the points lie on and near the regression line. This indicates that a strong correlation exists between the peak load and diameter of the rebar. This graph shows a positive strong correlation.

3.2 Results of Regression Analysis—Multiple Regression

As known, in multiple linear regression analysis, many variables together act as an independent variable. In this analysis, the independent variables are compression strength of the grout, the inner diameter of the sleeve, length of the sleeve, the thickness of the sleeve, anchorage length of the rebar in the connecting end, and the diameter of the rebar. The dependent variable is as same that of the simple linear regression. The dependent variable is the peak load.

Two multiple regressions are carried out, first is simply undergoing the regression analysis without considering the categorical response. The second regression analysis is done by considering the categorical response. A categorical predictor is an alphanumerical value that can be used in MINITAB software for analysis. The

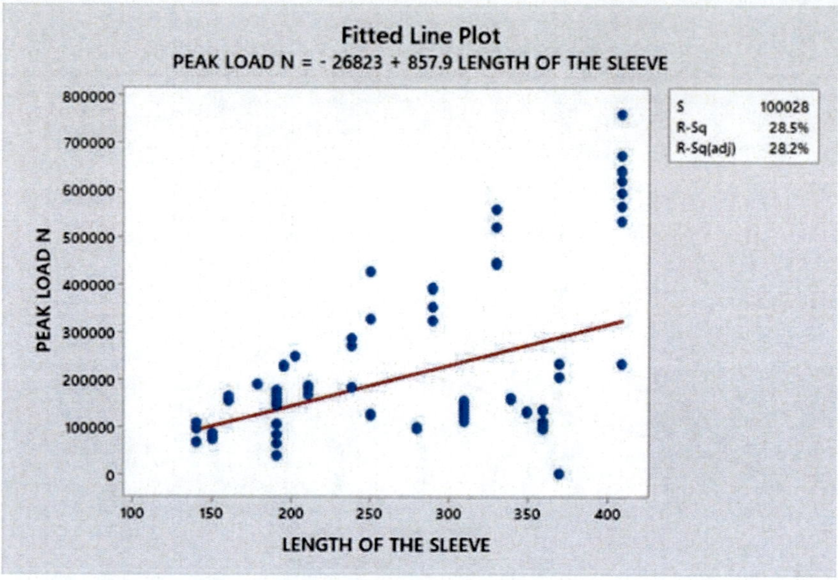

Fig. 13 Fitted line plot for peak load versus length of the sleeve

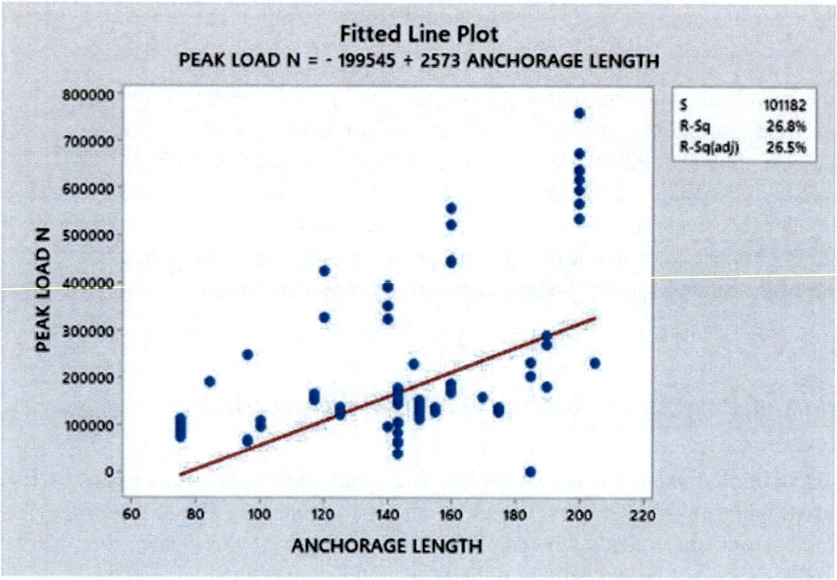

Fig. 14 Fitted line plot for peak load versus anchorage length of the rebar

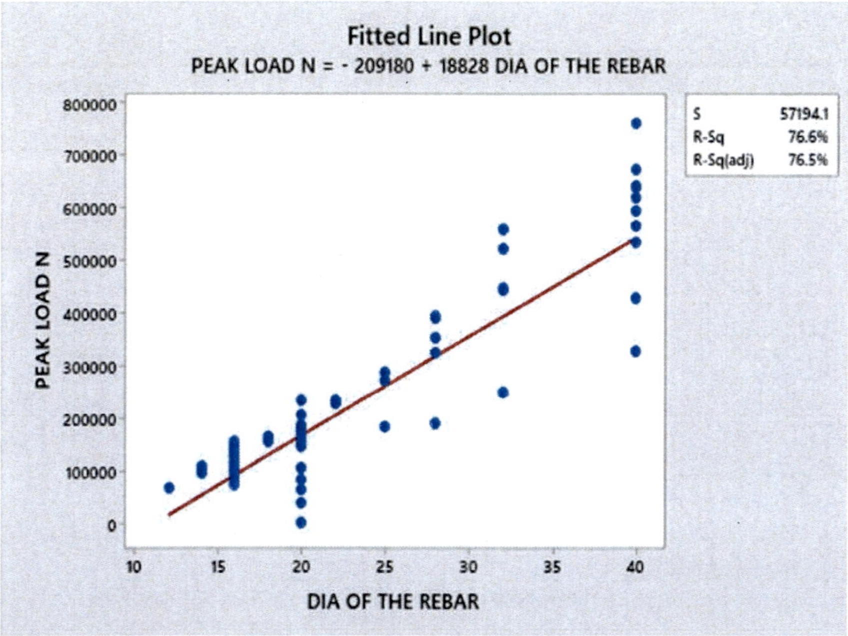

Fig. 15 Fitted line plot for peak load versus diameter of the rebar

categorical predictor used in the study is the mode of failure. From the data collected, there are only two types of failure modes in the journals, bond failure, and bar fracture. The mode of failure in the experimental study is a bond failure between the rebar and the grout within the sleeve.

3.2.1 Multiple Linear Regression Without Using the Categorical Response

In this multiple linear regression without using the categorical response peak load versus compression strength of the grout, inner diameter of the sleeve, length of the sleeve, the thickness of the sleeve, anchorage length of the rebar in the connecting end and the diameter of the rebar is predicted. The linear equation for this analysis is

$$\begin{aligned} \text{peak load} = & -219{,}069 - 346\ \text{compression strength n/mm}^2 \\ & + 508.2\ \text{length of the sleeve} - 1347\ \text{inner dia} \\ & + 7272\ \text{thickness of the sleeve} \\ & - 316\ \text{anchorage length} + 18{,}134\ \text{dia of the rebar} \end{aligned} \quad (6)$$

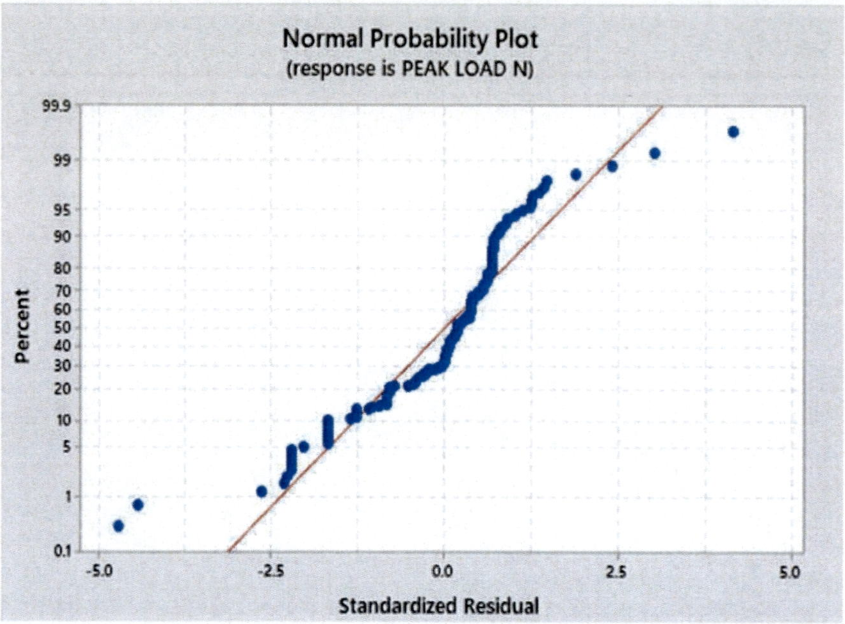

Fig. 16 Normal probability plot for the multiple linear regression without considering the categorical predictor

The value obtained from this equation is 132.05 kN. The confidence interval used in the analysis is 95%. The *R*-value is 0.91. This interprets that the correlation between these independent variables and peak load is very strong. The difference between the predicted and the experimental value is less. This shows that this equation can be used in predicting the values. A fitted line plot cannot be plotted for a multiple regression analysis hence a normal probability plot Fig. 16 is the plot with peak load as the main parameter. It has the percentage in its '*y*' axis and residual in the '*x*' axis. In the normal probability plot, the data is inferred as a set of good data if the data together form a straight line.

3.2.2 Multiple Linear Regression Using the Categorical Response

In this, multiple regression is performed by considering a categorical predictor. Peak load versus compression strength, length of the sleeve, inner dia, the thickness of the sleeve, anchorage length, dia of the rebar with failure mode as a categorical predictor is considered in this analysis. The mode of failure of the grouted sleeve couplers is chosen as the categorical predictor. The mode of failure is classified into two types bond failure and bar fracture. Two equations are obtained for these two categories. Since the mode of failure that occurred in the above experimental study is the bond failure between the grout and the rebar within the sleeve, only the equation

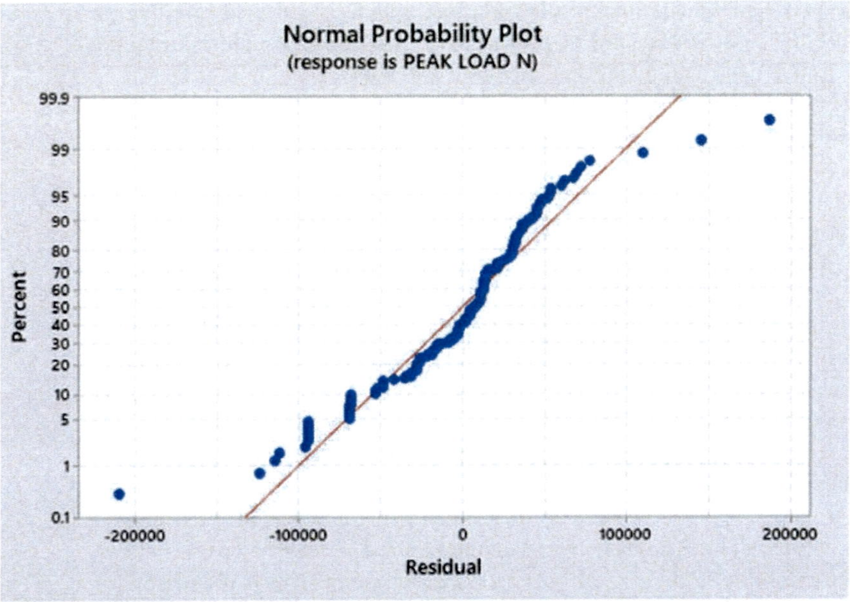

Fig. 17 Normal probability plot for the multiple linear regression considering the categorical predictor

corresponding to the bond fracture is considered for the comparison of the results. The regression equation is

$$\begin{aligned}\text{peak load} = & -219069 - 346\,\text{compression strength} \\ & + 508.2\,\text{length of the sleeve} - 1347\,\text{inner dia} \\ & + 7272\,\text{thickness of the sleeve} \\ & - 316\,\text{anchorage length} + 18{,}134\,\text{dia of the rebar}\end{aligned} \tag{7}$$

The value obtained from this equation is 135.771 kN. The confidence interval is maintained as 95%. The R-value for this equation is 0.93. This R-value indicates there exists a strong relationship between the independent variables and the dependent variable. A normal probability plot Fig. 17 was plot by considering the peak load, its percentage on the 'y' axis, and its residual on the 'x' axis.

4 Conclusion

- The experimental and numerical study on the behaviour of grouted sleeve couplers is studied in this paper.

- In the experimental study, the grouted sleeve coupler is subjected to a tension load until the specimen reached its failure point. The mode of failure for all the specimens is a bond failure between the grout and the rebar within the sleeve.
- The maximum load carried by the eccentrically placed specimens is lesser than that of the concentrically placed specimens. The major reason for this is the dispersion of the load which does not follow the same path of the load transfer mechanism.
- In the numerical study, linear regression analysis is carried out. Both simple and multiple linear regression analyses are carried out.
- In the equations obtained, only the equations which have a very strong correlation between the variables predicted the values that are nearer to the experimental value.
- Based on the R-value it is determined which equation is to be chosen for predicting the values.
- The peak load value obtained from the equations is not the same as that of the experimental value, because the data used to obtain the equations use a different type of grouted sleeve coupler such as full grouted sleeve coupler, threaded grouted sleeve coupler, welded grouted sleeve coupler, etc. The type of grouted sleeve coupler used in the study is a half-grouted sleeve coupler. This is the reason for the variation of the results.

Acknowledgements The authors wish to record their sincere and heartfelt thanks to Principal in Charge, Head of the Department, Former Head of the Department, Department of Civil Engineering, PSG College of Technology, Coimbatore.

References

1. Balineni H, Jagarapu D, Eluru A (2020) Analysis of dry and wet connections in the precast beam-column joint using ABAQUS software. Mater Today
2. Breccolotti M, Gentile S, Tommasini M (2016) Beam-column joints in continuous RC frames: comparison between cast-in-situ and precast solutions. Eng Struct 127:129–144
3. Cheng J, Wang Z, Liu Z, Ju S (2020) Experimental investigation of mechanical behaviour of rebar in steel half-grouted sleeve connection with defect in water/binder ratio. Structures 26:487–500
4. Deng M, Ma F, Song S (2020) Seismic performance of interior precast concrete beam-column connections with highly ductile fiber-reinforced concrete in the critical cast-in-place regions. Eng Struct 210:110360
5. Elliott SK (2002) Precast concrete structures, 1st edn.
6. Elsanadedy H, Almusallam T, Salloum Y, Abbas H (2017) Investigation of precast RC beam-column assemblies under column-loss scenario. Constr Build Mater 142:552–571
7. Guerrero H, Rodriguez V (2019) Experimental tests of precast reinforced concrete beam-column connections. Soil Dyn Earthq Eng 125:105743
8. Indian standard code IS 14858:2000. Compression testing machine used for testing of concrete and mortar—requirements

9. Indian standard code IS 1608:2005. Metallic materials—tensile testing at ambient temperature, 3rd rev
10. Liu C, Pan L, Liu H, Tong H, Yang Y, Chen W (2020) Experimental and numerical investigation on mechanical properties of grouted-sleeve splices. Constr Build Mater 260:120441
11. Ling J, Rahman A, Ibrahim I, Hamid Z (2016) Tensile capacity of grouted splice sleeves. Eng Struct 111:285–296
12. Parastesh H, Hajirasouliha I, Ramezani R (2014) A new ductile moment-resisting connection for precast concrete frames in seismic regions: an experimental investigation. Eng Struct 70:144–157
13. Wang Z, Zhu J, Wang J, Zhao G, Sun S, Zhang J (2021) Experimental study on a novel UHPC grout-filled pipe sleeve with mechanical interlocking for large-diameter deformed bars. Eng Struct 226:111358
14. Xu F, Wang K, Wang S, Li W, Liu W, Du D (2018) Experimental bond behaviour of deformed rebars in half-grouted sleeve connection with insufficient grouting defect. Constr Build Mater 185:264–274
15. Yan Q, Chen T, Xie Z (2018) Seismic experimental study on a precast concrete beam-column connection with grout sleeves. Eng Struct 155:330–344
16. Yuan H, Zhenggeng Z, Naito C, Weijian Y (2017) Tensile behaviour of half grouted sleeve connections: experimental materials 152:96–104
17. Zhanga Z, Dingb R, Nieb X, Fan J (2020) Seismic performance of a novel interior precast concrete beam-column joint using ultra-high-performance concrete. Eng Struct 222:111145
18. Zheng G, Kung Z, Xiao J, Pan Z (2021) Mechanical performance for defective and repaired grouted sleeve connections under uniaxial and cyclic loading. Constr Build Mater 233:117233

Comparative Study of Carbon Fiber-Reinforced Polymer Cabled Structure Subjected to Seismic Loads

Vaibhav Patel and Hanumant Magarpatil

1 Introduction

A cable structure is a structure that uses a cable or a set of cables as the visible and main load bearing structural element. When looking back at the history of cable constructions, it is clear that advancements in cable materials have a significant impact on their growth. Because of the existence of high-strength steel cables, long span cable systems, such as suspension bridges, can be installed. It also allows for the development of modern cable systems with new materials and shapes, such as cable roofs and cable facades.

Carbon fiber-reinforced polymer (CFRP) is a material made of nonmetallic composites composed of fibers of carbon reinforced with polymer resin. CFRP includes greater strength, fatigue behavior, corrosion resistance, and also lightweight. Therefore, CFRP has a big future as a cable and as a replacement for steel cables in cable systems, which will help it develop in this time frame. A comparative study is done on a mid-rise building which is being proposed by *X*-bracing and carbon fiber-reinforced polymer cables for different seismic parameters. Severe earthquake damages the structure with major fatal and economical losses in earthquake prone areas. Therefore, response spectrum analysis and time history analysis are done on ordinary rigid frame structure, *X*-bracing, and the structure with CFRP cables under major earthquake loads.

V. Patel (✉) · H. Magarpatil
School of Civil Engineering, Dr. Vishwanath Karad MIT World Peace University, Pune, India
e-mail: vaibhavpatel221998@gmail.com

H. Magarpatil
e-mail: hanumant.magarpatil@mitwpu.edu.in

A. Sil et al. (eds.), *Recent Trends in Civil Engineering*, Lecture Notes in Civil Engineering 274, https://doi.org/10.1007/978-981-19-4055-2_22

2 Literature Review

CABKOMA strand rod was created by Komatsu Seiten Fabric Laboratory in Japan, a new thermoplastic carbon fiber composite. It is light weighted and has higher strength than steel rod. Its first application was designed by Kengo Kuma for use on the façade of three story Komatsu seiten's head office building. For the project, the rods were inserted from the roof top to the ground around the outer side of the building. It is currently used as exhibition, workspace, and research facility. And it showed that it demonstrably improved behavior of the building under lateral forces. This results in an airy, organic façade that has a fabric-like esthetic while still offering the strength to resist ground motion during an earthquake event.

Eshelman presented a study which shows that strand rods provide greater strength with less weight. And also high-performance fiber-reinforced cementitious composite was tested in order to compare it with the strand rods in which composite panels add durability and ductility to beam-to-column and beam-to-shear wall connections, in order to strengthen the structure. It was seen that both methods were ease to use, cost effective and have minimal interruption [1].

Mochida and Imoto developed a carbon fiber-reinforced thermoplastic strand rod. Test like cavity ratio test, fiber content, alkali resistance test, tensile strength test, temperature dependency test, weather resistance test, etc., was performed on strand rods. It shows that it has a very high strength and can be used in seismic retrofit of the entire building [2].

In this research, carbon fiber-reinforced polymer (CFRP) cable is used as a replacement of the CABKOMA strand rod to make the structure strong enough to resist ground motion using same method.

Liu et al. present an overview of the use of CFRP cables in cable constructions with a historical review, the state of the art, and perspectives for the future. Properties of carbon fibers, the mechanical properties, and the structural shapes of CFRP cables are discussed which are included in the paper [3].

3 Structure Modeling

The ETABS software was used to model and analyze G+11 OMRF RCC frames in order to do a response spectrum and time history analysis for the building's seismic performance. Frames are designed optimally as per guidelines of IS 456:2000 [4], IS 1893:2000 (Part 1) [5], and IS 875:1987. The building has a rectangular shaped plan with dimensions 8.6 m * 28 m with a floor height of 3 m situated in seismic zone V in India on the medium soil type. Three structural models are built to carry analysis on earthquake loads, (i) rigid frame, (ii) structure with steel *X*-bracing frame, (iii) structure with CFRP cables attached around the building.

Structure general information and preliminary design considerations are shown in Tables below.

Table 1 General parameters of the structure

S. No.	Contents	Description
1	Structure type	RCC (OMRF)
2	Number of story	G+11
3	Materials	Concrete (M35) Steel reinforcement (Fe500)
4	Specific weight of RCC	25 kN/m^3

Table 2 Dimension values of structure

Model	Rigid	*X*-bracing	CFRP cables
Beam (mm)	450 * 300	450 * 300	450 * 300
Column (mm)	300 * 450	300 * 450	300 * 450
Slab (mm)	150	150	150
Height of each floor (m)	3	3	3
Total height of structure (m)	36.2	36.2	36.2

All the three structures were modeled with the general parameters and dimension values of the structure shown in Tables 1 and 2.

3.1 Rigid Structure

Structure is modeled with a shear wall with the above configurations of the building as mentioned in Tables 1 and 2 (Figs. 1 and 2).

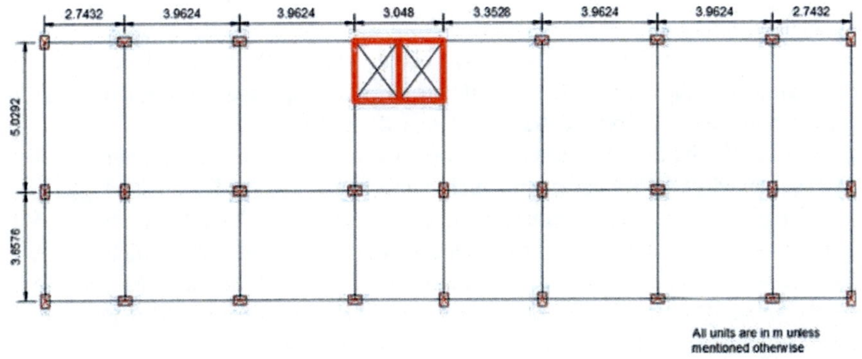

Fig. 1 Plan for rigid body and *X*-bracing

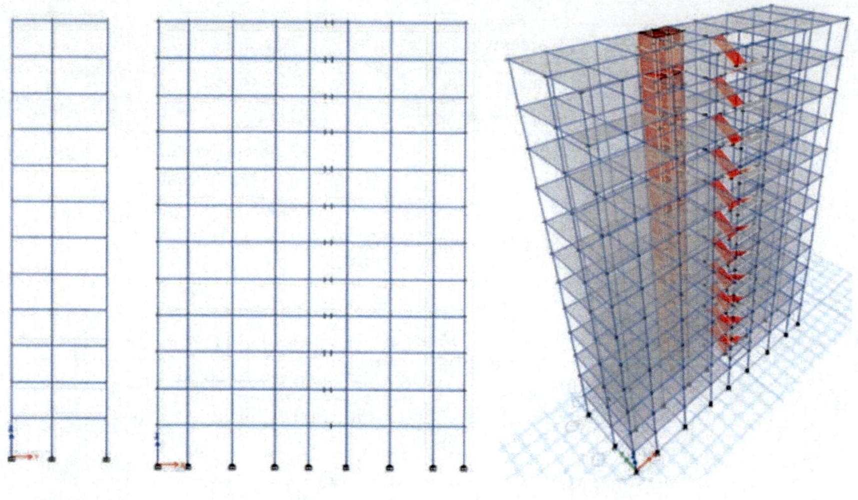
(a) (b) (c)

Fig. 2 **a**, **b** Elevation and **c** 3D model of rigid frame

3.2 *Braced Structure*

For braced structure, Indian standard medium weight beam is used for steel bracing in the structure which is ISMB 175 in order to build diagonal steel *X*-bracing system in *X*-direction of the building on both the sides (Fig. 3).

3.3 *CFRP Cabled Structure*

For CFRP cabled structure, cables were inserted from the roof top to the ground around the exterior of the building. Around 340 cables are attached around the building. The diameter of the cable is 40 mm, and around 4.5 m spacing is provided to pass a truck or vehicle easily into the building (Figs. 4 and 5).

As shown in Table 3, the density and elastic modulus of the steel rod are higher than the CFRP cable, while CFRP cable has much higher tensile strength than steel rods. Breaking length is an excellent parameter for demonstrating CFRP cable's high strength and lightweight. It is described as the greatest length of a hanging rod which can sustain its very own weight and may be computed using the formula $\sigma_u/(g)$, wherein g is the standard gravity value of 9.8 m/s^2. As shown in the table, CFRP cable breaking lengths are much better than steel rod breaking lengths.

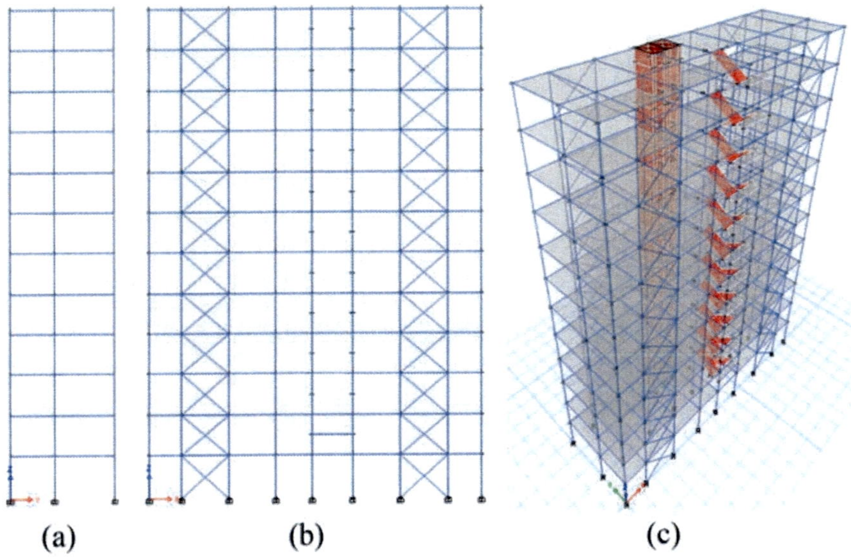

Fig. 3 **a**, **b** Elevation and **c** 3D model of braced frame

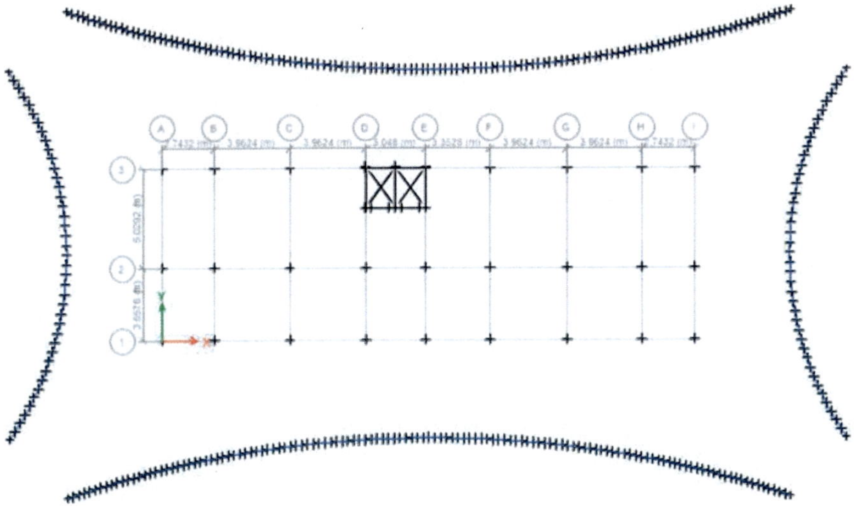

Fig. 4 Plan for CFRP cabled model

4 Structural Analysis

The analysis can be performed on the structures or structural material behavior and on the basis of external action. By using IS 1893-2002, earthquake loads and load combinations on the structure are considered for analysis.

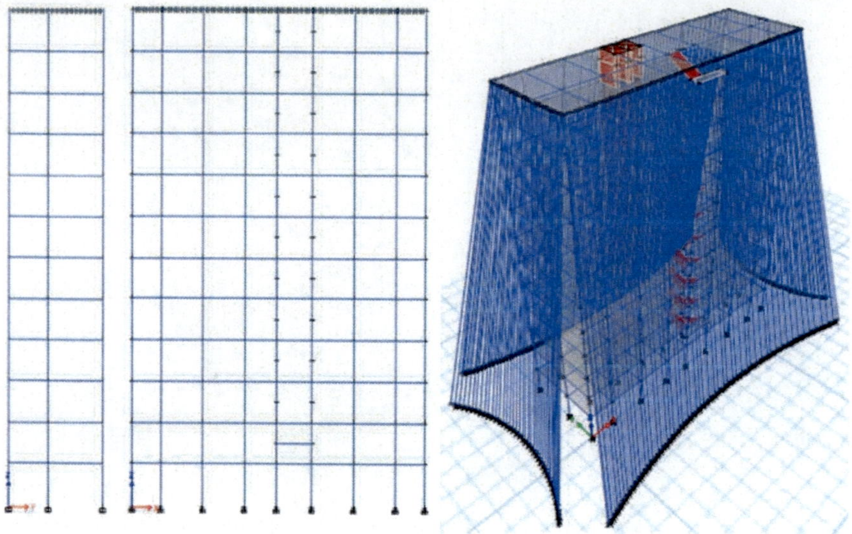

(a) (b) (c)

Fig. 5 **a**, **b** Elevation and **c** 3D model of CFRP cabled frame

Table 3 Comparing the properties of CFRP cable to steel cable

Material	Steel rod	CFRP cable
P—Density (kg/m^3)	7850	1500
σ_u—Tensile strength (Gpa)	1.5	2.1
E—Modulus of elasticity (Gpa)	160	137
$\sigma_u/(\rho g)$—Breaking length (km)	23	205

Three methods are used to see the performance of each structure.

a. Equivalent static method
b. Response spectrum method
c. Time history analysis.

Equivalent static method depends upon the fundamental period of structure defined by IS Code 1893:2002. Loads given in Table 4 are considered for carrying out the analysis.

Response spectrum analysis is a linear dynamic method used for determining a structure's maximum response when subjected to ground motion. It defines the base acceleration applied to each mode based on its period. It is carried out using IS:1893(Part 1)-2002 (Figs. 6, 7, 8; Table 5).

Time history analysis is a non-linear dynamic method under the loading which may differ according to specified time function. In this analysis, Chamoli earthquake 1999, Bishop round valley earthquake 1984, and N. Palm springs earthquake are taken into consideration.

Table 4 Loading parameters

S. No.	Contents	Description
1	Impose load	2–3 kN/m^2
2	Importance factor	1
3	Seismic zone	V
4	Zone factor	0.36
5	Response reduction factor	3

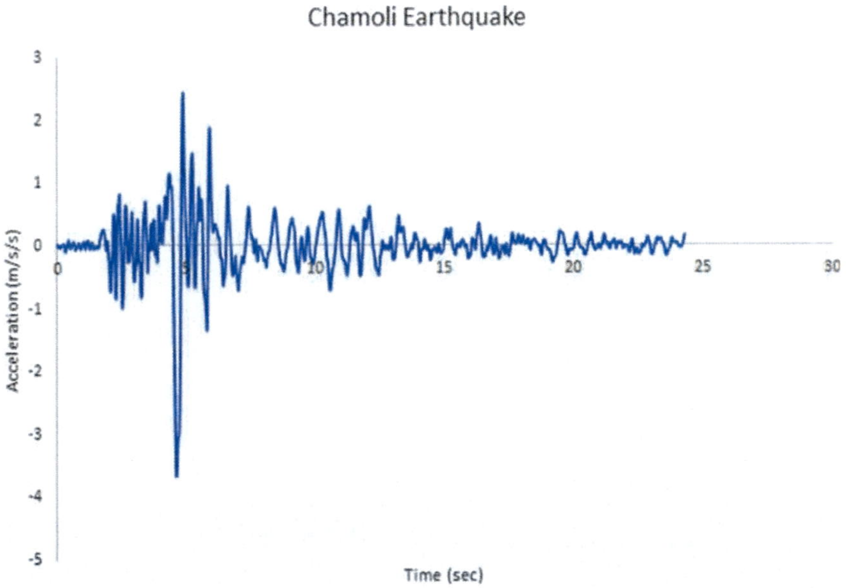

Fig. 6 Timeline of Chamoli earthquake

5 Results and Discussion

All the three models were analyzed under equivalent static method, response spectrum method, and time history analysis and are compared with story displacement in both *X*-direction and *Y*-direction.

5.1 Equivalent Static Method

As shown in Figs. 9 and 10, both braced and cabled structure show better results than normal rigid structure. Cabled structure perfomed 28.8% better than rigid structure and 11.45% better than braced structure in *Y*-direction. But in *X*-direction, braced

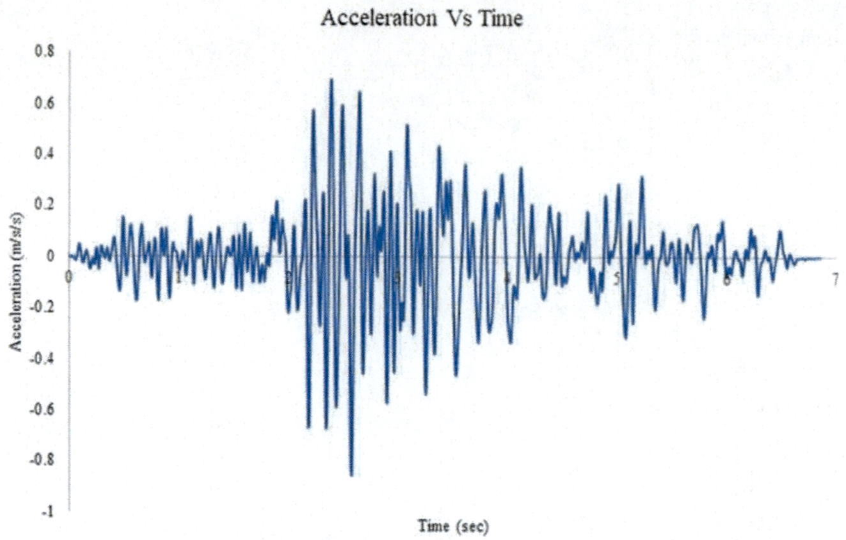

Fig. 7 Timeline of Bishop round valley earthquake

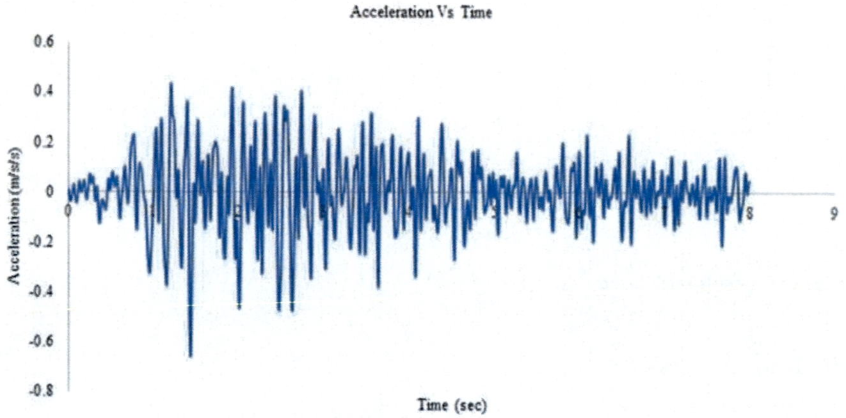

Fig. 8 Timeline of N. palm springs earthquake

structure performed 6.6% better than the cabled structure, whereas cabled structure shows 6.6% decrease in deflection than the rigid structure.

Table 5 Details of earthquake time histories considered in study

S. No.	Earthquake details	Recording station name	Epicentral distance (R) in kms
1	Chamoli earthquake, Uttarakhand, India; 29th March 1999; N20E component	IITR, Gopeshwar	17.3 (NF)
2	Bishop Round Valley; United States; 23rd November, 1984; NS component	McGee Creek surface	21.93 (FF)
3	N. Palm Springs; United States; 8th July, 1986; N162E component	Lake Matthews Dike Toe	79.23 (FF)

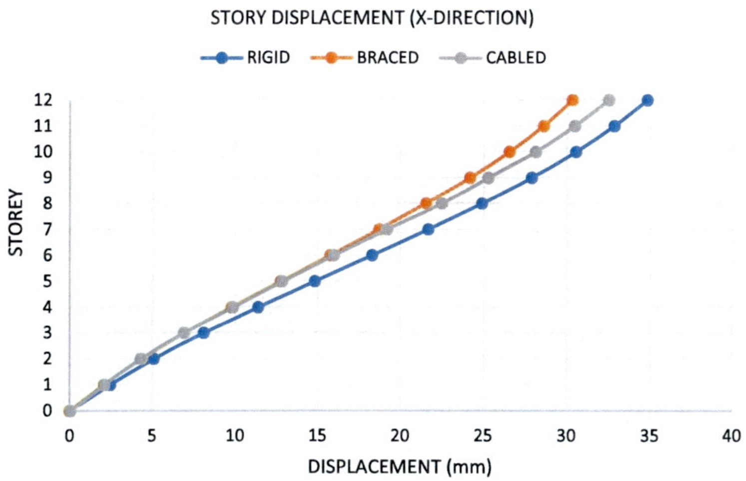

Fig. 9 Comparison of story displacement in X-direction for equivalent static method

5.2 *Response Spectrum Analysis*

Cabled structure performed pretty well in the response spectrum analysis in X-direction as well as Y-direction. As shown in Fig. 11, all the three structures show almost similar performance in X-direction but as compared to all the structure cabled structure shows less deflection than both the structures. Cabled structure performed 0.6% better than braced structure and 4.5% better than rigid structure. Whereas in Y-direction, braced structure showed poor performance than rigid and cabled structure as shown in Fig. 12. Around 24.4% less deflection is seen in cabled sturcture than the braced structure and 18% better performance of cabled structure with respest to rigid structure.

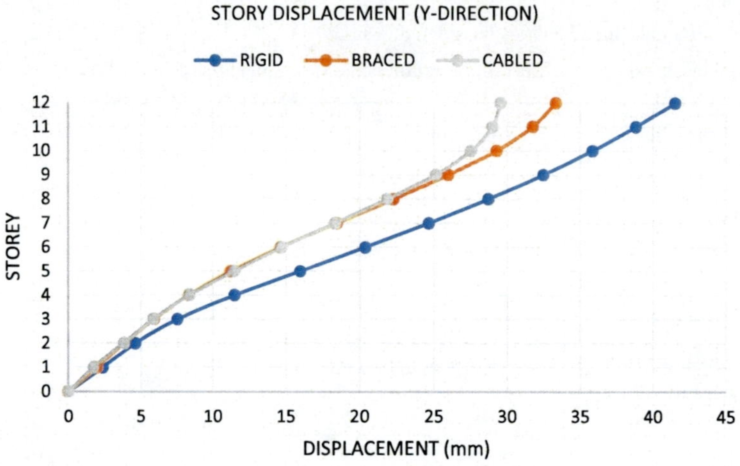

Fig. 10 Comparison of story displacement in *Y*-direction for equivalent static method

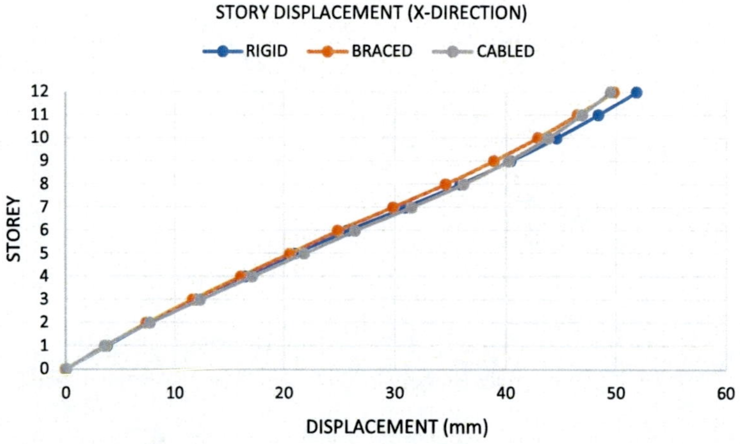

Fig. 11 Comparison of story displacement in *X*-direction for response spectrum method

5.3 *Time History Analysis*

For time history analysis, results of all the three earthquake records are compared together, and the earthquake with highest issue will be compared with all the three models.

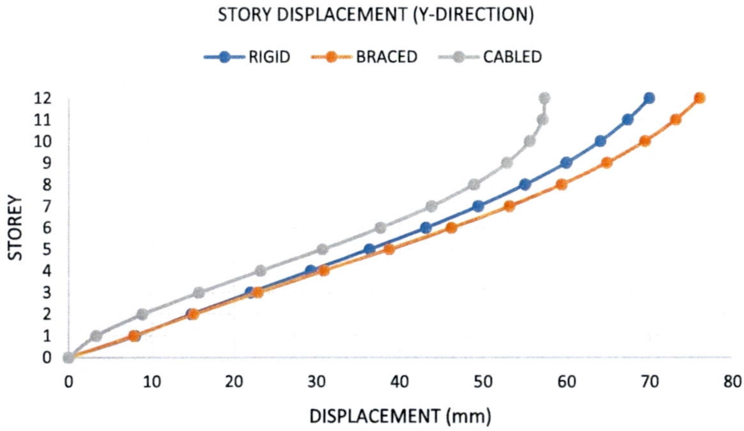

Fig. 12 Comparison of story displacement in *Y*-direction for response spectrum method

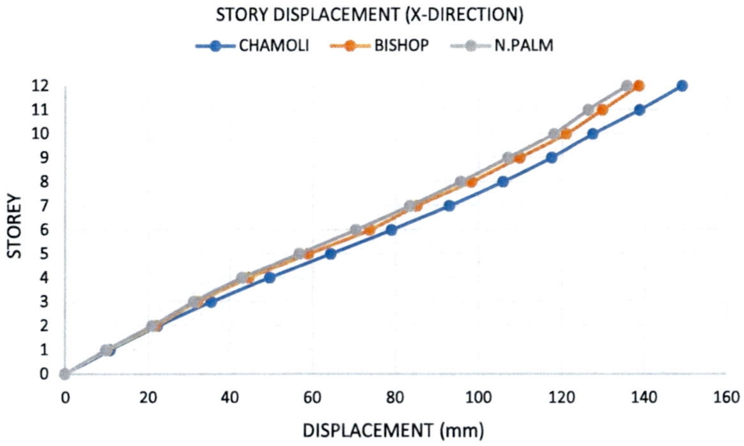

Fig. 13 Comparison of rigid frame in *X*-direction

5.3.1 Rigid Model

It has been seen from Fig. 13 as well as Fig. 14 that while comparing all the three time history records in rigid frame, Chamoli earthquake shows the highest diplacement in *X*-direction as well as *Y*-direction.

5.3.2 Braced Model

In the braced frame, Chamoli earthquake shows the highest displacement in both *X*-direction as well as *Y*-direction as shown in Figs. 15 and 16.

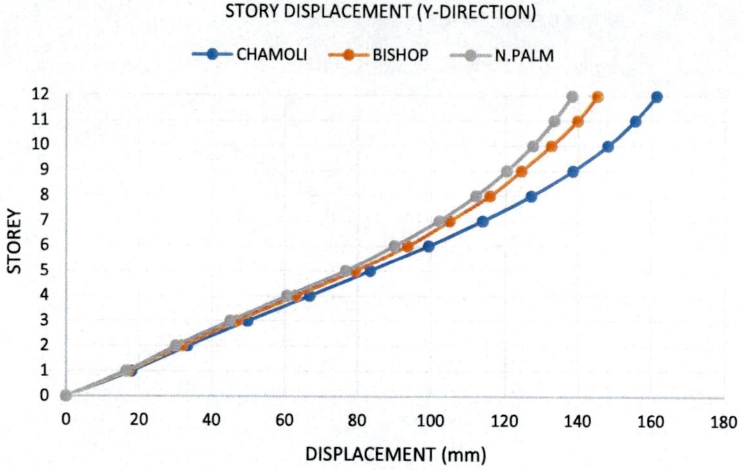

Fig. 14 Comparison of rigid frame in *Y*-direction

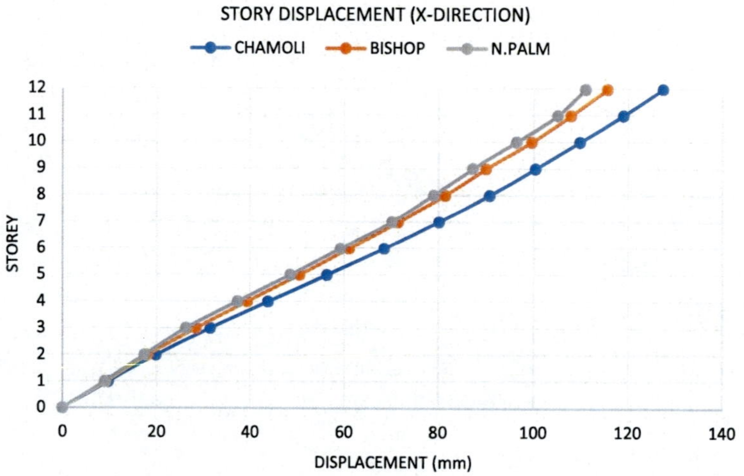

Fig. 15 Comparison of braced frame in *X*-direction

5.3.3 Cabled Model

It has been seen from Fig. 17 as well as Fig. 18 that while comparing all the three time history records in cabled frame, Chamoli earthquake shows the highest diplacement in *X*-direction as well as *Y*-direction.

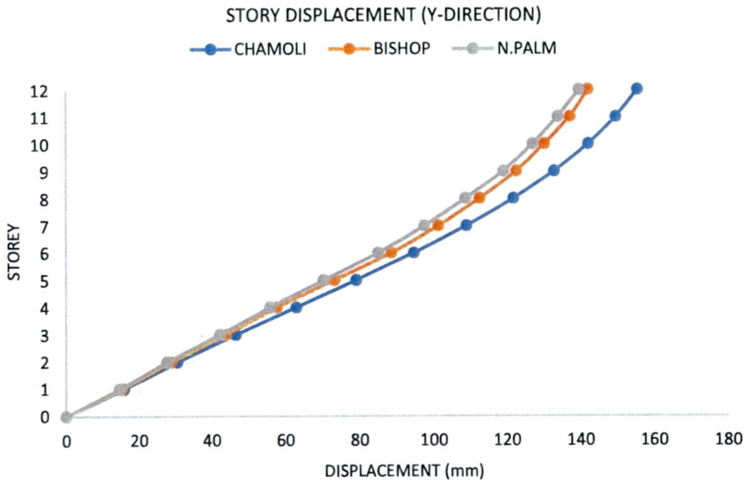

Fig. 16 Comparison of braced frame in *Y*-direction

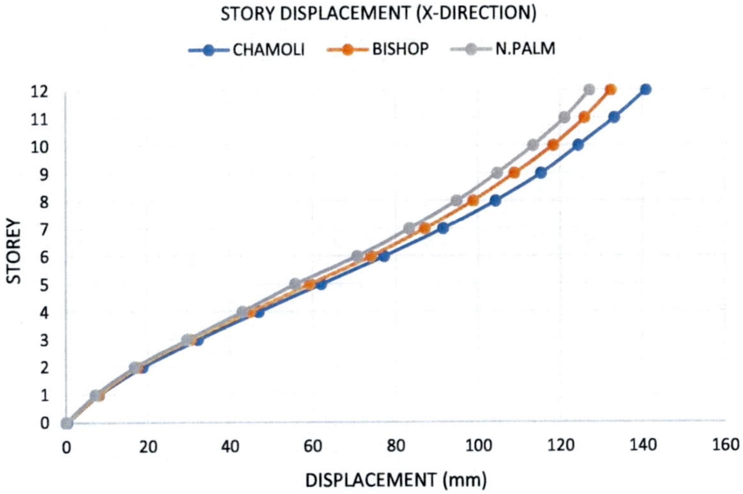

Fig. 17 Comparison of cabled frame in *X*-direction

5.3.4 Story Displacement Compared with All Models

As Chamoli earthquake got highest displacement in all the models, it is considered to compare the maximum displacement with rigid structure, braced structure, and cabled frame strcuture.

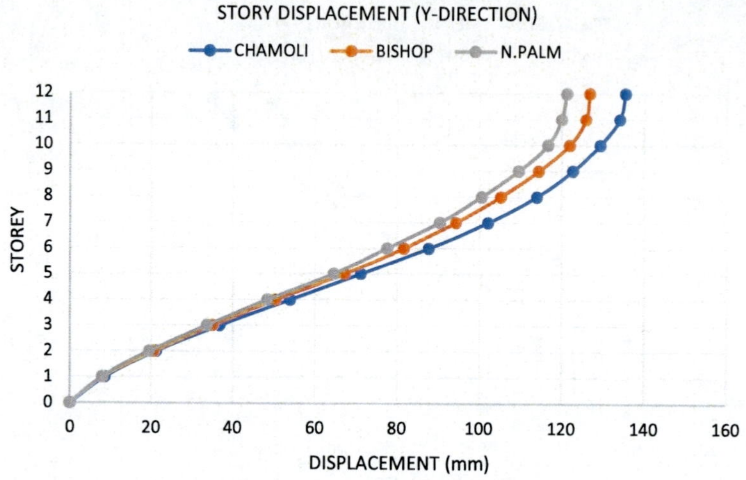

Fig. 18 Comparison of cabled frame in *Y*-direction

Cabled structure performed very well with good results in time history analysis. As shown in Fig. 19, cabled structure performed better than rigid structure in *X*-direction which is 5.49%. While braced structure shows 14% less displacement than rigid structure. Whereas in *Y*-direction, cabled structure shows 16% less deflection than rigid structure. And 12% better performance than braced structure as shown in Fig. 20.

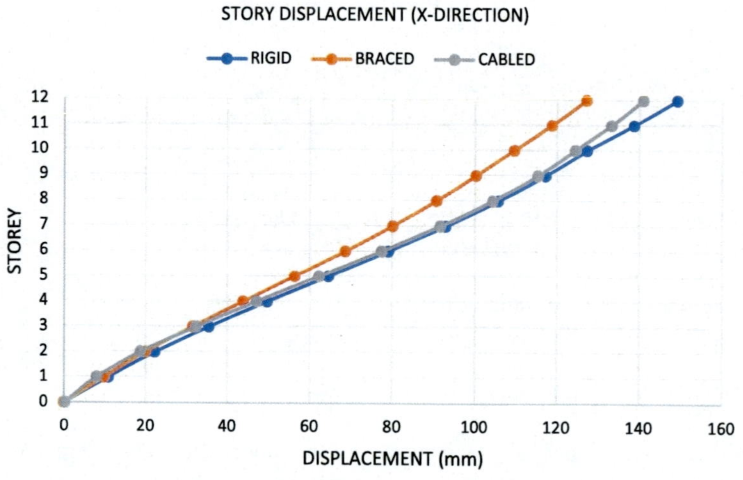

Fig. 19 Comparison of story displacement in *X*-direction for time history analysis

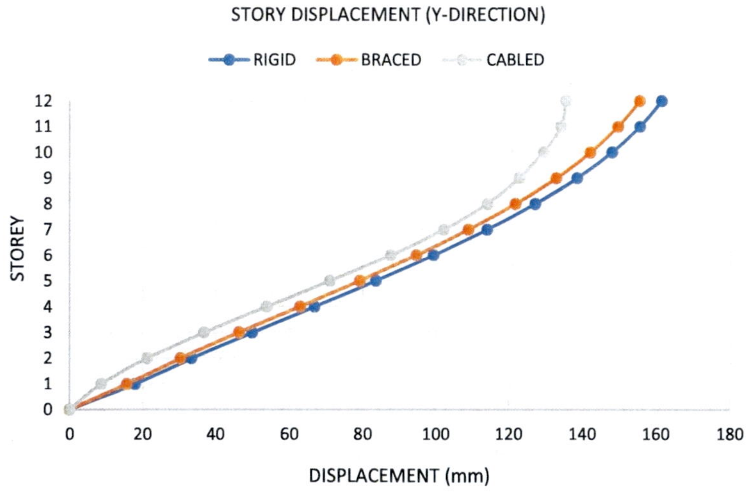

Fig. 20 Comparison of story displacement in *Y*-direction for time history analysis

5.4 *Time Period*

For comparison of time period of the structure, first three modes are taken into consideration. In which braced structure and cabled structure showed almost similar results with a difference of 0.014 s as shown in Fig. 21. In other words, braced structure performed 1.35% better than cabled structure. Also cabled structure performed 5.5% better compared to rigid structure.

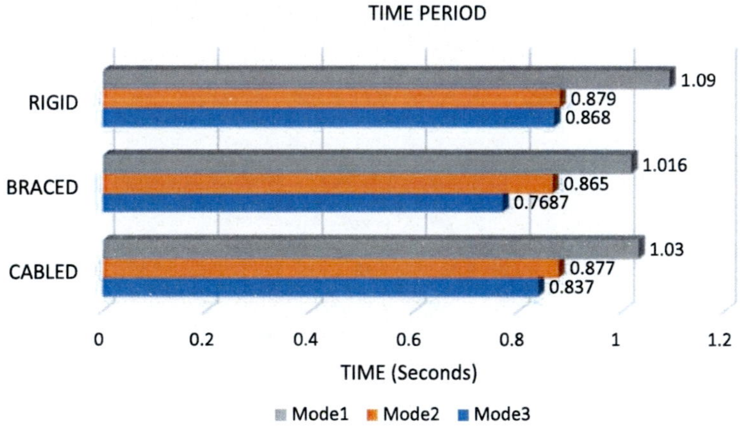

Fig. 21 Comparison of time period

6 Conclusion

In this work, analysis of all the three modeled structure has been carried out. Effect of CFRP cabled frame with steel X-braced frame and rigid frame under earthquake loads was studied, and it is concluded that:

- It has been observed that for CFRP cabled structures, the direction in which more cables are positioned, resulted in better output which is Y-direction.
- In cabled structure, there is 5–28% decrease in story displacement than rigid structure for equivalent static method, response spectrum method, and time history analysis.
- In time period, the braced structure outperformed the cabled structure by 1.35%.
- Cabled structure comparatively performed better than the other structures. So as a result, it can be used to improve the seismic stability of the structure.
- CFRP cables can be used for high strength and efficiency, but they are relatively costly than steel.

References

1. Eshelman LDP (2019) Strand rods and high-performance fiber-reinforced cementitious composites: alternative options for seismic retrofit of existing structures. Eng Struct 111:323–332
2. Mochida Y, Imoto Y (2019) Development of carbon fiber reinforced thermoplastic strand rod. Int J Geomate 16(57):109–115
3. Liu Y, Zwingmann B, Schlaich M (2015) Carbon fiber reinforced polymer for cable structures—a review. Polymers 7:2078–2099
4. IS 456:2000. Plain and Reinforced concrete—code of practice. Bureau of Indian Standards, New Delhi
5. IS 1893-2002 (Part-1). Criteria for earthquake resistant design of structures. General provisions and buildings. Bureau of Indian Standards, New Delhi

Structural Audit and Assessment of Application of External Post Tensioning Method on Bridge Girder

Rushikesh Hiralal Bicchal and H. R. Magarpatil

1 Introduction

In India, there large numbers of flyovers on highways and river-bridges are distress due to various factors. Fatigue damages and temperature stresses are additionally purposes behind these issues. They are many factors such as expansion in volume of traffic, low quality of development, and plan inadequacies, absence of grouting, lacking compressive strength of cement at move of prestress, consumption of links, relaxation of prestressing steel, permanent deformation due to declining of form works, non-working of bearings, damaged extension joints, and inappropriate drainage framework. It is accounted for that the US spent a few billion for fix and recovery of upset scaffolds. Same condition endures in European and different nations for the maintenance and updating of concrete structures. The distressed concrete structures have been showing major issues to bridge engineers.

Instead of reconstruction retrofitting of structure is the effective and economical answer for rehabilitation of bridges which are in bad condition and needed to be retrofitted. Existing constructions are retrofitted to satisfy different goals, for example, expanding flexural load conveying capacity, expanding the shear capacity, and so on. Post tensioning systems are further divided into two categories: (i) External Post tensioning (EPT) and (ii) Internal Post tensioning (IPT). The External Post tensioning system consist of anchoring tendons or cables to the external surfaces of the structure. The Internal Post tensioning consist of anchoring tendons or cables inside the structure. This is accomplished by either longitudinally or transitionally inside the structural member or by making a cavity along the structural member. An internal PT framework is utilized if the first cross section of the structure cannot be

R. H. Bicchal (✉) · H. R. Magarpatil
School of Civil Engineering, Dr. Vishwanath Karad MIT World Peace University, Pune, India
e-mail: rushikeshbicchal2@gmail.com

H. R. Magarpatil
e-mail: hanumant.magarpatil@mitwpu.edu.in

A. Sil et al. (eds.), *Recent Trends in Civil Engineering*, Lecture Notes in Civil Engineering 274, https://doi.org/10.1007/978-981-19-4055-2_23

changed. External Post tensioning has been used in retrofitting of existing structures since long time and has been considered as an effective and economical method for rehabilitation of distressed concrete bridges.

2 Methodology

The main objective of this paper is to perform structural audit of Mutha river-bridge by using Non-destructive tests such as Rebound Hammer test and Ultrasonic Pulse velocity test. After conducting NDTs to check the deflection of each deck by using institutionally developed displacement sensor. Using results from audit design retrofitting component for most critical girder by applying external post tensioning method in SAP2000 software.

2.1 Visual Inspection

While conducting the structural audit the initial step is visual investigation. It permits us to determine the current state of bridge, and defects in bridge. Also is gives rough idea about critical sections on which the further research is to be done. After visual examination of extension, numerous piers are in harmed condition. Numerous piers are having horizontal and vertical cracks. Some girders have multiple cracks and reinforcement is exposed (Figs. 1 and 2).

While observing condition of girders we found that many girders have horizontal and vertical cracks. Some girders have multiple cracks and reinforcement is exposed (Fig. 3).

Fig. 1 Pier of Mutha river-bridge

Fig. 2 Concrete condition of pier of Mutha river-bridge

Fig. 3 Girder of Mutha river-bridge

2.2 *Rebound Hammer Test*

After visual inspection Non-destructive tests are conducted. Initially the rebound hammer test is performed which determined the compressive strength of the concrete. The rebound hammer is also called as Schmidt hammer. Which consist of a spring controlled mass that slides on a plunger inside a rounded lodging. The Rebound hammer test depends on the principle that the rebound of an elastic mass relies upon the stiffness or hardness of the concrete surface against which the mass strikes.

Fig. 4 Rebound hammer

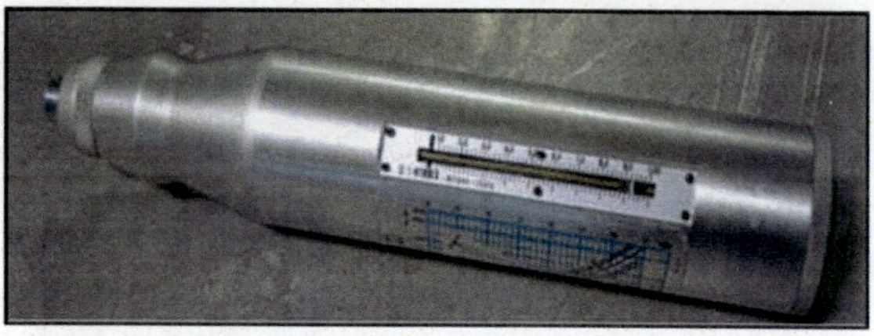

The rebound number is recorded form 1m height from the base of pier for each pier. Twenty-one readings are taken for each pier. By averaging the rebound number from 21 readings gave us quite accurate rebound number. And further process, to determine strength of concrete is calculated by referring the load-area chart which is mounted on rebound hammer (Fig. 4).

The rebound hammer test is most suitable method for determining the relation between compressive strength of concrete and rebound number is to test the concrete structures by using compression testing machine and simultaneously using rebound hammer (Table 1).

After performing the rebound hammer test it is observed that the pier number 7 and 8 have less strength as compared other piers.

2.3 *Ultrasonic Pulse Velocity Test*

After rebound hammer test, the ultrasonic pulse velocity test is conducted to check the quality of concrete of piers of Mutha river-bridge. Ultrasonic testing gear incorporates a transducer for changing electronic pulse into mechanical pulse, a pulse age circuit and comprising electronic circuit for creating pulses. The test is conducted by passing a pulse of ultrasonic through concrete to be tested and estimating the time taken by pulse to get past the design. Higher velocity shows great continuity and quality of concrete, while more slow velocity may demonstrate concrete with numerous breaks or voids (Fig. 5).

The ultrasonic test is performed for each pier, by taking three readings randomly on 350 and 750 mm, respectively, and averaging it. The test results are shown Table 2.

Interpretation of results: From Table 3 mentioned in IS:13311 (Part 1) the concrete quality (grading) is determined.

From the results of ultrasonic pulse velocity test most piers are in doubtful condition having poor concrete quality. Mainly Pier 1 and Pier 2 have the minimum velocity which indicates these piers are in doubtful condition.

Table 1 **a** Rebound hammer test results of abutment 1 and pier number 1, 2, 3, 4. **b** Rebound hammer test results of abutment 2 and pier number 6, 7, and 8

S. No.	Abutment 1	Pier 1	Pier 2	Pier 3	Pier 4
1	45	34	32	26	24
2	44	38	30	25	32
3	40	30	28	28	34
4	45	44	36	34	28
5	38	34	32	25	30
6	46	33	32	26	30
7	44	30	35	30	30
8	40	32	32	32	28
9	44	28	28	32	30
10	36	26	34	33	33
11	30	26	28	30	40
12	32	32	42	34	34
13	45	32	33	30	30
14	30	28	28	34	32
15	38	26	26	35	37
16	36	38	25	33	29
17	45	34	34	34	37
18	40	30	30	38	29
19	36	32	32	27	33
20	46	26	35	30	31
21	43	34	36	27	30
	843	667	668	643	661
Average	40.14	31.76	31.81	30.62	40.14
Strength (kg/cm^2)	352.86	233.71	234.57	216.19	352.86
Strength (N/mm^2)	34.6	22.92	23	21.2	22.42

S. No.	Pier 6	Pier 7	Pier 8	Abutment 2
1	25	27	42	41
2	24	26	45	50
3	26	27	46	26
4	27	31	40	28
5	27	26	41	43
6	20	26	30	38
7	25	38	40	42
8	23	28	34	28
9	21	37	28	30

(continued)

Table 1 (continued)

S. No.	Pier 6	Pier 7	Pier 8	Abutment 2
10	32	26	30	29
11	30	24	38	31
12	26	28	28	36
13	26	27	32	32
14	28	29	31	27
15	27	26	34	37
16	32	24	28	33
17	28	38	25	37
18	24	32	28	35
19	25	24	27	33
20	27	26	28	35
21	26	33	34	32
	549	603	709	723
Average	26.14	28.71	33.76	34.43
Strength (kg/cm^2)	158.99	187.14	257.62	268.57
Strength (N/mm^2)	15.59	18.35	25.26	26.34

Fig. 5 Ultrasonic pulse velocity test instrument

2.4 *Deflection Test by Institutionally Developed Sensor*

After conducting NDTs to determine the critical girder in bridge, we conducted the displacement test by using sensors and displacement meter test setup. The setup consists of three components:

Table 2 Ultrasonic pulse velocity test results

	Time (In μs)		Velocity (km/s)		Remark	
Length (in mm)	$L = 350$	$L = 750$	$L = 350$	$L = 750$	$L = 350$	$L = 750$
Abutment 1	154.83	299.83	2.33	2.51	Doubtful	Doubtful
Pier 1	160.50	334.00	2.20	2.25	Doubtful	Doubtful
Pier 2	144.67	317.83	2.46	2.37	Doubtful	Doubtful
Pier 3	145.00	240.83	2.45	3.13	Doubtful	Doubtful
Pier 4	104.83	227.50	3.35	3.31	Medium	Medium
Pier 6	171.33	263.67	2.06	2.85	Doubtful	Doubtful
Pier 7	157.67	266.33	2.24	2.82	Doubtful	Doubtful
Pier 8	144.17	269.00	2.44	2.79	Doubtful	Doubtful
Abutment 2	141.50	274.83	2.48	2.74	Doubtful	Doubtful

Table 3 Velocity criterion for concrete quality grading IS-13311 (part 1)

Pulse velocity (km/s)	Concrete quality (grading)
Above 4.5	Excellent
3.5–4.5	Good
3.0–3.5	Medium
Below 3.0	Doubtful

(i) The displacement sensor.
(ii) The recording instrument.
(iii) The laptop for interpretation of result.

The sensors can measure the slightest displacement variations of structure. The sensitivity of the sensors is up to two decimal points. They have mentioned the direction on them which show the direction in which the readings will be recorded. The sensors can also record acceleration, velocity and displacement in X, Y, and Z direction. These sensors are installed on the mid span of the deck platform and displacement readings are taken in Y direction (Table 4).

From the results of Deflection by Institutionally developed displacement meter is seems that the maximum deflection is at deck number 6. And also by visual inspection it is noticed that the girders of deck number 6 have major structural cracks. Hence we have taken same girder for analysis and further research purposes. The application of external post tensioning method is applied and analyzed by the means of SAP2000 software.

Table 4 Deflection test by displacement meter

Deck No.	Displacement recorded at mid span (in mm)
1	41.94
2	52.64
3	73.37
4	108.36
5	98.76
6	125.21
7	107.31
8	96.86
9	82.11

3 Analysis and Design

The initial analysis of bridge is done in Csi Bridge software. The material properties and various parameters required for design of bridge is considered form the results of audit. For example the minimum strength of concrete is 15 N/mm^2 for pier number 6 and maximum concrete strength is 34 for abutment 1. As an average strength 25 N/mm2 and for loading conditions IRC (Class-70R, Class-A, Class-AA) are considered for initial design of bridge in Csi Bridge. The purpose of this software analysis is to find the critical girder section which is suitable for application of external post tensioning (Figs. 6 and 7).

From the results, the maximum deflection is found in middle span of bridge, i.e., deck number 6.

The further research and design is carried out by using SAP2000 and application of external post tensioning is also analyzed. Initially the beam without application of EPT is analyzed. After that application of post tensioned tendon for same beam is applied and analyzed (Figs. 8 and 9).

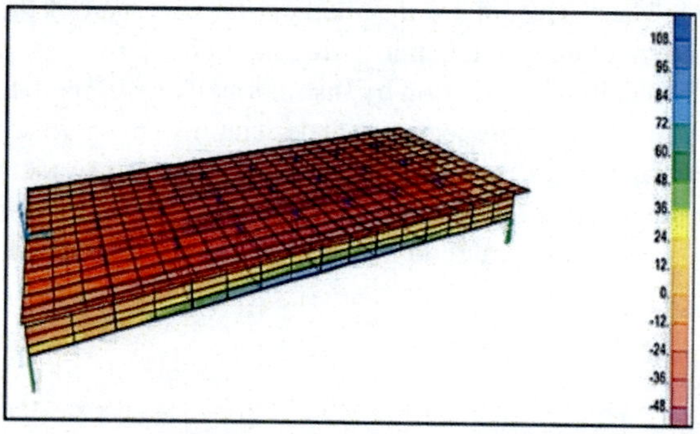

Fig. 6 Analysis of deck (by Csi bridge)

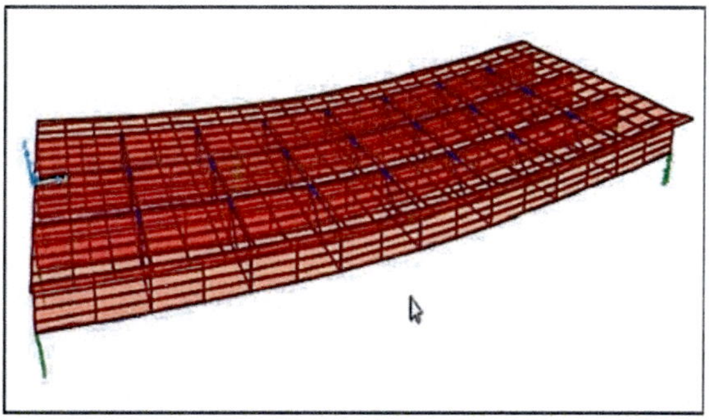

Fig. 7 Deflection of deck (by Csi bridge)

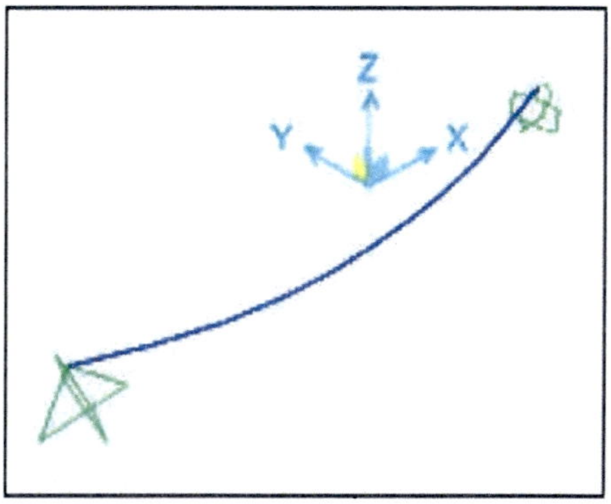

Fig. 8 Deflection of girder without application of EPT (by SAP2000)

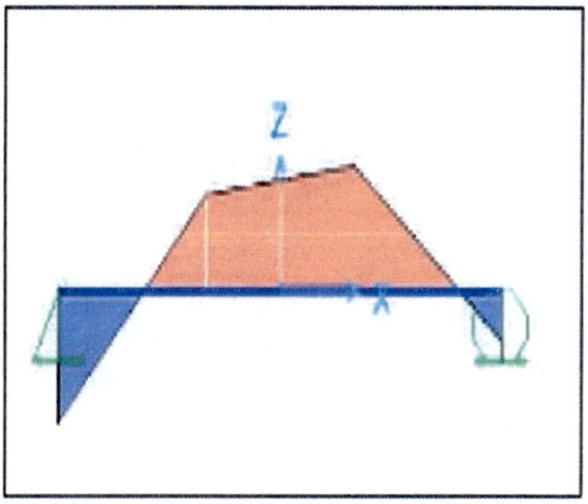

Fig. 9 Deflection and BMD of beam without application of EPT (by SAP2000)

Deformed beam before application of external post tensioning tendon. The moment carrying capacity of girder is 282.16 kNm before application of EPT (Figs. 10 and 11).

Deformation of girder after application of external post tensioning tendon. The moment carrying capacity of girder after the application of tendon is 478.96 kNm.

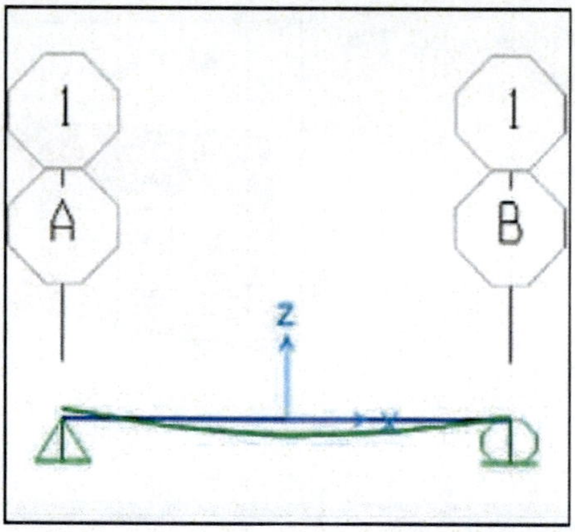

Fig. 10 FBD of girder with application of EPT (by SAP2000)

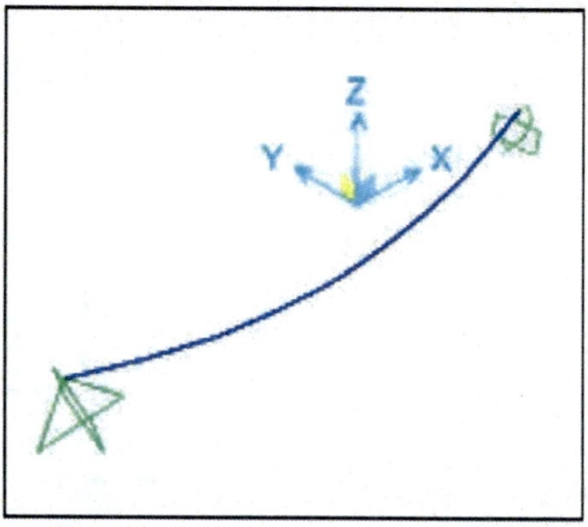

Fig. 11 Deformation of girder after application of external post tensioning same beam

4 Restrictions of External Post Tensioning

- The External post tensioning system can be applied only if the concrete is in relatively good condition means it should resist the stresses from the tendon.
- The anchorage can be difficult to install in case of existing bridge which requires desirable skills, and careful detailing is required to account for stress concentrations.
- There may be chances of corrosion in tendons due to weathering actions.
- The effect of traffic load applied on post tensioned structures may decrease the load carrying capacity of structure while fitting of system. Increasing the eccentricity of the tendons can improve the tensile strength contributing to strengthening the member.

5 Conclusion

From the both analysis, i.e., before application of EPT the moment demand was 282.16 kNm and after the application of EPT tendon the moment demand increased by 478.96 kNm. This states that by using External post tensioning method the moment carrying capacity of girder increases by 58%.

References

1. Manisekar R (2018) Effect of external post-tensioning in retrofitting of RC beams. The Institution of Engineers, India
2. Yuen TYP, Halder R. DFEM of a post-tensioned precast concrete segmental bridge with unbonded external tendons subjected to prestress changes. Elsevier. www.elsevier.com/locate/structures
3. Ghodke Y, Gandhe GR. Retrofitting of RCC structure WIH strengthening of shear wall with external post tensioning cables. Int J Eng Res 5(3):602–606
4. Maheri MR, Torabi A. Retrofitting external RC beam-column joints of an ordinary MRF through plastic hinge relocation using FRP laminates. Elsevier. www.elsevier.com/locate/structures
5. Gao D, Fang D. Flexural behavior of reinforced concrete one-way slabs strengthened via external post-tensioned FRP tendons. Elsevier. www.elsevier.com/locate/structures
6. Karthik MM, Terzioglu T. Magnetic flux leakage technique to detect loss in metallic area in external post-tensioning systems. Elsevier. www.elsevier.com/locate/structures
7. Rios AJ, O'Dwyer D. External post-tensioning system for the strengthening of historical stone masonry bridges. External post-tensioning system for the strengthening of historical stone masonry bridges, pp 602–606

Analysis of I-Section Composite Beam with Winding Wire and Metal Strap for Development of Effective Bond

Anand Kumar, Shailja Bawa, and M. Abdul Akbar

1 Introduction

Load and stress are transferred in reinforced cement concrete (RCC) structures through the steel–concrete bond. A clear bond stress-slip relationship is required to analyze the mechanical performance of key parts of RC structures in the elastic and plastic stage which include the beam-column joint, connection between beam and shear wall, connection between column and foundation [1]. Fiber-reinforced polymer material has been used by many researchers in their studies, and it is becoming a more popular material for concrete structures. The fiber-reinforced polymer material used in the structure has a higher strength and durability, particularly in harsh environments [2]. When a reinforced concrete beam is bonded with a carbon fiber-reinforced plastic (CFRP) laminate, the flexural behavior of the beam changes. As compared to steel, the advantages of carbon fiber-reinforced plastic composite are light weight, high strength, and strong resistance [3]. The composite beam and column can save money by having a smaller cross section and can handle heavier loads. Resistance to fire and corrosion, as well as high strength and ductility, is important characteristics of the composite. These composites are used in the construction of high-rise buildings and structures that carry a lot of weight. The ultimate axial load resistance capacity has improved due to structural steel in composite columns [4]. The key aim of using flexural behavior of carbon textile-reinforced concrete is to look at changes in the crack interface and failure mode. According to the experimental findings, the cracking

A. Kumar (✉) · S. Bawa · M. Abdul Akbar
Department of Civil Engineering, Dr. B. R. Ambedkar National Institute of Technology, G.T. Road, Amritsar Bypass, Jalandhar, Punjab 144011, India
e-mail: anand4897kumar2016@gmail.com

S. Bawa
e-mail: bawas@nitj.ac.in

M. Abdul Akbar
e-mail: akbarma@nitj.ac.in

A. Sil et al. (eds.), *Recent Trends in Civil Engineering*, Lecture Notes in Civil Engineering 274, https://doi.org/10.1007/978-981-19-4055-2_24

behavior has significantly improved, and it allows for significant cracking prior to failure. Textile-reinforced concrete (TRC) is a promising cement-based composite that is still relatively new [5].

The steel to concrete bond performance mainly focused on reinforced concrete structures. Similar to the bond between plain steel bars and concrete, the steel angle-to-concrete bond in steel-reinforced concrete (SRC) is primarily due to three mechanisms: chemical bonding, frictional resistance, and mechanical interlocking, with concrete strength, concrete cover thickness, transverse tie reinforcement ratio, anchorage length, shear connector, and surface condition being the main influencing factors. Steel angles have greater contact areas with the concrete than steel bars, and their cross-sectional forms are much more complex [6]. Several research and field experiments have shown the efficacy of iron-based form memory alloy (Fe-SMA, also known as "memory steel"). Because of the novelty of ribbed memory-steel reinforcing bars, only a small amount of study has been done on their use in structural engineering. Until being exposed to a pull-out load, memory-steel bars were mounted on concrete blocks and enabled by resistive heating. Based on obtained prestress, crack formation, slip, and bar strain, the effects of heating and thus prestressing, bar ductility, bond duration, and fixation rigidity were examined. A longitudinal splitting crack with negligible crack width occurred during the activation of the bars [7].

The pullout test on specimens of local bond studied the bond stress distribution between steel bar and steel fiber-reinforced concrete (SFRC) with separate volume fractions. The findings revealed that when steel fiber is added to high-strength concrete, the inhomogeneous distribution of bond stress between high-strength steel bar and SFRC increases. The slippage at the loaded end decreases as the volume fraction of steel fiber increases, while the bond stiffness at the loaded end increases dramatically [8]. High-strength wire (HSW) confined concrete columns were used in the experiments. The difference in confinement pressure was accomplished by adjusting the winding spacing of HSW in fifteen columns. The columns were studied for their stress–strain reaction, peak strength and strain, ultimate strain, and failure modes. Winding HSW is an inexpensive and reliable form of column jacketing that can greatly improve column strength and ductility, according to the results of the experiments [9].

An ABAQUS simulation of a concrete beam reinforced with I-section and flat straps on top and bottom and winding wire around the section is compared to another concrete beam reinforced with I-section in this article. Stress and strain patterns for composite beams with I-section steel and composite beams with I-section adding flat straps are calculated under evenly spaced load. The action of a composite beam with flat straps around the I-section and without flat straps around the I-section can be calculated using ABAQUS software and the finite element method (FEM). The geometric and material properties vary, and their results have been observed with respect to localized or uniformly distributed loads, as well as total beam deformation and path description, as seen in the figures.

The healthy functioning of a structure throughout its design life for the loads acting on it is the primary requirement of a structural engineer. This requirement satisfaction becomes more exhaustive in the case of composite materials wherein the load transfer between the different components of the structure has to be ensured. As discussed in the above literatures, several studies have been carried out on I-section composite beams. In this work, the effectiveness of using winding wire and flat strap on improving the bond between I-section and concrete is taken up. Finite element analysis is carried out using ABAQUS by varying the size and spacing of winding wire and flat strap.

2 Numerical Modeling

In this study, the effectiveness of bond stress developed is studied by analyzing three sets of models of I-section composite beams. One with I-section composite beam with winding wire around the I-section, second one with I-section composite beam with flat straps on the top and bottom of the steel section, and third one with I-section composite beam analyzed for optimal flat strap sizes for beam with increased fire rating in the form of increased clear cover. Rectangular beams are tested under the uniformly distributed load (UDL). The grade of concrete is M20. The modulus of elasticity of concrete is 22,360.675 MPa calculated from the given formula in IS 456:2000 [10] as per clause (6.2.3.1). The I-section modulus of elasticity is 200,000 MPa, and the Poison's ratios of concrete and steel are 0.2 and 0.3, respectively. The cross section of steel I-section (ISMB300) is taken from the IS 800:2007 Table (46) [11]. The beam is finalized for study has a length of 5000 mm, a width of 200 mm, and a depth of 360 mm.

A simplified view of the composite beam subjected to direct uniformly distributed load (UDL) over the span of the beam is shown in Fig. 1a, b. The beam is a simply supported (both sides pinned or hinged) applied with a UDL of 100kN acting along the length of the beam. Detailed finite element analysis of the I-section composite beam was carried out using ABAQUS.

The parameters used for analysis of the composite beam are given in Table 1. These parameters are used in all of the models in this analysis.

For the composite beam, cross section of 200 mm × 360 mm with winding wire ties around the I-section throughout the span of beam as shown in Fig. 2. Seven models were created for the study of winding wire behavior, each with a different wire diameter. The diameter of winding wire is specified by IS 280:2006 as per clause (7.1) [12]. The diameter of the winding wire used in this analysis was selected from a selection at a specific interval (0.125, 0.224, 0.4, 0.710, 1.25, 2.5, and 5 mm). The flat strap tie on the I-section was examined as another composite. Sixteen models were created using the IS 1730:1989 [13]. Figure 3 shows a concrete beam with flat strap ties at the top and bottom of the I-section over the span of the beam. The flat strap's length is equal to the width of I-section flange. The size of flat strap used in this analysis was chosen at some interval from the table with increasing flat strap width and thickness. The widths of flat straps are 20, 30, 40, and 45 mm. Each flat strap is made with a thickness of 3, 6, 10, and 12 mm.

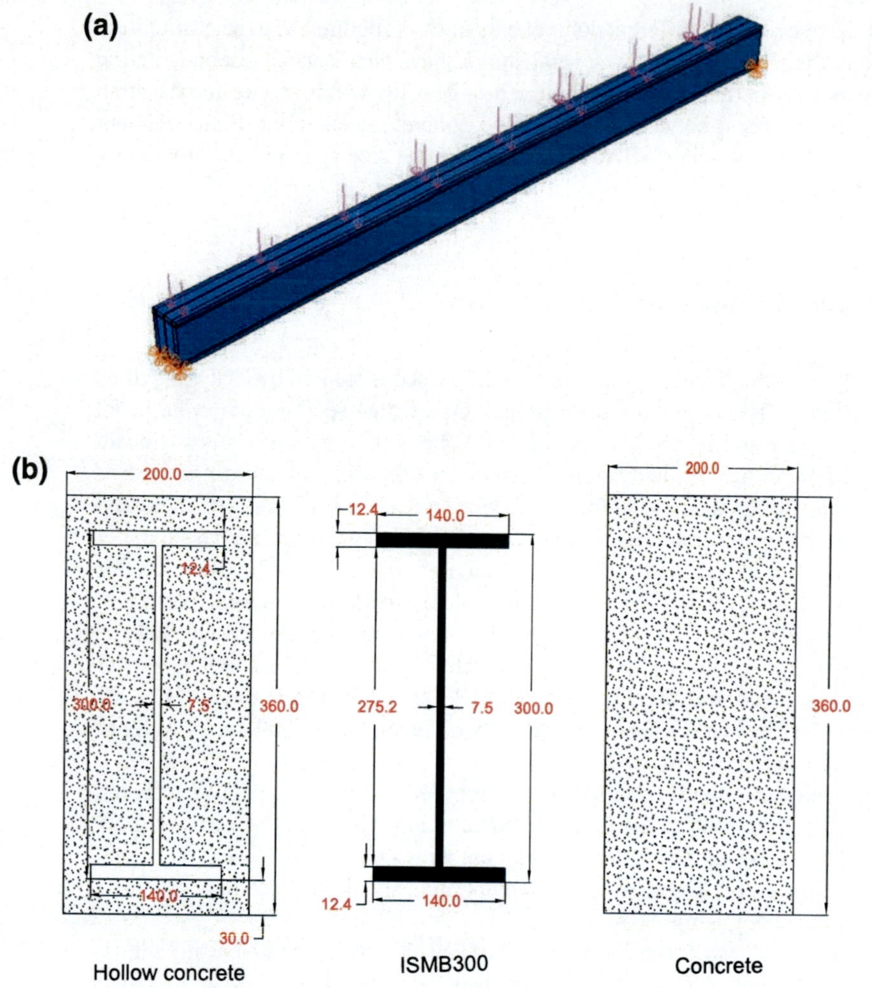

Fig. 1 **a** Beam modeled in ABAQUS. **b** Cross section of beam (all dimension in mm)

Mesh convergence study was carried out by assuming boundary condition as pinned or hinged support at the edge of the beam end by varying the mesh size. Global seeding was used. After convergence study, a global seeding of 15 mm is finalized as shown in Fig. 4. A converged mesh size was used in all model. The boundary condition used is same as that used in case of mesh convergence study, and 0.1 N/mm^2 uniform pressure load is applied (i.e., equivalent to 100 kN point load on the top of the beam).

It is necessary to bind all nodes with elements of all parts in the model in order to pass load and stress. The model in this study was complex, and meshing the portion using global seeding was difficult and bad meshing. For partitioning of complex

Table 1 Parameters used in modeling

Parameter	Value
Method of analysis	Static, general
Total load	100 kN
Boundary condition	Hinged support
Length of span	5000 mm
Density of steel	77.005 kN/mm^3
Density of concrete	24.991 kN/mm^3
Element types	C3D8R, an 8-node linear, a 2-node linear, 3D truss element for winding wire

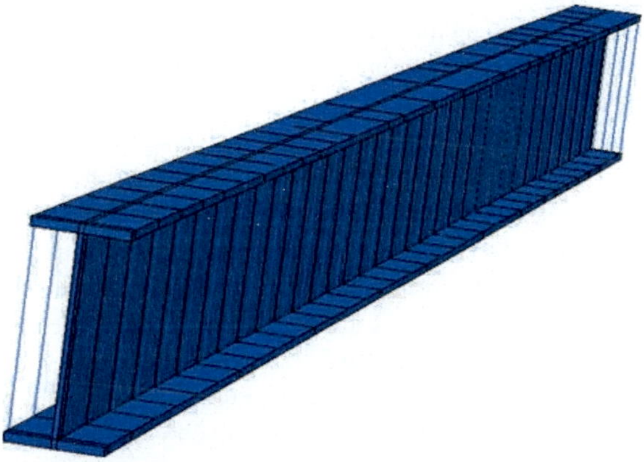

Fig. 2 I-section with winding wire

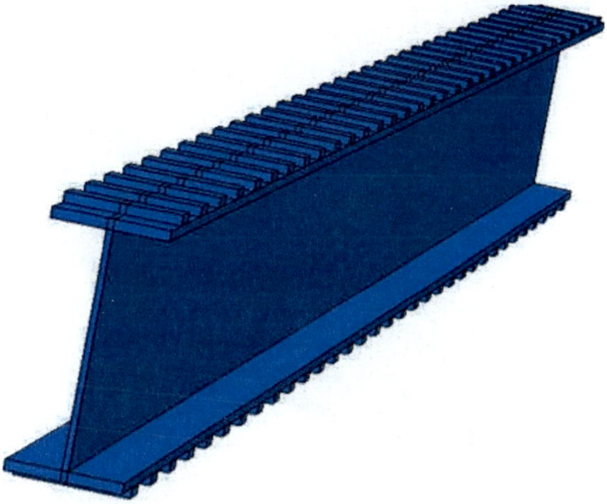

Fig. 3 I-section with flat strap

Fig. 4 Global meshing of beam

geometry, the partition technique was used. As a result, some of the mesh sizes in this model are smaller than others due to partition of model. The development of a clean, uniform mesh for structural analysis was facilitated by improved meshing algorithms and software. The mesh size varies, but the average is 15 mm. The analysis of the model's outputs and representation of the value in graphs, as well as the analysis of results, do not reveal a consistent trend or pattern. The path is required for analyzing the model's outputs in order to provide a consistent pattern. The path was taken on I-section bottom face excluding all part of beam from one end to another end. The aim of selecting the direction at the bottom face of the I-section was to analyze the stress in the tension region. Figure 5.

Defining the property of the material and section is an important step in the finite element modeling process. Embedded and tie constraints were used to assemble the parts together in this study. The structured mess was chosen, and the partition tool

Fig. 5 Direction of path

was used to separate the complex geometry into geometry that could be used in the structured mess.

3 Results and Discussion

A detailed finite element analysis of a beam with various composites was performed. In this study, several models were made, including a base model (concrete reinforced with I-section steel) and seven models with winding wire of various diameters, as well as sixteen models with flat straps of varying width and thickness. ABAQUS was used to run the static linear analysis. Vertical displacement (U2), bending tensile stress (S33), and shear stress (S23) for the total beam and path along the span of the beam on the bottom face of the I-section were calculated for all models. The shear stress (S23) illustrates the bond stress effect.

The minimal shear stress value was calculated using flat straps of 30 mm × 10 mm and 30 mm × 12 mm with spacing of 75 and 150 mm. It is close to the value provided in the code for simple and deformed bar design bond stress [IS 456:2000 (clause 26.2.1.1) and SP: 34 (S&T)-1987 as per clause (4.2.2.1)]. Variation of shear stress corresponding to different sizes of flat strap is presented in Figs. 6 and 7. Although flat strap spacing of 75 mm is smaller, the value of shear stress at this spacing is the smallest; it may be avoided as lesser spacing mean higher cost of project and assembly time.

The distance between the plates that were used differed. The various sizes available are 75, 150, 225, and 300 mm. It was observed that 150 and 225 mm are optimal spacing considering the least value obtained for S23 as seen in Fig. 8.

The model was created with various sizes of flat straps in this analysis. It was concluded that stretching the plate's size beyond 30 mm width and having a reasonable result was not worth it. The data from the path study of the beam was compared

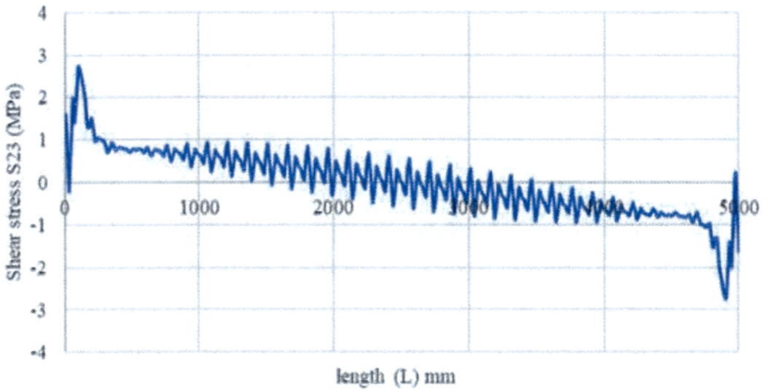

Fig. 6 Variation of shear stress of beam with flat strap 30 mm × 12 mm

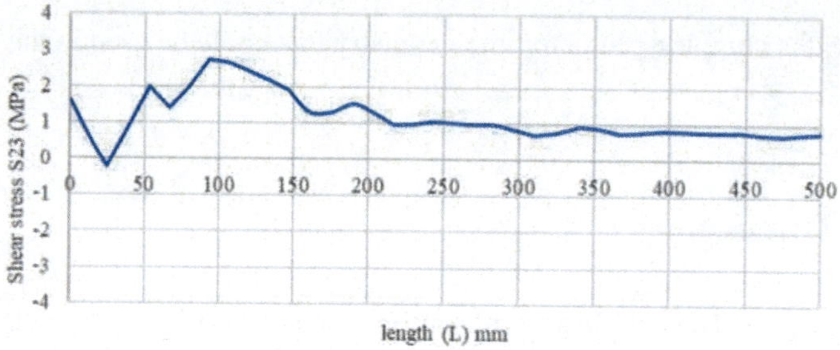

Fig. 7 Optimal shear stress in beam with flat strap 30 mm × 12 mm

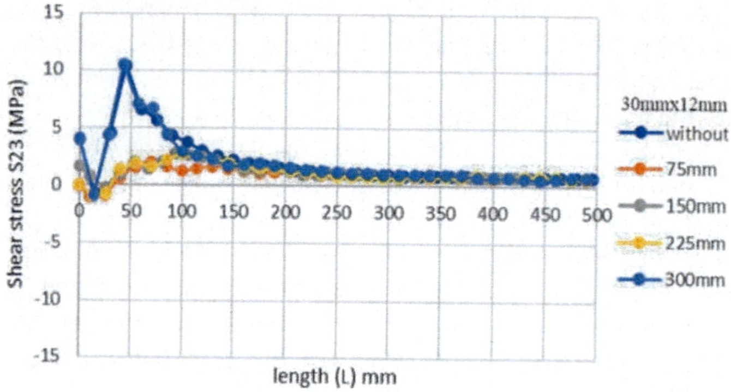

Fig. 8 Variation of shear stress with spacing

to various sizes of flat strap. After the 30 mm width, the shear stress increases. As seen in Figs. 9 and 10, it could have a local influence.

It is observed that for flat strap sizes of 30 mm × 10 mm and 30 mm × 12 mm, shear stress was lesser at spacings of 75, 150, and 225 mm flat straps. Flat strap was attached or joined with I-section by welding and varying the spacing (75, 150, 225,

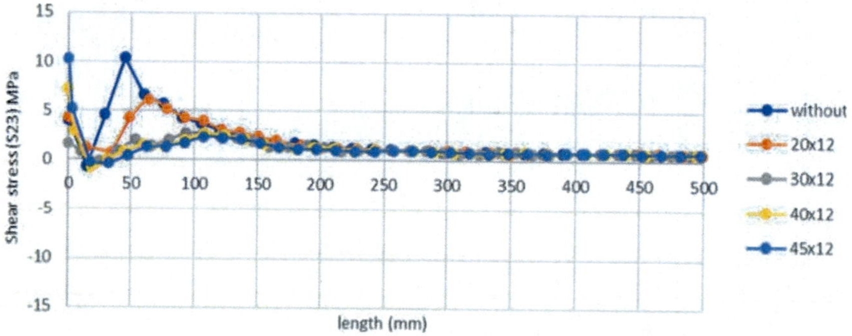

Fig. 9 Variation of shear stress with width

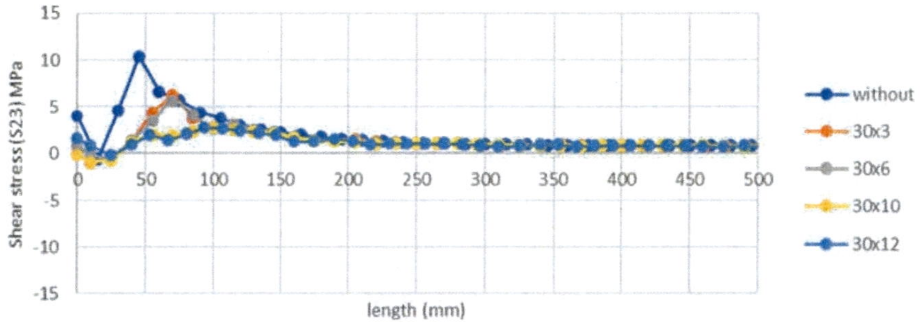

Fig. 10 Variation of shear stress with thickness

and 300 mm). Another parameter considered in this study is the displacement, and it was observed that displacement of the beam increases with the increase in the size of flat strap as shown in Figs. 11 and 12. As the size of the plate is increased, the deflection of the beam and the weight of the flat strap increase.

In the analysis of winding wire, variation was not much in displacement (U2), bending tensile stress (S33), and shear stress (S23) as compared to the composite beam I-section. The results of models with winding wire data are shown in Table 2. The variation was observed in displacement, bending stress, and shear stress with the addition of flat strap, and hence, it is effective for improving the bond strength of I-section against slip and flexure. The variation of bending tensile stress obtained was compared with base model composite beam without flat plate as shown in Figs. 13

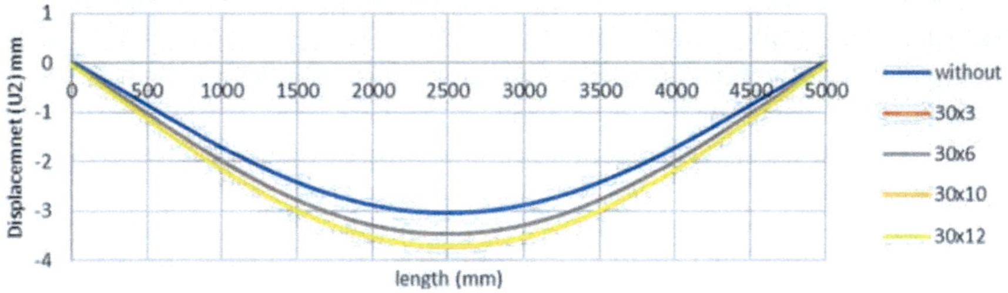

Fig. 11 Variation of displacement with thickness

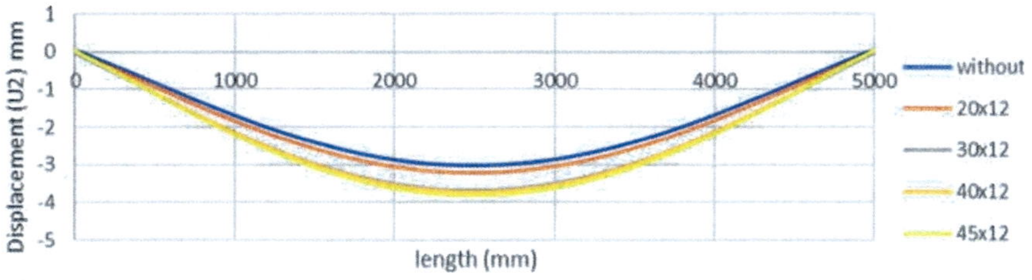

Fig. 12 Variation of displacement with width

Table 2 Parameters observed in winding wire composite

Composite beam with winding wire and I-section					
Diameter (mm)	Area (mm^2)	Displacement U2 (mm)	Bending tensile stress S33 (MPa)	Shear stress S23 (MPa)	Spacing (mm)
0.125	0.012266	3.042	39.28	26.77	150
0.224	0.039388	3.042	39.28	26.77	150
0.4	0.1256	3.042	39.28	26.77	150
0.71	0.395719	3.042	39.29	26.77	150
1.25	1.226563	3.042	39.3	26.76	150
2.5	4.90625	3.042	39.32	26.74	150
5	19.625	3.041	39.34	26.73	150

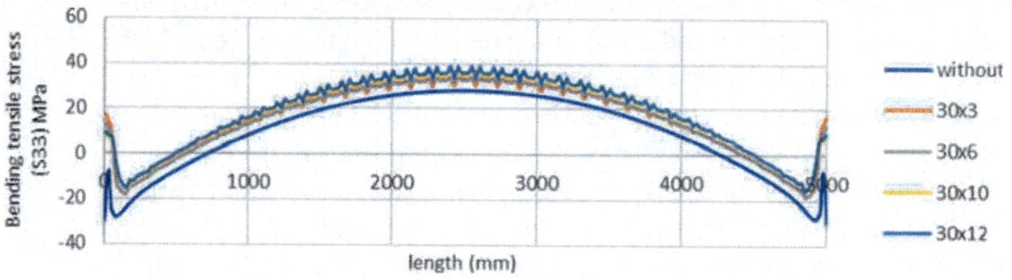

Fig. 13 Variation of bending stress with thickness

and 14. In this figure, clear variation was observed in bending tensile stress increasing with increase in the thickness and width of flat strap. The obtained value of shear stress from the comparison study between flat strap and without flat strap with I-section is shown in Fig. 15.

Analysis of composite beam fire resistance as the cover of the beam around the steel portion is increased. Since steel can melt at high temperatures, a fire rating study was conducted for the protection of the construction at a certain temperature.

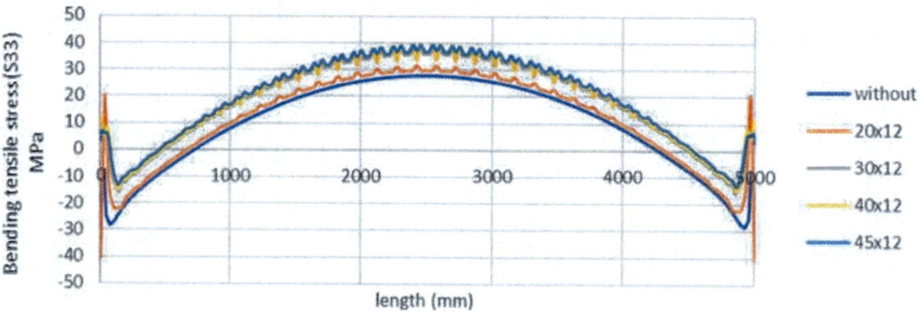

Fig. 14 Variation of bending stress with width

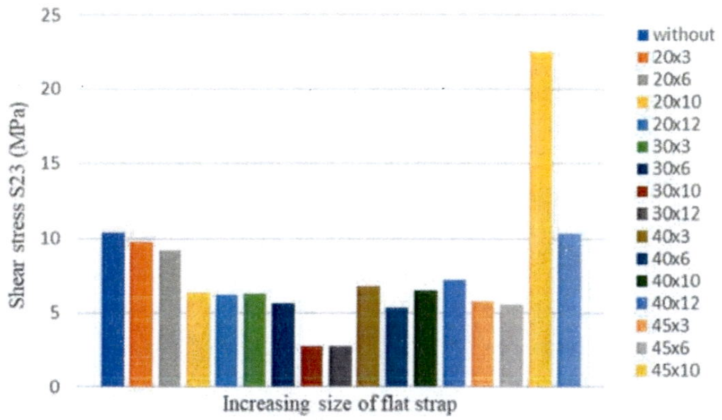

Fig. 15 Variation of shear stress with 150 mm spacing

Nominal cover is specified period of fire resistance in the IS-code 456:2000 as per table 16(A) for simply supported beam. Variation was observed in the study of fire resistance as seen in Table 3. Ten models were modeled for the fire resistance analysis using flat straps of size 30 mm × 12 mm with spacing 150 mm and 30 mm × 10 mm with spacing 225 mm by varying the cover (Table 3). These flat strap composites were chosen because they generated the minimum shear stress as seen in the previous study.

Variations in deflection and bending tensile stress as well as shear stress were observed. As seen in Fig. 16, deflection (U2), bending tensile stress (S33), and shear stress (S23) decrease as the cover of the beam (fire rating) increases. While the cross-section area of the beam changes as the fire rating is increased, the average load remains consistent for all models. Since the load distribution area was expanded,

Table 3 Results for variation in cover

Fire resistance for composite beam with flat strap and I-section							
Cover (mm)	Thickness (mm)	Width (mm)	U3 (mm)	S22 (MPa)	S11 (MPa)	U11 (mm)	Spacing (mm)
20	10	30	0.7992	11.77	52.23	0.07103	225
30	10	30	0.6155	12.04	37.91	0.05224	225
40	10	30	0.3928	18.23	12.27	0.03665	225
60	10	30	0.3004	8.595	9.613	0.02716	225
70	10	30	0.262	7.047	6.358	0.02221	225
20	12	30	0.7925	15.33	51.51	0.07071	150
30	12	30	0.6011	12.13	43.36	0.05229	150
40	12	30	0.395	18.68	12.76	0.03667	150
60	12	30	0.3016	7.827	9.715	0.0271	150
70	12	30	0.2647	6.851	6.503	0.02236	150

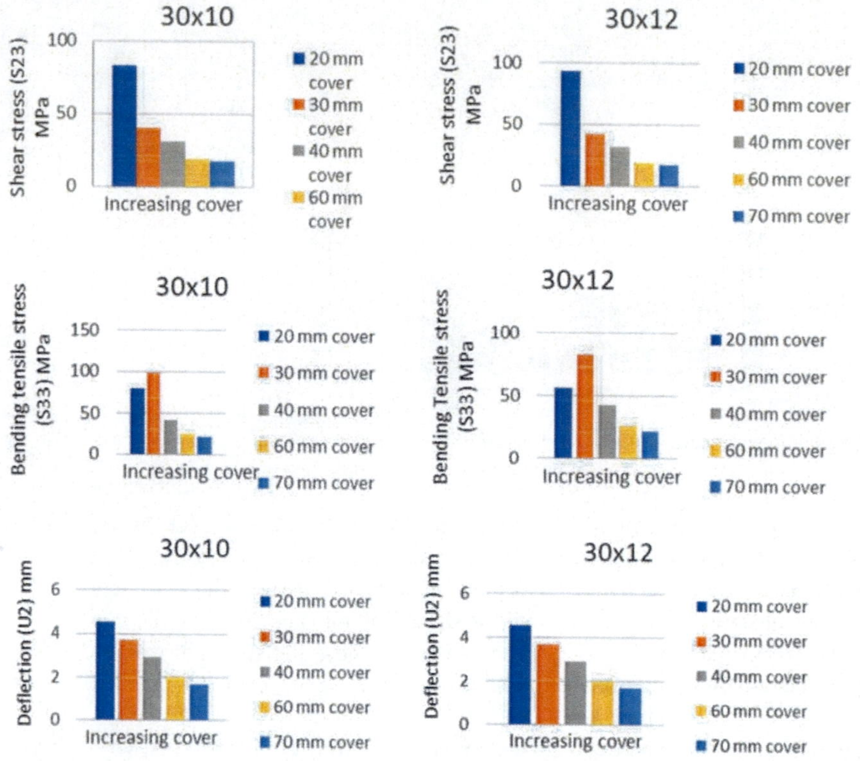

Fig. 16 Variation in fire rating of beam

the load intensity was reduced. Here, U1, U2, and U3 are the displacement of the beam along directions of axis (x, y, z), and S11, S22, and S33 are the tensile stresses of the beam along the respective directions.

Out of all models, the minimum shear stress (S23) of 2.003 MPa for 30mmx12mm spacing 75 mm and 2.740 MPa for 30 mm × 12 mm spacing 150 mm was observed in flat strap model. There was a sharp decrease in the value of shear stress for the width of 20 and 30 mm of the plate after the shear stress increased due to local effect. The displacement of the beam is greatest along the y-axis as seen in Fig. 17, and it is found that as the flat strap size is increased, the displacement of the beam increases, but the increment is small, while the strength of the beam increases significantly in the form of increased bending tensile stress and shear stress. The results taken from the analysis correspond to the path (Fig. 5).

The study of a concrete beam reinforced with I-section steel by welding flat straps on the top and bottom of the I-section flange. After the load was applied, the beam deflected. The composite beam's deflected outline shows some deformation in the flat strap's shape. The shape of the flat strap in the middle and near the beam's supports is out of shape. In these locations, it is easily recognizable as shown in Fig. 18.

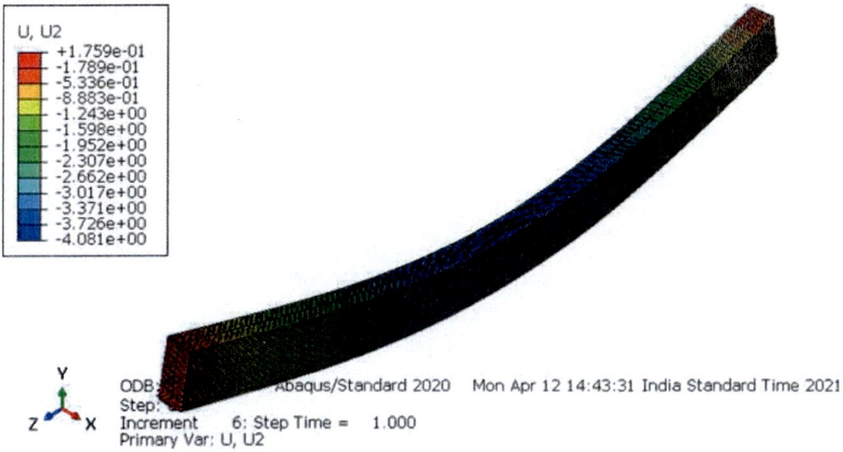

Fig. 17 Displacement contour of beam (flat strap 30 mm × 10 mm with 150 spacing)

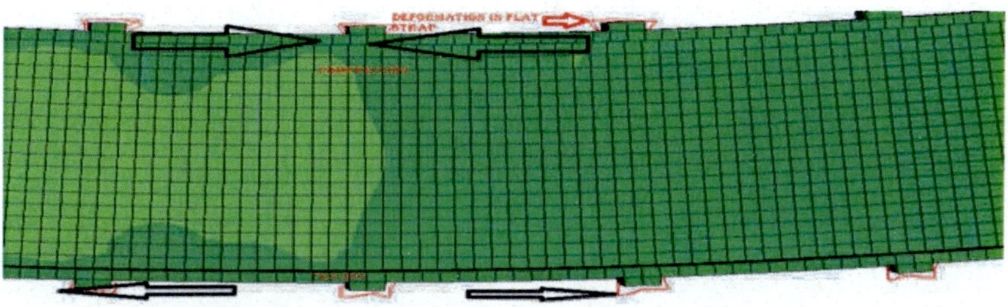

Fig. 18 Deformed shape of flat strap

In all of the models investigated, the shear stress was found to be higher near the support, as seen in Fig. 19. As the thickness of the metal strap was extended up to 30 mm, the beam's shear stress decreased. The use of a flat strap is useful in improving bond between the I-section and concrete.

As per this study, the use of winding wire did not improve the bond stress significantly. This is attributed to the smaller area of the winding wire compared to the flat strap. In a deflected beam, the winding wire outline is distorted for all sizes as seen in Fig. 20.

4 Conclusion

This paper presents the finite element analysis of composite beam (concrete reinforced with I-section steel and adding flat strap and winding wire) subjected to static load 100 kN over the span (uniformly distributed load). Based on the analysis

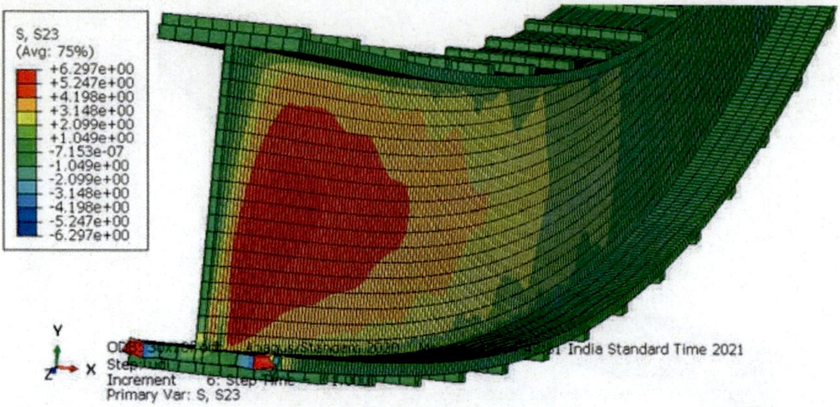

Fig. 19 Shear stress contour of beam (S23)

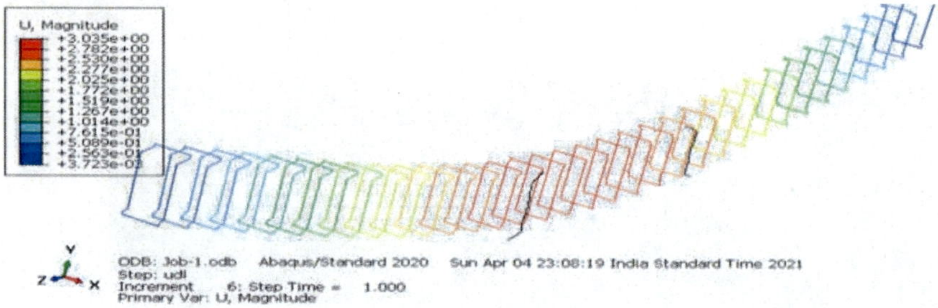

Fig. 20 Deflected shape of winding wire

performed for winding wire and flat strap with different size, spacing, and concrete cover, the following conclusion were made.

- After comparing the effects of concrete reinforced with I-section adding winding wire and flat strap, it was observed that using flat strap is beneficial to the beam and increases its bond strength.
- Shear stresses were higher in concrete reinforced with I-section steel, as predicted. As a result, shear stress is closer to the support in the model without the flat strap than in the model with the flat strap.
- Optimal results were observed for size of flat strap 30mmx10mm with spacing of 225 mm, and corresponding value of stress is 2.710 MPa.
- When comparing the results of the model with the flat strap, it is clear that the bending tensile stress increases as the flat strap size increases.
- After examining all models of flat straps, it is observed that the relationship between concrete and I-section will be improved if flat straps were included. The optimal sizes of flat strap were observed 30 mm × 10 mm and 30 mm × 12 mm.

- By increasing the cover to meet different fire rating requirement, it is observed that as the thickness of the concrete cover increases, the strength of the beam decreases.
- It has been observed that for the flat strap size of 30 mm × 10 mm and 30 mm × 12 mm, the displacement of the beam increases, while the displacement of the beam reduces as the concrete cover increases.

References

1. Gao X, Li N, Ren X (2019) Analytic solution for the bond stress-slip relationship between rebar and concrete. Constr Build Mater 197:385–397. https://doi.org/10.1016/j.conbuildmat.2018.11.206
2. Kapildev P, Kumaran G (2020) Theoretical and experimental on the behaviour of rectangular concrete beams reinforced internally with hfrp reinforcements under pure torsion. Int J Adv Res Eng Technol 11(3):29–47. https://doi.org/10.34218/IJARET.11.3.2020.004
3. Shahawy MA, Arockiasamy M, Beitelman T, Sowrirajan R (1996) Reinforced concrete rectangular beams strengthened with CFRP laminates. Compos Part B Eng 27(3–4):225–233. https://doi.org/10.1016/1359-8368(95)00044-5
4. Kartheek T, Das TV (2020) 3D Modelling and analysis of encased steel-concrete composite column using abaqus. Mater Today Proc 27:1545–1554. https://doi.org/10.1016/j.matpr.2020.03.200
5. Botelho Goliath K, Daniel DC, Silva FA (2021) Flexural behavior of carbon-textile-reinforced concrete I-section beams. Compos Struct 260:113540. https://doi.org/10.1016/j.compstruct.2021.113540
6. Zhang D, Zeng Y, Chen J, Gong F, Liu C (2021) Bond properties of steel angles in concrete filled steel tubes. J Constr Steel Res 182:106653. https://doi.org/10.1016/j.jcsr.2021.106653
7. Schranz B, Czaderski C, Vogel T, Shahverdi M (2020) Bond investigations of prestressed, near-surface-mounted, ribbed memory-steel bars with full bond length. Mater Des 196:109145. https://doi.org/10.1016/j.matdes.2020.109145
8. Gao D, Huang Y, Chen G, Yang L (2020) Bond stress distribution analysis between steel bar and steel fiber reinforced concrete using midpoint stress interpolation method. Constr Build Mater 260:119866. https://doi.org/10.1016/j.conbuildmat.2020.119866
9. Wei Y, Wu Y (2014) Compression behavior of concrete columns confined by high strength steel wire. Constr Build Mater 54:443–453. https://doi.org/10.1016/j.conbuildmat.2013.12.083
10. Indian standard plain and reinforced concrete—code of practice. IS 456:2000, Bureau of Indian Standard, New Delhi, India
11. Indian standard general construction in steel—code of practice. IS 800:2007, Bureau of Indian Standard, New Delhi, India
12. Indian standard mild steel wire for general engineering purposes. IS 280:2006, Bureau of Indian Standard, New Delhi, India
13. Indian standard flats for structural and general steel plates, sheets, strips and engineering purposes. IS 1730:1989, Bureau of Indian Standard, New Delhi, India

Comparison of Response on Building Due to Wind Load as Per Wind Codes [IS 875—(Part 3)—2015] and [AS/NZ1170.2-2011]

Pradeep K. Goyal and Naveen Suthar

1 Introduction

With the advancement of computational methods and design procedure, tall buildings are widely constructed which are slender and lightweight. Construction of tall buildings is required due to scarcity of lands especially commercial areas. Wind effect needs to be considered while designing of high-rise structures. Wind codes, namely Indian standard code [1], combined code of Australia and New Zealand [2] are available for the design of high-rise building. Effect of wind on high-rise structure depends on terrain category and topography conditions and also depends on shape. Wind loads on tall building are obtained as per procedure given in the code. Wind pressure and wind force coefficient values for a particular building have to be taken from relevant countries' code during designing of tall building. The IS code [1] gives the procedure for calculating wind force using pressure coefficient method and force coefficient method. In this study, the force coefficient method is taken to design the high-rise structure for wind loads. The Indian standard code [1] remains silent about pressure coefficient of building having various side ratio and height and various shapes of structures. Kumar and Swami [3] presented a static and dynamic analysis for the high-rise structure up to 100 floors using IS:875(Part1)-1987 and compared the results. Effect of wind on coastal region and interior region of the country was also carried out. Kwon and Kareem [4] conducted a comparative study on wind loads on tall buildings and their effects using eight major international codes/standards. Reddy and Tupat [5] carried out the comparison of wind and earthquake load. It was found that the wind effect is more severe on high-rise structure than the earthquake effect. Jayasinghe and Weerasuriya [6] conducted a wind analysis of a tall building of height 182 m using five wind codes. In building acceleration calculation, 5-year return period wind speeds were utilized. Badria et al. [7] conducted a wind

P. K. Goyal · N. Suthar (✉)
Department of Civil Engineering, Delhi Technological University, New Delhi, Delhi, India
e-mail: naveensuthar90@gmail.com

A. Sil et al. (eds.), *Recent Trends in Civil Engineering*, Lecture Notes in Civil Engineering 274, https://doi.org/10.1007/978-981-19-4055-2_25

tunnel test for comparative study for calculating wind loads on tall building using different codes. Analytical results were compared with experimental results. Daniel et al. [8] found that failure of many tall buildings in India has occurred due to wind. Comparative analysis was carried out by using Indian code (IS 875:1987 Part 3) and second American code (ASCE-7:2002). It has been found that American codes have less chances of failure of structures as compared to Indian code. Shams and Mandal [9] carried out a comparative study using five countries' wind code. The codes and standards of wind loads, namely ASCE-7-98, AS1170.2, NBC-1995, RLB-AIJ-1993 and Eurocode 1-4 (1993), were used in this study. Alinejad et al. [10] conducted a study on the high-rise building for calculating wind load using ASCE-16-2017 and compared with Korean Building Code KBC-2016. Lee et al. [11] consider the effect of wind on high-rise tower using four different countries' wind codes, namely ASCE-7-10, KBC2009, AIJ2004 and EUROCODE. It was found that design wind velocity obtained different codes are slightly differed. In this paper, compression study of response on building due to two different wind codes AS/NZS 1170 (Part 2)-2011 and Indian (IS 875 (2015)-part 3) is carried out. Design wind pressure, wind load, lateral deflection and storey drift are computed using STAAD, and values obtained from both the codes are also compared. Suthar and Goyal [12] conducted a comparative study for calculating wind load on G+11 building using IS875-part 3-1987 and 2015. The comparison of deflection and storey drift also was studied.

This paper presents the analysis of wind loads on tall building using two different wind codes AS/NZS 1170 (Part 2)-2011 and Indian (IS 875 (2015)-part 3) and compares the result. In this study, design wind pressure, wind load, lateral deflection, storey drift, storey shear and storey moment are also compared.

2 Methodology

For calculation of wind forces on high-rise structures, two different countries' wind codes are considered in this study. Comparison of design wind speed, design wind pressure, wind load as per IS 875 part 3 (2015) and Australian/New Zealand standard AS/NZS 1170 part 2 (2011) is shown in Table 1. The table shows the variation of parameters like design wind speed, design wind pressures and pressure coefficients.

The wind speed varies with height and is also dependent on terrain category. Magnitude of wind speed at a given height is lesser in rougher terrain as compared to smoother terrains. In this study, terrain category 3 is considered for the comparison of wind load. For the calculation of wind loads, partial safety factor for loads is taken as 1.5 [IS-456-2000]. Partial factor of safety is considered as 1.15 and 1.5 for steel and concrete, respectively. Dead load (D.L.), impose load (L.L.) and wind load (W.L.) are taken for the design of building. Following load combinations, which are considered for the calculation of the wind load as per IS 456-2000 given below:

(i) $1.5 \times$ (D.L. + L.L.).
(ii) $1.2(\text{D. L} + \text{L.L} + W_{\text{in }X\text{+ve}})$, in X direction.

Table 1 Comparison of design parameters as per IS 875 (part 3)-2015 and AS/NZS 1170 (part 2)-2011

IS 875:2015 (part 3)	AS/NZ 1170.2 2011
The design wind speed (Vz) is given by $V_z = V_b K_1 K_2 K_3 K_4$ where V_z is the design wind speed at any height z in m/s V_b is the basic wind speed in m/s, K_1 is the probability factor K_2 is terrain height and structure size factor K_3 is the topography factor, and K_4 is the importance factor for the cyclonic region	The site wind speed is calculated as follows $V_{sit,\beta} = V_R M_d M_{Z,cat} M_s M_t$ where V_{sit} is the site wind speeds, V_R is the regional 3 s gust wind speed, in m/s for annual probability M_d is the wind directional multipliers $M_{z,cat}$ is the terrain/height multiplier M_s is the shielding multiplier, and M_t is the topographic multiplier
The basic wind pressure is obtained at height z as $P_z = 0.6\ Vz^2$ P_z is the design wind pressure in N/m^2 at height z V_z is the design wind speed at any height z in m/s $P_d = K_d K_a K_c P_z$ where P_d is the design wind pressure in N/m^2 at height z, K_d is the wind directionally factor K_a is the area averaging factor, and K_c is the combination factor	The wind pressures (P) are often determined as $P = 0.5\ [V_{des,\theta}]^2 \rho_{air} C_{fig} C_{dyn}$ P is design wind pressure ρ_{air} is the density of air taken as 1.2 kg/m^3 $V_{des,\theta}$ is the building orthogonal design wind speeds C_{fig} is the aerodynamic shape factor, and C_{dyn} is the dynamic factor
The wind load (F) on building or structure is determined by $F = C_f A_e P_d$ where A_e is the effective frontal area P_z is the design wind pressure in N/m^2 C_f is the force coefficient which depends upon shape of element plan size and wind dir.	Design wind force derived from force coefficients $F = 0.5\ [V_{des,\theta}]^2\ \rho_{air}\ C_{fig}\ C_{dyn}\ A_Z$ $F = pA_Z$ where A_z is projected area with drag force coefficient (C_d)

(iii) 1.2(D. L + L.L + $W_{\text{in Z+ve}}$), in Z direction.

2.1 *Analysis Steps*

- Nodal points are created as per the dimensions and plan in STAAD pro software.
- Structure is created by wizard structure command in STAAD.
- Fixed supports are assigned at the bottom of the structure. Beam and column dimensions are also assigned.
- Wind loads are obtained according to IS 875 PART 3 2015 and AS/NZ 1170.2 and added in + X and + Z directions.

- D.L. and L.L. are calculated as per IS 875 (part 1) and IS 875 (part 2), respectively. L.L. is taken for each storey as 3 kN/m^2.
- The load combination as described above is taken for the calculation of wind loads.
- Analysis is performed after all steps and checked for errors.
- As per IS 456:2000 and IS 800:2007 RCC frame design is performed and assign suitable command for each structure's member.

3 Numerical Study

Study was carried out on a high-rise building which is a RCC frame structure residential type building assumed to be located in a Delhi, and there are no vertical irregularities. Topography is assumed as flat in all directions. Building is rectangular in cross Sect. (50 m × 70 m) having height above ground surface that is 150 m and flat roof. Wind direction is normal to the 70 m wall face of the building. Basic wind speed is 47 m for Delhi region (Zone IV) as per IS875 (part3)-2015. Terrain category 3 is assumed for the region where building is situated in both directions. In order to evaluate the storey displacement and lateral displacement, analytical study is done in STAAD pro software. Design properties of building are shown in Table 2.

The rectangular building of dimension and properties as given above is modelled in STAAD pro. The 3D view of rectangular building and rending view of building are shown in Figs. 1 and 2.

Wind load on building is applied in *X* and *Z* directions which are considered for the wind load calculation on the building as per load combination considered in this study. Wind load in *X* direction is shown in Fig. 3 and in *Z* direction in Fig. 4.

Table 2 Design parameters of 150 m height building

No. of storey	50
Column size	1 m × 1m
Beam size	300 mm × 700 mm
Size of slabs (m)	0.150
L.L. on slab (kN/m^2)	3
Floor finish at each storey (kN/m^2)	3
F_{ck} in column	M 40
F_{ck} in beam	M 40
Fe of steel	Fe 500
Height of building (m)	150
Ground storey height (mm)	3000
Height of each storey (mm)	3000
One frame length and width (m)	5
Size of wall (mm)	150

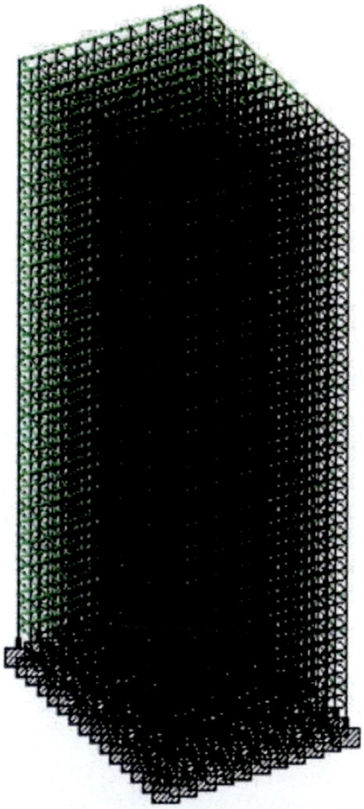

Fig. 1 3D view of framed structure

Fig. 2 Rendering view of framed structure

Fig. 3 Wind load acting in x direction

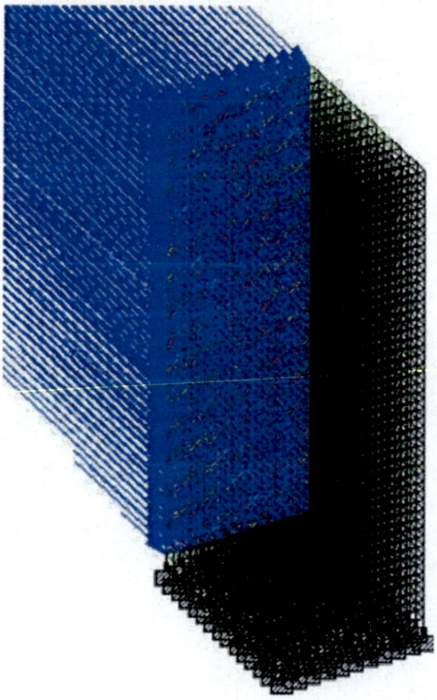

Fig. 4 Wind load acting in z direction

4 Results and Discussion

The comparison of wind load on the tall building for rectangular shape of the building due to different codes is studied. The storey response was checked in the STAAD, and all the storey is passed against the wind excitation. The effect of height on wind intensity is studied. The comparison of variation of wind intensity with height as per IS875 (part3) 2015 and AS/NZ 1170.2-2011 is shown in Fig. 5. It is clear from the figure that wind intensity is much higher in case of IS875 (part3) 2015 as compared to AS/NZ 1170.2-2011.

The comparison of variation of wind load with height as per IS875 (part 3) 2015 and AS/NZ 1170.2-2011 is shown in Fig. 6. It is shown from the figure that wind load increases gradually with height. It is also clear from the figure that AS/NZ1170.2-2011 gives more wind load as compared to IS875 (PART 3)-2015.

The comparison of lateral displacement with height is shown in Fig. 7. It is clear from the figure that the lateral displacement will be different at each storey of building.

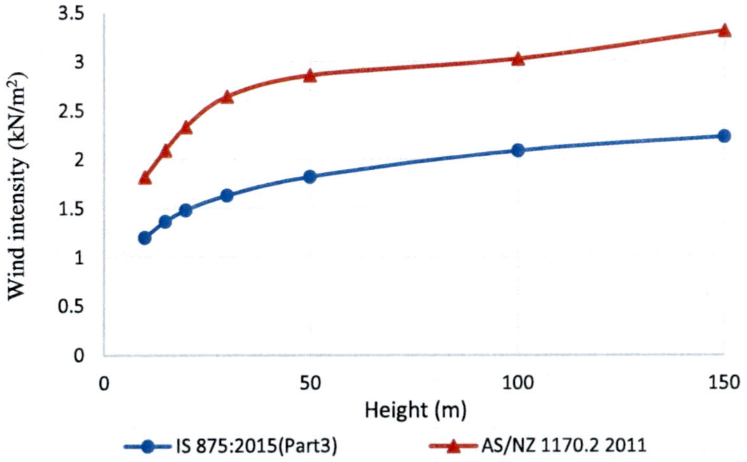

Fig. 5 Comparison of wind intensity with height

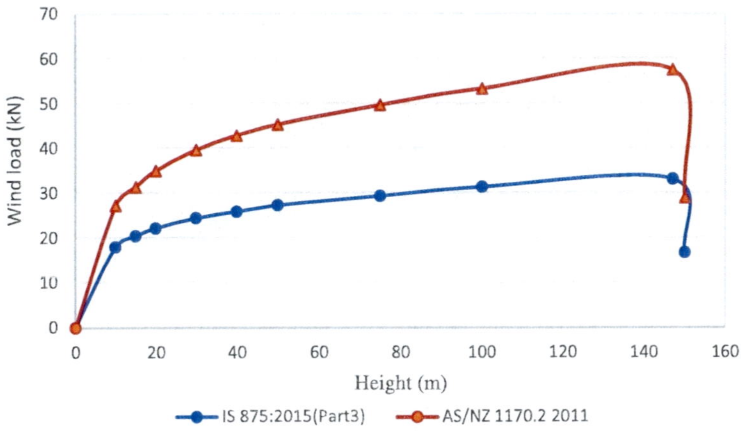

Fig. 6 Comparison of wind load in kN

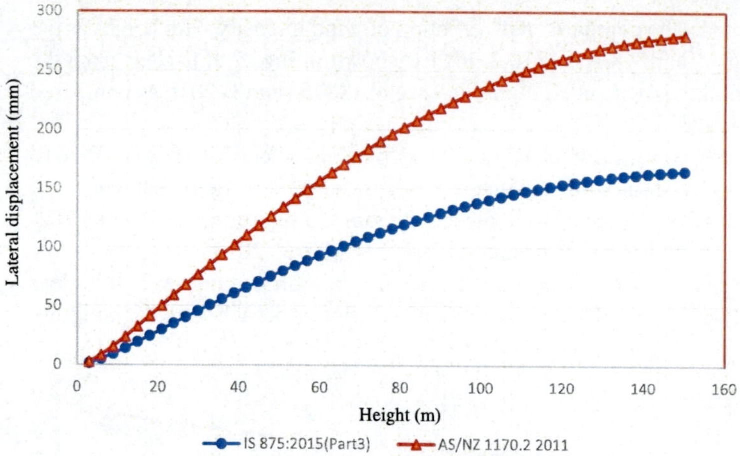

Fig. 7 Comparison of deflection in mm at different storeys of building in *x* dir.

There is no much difference of lateral displacement obtained from IS875 part 3-2015 and AS/NZ1170.2-2011 up to the height 50 m. There is significant difference of lateral displacement obtained from IS875 part 3-2015 and AS/NZ1170.2-2011 beyond 50 m height.

The comparison of storey drift with height is shown in Fig. 8. It is seen from the figure that there is significant difference in the storey drift as calculated by IS875 (part 3)-2015 and AS/NZ1170.2-2011 up to the height of 50 m.

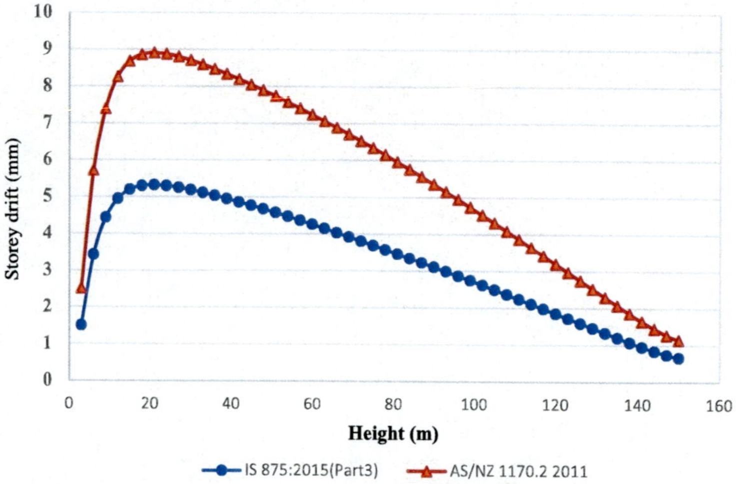

Fig. 8 Comparison of storey drift in mm at different storeys of building in *x* dir.

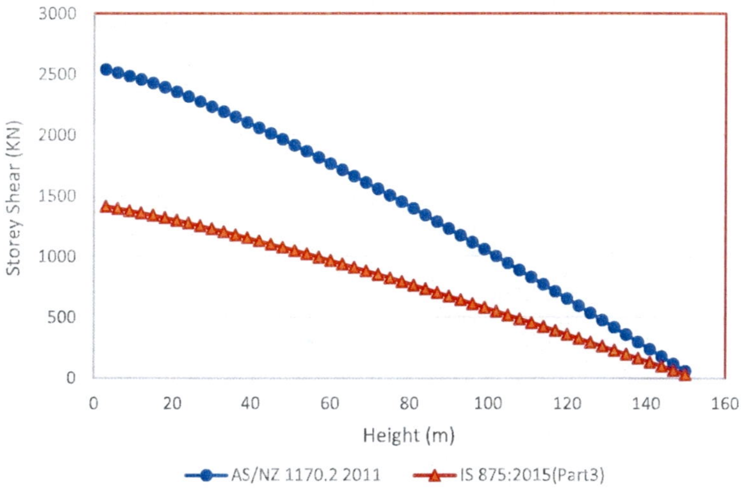

Fig. 9 Comparison of base shear at different storeys of building in *x* dir.

The comparison of storey shear with height is shown in Fig. 9. It is seen from the figure that there is significant difference at the bottom of building as calculated by IS875 (part 3)-2015 and AS/NZ1170.2-2011. Maximum storey shear is much higher at the bottom, and it is decreasing with height significantly.

The comparison of storey moment with height is shown in Fig. 10. It is seen from the figure that there is significant difference at the bottom of building as calculated by IS875 (part 3)-2015 and AS/NZ1170.2-2011. There is no significant difference after 80 storey. Maximum storey moment is much higher at the bottom and it is decreasing with height significantly.

5 Conclusion

From the study, it is concluded that analysis using the AS/NZ 1170.2 2011 code gives the higher defection, storey drift and wind load as compared to IS: 875 (Part 3)-2015. Therefore, building will be designed for higher values of wind load and deflection. That will give more safety to structures. Following specific conclusions are made from this study.

- Storey drift limitation should not be more than 0.002 times the storey height or H/500 in both the cases. Storey drift is much higher as calculated by AS/NZ1170.2-2011as compared to IS875 (part 3)-2015 up to the height of 50 m.
- The maximum lateral deflection 145.36 mm in the terrace level for tall building which is analysed as per IS:875 (Part 3) 2015 and 246.32 mm in tall building

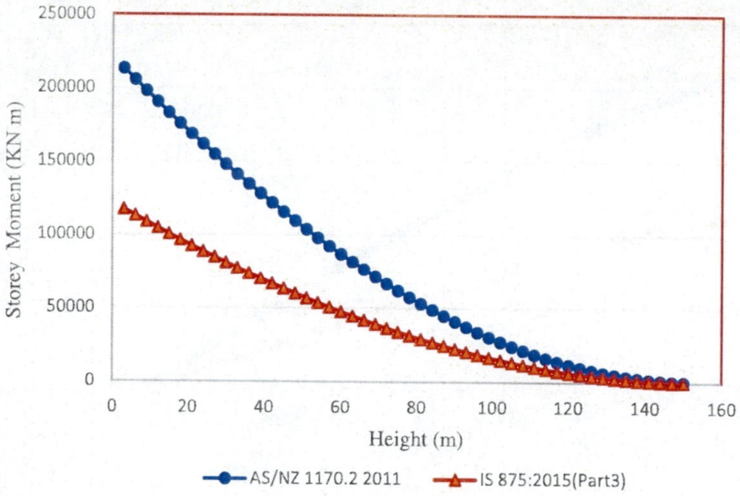

Fig. 10 Comparison of storey moment at different storeys of building in x dir.

which is analysed as per AS/NZ 1170.2 2011 in x direction which is less than permissible limit 300 mm for 150 m height building.

- As per the IS:875 (Part 3)-2015, wind load on high-rise structure is decreased as compared to AS/NZ 1170.2 2011 and the decrease in percentage of wind load is 42.27% on the topmost storey along X direction.
- As per 875 (Part 3)-2015, lateral defection at the terrace level of 50-storey building is decreased as compared to AS/NZ 1170.2 2011 and decrease in percentage of lateral deflection is 40.90% along X direction.
- As per 875 (Part 3)-2015, storey drift at the terrace level of 50-storey building is decreased as compared to AS/NZ 1170.2 2011 and decrease in percentage of lateral deflection is 48.56% along X direction.
- As per 875 (Part 3)-2015, storey shear at the bottom is lesser approximately 44.2% as compared to AS/NZ 1170.2 2011.
- As per 875 (Part 3)-2015, storey moment at the bottom is lesser approximately 44.95% as compared to AS/NZ 1170.2 2011.

References

1. IS875-part 3 (2015) Indian standards code of practice for design loads for buildings and structures part 3—wind loads. Bureau of Indian Standards, India
2. Australian/New Zealand Standard AS/NZS 1170.2 (2011) Structural design actions part 2—wind actions
3. Kumar BD, Swami BLP (2012) Critical gust pressures on tall building frames-review of Codal provisions. Int J Adv Technol Civ Eng 1(2):66–66

4. Kwon DK, Kareem A (2013) Comparative study of major international wind codes and standards for wind effects on tall buildings. Eng Struct 51:23–35
5. Reddy KRC, Sandip A, Tupat SA (2014) The effect of zone factors on wind and earthquake loads of high-rise structures. India IOSR J Mech Civ Eng (IOSR-JMCE):53–58
6. Jayasinghe MTR, Weerasuriya AU (2014) Wind loads on high-rise buildings by using five major international wind codes and standards. Engineer 18(3):13–25
7. Badria AA, Hussein MM, Attiab WA (2015) Study of wind tunnel test results of high-rise buildings compared to different design codes. Wind Struct 20(5):623–642
8. Daniel C, Daniel L, Shelton J, Raj A, Vincent Sam Jebadurai S, Hemalata G (2017) influence of wind loads design of Indian code compared with American code. Int J Civ Eng Technol 8(5):676–685
9. Shams A, Mandal S (2017) Comparative Study of along-wind response of major international codes with Indian code. Int J Eng Res Technol 6(11):256–260
10. Alinejad H, Jeong SY, Kang THK (2020) Comparative assessment of ASCE 7-16 and KBC 2016 for determination of design wind loads for tall buildings. Wind Struct 3(6):575–591
11. Lee S, Kim S, Kwon D-S (2021) A comparative study of wind tunnel test and design codes in evaluation of along-wind effect on towers. Shick Vibr 2021:13. https://doi.org/10.1155/2021/8818057
12. Suthar N, Goyal PK (2021) Comparison of response of building against wind load as per wind codes [IS 875—(Part 3)—1987] and [IS 875—(Part 3)—2015]. IOP Conf Ser Earth Environ Sci 796:012007

Strengthening Measures for Reinforced Concrete Column Against Blast Loading—A Review

Atul Pandey and Hari Krishan Sharma

1 Introduction

The past two to three decades have witnessed a gradual shift in the focus of the scientific community from understanding the effects of gravity, seismic and wind loads on structures to more unconventional loads of which blast loads are the most prominent. The reason for this switch is not lost on anyone since there is an increased level of threat to Buildings and other infrastructures due to geopolitical instability, proliferation of weapons and unprecedented rise in the number of terrorist outfits. The fatalities arising due to direct or indirect (because of flying fragments and failure of infrastructure) effects of blast cannot just be blamed on deliberate targeting but there are also numerous cases of human negligence in the handling of flammable and detonative materials which have resulted in fatal situations. The most current example being the 2020 Beirut explosion in which an estimated amount of 2750 tonnes of ammonium nitrate stored at a warehouse in the port city of Beirut [1], the capital of Lebanon exploded leading to at least 210 deaths and property damage of 15 billion USD. Thus, it has become more important than ever for researchers and engineers to study the behaviour of RC (because of it being the most widely used construction material in the world) structural members when subjected to blast loading and recommend resistant measures for the same especially RC columns because, in most explosion scenarios, the blast overpressure acts in a lateral direction of a structure, thereby causing more damage to vertical members such as columns and walls than slabs or beams.

The primary objective of this review paper is to discuss the dynamic response and failure modes of an RC column due to both contact and non-contact explosion as well as various strengthening/retrofit solutions available in the literature to safeguard against its detrimental effect. Moreover, the strength enhancement, effectiveness and

A. Pandey (✉) · H. K. Sharma
Department of Civil Engineering, National Institute of Technology, Kurukshetra 136119, India
e-mail: atul_31902305@nitkkr.ac.in

A. Sil et al. (eds.), *Recent Trends in Civil Engineering*, Lecture Notes in Civil Engineering 274, https://doi.org/10.1007/978-981-19-4055-2_26

applicability of each retrofit solution are also discussed along with the variation in blast resistance with different parameters of the column as well as the retrofit method.

A blast event occurs due to detonation of explosive materials which may be solid, liquid or gaseous in their physical states and typically results in a sudden increase in the pressure (called peak positive incident overpressure) and temperature of the air surrounding the explosive. The surrounding air particles get displaced and form a shock front or blast wave which travels at a very high velocity simultaneously creating high-pressure zones in the regions it passes. Although after a blast wave has passed a region, its pressure exponentially falls below the ambient atmospheric pressure (known as negative phase) before reaching pre-blast level pressure, the problem arises when there is an obstruction or target such as a building in the way of the blast wave which gives rise to peak positively reflected overpressure which can be 2–8 times in magnitude as compared to the incident one [2]. This happens because the air particles in between the target structure and blast wave take more time in clearing as they have to go around the target and are thus compressed against the structure by the incoming blast wave. A general pressure–time history curve for a region under contact of a blast wave is given in Fig. 1 which is mathematically expressed using Modified Friedlander equation given in Eq. 1.

It is also crucial to note that the negative pressure phase results in formation of a vacuum in which air gets filled consequently at a rapid rate due to which pressure acts in the opposite direction which can lead to moment reversal in case the blast wave impinges on a target structure. The negative phase is often neglected because of its small magnitude to simplify the analysis.

$$P(t) = P_0 + P_{so^+}\left(1 - \frac{t}{t_{pos}}\right)e^{-b\frac{t}{t_{pos}}} \tag{1}$$

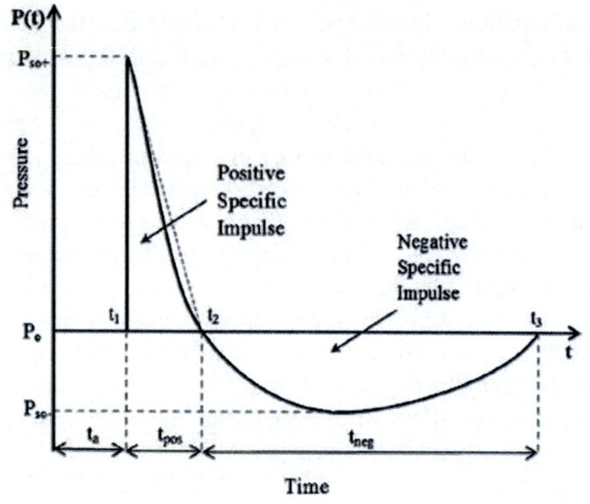

Fig. 1 Pressure–time history curve for a Blastwave [3]

where $P(t)$ is the incident pressure at any time (t), (P_0) is the ambient atmospheric pressure of the region, (P_{so+}) is the peak positive incident overpressure as discussed before, (t_{pos}) is the duration of the positive phase and (b) is the decay parameter.

While an explosion can be categorized on the basis of type of explosive material, confinement of charge and shape of charge, the studies related to response and blast performance enhancement of structural elements (Column, beam, slabs and walls) have been mostly done considering the classification based on scaled distance (Z) of the explosive which is given by Eq. 2.

$$Z = \frac{R}{W^{1/3}} \tag{2}$$

where (R) is the distance between the point of detonation and the target specimen (termed as standoff distance) in metres and (W) is the equivalent mass of TNT in kilograms. On the basis of scaled distance an explosion can be termed as far-range for ($Z \geq 1.2$ m/kg$^{1/3}$), close-in for ($Z < 1.2$ m/kg$^{1/3}$) and contact explosion for ($Z = 0$). Far-range and close-in explosions can be grouped together as non-contact explosions.

2 Response of Conventional RC Column to Blast Loading

To obtain the response and behaviour of typical RC columns against blast overpressure, analysis in the literature has been done experimentally through controlled on-field explosions, analytically by equivalent SDOF spring-mass systems and numerically by finite element modelling in LS-DYNA, AUTODYN and ABAQUS softwares. The failure criterion for RC column is quite different for contact blast when compared to non-contact blast.

In the case of non-contact explosions which have a non-zero standoff distance, the failure in column is mainly due to excessive plastic flexural deformation accompanied by crushing of concrete at the blast facing or compression face and spalling at the tensile face. Over stiff columns are more susceptible to fail in direct shear or diagonal shear at the supports or midspan. Of course, the degree of damage in column due to an explosion is very much dependent on the scaled distance, grade of rebar and concrete, column width and depth, reinforcement ratio and boundary conditions. Post-blast damage profiles of bare RC column specimens subjected to three different close-in blasts with varying scaled distances by Yan et al. [4] is given in Fig. 2.

It can be concluded from Fig. 2 that the extent of damage increases with a decrease in scaled distance. The midspan displacement, length of spall and crushing as well as the support rotations increased in successive cases. Shear cracks can also be seen on the right support in cases (b) and (c) which can be attributed to greater reflected impulse acting on the front face of the column due to reduced scaled distance.

Damage and failure in column due to contact blast ($Z = 0$) however is much more severe and catastrophic than non-contact blast because of considerable material failure and material loss, this happens because of internal blast wave reflections which

(a) Z = 0.54m/kg$^{1/3}$

(b) Z = 0.45m/kg$^{1/3}$

(c) Z = 0.40m/kg$^{1/3}$

Fig. 2 Comparison of damage modes of bare RC column for different scaled distances [4]

cause local material damage. In this case, the energy transferred by the blast to the column is not dissipated through flexural deformation but rather by causing heavy crushing of concrete which is limited to some length above and below the explosive location before detonation. This results in reduction of cross-section of the column which causes significant drop in the axial capacity of column post-blast. Depending upon the cross-section and mass of the explosive the damage due to contact blast can vary from crater-spall failure in which a crater is formed at the contact face which merges with the side spall to total crushing of material core with spalling along the length of column on all sides. Post-blast damage profiles of square RC column specimens subjected to three different contact blasts with varying mass of explosive by Dua et al. [5] is given in Fig. 3. It can be seen that there is a small crater-spall failure in column for 115 g TNT explosion, followed by partial and complete crushing of concrete core for 0.5 kg and 1 kg TNT explosion, respectively.

Fig. 3 Comparison of damage modes for 300 × 300 mm column subjected to contact blast from 115 g, 0.5 kg and 1 kg TNT, respectively (from left to right) [5]

3 Strengthening of RC Columns for Non-contact Blast

The traditional approach to make an RC column more blast resistant is to increase the mass or section of the column by using higher grade concrete such as Ultra-High-Performance Concrete (UHPC) or Ultra High-Performance Fibre Reinforced Concrete (UHPFRC) instead of Normal Strength Concrete (NSC) but it is only feasible for upcoming structures and not existing ones and also the initial cost associated with it is very high. Increasing the section of the column on the other hand by concrete jacketing has its own downsides since if the new concrete is not properly bonded to the old concrete, it will play no role in the load transfer and will instead act as additional dead weight. Therefore, the more recent and advanced methods to retrofit columns and other RC members against detrimental effects of explosion has been to use composite wraps with high strength to weight ratios, high failure strains and sufficient ductility, stiffness and energy absorption capacities such as Fibre Reinforced Polymers (FRP's), Strain Hardening cementitious Composites (SHCC), Sprayed-On Elastomeric Polymers (EP), Polyurethane (PU) Bricks, Metallic foams such as aluminium, Steel Jackets, etc. Since a single composite material may not have all the desirable properties stated above, Hybrid composite laminates are also externally bonded to RC members such as Hybrid FRP-EP in which elastomeric polymer is sprayed over FRP laminate to compensate for the low failure strain of the latter. From the previously stated composite materials only Carbon Fibre Reinforced Polymer (CFRP), Glass Fibre Reinforced Polymer (GFRP), Steel Jacket and PU bricks have been explored as retrofitting options for columns. The applicability and effectiveness of other composite materials for blast retrofitting of RC column is

Table 1 Mechanical properties of blast retrofitting composite materials [3]

Material	Young's modulus (GPa)	Tensile strength (MPa)	Failure strain (%)	Strain rate sensitivity
Aluminium foam	0.1063	3.94	6.3	NA
AFRP*	70–124	3000–3600	2.4–3.6	NA
CFRP	150–250	3000	1.5–1.9	Noticeable for rates > 50 s^{-1}
GFRP	17–72	300–2300	4.5–4.9	Yes
Polyurea (EP)	Bond strength 2.1 MPa	20	310	Yes
SHCC	–	5	4	Yes
SFRP*	36	514	1.63	–

* *AFRP* aramid fibre reinforced polymer, *SFRP* steel fibre reinforced polymer

a subject for future research. Mechanical properties of some of the composites are listed in Table 1.

Muszynski and Purcell [6] in order to establish the effectiveness of CFRP and GFRP laminates for blast resistance subjected two full-fledged test structures (5.67 $\times$ 2.53 m^2) comprising different combinations of retrofitted (also type of retrofit laminate was varied for different members) and unretrofitted RC walls and columns to a detonation of 860 kg TNT charge at standoff's varying from 13.5 to 15.9 m. The structure in which the retrofitted walls or columns were facing the blast remained intact and only suffered lateral displacements and spalling as opposed to the unmodified one which failed catastrophically. They, however, did not quantify the damage and displacements of individual retrofitted and unretrofitted walls and columns and since both structures were laminated by CFRP as well as GFRP, it did not become clear as to which one among the two FRPs has greater blast resistant capacity, although theoretically CFRP because of a much larger elastic modulus and tensile strength than GFRP should be more effective.

Mutalib et al. [7] developed pressure-impulse (P-I) diagrams through numerical modelling for CFRP retrofitted (wrapped in longitudinal strips, transverse wrap and combination of both) and bare control RC columns subjected to close-in and far-range blast loads and concluded that the columns retrofitted with longitudinal strips had a greater pressure asymptote (P_O) for a particular damage degree (based upon residual axial capacity on a scale of 0–1) indicating higher flexural resistance than control columns, columns retrofitted with transverse wraps had a greater impulse asymptote (I_O) indicating higher shear resistance than control column and columns wrapped with combination of both had a high flexural as well as high shear resistance and thus it can resist a greater blast load without failure. They also did a parametric analysis to study the effect of column width, height, depth, CFRP strength, thickness of CFRP, longitudinal and transverse steel reinforcement ratios and grade of concrete on blast resistant efficacy of RC column which is summarized in Table 2.

Table 2 Effect of various parameters on blast resistance capacity of RC column

Parameter	Effect on blast resistant capacity of column	
	Impulse asymptote/shear strength	Pressure asymptote/flexural strength
CFRP strength	Considerable increase	Nominal increase
CFRP thickness	Tremendous increase	Nominal increase
Column depth	Tremendous increase	Considerable increase
Column width	Nominal decrease	Considerable decrease
Column height	Considerable decrease	Considerable decrease
Concrete strength	Considerable increase	Nominal increase
Longitudinal reinforcement ratio	Nominal increase	Considerable increase
Transverse reinforcement ratio	Considerable increase	Nominal increase

Codina et al. [8] conducted experiments to study the dynamic response of reinforced concrete (RC) columns retrofitted with classical steel jacketing (completely wrapped) and reinforced polyurethane brick sacrificial layer subjected to close-in blast. They recorded a 60% reduction in maximum final displacement of steel jacketed column and 20% reduction in maximum final displacement of column provided with a sacrificial layer of polyurethane bricks on the blast facing side for scaled distance (Z) of 0.3 m/kg$^{1/3}$. The adopted protection measures also managed to reduce the localized crushing and spalling of concrete indicating a larger residual axial capacity than the control column. From the experimental results, they also concluded that it is more advantageous to increase the strength of column in a blast case scenario (as done by steel jacketing) than by dissipating energy through sacrificial claddings (as done by PU bricks).

Omran et al. [9] studied the effect of configuration of steel jackets on the improvement in axial strength and blast resistance of RC column through numerical modelling in ABAQUS. They subjected 6 RC columns of same cross-section (200 × 200 mm^2) and rebars, first one being unretrofitted, 2nd and 3rd being jacketed with four angles plus 3 and 6 connecting plates on all four sides, respectively, 4th and 5th being jacketed with two channels plus 3 and 6 connecting plates on all four sides, respectively and the 6th column being completely wrapped by steel plates to a fixed blast pressure. The results of their investigation are summarized in Table 3.

From Table 3 it can be concluded that the jacketing with channels is the most successful configuration followed by complete steel plate wrapping followed by

Table 3 Maximum midspan displacement and residual axial capacity of RC columns

Column specimen	Maximum midspan displacement (mm)	Axial capacity (kN)	Residual axial capacity (kN)
Col-0	60	1231	–
Col-1	23	1897	280
Col-2	16	1684	791
Col-3	4	1627	1506
Col-4	4	1873	1762
Col-5	10	1524	990

angle sections in a non-contact blast case scenario because of maximum reduction in lateral displacement and considerable improvement in post-blast axial capacity of column. It can also be seen that both the nominal and residual axial capacity of the channel retrofitted RC column increased with an increase in the number of connecting plates.

The more recent analysis of CFRP strengthened RC columns subjected to close-in blast loading was done by Yan et al. [4] They established through numerical simulation that in addition to considerable reduction in maximum and residual midspan displacement of RC columns, external CFRP bonding also led to reduced stresses in rebars and internal energy of concrete thus indicating energy absorption properties of CFRP. They also discussed the pros and cons of retrofitting both the tension as well as compression face of the column and concluded that although laminating the column on both sides reduced the peak displacement and the possibility of column failure due to moment reversal during the negative phase of blast pressure, it made the column over stiff thereby resulting in shear failure for blast loads characterized by large amplitude and small duration. They also did a parametric study to understand the effects of bonding strength between CFRP and Concrete, anchorage systems and pre-blast axial load on the strength enhancement of RC columns against close-in blasts and concluded that the increase in bonding strength up to 10 MPa reduced the maximum and residual midspan displacement, any increase beyond 10 MPa did not improve the resistance, anchorage systems are effective only when its bonding strength is less than the tensile strength of concrete and application of pre-blast axial loads results in decrease of lateral displacements.

4 Strengthening of RC Columns for Contact Blast

One of the methods proposed to prevent failure of RC columns due to surface or contact blast which is both experimentally and numerically validated by Dua et al. [5] is to increase the width of the blast facing side w.r.t depth such that the damage mode changes from crushing and side spalling for square column to small craters and minor cracks for rectangular columns with more aspect ratio. In addition to reduced damage,

Table 4 Residual axial capacity of equivalent square and rectangular RC columns [5]

Cross-section of square/rectangular column (mm × mm)	Mass of TNT (g)	Nominal axial capacity (kN)	Residual axial capacity of square/rectangular column (kN)
390 × 390/300 × 500	500	3044	923/895
460 × 460/300 × 700	500	4028	3060/4028
520 × 520/300 × 900	2500	5206	665/5206

the rectangular columns having equivalent cross-sectional area and nominal axial capacity as that of square columns exhibited a much higher residual axial capacity than its square counterparts which is presented in Table 4. The suggested method can however only be employed in case of upcoming structures. Also, the increase in the width of the blast facing side may prevent damage of columns for contact explosion but it results in a decrease of flexural strength of rectangular column of equal cross-sectional area as the equivalent square column for out of plane bending which makes it susceptible to failure due to flexural yield on application of blast overpressure from far-field or close-in explosives or other natural and accidental lateral forces.

Liu et al. did [10] a numerical and experimental study of CFRP protective circular RC piers under contact explosion and concluded that CFRP can provide effective protection for RC piers under contact explosion by virtue of decrease in peak and residual internal energy and axial stresses in concrete and reinforcement of retrofitted piers. They stated that the concrete can absorb external energy and translate it into internal energy by concrete deformation and fracture. From the internal energy time history, the peak internal energy and residual internal energy was greater for bare RC pier than for CFRP retrofitted piers and the residual internal energy of bare RC pier was nearly ten times greater than that of the other laminated piers, as was interpreted by more concrete deformation and fracture leading to greater residual internal energy of concrete. Therefore, the damage extent of unprotected piers was more serious in accordance with concrete damage. They subjected four piers to a contact explosion of 1 kg TNT at a height of 330 mm from the footing. One of them was a control pier and the other three were retrofitted with one, two and three layers of 0.163 mm thick CFRP sheet, respectively. Figure 4 shows the height of crushing on each pier.

It can be seen that the damage height and depth reduced significantly with the number of CFRP layers. They further performed numerical simulations to determine the reduction in peak and residual internal energy of concrete and axial stresses in the longitudinal and transverse reinforcements due to CFRP lamination which is tabulated in Table 5. The remaining axial capacity of the piers however was not determined.

Fig. 4 Comparison of damage of bare RC pier and pier retrofitted with one, two and three layers of CFRP, respectively subjected to contact explosion from 1 kg TNT (from left to right) [10]

Table 5 Peak and residual internal energy and reinforcement stresses of bare and CFRP-wrapped RC piers

Pier specimen	Peak/residual internal energy of concrete (kJ)	Peak/residual axial stress in blast facing longitudinal rebar (MPa)	Peak/residual axial stress in blast facing transverse rebar (MPa)	Peak/residual axial stress in back face longitudinal rebar (MPa)	Peak/residual axial stress in back face transverse rebar (MPa)
Control pier	67/20	500/150	780/75	650/125	700/150
CFRP-wrapped (1 layer)	2.7/2.3	350/0	450/300	200/100	375/75
CFRP-wrapped (2 layers)	3.3/2.7	350/0	400/300	200/100	300/75
CFRP-wrapped (3 layers)	3.3/2.7	325/0	300/200	225/100	275/75

5 Conclusion

From the undertaken review it can be concluded that majority of studies in the field of blast protection of RC columns concern itself with the use of CFRP and steel jacketing as an effective blast retrofit solution even though there are high costs associated their application. Therefore, there is a need to study the efficacy of other economical composite materials such as SHCC, AFRP, SFRP, Metallic foams and elastomeric polymers which have shown agreeable results in case of other RC members especially slabs [11–14] for blast protection of columns. It can also be seen that research related to the behaviour and retrofitting schemes for RC column against contact explosion is in a very nascent stage and requires more focus and investigation because the threat level posed by it is very much severe than far-range or close-in explosion. The paper also highlights the effective application of FRP since large thickness or number of layers does not guarantee greater blast resistance but can in turn lead to pre-mature failure by making the column stiffer.

References

1. How ship's deadly cargo ended up at Beirut port. BBC News. 6 Aug 2020
2. Trajkovski J, Kunc R, Parenda J, Prebil I (2014) Minimum mesh design criteria for blast wave development and structural response—MM-ALE method. Latin Am J Sol Struct 11:1999–2017
3. Goswami A, Adhikary SD (2019) Retrofitting materials for enhanced blast performance of structures: recent advancement and challenges ahead. Constr Build Mater 204:224–243
4. Yan J, Liu Y, Xu Z, Li Z, Huang F (2020) Experimental and numerical analysis of CFRP strengthened RC columns subjected to close-in blast loading. Int J Impact Eng 146:103720
5. Dua A, Braimah A, Kumar M (2020) Experimental and numerical investigation of rectangular reinforced concrete columns under contact explosion effects. Eng Struct 205:109891
6. Muszynski LC, Purcell MR (2003) Composite reinforcement to strengthen existing concrete structures against air blast. J Compos Constr 7:93–97
7. Mutalib AA, Hao H (2011) Development P-I diagrams for FRP strengthened RC columns. Int J Impact Eng 38:290–304
8. Codina R, Ambrosini D, Borbón FD (2016) Alternatives to prevent the failure of RC members under close-in blast loadings. Eng Fail Anal 60:96–106
9. Omran ME, Mollaei S (2017) Investigation of axial strengthened reinforced concrete columns under lateral blast loading. Shock Vib 2017:3252543
10. Liu L, Zong Z, Gao C, Yuan S, Lou F (2020) Experimental and numerical study of CFRP protective RC piers under contact explosion. Compos Struct 234:111658
11. Adhikary SD, Chandra LD, Christian A, Ong KCG (2018) SHCC-strengthened RC panels under near-field explosions. Constr Build Mater 183:675–692
12. Wu C, Huang L, Oehlers DJ (2011) Blast testing of aluminum foam-protected reinforced concrete slabs. J Perform Constr Facil 25:464–474
13. Silva PF, Lu B (2007) Improving the blast resistance capacity of RC slabs with innovative composite materials. Compos Part B Eng 38:523–534
14. Kong X, Qi X, Gu Y, Lawan IA, Qu Y (2018) Numerical evaluation of blast resistance of RC slab strengthened with AFRP. Constr Build Mater 178:244–253
15. Raman SN, Ngo T, Mendis P, Pham T (2012) Elastomeric polymers for retrofitting of reinforced concrete structures against the explosive effects of blast. Adv Mater Sci Eng 2012:754142

Variation of Progressive Collapse Resistance of Diagrid Structures Based on Plan Geometry and Section

Nipun B. Harigopal and Mini Remanan

1 Introduction

Lateral load resistance is of high priority in high-rise buildings along with resistance to vertical loads. Some of the structural systems used for high-rise buildings are braced frame structural system, rigid frame structural system, wall frame system, shear wall system, core and outrigger structural system, infilled frame structural system, flat plate and flat slab structural system, tube structural system, coupled wall system, and diagrid structural system. In diagrid structures, almost all the vertical columns are eliminated at the exterior and replaced by triangulated exterior members. The diagonal members in diagrid structures resist vertical loads and lateral loads, whereas the diagonals in braced frame type structures resist only lateral loads. Diagrid structures tend to have better resistance to seismic loads as the elements produce a structure that has a better lateral stiffness when compared to other tubular structural systems. They are also more architecturally pleasing than other structural systems.

Progressive collapse is the spread of an initial local failure from element to element, resulting eventually in the collapse of an entire structure or a disproportionately large part of it. (ASCE 7-16). These events can become chaotic as the collapse can cause multiple injuries and destruction of property in and at the immediate vicinity of the building. Progressive collapse is a topic of interest as the scale of damage may be very large compared to the initial failure. Progressive collapse resistance of a structure depends on the ability of the structure to redirect or redistribute the load through a different path to the ground. The response of a building due to local damage in the load-bearing elements leads to nonlinear deformation of

N. B. Harigopal (✉) · M. Remanan
Department of Civil Engineering, National Institute of Technology Calicut, Calicut, Kerala 673601, India
e-mail: nipunharigopal@gmail.com

M. Remanan
e-mail: mini@nitc.ac.in

A. Sil et al. (eds.), *Recent Trends in Civil Engineering*, Lecture Notes in Civil Engineering 274, https://doi.org/10.1007/978-981-19-4055-2_27

the structural elements prior to failure. Therefore, progressive collapse is a nonlinear dynamic phenomenon.

The analysis for determining progressive collapse resisting capacity is done by arbitrary column removal scenario. The guidelines for design and analysis for progressive collapse are given by General Services Administration (GSA) and Department of Defence (DoD) in the Unified Facilities Criteria (UFC), UFC 4-023-03. Diagrid structures tend to have a good progressive collapse resisting capacity as compared to other structures due to the presence of multiple alternate paths for load transfer because of the triangulated exterior elements.

Moon [1] developed a preliminary design method for determining the sizes of diagonals and studied the influence of the diagonal angle on the behaviour of diagrid structures. Kim and Lee [2] compared the progressive collapse resisting capacity of diagrid structures to conventional tubular structures with exterior moment resisting frame based on GSA (2003) and UFC 4-023-03 (2005) guidelines by arbitrary column removal scenario and found that the structures had similar progressive collapse resistance.

Kwon and Kim [3] studied the progressive collapse and seismic load resisting capacities of twisted diagrid buildings based on the ATC 63 (Applied Technology Council) code. As the twisting angle increased, the progressive collapse resisting capacity increased and so did the seismic load resistance. Linear analysis takes into account linear behaviour of material, whereas nonlinear analysis considers the geometric and material nonlinearity. Study by Kwon et al. [4] on two existing buildings compared the various analysis procedure for calculating the progressive collapse resisting capacity based on the GSA (2003) and the UFC 4-023-03 (2005) guidelines. Their study (based on arbitrary column removal scenario) concludes that the linear static method is conservative in the prediction of progressive collapse resisting capacity of building structure when compared with nonlinear analysis procedures. A study on the dynamic increase factor (DIF) was conducted by Mashhadiali et al. [5], and they proposed a methodology for determining the DIF for a structure based on GSA [6] and UFC 4-023-03 [7] guidelines.

Linear static procedure is recommended for buildings of 10 storeys or less. The use of the procedure is subject to limitations—demand-capacity ratio (DCR) and irregularities defined in GSA [6]. In the nonlinear static analysis, the plastic hinges are included. There are no DCR or geometric limitations to the application of this analysis. A DIF factor is used to account for the dynamic effects. An increased gravity load combination is applied on the bays immediately adjacent to the removed element and at all floors above the removed element as given in Eqs. 1 and 2 for linear static analysis and nonlinear static analysis, respectively;

$$G_{\mathrm{L}} = \Omega_{\mathrm{L}}[1.2D + (0.5L \text{ or } 0.2S)] \tag{1}$$

$$G_N = \Omega_N[1.2D + (0.5L \text{ or } 0.2S)] \tag{2}$$

G_{L} Increased gravity loads for linear static analysis.

D Dead load.
L Live load.
S Snow load.
Ω_L Load increase factor for linear static analysis.
G_N Increased gravity loads for nonlinear static analysis.
Ω_N Dynamic increase factor for nonlinear static analysis.

In nonlinear dynamic procedure, the dynamic effects of progressive collapse and the plastic hinge formation are included. This method is the most accurate in prediction of progressive collapse even though the time required for this procedure is longer. There are no DCR or geometric irregularity limitations on the use of the nonlinear dynamic procedure. There is no DIF that is to be multiplied to the load. The load combination to be applied on the entire area of the floor for this analysis is given in Eq. 3. In the current study, nonlinear dynamic analysis is done.

$$G_{\mathrm{ND}} = 1.2D + (0.5L \text{ or } 0.2S) \tag{3}$$

G_{ND} Gravity loads for nonlinear dynamic analysis.

In most of the studies, two diagonal elements from the triangulated exterior elements were removed at ground storey level [2, 8]. In the current study, the same procedure is adopted. UFC 4-023-03 (2009) gives a lateral load that can be used in combination with the vertical load for estimating the progressive collapse resistance of various structures. The lateral load to be used along with Eqs. (1–3) is given in Eq. (4). The lateral load is to be applied on the structure along in each direction, one at a time, in the horizontal plane to estimate the progressive collapse resistance of the structures.

$$L_{\mathrm{LAT}} = 0.002\Sigma P \tag{4}$$

G_{ND} Gravity loads for nonlinear dynamic analysis.
L_{LAT} Lateral load.
ΣP Sum of dead and live loads acting only on that floor.

The use of lateral load was removed from UFC 4-023-03 c2 (2013) onwards as the initial and primary damage is limited to the element removal location, with the rest of the structure being intact.

2 Modelling and Analysis

The structures are modelled in such a manner that the sum of the floor areas is the same for all twelve structures. The modelling is done in Rhinoceros 3D using grasshopper plugin. The models are then designed in SAP2000 as per IS 800:2007 design provisions and are checked against AISC 360-16. All model structures are 40 storey buildings with floor height of 4 m. The dead and live loads are assumed to be 5 kN/m^2 and 2.5 kN/m^2, respectively, and the structure is assumed to be located in seismic zone 4 as per IS 1893:2016. The (straight, convex, and concave) elevations of the structures with square, hexagonal, octagonal, and circular plan are shown in Fig. 1a–d, respectively. The plan and a module (with two storeys from the ground) of the structure with square plan and straight elevation are shown in Figs. 2 and 3, respectively. The core structure of all the models is square in shape with sides of 10 m as shown in Fig. 2. A module of a structure consists of 2 storeys through which the diagrid elements are connected. The models were designed as per IS 800:2007 with seismic loads as per IS 1893:2016. Then, the models were checked to conform to AISC 360-16 specifications. Pipe sections are used for the diagrid elements, and I-sections are used for the remaining load carrying elements. The diagrid elements are designed to resist the gravity loads as well as the lateral loads, whereas the core structure is designed to resist only the dead and live loads. No lateral load is transferred to the core structure.

2.1 *Nonlinear Dynamic Analysis*

Nonlinear dynamic analysis is done in SAP 2000 by simulating a sudden removal of a pair of diagonal elements according to the steps and specifications provided in UFC 4-023-03 [7]. A structure with a pair of diagonal elements removed from the ground storey is shown in Fig. 4. Initially after modelling the structure, a nonlinear static analysis is carried out with the load combination given in Eq. (3). This analysis is done to find the axial loads carried by the pair of diagonal elements that has to be removed. Later, in the model, the pair of diagonal elements is replaced with their corresponding axial loads at the joint from where it is removed. Nonlinear dynamic analysis is carried out in such a manner that the axial loads applied at the joints are removed after 10 s by using a user defined time history function. The maximum vertical displacement of the joint from which the diagonal elements are removed is found to understand the progressive collapse resistance of the structures.

Nonlinear dynamic analysis is also carried out with the load combination including lateral loads as given in Eq. (4). The axial loads carried by the pair of diagonal elements are found again and are applied at the joints where it is removed, and the maximum vertical displacements of the joints are obtained for the case with lateral loads. The plastic hinge distribution and the elements within the structure that are active in resisting progressive collapse are found for all the structures.

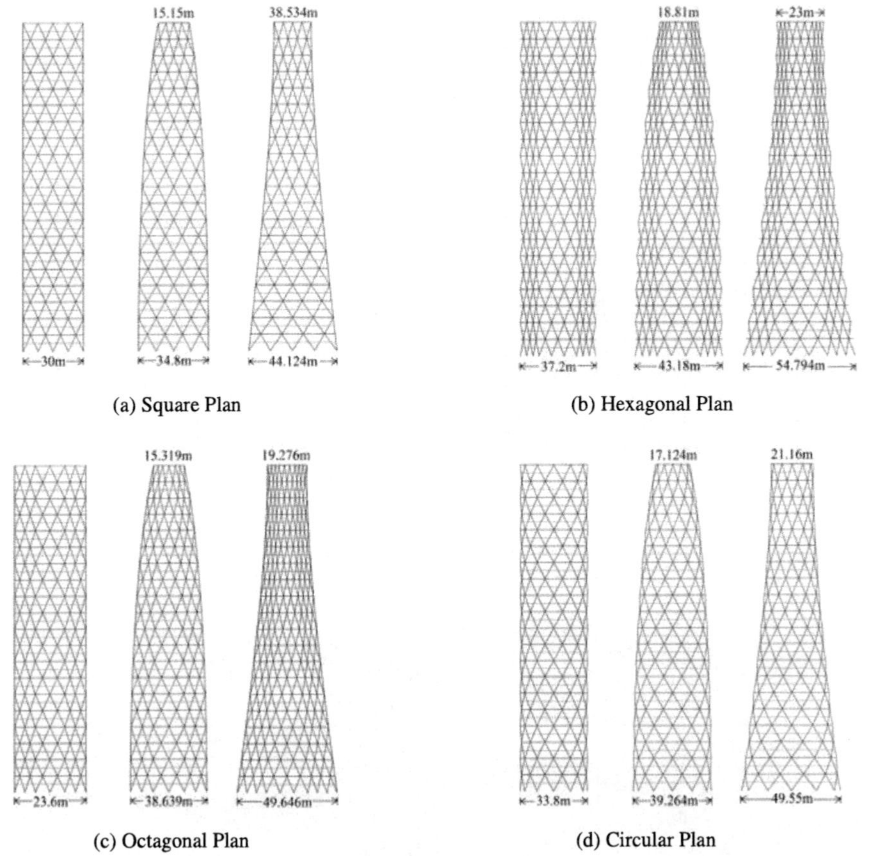

Fig. 1 Analysis models with **a** square plan, **b** hexagonal plan, **c** octagonal plan, and **d** circular plan

3 Nonlinear Dynamic Analysis Results

Since nonlinear dynamic analysis considers the geometric and material nonlinearity of structures along with the dynamic nature of progressive collapse, it yields much better results when compared to linear static and nonlinear static procedures. Nonlinear dynamic analysis was first done with the vertical load combination alone and was repeated with vertical and lateral load combination.

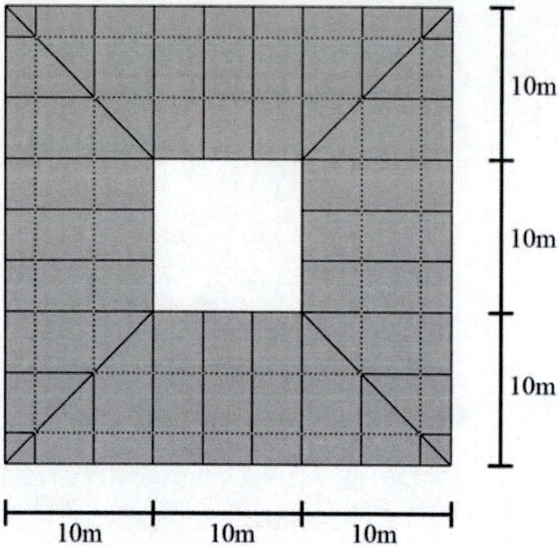

Fig. 2 Plan of a diagrid structure with square plan and straight elevation

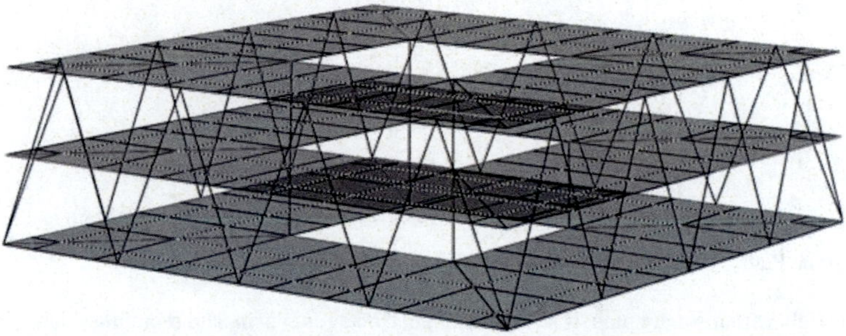

Fig. 3 Module of a diagrid structure with square plan and straight elevation

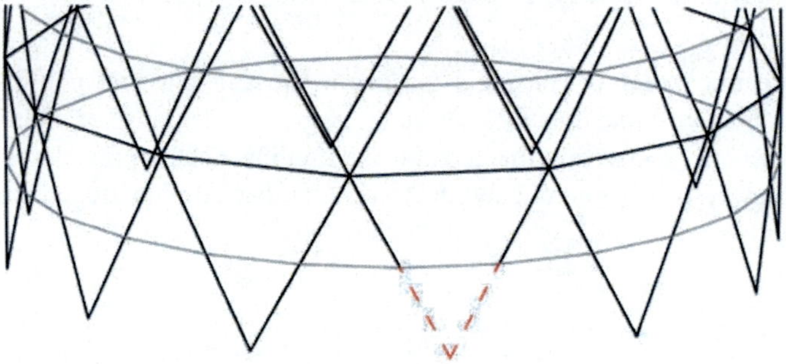

Fig. 4 Structure with circular plan and straight elevation with a pair of diagrid elements removed from the ground floor level

3.1 Vertical Displacement Time History Curves

The maximum vertical displacement that each structure undergoes due to the sudden removal of a pair of load-bearing elements represents the progressive collapse resistance of the structure. Higher vertical displacement indicates a lower resistance to progressive collapse. In the analysis, the suggested load combination as given in Eq. 3 is sustained for 10 s on the structures to attain stability, and then, the elements are removed suddenly to simulate damage. The vertical displacement time history curves of (the straight, convex, and concave) structures with square, hexagonal, octagonal, and circular plan under vertical load combination are shown in Fig. 5a–d, respectively, and that under vertical and lateral load combination are shown in Fig. 6a–d, respectively. The vertical displacement time histories of all the structures indicate that they are stable even after the sudden removal of the pair of diagrid elements from the ground floor level. The maximum vertical displacement of all the twelve structures under the two load combinations is given in Table 1.

For structures with the same plan shape, the maximum vertical displacement is for the structure with concave elevation, and the minimum vertical displacement is for the structure with convex elevation. The vertical displacement time history of a structure subjected to lateral load in addition to vertical load combination shows a small increase for all the cases. The maximum percentage increase in the vertical displacement when the lateral loads are also applied is 4.87% as shown in Table 1.

Out of all the twelve structures, the maximum vertical displacement is for the structure with square plan and concave elevation, whereas the minimum vertical displacement is for the structure with circular plan and convex elevation for both the load cases. In structures with the same elevation, the vertical displacement varies with the variation of the plan shape. For all the structures with the same elevation, the maximum vertical displacement is for structures with square plan, and the minimum vertical displacement is for structures with circular plan. Of all the twelve structures, the least maximum vertical displacement is shown by structures with circular plan with convex elevation, which implies that this structure gives the best performance in resisting progressive collapse compared to the other structures considered in the study.

3.2 Plastic Hinge Distribution

The plastic hinge formation in an element indicates the participation of the element in resisting collapse. The plastic hinge distribution in the diagrid elements of (the straight, convex, and concave) structures with square, hexagonal, octagonal, and circular plan under vertical load combination is shown in Fig. 7a–d, respectively, and that under vertical and lateral load combination are shown in Fig. 8a–d, respectively. The plastic hinge distribution of structures with concave elevations (in all plan shapes) is less spread out towards the higher storeys and more concentrated towards the lower

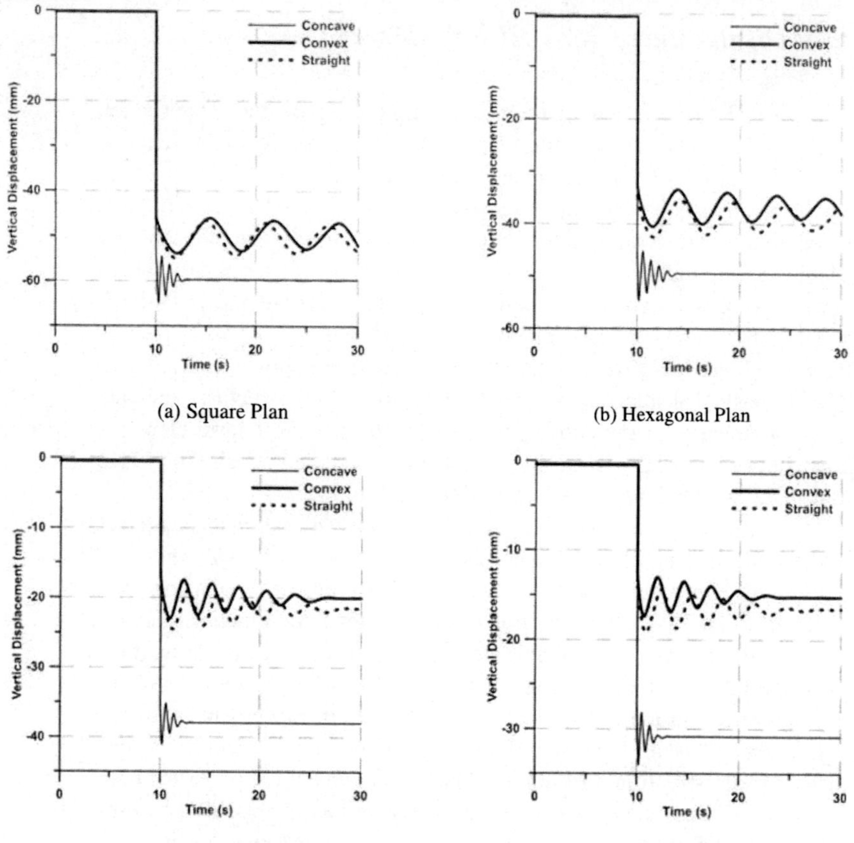

(a) Square Plan
(b) Hexagonal Plan
(c) Octagonal Plan
(d) Circular Plan

Fig. 5 Vertical displacement time histories of straight, convex, and concave structures under vertical load combination with **a** square plan, **b** hexagonal plan, **c** octagonal plan, and **d** circular plan

storeys when compared to that in structures with convex and straight elevations. This implies that less number of structural members participate in the collapse resistance of structures with concave elevation when compared to the structures with convex and straight elevations. In the structures with convex and straight elevations, the plastic hinges are distributed to the higher storeys and to more number of elements, implying that these structures are better at resisting progressive collapse. The plastic hinge distribution in structures with the same plan geometries having convex and straight elevations is similar. For structures subjected to combination of vertical and lateral loads, the plastic hinge distribution was similar to when the structures were subjected to vertical load combination alone. There is only a difference in the location of the plastic hinges, and they are spread out vertically and horizontally through the structure in similar manner.

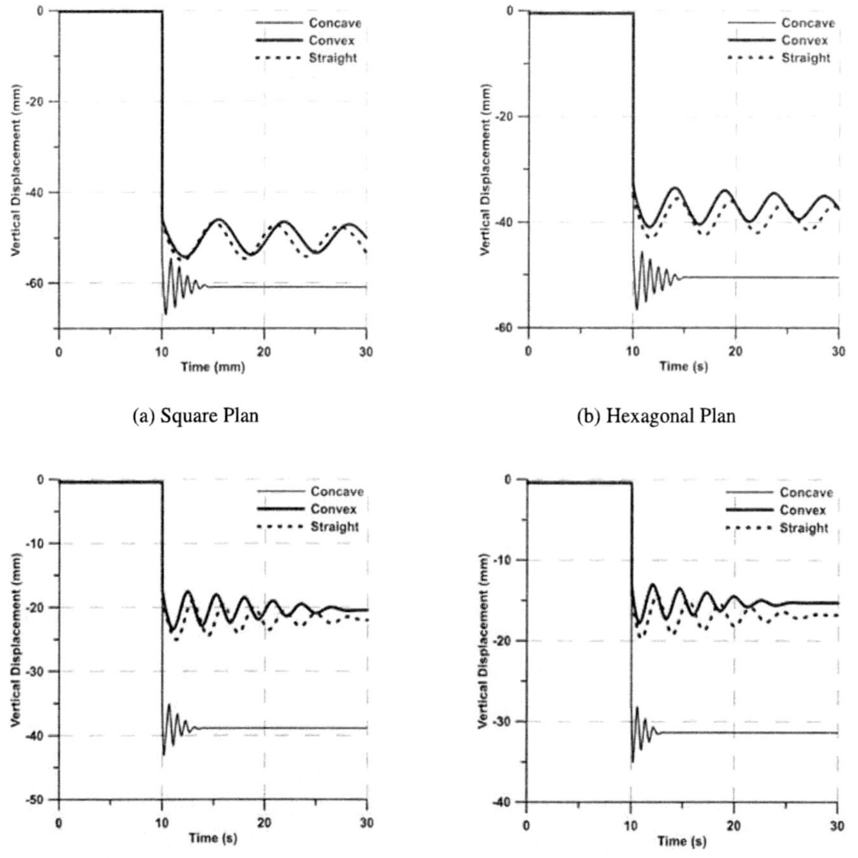

(a) Square Plan

(b) Hexagonal Plan

(c) Octagonal Plan

(d) Circular Plan

Fig. 6 Vertical displacement time histories of straight, convex, and concave structures under vertical and lateral load combination with **a** square plan, **b** hexagonal plan, **c** octagonal plan, and **d** circular plan

4 Conclusions and Future Work

4.1 Conclusions

The following conclusions were drawn from the study. The nonlinear dynamic analysis of the structure showed that progressive collapse resistance of diagrid structures is dependent on the variation of the plan shape and elevation of the structure. The analysis also showed that all the twelve structures were stable even after the sudden removal of a single pair of diagonal elements irrespective of the variation in plan shape and elevation. Circular structures are comparatively more resistant to

Table 1 Maximum vertical displacements in the structures

Plan shape	Elevation shape	Max. displacement (mm)		Percentage increase in displacement due to lateral load (%)
		Vertical load combination	Vertical and lateral load combination	
Circle	Straight	19.5	19.6	0.51
Circle	Convex	17.5	17.8	1.71
Circle	Concave	34.0	35.0	2.94
Octagon	Straight	24.0	25.0	4.16
Octagon	Convex	23.0	23.5	2.17
Octagon	Concave	41.0	43.0	4.87
Hexagon	Straight	42.5	43.0	1.17
Hexagon	Convex	40.9	41.0	0.24
Hexagon	Concave	55.0	56.5	2.72
Square	Straight	54.5	55.2	1.28
Square	Convex	53.0	54.3	2.45
Square	Concave	65.0	67.0	3.07

progressive collapse amongst all the selected plan geometries. Of the three variations in elevations considered, those structures with concave elevations are more affected by the sudden removal of elements as compared to their convex and straight counterparts. Plastic hinges formed in the concave structures indicate that lower storey diagrid elements are more active in resisting the collapse of the structure than elements in higher storeys. Plastic hinges formed in convex structures indicate that the higher storey diagrid elements are also active in resisting progressive collapse when compared to their concave counterparts. From the results obtained, we can conclude that structures with concave elevations should be avoided and those with circular plan and convex elevation should be preferred. Also, there is only less than 5% difference in the vertical displacement of the structures due to the lateral loads. This substantiates the removal of the lateral load from the load combination in the revised guidelines in UFC 4-023-03 c2 (2013) in determining the progressive collapse resistance of the structures.

4.2 Future Works

The current study only considers the removal of a single pair of diagrid element from the structures. This can be extended to the removal of more elements and also from different sides of the structure in succession or simultaneously. In this study, a square

Fig. 7 Plastic hinge distribution of straight, convex, and concave structures under vertical load combination with **a** square plan, **b** hexagonal plan, **c** octagonal plan, and **d** circular plan

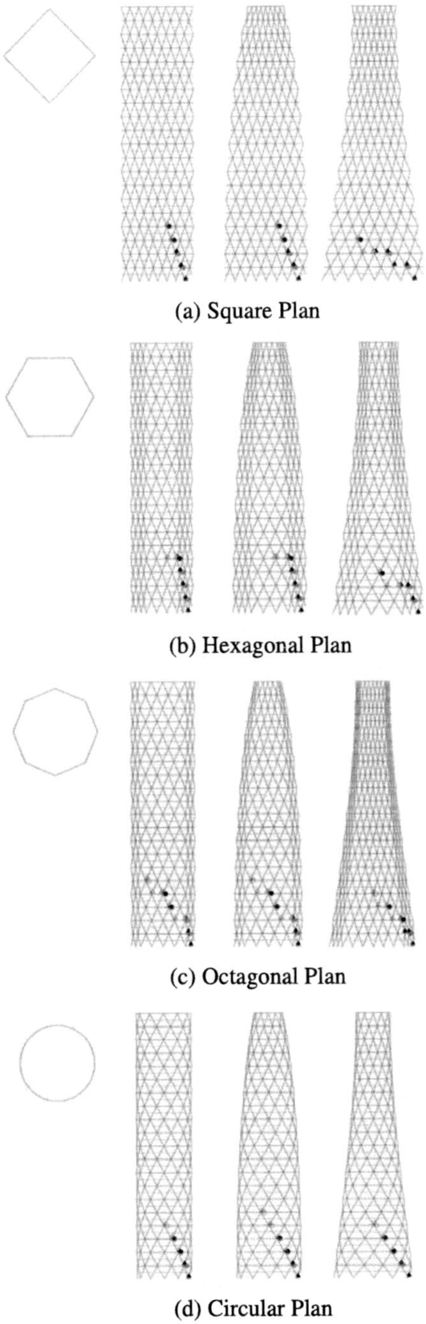

Fig. 8 Plastic hinge distribution of straight, convex, and concave structures under vertical and lateral load combination with **a** square plan, **b** hexagonal plan, **c** octagonal plan, and **d** circular plan

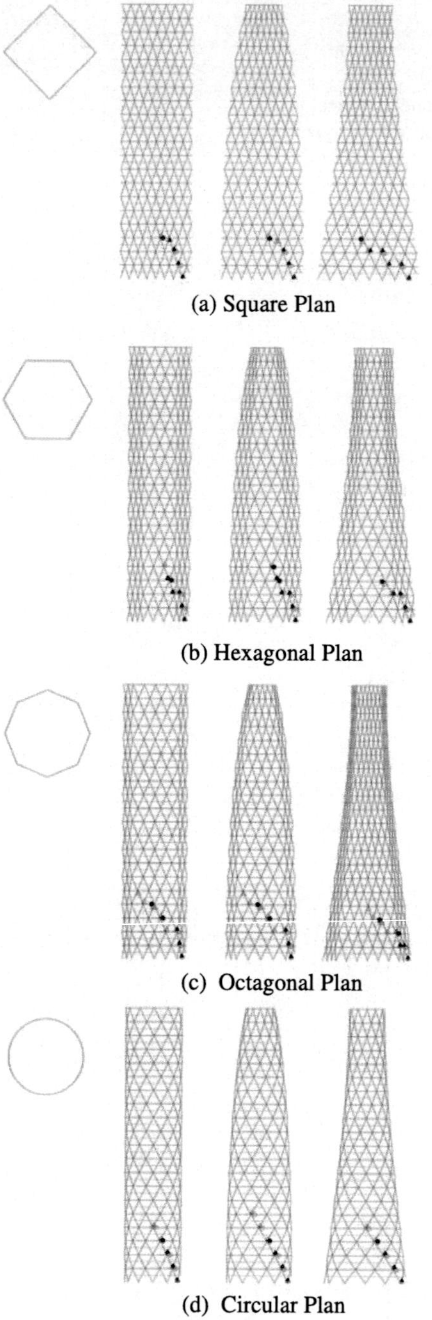

core was considered. A future study could focus on the influence of core geometry in the progressive collapse of diagrid structures.

References

1. Moon KS, Connor JJ, Fernandez JE (2007) Diagrid structural systems for tall buildings: characteristics and methodology for preliminary design. Struct Design Tall Spec Build 16:205–230. https://doi.org/10.1002/tal
2. Kim J, Lee Y (2010) Progressive collapse resisting capacity of tube-type structures. Struct Des Tall Special Build 19:761–777. https://doi.org/10.1002/tal.512
3. Kwon K, Kim J (2014) Progressive collapse and seismic performance of twisted diagrid buildings. Int J High-Rise Build 3(3):223–230
4. Kwon K, Park S, Kim J (2012) Evaluation of progressive collapse resisting capacity of tall buildings. Int J High-Rise Build 1(3):229–235
5. Mashhadiali N, Kheyroddin A, Zahiri-Hashemi R (2016) Dynamic increase factor for investigation of progressive collapse potential in tall tube-type buildings. J Perform Const Facilities:1–9. https://doi.org/10.1061/(ASCE)CF.1943-5509.0000905
6. General Services Administration (GSA) (2016) Alternate path analysis & design guidelines for progressive collapse resistance
7. Department of Defense (DoD), Unified Facilities Criteria (UFC) UFC 4-023-03 (2016) Design of buildings to resist progressive collapse
8. Kim J, Kong J (2012) Progressive collapse behavior of rotor type diagrid buildings. Struct Design Tall Spec Build. https://doi.org/10.1002/tal.762

Influence of Side Ratio on Wind Induced Responses of U Plan Shape Tall Building

Shanku Mandal, Sujit Kumar Dalui, and Soumya Bhattacharjya

1 Introduction

The evaluation of wind load is one of the most essential factors for the construction of tall buildings. The wind response depends on several factors like height, shape, location of the building and wind direction, etc. Various researchers took those aspects into consideration to determine the wind impact. Kawai [1] investigated the characteristics of vortex and pressure variation around the square building in different wind directions. Choi and kwon [2] studied the Strouhal number fluctuation on the rectangular shaped building with corner cuts. Gu and Quan [3] presented the across wind responses of rectangular, square, bevel and concave corner square models of different aspect ratios. Lin et al. [4] discussed the effect of elevation, side ratio and aspect ratio on wind forces of a square tall building. Gomes et al. [5] utilized wind tunnel and computational fluid dynamics (CFD) to measure the wind pressure on U and L shape buildings. Matsumoto et al. [6] compared the flutter response between rectangular and H shaped buildings. Memon et al. [7] explained the dependency of air temperature variation on aspect ratio and wind speed. Keast et al. [8] calculated the correlation of different wind responses of medium-rise two rectangular and one L shape model. Huang et al. [9] studied the wind interaction on a stepped tall building at high turbulence conditions. Peng et al. [10] evaluated the peak wind pressure on the building roof. Tecle et al. [11] studied the internal pressure variation due to compartmentalization and opening in low rise structures. Iqbal and chan [12] concluded that the proper aspect ratio and orientation of the building can considerably improve the wind circulation at building corners. Ierimonti et al. [13] proposed the life-cycle

S. Mandal (✉) · S. K. Dalui · S. Bhattacharjya
Department of Civil Engineering, Indian Institute of Engineering Science and Technology, Shibpur, Howrah, India
e-mail: shankumandalce@gmail.com

S. Bhattacharjya
e-mail: soumya@civil.iiests.ac.in

A. Sil et al. (eds.), *Recent Trends in Civil Engineering*, Lecture Notes in Civil Engineering 274, https://doi.org/10.1007/978-981-19-4055-2_28

cost analysis technique to understand the structural damage due to wind flow. The aerodynamic response of geometric alterations on rectangular buildings is investigated by Alminhana et al. [14]. Daemei and Eghbali [15] investigated the wake flow characteristics of triangular and square shapes having sharp, recessed, rounded and chamfered corners. Sharma et al. [16] suggested the set-back technique for high rise buildings over tapering since this technique reduces more wind force. Zheng et al. [17] measured the wind pressure to understand the ventilation efficiency of external shading louvers. Li et al. [18] considered the L shape with different side and aspect ratios to measure the pressure in four terrain categories. Sanyal and Dalui [19] explained the wind impact of Y shaped building with varying side ratios. Noormohamadian and Salajegheh [20] analyzed the aerodynamic performance of triangular shaped buildings using the surrogate modeling technique. Weerasuriya et al. [21] explored the wind speed variation near the pedestrian zone of a rectangular shaped building.

In modern construction practices, the selection of alphabetic shapes for building construction is the popular choice. Some researchers have evaluated the wind effects on alphabetic shapes in different wind aspects but the influence of side ratio in irregular alphabetic shapes is not well explored. The novelty of this study is that the wind responses in varying side ratios of U alphabetic shape tall buildings are investigated in various wind angles by numerical analysis using CFD. The fluctuations in mean pressure, pressure contour, flow pattern and force coefficients has been indicated.

2 Details of the Building Models

The detailed dimensions of building models have been tabulated in Table 1.

The building models (see Fig. 1) for numerical analysis have been prepared at 1:300 geometric scale. The wind effects on the U alphabetic shape building with different side ratios (2, 3, 4, 5 and 6) have been evaluated at 0°, 30°, 60° and 90° wind angle keeping the same height (H = 500 mm) and approx. same plan area in each case.

Table 1 Dimensions of different building models

Model	Width (W) (mm)	Length (L) (mm)	Limb Width (W') (mm)	Side Ratio (L/W')
MU1	244	122	61	2
MU2	250	150	50	3
MU3	258	172	43	4
MU4	273	195	39	5
MU5	280	210	35	6

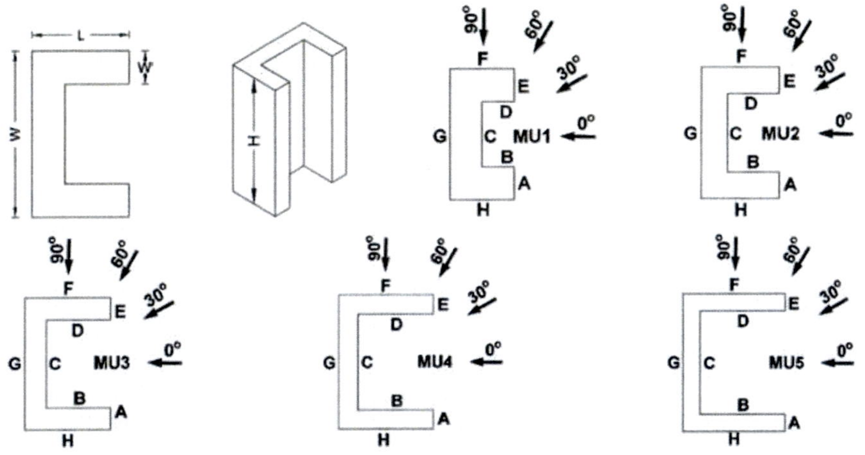

Fig. 1 Details of building models with the angle of wind and face name

3 Solution Methodology

The numerical analysis using CFD has been executed in Ansys CFX. A CFD analysis follows some steps which include the selection of suitable domain, meshing technique and set up of boundary conditions.

3.1 Setup of Computational Domain and Boundary Conditions

A suitable computational domain (Fig. 2) has been adopted where a significant place is available for wind flow simulation. The outlet, inlet and side wall (free-slip wall) boundaries of the domain are kept at a distance of 15, 5 and 5H from the model faces (no-slip wall). The turbulence has been created using the *k*-ε turbulence model. The boundary layer flow is simulated by the power-law equation [$U/U_0 = (H/H_0)^{\alpha}$]. The reference height and any elevation height from the ground are noted as 'H_0' and 'H.' The wind speed at heights H_0 and H is denoted as 'U_0' and 'U.' The values of power-law exponent (α), Inlet velocity and reference height (H_0) are adopted 0.133, 10 m/s and 1 m.

3.2 Generation of Mesh

The tetrahedral elements (Sanyal and Dalui [19]) are adopted for meshing the entire domain but for the higher accuracy of wind responses evaluation; few fine layers of

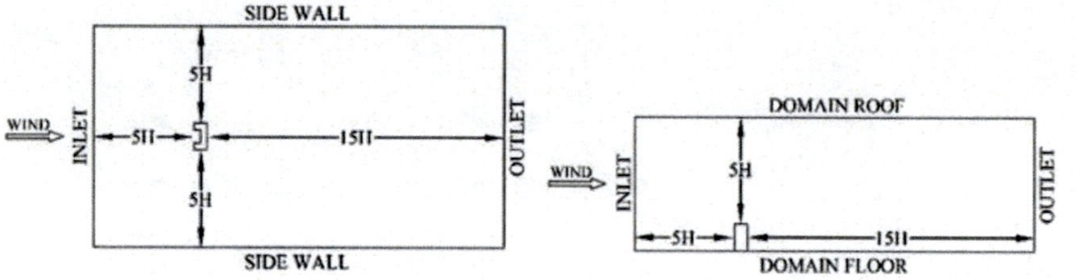

Fig. 2 Plan and elevation view of CFD domain

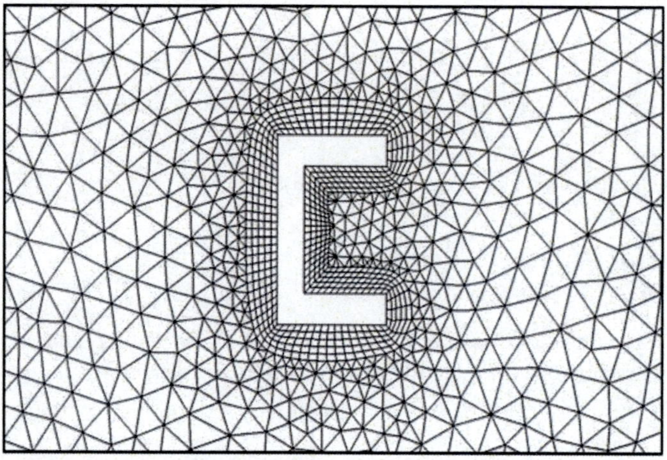

Fig. 3 Mesh pattern of the CFD model

hexahedron elements are taken near the model faces (Fig. 3). This meshing technique is often used in CFD simulation since it is capable of generating accurate wind flow around the model and also takes less computational resources.

3.3 Mesh Sensitivity Study

The mesh sensitivity study has been performed for the MU1 model at 0° wind angle to measure the suitable mesh element size. The drag coefficient has been measured in the different trial (T1–T7), where the mesh size has been changed from very coarse to very fine until the value converge (Fig. 4). The total elements [1284571(T1), 5247965(T2), 8467493(T3), 15214796(T4), 23572414(T5), 30247647(T6) and 38745769(T7)] are calculated for each trial which demonstrate that the mesh size has been gradual changes from very coarse to very fine. T6 mesh size trial is selected for the current study since the differences in the drag coefficient value between T6

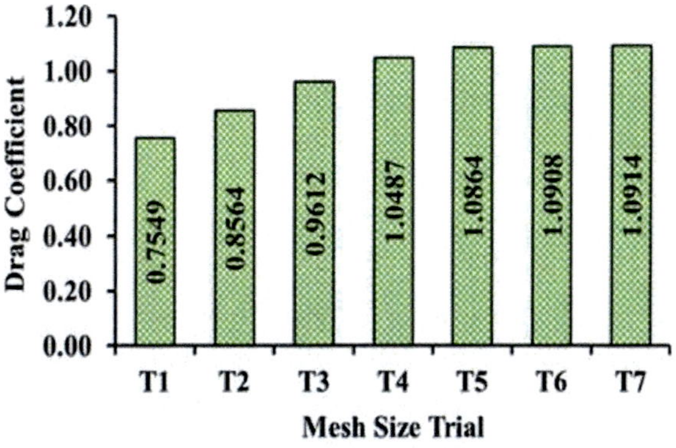

Fig. 4 Mesh sensitivity study for the MU1 model

and T7 is almost negligible and this indicates that the value converged in this mesh size.

3.4 Validation of Numerical Analysis

The present CFD analysis has been validated with the research article of Gomes et al. [5]. A similar U shape model as per the article has been prepared in Ansys CFX and the pressure coefficient of face D through the vertical line at 0° angle has been compared (Fig. 5). Almost negligible fluctuations in the pressure coefficient have been noted from the comparison. Hence, the CFD study in Ansys CFX is reliable.

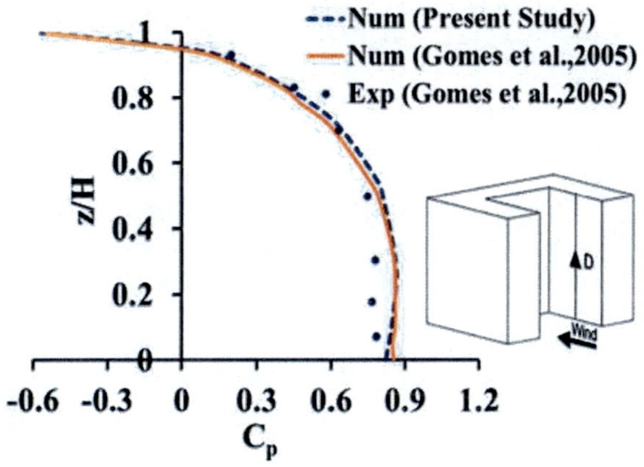

Fig. 5 Validation study of current numerical analysis

4 Results and Discussion

The influence of side ratio modification and wind direction has been indicated by comparing the velocity streamline, wind force, local pressure and pressure contour profile.

4.1 Velocity Streamline of Wind Flow

The velocity streamline indicates the wind velocity fluctuation around the models (Fig. 6). The velocity fluctuates when the flow of wind is obstructed by the building models and also the flow pattern varies with the outer shape of the model and wind angle. The side ratio greatly influences the area of the wake zone. At 0° wind angle, the symmetric flow pattern is observed in both sides of the U shape model but the wake zone extended when the side ratio increases. The two limbs of this shape modified the flow pattern when the wind angle shifted. In between the limbs, the location of low velocity regions decreased when the wind angle shifted to 30° from 0°. It is visible that the location of low velocity regions almost vanished at 30° angle when the side ratio increases. At 60° angle, almost all the locations between the limbs of smaller side ratio models exerted low velocity due to vortex formation. With the side ratio increment, the low velocity regions decreased and shifted to the limb corner. At 90° angle, almost all the regions in between the limbs of the building models experiencing low velocity. These velocity variations influence the wind responses of the buildings.

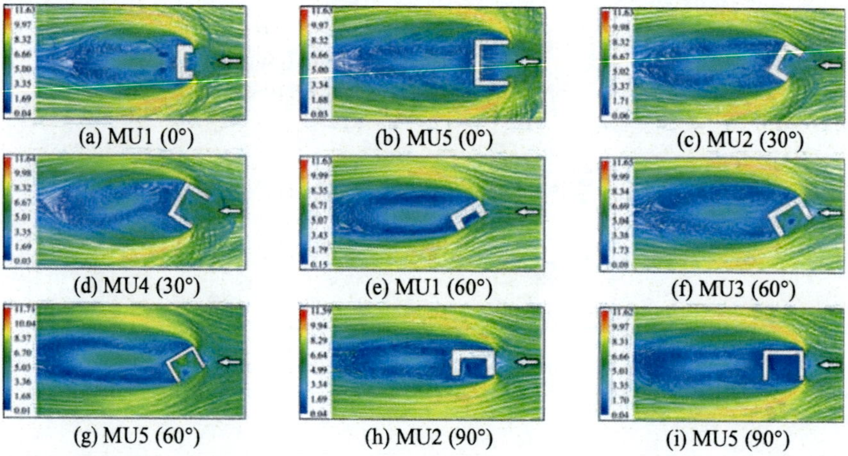

Fig. 6 Velocity streamline of wind around the different models at the various wind angle

Table 2 Comparison of force coefficients at the various wind angle

Model	0°		30°		60°		90°	
	C_{fx}	C_{fy}	C_{fx}	C_{fy}	C_{fx}	C_{fy}	C_{fx}	C_{fy}
MU1	1.091	0.000	1.025	−0.432	0.473	−1.057	−0.038	−0.895
MU2	1.096	0.000	1.029	−0.571	0.489	−1.133	−0.074	−0.911
MU3	1.117	0.000	1.038	−0.670	0.492	−1.174	−0.083	−0.919
MU4	1.128	0.000	1.055	−0.781	0.515	−1.183	−0.055	−0.921
MU5	1.157	0.000	1.059	−0.864	0.386	−1.163	−0.019	−0.954

4.2 Comparison of Force Coefficients

The comparison of force coefficients of various models has been tabulated in Table 2.

The drag coefficient (C_{fx}) rises with the side ratio at 0° and 30° angles. When the wind angle shifted to 60° and 90°, the gradual increment in C_{fx} is noted upto the side ratio 5 but at side ratio 6, the decrement in C_{fx} is observed. The C_{fx} is also decreasing when the wind angle changes. The lift coefficient (C_{fy}) also follows a similar trend of drag coefficient but the decrement in C_{fy} at side ratio 6 is noted only at 60° angle. For all models, C_{fy} is maximum when the angle is 60°. Further shifting the angle to 90° from 60° draws less C_{fy} compared to the previous case. The maximum C_{fx} (1.157) has been observed for the MU5 (side ratio 6) model at 0° angle, whereas the maximum C_{fy} (−1.183) has been noted for the MU4 (side ratio 5) model at 60° angle. This force coefficient variation trend manifest that the higher side ratio attracts greater wind force.

4.3 Comparison of Mean Pressure Coefficients

The velocity variations due to change in side ratio and wind angle influence the face pressure coefficient (C_p). The comparison of C_p at a different angle has been demonstrated in Fig. 7. At 0° angle, gradual positive pressure drop has been observed at face A and E with the increase in side ratio. But no such significant variation is observed in other faces. The negative C_p is noted in faces F, G and H due to vortex formation in those regions. The significant C_p variation is measured in faces A, D and F when the angle changes to 30° from 0°. At face D, the positive C_p decreases when the side ratio increases. The face E shows a huge increment in positive C_p on this angle. Further angle changes to 60° generate a gradual dip in positive C_p at faces B, C, D and E with the side ratio increment. All the faces other than face F experiencing massive negative C_p at 90° angle. The positive C_p in face A and negative C_p in face F gradually converted to negative C_p and positive C_p, respectively, when the angle

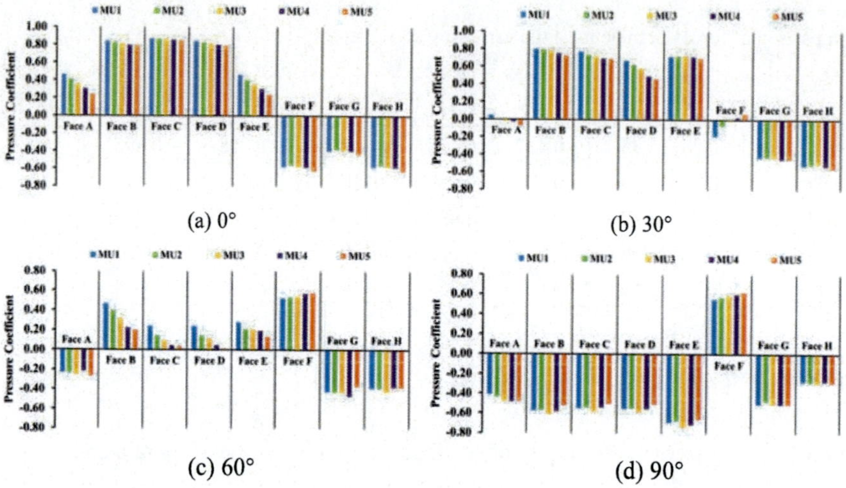

Fig. 7 Comparison of face pressure coefficient between various models at the different wind angle

changes. The comparison indicates that the flow separation region changes with the angle and those corners show remarkable fluctuation in pressure.

4.4 Comparison of Pressure Contour Plots

The pressure contour plots represent all regions of positive and negative pressure on the faces. The contours of the MU1 model for different angel has been demonstrated in Fig. 8. The identical pressure fluctuations are noted between face A and E, face B and D and face F and H at 0° angle since those faces are the axisymmetric face of U shape. Almost all regions of faces B, C and D experiencing immense positive pressure. The corners of the face A and E have positive pressure. Most of the surfaces of faces F, G and H show negative pressure regions. When the angle changes, pressure plots showing remarkable variation. At 30° angle, although most of the surfaces of faces B, C and D have positive pressure but the regions of huge positive pressure shrink. The enormous positive pressure is visible at the upper corner of face E but the surface of face A showing both positive and negative pressure regions. Face F's corner surface has encountered tremendous positive pressure at 60° angle but the comparatively low intensity of positive pressure is measured in faces B, C, D and E. At 90° angle, only the face F is showing the positive pressure. A small strip of huge negative pressure region is observed at faces E and G. The comparison of pressure contour plots indicates the critical surface of the different faces.

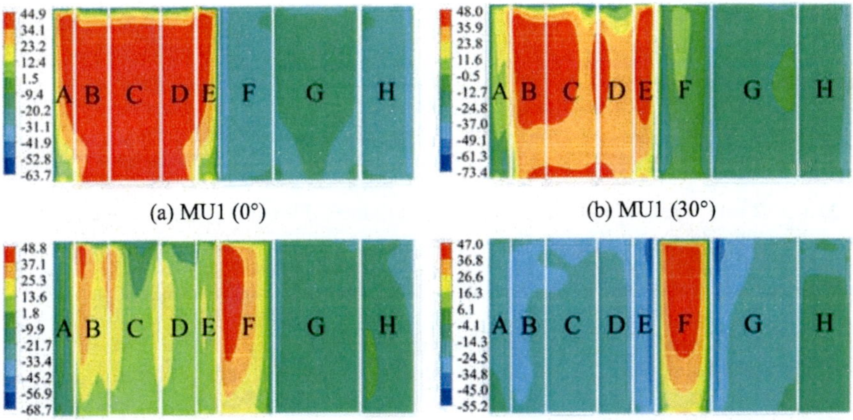

(a) MU1 (0°) (b) MU1 (30°)
(c) MU1 (60°) (d) MU1 (90°)

Fig. 8 Pressure contour plot of all the faces of MU1 type building

5 Conclusion

This study illustrated the influence of side ratio on variation in wind induced responses of U shape building for 0°, 30°, 60° and 90° angles of the wind. The shape irregularity in the horizontal direction has been varied with side ratio, which generates tremendous impact in wind responses. From the observations of velocity streamline and the comparison of force coefficient, pressure coefficient and pressure contour plots, some critical findings can be summarized as follows.

- The pattern and velocity of the flow vary with side ratio and wind angle. The increasing area of wake region for high side ratio building indicates that if a building with a high side ratio has been constructed, this may also influence the wind responses of neighboring structures.
- In most cases, the drag and lift force of the building is more for high side ratio buildings. So, high strength structural elements are required for those buildings.
- The flow separation corner regions have greater importance in cladding design since the pressure variation is critical in those regions.
- The contour plots demonstrated the critical region of fluctuated pressure. The clad material should be selected based on this so that these regions can resist the extra pressure.

The selection of proper side ratios for U shape buildings depends on various factors. The high side ratio attracts more wind force but on the other hand, a high value of side ratio is desirable from air circulation and light distribution point of view. The selection of clad elements is also an important factor irrespective of the selected side ratio. The critical observations from this study will guide the engineers to choose the suitable side ratio for U shape buildings.

References

1. Kawai H (2002) Local peak pressure and conical vortex on building. J Wind Eng Ind Aerodyn 90:251–263. https://doi.org/10.1016/S0167-6105(01)00218-5
2. Choi CK, Kwon DK (2003) Effects of corner cuts and angles of attack on the Strouhal number of rectangular cylinders. Wind Struct An Int J 6:127–140. https://doi.org/10.12989/was.2003.6.2.127
3. Gu M, Quan Y (2004) Across-wind loads of typical tall buildings. J Wind Eng Ind Aerodyn 92:1147–1165. https://doi.org/10.1016/j.jweia.2004.06.004
4. Lin N, Letchford C, Tamura Y, Liang B, Nakamura O (2005) Characteristics of wind forces acting on tall buildings. J Wind Eng Ind Aerodyn 93:217–242. https://doi.org/10.1016/j.jweia.2004.12.001
5. Gomes MG, Moret Rodrigues A, Mendes P (2005) Experimental and numerical study of wind pressures on irregular-plan shapes. J Wind Eng Ind Aerodyn 93:741–756. https://doi.org/10.1016/j.jweia.2005.08.008
6. Matsumoto M, Shirato H, Mizuno K (2008) Flutter characteristics of H-shaped cylinders with various side-ratios and comparisons with characteristics of rectangular cylinders. J Wind Eng Ind Aerodyn 96:963–970. https://doi.org/10.1016/j.jweia.2007.06.022
7. Memon RA, Leung DYC, Liu CH (2010) Effects of building aspect ratio and wind speed on air temperatures in urban-like street canyons. Build Environ 45:176–188. https://doi.org/10.1016/j.buildenv.2009.05.015
8. Keast DC, Barbagallo A, Wood GS (2012) Correlation of wind load combinations including torsion on medium-rise buildings. Wind Struct An Int J 15:423–439. https://doi.org/10.12989/was.2012.15.5.423
9. Huang S, Li R, Li QS (2013) Numerical simulation on fluid-structure interaction of wind around super-tall building at high reynolds number conditions. Struct Eng Mech 46:197–212. https://doi.org/10.12989/sem.2013.46.2.197
10. Peng X, Yang L, Gavanski E, Gurley K, Prevatt D (2014) Journal of wind engineering a comparison of methods to estimate peak wind loads on buildings. J Wind Eng Ind Aerodyn 126:11–23. https://doi.org/10.1016/j.jweia.2013.12.013
11. Tecle AS, Bitsuamlak GT, Chowdhury AG (2015) Opening and compartmentalization effects of internal pressure in low-rise buildings with gable and hip roofs. J Archit Eng 21(04014002):1–14. https://doi.org/10.1061/(asce)ae.1943-5568.0000101
12. Iqbal QMZ, Chan ALS (2016) Systematic influence of wind incident directions on wind circulation in the re-entrant corners of high-rise buildings. Wind Struct An Int J 22:409–428. https://doi.org/10.12989/was.2016.22.4.409
13. Ierimonti L, Caracoglia L, Venanzi I, Materazzi AL (2017) Investigation on life-cycle damage cost of wind-excited tall buildings considering directionality effects. J Wind Eng Ind Aerodyn 171:207–218. https://doi.org/10.1016/j.jweia.2017.09.020
14. Alminhana GW, Braun AL (2018) A numerical-experimental investigation on the aerodynamic performance of CAARC building models with geometric modifications. J Wind Eng Ind Aerodyn 180:34–48. https://doi.org/10.1016/j.jweia.2018.07.001
15. Baghaei Daemei A, Rahman Eghbali S (2019) Study on aerodynamic shape optimization of tall buildings using architectural modifications in order to reduce wake region. Wind Struct An Int J 29:139–147 https://doi.org/10.12989/was.2019.29.2.139
16. Sharma A, Mittal H, Gairola A (2019) Aerodynamics of tapered and set-back buildings using detached-eddy simulation. Wind Struct An Int J 29:111–127. https://doi.org/10.12989/was.2019.29.2.111
17. Zheng J, Tao Q, Li L (2020) Wind pressure coefficient on a multi-storey building with external shading louvers. Appl Sci 10(1128):1–25. https://doi.org/10.3390/app10031128
18. Li Y, Duan RB, Li QS, Li YG, Li C (2020) Research on the characteristics of wind pressures on L-shaped tall buildings. Adv Struct Eng 23:2070–2085. https://doi.org/10.1177/1369433220906934

19. Sanyal P, Dalui SK (2021) Effects of side ratio for 'Y' plan shaped tall building under wind load. Build Simul 14:1221–1236. https://doi.org/10.1007/s12273-020-0731-1
20. Noormohamadian M, Salajegheh E (2021) Evaluation and minimization of moment coefficient of tall buildings with trilateral cross-section via a surrogate model. SN Appl Sci 3(234):1–14. https://doi.org/10.1007/s42452-020-04128-5
21. Weerasuriya AU, Zhang X, Lu B, Tse KT, Liu CH (2021) A Gaussian process-based emulator for modeling pedestrian-level wind field. Build Environ 188(107500):1–15. https://doi.org/10.1016/j.buildenv.2020.107500

Comparative Study of Time History Analysis of Outrigger and Tube-In-Tube Structures Using ETABS

Makam Jeevan Kumar and S. M. Gupta

1 Introduction

India which is a developing nation is undergoing massive urbanization. The main reasons for urbanization are to improve the infrastructure of the country and to provide all the residential and commercial facilities available for the people. In these days, the pandemic had shaken the world with its wide spread, and there is shortage of facilities. Especially in case of urban areas, the density of population has been rising immensely because of migration of people to cities, and there is a great shortage of land. Because of rise in population, the accommodation is very difficult in horizontal direction due to shortage of land. Hence, the only possibility to accommodate is to move vertically, i.e., to go for the construction of tall buildings. Tall buildings also serve as iconic status of the country which also increase the tourism.

Tall buildings are more useful in case of urban areas because the total floor area of the building can be increased to a great extent with in a very small ground area. The high-rise structures are useful for residential as well as for commercial purposes. But the design and construction of these tall buildings are very difficult. The design and the construction must be accurate as small error leads to huge damage which in turn also leads to the risk of human lives. Hence, it became a complex task for the engineering field.

Tall buildings are subjected to gravity loads and transverse loads like wind load and earthquake load. Tall buildings are very much sensitive to lateral loads, and they tend to sway when it experiences lateral loads. The tall buildings are becoming slenderer which increases the possibility of extra sway to the structure. Therefore, in order to withstand these transverse loads, a transverse load resisting entity must be introduced. These tall buildings must be verified in all the aspects.

M. J. Kumar (✉) · S. M. Gupta
Department of Civil Engineering, NIT Kurukshetra, Kurukshetra, Haryana 136119, India
e-mail: jeevankumar2701@gmail.com

A. Sil et al. (eds.), *Recent Trends in Civil Engineering*, Lecture Notes in Civil Engineering 274, https://doi.org/10.1007/978-981-19-4055-2_29

Nowadays, there are many software's available for the analysis and the design which makes it easy for the engineers. These software show the highly stressed areas with maximum bending moment, shear force, etc.; hence, we can take suitable measures to resist them by using different techniques. These software are used for simulation purpose, and they can impose earthquake on the structures and analyze the structure for different parameters.

Many techniques are available for resisting the lateral loads.

General Types of Structural Systems in Tall Buildings.

- Rigid frame system
- Tubular system
- Braced frame system
- Shear wall system
- Coupled wall system.

1.1 Rigid Frame System

A structural system consisting of beams, columns, slab, and walls which are used to resist vertical and lateral loads.

1.1.1 Outrigger Structural System

The structural entity in which there is a central core element and the peripheral columns withstanding the transverse loads and self-weight. The central core is united with chosen peripheral columns by horizontal members are called as outriggers at various floors through the extent upward of the structure. The outrigger structure is an addition of the central core to improvise the rigidity of the high-rise buildings.

The components which are linked from the perimeter columns to the central structure in two sides to take hold of the structure to withstand the transverse loads are known as outriggers. The central shear walls are adopted to receive the weights and distribute the weights which help to withstand the loads against external loads. This entity gives more rigidity to the tower than conservative frame structure. It also acts as overhangs which are used to balance the structure against the transverse loads.

The outrigger entity is used from a long time ago, and it has been used to balance the structures from the transverse loads which are acting on the structure. The outrigger concept is used in the steel buildings as well as in case of the reinforced concrete structures. This entity is very much capable of withstanding the earthquake loads as well as the wind loads in case of high-rise structures. Outrigger entity is used for withstanding both the transverse loads as well as the self-weight (Fig. 1).

There are many software's available like ETABS, STAAD.Pro, etc., which are used for the analysis and design of the tall outrigger structures. An outrigger entity

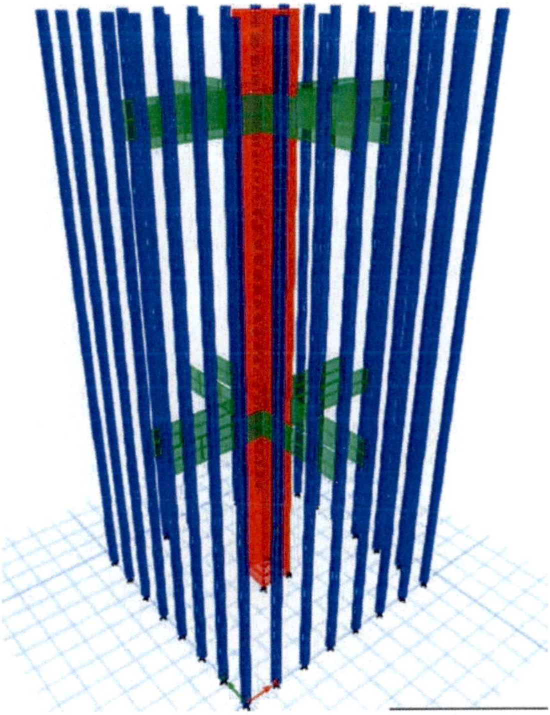

Fig. 1 Outrigger structural system

is created in these software, and the loads are imposed on the structure. The best position of the outrigger shows the least values of displacement, story drift.

The outrigger structure stands along with the structure and provides transverse stiffness to the structure. Outrigger structure with belt truss entity is more capable in withstanding the transverse loads. It helps the structure in withstanding the overturning moments and twisting moments and rotation of the structure.

1.2 Tubular Structures

The structural system in which the structural elements are arranged in such a way that it can resist the external loads on the structure efficiently mainly in case of the lateral loads. The tubular system consists of various elements like beams, slab, columns, and girders. The core and the walls are used to resist the lateral loads.

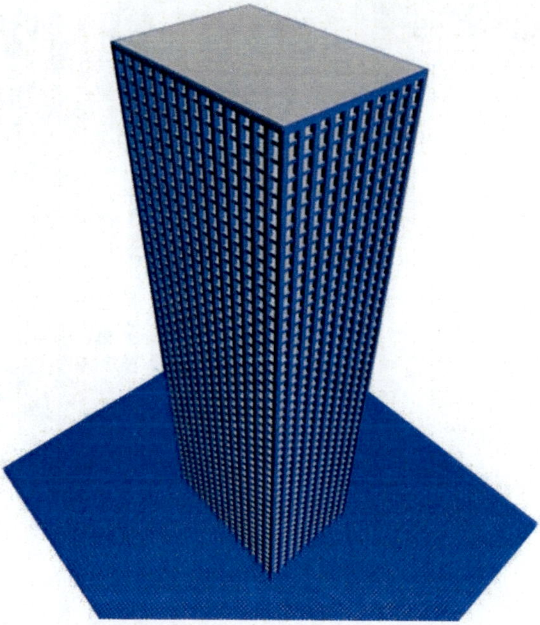

Fig. 2 Framed tube structural system

Types of Tubular Systems

- Framed tube structural system
- Tube-in-tube structural system
- Bundled structural system
- Braced tube structural system.

1.2.1 Framed Tube Structural System

A structural entity in which the spacing between the adjacent columns in the outer frame is less, and the deep beams are used to connect the external columns to each other for better tube action is known as framed tube structural system. The internal vertical elements which consist of the central core or columns are used to withstand self-weight of the structure (Fig. 2).

1.2.2 Tube-In-Tube Structural System

A structural entity in which there is an internal tube comprising of core element and an outer periphery known as the tube-in-tube structure. The outer perimeter is connected to internal perimeter via beams, slab.

The tube-in-tube structure is capable of withstanding the transverse loads and self-weight. The tube-in-tube structure resists the bending moments as well as the

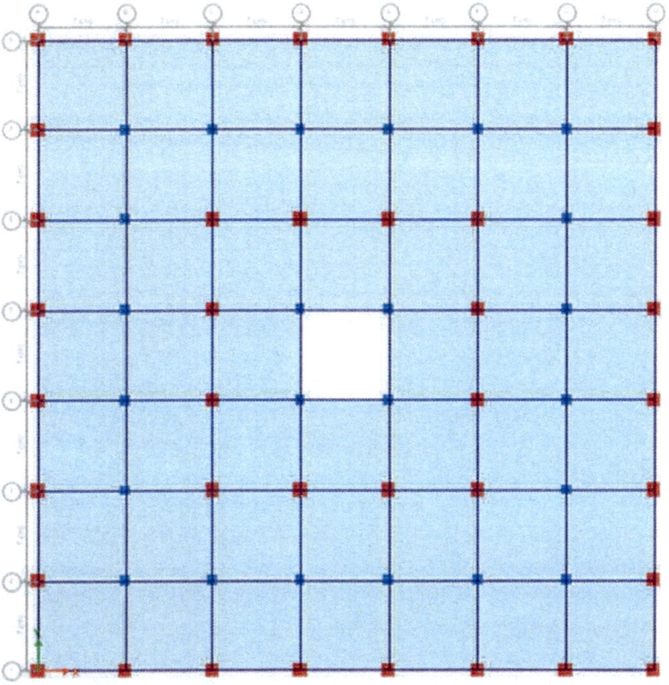

Fig. 3 Tube-in-tube structural system

twisting moments. These structures distribute the loads between the inner and the outer tubes. Both the inner and outer tubes are joined together by beams or truss. The transverse loads like wind and earthquake loads are taken initially by the outer tube and transfer them to the inner tube, whereas the gravity loads are taken by the inner tube itself.

The tube-in-tube structures can be either of steel or concrete or the combination of both the materials (Fig. 3).

1.2.3 Bundled Tube Structural System

A structural entity which is formed by the assembling of individual tubes is known as bundled tube structure. This structural entity is used for extreme heights of the structure and also comprises of huge floor area. Burj Khalifa which is the tallest structure in the world is of bundled tube structure (Fig. 4).

The shear lag can be reduced in the flange beams when the internal Webs are introduced in this entity. Hence, the columns are stressed evenly when compared with single tube structures, and therefore, the transverse stiffness of the structure also increases.

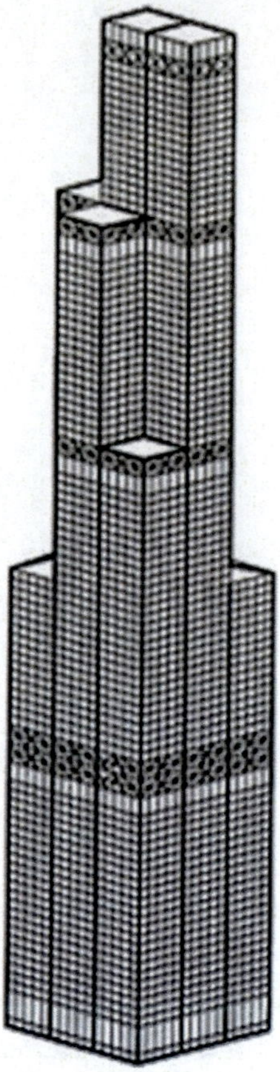

Fig. 4 Bundled tube structure [10]

1.2.4 Braced Tube Structural System

The tubular structure consists of inner and external tubes connecting with each other. In braced tube structure, the outer peripheral columns are connected with the braces. With the usage of these braces, the spacing of the outer peripheral columns is increased, and the depth of the deep beams is reduced which makes it more efficient and economical (Fig. 5).

2 Objectives

1. The effect of transverse loads on the structures with outrigger and tube-in-tube structures is determined.
2. To examine the transverse story displacement, story drift, and base shear in both the structures.

Fig. 5 Braced tube structural system [10]

3. To identify the highly vulnerable structure considered under seismic action.
4. To reduce the transverse displacement of the high-rise structure.

3 Methodology

1. A 3-dimensional-reinforced concrete structure with forty stories and each story height is 3 m of a total height of 120 m is considered for the analysis.
2. A bare frame model is modeled in the ETABS software as a reference.
3. Outrigger and tube-in-tube structures are modeled by considering the bare frame.
4. The material properties as well as the sectional properties of bare frame, outrigger, and tube-in-tube structures are kept same.
5. The distance from one floor to another floor is also kept equal in all the models.
6. In case of the outrigger, the analysis is performed by finding the best position of the outrigger.
7. A non-linear dynamic (time history analysis) analysis is done on all the models, and a comparison is made in between them.
8. From the observation of the results of various parameters, a conclusion is made.

4 Model Geometry

No. of stories: G + 39

Plan dimensions: 49m × 49m

No. of bays in X and Y-direction: 7 × 7

Story height: 3 m

Grade of concrete: M40

Grade of steel: Fe550

Beam size: 0.45 m × 0.65 m

Column size: 1.1 × 1.1 m, 0.8 × 0.8 m

Slab thickness: 150 mm

Shear wall thickness: 250 mm

Seismic zone: IV

Soil type: II

Importance factor(I): 1.5

Reduction factor(R): 5

LL: 4 kN/m^2

DL: 1.5 kN/m^2

FL: 1 kN/m^2.

5 Material Properties

Modulus of elasticity of steel = 200,000 N/mm^2

Modulus of elasticity of concrete = 31,622 N/mm^2

Characteristic strength of concrete = 40 N/mm^2

Yield stress of steel = 550 N/mm^2

Ultimate tensile strain in bending = 0.0035.

6 Analysis

In the analysis, a three-dimensional 40 story building with 7 bays in x-direction and 7 bays in y-direction, each bay is of 7 m is analyzed. Three type of models is prepared bare frame system, outrigger structural system, and tube-in-tube structural system. These models have similar geometry like number of floors, floor height, beam and column dimensions, plan dimensions, etc. The basic difference between these models is the structural system employed in it. Bare frame model does not have any additional structural system. Loads are applied on the structure such as dead load, live load, floor finish, and lateral load like earthquake load.

In the outrigger structure, the optimum location of outrigger is found out, and the depth of each outrigger is of two floors each. Two outriggers are placed in the structure. In case of tube in tube, the optimum location of the internal tube is found out by performing various trails.

Time history analysis is performed on outrigger, tube in tube as well as in bare frame in the ETABS software, and the results are studied in comparison with those of the bare frame (Figs. 6, 7 and 8).

Model 1: Bare frame structure

Model 2: Outrigger structure

Model 3: Tube-in-tube structure

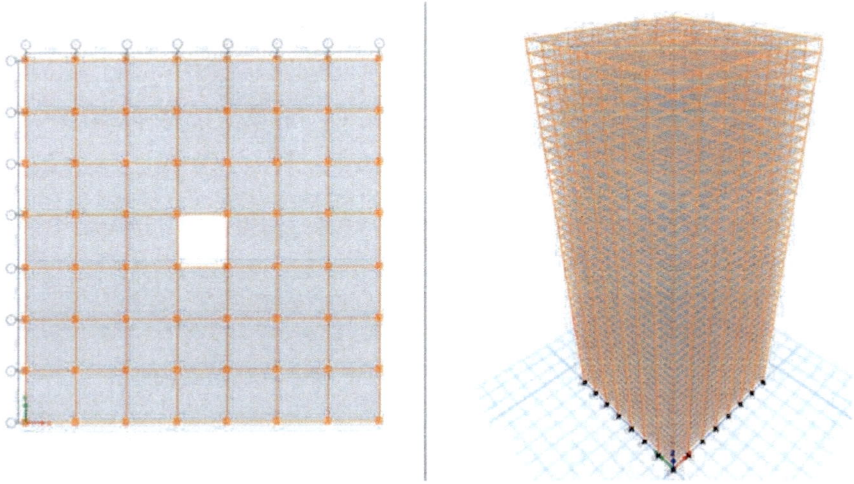

Fig. 6 Plan view of modeled bare frame structural system

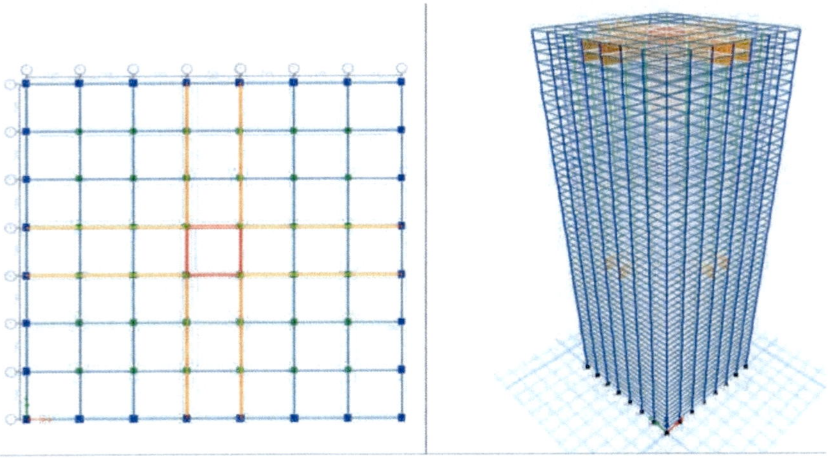

Fig. 7 Plan view of modeled outrigger structural system

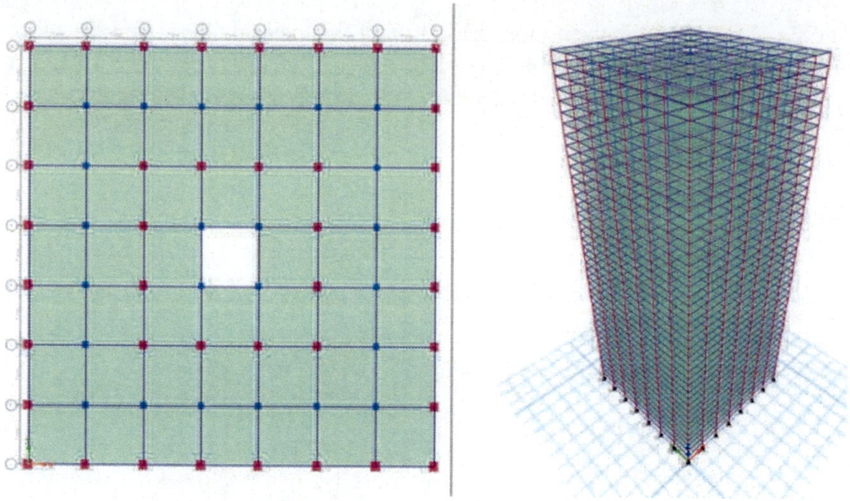

Fig. 8 Plan view of modeled tube-in-tube structural system

7 Results and Discussion

From the time history analysis of bare frame, outrigger, and tube-in-tube structures, the results obtained are shown and compared with that of bare frame.

	Bare frame structure	Outrigger structure	Tube-in-tube structure
Max story displacement (mm)	519	268	369
Base shear (kN)	42,524	69,405.85	44,592.49
Max story drift	0.006742	0.003545	0.004731

7.1 Max Story Displacement

The 3-dimensional 40 story building when undergone time history analysis (non-linear dynamic analysis) in ETABS is shown in the following figure. From the comparison of results of bare frame, outrigger, and tube-in-tube structures, it is observed that bare frame shows larger value of displacement and the displacement is reduced in tube-in-tube and outrigger structures. Outrigger structure has least value of displacement of 268 mm, bare frame shows maximum displacement of 519 mm and that of tube in tube shows 369 mm (Fig. 9).

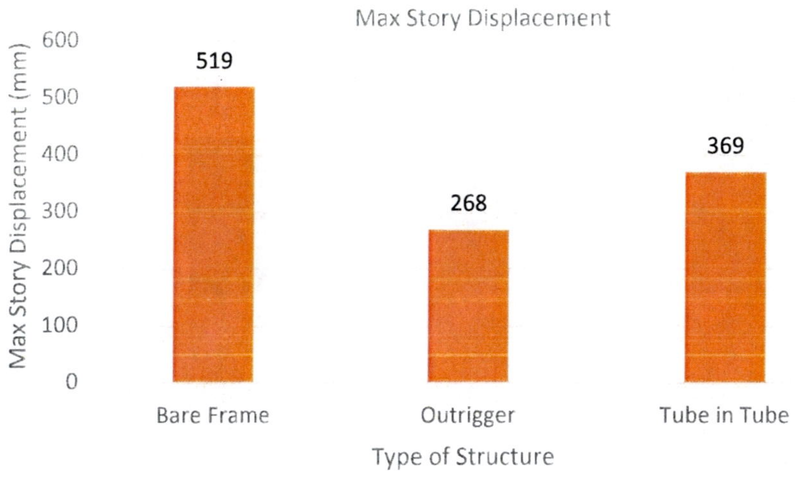

Fig. 9 Max story displacement

7.2 *Base Shear*

The 3-dimensional 40 story building when undergone time history analysis (non-linear dynamic analysis) in ETABS with base shear is shown in the following figure. From the comparison of results of bare frame, outrigger, and tube-in-tube structures, it is observed that bare frame shows least value of base shear and that outrigger structure shows maximum value (Fig. 10).

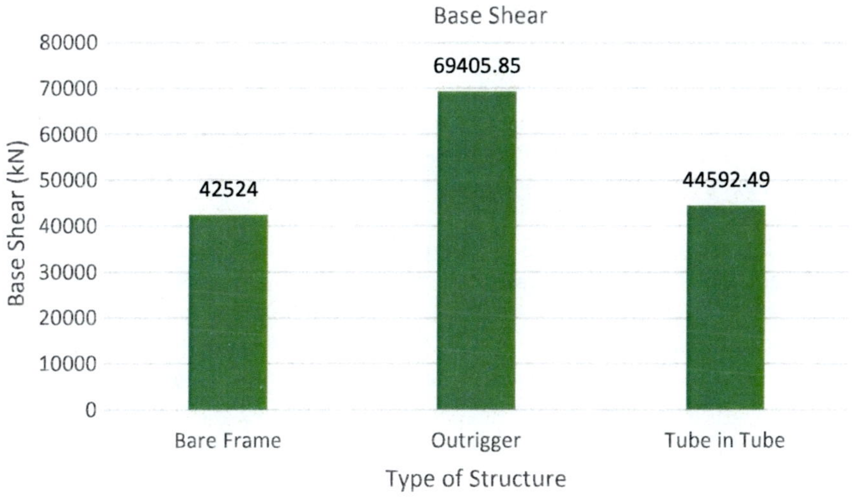

Fig. 10 Base shear

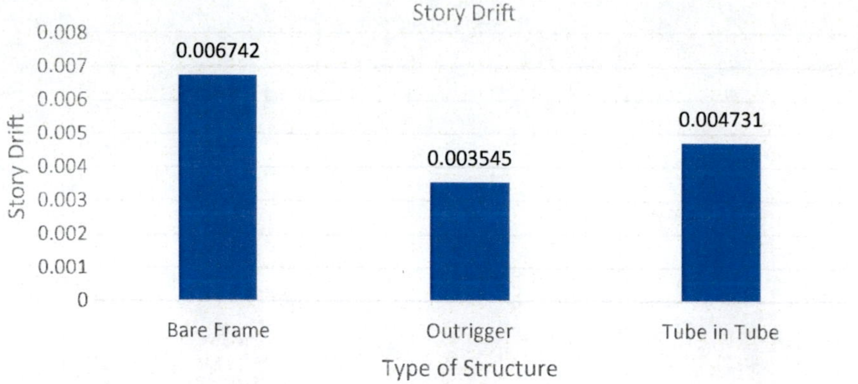

Fig. 11 Story drift

7.3 *Story Drift*

The 3-dimensional 40 story building when undergone time history analysis (non-linear dynamic analysis) in ETABS with story drift is shown in the following figure. From the comparison of results of bare frame, outrigger, and tube-in-tube structures, it is observed that bare frame shows larger value of story drift and the story drift is reduced in tube-in-tube and outrigger structures. Outrigger structure has least value of story drift of 0.003545, whereas bare frame has maximum value of 0.006742 and that of tube in tube has 0.004731 (Fig. 11).

8 Conclusion

1. The outrigger and tube-in-tube structures provide the strength and rigidity to the structure by resisting the transverse displacement and story drift.
2. The story drift is minimum at the position of the outrigger.
3. The best position of the outriggers for the 40 story structure is at 15, 16 floors and 35, 36 floors.
4. The story displacement is minimum in case of outrigger structure in comparison with tube-in-tube structure and bare frame. The story displacement is 268 mm in case of outrigger structure and 369 mm, 519 mm in case of tube-in-tube structure and bare frame, respectively.
5. The story drift is also minimum in case of outrigger structure in comparison with tube-in-tube structure and bare frame. The story drift is 0.003545 in case of outrigger structure and 0.004731, 0.006742 in case of tube-in-tube structure and bare frame, respectively.
6. Base shear is maximum in case of outrigger structure when compared to tube-in-tube structure and bare frame. The base shear is 69,405.85 kN in case of outrigger

structure and 44,592.49 kN, 42,524 kN in case of tube-in-tube structure and bare frame, respectively.

7. Outrigger structure resists more load than tube-in-tube structure and bare frame.
8. The lateral loads are resisted by the perimeter columns both in outrigger and tube-in-tube structures.
9. The outrigger structure performs better as compared to tube-in-tube structure, and bare frame when time history analysis is performed.

References

1. Mulla AK, Srinivas BN (2015) A study on outrigger system in a tall R.C structure with steel bracing. Int J Eng Research & Technol (IJERT) 4(7), ISSN: 2278-0181
2. Archana J, Reshmi PR (2016) Comparative study on tube-in-tube structures and tubed mega frames. Int J Innov Res Sci (IJIRS) 5(8)
3. Dangi A, Jamle S (2018) Determination of seismic parameters of R.C.C. building using shear core outrigger, wall belt and truss belt systems. Int J Adv Eng Res Sci (IJAERS) 5(9), ISSN: 2349-6495(P), 2456-1908(O). https://doi.org/10.22161/ijaers.5.9.36
4. Dangi A, Jamle S (2019) Stability enhancement of optimum outriggers and belt truss structural system. Int Res J Eng Technol (IRJET) 6(2), e-ISSN: 2395-0056, p-ISSN: 2395-0072
5. Naik BH, Chandra BSS (2017) Comparative analysis between tube in tube structure and conventional moment resisting frame, Int Res J Eng Technol (IRJET) 4(10), e-ISSN: 2395-0056, p-ISSN: 2395-0072
6. Herath N, Haritos N, Ngo T, Mendis P (2009) Behaviour of outrigger beams in high rise buildings under earthquake loads. Australian earthquake engineering society (2009) conference
7. Jahanshahi MR, Rahgozar R, Malekinejad M (2012) A simple approach to static analysis of tall buildings with a combined tube in tube and outrigger-belt truss system subjected to lateral loading. Int J Eng (IJE). https://doi.org/10.5829/idosi.ije.2012.25.03a.10
8. Lee KK, Loo YC, Guan H (2001) Simple analysis of framed-tube structures with multiple internal tubes. J Struct Eng (JSE) 127(4), ISSN:0733-9445/01/0004-0450–0460, Paper No. 21915
9. Scaria KJ, Karmakar A (2018) Comparative analysis of behaviour of RCC pentagonal and hexagonal tube in tube structural system and the conventional RCC pentagonal and hexagonal structure subjected to lateral loads in different seismic zones. Int J Sci Eng Res (IJSER) 9(11), ISSN 2229-5518
10. Kamath K, Divya N, Rao AU (2012) A study on static and dynamic behavior of outrigger structural system for tall buildings. Bonfring Int J Ind Eng Manage Sci 2(4), ISSN 2277-5056
11. Lavanya T, Sridhar RS (2017) Dynamic Analysis of Tube-In-Tube Tall Buildings. Int Res J Eng Technol (IRJET), 4(4), e-ISSN: 2395–0056, p-ISSN: 2395-0072
12. Shadulla MT, Kiran KM (2018) Analysis of tube in tube structures with different size of inner tube. Int J Tech Innov Modern Eng Sci (IJTIMES) 4(10), e-ISSN: 2455-2585
13. Mohan KT, Rahul Y, Kumara KNV (2017) Analysis of different forms of tube in tube structures subjected to lateral loads" Int J Innov Res Technol (IJIRT) 4(2), ISSN: 2349-6002
14. Kawade MP, Bangde VS, Sawai GH (2020) Seismic analysis of tall building with central core as tube structure. Int Res J Modernization Eng Technol Sci 2(9), e-ISSN: 2582-5208
15. Dileep N, Renjith R (2015) Analytical investigation on the performance of tube-in-tube structures subjected to lateral loads. Int J Tech Res Appl (IJTRA) 3(4):284-288, e-ISSN: 2320-8163
16. Kumar PK, Patnaikuni CK, Balaji KVGD, Santosh Kumar B (2020) Seismic performance of bundled tube structures in seismic zone Iv & Zone V of India. Int J Adv Res Eng Technol (IJARET) 11(6):328-337, Article ID: IJARET_11_06_029, ISSN Print: 0976-6480

17. Chang PC (1985) Analytical modeling of tube-in-tube structure J Struct Eng (JSE), Engineering 111(6), Paper No. 19808, ISSN 0733-9445/85/0006-1326
18. Kian PS, Siahaan FT (2001) Outrigger and belt truss system for high-rise concrete buildings. Dimensi Teknik Sipil, DTS 3(1), Maret, Hal. 36–41, ISSN 1410-9530
19. Biradar PN, Bhandiwad MS (2015) A performance based study on static and dynamic behaviour of outrigger structural system for tall buildings. Int Res J Eng Technol (IRJET) 2(5), e-ISSN: 2395-0056, p-ISSN: 2395-0072
20. Kafina R, Sagaseta J (2018) Analysis of outrigger-braced reinforced concrete supertall buildings: core-supported and tube-in-tube lateral systems. Struct Design Tall Spec Build, e1567. https://doi.org/10.1002/tal.1567
21. Tavakoli R, Kamgar R, Rahgozar R (2019) Seismic performance of outrigger–belt truss system considering soil–structure interaction. Int J Adv Struct Eng (IJASE). https://doi.org/10.1007/s40091-019-0215-7
22. Sarath BN, Claudiajeyapushpa D (2015) Comparative seismic analysis of an irregular building with a shear wall and frame tube system of various sizes. Int J Eng Comput Sci (IJECS) 4(4):11687–11697, ISSN:2319-7242
23. Modi SA, Agrawal VV, Arekar VA (2017) Parametric Study of Various Tube in Tube Structures. Int J Adv Res Eng Sci Technol (IJAREST) 4(5), e-ISSN: 2393-9877, print-ISSN: 2394-2444
24. Balakrishnan S, James RM (2019) Comparative study on tube in tube and tubed mega frames on different building geometry using ETABS. Int J Appl Eng Res (IJAER) 14(12), ISSN 0973-4562
25. Shivacharan K, Chandrakala S, Karthik NM (2015) Optimum position of outrigger system for tall vertical irregularity structures. IOSR J Mech Civil Eng (IOSR-JMCE) 12(2):54–63, e-ISSN: 2278-1684, p-ISSN: 2320-334X
26. Kogilgeri SS, Shanthapriya B (2015) A study on behaviour of outrigger system on high rise steel structure by varying outrigger depth. Int J Res Eng Technol (IJRET) 4(7), e-ISSN: 2319-1163|p-ISSN: 2321-7308
27. Wadagule SR, Charantimath VR (2019) Comparative study of tube in tube structure and frame tube structure. Int Res J Eng Technol (IRJET) 6(7), e-ISSN: 2395-0056, p-ISSN: 2395-0072
28. Singh S, Lata N, Nagar B, Kumar R (2019) Comparative analysis of RCC structure and tube-in-tube structure. Int Res J Eng Technol (IRJET) 6(7), e-ISSN: 2395-0056, p-ISSN: 2395-0072

Evaluation of the Behavior of Gusset Plates in Concentrically Braced Steel Frames Under Cyclic Loading

Mehdi Ebadi-Jamkhaneh, Denise-Penelope N. Kontoni, and Masoud Ahmadi

1 Introduction

In steel braced frame structures, the connection of the steel brace member to the boundary element must be provided in such a way as to form a truss shape. Therefore, in all stages of loading, the axial forces are generated in the steel brace members. The idea of using a brace connection plate to connect the brace members to the steel frame was the best idea that was introduced in this regard and is still the main method of connecting the brace to the main frame. The gusset plate (GP) is a plate element that transmits braces' force to the main frame and is usually used in rectangular or trapezoidal shapes, which is more common in conventional steel structures due to its simplicity of execution. The development of bracing systems began in the 1960s and is still being investigated. Shibata et al. [1] performed an investigation of the concentrically braced frames. The results of these experiments showed that, although the initial compressive strength of the member is approximately equal to its initial tensile strength, after the first period of compressive loading and occurrence of buckling, the compressive strength decreases significantly. The tests clearly demonstrated the importance of the brace's slenderness under cyclic loads and reducing its

M. Ebadi-Jamkhaneh (✉)
Department of Civil Engineering, School of Engineering, Damghan University, Damghan, Iran
e-mail: m.ebadi@du.ac.ir

D.-P. N. Kontoni
Department of Civil Engineering, School of Engineering, University of the Peloponnese, 26334 Patras, Greece
e-mail: kontoni@uop.gr; kontoni.denise@ac.eap.gr

School of Science and Technology, Hellenic Open University, 26335 Patras, Greece

M. Ahmadi
Department of Civil Engineering, School of Engineering, Ayatollah Boroujerdi University, Boroujerd, Iran
e-mail: masoud.ahmadi@abru.ac.ir

A. Sil et al. (eds.), *Recent Trends in Civil Engineering*, Lecture Notes in Civil Engineering 274, https://doi.org/10.1007/978-981-19-4055-2_30

compressive strength at each stage of loading and unloading. Jain et al. [2] studied the effect of braces' slenderness on their seismic behavior. These studies revealed that the phenomenon of decreasing strength and compressive stiffness increases with increasing slenderness coefficient. Experimental research was conducted by Black et al. [3] to express the effect of slenderness coefficient, boundary end conditions, and cross-sectional shape of the brace on the cyclic behavior of the braces. Its behavior shows that the reduction of strength and compressive stiffness increases with the increasing of slenderness coefficient. The use of thin-walled tubular closed-sections has better performance than thin-walled open-sections. That is why the closed form of the thin-walled section is more common. The test results of the studies conducted by Zayas et al. [4] revealed that braces with a circular cross section with a low ratio of diameter to thickness have a proper performance. Braces made with these sections have more expand hysteresis loops and less reduction of resistance. Ikeda and Mahin [5] proposed the use of large braces instead of slim braces to reduce the risk of local buckling in brace members. Walbridge et al. [6] conducted a study on GPs under monotonic loading. The examined parameters included the interaction between the brace and the gusset plate and the thickness of the GP. They concluded that a strong braced system and a weak GP show slight pinching in the hysteresis diagrams. Lutz et al. [7] experimentally tested the behavior of perforated GPs at compressive loading. One of the most important parameters in this study is the effect of thickness, width, and length of the perforated plate. The most important result of this research is to provide an analytical model for buckling of the perforated plates. Ling et al. [8] studied the shear failure of GPs experimentally. The main focus of that study is to investigate shear failure in GPs with high yield strength (1350 MPa). In that study, they modified the reliability coefficients related to the design methods of high-strength connection plates. Zhang et al. [9] examined the cyclic behavior of the concentric brace (Chevron) and its connection plates. One of their most important results is to investigate the ductility of GP. They conducted an experimental study of the connection plates under compressive loading and their buckling modes. In this regard, the Abaqus finite element (FE) software was used for nonlinear analysis. By investigation of the out-of-plane deformation obtained from the experimental and numerical results, they showed that the final compressive load of the GPs can be predicted to a large extent by the Whitmore method. A series of experimental studies was carried out to examine the low yield point (LYP) steel gusset plates under cyclic loads by Chen and Chang [10]. They found that adding slot-type restrainers to the LYP steel GP greatly enhances the seismic resistance of the GP. Fang et al. [11] performed a numerical study of connection plates in compression. Their results are presented in terms of the buckling resistance in the connection plates and behavior of the connection plates with the characteristics of material properties, strength, and different geometry. Salawdeh et al. [12] investigated the performance of different brace connection configurations, and GP designs were examined using shake-table testing. Ebrahimi et al. [13] proposed a new procedure to determine the dimensions of a GP. Fifteen one-bay, one-story braced frames with the different brace angles are subjected to monotonic loading. The results revealed that the proposed procedure presents a more accurate distribution of forces at GP interfaces than other methods.

Ebrahimi et al. [14] tested two full-scale special concentrically braced frames with a new GP connection passed through a hollow structural section column. The cyclic results of specimens showed that beam-to-column moment connection by through-plate increased strength, ductility, and energy dissipation capacity for approximately 41%, 79%, and 110%, respectively.

In this study, to investigate the behavior of gusset plates with different dimensions of the plate and various types of stiffeners, including edge, longitudinal, and internal stiffeners, several parametric studies are performed with the three-dimensional (3D) finite element (FE) method.

2 Numerical Modeling

2.1 *Materials Characteristics*

To model the properties of the steel materials in the joint components, the von Mises yield criterion was used to investigate their behavior. Nominal stress–strain relations were found from coupon tests and were then converted into true stress–strain relations based on Hill [15] to represent materials nonlinearity as the following:

$$\varepsilon = \frac{\sigma}{E} + 0.002\left(\frac{\sigma}{\sigma_y}\right)^n \tag{1}$$

where σ and σ_y are the true stress and yield stress, respectively. E is the modulus of elasticity, and ε is the nominal strain. In order to simulate the post-yield behaviors of the materials, the kinematic hardening was used. The stress–strain relationship of materials is shown in Fig. 1.

2.2 *Element Features and Interaction Between the Different Parts*

A 3D model was generated based on the FE method in Abaqus [16] software for parametric studies. The material property and nonlinear geometry were considered in numerical models. S4R element was utilized for numerical models. S4R element is a four-node doubly curved thin or thick shell, reduced integration, hourglass control, and finite membrane strains. Nonlinear static analysis was conducted considering the effect of large deformations. Tie constraints are used for the connections of the steel column and beam elements and for the connections of the GP to the beam-column joint [17].

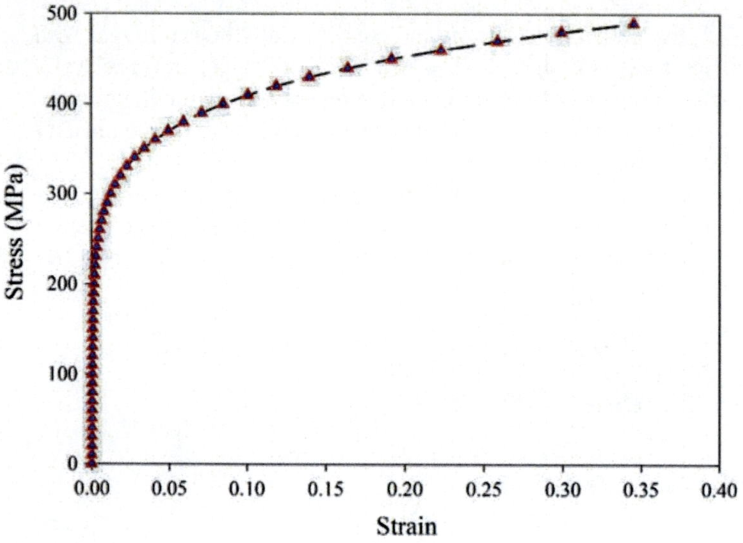

Fig. 1 Stress–strain curves of the different parts

2.3 *Loading Protocol and Boundary Conditions*

After assembling the steel components, the members are subjected to cyclic loading in accordance with the instructions of ATC-24 [18]. This protocol requires a number of elastic cycles to determine the stiffness and yield strength, and then cycles involving inelastic deformation are repeated by increasing the load amplitude to evaluate the cyclic inelastic performance. The loading is done by applying a displacement to the connection of the beam and the column at the top corner of the frame. Due to the fact that displacement loading control in the Abaqus software has better convergence than force control, and also in most regulations of the world, the displacement loading model is used to study the ductility of structural members. In this study, loading has been applied in the form of displacement control. The ATC-24 [18] instruction has also been used to take into account the nature of dynamic load cycles and reduce computational errors. Figure 2 shows the loading history.

According to Fig. 3, the lower part of the columns has roller supports, and the top section of the boundary columns is prevented from moving out of plane. For this purpose, lateral support has been introduced in the model. Also, to prevent out-of-plane buckling and deformation, the bottom beam and the two ends of the column are restrained in a direction perpendicular to the axis of the frame (Z-axis).

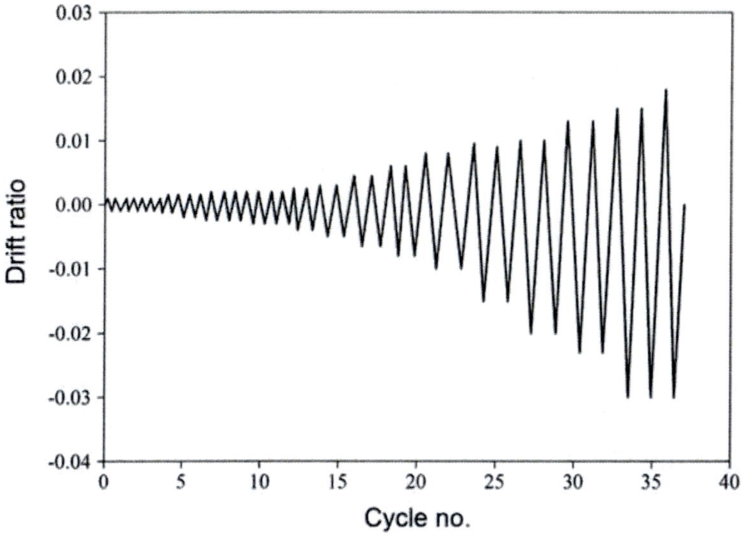

Fig. 2 Loading protocol scheme

3 Numerical Validation

The numerical model was validated by an experimental study which was done by Yoo et al. [19]. The one-story steel braced frame with a width and height of 3670 mm was selected. The cross section of steel brace was *HSS* $5 \times 5 \times 3/8$. $HSS5 \times 5 \times 3/8$. Mechanical properties of materials and geometrical characteristics of beam and column elements are shown in Tables 1 and 2, respectively.

Figure 4 shows the results of the experimental specimen [19] and the numerical model of the present study. A comparison between the inelastic response of the numerical cycles in the form of general and local failure mode with the experimental specimen is presented. The general response includes the stiffness of the frame and its strength close to each other. There is a difference of less than 7% between the final strength obtained from the numerical model and the experimental specimen. It should also be noted that the inelastic FE analysis correctly is able to investigate the behavior of the simulated model. This is because the FE model is able to show the position and development of the place of yielding. Specifically, the findings of the numerical and test models are mentioned below.

(1) Local failure and final fractures in the center of the brace due to local damage, which is the result of out-of-plane deformation and buckling in both numerical model and experimental specimen [19], are well known.
(2) Local yielding in the GP and limited fracture of the welds of this plate is due to large deformations of the plate and the rotation at the end of the brace. This behavior is shown in numerical observations.

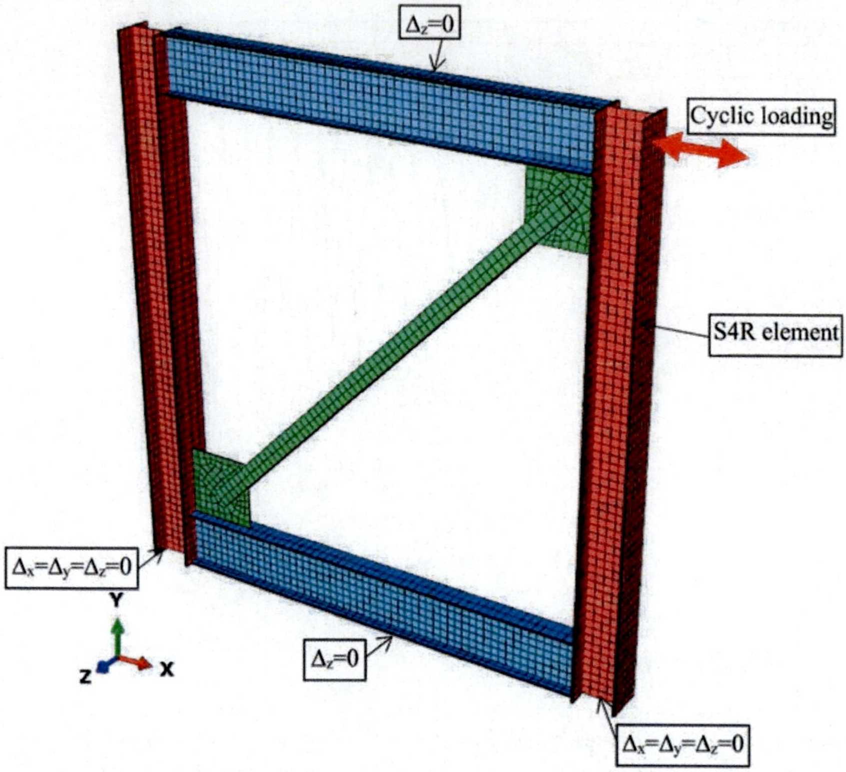

Fig. 3 Boundary conditions

Table 1 Mechanical properties of steel materials [19]

Gusset plate		Brace		Column		Beam	
Yield stress (MPa)	Ultimate stress (MPa)	Yield stress (MPa)	Ultimate stress (MPa)	Yield stress (MPa)	Ultimate stress (MPa)	Yield stress (MPa)	Ultimate stress (MPa)
447	602	505	549	409	522	395	501

Table 2 Geometrical features of beam and column members [19]

Member	Section	Area (mm^2)	Depth (mm)	Flange width (mm)	Flange thickness (mm)	Web thickness (mm)
Column	W 12 × 72	13,613.00	311.15	305.82	17.02	10.92
Beam	W 16 × 45	8580.63	409.70	178.70	14.35	8.76

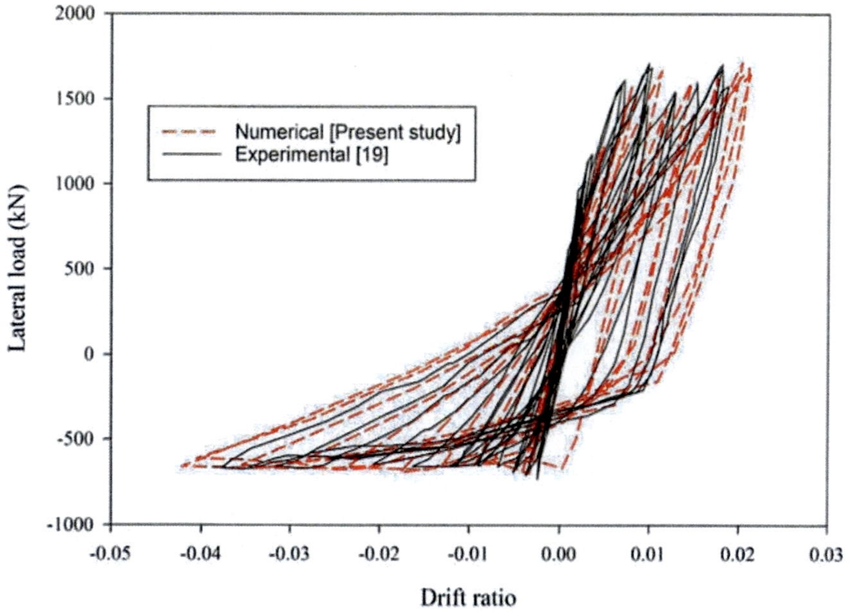

(a) Comparision of the "lateral force - drift ratio" relationships between the experimental specimen and the numerical model.

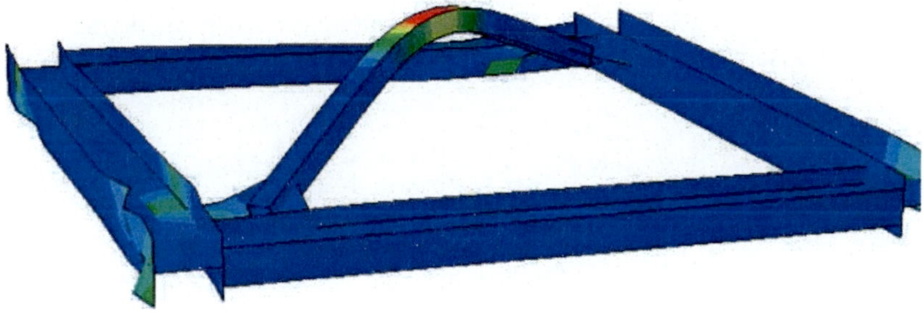

(b) Buckling mode of the numerical brace.

Fig. 4 Comparison of the results between the experimental specimen and the numerical model

4 Definition of the Numerical Models

Numerical models with changes in the dimensions and thickness of the gusset plate (GP) and the use of three stiffener plates (edge, longitudinal, and internal stiffeners) are described in Table 3. The letter L_{GP} indicates the length of the GP, t_{GP} indicates the thickness of the GP, S_e indicates the presence of edge stiffener, S_m indicates the presence of the middle stiffener, and S_L indicates the presence of the longitudinal stiffener. Figure 5 shows the position of the stiffener plates on the GP. The width

Table 3 Definition of the numerical models

Model name	S_L	S_m	S_e	t_{GP} (mm)	L_{GP} (mm)
G1	–	–	–	8	400
G2	–	–		8	400
G3	–		–	8	400
G4		–	–	8	400
G5	–	–	–	16	400
G6	–	–		16	400
G7	–		–	16	400
G8		–	–	16	400
G9	–	–	–	8	650
G10	–	–		8	650
G11	–		–	8	650
G12		–	–	8	650
G13	–	–	–	16	650
G14	–	–		16	650
G15	–		–	16	650
G16		–	–	16	650

of all three stiffener plates is equal to 100 mm. The material and thickness of these sheets are the same as the connection plate. The loading pattern used in the modeling is in accordance with the cyclic loading pattern in the ATC-24 [18] code.

5 Results and Discussion

5.1 Gusset Plate with a Width of 400 mm

The G1 model is subjected to cyclic loading. In this model, the dimension of the connection plate is equal to 400 mm, and its thickness is 8 mm. The final deformation of the frame is shown in Fig. 6. Due to the smaller thickness of the connection plate than the brace plate, the GP enters to the plastic area earlier than the brace, and buckling occurs in the connection plate. As can be seen, the brace does not buckle under the applied loads even in high cycles. The failure mechanism has occurred in the gusset plate, and then, the heel of the column (column flange near the support) is buckled. In the high cycles, the upper beam also becomes distorted and deviates from the plane. Figure 7 shows the force–drift ratio behavior of the brace.

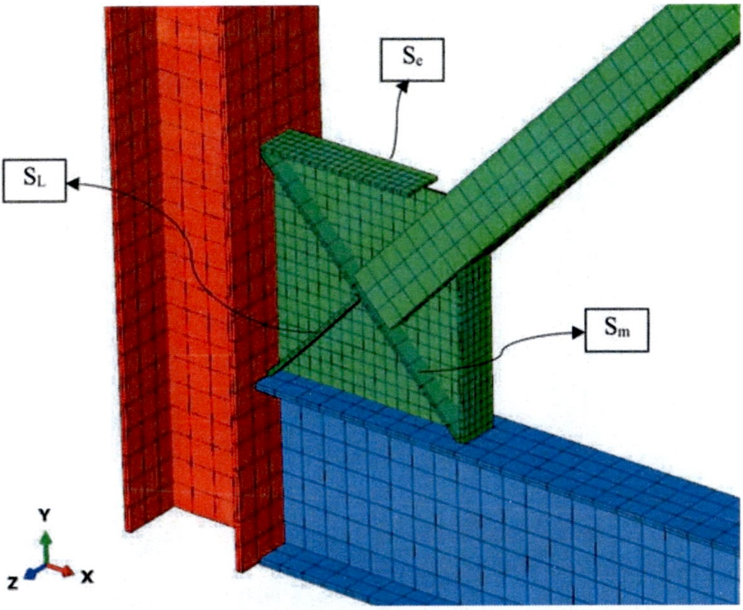

Fig. 5 Position of the stiffener plates on the GP

The G5 model, which is the only difference from the G1 model in the thickness of the gusset plate, was subjected to cyclic loading. By doubling the thickness of the G1 connection plate from 8 to 16 mm, it is observed that, unlike the G1 model, the failure mode does not start from buckling and yielding of the gusset plate. Also, in this model, the brace buckles, which was not the case in the G1 model. Figure 8 shows the final deformation in the final loading cycle. Figure 9 shows the force–drift ratio curve of the model.

By comparing the two models (G1 and G5), it is found that the model without stiffener and only by increasing the thickness of the gusset plate by a double value changes the mechanism of failure from the buckling of the GP to the buckling of the brace. Also, the maximum amount of lateral force has increased from 1625 kN in the G1 model to 1905 kN in the G5 model (17% increase in load-bearing capacity).

In Fig. 10, a comparison is made between different models with different thicknesses. As can be seen, in the case of the 400 mm gusset plate with two different thicknesses, the use of stiffeners has increased the load-bearing capacity and ductility of the connection plate. For example, in the G2 model, which uses an edge stiffener, the bearing capacity is increased by 10% just by doubling the thickness of the GP. The G7 model has an 18% increase in load-bearing capacity compared to the G3 model, and the G8 model has an approximately 38% higher load-bearing capacity than the G4 model.

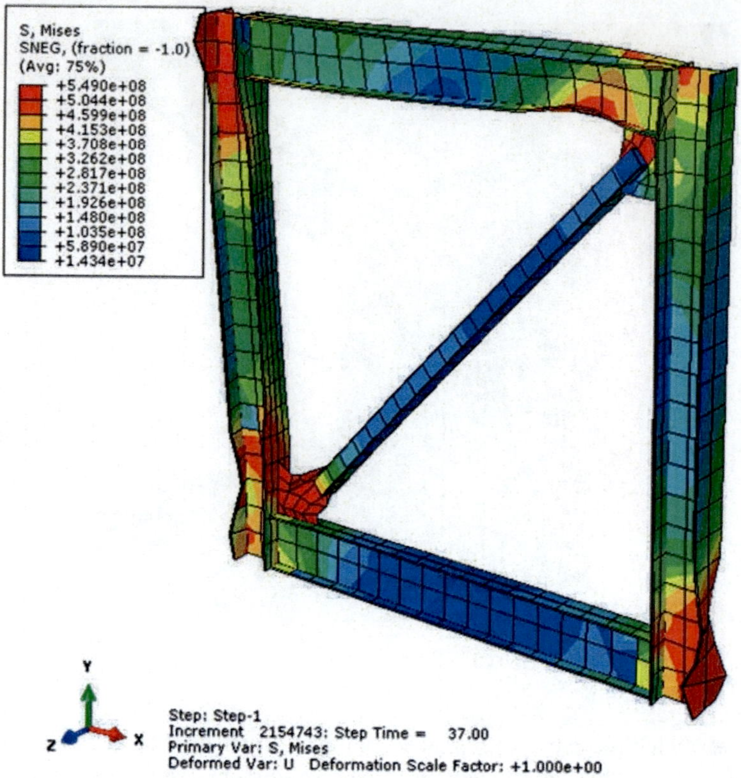

Fig. 6 Stress distribution of G1 model

In Fig. 11, a comparison is made between the final deformation of the models. Comparison of the buckling mode of the model according to Fig. 11 shows that due to the inherent weakness of the edges of the joint plate, if the plate's edges are not restrained, premature buckling occurs at the edge. Buckling occurs at the edge due to lack of control or restraint of the edges. However, due to the increase in stiffness in the connection with the longitudinal stiffeners, buckling occurs in a local region and prevents its global buckling. In the model with longitudinal stiffener, the occurrence of edge buckling prevents the influence of other factors affecting the compressive behavior of the joint. On the other hand, due to the deformation at the end of the member under the influence of edge buckling, the middle stiffener induces its effect to some extent.

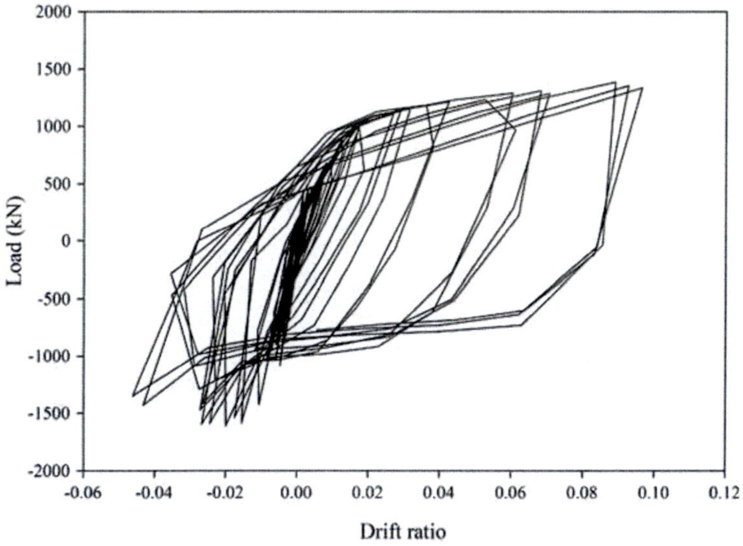

Fig. 7 Force–drift ratio curve of G1 model

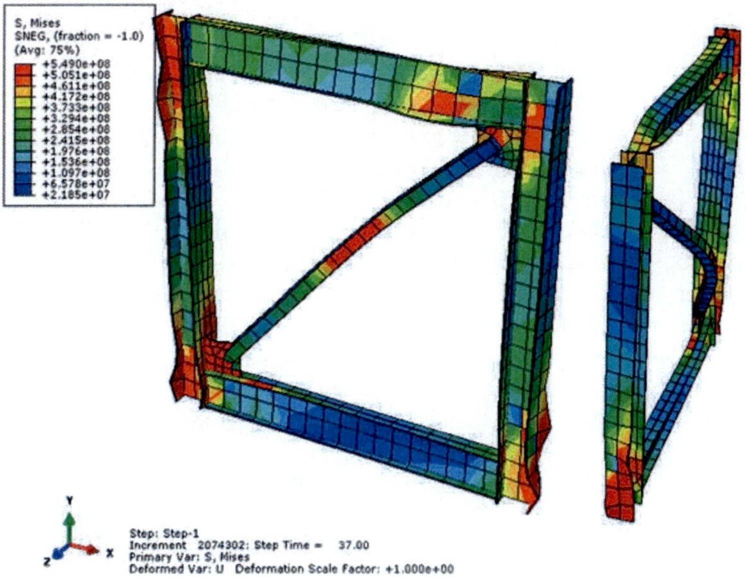

Fig. 8 Stress distribution of G5 model

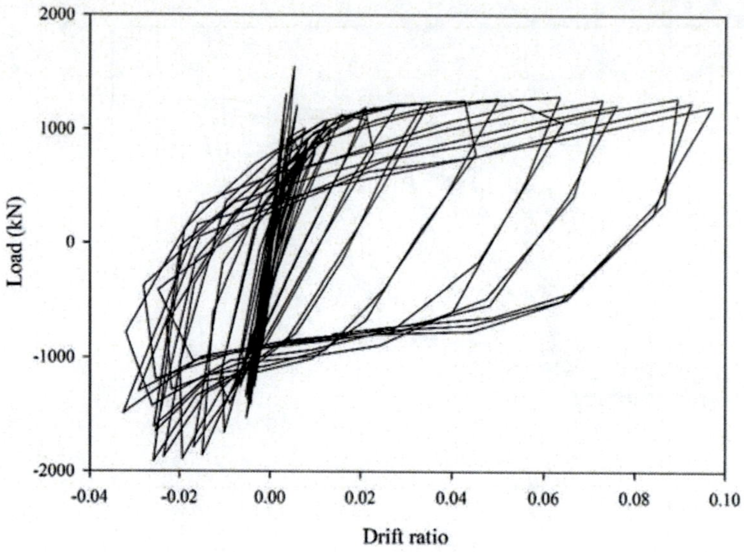

Fig. 9 Force–drift ratio curve of G5 model

5.2 *Gusset Plate with a Width of 650 mm*

In this section, the previous FE models were subjected to cyclic loading only by changing the dimensions of the gusset plate to the beam and column to evaluate the effect of enlarging the GP dimension on buckling modes and load-bearing capacity. Figure 12 shows the force–drift ratio curve of the models.

By adding the edge stiffener, the load-bearing capacity increases by 10% compared to the model without stiffener, and by adding the longitudinal stiffener, the load-bearing capacity increases by approximately 37%. The longitudinal stiffener significantly increases the load-bearing capacity and ductility rather than the edge stiffener due to the control of the critical area of the connection plate at the end of the joint member. The longitudinal stiffener does not cause a noticeable change in the buckling mode of the gusset plate but improves the compressive strength and ductility of the GP.

In Fig. 13, a comparison between the final deformation modes of the models is performed.

It is noteworthy that by comparing similar situations in two dimensions of the connection plate, it is determined that with the enlargement of the connection plate when using a plate with a thickness of 8 mm, the general buckling mode of the brace prevails. However, this did not happen on plates with a dimension of 400 mm and a thickness of 8 mm.

Fig. 10 Comparison of the results of numerical models with different types of stiffeners

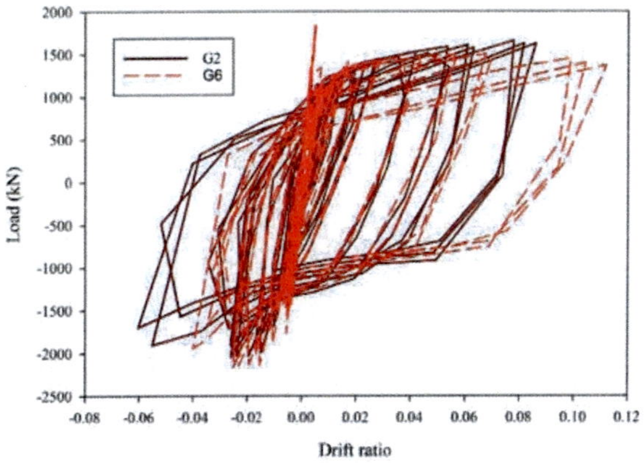

(a) Models with edge stiffeners.

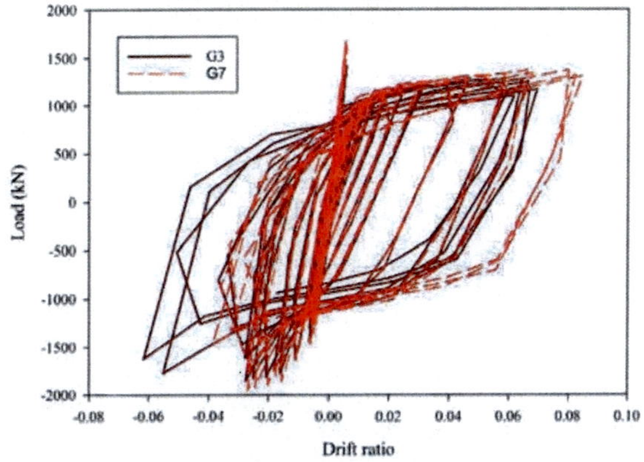

(b) Models with internal stiffeners.

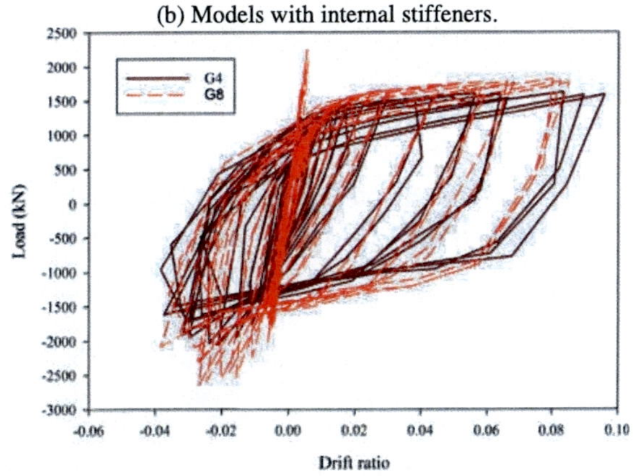

(c) Models with longitudinal stiffeners.

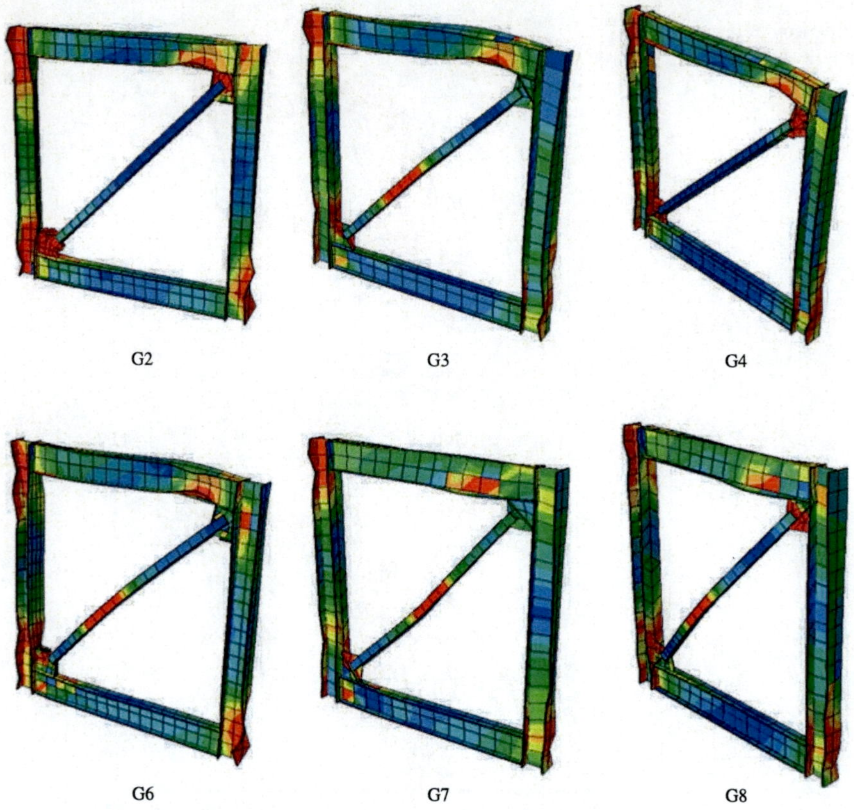

Fig. 11 Comparison of final deformation of the models

6 Conclusions

In this study, 16 numerical FE models were subjected to cyclic loading by changing the dimensions of the connection plate and its thickness along with three types of stiffeners in a one-span frame. The results are summarized below.

- The edge stiffener changes the buckling mode and increases the compressive capacity of the model.
- The longitudinal stiffener does not cause a noticeable change in the buckling mode of the joint but improves the compressive capacity and ductility of the joint.
- The edge stiffener has a greater effect on improving the behavior of the joint plate than the middle stiffener and increases ductility, stiffness, load-bearing capacity, and energy dissipation.
- The edge stiffeners do not have a significant effect on the post-buckling strength

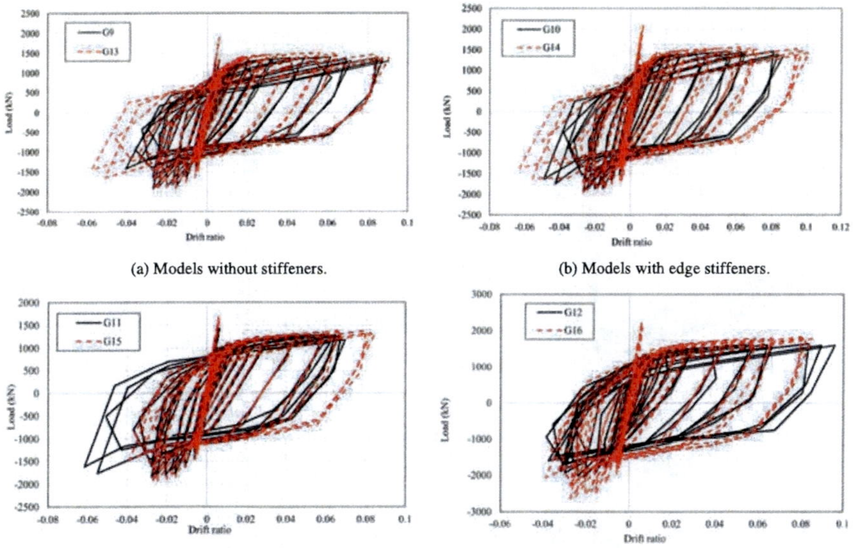

(a) Models without stiffeners.

(b) Models with edge stiffeners.

(c) Models with middle stiffeners.

(d) Models with longitudinal stiffeners.

Fig. 12 Force–drift ratio relationships of the models with a dimension of 650 mm of GP

of the connection, but contribute to the stability of the connection in the post-buckling area, which leads to higher post-buckling strength. The effect of edge stiffeners on increasing the buckling strength seems to depend on the thickness of the gusset plate.

- It is suggested that in future research, the behavior of the gusset plate with high-strength bolts under tensile force should be considered.

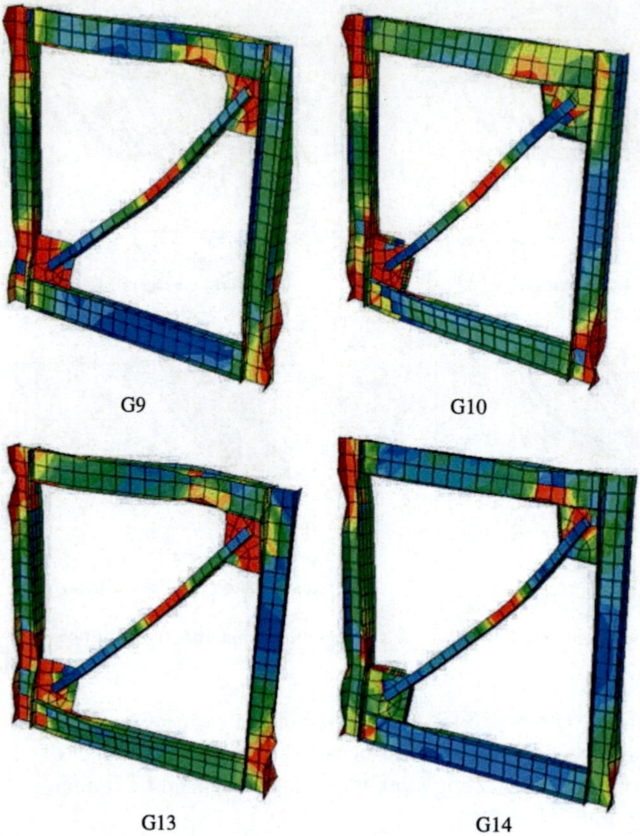

Fig. 13 Final deformation of the FE models

References

1. Shibata M, Nakamura T, Yoshida N, Morino S, Nonaka T, Wakabayashi M (1974) Elastic-plastic behavior of steel braces under repeated axial loading. In: Proceedings of the 5th world conference on earthquake engineering, Rome, Italy, vol 1, pp 845–848
2. Jain AK (1978) Hysteresis behavior of bracing members and seismic response of braced frames with different proportions. Ph.D. Dissertation, University of Michigan, Dept. of Civil Engineering. http://hdl.handle.net/2027.42/127507
3. Black RG, Wenger WAB, Popov EP (1980) Inelastic buckling of steel struts under cyclic load reversals. University of California, Berkeley, California, National Science Foundation American Iron and Steel Institute
4. Zayas VA, Popov EP, Mahin SA (1979) Cyclic inelastic buckling of tubular steel braces. Report No. UCB/EERC-80/16. Berkeley (CA): Earthquake Engineering Research Center, Univ. of California
5. Ikeda K, Mahin SA (1984) A refined physical theory model for predicting the seismic behavior of braced steel frames. UCB-EERC 84–12 Earthquake Engineering Research Center, Berkeley, CA. https://nehrpsearch.nist.gov/article/PB85-191450/XAB

6. Walbridge S, Grondin G, Cheng JJ (2005) Gusset plate connections under monotonic and cyclic loading. Can J Civ Eng 32(5):981–995. https://doi.org/10.1139/l05-045
7. Lutz D, LaBoube R (2005) Behavior of thin gusset plates in compression. Thin-Walled Struct 43(5):861–875. https://doi.org/10.1016/j.tws.2004.10.007
8. Ling TW, Zhao XL, Al-Mahaidi R, Packer JA (2007) Investigation of block shear tear-out failure in gusset-plate welded connections in structural steel hollow sections and very high strength tubes. Eng Struct 29(4):469–482. https://doi.org/10.1016/j.engstruct.2006.05.015
9. Zhang W, Huang M, Zhang Y, Sun Y (2011) Cyclic behavior studies on I-section inverted V-braces and their gusset plate connections. J Constr Steel Res 67(3):407–420. https://doi.org/10.1016/j.jcsr.2010.09.012
10. Chen S-J, Chang C-C (2012) Experimental study of low yield point steel gusset plate connections. Thin-Walled Struct 57:62–69. https://doi.org/10.1016/j.tws.2012.03.014
11. Fang C, Yam MC, Cheng JR, Zhang Y (2015) Compressive strength and behaviour of gusset plate connections with single-sided splice members. J Constr Steel Res 106:166–183. https://doi.org/10.1016/j.jcsr.2014.12.009
12. Salawdeh S, English J, Goggins J, Elghazouli AY, Hunt A, Broderick BM (2017) Shake table assessment of gusset plate connection behaviour in concentrically braced frames. J Constr Steel Res 138:432–448. https://doi.org/10.1016/j.jcsr.2017.07.022
13. Ebrahimi S, Mirghaderi SR, Zahrai SM (2019) Proposed design procedure for gusset plate dimensions and force distribution at its interfaces to beam and column. Eng Struct 178:554–572. https://doi.org/10.1016/j.engstruct.2018.09.066
14. Ebrahimi S, Mirghaderi SR, Zahrai SM, Najafi A, Otaghsaraie SMS (2021) Experimental study on brace to HSS column connection using through-gusset plate. Eng Struct 234:111948. https://doi.org/10.1016/j.engstruct.2021.111948
15. Hill H (1944) Determination of stress-strain relations from "offset" yield strength values. NACA-TN-927, NACA Technical Notes. http://hdl.handle.net/2060/19930081673
16. Abaqus, version 6.14: Documentation. Dassault Systemes Simulia Corporation, Providence, RI, USA (2014). http://130.149.89.49:2080/v6.14/
17. Salimi SM, Rahimi S, Hoseinzadeh M, Kontoni DP, Ebadi-Jamkhaneh M (2021) Numerical 3D finite element assessment of bending moment-resisting frame equipped with semi-disconnected steel plate shear wall and yielding plate connection. Metals 11(4):604. https://doi.org/10.3390/met11040604
18. ATC-24 (1992) Guidelines for cyclic seismic testing of components of steel structures. Applied Technology Council (ATC)
19. Yoo J, Lehman D, Roeder C (2008) Influence of gusset plate parameters on the seismic resistance of braced frames. J Constr Steel Res 64:607–623. https://doi.org/10.1016/j.jcsr.2007.11.005

Modeling of Soil-Structure Interaction in Liquefiable Soils Using an Equivalent Linear Approach Including Shear Modulus Updating

Mohammad Iman Khodakarami, Marzieh Dehghan, and Denise-Penelope N. Kontoni

1 Introduction

One of the criteria that can affect the design of a structure is the correct understanding of the dynamic response of structures under seismic loading. Soil-structure interaction (SSI) is one of the important factors that can significantly affect the dynamic response of structures [1–6]. In the past, in the analysis of structures, soil and structure were considered separately from each other, which resulted in a difference between the analysis results and the actual results. Soil and structure interaction (SSI) is one of the factors that can have a significant impact on the dynamic behavior of a structure.

There are two approaches to consider the SSI: the direct method and the substructure method. The direct method is based on numerical methods in which soil and structures are completely modeled in software. There are various numerical methods to model the propagation of waves in soil layers, such as the boundary element method (BEM) [7], coupled finite element method (FEM) and BEM [8], viscous boundaries [9], transmitting boundaries [10–12], and infinite elements [13–15], while the nonlinear behavior of the soil and structure can be used in these methods [16–20]. These methods have some drawbacks but still provide the most accurate results.

M. I. Khodakarami · M. Dehghan
Faculty of Civil Engineering, Semnan University, Semnan, Iran
e-mail: khodakarami@semnan.ac.ir

M. Dehghan
e-mail: marziehdehghan.73@yahoo.com

D.-P. N. Kontoni (✉)
Department of Civil Engineering, School of Engineering, University of the Peloponnese, 26334 Patras, Greece
e-mail: kontoni@uop.gr; kontoni.denise@ac.eap.gr

School of Science and Technology, Hellenic Open University, 26335 Patras, Greece

A. Sil et al. (eds.), *Recent Trends in Civil Engineering*, Lecture Notes in Civil Engineering 274, https://doi.org/10.1007/978-981-19-4055-2_31

These methods have drawbacks. For example, the BEM is not successfully applied to nonlinear fluid flow problems. The mathematics used in BEM is unfamiliar to the engineering community.

One of the main drawbacks of the FEM is that the finite elements associated with a crack must conform to crack faces. Furthermore, remeshing techniques are required to investigate and follow crack propagation patterns [21].

Analytical methods are much simpler than numerical methods and have two separate systems of soil and structure. Analytical methods are used in the substructure methods. In the substructure method, the soil and the structure are considered as distinct systems. The behavior of the foundation and soil is represented by stiffness and damping coefficients, and the effect of interaction between the soil and foundation is transmitted to the structure by means of dashpots and springs. In the substructure method, the soil is replaced by springs and dampers such that these springs and dampers behave similarly to the original soil. In the substructure method, the seismic behavior is modeled linearly because the stiffness and damping properties are considered linear. This method also has limitations as it is unable to model the nonlinear behavior of the soil.

Laboratory studies have shown that earthquakes significantly reduce the resistance of sandy soils to liquefaction [22, 23]. The phenomenon of liquefaction causes serious damage to infrastructure, for example, the tilting of buildings and the rise of buried structures [24, 25]. Displacement due to soil deformation after liquefaction can be large in structures such as dams and embankments [26–28]. It is necessary to mention that the phenomenon of liquefaction can occur multiple times in an area [29]. There are different methods for modeling the liquefaction phenomenon, including the direct element method [30–33].

There are many examples of structural damage in the world where the main reason for the instability is the increase in pore water pressure during an earthquake. An example is the failure of bearing capacity due to liquefaction in the Kwagishi-Chu buildings during the 1964 Niigata earthquake [34]. Previous studies have also shown that increasing pore water pressure can increase soil pressure [35, 36]. Research shows that increasing pore water pressure reduces the effective stress, and thus the shear modulus during earthquakes is reduced. The liquefaction phenomenon can significantly change the response of the soil system. As a result, considering the effect of pore water pressure in the substructure method, it can give more suitable results.

In this paper, at each time step, the parameters of the cumulative absolute velocity (*CAV*) and the pore water pressure ratio (r_u) are obtained linearly, and with the help of these mentioned parameters, a new shear modulus is obtained at each step. According to the above, this shear modulus decreases during the earthquake and approaches zero during liquefaction. To calculate the seismic response of soil and structure, a multiple degree of freedom (MDOF) system has been used, and in order to solve the equations obtained, the Newmark-beta method has been used. All of the above procedure is coded in MATLAB [37] and solved automatically. It should be noted that in this article, two structures with different heights and weights, two different soil types, and a seismic input with low risk level are considered. Finally, the displacement of the structures (in three directions of horizontal, vertical, and rotational) is presented as output.

2 Methodology

When an earthquake occurs in saturated granular soil, the pore water pressure in the soil increases, resulting in a reduction in the effective stress. Also, the strength and stiffness of the soil change significantly at any time, according to the Mohr–Coulomb model. This decrease in the strength increases the response of the structure. The calculation of the pore water pressure ratio (r_u) is of special importance in the liquefaction phenomenon. It should be noted that the type of soil, the height of soil layers, as well as the parameters affected by the earthquake can be effective in increasing the pressure of pore water. It is effectively related to the cumulative absolute velocity (*CAV*). The cumulative absolute velocity parameter is introduced as the absolute value integral of the acceleration of earthquake time series [38]. The cumulative absolute velocity (*CAV*) is mathematically represented by the following equation:

$$CAV(t) = \int_0^t |\ddot{u}_g(t)| \mathrm{d}t \tag{1}$$

where $\left|\ddot{u}_g(t)\right|$ is the absolute of the acceleration time series. Equation (1) can calculate the *CAV* at the time domain.

Based on the previous studies [38], the seismic (earthquake-induced) excess pore water pressure ratio (r_u) is calculated as a function of *CAV*, as follows:

$$r_u(t) = 1 - \exp[-\alpha CAV(t)] \tag{2}$$

where α is a constant related to the soil parameters and seismic conditions (the α is assumed to be 0.01), and *CAV* is the cumulative absolute velocity (cm/s).

The stiffness of the soil reduces during the seismic loading due to the increase in nonlinear soil pressure, and it can be calculated at each step based on the equation between the pore water pressure ratio and shear modulus. According to a simple equation, the reduced shear modulus is as follows:

$$G(t) = G_{\max}[1 - r_u(t)] \tag{3}$$

where $G_{\max}$ is the maximum shear modulus before liquefaction ($G_{\max} = \rho {V_s}^2$).

With the help of the equations mentioned in the NIST GCR 12-917-21 [39] regulations, it is possible to calculate the stiffness and damping of the liquefied soil model at each earthquake step. It should be noted that according to the equations introduced in these regulations [39], the stiffness and damping are calculated approximately. In

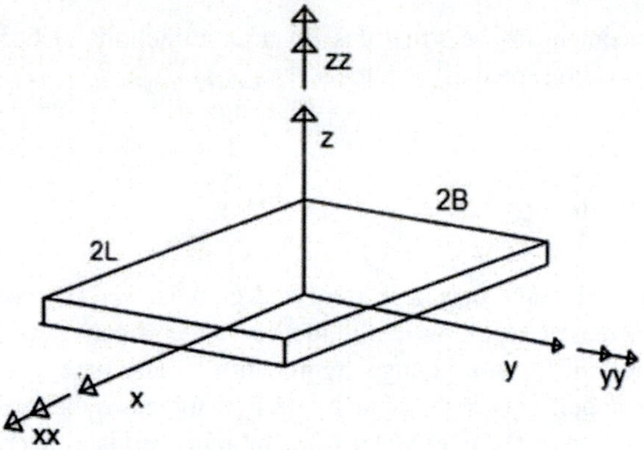

Fig. 1 Coordinate axis directions of the foundation

this article, the soil is considered as three degrees of freedom in horizontal, vertical, and rotational directions. The equations extracted from the regulations for stiffness and damping are given in Eqs. (4)–(12), where these equations were proposed by Pais and Kausel [40]. Figure 1 shows the coordinate axis directions of the foundation.

The equations related to the horizontal degree of freedom (translation along horizontal y-axis) are as follows [39, 40]:

$$K_y = \frac{\mathrm{GB}}{2-\nu}\left[6.8\left(\frac{L}{B}\right)^{0.65} + 0.8\left(\frac{L}{B}\right) + 1.6\right] \quad (4)$$

$$B_y = \left[\frac{4 \times \left({}^{L}/{}_{B}\right)}{{}^{K_y}/{}_{GB}}\right] \times \left[\frac{a_0}{2a_y}\right] \longrightarrow \begin{cases} a_y = 1 \\ a_0 = \frac{\omega B}{V_s} \end{cases} \quad (5)$$

$$C_y = \frac{2K_y \times B_y}{\omega} \quad (6)$$

The equations related to the vertical degree of freedom (translation along vertical z-axis) are as follows [39, 40]:

$$K_z = \frac{GB}{1-\nu}\left[3.1\left(\frac{L}{B}\right)^{0.75} + 1.6\right] \quad (7)$$

$$B_z = \left[\frac{4\psi \times (\mathrm{L/B})}{K_z/GB}\right] \times \left[\frac{a_0}{2a_z}\right] \rightarrow \begin{cases} a_z = 1 - \left[\dfrac{\left(0.4 + \frac{0.2}{\mathrm{L/B}}\right)a_0^2}{\left(\frac{10}{1+3(\mathrm{L/B}-1)}\right) + a_0^2}\right] \\ a_0 = \dfrac{\omega B}{V_s} \\ \psi = \sqrt{2(1-\nu)/(1-2\nu)} \\ \psi \le 2.5 \end{cases} \quad (8)$$

$$C_z = \frac{2K_z \times (B_z)}{\omega} \tag{9}$$

The equations related to the rotational degree of freedom (rocking about x-axis) are as follows [39, 40]:

$$K_{xx} = \frac{GB^3}{1-\nu}\left[3.2\left(\frac{L}{B}\right)^{0.75} + 0.8\right] \tag{10}$$

$$B_{xx} = \left[\frac{(4\psi/3) \times (L/B)a_0^2}{(K_{xx}/GB^3)\left[\left(2.2 - \frac{0.4}{(L/B)^3}\right) + a_0^2\right]}\right] \times \left[\frac{a_0}{2a_{xx}}\right] \rightarrow \begin{cases} a_{xx} = 1 - \left[\dfrac{\left(0.55 + 0.01\sqrt{L/B - 1}\right)a_0^2}{\left(2.4 - \frac{0.4}{(L/B)^3}\right) + a_0^2}\right] \\ a_0 = \dfrac{\omega B}{V_s} \\ \psi = \sqrt{2(1-\nu)/(1-2\nu)} \\ \psi \le 2.5 \end{cases} \tag{11}$$

$$C_{xx} = \frac{2K_{xx} \times (B_{xx})}{\omega} \tag{12}$$

In the above equations:

K_y is the equivalent soil stiffness in y direction.
B_y is the damping coefficient in y direction.
C_y is the equivalent damping coefficient in y direction.
K_z is the equivalent soil stiffness in z direction.
B_z is the damping coefficient in z direction.
C_z is the equivalent damping coefficient in z direction.
K_{xx} is the equivalent soil stiffness related to rotation around the x-axis.
B_{xx} is the soil damping coefficient related to rotation around the x-axis.
C_{xx} is the soil equivalent damping coefficient related to rotation around the x-axis.
B is the half of the foundation dimension in the y direction.
L is the half of the foundation dimension in the x direction.
G is the shear modulus.
a_z is a constant, ω is the circular frequency, a_0 is a dimensionless frequency, ψ is a constant, and ν is the Poisson ratio of the soil.

Figure 2 shows a diagram of the SSI problem, where subscripts s and f refer to the structure and foundation, respectively, K_h, K_v, and K_r refer to the horizontal, vertical, and rotational stiffness of the soil, and C_h, C_v, and C_r refer to the horizontal, vertical, and rotational damping of the soil.

Fig. 2 SSI problem

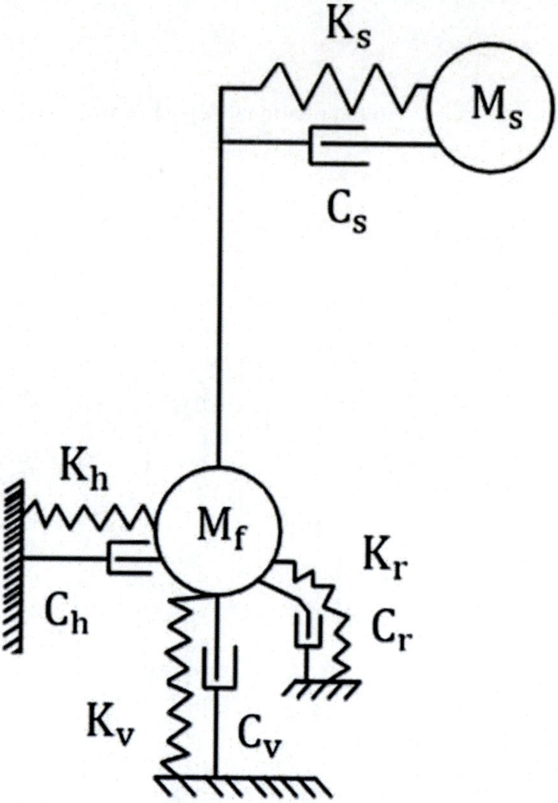

Finally, by solving the equilibrium equations by the numerical Newmark-beta method, the displacement of the soil and structure in different degrees of freedom is obtained.

In this paper, at each time step, the parameters of the cumulative absolute velocity (*CAV*) and the pore water pressure ratio (r_u) are obtained linearly, and with the help of the above-mentioned parameters, a new shear modulus is obtained at each time step. According to the above, this shear modulus decreases during the earthquake and approaches zero during liquefaction. To calculate the seismic response of the soil and structure, a multiple degree of freedom (MDOF) system has been used, and in order to solve the equations obtained, the Newmark-beta method has been used. All the above procedure is coded in MATLAB [37] and solved. Finally, the displacements (horizontal, vertical, and rotational) of the building structures are given as output.

3 Validation

The article of Mehrzad et al. [41] has been used for validation, where a nine-floor structure was assumed with foundation dimensions of 9.6 by 6.4 m and foundation static contact pressure of 88.5 kPa. Input excitation is in the form of a sinusoidal wave with an amplitude of 0.18 g and frequency of 2 Hz.

Validation results are presented in Figs. 3, 4, 5, 6 and 7. According to Fig. 3, at a depth of 2.4 m, the error in the worst case is 10%. However, at a depth of 7.2 m, this error becomes 11% in the worst case (Fig. 4). At depths of 12, 17.6, and 24 m, there is the best similarity between the outputs of that article [41] and the present article, where in the worst case, only a 2% error is observed (Figs. 5, 6 and 7).

Thus, the comparison between the results of the proposed technique and the simulation using a finite difference software [41] shows that the present technique has a good accuracy.

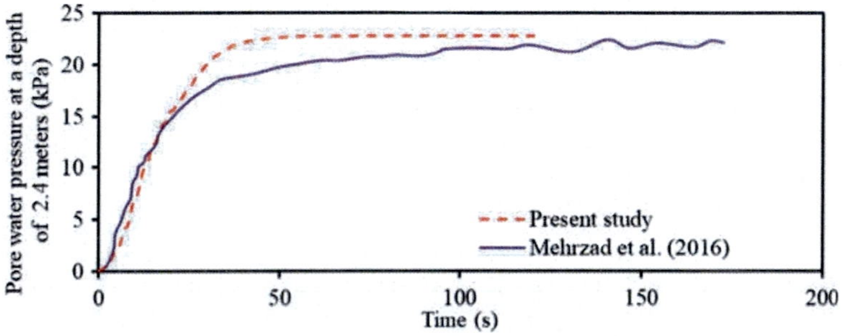

Fig. 3 Validation results at a depth of 2.4 m

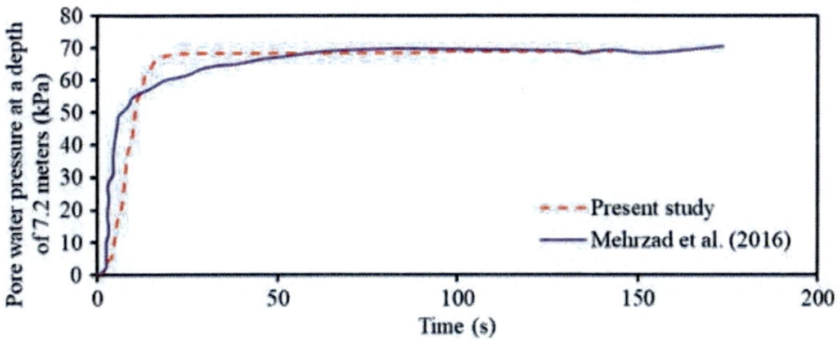

Fig. 4 Validation results at a depth of 7.2 m

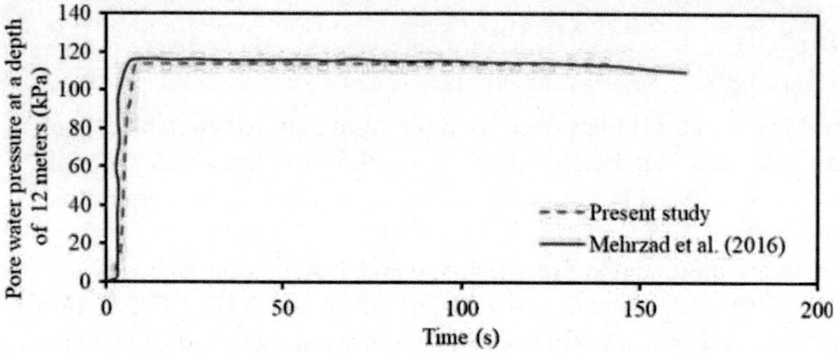

Fig. 5 Validation results at a depth of 12 m

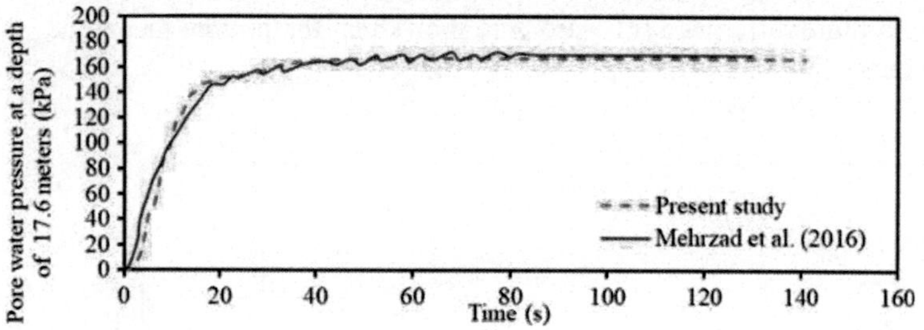

Fig. 6 Validation results at a depth of 17.6 m

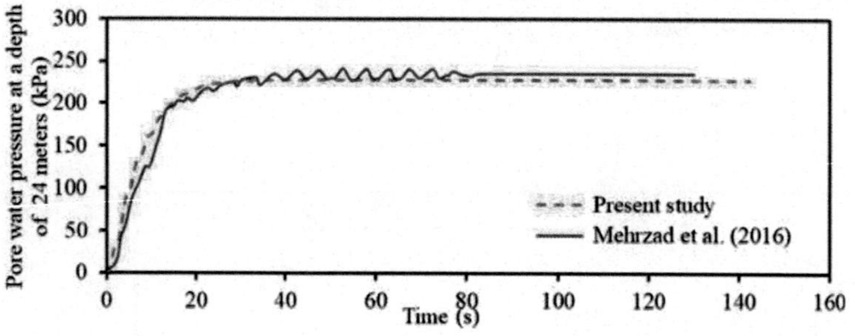

Fig. 7 Validation results at a depth of 24 m

4 Results and Discussion

In this article, two building structures with different heights and weights, two different soil types, and a seismic input with low risk level were considered.

Herein, the foundation dimensions are 8 m × 8 m, and two building structures with a different number of floors and weight are considered. The characteristics of the two building structures are presented in Table 1.

Table 1 Characteristics of the two building structures

Number of floors	Height (m)	Mass (ton)
15	45	1920
35	105	4480

Table 2 Dynamic characteristics of soils

Soil type	ξ_g	E (kN/m^2)	G (kN/m^2)	γ (kg/m^3)	ν	V_s (m/s)
S1	0.05	7,000,000	2,692,310	2000	0.30	1149.10
S2	0.05	500,000	192,310	1900	0.35	309.22

Table 3 Specifications of the earthquake record used

Earthquake record name	Year	PGA (g)	Magnitude (R)	Effective duration (s)	Seismic hazard level
Tabas	1978	0.047	7.35	24.2	Low

The mass of each structure is assumed to be 2000 kg/m^2, and the height of each floor is 3 m. The density of concrete used in the foundation is considered to be 2400 kg/m^3.

Also, the characteristics of the two soils and the earthquake used are given in Tables 2 and 3, respectively.

Herein, an earthquake with low risk level has been considered. Earthquakes with higher risk levels (all levels of seismic hazard) will be considered in in a future article.

According to the algorithm proposed in Sect. 3, four models with two building structures, two soil types, and one earthquake were examined, and the results of which are presented in Figs. 8, 9, 10, 11, 12, 13 and 14.

From the displacement results of the two building structures on soil type 1, it is concluded that the higher the height of the structure, the higher the displacements of the structures. This conclusion can also be seen for the building structures located on soil type 2.

Moreover, from Figs. 8 and 9, it can be seen that as the soil loosens and the shear wave velocity decreases, the displacements of the foundations decrease.

Figures 12, 13 and 14 show that during an earthquake, the pore water pressure ratio (r_u) increases, and the looser the soil, the higher the slope of the r_u curve. It is known that when the pore water pressure ratio reaches 1, the soil actually liquefies. Therefore, according to the numbers in Figs. 12, 13 and 14, the two soils considered are not degraded by the Tabas earthquake.

Figures 12, 13 and 14 also show that the soil stiffness and damping decrease during an earthquake. This result can be seen in both types of soil. It is also noteworthy that with decreasing soil shear wave velocity and soil loosening, the soil stiffness also decreases. At the time of liquefaction, the soil stiffness goes to zero. As a result, from

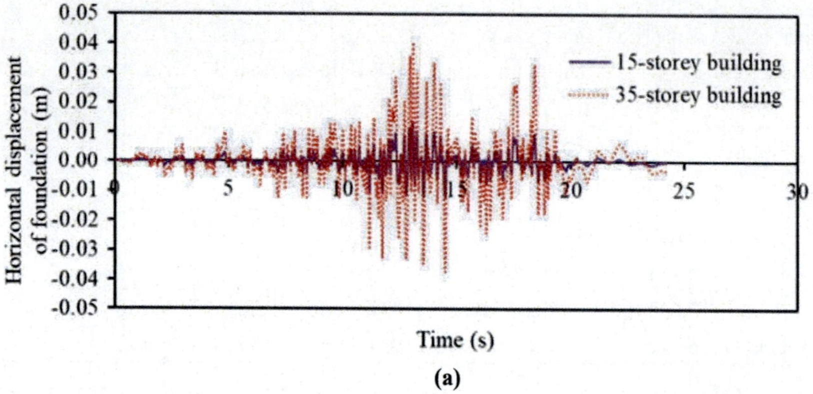

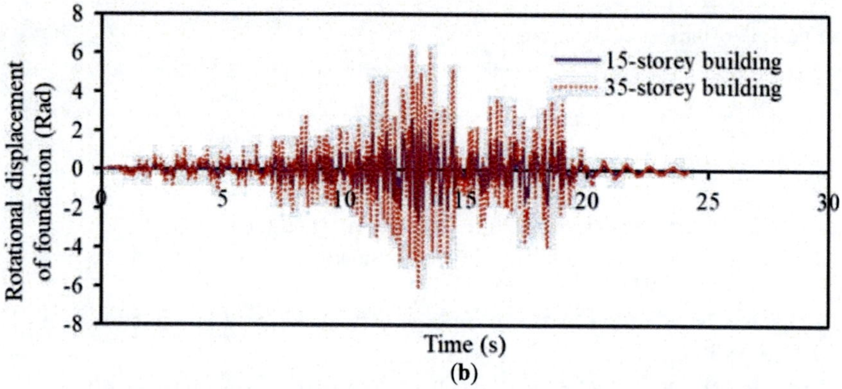

Fig. 8 Horizontal and rotational displacement of the foundations of the 15- and 35-story building on Type 1 soil under the Tabas earthquake: **a** Horizontal displacement, **b** Rotational displacement

the non-zero stiffness of the soils, the complete absence of liquefaction can also be justified.

5 Conclusions

In the present paper, a simplified approach using the substructure method to model the nonlinearity of the liquefiable soil in a system of soil-structure interaction was proposed. In this paper, two building structures with 15 and 35 stories are placed on Type 1 and Type 2 soils.

The occurrence of the liquefaction phenomenon, as well as the amount of displacement of the foundations in the three directions and the displacement of the structure

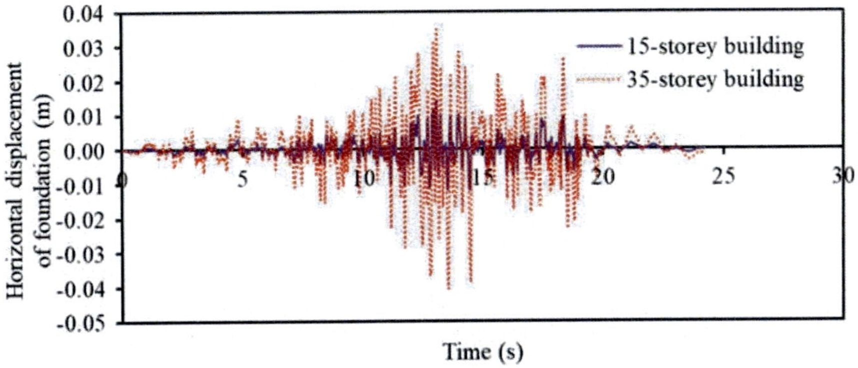

(a)

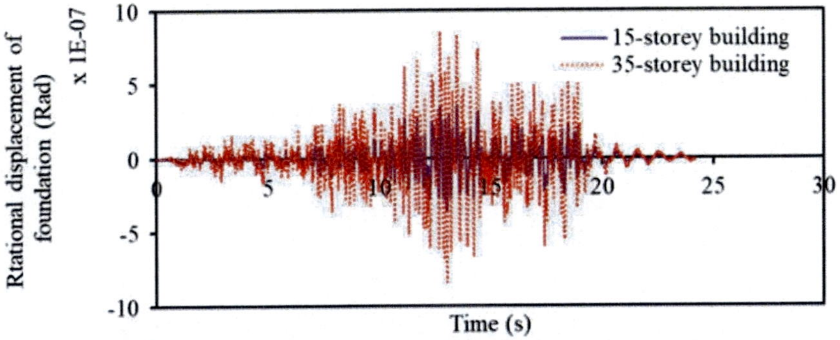

(b)

Fig. 9 Horizontal and rotational displacement of the foundations of the 15- and 35-story building on Type 2 soil under the Tabas earthquake: **a** Horizontal displacement, **b** Rotational displacement

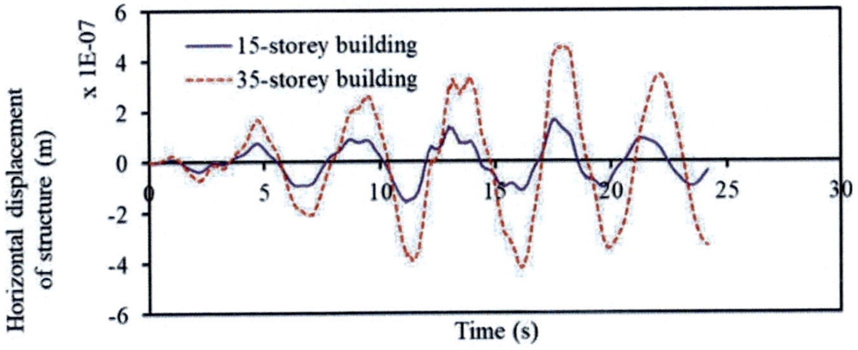

Fig. 10 Horizontal displacement of the 15-, 35-story buildings on Type 1 soil under the Tabas earthquake

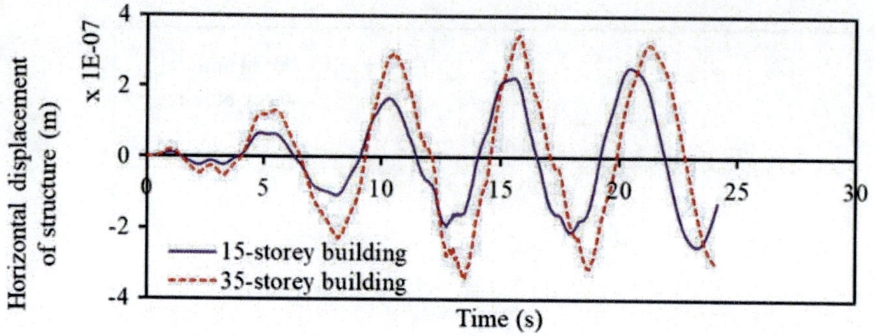

Fig. 11 Horizontal displacement of the 15-, 35-story buildings on Type 2 soil under the Tabas earthquake

in the horizontal direction, have been investigated. The conclusions of the proposed method and the outputs of MATLAB code are stated below.

Comparing the results of this article with the article of Mehrzad et al., it can be seen that the proposed method has a good accuracy (error in the worst case between 2 to 11%). Simple modeling, short run time, and appropriate accuracy are among the advantages of this method.

The Tabas earthquake increased the pore water pressure ratio (r_u), but it did not cause liquefaction of the two considered soils.

It can be concluded that the higher the building structure, the higher the displacements of the structure.

It can also be concluded that the looser the soil, the lower the displacements of the foundation and the structure. It must be pointed out that the rotation of the foundation is the most dangerous, especially for high-rise buildings.

Soil stiffness in the rotational direction is more than the vertical and horizontal stiffness of soils. This means that the soil is more resistant to rotation than to horizontal and vertical displacement. This trend is also observed in soil damping, considering the rotational, horizontal, and vertical degrees of freedom.

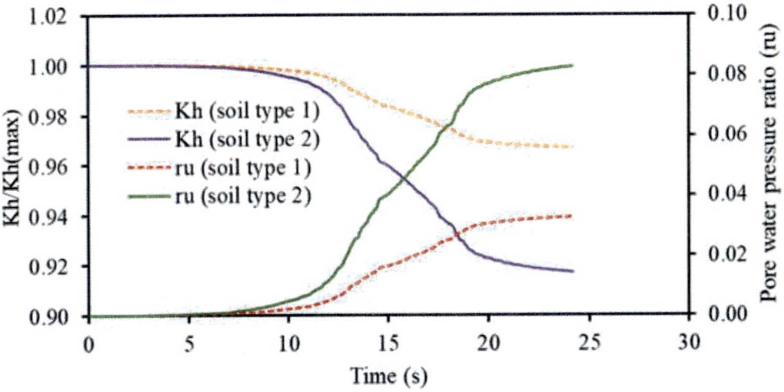

(a)

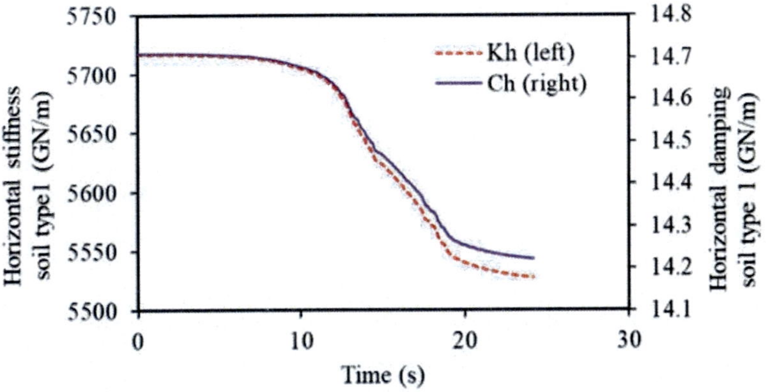

(b)

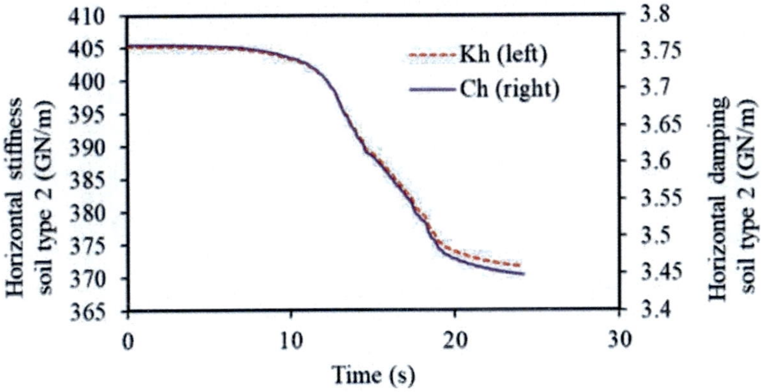

(c)

Fig. 12 Pore water pressure ratio and horizontal stiffness and damping of soil type 1 and soil type 2 under the Tabas earthquake

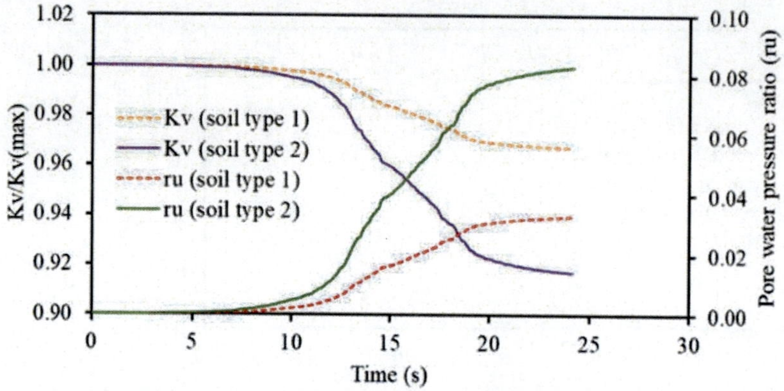

(a)

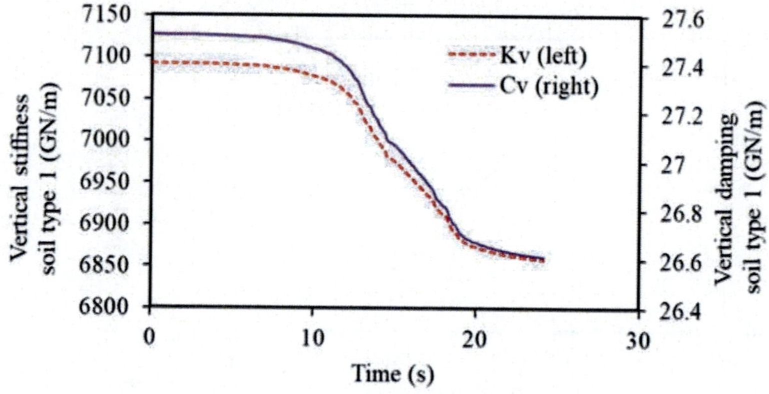

(b)

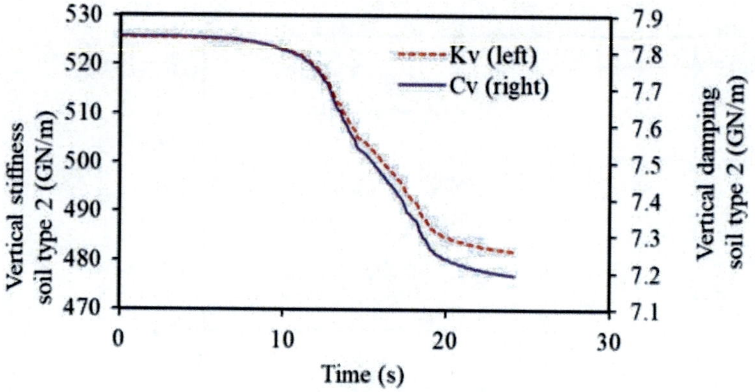

(c)

Fig. 13 Pore water pressure ratio and vertical stiffness and damping of soil type 1 and soil type 2 under the Tabas earthquake

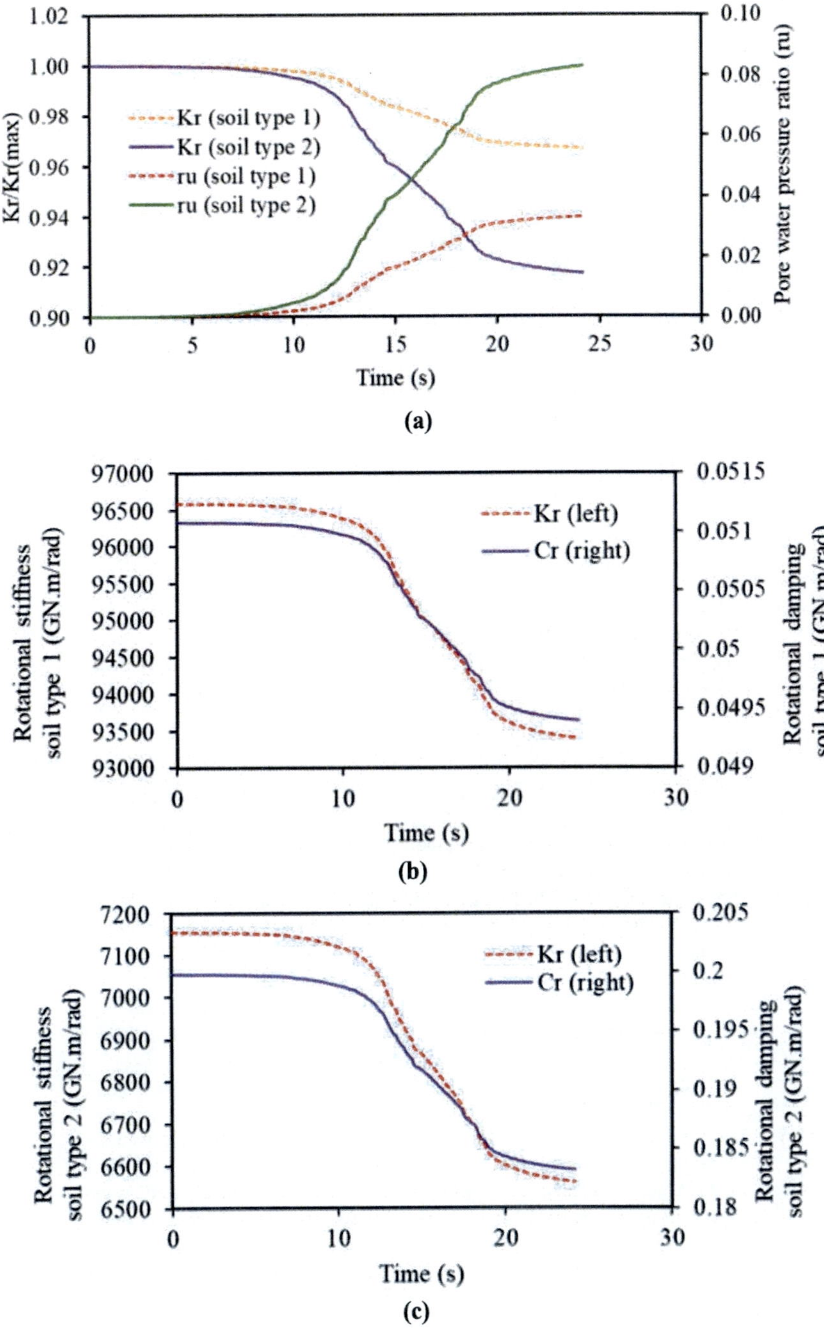

Fig. 14 Pore water pressure ratio and rotational stiffness and damping of soil type 1 and soil type 2 under the Tabas earthquake

References

1. Wolf JP (1985) Dynamic soil-structure-interaction. Prentice Hall, Englewood Cliffs, New Jersey
2. Stewart JP, Seed RB, Fenves GL (1999) Seismic soil-structure-interaction in buildings II: empirical findings. J Geotech Geoenviron Eng 125(1):38–48. https://doi.org/10.1061/(ASCE)1090-0241(1999)125:1(38)
3. Pitilakis D, Clouteau D (2010) Equivalent linear substructure approximation of soil-foundation-structure interaction: model presentation and validation. Bull Earthq Eng 8(2):257–282. https://doi.org/10.1007/s10518-009-9128-3
4. Maleki M, Khodakarami MI (2017) Feasibility analysis of using MetaSoil scatterers on the attenuation of seismic amplification in a site with triangular hill due to SV-waves. Soil Dyn Earthq Eng 100:169–182. https://doi.org/10.1016/j.soildyn.2017.05.036
5. Gouasmia A, Djeghaba K (2007) Non-linear seismic soil-structure interaction analysis of structures based on the substructure method. Asian J Civil Eng (Build Hous) 8(2):183–201
6. Farajian M, Khodakarami MI, Kontoni D-PN (2017) Evaluation of soil-structure interaction on the seismic response of liquid storage tanks under earthquake ground motions. Computation 5(1):17. 1–12. https://doi.org/10.3390/computation5010017
7. Hall WS, Oliveto G (eds) (2003) Boundary element methods for soil-structure interaction. Springer—Kluwer Academic Publishers, Dordrecht, Netherlands
8. Von Estorff O, Antes H (1991) Dynamic response of tunnel structures. European conference on structural dynamics, EUDODYN'90, pp 803–810 Balkema AA, Rotterdam, Netherlands
9. Lysmer J, Kuhlemeyer RL (1969) Finite dynamic model for infinite media. J Eng Mech Div 95(4):859–877. https://doi.org/10.1061/JMCEA3.0001144
10. Lysmer J, Drake LA (1971) The propagation of love waves across nonhorizontally layered structures. Bull Seismol Soc Am 61(5):1233–1251. https://doi.org/10.1061/JMCEA3.0001144
11. Lysmer J, Waas G (1972) Shear waves in plane infinite structures. J Eng Mech Div 98(1):85–105. https://doi.org/10.1061/JMCEA3.0001583
12. Kausel Bolt EAM (1974) Forced vibrations of circular foundations on layered media. Doctoral dissertation, Dept. of Civil Engineering, Massachusetts Institute of Technology. http://hdl.handle.net/1721.1/13842
13. Kim DK, Yun CB (2000) Time-domain soil–structure interaction analysis in two-dimensional medium based on analytical frequency-dependent infinite elements. Int J Numer Meth Eng 47(7):1241–1261. https://doi.org/10.1002/(SICI)1097-0207(20000310)47
14. Yun CB, Kim DK, Kim JM (2000) Analytical frequency-dependent infinite elements for soil-structure interaction analysis in two-dimensional medium. Eng Struct 22(3):258–271. https://doi.org/10.1016/S0141-0296(98)00070-4
15. Yang YB, Hung HH, Lin KC, Cheng KW (2015) Dynamic response of elastic half-space with cavity subjected to P and SV waves by finite/infinite element approach. Int J Struct Stab Dyn 15(07):1540009. https://doi.org/10.1142/S021945541540009X
16. Lysmer J (1969) Finite dynamic model for infinite media. J Eng Mech Div 95(4):859–877. https://doi.org/10.1061/JMCEA3.0001144
17. Tohidvand HR, Hajialilue-Bonab M (2014) Seismic soil-structure interaction analysis using an effective scaled boundary spectral element approach. Asian J Civil Eng (BHRC) 15(4):501–516
18. Liao ZP, Wong HL (1984) A transmitting boundary for the numerical simulation of elastic wave propagation. Soil Dyn Earthq Eng 3(4):174–183. https://doi.org/10.1016/0261-7277(84)90033-0
19. Khodakarami MI, Khaji N (2014) Wave propagation in semi-infinite media with topographical irregularities using decoupled equations method. Soil Dyn Earthq Eng 65:102–112. https://doi.org/10.1016/j.soildyn.2014.06.006
20. Mirzajani M, Khaji N, Khodakarami MI (2016) A new global nonreflecting boundary condition with diagonal coefficient matrices for analysis of unbounded media. Appl Math Model 40(4):2845–2874. https://doi.org/10.1016/j.apm.2015.09.083

21. Belytschko T, Black T (1999) Elastic crack growth in finite elements with minimal remeshing. Int J Numer Meth Eng 45:601–620. https://doi.org/10.1002/(SICI)1097-0207(19990620)45:5%3C601::AID-NME598%3E3.0.CO;2-S
22. Price AB, DeJong JT, Boulanger RW (2017) Cyclic loading response of silt with multiple loading events. J Geotech Geoenviron Eng 143(10):04017080. 1–13. https://doi.org/10.1061/(ASCE)GT.1943-5606.0001759
23. Toyota H, Takada S (2017) Variation of liquefaction strength induced by monotonic and cyclic loading histories. J Geotech Geoenviron Eng 143(4):04016120. 1–11. https://doi.org/10.1061/(ASCE)GT.1943-5606.0001634
24. Ishihara K, Koga Y (1981) Case studies of liquefaction in the 1964 Niigata earthquake. Soils Found 21(3):35–52. https://doi.org/10.3208/sandf1972.21.3_35
25. Koseki J, Matsuo O, Ninomiya Y, Yoshida T (1997) Uplift of sewer manholes during the 1993 Kushiro-Oki earthquake. Soils Found 37(1):109–121. https://doi.org/10.3208/sandf.37.109
26. Cubrinovski M, Henderson D, Bradley BA (2012) Liquefaction impacts in residential areas in the 2010–2011 Christchurch earthquakes. University of Canterbury, Civil and Natural Resources Engineering, pp 1–14. https://ir.canterbury.ac.nz/handle/10092/6712
27. Yasuda S, Harada K, Ishikawa K, Kanemaru Y (2012) Characteristics of liquefaction in Tokyo Bay area by the 2011 Great East Japan earthquake. Soils Found 52(5):793–810. https://doi.org/10.1016/j.sandf.2012.11.004
28. Ishikawa A, Zhou YG, Shamoto Y, Mano H, Chen YM, Ling DS (2015) Observation of post-liquefaction progressive failure of shallow foundation in centrifuge model tests. Soils Found 55(6):1501–1511. https://doi.org/10.1016/j.sandf.2015.10.014
29. Kajihara K, Okuda H, Kiyota T, Konagai K (2020) Mapping of liquefaction risk on road network based on relationship between liquefaction potential and liquefaction-induced road subsidence. Soils Found 60(5):1202–1214. https://doi.org/10.1016/j.sandf.2020.07.007
30. Wang R, Fu P, Zhang JM, Dafalias YF (2016) DEM study of fabric features governing undrained post-liquefaction shear deformation of sand. Acta Geotech 11(6):1321–1337. https://doi.org/10.1007/s11440-016-0499-8
31. Wang R, Dafalias YF, Fu P, Zhang JM (2020) Fabric evolution and dilatancy within anisotropic critical state theory guided and validated by DEM. Int J Solids Struct 188:210–222. https://doi.org/10.1016/j.ijsolstr.2019.10.013
32. Wei J, Huang D, Wang G (2020) Fabric evolution of granular soils under multidirectional cyclic loading. Acta Geotech 15(9):2529–2543. https://doi.org/10.1007/s11440-020-00942-8
33. Huang X, Hanley KJ, Zhang Z, Kwok C, Xu M (2019) Jamming analysis on the behaviours of liquefied sand and virgin sand subject to monotonic undrained shearing. Comput Geotech 111:112–125. https://doi.org/10.1016/j.compgeo.2019.03.008
34. Kramer SL (1996) Geotechnical earthquake engineering. Prentice-Hall international series in civil engineering and engineering mechanics. Prentice-Hall, New Jersey
35. Bray JD, Dashti S (2014) Liquefaction-induced building movements. Bull Earthq Eng 12(3):1129–1156. https://doi.org/10.1007/s10518-014-9619-8
36. Karimi Z, Dashti S (2015) Numerical and centrifuge modeling of seismic soil-foundation-structure-interaction on liquefiable ground. J Geotech Geoenviron Eng 142(1):04015061. 1–14. https://doi.org/10.1061/(ASCE)GT.1943-5606.0001346
37. MATLAB version R2017b (9.3.0) (2017) The MathWorks Inc., Natick, Massachusetts
38. Khodakarami MI, Lashgari A (2018) An equivalent linear substructure approximation for the analysis of the liquefaction effects on the dynamic soil–structure interaction. Asian Journal of Civil Engineering 19:67–78. https://doi.org/10.1007/s42107-018-0008-5
39. NIST: Soil-Structure-Interaction for Building Structures (NIST GCR 12–917–21). National Institute of Standards and Technology, Gaithersburg, MD (2012). https://tsapps.nist.gov/publication/get_pdf.cfm?pub_id=915495

40. Pais A, Kausel E (1988) Approximate formulas for dynamic stiffnesses of rigid foundations. Soil Dyn Earthq Eng 7(4):213–227. https://doi.org/10.1016/S0267-7261(88)80005-8
41. Mehrzad B, Haddad A, Jafarian Y (2016) Centrifuge and Numerical Models to Investigate Liquefaction-Induced Response of Shallow Foundations with Different Contact Pressures. International Journal of Civil Engineering 14:117–131. https://doi.org/10.1007/s40999-016-0014-5

Geotechnical Engineering

Ground-Granulated Blast Furnace Slag-Based Geopolymer-Treated Fibrous Peat

Suhail Ahmad Khanday, Monowar Hussain, and Amit Kumar Das

1 Introduction

Fibrous peat is made up of loosely arranged unconsolidated organic matter whose particles are not cemented with each other. The peat is mainly found in submerged or wetlands in acidic and anaerobic conditions [1]. Approximately, the fibrous peat constitutes about 8% of total global land [2]. In order to catergorise fibrous peat, the fiber content present in the soil must be greater than 66% [3]. This type of soil is highly problematic in its natural conditions because of low shear strength (5–20 kPa) and high compressibility due to the presence of fresh fiber and microorganisms [2, 4]. The other geotechnical properties of fibrous peat include the high rate of creep, high permeability and high-water holding capacity (up to 850%), which is mainly because of the presence of a high void ratio in the range of 7.5–30 [5]. Therefore, to use it as a sustainable geomaterial for the foundation of any infrastructure, its stabilization is inevitable.

The chemical stabilization involves the mixing of the binder with soil to form a blend, which upon reaction forms a hardened matrix [6, 7]. The GGBS used in chemical stabilization is obtained from the steel industry as a by-product and is highly non-crystalline in nature [8]. The GGBS after rapidly cooled is grounded. The main constituents are Si, Al, Ca, O and Mg. The rate of hydration is slow after combining with water; however, the rate of hydration can be increased by increasing the pH of the solution. Therefore, alkalis like sodium hydroxide, magnesium oxide and sodium silicate are used to accelerate the hydration rate [9].

Geopolymer is formed from the dissolution of Si and Al from the binder in alkaline solution [10]. The GGBS-based geopolymer has been successfully used

S. A. Khanday (✉)
Department of Civil Engineering, National Institute of Technology, Silchar 788010, India
e-mail: khandaysuhail2013@gmail.com

M. Hussain · A. K. Das
National Institute of Technology, Silchar 788010, India

A. Sil et al. (eds.), *Recent Trends in Civil Engineering*, Lecture Notes in Civil Engineering 274, https://doi.org/10.1007/978-981-19-4055-2_32

in concrete technology and recently, its use has been extended as a clean stabilizer for the stabilization of the weak soils. The use of GGBS-based geopolymer has resulted in enhancing the properties of weak soil like high resistance to sulfate attack and wetting to drying, high strength and lower permeability [11–13]. Previous literature mainly focuses on the stabilization of inorganic soil with GGBS-based geopolymer. Therefore, in this study, an attempt has been made to explore the effect of GGBS-based geopolymer on fibrous peat. A series of tests, including pH, electrical conductivity (EC), and unconfined compressive strength (UCS), is conducted to investigate the properties of stabilized peat.

2 Materials and Methodology

2.1 *Materials*

2.1.1 Peat

Fibrous peat was collected from tea estate rich in tea plant cultivation in the Hailakandi district of state Assam, India. The peat layer was found around 4.5 m thick ranging from 1.5 to 6 m below ground level. The peat layer is sandwiched between blackish organic soil on the top and silty clay on the bottom. The physicochemical characteristics of the peat are shown in Table 1.

Table 1 Physiochemical properties of fibrous peat

Properties	Values
Organic content (%)	34
Ash content (%)	66
Fiber content (%)	73
pH	5.2
Electrical conductivity (μS/cm)	264
Specific gravity	1.73
Water content (%)	431
Bulk unit weight (kN/m^3)	9.8
Liquid limit (%)	75
Plastic limit (%)	63
Degree of humification	H1–H4

Table 2 Physical properties of GGBS

Properties	Values
Specific gravity	2.85
Bulk density (gm/cc)	2.9
Particle finer than 45 microns (%)	97
Specific surface area (m^2/kg)	410
pH	8.5

2.1.2 Ground-Granulated Blast Furnace Slag

GGBS is obtained from Astrra Chemicals (Chennai, India) in fine powder form. The physical composition of GGBS is shown in Table 2.

2.1.3 Sodium Hydroxide

The 98% pure sodium hydroxide in pellet form with a specific gravity of 2.13 was used in this study as an activator. It has shown better dissolution capabilities than sodium silicate activator [11]. The solution of NaOH with the required molarity was prepared one day before its use.

2.2 *Methodology*

The natural water content is obtained by drying the collected peat at a temperature of 105 °C as per IS 2720 Part 2 (1973) [14]. The density bottle method and core cutter method were used to find out specific gravity and bulk density (i.e., in situ density) as per IS 2720 Part 3 (1980) and IS 2720 Part 29 (1975), respectively [15, 16]. The conductivity meter (Model CM 180) and digital pH meter (Model L1 120) were used for determining electrical conductivity and pH of the soil, respectively, as per ASTM D2976 (2015) [17]. The Atterberg limits were found out as per IS 2720 Part 5 (1985) [18]. The muffle furnace was used to find out organic content as per ASTM D2974 (2014) [19]. The fiber content was calculated by weighing the dry fiber retained on a 150 μm sieve as per ASTM D1997 (2013) [20]. The UCS of the untreated and treated test was performed as per IS 2720 Part 10 (1991) [21]. The final test value was taken as the mean value of three samples.

2.2.1 Specimen Preparation

The GGBS as a binder in the percentages of 10, 20, and 30% by dry weight of fibrous peat along with NaOH as an activator was taken. The three concentrations/molarities (i.e., 6, 9, and 12) and three alkali to binder ratio (A/B) (i.e., 0.5, 0.7, and 0.9) were chosen as per [9].

The required quantity of oven-dried peat and binder was mixed thoroughly for 3–5 min. Then the required amount of alkali was added to form a homogenous mixture. The specimens were prepared at the liquid limit to increase workability and induce through mixing [11, 22]. The UCS sampler of height 76 mm and diameter 38 mm was then filled with the mixture of the soil-geopolymer blend, compacted manually in three layers to release any air if present from the voids. The curing was achieved at room temperature for 7, 14, and 28 days. After desired curing periods, the samples were subjected to UCS testing. The pH and EC were calculated from post-tested samples of UCS.

3 Results and Discussion

3.1 *Role of pH on UCS*

The leaching of silica and alumina from the binder mainly depends on the pH of the alkaline solution. The pozzolanic reactions get encouraged when the pH of the solution is equal to or greater than 10.5 [23]. Figure 1 shows the variation of pH with molarity at 7 and 14 days of curing period. It is found that with the increase in the concentration of NaOH solution, the alkalinity increases in the pore solution. This is due to the increase in hydroxide ions released in the solution. This high alkalinity will help in the dissolution of silica and alumina from the GGBS, leading to the formation of a geopolymer that is responsible for the strength gain in treated peat.

Further, with an increase in the curing period from 7 to 14 days, the pH got decreased. However, the pH shows very little attenuation after 14 days of curing period which is also in good agreement with published literature study [9]. This happens because the hydroxide ions responsible for alkalinity get used up in the process of geopolymerization. Therefore, to keep the pozzolanic reactions going, it is necessary that the pH of the pore solution should be greater than 10.5.

Figure 2 shows the variation of EC with molarity after 7 and 14 days of curing. The results show the increasing trend of EC with the increase in the concentration of the solution. This is because of the coming of more and more ions in the solution due to the leaching of silica and alumina from binder and hydroxide ion liberated from NaOH. However, with the increase in the curing period, the EC shows a decreasing trend. The similar trends are observed by Khanday et al. [9] for 28 days of curing period [9]. This is attributed to the blocking of movement of ions through the pores, which gets filled up by cementitious gel, and also because the ions get eaten up in the geopolymerization process.

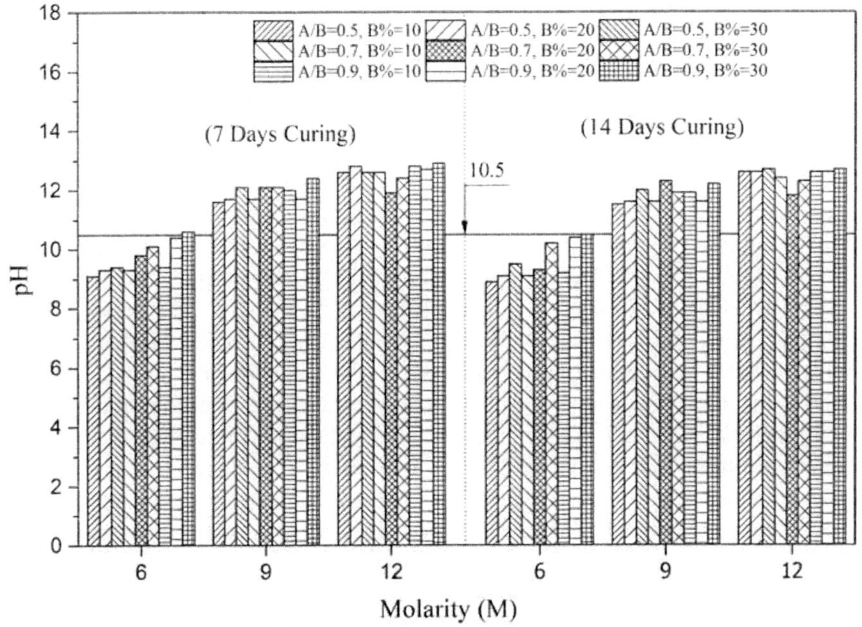

Fig. 1 Variation of pH with molarity at different A/B ratio and binder content after 7 and 14 days of curing

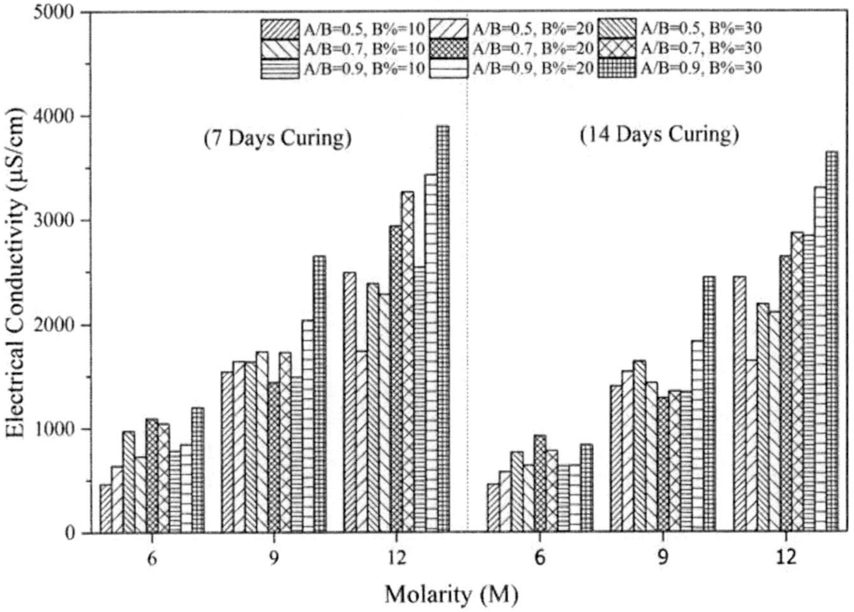

Fig. 2 Variation of electrical conductivity with molarity at different A/B ratio and binder content after 7 and 14 days of curing

3.2 *Effect of Binder Content, A/B Ratio, Molarity and Curing Period on UCS*

Figure 3a, b and c shows the unconfined compressive strength of fibrous peat with respect to binder content at different A/B ratios and molarities after 7, 14 and 28 days of curing, respectively. The results show that the UCS increases with the increase in binder content. This is because as more the binder content more is silica and alumina present for the geopolymerization. It was also found that the UCS increases up to a certain limit of A/B ratio (i.e., 0.7) and then decreases. This is because at a low ratio, the amount of alkali required to leach the silica and alumina from the binder is less, and at a higher A/B ratio, the excess hydroxide ions cause the crystallization stage of geopolymerization before the condensation stage [24]. Further, it is clear from Fig. 3 that the UCS is maximum at a molarity of 9. This is because, at a lower concentration of 6 M, the pH-induced is not sufficient to liberate silica and alumina ions from the binder, therefore inhibit the geopolymerization. Further, at a higher concentration, a semi-plastic mixture is formed, which leads to lower strength. However, at an optimum NaOH concentration of 9 M, the pH was sufficient, and samples were easily moldable, leading to higher UCS.

From Fig. 3, it is evident that the UCS increases with the increase in the curing period. However, the rate of increment of UCS was greater for 7 days, then it

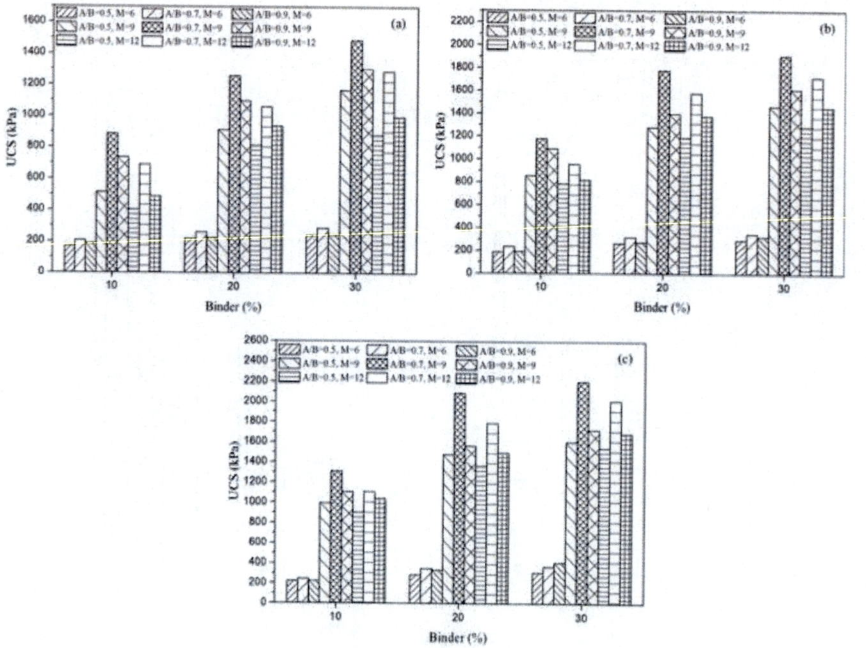

Fig. 3 Variation of UCS with binder content at different A/B ratio and molarity after **a** 7 days, **b** 14 days, and **c** 28 days of curing

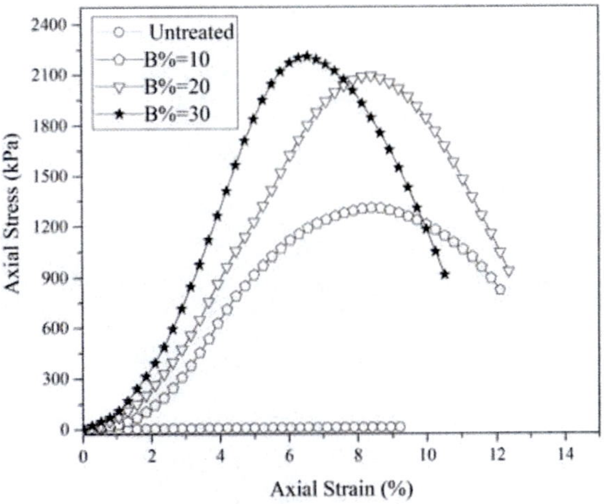

Fig. 4 Stress–strain response of stabilized peat at A/B ratio of 0.7 and 6 M with different binder content

decreased. This is because the geopolymer reactions take only a few weeks to complete, unlike cement, where pozzolanic reactions take place for a longer duration. The maximum UCS value of 2207 kPa was achieved at the binder content of 30%, A/B ratio of 0.7, and molarity of 9 after 28 days of curing.

3.3 Stress–strain Behavior of GGBS-Based Geopolymer Stabilized

Figure 4 shows the stress–strain behavior of various percentages of GGBS-based geopolymer stabilized peat at 9 M and A/B ratios of 0.7. With the addition of binder, there is a decrease in axial strain and an increase in the slope of the stress–strain curve. Further, it is evident that with the increase in the binder, there is a shift of curve toward the left-hand side and decrement in failure strain. This means there is an increase in elastic modulus of GGBS-based geopolymer stabilized peat specimens. Furthermore, with the increase in binder content, there is an obvious transition of the specimen from ductile to brittle failure, as shown in Fig. 5. This is due to the cementitious product formation by geopolymerization.

3.4 Role of Na/Al on Peat-GGBS Geopolymer

From Fig. 3, it is evident that high UCS may not always be yielded at high molarity and A/B ratios. The highest UCS would be obtained at a particular combination of all constituents of the soil-geopolymer mix. The UCS of geopolymer stabilized soil is governed by the combination of atomic ratios of Na to Al and Si to Al ratios which

Fig. 5 Failure pattern of **a** untreated peat and **b** 10% **c** 20% **d** 30% GGBS-based geopolymer stabilized peat

controls the kinetics of reaction [24, 25]. NaOH is the source for Na, while the source of Al is GGBS. Therefore, the ratio of Na/Al increases with the increase in molarity and A/B ratio and is independent of binder percentage in mix.

Further, Fig. 6 shows the variation of UCS with Na/Al ratios at constant Si/Al ratio (2.2) and binder content of 30% with different curing periods. It was noticed that the improvement of UCS was low at Na/Al equal to 0.9 and increased sharply beyond 0.9 up to 1.8, and then there was a decrease in UCS. The improvement of UCS with Na/Al ratio may be due to the increase in Na content in the sample, which increases adhesion strength [26]. Therefore, the optimum Na/Al ratio is 1.8. However, Singhi et al. [27] found Na/Al ratio equal to 1.2 to be optimum for inorganic soil stabilized with GGBS-based geopolymer [27]. This difference in optimum values may be due to the effect of high organic content on cation exchange capacity.

4 Conclusion

- Peat is made up of fully or partially decomposed animal or plant remains. This highly organic soil is geotechnically problematic because of high natural moisture content (431%), high liquid limit (75%), low pH (5.2), low specific gravity (1.73), and low bulk unit weight (9.8 kN/m^3).

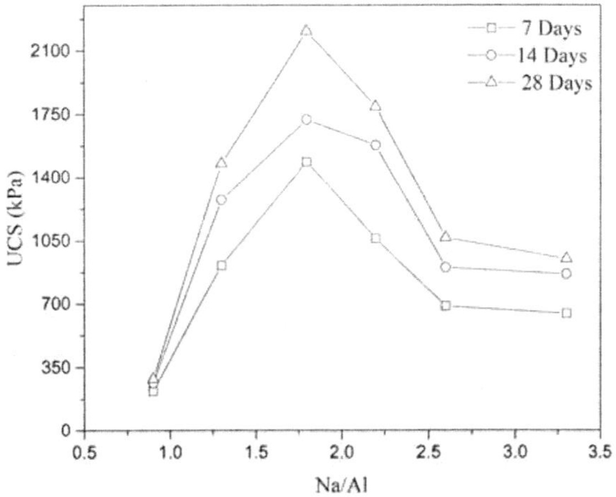

Fig. 6 Variation of UCS with Na/Al ratio at Si/Al = 2.2 and B = 30% for GGBS-based geopolymer-treated fibric peat

- There is an high improvement of UCS at A/B ratio of 0.7 and NaoH concentration of 9 molarity; below and above it, there is an decrease in UCS. The low alkaline environment cannot dissolve Si and Al sufficiently to form a geopolymer, while a high alkaline environment hinders strength gain by causing early crystallization of geopolymer.
- The highest value of UCS was obtained at GGBS content of 30%. It has been found that the lower amount of GGBS-geopolymer was not good enough to hold the peat particles together. Further, the UCS has increased rapidly first 7 days and after that the increament was gradual.
- The maximum UCS for GGBS-based geopolymer is achieved for GGBS of 30% as binder content, A/B and Na/Al ratio of 0.7 and 1.8, respectively, and the molarity of 9. The UCS of treated peat in this combination was found to be 208 times more than untreated peat (9.8 kPa). Hence, the application of GGBS-geopolymer to enhance the strength of fibrous peat is found to be a good alternative for cement.

References

1. Huat BBK, Prasad A, Asadi A, Kazemian S (2019) Geotechnics of organic soils and peat. CRC press
2. Dehghanbanadaki A, Arefnia A, Keshtkarbanaeemoghadam A, Ahmad K, Motamedi S, Hashim R (2017) Evaluating the compression index of fibrous peat treated with different binders. Bull Eng Geol Environ 76:575–586
3. Mesri G, Ajlouni M (2007) Engineering properties of fibrous peats. J Geotech Geoenviron Eng 133:850–866
4. Paul A, Hussain M (2019) Geotechnical properties and microstructural characteristics of Northeast Indian Peats Mires Peat 24
5. Moayedi H, Kassim KA, Kazemian S, Raftari M, Mokhberi M (2014) Improvement of peat using Portland cement and electrokinetic injection technique. Arab J Sci Eng 39:6851–6862

6. Khanday SA, Hussain M, Das AK (2021) A review on chemical stabilization of peat. Geotech Geol Eng, 1–15
7. Moayedi H, Nazir R (2018) Malaysian experiences of peat stabilization, state of the art. Geotech Geol Eng 36:1–11
8. Jiang N-J, Du Y-J, Liu K (2018) Durability of lightweight alkali-activated ground granulated blast furnace slag (GGBS) stabilized clayey soils subjected to sulfate attack. Appl Clay Sci 161:70–75
9. Khanday SA, Hussain M, Das AK (2021) Stabilization of Indian peat using alkali-activated ground granulated blast furnace slag. Bull Eng Geol Environ 1–13
10. Xu H, Van Deventer JSJ (2000) The geopolymerisation of alumino-silicate minerals. Int J Miner Process 59:247–266
11. Singhi B, Laskar AI, Ahmed MA (2017) Mechanical behavior and sulfate resistance of alkali activated stabilized clayey soil. Geotech Geol Eng 35:1907–1920
12. Sargent P, Hughes PN, Rouainia M (2016) A new low carbon cementitious binder for stabilising weak ground conditions through deep soil mixing. Soils Found 56:1021–1034
13. Du Y-J, Wu J, Bo Y-L, Jiang N-J (2020) Effects of acid rain on physical, mechanical and chemical properties of GGBS–MgO-solidified/stabilized Pb-contaminated clayey soil. Acta Geotech 15:923–932
14. IS 2720 Part 2 (1973) Method of test for soils: Determination of water content of soils. BIS, New Delhi. (1973)
15. IS 2720 Part 3 (1980) Method of test for soils: determination of specific gravity of fine grained soils. BIS, New Delhi
16. IS 2720 Part 29 (1975) Method of test for soils: determination of dry density of soils in-place by the core-cutter method. BIS, New Delhi
17. ASTM D2976 (2015) Standard test method for pH of peat materials. ASTM Int
18. IS 2720 Part 5 (1985) Method of test for soils: determination of atterberg limits. BIS, New Delhi
19. ASTM D2974 (2014) Standard test methods for moisture, ash, and organic matter of peat and other organic soils. ASTM Int
20. ASTM D1997 (2013) Standard test method for laboratory determination of the fibre content of peat samples by dry mass. ASTM Int
21. IS 2720 Part 10 (1991) Method of test for soils: determination of unconfined compressive strength of soils. BIS, New Delhi
22. Yi Y, Li C, Liu S (2014) Alkali-activated ground-granulated blast furnace slag for stabilization of marine soft clay. J Mater Civ Eng 27:4014146
23. Davidson LK, Demirel T, Handy RL (1965) Soil pulveration and lime migration in soil-lime stabilization. Highw Res Rec (1965)
24. Khale D, Chaudhary R (2007) Mechanism of geopolymerization and factors influencing its development: a review. J Mater Sci 42:729–746
25. Mozumder RA, Laskar AI (2015) Prediction of unconfined compressive strength of geopolymer stabilized clayey soil using artificial neural network. Comput Geotech 69:291–300
26. Khan MI, Azizi K, Sufian S, Man Z (2014) Effect of Na/Al and Si/Al ratios on adhesion strength of geopolymers as coating material. In: Applied Mechanics and Materials. pp 85–89. Trans Tech Publ
27. Singhi B, Laskar AI, Ahmed MA (2016) Investigation on soil–geopolymer with slag, fly ash and their blending. Arab J Sci Eng 41:393–400

Finite Element Modelling of RC Silo subjected to Thermal Loads

Zaid Mohammad, Mohd. Anwar, Saad Shamim Ansari, and Abdul Baqi

1 Introduction

Silos are commonly used for storage and subjected to static and dynamic loading conditions such as gravity load, granular infill pressure, thermal load due to storage material, wind as well as earthquake loads. The design of silos is primarily governed by the properties and physical condition of stored material. As the density, flow and frictional properties of grains, cement, coal, carbon black, etc. vary widely, failure stresses may vary in different parts of silo. Silo structures are also prone to buckling due to external pressures caused by wind force. Empty silos have been found to be more vulnerable to wind forces. Further, the occupancy levels inside the silos play major role in defining the variable stress development inside the concrete walls. The values of wall stresses and the critical locations have been observed to vary with the amount of infill material inside the silo structure. Many researchers have investigated the performance of concrete silos under wind loads, earthquake loads and thermal loads due to seasonal temperature variations. However, the effect of temperature along with the pressure exerted on the structure due to infill materials and its cracking behaviour at critical locations of extreme stresses has yet to be explored.

The state-of-the-art studies [1–11] have provided various empirical and analytical models to ascertain the internal pressure inside the silo and analysed the vulnerability of the steel as well as concrete silo structure against earthquake and wind loads. Mohammad et al. [12] reviewed prominent research studies conducted on steel as well as concrete silos, in which numerical as well as experimental investigations [3–7, 9–11] were conducted to compare the previously developed closed-form solutions proposed by Jansen [1], Airy [2] and Reimbert [8].

Z. Mohammad (✉) · Mohd. Anwar · S. S. Ansari · A. Baqi
Department of Civil Engineering, Z.H.C.E.T, Aligarh Muslim University, Aligarh 202002, India
e-mail: zmohammad_co@myamu.ac.in

A. Sil et al. (eds.), *Recent Trends in Civil Engineering*, Lecture Notes in Civil Engineering 274, https://doi.org/10.1007/978-981-19-4055-2_33

A very few research studies had explored the structural performance of RC or steel silo under thermal loads due to seasonal climatic variation or infill materials. Orosz [13] surveyed fundamental formulations for the determination of thermal stresses as proposed by previous studies and developed an approximate method for the design of silos. It was observed that the variation in outer temperature causes deformation in the silo wall and the stored material exposed to expansion and contraction cycles follows the wall movements, becomes increasingly elastic upon compaction, resulting in force effects acting as tensile forces in the cell wall of silo. Cha et al. [14] presented thermal stress analysis of a silo in a waste repository and risk analysis was performed using numerical analysis. It was concluded that the magnitude of the thermal stresses in the silo varied according to the construction conditions. Chen et al. [15] conducted an experimental study on a concrete silo with 136.5 m in diameter and 19.35 m in height under different circumstances of temperature patterns due to sunshine and yearly climatic conditions. The results agree well with the numerical simulation results. Khalifa and El-Kashif [16, 17] presented a comparative study in which design formulation for the internal pressure in the RC silo was discussed as stated by international code of practices, viz. ACI, Euro and Poland. The empirical relationships were then compared with a numerical study. Linear and nonlinear analyses were used to evaluate the stresses and displacements for different silos configurations and different loading combinations. It was observed that the nonlinear analysis of FEM has good matching with the corresponding values in ACI, leading to the conclusion that nonlinear analysis is more accurate than linear analysis. Moreover, the study results of hoop forces showed a distinct pattern with the temperature difference, silo radii and insignificant silo wall thickness for each of FEM, EU and Poland codes. Maraveas [18] published a review article discussing various studies concerning failure, design issues, repair and strengthening of reinforced concrete (RC) silos, used in agricultural industry. It was observed that high-grade steel used in silo wall reinforcement is susceptible to corrosion. In contrast, fibre-reinforced polymers (FRP) provide better mechanical properties and are resistant to corrosion. Additional challenges include design code limitations, poor design practice and failure due to periodical loading and unloading of infill materials.

Previous research works have explored the behaviour of infill material on the lateral pressure and proposed various analytical closed-form solutions as well as empirical relationships to calculate the infill pressure. However, the effect of temperature variation due to the infill material on the development of critical stresses inside silo structures is very much needed to be investigated to effectively design the RC cement silo structures. Thus, in the present study effect of gravity load, internal pressure load due to different occupancy levels was observed on the reinforced concrete silo by using Jansen's lateral pressure theory. Further, the effect of temperature variation of infill material on the structural behaviour of silo was observed by applying thermal loads on the structure. Numerical analysis of a three-dimensional quarter finite element model of RC silo was carried out in ABAQUS software. The concrete structural components were modelled using solid brick elements (C3D8R), whereas, truss elements (T3D2) were used to model the internal reinforcement. The Concrete Damage Plasticity material model was used to simulate the material behaviour of the concrete. Thermal properties of the concrete material were undertaken from a previously published study [10]. The results were obtained and discussed in terms of critical stresses and lateral deflection in the silo wall along the height, with different occupancy levels of infill material. At last appropriate solutions that improve the design of reinforced concrete silo have been suggested.

2 Modelling and Analysis

The present study investigates the effect of lateral pressure and thermal temperature variation (from normal to 120 °C) due to infill material at various occupancy levels, viz. 33, 50, 66 and 100%, on the structural performance of an RC silo. A reinforced concrete silo taken from a practical case study was undertaken. The dimensions and reinforcement details of the silo structure have been discussed in previously published study conducted by Mohammad et al. [12] (see Fig. 1). The three-dimensional finite element model of the silo was modelled in ABAQUS/CAE software and material properties and boundary conditions were assigned in the provided modules. In addition to the self-weight of the structure, the primary loads, viz. Horizontal pressure (p_h) and vertical shear stress (p_v) were applied inside the silo according to the infill occupancy level. These values were determined from the equations proposed by Janssen [1, 12].

The Concrete Damage Plasticity material model was used to simulate the material behaviour of the concrete. The material parameters for the concrete and rebar were taken from previously published study [10, 12]. The concrete structural components

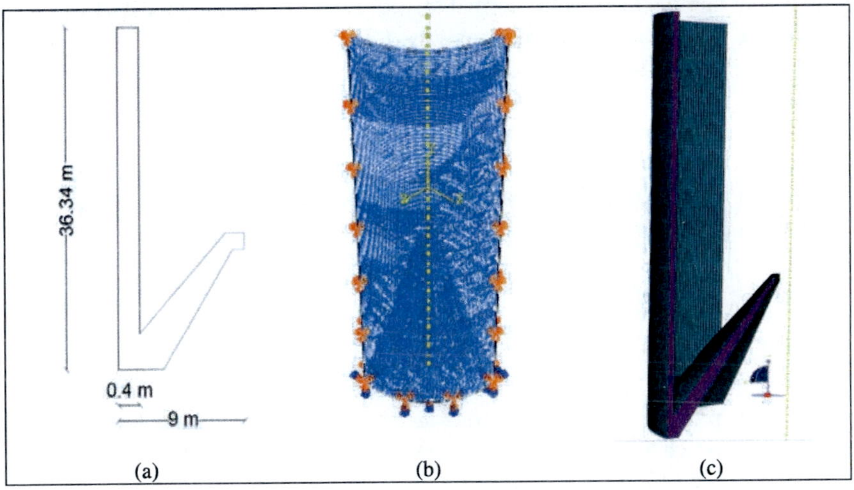

Fig. 1 Finite element model of RC silo: **a** geometric dimensions, **b** boundary condition and **c** mesh details

Table 1 Mesh properties of RC silo

Part/Instance	Mesh/element size (mm)	Element type
Concrete wall	200	C3D8R
Reinforcement	8	T3D2

were modelled using solid brick elements (C3D8R), whereas, truss elements (T3D2) were used to model the internal reinforcement. The rebar elements were embedded in the concrete material and required contact properties between the structure and filler material have been assigned. The boundary condition at the foundation of silo was assigned as fixed or *ENCASTRE*. Since the RC silo has been modelled as a quarter model, the walls have been constraint to move in the circumferential direction of the silo structure, however outward or inward translation or rotation have been allowed (see Fig. 1). A mesh sensitivity analysis study was also conducted for obtaining the optimum element size of the mesh. Reduced integration was used to reduce the computational time. Meshing was separately applied on every single component because of the difference of geometry and size of components [12] (see Table 1).

3 Results and Discussion

A three-dimensional model of RC cement silo was created in ABAQUS software and finite element analysis was performed. In addition to the gravity loads, the effect of internal pressure load due to different occupancy levels was observed on structural performance of the silo using Jansen's lateral pressure equations. Further, the effect of temperature variation of infill material on the structural behaviour of silo was observed by applying thermal loads from normal to 120 °C, with each increment of 5 °C on the structure. The silo models were analysed for different load conditions, viz. Case 1; for the silo filled with infill material at 33, 50, 66 and 100% of interior volume of the silo and Case 2; for the silo filled with cement with different levels of filled conditions at elevated temperature of 120 °C. The Concrete Damage Plasticity material model was used to simulate the material behaviour of the concrete. The results were obtained and discussed in terms of critical stresses and lateral deflection in the silo wall along the height.

3.1 Behaviour of RC Silo Under Thermal Loads

Previous studies have proposed that the concrete silos are vulnerable to structural cracks in the silo wall, which were developed due to seasonal temperature variations. However, the temperature of infill material in the silo may also affect its structural response. Thus, in the present study, the behaviour of concrete silo under thermal

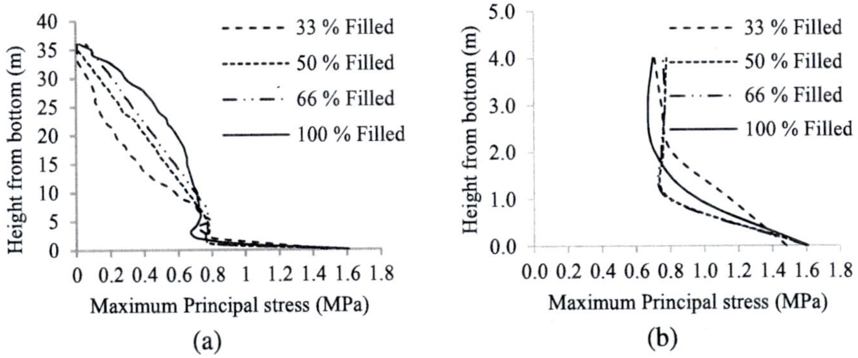

Fig. 2 Maximum principal stress variation in filled silo at elevated temperature

loads was investigated to ascertain its structural performance at elevated temperatures. The structural parameters obtained from the numerical study were compared with the structural behaviour of silo at normal as well as at elevated temperatures.

The structural performance of concrete silo was observed under gravity, lateral internal pressure due to infill material at normal as well as elevated temperature of 120 °C. The pressure distribution along the height was calculated from Jansen's theory of lateral pressure and applied accordingly.

Figure 2a shows the variation of maximum principal stress due to gravity and internal pressure load of infill at different levels of filling of silo at elevated temperature. The maximum principal stress was observed to be increased with decrease in silo height. However, stress increases drastically at nearly most bottom part of the silo (Fig. 2b). Further, the stress values were observed to be increased with increase in the volume of infill material, while at the bottom of the silo, this variation was found to be disturbed.

The critical locations at which the value of stress was found to be maximum, have been shown in Fig. 3. The critical stress values were calculated at various occupancy levels and have been compared in Table 2. The height of critical location of maximum stress from bottom firstly decreased on increasing volume of infill material, then the height was increased at 66% of infill volume. The maximum stress value of 0.79 MPa was observed at the occupancy level of 50% of the total volume.

Figure 4 describes the variation of lateral deflection of silo wall along the height under gravity, pressure and thermal loads at different occupancy levels. It was observed that with increase in the height of silo from the bottom; deflection of silo also increased to approximately 87×10^{-5} m at 9 m, then deflection was decreased with further increase in height of the silo wall. The maximum value of the deflection was observed at 66% occupancy of infill material, as compared with 33, 50 and 100% capacity (Fig. 5). Table 3 shows the values of maximum deflection of silo wall, it was noted that there were not much variations in deflections and critical height locations. At the first, the height of critical locations was decreased as occupancy in the silo was increased. After that, at 66% occupancy level, the height of the critical location was

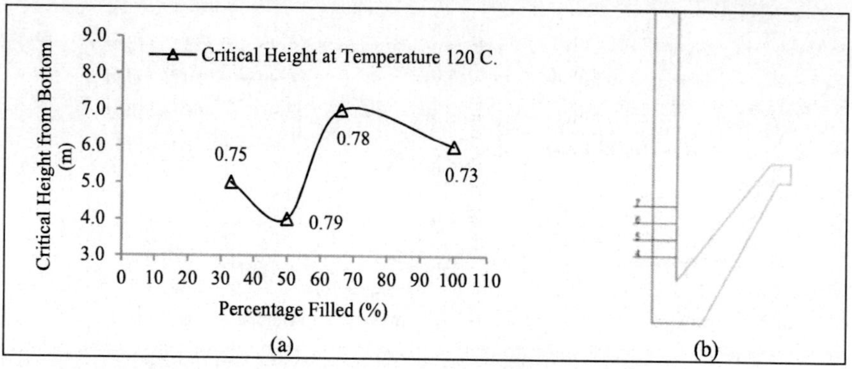

Fig. 3 Critical locations for maximum principal stress in filled RC silo

Table 2 Maximum principal stress (MPa) at different filled conditions at 120 °C

Critical location from bottom (m)	Filled conditions			
	33%	50%	66%	100%
4	0.70	**0.79**	0.71	0.67
5	**0.75**	0.75	0.78	0.72
6	0.65	0.78	0.77	**0.73**
7	0.64	0.78	**0.78**	0.72
At cone 12.36	0.61	0.62	0.62	0.62

found to be increased, then it again went down at 100% occupancy level as shown in Fig. 5.

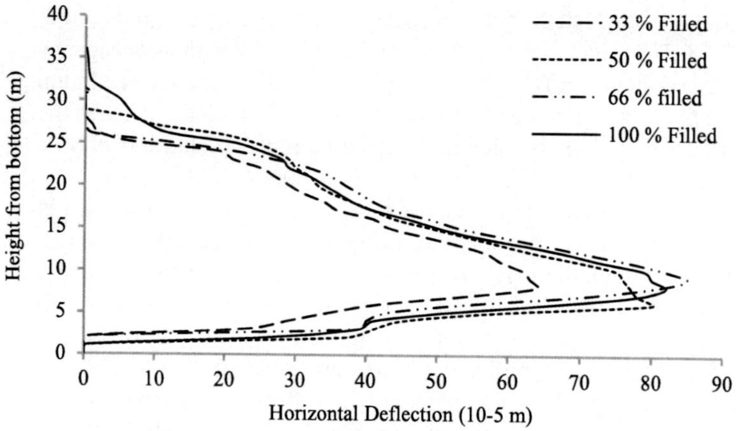

Fig. 4 Lateral deflection variation in filled silo

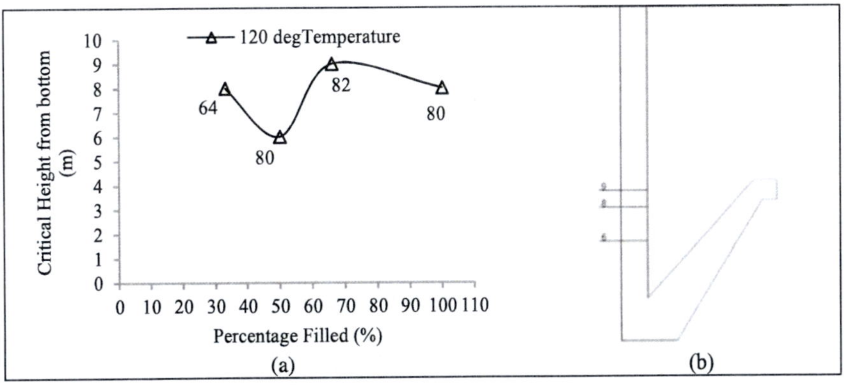

Fig. 5 Critical locations for lateral deflection in filled silo

Table 3 Maximum deflection (10^{-5} m) at different filled conditions at 120 °C

Critical Location From bottom (m)	Filled conditions			
	33%	50%	66%	100%
6	42	**80**	54	66
8	**64**	77	80	**82**
9	63	76	**85**	80

The comparison of maximum principal stress values in silo wall obtained at normal and at 120 °C temperatures, has been shown in Fig. 6. Stress values at elevated temperatures were found to be lower than stress at normal temperatures at various critical locations. Further, increase in the infill material results in the reduction in this variation and at 66 and 100%, maximum principal stress values were found to be approximately similar.

Figure 7 describe the comparison of maximum horizontal deflection in the RC silo wall at normal as well as at elevated temperature of 120 °C. Horizontal deflection at normal temperature was observed to be less as compared with the deflection values at 120 °C. Also, further increase in the occupancy of the silo results in higher variation in the values of deflection.

4 Conclusion

The present study investigates the influence of lateral pressure at different occupancy levels of infill material on the structural performance of RC silo. Further, the effect of temperature variation of infill material on the structural behaviour of silo was observed by applying thermal loads from normal (20 °C) to 120 °C, with increment of 5 °C on the structure. Numerical study was carried out on the RC silo structure

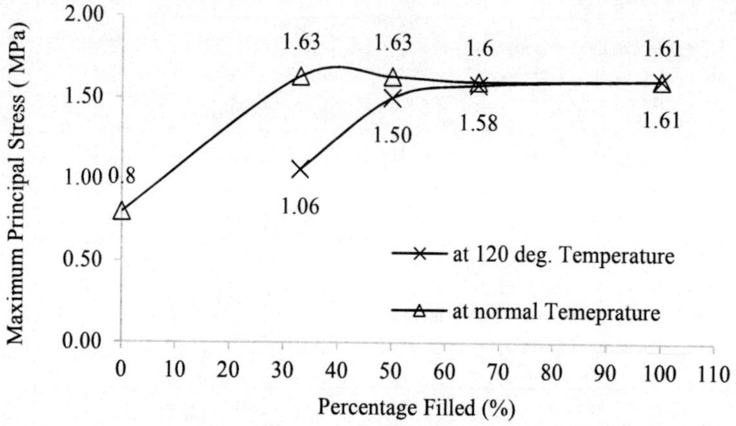

Fig. 6 Maximum principal stress values at different occupancy levels

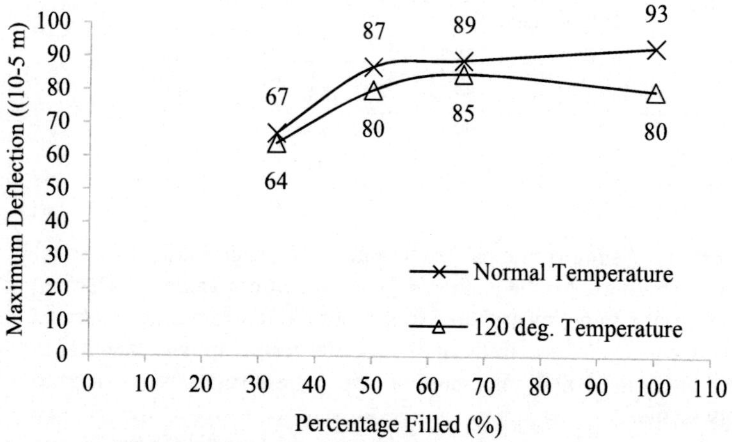

Fig. 7 Maximum lateral deflection values at different occupancy levels

in a finite element software. The Concrete Damage Plasticity material model was employed to simulate the material behaviour of the concrete.

The effect of lateral pressure due to infill material was observed at various occupancy levels, viz. 33, 50, 66 and 100%. It was observed that the maximum stress values obtained from the numerical study corroborated with Jansen's theoretical model.

The variation of lateral deflection of silo wall along the height under gravity, internal pressure and thermal loads at different occupancy levels, was evaluated. It was observed that with increase in the height of silo from the bottom; deflection of silo also increased to approximately 87×10^{-5} m at 9 m, then deflection was decreased with further increase in height of the silo wall. The maximum value of the deflection was observed at 66% occupancy of infill material, as compared with 33, 50 and 100% capacity.

The effect of temperature on the structural behaviour revealed that the stress values at elevated temperatures were found to be lower than that at normal temperatures. However, at higher occupancy levels of infill material, i.e. 66 and 100%, the effect of temperature was found to be negligible. The lateral deflection at critical locations was also found to be lower than that at normal temperature. Thus it is concluded that the effect of temperature variation of the infill material on the performance of concrete is a major concern and should be analysed and incorporated in the design properly.

References

1. Janssen HA (1885) Versuche uber getreidedruck in silozellen. VDI Zeitschrift Dusseldorf 39:1045–1049
2. Airy W (1897) The pressure of grain. Minutes Proc Inst Civil Eng London 131:458–465
3. Faber J, Mead F (1961) Reinforced concrete. E. & F. N. Spon, London
4. Ooi JY, Rotter JM (1990) Wall pressures in squat steel silos from simple finite element analysis. Comput Struct 37(4):361–374
5. Nielsen J (1998) Pressures from flowing granular solids in silos. Phil Trans R Soc A 356:2667–2684
6. Khouri MF (2005) Comparison of various methods used in the analysis of silos without wall friction. WIT Trans Model Simul WIT Press 41:425–441
7. Khouri MF (2010) Analysis of silo walls without friction. AES Tech Rev Int J Part B IJAMAIM 1(1):69–79
8. Reimbert M, Reimbert A (1976) Silos-theory and practice. Trans Tech Publication, 1st Edition, Claustal, Germany
9. Dogangun A, Karaca Z, Durmus A, Sezen H (2009) Cause of damage and failures in silo structures. J Perform Constr Facil 23(2):65–71
10. Nateghi F, Yakhchalian M (2011) Seismic behaviour of reinforced concrete silos considering granular material-structure interaction. Procedia Eng 14:3050–3058
11. Raeesi A, Ghaednia H, Zohrehheydariha J, Das S (2017) Failure analysis of steel silos subject to wind load. Eng Fail Anal 79:749–761
12. Mohammad Z, Ansari SS, Anwar M, Baqi A (2021) Structural behaviour of RC silo under gravity and infill loads. Recent Trends in Civil Engineering, Lecture Notes in Civil Engineering 274. https://doi.org/10.1007/978-981-19-4055-2_20
13. Orosz A (1977) Thermal effects in reinforced concrete silos. Periodica Polytechnica Civil 22(3):125–131
14. Cha SL, Lee Y, Kim JK (2017) Thermal stress analysis of silo in radioactive waste repository considering construction conditions. Nucl Eng Des 322:412–426
15. Chen Z, Li X, Yang Y, Zhao S, Fu Z (2018) Experimental and numerical investigation of the effect of temperature patterns on behaviour of large-scale silo. Eng Fail Anal 91:543–553
16. Khalifa WMA, El-Kashif KFO (2019) Evaluation of design of RC silos subjected to thermal loads. IOSR J Mech Civil Eng 16(1):1–11
17. Khalifa WMA, El-Kashif KFO (2019) Computational model for the evaluation of reinforced concrete silos subjected to thermal load. Eng Technol Appl Sci Res 9(4):4411–4418
18. Maraveas C (2020) Concrete silos: failures, design issues and repair/strengthening methods. Appl Sci 10:3938

Utilization of GGBS-Based Geopolymer Lateritic Soils for Sustainable Pavements

T. V. Nagaraju, B. M. Sunil, and M. Venkata Rao

1 Introduction

Lateritic soils were formed with highly weathering and having special properties based on their minerals such as quartz and secondary oxides of iron [1]. The process of weathering involves leaching of silica, the formation of colloidal sesquioxides and crystallization. Crystallization leads to the formation of iron oxides and aluminium oxides [1, 2]. Dehydration of colloidal hydrated oxides plays a significant role in the formation of laterites and also mineralogy of laterites varies with the rainfall intensity [3]. Index properties of lateritic soils are extremely varied from region to region and it is difficult to derive an acceptable generalization on plasticity and gradation of lateritic soils [4, 5]. Hence, Lohnes and Demirel [7] maintain that specific gravity may be a reliable indicator of engineering behaviour of lateritic soils. In lateritic soils both cohesion and frictional between in the grains vary from moderate to very high. This can be attributed due to the binding forces offered by the minerals sesquioxides [6]. In general most of the soil's strength characteristics were affected by the moisture content; in lateritic clay, it is much more dependent. Moreover, low amount of moisture content improves both cohesion and friction between the particles [10].

As lateritic soils are an important construction material used in various civil engineering infrastructures such pavement bases, subgrades, embankments and retaining walls, it is highly significant that their index properties, engineering properties and California Bearing Ratio (CBR) are studied. Moreover, soils such as lateritic soils,

T. V. Nagaraju (✉) · M. V. Rao
Department of Civil Engineering, SRKR Engineering College, Bhimavaram 534204, India
e-mail: varshith.varma@gmail.com

B. M. Sunil
Department of Civil Engineering, National Institute of Technology Karnataka, Mangalore 575025, India

A. Sil et al. (eds.), *Recent Trends in Civil Engineering*, Lecture Notes in Civil Engineering 274, https://doi.org/10.1007/978-981-19-4055-2_34

expansive clays and sands altering with cement and lime give promising results in terms of strength and stiffness [7, 8].

In India, fly ash and ground granulated blast furnace slag were two major by-products coming from the industries in million tonnes [9, 10]. Moreover, fly ash which has, of late, occupied an important place among siliceous pozzolanic materials that are added to treat problematic soils for reducing plasticity and increasing strength [11]. In field, fly ash is widely used as a material in subgrade, backfill, embankment fill and concrete floors and slabs [12]. GGBS, another industrial by-product, is also a siliceous as well as aluminous material used in clays and lateritic soils as a pozzolanic additive. It improves plasticity, compressibility and shear strength of clayey soils [13, 14]. By comparison, GGBS with fly ash, limited research has been done on GGBS blended lateritic soils. On other hand, GGBS-based geopolymer soils with addition of precursor as GGBS with the variation of $NaOH/Na_2SiO_3$, precursor and curing period. It was reported that for green pavement bases with low carbon content, fly ash and GGBS-based geopolymer is a promising material [15].

This paper investigates the effect of GGBS-based geopolymers on strength characteristics of blended lateritic soils. GGBS contents were varied with an increment of 5% up to 30% by dry weight of the soil. Stress–strain characteristics, strength characteristics and structural development of the blends were studied.

2 Experimental Investigation

The influence of GGBS-based geopolymers on the lateritic soil was examined using a strength tests and micro-structural analysis. First, an experimental investigation was conducted on the raw materials used in this study such as lateritic soil, ground granulated blast furnace slag (GGBS) and alkali-activated solution. The alkali-activated solution is prepared based on the mixture of sodium silicate gel (Na_2SiO_3) and sodium hydroxide pallets (NaOH), which were collected from the local market. The solution was maintained 7 molar concentration of NaOH and the ratio of $Na_2SiO_3/NaOH$ as 2.5. The geotechnical properties of lateritic soil were shown in Table 1.

2.1 *Compaction and Strength Characteristics*

Proctor compaction tests (according to ASTM standards) were determined on the lateritic soil blended with GGBS content. Further, at their respective optimum moisture content and maximum dry densities of the GGBS-soil blends, unconfined compressive strength (UCS) tests were conducted for GGBS-soil blends (alkali-activated solution is used instead of water for preparation of UCS samples). The geopolymer soil blends were prepared by varied percentages of precursor (GGBS) content as an increment of 5% up to 30% by dry weight of oven-dried soil. The mixing of the geopolymer soil blends were done manually and proper care was taken to mix

Table 1 Index and engineering properties of lateritic soil

Property	Soil
Specific gravity, G	2.68
Gravel (%)	5
Sand (%)	37
Silt (%)	6
Clay (%)	52
Liquid limit, LL (%)	51.5
Plastic limit, PL (%)	21
Plasticity index, PI (%)	30.5
USCS classification	CH
Optimum moisture content, OMC (%)	14.5
Maximum dry density, MDD (kN/m^3)	22.1

alkali-activated solution to the GGBS-soil homogeneously. Unconfined compressive strength of the untreated soil and treated soil with geopolymers as mentioned were prepared. Cylindrical geopolymer samples (38 mm in diameter and 76 mm in length) were then subjected to an axial loading in the compressive testing machine. The UCS samples were examined at different curing conditions; curing in ambient conditions for 14-days, 40-days, 60-days and 90-days and curing the samples for 5 h at 250 °C and for 7-days at 100 °C, respectively.

2.2 Scanning Electron Microscope (SEM)

The micro-structural changes of the geopolymer lateritic soil blends were investigated to elucidate the geopolymerization mechanisms. For SEM analysis, samples were taken from tested blends that had been fractured during compressive strength testing; the samples were coated with gold–palladium alloy. The beam used in the analysis was applied at 10 keV. The whole process and testing were conducted in the applied analytical laboratory Andhra University, Visakhapatnam. The SEM micrographs were helpful to analyze the morphology of the geopolymer lateritic soils.

Table 2 Effect of GGBS-based geopolymers on compaction characteristics and unconfined compressive strength

GGBS content (%)	Compaction characteristics		Unconfined compressive strength (kPa) for curing periods of					
	OMC (%)	MDD (kN/m^3)	14 days	40 days	60 days	90 days	5 h	7 days
			Ambient curing				250 °C	100 °C
0	14.5	22.1	550	700	1150	1050	1940	1589
5	14.1	22.4	1040	1130	1340	1940	3513	2585
10	13.7	22.7	1530	2340	2740	4280	5684	5553
15	13.2	23.3	2520	4360	6160	7390	8529	7743
20	12.5	23.8	3840	6070	8420	10090	11166	8924
25	12.0	24	5000	8050	10150	12200	12818	11582
30	11.6	24.2	5930	10030	11480	12520	13531	12888

3 Discussion of Test Results

3.1 Variation of Compaction Characteristics with GGBS Content

Table 2 shows the test results of influence of additive content in the geopolymer soils. Compaction properties of the lateritic soil improved with an increment of GGBS content. This can be due to the lateritic soil structures were flocculated with addition of GGBS content. Further, flocculated particles were dominant in gravity forces than surface forces. Maximum dry density (MDD) improved from 22.1 to 24.2 kN/m^3 with a percentage increment of 9.5 when additive content increased from 0 to 30%. Phanikumar and Nagaraju [11] also reported the similar phenomena in expansive clays blended with GGBS content. Whereas, for alkali-activated clay blends, the OMC may not be decreased with addition of precursor content due to the physical and chemical properties of the alkali solution is different to water.

3.2 Variation of Strength Characteristics with GGBS Content

Strength characteristics of alkali-activated GGBS-soil blends varied with varying GGBS content. Stress–strain patterns of the unblended soil and blended soil with precursor content were shown in Figs. 1 and 2 for the ambient curing exposure of 60-days and 90-days, respectively. The stress–strain behaviour clearly indicates that with a given strain, the compressive axial stress improved with increasing precursor content in the blends. Moreover, the stress–strain peaks of the geopolymer blends showed the brittle nature. Whereas, in the case of lateritic soils treated with GGBS content without any geopolymers possess low to moderate brittle nature (Vamsi

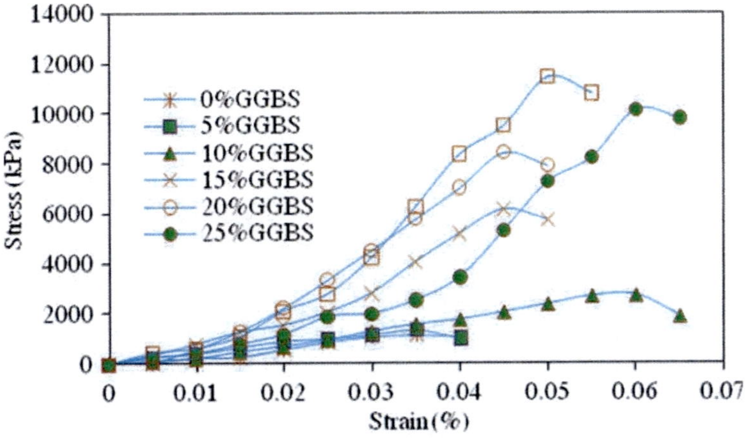

Fig. 1 Stress—strain curves of geopolymer soils for 60 days ambient curing

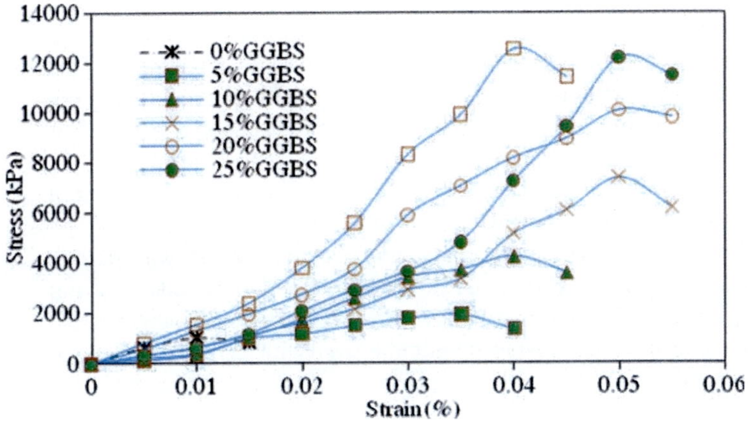

Fig. 2 Stress—strain curves of geopolymer soils for 90 days ambient curing

et al. 2018). The increase in strength of geopolymer with increasing GGBS content is due to the variation of Na/Al and Si/Al ratios. The Na/Al ratios of the geopolymer blends was 0.470, 0.475, 0.490, 0.498, 0.504, 0.512 and 0.519 for blends of 0%, 5%, 10%, 15%, 20%, 25% and 30% GGBS content, respectively. The Na/Al ratio of geopolymer blends was significantly increased with increasing precursor content. The higher Na/Al ratio of the geopolymer blend attains higher strength due to the effective condensation process occurs with the dilution of Al ions with Na^{+2} ions resulted in dense geopolymer matrix. In similar fashion, there is a variation of Si/Al ratios with increasing GGBS content. The Si/Al ratios of the geopolymer blends was 2.581, 2.578, 2.575, 2.571, 2.567, 2.564 and 2.560 for blends of 0%, 5%, 10%, 15%, 20%, 25% and 30% GGBS content, respectively.

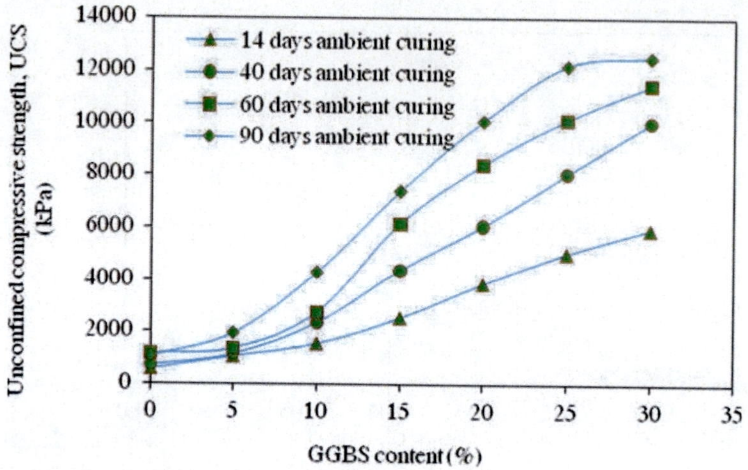

Fig. 3 Effect of additive content and curing period on UCS

Figure 3 shows, by comparison, the effect of different exposure conditions on the strength characteristics of the tested geopolymer blends. Compressive strength of the geopolymer blends significantly improved with increasing curing period for a given precursor content. As similar to the cementitious materials, geopolymer materials also become dense phases with increasing curing periods [16, 17]. For example, 30% GGBS blended geopolymer soils, the strength improvement shows an increment of 111% when the ambient curing exposure is from 14 to 90 days.

Figure 4 and 5 shows the stress–strain behaviour of geopolymer soils with varying precursor content (0, 5, 10, 15, 20, 25 and 30%) for the curing periods of 5 h at 250 °C and 7-days at 100 °C. Moreover, at any given percentage of strain, the axial compressive resistance of the geopolymer blends improved with increasing precursor and also with temperature. The maturity of strength development in the geopolymer products also depends upon the temperature. Even though for high temperature of 250 °C, the micro-cracks or flaws are clearly observed on the surface texture (vide Fig. 8), the specimen exhibits higher strength. However, the higher temperature and heat duration is not always positive, especially in geopolymer blends, because of presence chemically bound water in the soil structure or adsorbed water in the voids leads to shrinkage at various stages of thermal conditions.

For any case, ambient condition or at higher temperature curing, the unconfined strength of the geopolymer soils were increased with increasing GGBS content up to 30%. On the other hand without alkali activation, GGBS stabilized soils resulted optimum content in the range of 8–15% [8, 13, 14].

Figures 1, 4 and 5 show, by comparison, the stress peaks for a given strain were influenced by the processing temperature. The geopolymer blends were losing ductile behaviour and behaving as semi-rigid blends when the temperature increasing. However, at higher temperatures, the formation of geopolymer products was poor

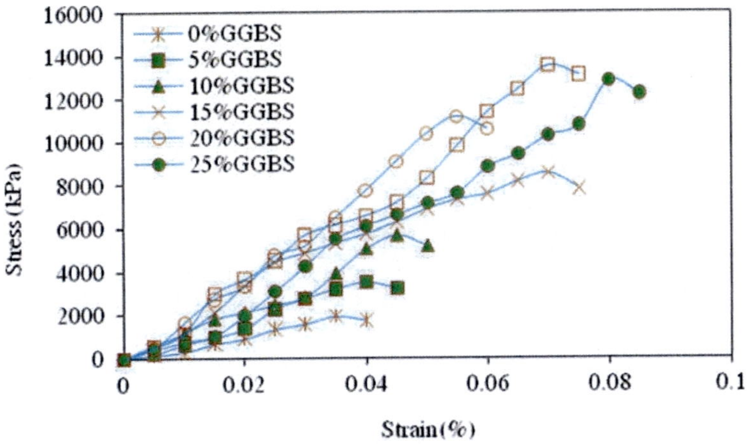

Fig. 4 Stress–strain curves of geopolymer soils for 3 h curing at 250 °C

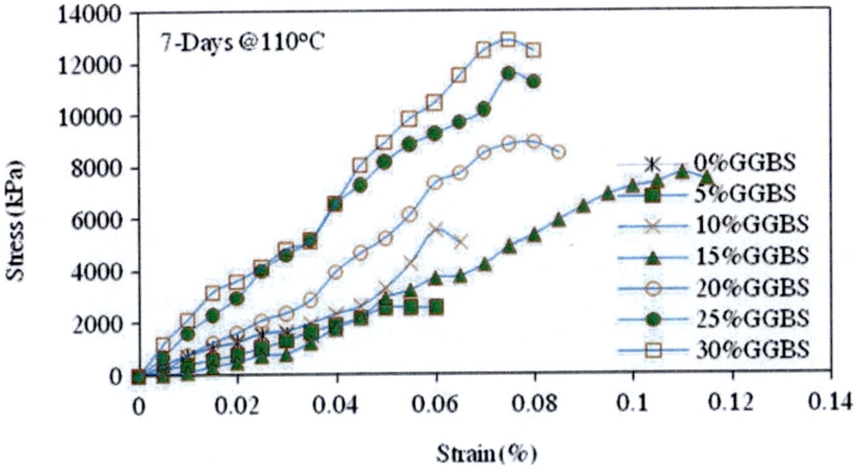

Fig. 5 Stress–strain curves of geopolymer soils for 7 days at 110 °C

due to the inefficient dissolution of Si and Al in the alkaline solution. Further, it leads to the poor bond between the soil and geopolymers.

3.3 Scanning Electron Microscopy Studies

SEM analysis was conducted for the untreated soil, precursor and the geopolymer soil blends for assessing the micro-structural behaviour and chemical reactions. Micrographs of the untreated lateritic soil and precursor were shown in Figs. 6a and 6b,

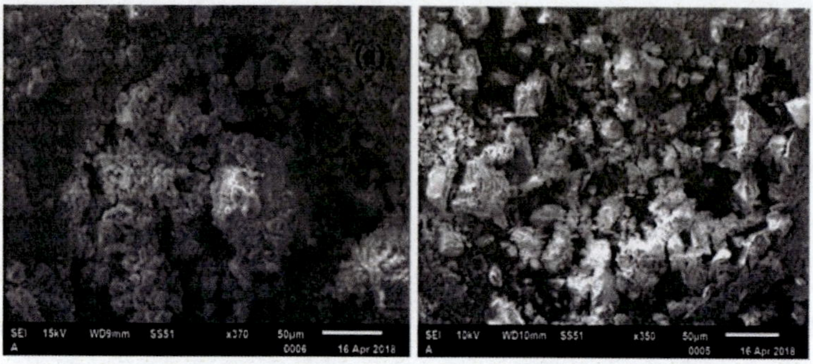

Fig. 6 SEM micrograph of **a** lateritic soil, **b** GGBS

respectively both were exhibiting open textures on their surface and non-uniform or random particle distribution.

Figures 7, 8 and 9 illustrate the influence of GGBS content (10%, 20% and 30%, respectively) on the micro-structural development of the blends. The SEM micrographs of alkali-treated blends were clearly indicating that the open textures on the raw materials (lateritic soil and GGBS) were closely bound by the alkali activation and seemingly filled pores exhibits dense phase. For alkali-activated lateritic soil blends without using precursor (GGBS) in the blends witness the pockets left on the surface of the blends, later with the addition of GGBS content to the blends pore spaces were reduced due to the pozzolanic reaction (Fig. 7). As observed from Fig. 8 blend surface was dense pack with random micro-cracks, which affects the bonding between the grains. However, this behaviour is not significant in case of the geopolymer products due to the heterogeneous products with the both crystalline and amorphous phases which acts as a composite dense phase. Further these micro flaws not effects on the strength of the geopolymer blends [17]. On the other side, at processing temperatures and higher precursor content, etched holes are witnessed on the surface (Fig. 9). This can be contributed due to the leaching of minerals (silica and alumina oxides) present in the precursor with the alkaline solution. The leached minerals react with the calcium and sodium ions present in the blend and helpful in geopolymer products (Na-Al-S-H and Na-C-S-H). The similar observations were seen in the fly ash-based geopolymer blends [18, 19].

4 Conclusions

This work highlights the use of GGBS to improve its properties of lateritic clay. Following conclusions drawn from the above study.

(1) Compaction characteristics showed improvement as maximum dry density (MDD) from initial value of 22.1 to 24.2 kN/m^3 when GGBS content increased from 0 to 30%. And optimum moisture content (OMC) decreased from 14.5 to

Fig. 7 SEM micrographs of geopolymer soils at 60 days of ambient curing having GGBS content as 10%

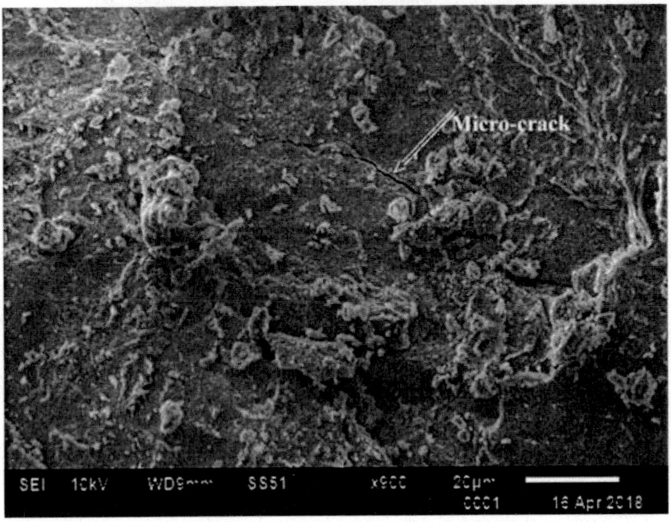

Fig. 8 SEM micrographs of geopolymer soils at 60 days of ambient curing having GGBS content as 20%

11.6% when the GGBS content from 0 to 30%. The percentage improvement in MDD and OMC was 8% and 20%, respectively.

(2) Strength characteristics of the geopolymer blends improved with increasing precursor and curing temperature. The UCS values of heated alkali-activated GGBS-soil blends are much higher than the blended samples cured at ambient conditions, which might be due to quick reactions of GGBS-based geopolymer precursors.

(3) SEM results showed the micro-structure of the geopolymer blends improved with increasing precursor content. The improvement in strength characteristics

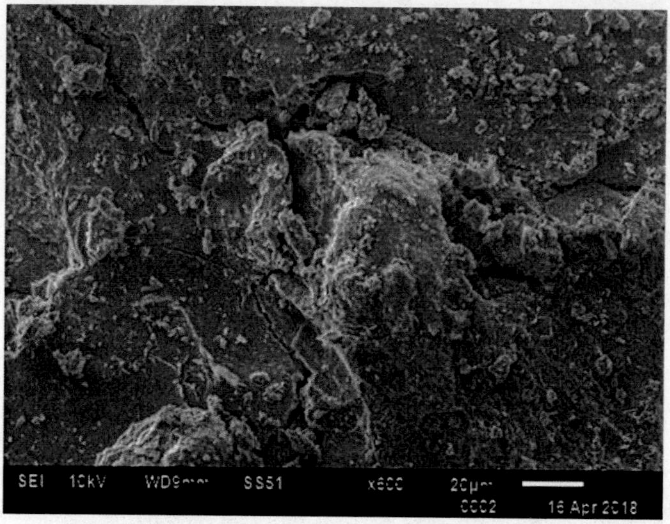

Fig. 9 SEM micrographs of geopolymer soils at 60 days of ambient curing having GGBS content as 30%

of the geopolymer blends was mainly due to the higher Si/Al ratio (at 30% GGBS content).

(4) The study showed that chemical amelioration of lateritic clays by geopolymer improves soil properties and its usage in construction industry can reduce the carbon footprint with large-scale utilization of GGBS in the pavement subgrades (up to 30% GGBS content).

References

1. Amadi AA, Okeiyi A (2017) Use of quick and hydrated lime in stabilization of lateritic soil: a comparative analysis of laboratory study. Int J Geo-Eng 8:3
2. Amadi A (2010) Evaluation of changes in index properties of lateritic soil stabilized with fly ash. Leonardo Electron J Practices Technol 9(17):69–78
3. Arca MN, Weed SB (1966) Soil aggregation and porosity in relation to contents of free iron oxide and clay. Soil Sci 101:164–170
4. Chatterjee AK (2011) Indian fly ashes: Their characteristics and potential for mechanochemical activation for enhanced usability. J Mater Civ Eng 23(6):783–788
5. Higgins DD (2005) Soil stabilisation with ground granulated blastfurnace slag. UK cementitious slag makers association (CSMA), pp 1–15
6. Phummiphan I, Suksun H, Tanakorn P, Arul A, Shui-Long S (2016) Marginal lateritic soil stabilized with calcium carbide residue and fly ash geopolymers as a sustainable pavement base material. J Mater Civ Eng ASCE 29(2):04016195
7. Lohnes RA, Fish RO, Demirel T (1971) Geotechnical properties of selected Puerto Rican soils in relation to climate and parent rock. Geol Soc, Am 82:2617–2624
8. Lohnes RA, Demirel T (1973) Strength and structure of laterities and lateritic soils. Eng Geol 7:13–33

9. Patrick NL, Achille BM, Elie K, Uphie CM, Marie PD, Hubert R (2014) Influence of the processing temperature on the compressive strength of Na activated lateritic soil for building applications. Constr Build Mater 65:60–66
10. Paulson SK (1975) The strength, structure and mineralogy of selected Hawaiian lateritic soils. Unpublished M.S. thesis, Iowa State University, Ames, Iowa
11. Phanikumar BR, Nagaraju TV (2018) Engineering behaviour of expansive clays blended with cement and GGBS. Ground Improv. https://doi.org/10.1680/jgrim.17.00054
12. Phoo-ngernkham T, Chindaprasirt P, Sata V, Pangdaeng S, Sinsiri T (2013) Properties of high calcium fly ash geopolymer pastes with Portland cement as an additive. Int J Miner Metall Mater 20(2):214–220
13. Phummiphan I, Suksun H, Runglawan R, Arul A, Shui Long S, Prinya C (2018) High calcium fly ash geopolymer stabilized lateritic soil and granulated blast furnace slag blends as a pavement base material. J Hazard Mater 341:257–267
14. Shaymaa A, Nima F, Afshin A, Bujang KM (2017) Collapsibility potential of gypseous soil stabilized with fly ash geopolymer; characterization and assessment. Constr Build Mater 137:390–409
15. Sherman GD (1949) Factors influencing the development of lateritic and laterite soils in the Hawaiian Islands. Pacific Sci 3(4):307–314.120
16. Sherman GD (1950) The genesis and morphology of Hawaiian ferruginous laterite crusts. Pac Sci 4:315–322
17. Winterkorn HF, Chandrashekharan EC (1951) Laterite soils and their stabilization. Highw Res Board Bull 44:10–29
18. Yadu L, Tripathi RK (2013) Effects of granulated blast furnace slag in the engineering behaviour of stabilized soft soil. Procedia Eng 51:125–131
19. Yao ZT, Ji XS, Sarker PK, Tang JH, Ge LQ, Xia MS, Xi YQ (2015) A comprehensive review on the applications of coal fly ash. Earth Sci Rev 141:105–121

Parametrical Study on the Seismic Performance of Post-installed Anchors Using Injectable Epoxy Mortars

Ali Kheyroddin, Hamed Arshadi, Mehran Gudarzi, and Mahdi Kioumarsi

1 Introduction

During the last century, reinforced concrete (RC) materials were the most common materials being utilized all over the world [1, 2]. However, their production process is not environmentally friendly due to vast CO_2 emissions [3, 4]. One of the most visible consequences of vast CO_2 emissions is global warming which increases the intensity and frequency of extreme weather events [5, 6]. Therefore, to satisfy global and local climate strategies, the material consumption of RC structures must be decreased while their strength, ductility, and resistance are maintained [7–9]. Recent technological proceedings facilitated the production of high-strength materials (HSMs), which led to decreases in the dimension of structural elements, consuming much fewer materials and time to build RC structures [10, 11]. On the other hand, using HSMs can have repercussions, e.g., a possible reduction in ductility and energy absorption of RC structures [8, 12]. It must be reminded that using fibers also increases the tension and shear strength of concrete and restricts cracking in concrete. Moreover, the other environmentally friendly strategy is to keep existent obsolete RC structures under service by rehabilitation and strengthening methods rather than their destruction and substitution with new ones. This issue is not only a time-and money-saving process, but it is not also harmful to the environment. There are several strategies to rehabilitate and strengthen RC structures such as using steel and concrete jacketing, fiber-reinforced polymer (FRP) sheaths and sheets, steel bracing, shear walls, and base isolators. Each approach has its unique consequences and effects on the

A. Kheyroddin · H. Arshadi · M. Gudarzi
Department of Civil Engineering, Semnan University, Semnan, Iran

M. Kioumarsi (✉)
Department of Civil Engineering and Energy Technology, OsloMet—Oslo Metropolitan University, Pilestredet 35, 0166 Oslo, Norway
e-mail: mahdik@oslomet.no

Department of Engineering, Østfold University College, 1757 Halden, Norway

A. Sil et al. (eds.), *Recent Trends in Civil Engineering*, Lecture Notes in Civil Engineering 274, https://doi.org/10.1007/978-981-19-4055-2_35

behavior of RC structures. The strengthening performance of these methods can be evaluated by experimental and analytical methods (such as incremental dynamic analysis, endurance time method, and time-history analysis.) [13].

Rehabilitation by installing anchors or bars plays a significant role in rehabilitating RC structures. The post-installed anchors are used to add shear walls, concrete jackets, the thickness of slabs, and structural and non-structural elements to structures. There are different anchor types with unique tension load transfer mechanisms such as undercut anchor (UA), screw anchor (SA), expansion anchor—sleeve-type (EAS), expansion anchor–bolt-type (EAB), bonded anchor (BA—with cement-based mortar or epoxy adhesives mortars). However, using bonded anchors is more user-friendly due to their simple operation. The method of installing anchors using epoxy-based materials is similar to that of installing them with cement-based materials, in which epoxy resin or bar installation adhesive are used instead of cement with anchors. Due to the extremely high adhesion of epoxy adhesives, it is obvious that the hole diameter and embedded length will be smaller, thus making the drilling operation easier. But, the price of epoxy adhesive materials is more expensive than cement-based materials. In addition, epoxy adhesives mortars have excellent adhesion to concrete and steel surfaces. Therefore, they cannot be packed and transported in a complete combination, because they are quickly caught and spoiled.

During seismic excitations, anchors that are utilized to join structural and non-structural elements to the concrete are exposed to cyclic loading. Imperfectly designed or insufficiently qualified anchors have produced severe damage and casualties in the past. To guarantee safe anchorage, the American Concrete Institute design code ACI 318 Appendix D necessitates anchor products to satisfy provisions of ACI 355.2 for mechanical anchors and ACI 355.4 for adhesive ones [14]. In recent years, researchers have inadequately studied the performance of post-installed bars and anchors under seismic and static loading [15]. Razaqpur et al. experimentally investigated the seismic response of three types of anchors (ribbed ones, expansion sheath, and undercut) installed in concrete under three types of loading (tension, shear, and tension-shear combination) [16]. They indicated that the load-bearing capacity of all three anchors under the combined static load of tension and shear was less than the computed ones using linear relations suggested by ACI 355.2. The effect of loading rate on anchor load-bearing behavior using experiments on adhesive anchors, expansion anchors, and torque-controlled adhesive anchors in cracked concrete exposed to rapid tension loading was studied by Hoehler et al. [17]. They reported that tests at fast load rates were not essential for the qualification of anchors for seismic applications. The load–displacement behavior of ordinary types of anchors installed in cracked concrete and exposed to cyclic tension was investigated by Mahrenholtz et al. [18]. They confirmed strong behavior for anchors loaded in cyclic tension, even along with the existence of crack widths in the anchorage material larger than presently required by ACI 355.2. Bhoir and Patil studied the influence of bar depth installed in the RC joints using finite element analyses (FEA) [19]. They resulted that the displacement of installed bars in the joints at an installation depth of 378 mm was less than the displacement of the anchors installed at installation depths of 10, 15, and 20 times the anchor diameter. The stresses in bars with a depth of ten times the bar's diameter in the joints were more than twice the stresses of the bars in the structural models.

Since the chemical (bonding) method of anchor installation is easier and faster than other methods of anchor installation [20], studying its different governing aspects is important. This research aimed to study the effects of governing parameters on the behavior of post-installed anchors in concrete using adhesive epoxy mortars to resist tension load and develop a logical principle for their design and application. Therefore, the seismic and static performance of post-installed bars (anchors) using the chemical method (epoxy adhesive mortar-Hilti HIT-RE 500 V3) in cracked and non-cracked reinforced concrete was analyzed using Hilti International software. For this purpose, 52 RC specimens with different anchor diameters, bar installation' depth, and edge distances were modeled and analyzed under static and seismic tension loading. Furthermore, the failure modes of the specimens (concrete break out, pull out, etc.) were studied. To verify the results of the Hilti software (Hilti PROFIS Anchor soft.), one slab in which three anchors with a diameter of 13 mm and installed depth of 80 mm were post-installed was constructed and subjected to cyclic tension loading. The experimental results complied with the analytical results, which show the accuracy of the software. The analytical results demonstrated that the presence of cracks in the concrete and applying the seismic load to the installed anchors decreased the capacity of the anchoring system by about 35% and 50%, respectively. In addition, low edge distances reduced the anchoring capacity by up to 50%. Finally, the decreases in anchor diameter decreased the adhesion capacity and caused adhesion failure rather than the concrete breakout or anchor failure.

2 Method and Modeling

In this analytical campaign, 52 cracked and uncracked concrete samples with the compressive strength of 20 MPa at 28 days, with different bar diameters (10, 13 and 19 mm), installation depths (60, 80, 140, 180 and 250 mm), and edge distances (80, 150, 280 and 300 mm) installed in concrete were modeled and analyzed. Moreover, three anchors with a diameter of 13 mm, installation depth of 80 mm and edge distance of 300 mm were constructed and tested under an unconfined tension test to verify the results of the Hilti PROFIS Anchor software.

2.1 Materials

2.1.1 Adhesive Epoxy Mortar HIT-RE 500 V3

Epoxy adhesive mortars as crack injection resins have widespread use throughout the world. Various producers are manufacturing different types of epoxy mortars for the post-installation of anchors. The use of these products is increasing widely,

Table 1 The physical and chemical characteristics of the HIT-RE 500 V3 Epoxy Mortar

Physical state	Solid
Colour	Red
pH	11.5
Density (g/cm^3)	1.31
Solubility	Insoluble in water
Viscosity, dynamic (Pa·s)	50–70

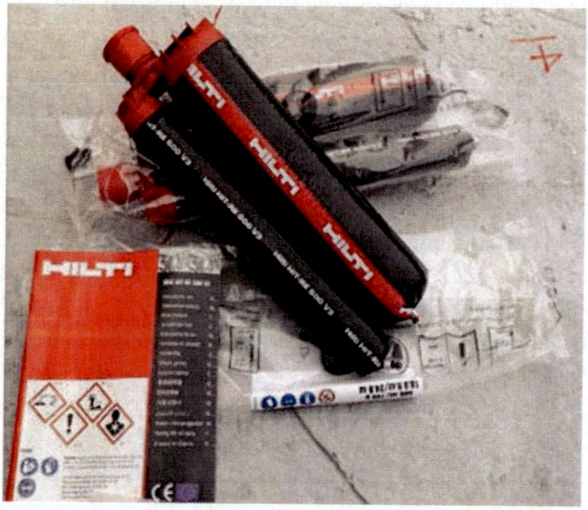

Fig. 1 Hilti's HIT-RE 500 V3 epoxy

and there is an important necessity to study their properties, performance and effects by testing and analyzing. One of the novel and efficient adhesive epoxy mortars is HIT-RE 500 V3 produced by HILTI Corp. Table 1 shows the specifications of this product. Figure 1 also shows one of the product tubes used in this study.

2.2 Test Process

Figure 2a and b indicate an example of an unconfined tension test setup for adhesive anchors and a cracked concrete slab for testing. Unconfined tension tests allow the conical failure of concrete to take different forms. The loading capacity of confined tension tests is higher than the unconfined ones. In confined tension tests, the conical failure of concrete is eliminated due to the reaction of the forces around the anchor within the concrete. ACI 355.4 M suggested that anchors be installed in cracked RC slabs to make the experiments more similar to the real seismic loading situation [21]. The cracking of concrete under the cyclic loads will create special conditions for the installed anchors [22]. Therefore, the need for research and testing on the opening and closing of cracks and the effect of impact resulting from the behavior of the

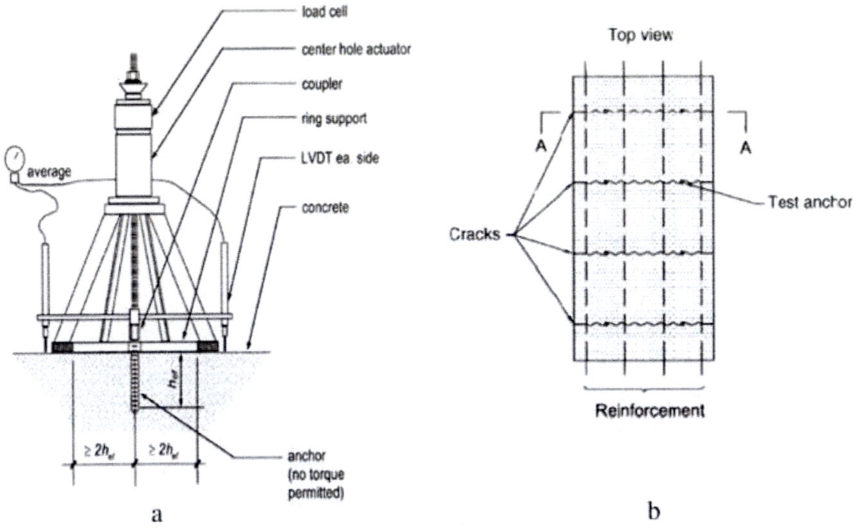

Fig. 2 **a** Example of unconfined tension test setup for adhesive anchors, and **b** cracked concrete slab for testing [21]

installed anchor under cyclical loading is essential. Although ACI 355.4 M specified the minimum installation depth (as six times the anchor's diameter) [21], the effects of the other installation depths were investigated in this study. It must be reminded that different failure modes such as concrete break-out, pull-out, sustained pull-out, shear failure, and conical shape failure can happen under tension loading.

Loading should be applied in 140 alternating sinusoidal cycles according to Fig. 3, and with a frequency between 0.1 to 2 Hz.

Where N_{eq} is the maximum seismic tension test load, equal to 50% of the mean tension capacity in cracked concrete from reference tests, N_m is one-fourth of the mean tension capacity in cracked concrete from reference tests; N_i is calculated using Eq. 1 and Table 2.

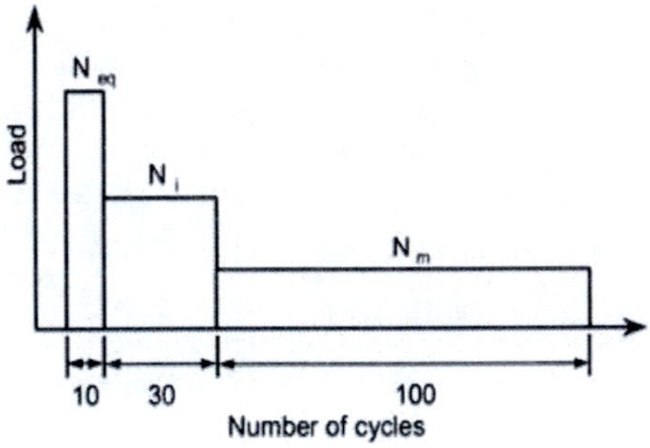

Fig. 3 Loading pattern for simulated seismic-tension test ACI 355.4 M [21]

Table 2 Required history of seismic tension load

Load level	N_{eq}	N_i	N_m
Number of cycles	10	30	100

$$N_{\mathrm{i}} = (Neq + N_m)/2 \quad (1)$$

2.3 Setup Device

To verify the results of the Hilti software (Hilti PROFIS Anchor soft.), one RC slab with dimensions of 1500 mm × 800 mm × 300 mm, (in which three anchors with diameters of 13 mm and embedded depth of 80 mm were post-installed) was constructed and subjected to unconfined tension testing. The compressive strength of concrete was 20 MPa. Figures 4a and b show the post-installation process of anchors using HIT-RE 500 V3 epoxy mortar and the unconfined tension test setup. It must be mentioned that the slabs were cracked by being subjected to two-point loading, and then the anchors were installed on the crack lines (the axis of the anchor coincided with the crack plane) based on the suggestions of ACI 355.4 M [21]. The crack widths were controlled by means of embedded longitudinal reinforcing bars positioned perpendicular to the intended crack plane and situated symmetrically over the test member cross-section. The ratio of tension longitudinal bars for both the top and bottom layers to the area of the crack plane was nearly 1%. The slab was fixed to the strong floor using supports surrounding the slab. These supports were also fixed to the rigid frame by ultra-high-strength bolts. The displacement of the steel anchors and

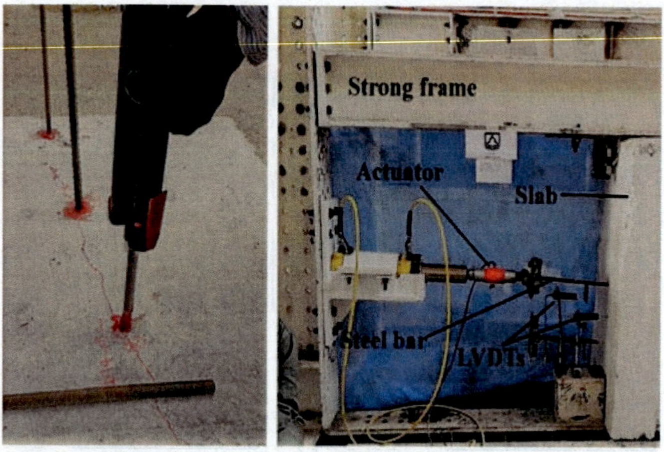

a b

Fig. 4 **a** Post-installation process of anchors, and **b** test setup the slab was monitored using two different linear variable differential transformers (LVDTs). A steel plate was welded to each bar to measure the displacement of the steel anchor to join the LVDT. The cyclic load was applied by a hydraulic actuator with a load capacity of 200 kN in both the tension and compression.

3 Results

3.1 Effects of Bar Diameter and Post-installation Depth

Figure 5 presents the medium and settings of the Hilti PROFIS Anchor software. This software is a product of Hilti International Company, which has been produced for the simulation and design of anchors installed in concrete. In this software, the type and size of anchors can be selected through the list provided by the software. Because this software is Hilti's proprietary software, it only provides the user with a list of types of Hilti adhesives. The compressive strength of concrete, being cracked concrete or uncracked one, the types of loads (shear, tension and the combination of them), regulation codes and edge distances can be chosen in the software. This software calculates the amount of loading capacity of anchor, adhesive mortar and concrete according to the regulation codes specified by the user, and compares it with the applied load. It also calculates the percent ratio of applied load to the loading capacity of anchor, adhesive mortar, and concrete. By applying seismic and static tension load on cracked and uncracked concrete, the influences of effective parameters such as loading type, installation depth, anchor diameter, edge distance and the presence of cracks in concrete were studied analytically in the software. It

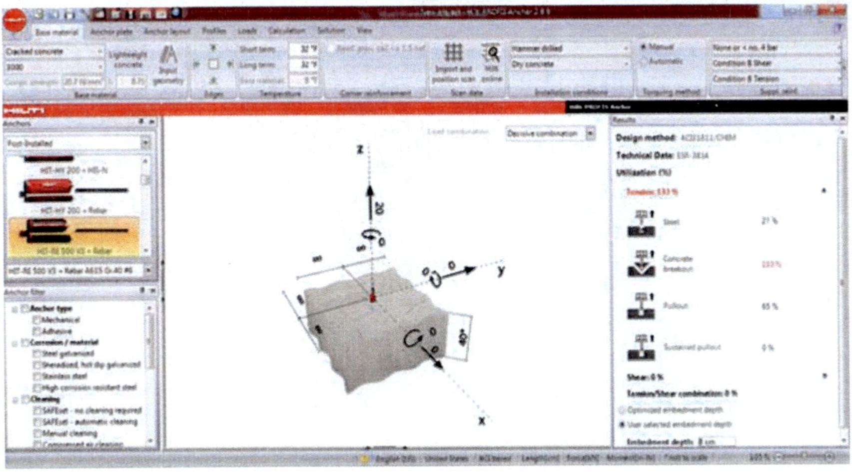

Fig. 5 The imports and exports of the Hilti PROFIS anchor software

should be noted that the strength of concrete was 20 MPa, the analysis was based on ACI 318.11, and the adhesive epoxy resin used was HIT-RE 500 V3. It must be reminded that the results of the experiments were in accordance with those of the software. Figures 6a and b indicate the pulled-out anchor and failure observed in the concrete slab. The specimens were named by a common rule with the first letter (D) standing for the anchor's diameter and the following number for its amount (mm), the second letter (H) for embedded length, and the following number for its amount (mm). The third letter (C or U) stands for cracked or uncracked concrete. The fourth letter (L) presents for the edge distance which the following number is its amount (mm). Figure 7 demonstrates the effects of applying 70 kPa force to the 19 mm anchor with a depth of 180 mm in different cases of having cracks or not, static and seismic analyses. Table 3 shows the loading capacity ratios to the maximum applied load of the specimens in different cases of bar's diameter, installation depth, and in different failure scenarios.

Analytical results indicated that the failure mode at the lowest installation depths (6 times the anchor's diameter) for all three diameters of anchors (10, 13, and 19 mm) was in the form of concrete breakout (conical shape). By increasing the installation depth to 10 times anchor diameter, the failure mode tended to pull out through adhesion failure. It was observed that if the strength of concrete was low, the failure would be through the formation of the concrete cone, and if the anchor installation principles were not applied correctly, the failure mode would be pulled out through adhesion failure. By increasing the installation depth to 15 times anchor diameter, the failure mode got closer to the combination of concrete breakout and adhesion failure (pulling out). These installation depths caused the anchors to be yielded. Finally, the

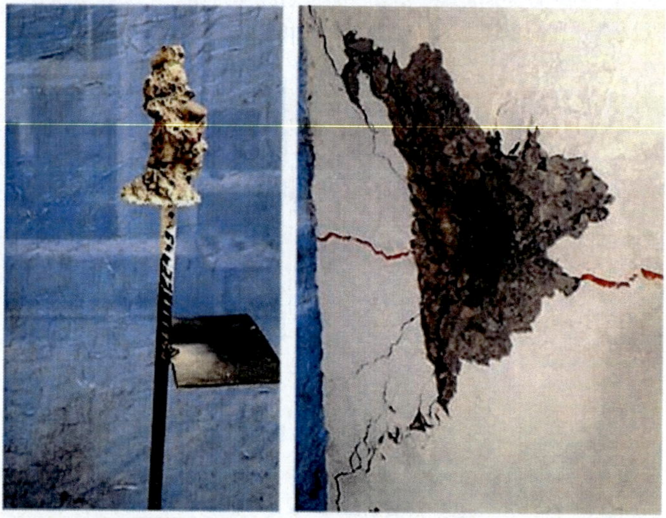

a b

Fig. 6 **a** Pulled-out anchor and **b** conical failure of concrete

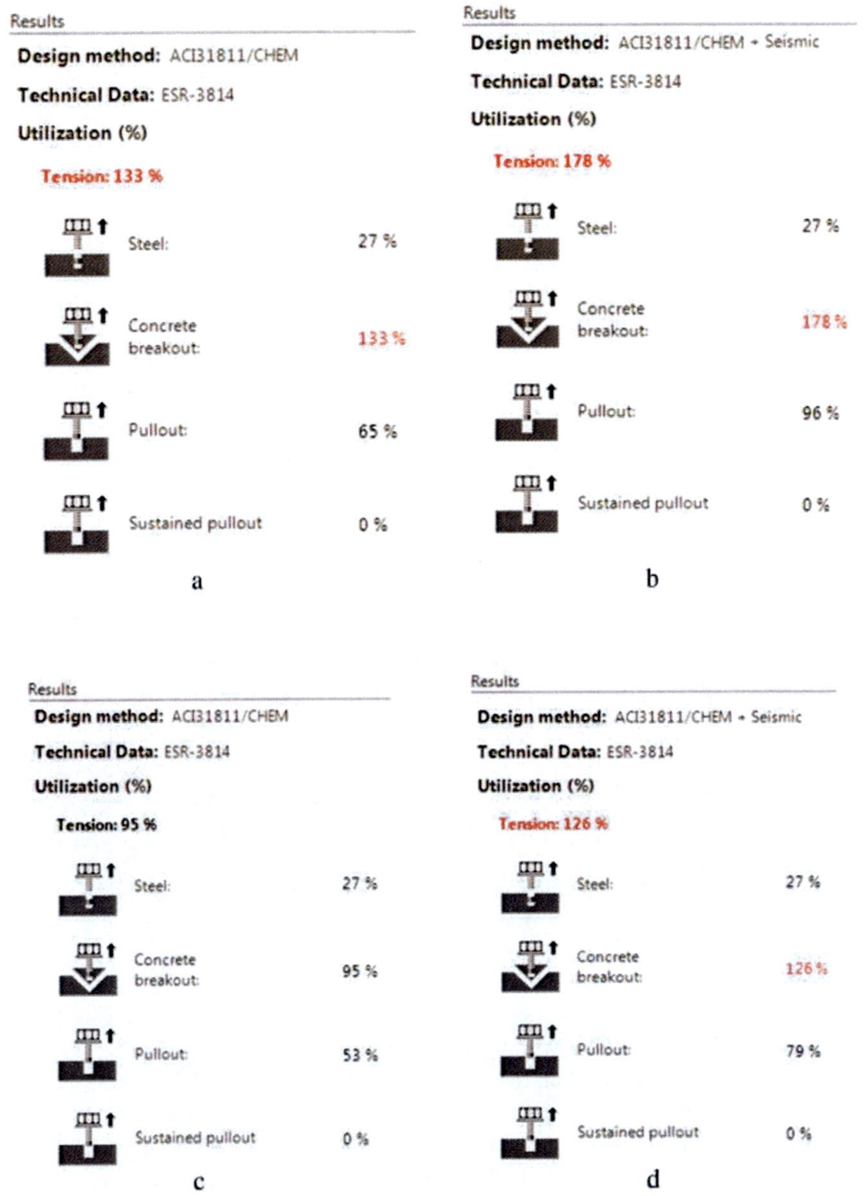

Fig. 7 The effects of applying 70 kPa force to the 19 mm anchor with a depth of 180 mm in the case of: **a** cracked concrete and seismic loading, **b** cracked concrete and static loading, **c** uncracked concrete and seismic loading, and **d** uncracked concrete and static loading

Table 3 The loading capacity ratios to the maximum applied load of the specimens in different cases of bar's diameter, installation depth, and in different failure scenarios

Specimen name	Static loading			Seismic loading			Maximum applied load (kN)
	Steel (%)	Concrete breakout (%)	Pull-out (%)	Steel (%)	Concrete breakout (%)	Pull-out (%)	
D19H8CL30	27	178	96	27	133	65	20
D19H8UL30	27	126	79	27	95	53	20
D19H18CL30	92	184	149	92	138	101	70
D19H18UL30	92	131	122	92	98	82	70
D19H25CL30	132	161	153	132	121	103	100
D19H25UL30	132	114	125	132	86	85	100
D13H8CL30	58	178	149	58	133	101	20
D13H8UL30	58	126	114	58	95	77	20
D13H14CL30	145	192	212	145	144	143	50
D13H14UL30	145	136	163	145	102	110	50
D13H18CL30	202	138	156	202	184	231	70
D13H18UL30	202	131	177	202	98	120	70
D10H6CL30	79	205	200	79	154	135	15
D10H6UL30	79	145	150	79	109	101	15
D10H10CL30	132	159	200	132	119	135	25
D10H10UL30	132	113	150	132	85	101	25
D10H14CL30	184	134	200	184	101	135	35
D10H14UL30	184	95	150	184	72	101	35

results showed that the more the installation depth, the more probable the anchor failure. As for the effects of the anchor diameter, they provide the cross-sectional bonding area between the adhesive mortar, anchor, and concrete. Increasing the diameter of the anchor increases the bonding strength of the adhesion and thus reduces the probability of adhesion failure. The results confirmed this issue and showed that the adhesion failure (poll out) for the lower diameters of the anchor was more likely. By increasing the diameter of the anchor, the failure mode changed from the failure of the adhesion to the combinational failure of the concrete breakout and adhesion, or the failure of the installed anchor despite having sufficient installation depth. Moreover, the results demonstrated that cracking significantly reduced the tension capacity of specimens. The reason behind the conspicuous influences of cracking on the seismic behavior of post-installed anchors during higher seismic excitations is the phenomenon of opening and closing of the cracks during these excitations. Under static loads, the loading capacity of cracked concrete was reduced by about 35% compared to uncracked concrete, and also, the adhesive loading capacity of cracked concrete under static load was reduced by about 20% compared to the uncracked one. Finally, it was observed that the effect of the type of loading (static or seismic) on the failure capacity of concrete breakout and poll-out failure of the adhesive was significant. By employing the seismic conditions in the software, it was observed that the concrete failure capacity and poll-out failure were reduced by about 50%. One of the reasons for the decrease in the capacity of the installed anchor was that the concrete had been cracked in seismic conditions, and another reason was the type of loading. Alternating loading caused cracks to open and close, leading to reducing concrete and bonding capacity.

3.2 *The Effects of Edge Distance*

The minimum edge distance is considered to be equal to the installation depth to supply enough space around the embedded anchor to form a cone. Comparing the edge distances of 80, 150, 280, and 300 mm for an anchor with a diameter of 13 mm and an installation depth of 14 cm, it was observed that reducing the edge distance of the anchor significantly reduced the bonding capacity of adhesion and concrete capacity. At a distance twice the depth of the installed anchor from the edge, the edge distance had no considerable effect on the loading capacity of the anchoring system (concrete, anchor, and adhesive mortar). Comparing the software analyses at edge distances of 80 and 280 mm, it was observed that at a distance of 80 mm, failure capacity was reduced by 50% compared to a distance of 280 mm. The adhesion failure capacity had also been reduced by 57%. Figure 8 shows the effects of different edge distances (80, 150, 280, and 300 mm) in the case of applying 35 kN to an anchor with a diameter of 13 mm and an installation depth of 140 mm. Table 4 indicates The loading capacity ratios to the maximum applied load of each specimen with different edge distances in different failure mode scenarios.

4 Conclusion

This study analyzed the seismic and static performance of post-installed bars (anchors) using a chemical method (epoxy adhesive mortar) in cracked and non-cracked reinforced concrete using Hilti International software. For this purpose, 52 RC specimens with different anchor diameters, bar installation' depth, and edge distances were modeled to study their effects on the performance of anchors. First, to verify the software's results three anchors with diameters of 13 mm, and installation depths of 80 mm were installed in a slab and subjected to unconfined tension tests. The analytical results showed that the presence of cracks in the concrete reduced the capacity of the anchor system by about 35%. Moreover, applying the seismic load to the installed anchor decreased the capacity of the anchoring system by about 50%. Furthermore, low distances from the edge reduced the capacity of the anchoring system by up to 50%. The observations showed that the failure mode at the installation depths below 100 mm was the concrete breakout, for the depths from 10 to

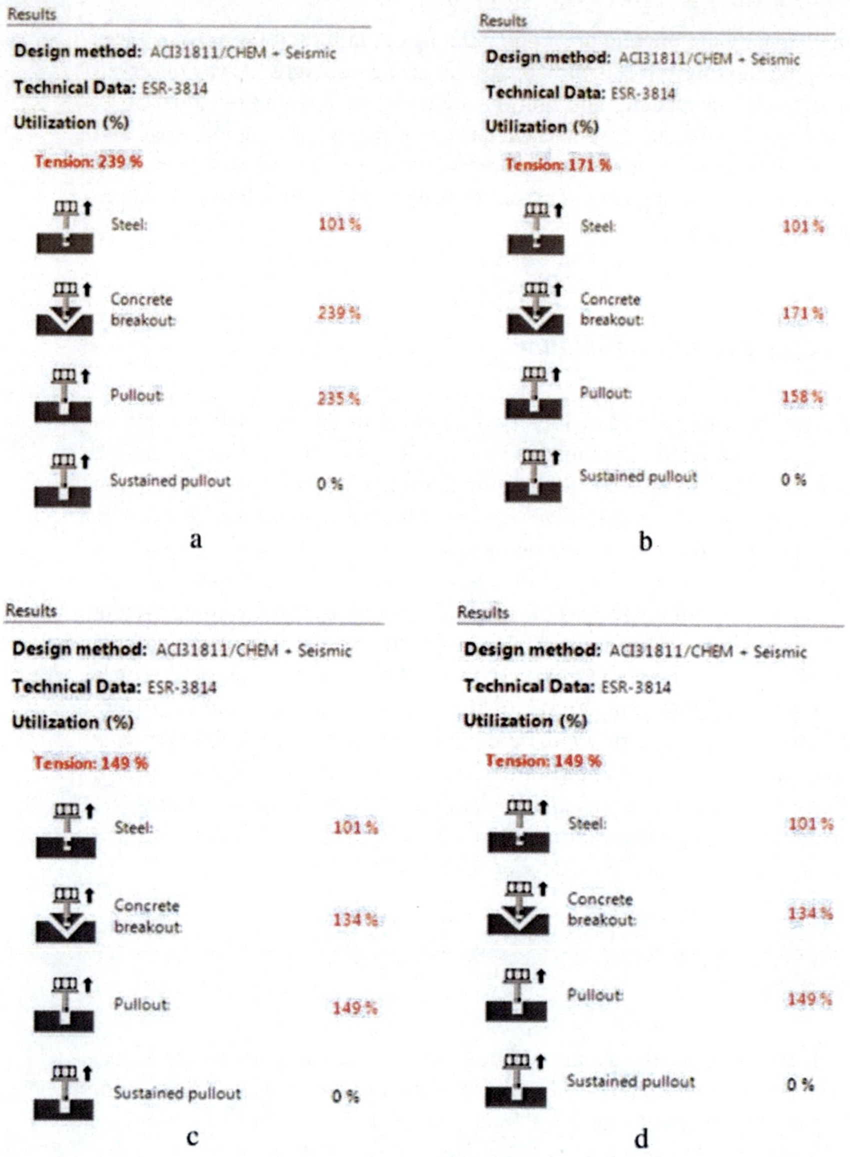

Fig. 8 The effects of applying 35 kPa load to the 13 mm anchor with a depth of 140 mm in different edge distances: **a** 80 mm, **b** 150 mm, **c** 280 mm, and **d** 300 mm

Table 4 The loading capacity ratios to the maximum applied load of specimens with different edge distances in different failure mode scenarios

Specimen name	Static loading			Seismic loading			Maximum applied load (kN)
	Steel (%)	Concrete breakout (%)	Pull-out (%)	Steel (%)	Concrete breakout (%)	Pull-out (%)	
D13H14CL8	101	179	159	101	239	235	35
D13H14CL15	101	129	107	101	171	158	35
D13H14CL28	101	101	101	101	134	149	35
D13H14CL30	101	101	101	101	134	149	35
D13H14UL8	101	142	176	101	189	260	35
D13H14UL15	101	102	119	101	136	175	35
D13H14UL28	101	72	77	101	95	114	35
D13H14UL30	101	72	77	101	95	114	35

20 cm, the failure mode was adhesion failure or the combination of concrete breakout and adhesion failure, and for the depths greater than 20 cm the failure mode got closer to anchor failure (yielding). Finally, the anchor diameter was influential in the adhesion capacity of the adhesive mortar, and low diameters of anchors caused adhesion failure. The printing area is 122 mm × 193 mm. The text should be justified to occupy the full line width, so that the right margin is not ragged, with words hyphenated as appropriate. Please fill pages so that the length of the text is no less than 180 mm, if possible.

Acknowledgements The authors would like to express our special appreciation to Mr. Bakhshaie (head of the structural laboratory of Semnan University) who helped us do these experiments with his advice and guidance.

References

1. Bolooki Poorsaheli H, Behravan A, Tabatabaei Aghda ST, Gholami A (2019) A study on the durability parameters of concrete structures reinforced with synthetic fibers in high chloride concentrated shorelines. Constr Build Mater 200:578–585. https://doi.org/10.1016/j.conbuildmat.2018.12.155
2. Rautenberg JM, Pujol S, Tavallali H, Lepage A (2013) Drift capacity of concrete columns reinforced with high-strength steel. ACI Struct J. https://doi.org/10.14359/51684410
3. Kheyroddin A, Arshadi H, Salehzadeh J (2021) Impact behavior of fiber-reinforced concrete with polypropylene fibers and carbon fiber reinforced polymers. ASTM J Test Eval 49(6). https://doi.org/10.1520/JTE20200670
4. Chang C-J (2017) High-strength reinforcement in exterior beam-column joints under cyclic loading. ACI Struct J 114(5). https://doi.org/10.14359/51700788
5. Kheyroddin A, Arshadi H, Ahadi MR, Taban G (2021) The impact resistance of fibre-reinforced concrete with polypropylene and GFRP wrapping. In: 2nd international conference on aspects

of materials science and engineering. Panjab: materials today: proceedings, Elsevier. https://doi.org/10.1016/j.matpr.2021.02.116
6. Kheyroddin A, Arshadi H, Khedri J (2021) Resistance of fiber-reinforced concrete with steel fibers and CFRP to drop-weight impact. Periodica Polytechnica Civil Eng 65(2):666–676. https://doi.org/10.3311/PPci.17477
7. Khaksefidi S, Ghalenove M, De Brito J (2021) Bond behaviour of high-strength steel rebars in normal (NSC) and ultra-high performance concrete (UHPC). J Building Eng 33. https://doi.org/10.1016/j.jobe.2020.101592
8. Arshadi H, Kheyroddin A, Naderpour H (2017) An investigation into the behavior of special moment frames with high-strength reinforcement subjected to cyclic loading. J Build Eng 26. https://doi.org/10.1016/j.jobe.2019.100905
9. Arshadi H, Kheyroddin A, Naderpour H, Kioumarsi M (2021) Experimental study on the failure traits of frames with high-strength reinforcements. Int J Mater Sci Eng (IJMSE) 9(1):1–7. https://doi.org/10.17706/ijmse.2021.9.1.8-14
10. Mahboubi S, Kioumarsi M (2021) Damage assessment of RC bridges considering joint impact of corrosion and seismic loads: a systematic literature review. Constr Build Mater 295:123662. https://doi.org/10.1016/j.conbuildmat.2021.123662
11. Arshadi H, Kheyroddin A, Naderpour H, Haji M (2020) Study of the damage indices of concrete members reinforced with high-strength steel. Struct Build, (in press). https://doi.org/10.1680/jstbu.20.00165
12. Yang C (2010) Development of High Strength Construction Rebars. In: Proceedings of international seminar on production and application of high strength seismic grade rebar containing vanadium
13. Arshadi H (2016) An overview on the concepts and methodologies of incremental dynamic analysis IDA (with a single record and multiple records). In: world Congress on Jeju Island, Korea
14. American Concrete Institute (ACI) (2019) ACI 318M-19: Building code requirements for structural concrete
15. Andrew JJ, Srinivasan SM, Arockiarajan A, Dhakal HN (2019) Parameters influencing the impact response of fiber-reinforced polymer matrix composite materials: a critical review. Compos Struct 224. https://doi.org/10.1016/j.compstruct.2019.111007
16. Razaqpur AG, Foo S, Mostafa A, Saatcioglu M (2008) Seismic response of three types of post-installed anchors in concrete. In: The 14 th world conference on earthquake engineering. Beijing, China
17. Hoehler MS, Mahrenholtz P, Eligehausen R (2011) Behavior of anchors in concrete at seismic-relevant loading rates. ACI Struct J 108(2):238–247
18. Mahrenholtz P, Eligehausen R, Hutchinson TC, Hoehler MS (2016) Behavior of post-installed anchors tested by stepwise increasing cyclic load protocols. ACI Struct J 113(5):997–1008. https://doi.org/10.14359/51689023.BEHAVIOR
19. Bhoir AS, Patil SP (2019) Effect of post-installed rebar connection on concrete structure. Int J Sci Technol Res 8(9)
20. Kheyroddin A, Arshadi H, Abdollahi M (2021) Bond behavior of headed and straight NSS and HSS bars in concrete with steel fibers. In: Proceedings of the institution of civil engineers-structures and buildings, (in press)
21. ACI Committee 355 (2011) Qualification of post-installed adhesive anchors in concrete (ACI 355 . 4M-11) Farmington Hills, MI.: american concrete institute
22. Kioumarsi M, Benenato A, Ferracuti B, Imperatore S (2021) Residual flexural capacity of corroded prestressed reinforced concrete beams. Metals 11(3):442

Stability Analysis of Non-homogeneous Slopes and Assessment of Failure Characteristics Using HYRCAN

G. Sreelakshmi, K. Lini Dev, N. Nithesh, and H. S. Harshan

1 Introduction

Landslides affected 12% of Indian terrain, and from 2004 to 2020, the country recorded 18% of all landslide fatalities. While landslides are caused by monsoon rains every year, economic activities and climate crises have increased the risk, especially in the Himalayas and the Western Ghats regions. One significant contributor to slope failure-related landslides is water-level fluctuations along sloping surfaces of mountainous regions. The main reasons for water accumulation near sloping surfaces are improper seepage of surface runoff, impermeable soil layers near base level, change in river course (artificial or natural) for irrigational purposes and power generation.

Researchers have carried several studies to investigate the influence of rainfall, cloud busts, and the effect of vegetative cover, but only a few studies concentrated on the impact of the ponded water table near the sloping surface. Griffiths and Lu [1] adopted numerical analysis based on the elastoplastic method to investigate stability analysis of unsaturated slopes and concluded that the location of slopes influences the water infiltration rate. Huat et al. [2] performed experimental and model analyses to estimate the influence of water seepage on the slope with different types of vegetative covers. They concluded that the safety of slope directly depended on permeability characteristics of surface materials. Zhu [3] adopted the Bishop method to determine

G. Sreelakshmi (✉) · N. Nithesh · H. S. Harshan
Department of Civil Engineering, CMR Institute of Technology, Bangalore 560 037, India
e-mail: engr.lakshmi@gmail.com

N. Nithesh
e-mail: nitn17cv@cmrit.ac.in

H. S. Harshan
e-mail: hahs17cv@cmrit.ac.in

K. Lini Dev
Department of Civil Engineering, National Institute of Technology Patna, Patna, India
e-mail: linisubheesh@nitp.ac.in

A. Sil et al. (eds.), *Recent Trends in Civil Engineering*, Lecture Notes in Civil Engineering 274, https://doi.org/10.1007/978-981-19-4055-2_36

the factor of safety of slopes, and it has proved effective in satisfying all kinds of equilibrium conditions compared to other stability analysis methods. Gavin and Xue [4] adopted a finite element approach to investigate the effect of stability of saturated soil and concluded that soil characteristics influenced the saturation capacities of the wetted sloping surface. A numerical investigation of an unstable slope with a rainfall model was investigated by Muntohar and Liao [5]. They suggested that the model could predict the mode and occurrence of rainfall-induced landslides. The prediction of pore water pressure distribution along an unstable slope is always a matter of research. Ji-Cheng et al. [6] stated that the magnitude and direction of pore water along the sloping surface changes soil's physical and engineering properties. Studies conducted by Johansson [7] focused on the stability of waterfront slopes subjected to varying pore water pressure. They concluded that pore pressure's long-term effects significantly influenced the slope, generating seepage during the landslide process. Later [8] adopted statistical techniques to investigate the stability of the slope with an existing water table. Researchers like [9, 10] investigated the effect of surcharge load and concluded that slope stability decreased with increasing surcharge load and slope angle. The slopes with higher surcharge load also had a significant settlement and swelling effects. The analysis results revealed that the presence of water table reduced the safety of slopes and undermined the significance of groundwater level instability problems associated with landslides. The studies revealed that the effect of surface or ponded water along the waterfront slopes and surcharge load significantly influenced the stability of the non-homogenous slope. So, following objectives have arrived for the present study:

- Investigate the influence of non-homogeneous soil profile on the stability of finite slopes
- Estimate the variation of a factor of safety under surface water and surcharge load on the front end of non-homogenous finite slopes
- Compare the variation of pore water pressure, shear force and normal force for two-layered and multi-layered sloping profiles.

2 Methodology

The steps adopted in the stability study of non-homogenous finite slopes are detailed here, with a flowchart depicting the stages in order (Fig. 1).

- Generate the profile of the slope geometry
- Fix the direction of slope failure (left to right or right to left)
- Specify the physical properties of the infill materials
- Assign the properties to the slope geometry
- Establish the level of the ponded water table at the sloping surface
- Define the limits of the section near the external slope geometry
- Perform Bishop simplified analysis for specific sloping surface

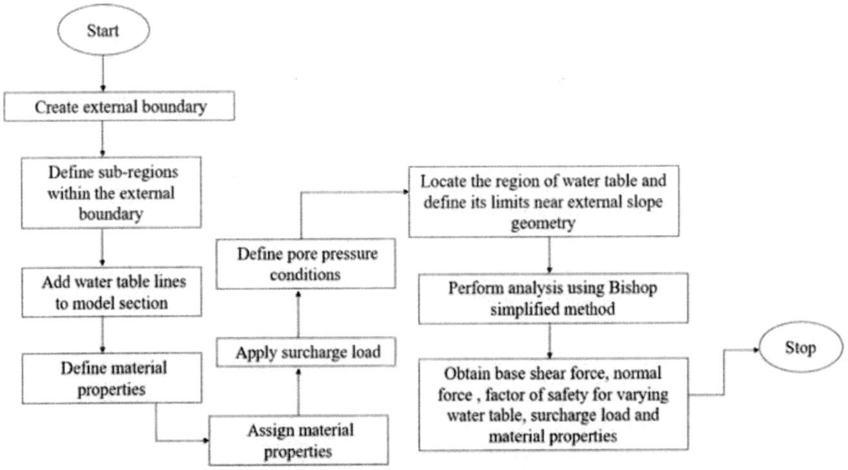

Fig. 1 Steps involved in the stability analysis of non-homogenous finite slopes

- Evaluate the base shear, normal force, and factor of safety for varying water table, surcharge load, and material properties.

3 Model Setup

The following section discusses the software, model generation, physical characteristics of infill materials used, and location of the water table adopted in analytical modeling of finite slopes.

3.1 HYRCAN Software

HYRCAN is a two-dimensional slope stability software developed by Roozbeh Geraili Mikola in April 2021. This open-source software evaluates factor of safety, failure profile of slip surface in soil and rocky slopes under varying external loading, groundwater, and support conditions. The software HYRCAN based on Vatti's clipping algorithm, an open-source freeware library, enables the user to specify boundaries, geometry, material properties, and supports elements through a CAD-type graphical user interface (GUI).

3.2 Material Properties of Soil Mass Used in the Study

Two different extreme soil masses like sand and clay are selected as the embedded medium for studies to simulate actual site conditions. The density of sand is fixed through pluviation method using two different heights of fall, viz., 5 and 15 cm. Accordingly, infill densities of sand obtained are 1740 and 1820 kg/m^3. A maximum dry density of 1400 kg/m^3 is obtained through a compaction test, and accordingly, two moisture content levels: 10 and 12%, are selected to achieve the density of clayey soil. The shear strength parameters of infill, such as cohesion and friction angle, are obtained through direct shear test results. The HYRCAN software creates a model section's material boundary, external boundary, and subregions within the external boundary and defines the water table's range. In the present study, two different stratified soil conditions were chosen to simulate the slope section's non-homogeneity. The two cases adopted for the study are: (i) Case 1: Two-layered soil profile consisting of sandy soil (first layer) with an infill density of 1740 kg/m^3 until a depth of 5 m, followed by sandy soil (second layer) with a dry density of 1820 kg/m^3 until a depth of 9 m. Table 1 lists the physical characteristics of the infill material used in the analytical model, and Fig. 2 shows a two-layered non-homogenous slope profile.

(ii) Case 2: Four-layered soil profile consisting of sandy soil (first layer) with an infill density of 1740 kg/m^3 until a depth of 2 m, followed by sandy soil (second

Table 1 Physical properties of infill material used in the analytical model

Material	Infill density (kg/m^3)	Cohesion (Pa)	Friction Angle (degrees)	Specific gravity	Coefficient of curvature	Uniformity coefficient	Soil classification
Sandy soil (First layer)	1740	4.5	20	2.62	2.23	1.19	SP
Sandy soil (Second layer)	1820	7.5	25	2.62	2.23	1.19	SP

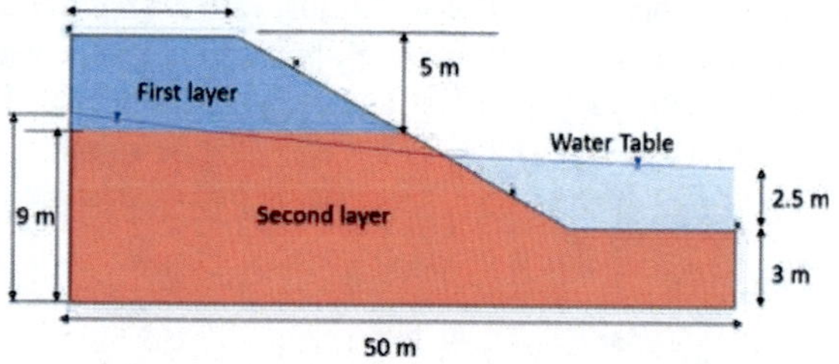

Fig. 2 Two-layered non-homogenous slope with the water table

Table 2 Physical properties of infill material for layered slope

Material	Infill density (kg/m^3)	Cohesion (Pa)	Friction Angle (degrees)	Specific gravity	Soil classification
Sandy soil (First layer)	1740	4.5	20	2.62	SP
Sandy soil (Second layer)	1820	7.5	25	2.62	SP
Clayey soil (Third layer)	1350	45	5	2.79	CL
Clayey soil (Fourth layer)	1250	35	0	2.79	CL

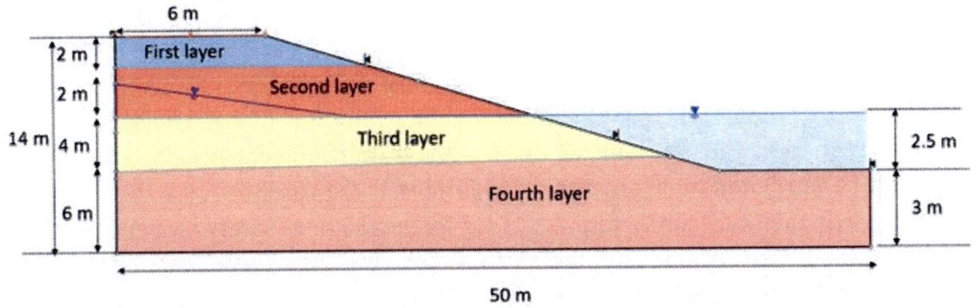

Fig. 3 Four-layered non-homogenous slopes with the water table

layer) with a dry density of 1820 kg/m^3 until a depth of 2 m. The third layer consists of clayey soil with an infill density of 1350 kg/m^3 till a depth of 4 m, and the fourth layer of clayey soil with an infill density of 1250 kg/m^3 till 6 m. Table 2 summarizes the physical characteristics for soils adopted in the analytical model study, and Fig. 3 shows a four-layered non-homogenous slope with the water table at a depth of 2.5 m.

The stability of the non-homogeneous slopes was investigated by varying surcharge load, depth of water table, and stratification of infill medium. Table 3 summarizes the parameters used for the study.

Table 3 Summary of parameters adopted for the study

	Surcharge load (kPa)	Depth of water table from the bed level (m)		Surcharge load (kPa)	Depth of water table from the bed level (m)
Two-layered	50	2.5	Multi-layered	50	2.5
	100	2.5		100	2.5
	50	1.5		50	1.5
	100	1.5		100	1.5

4 Result and Discussions

The following section discusses the influence of the water table and surcharge load on the stability of non-homogenous slope with the two-layered and multi-layered soil profiles.

4.1 Stability Analysis of Non-Homogenous Slopes with Surcharge Load of 50 kPa Under a Varying Water Table of 2.5 and 1.5 m

Figure 4 shows the variation of a factor of safety of two-layered and multi-layered slopes under a varying water table of 2.5 m for a surcharge load of 50 kPa. The factor of safety of the slope reduced from 1.488 to 1.475 for a water depth of 2.5 m. The non-homogeneity of the ground profile influenced this variation. The decreased cohesion and lesser frictional angle value of the third and fourth layers (weak soil near the toe region) contributed to an enlarged slip surface at the base.

With the decrease in the water table from 2.5 to 1.5 m, the factor of safety reduced from 1.488 to 1.484 in two-layered and 1.475 to 1.381 in multi-layered slopes, respectively. The presence of fully saturated clayey soil beneath the toe has resulted in a severe reduction in the factor of safety. In this case, the chances of slope failure are greater as less shear strength is developed along the base of the critical failure surface. Figure 5 shows the variation of a factor of safety of two-layered and multi-layered slopes under a varying water table of 1.5 m for a surcharge load of 50 kPa.

Since the modified Bishop's method is adopted in the present analysis, it is necessary to investigate vertical equilibrium, moment equilibrium, and horizontal forces on slices. Table 4 presents the comparative analysis of pore pressure, base shear force, and normal base force for two-layered and multi-layered slopes for a mid-slice (slice 13). It is seen that pore pressure decreased from 84.29 to 60.83 kPa for multi-layered slopes due to a decrease in water level from 2.5 to 1.5 m. By comparing the two cases, for a water depth of 2.5 m, base shear force decreased from 43.89 to 34.7 kN, and normal force decreased from 194.77 to 188.56 kN due to decreased shear strength offered by saturated clay layers. The surface adsorption phenomenon that occurred between clay mineral grains and pore water resulted in a loss of cohesion, thus reducing the strength of the soil. Since the base shear developed depends on cohesion, friction angle, and normal force, a decrease in friction angle has reduced shear force values in multi-layered slopes. Also, the factor of safety (FOS) of two-layered slopes did not reduce drastically as the normal force component reduced the tendency of sliding failure.

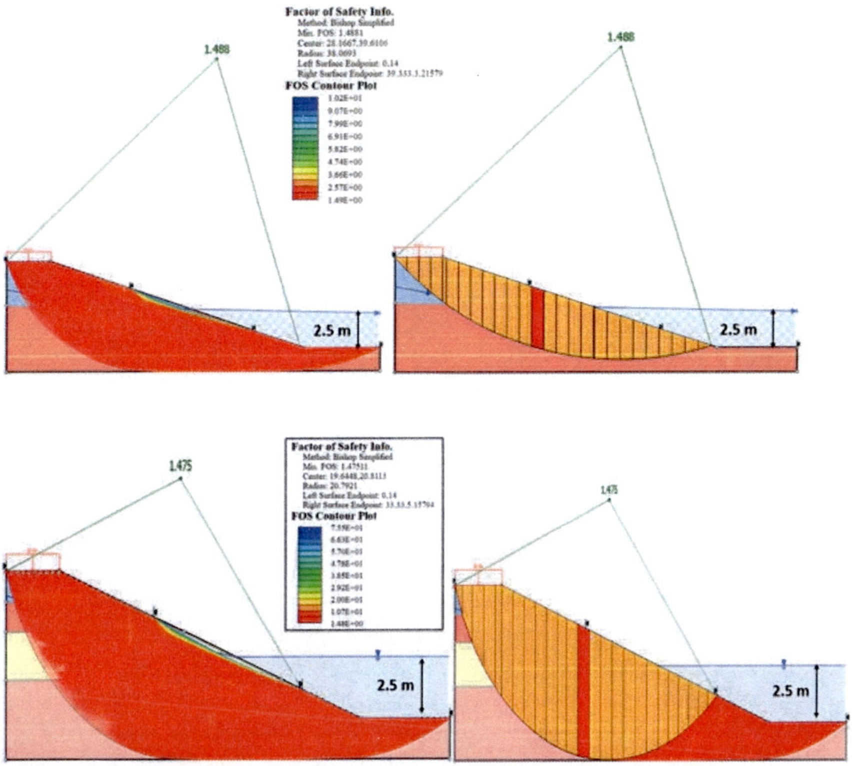

Fig. 4 Variation of the factor of safety of two-layered and multi-layered slopes for a water depth of 2.5 m

4.2 Stability Analysis of Non-Homogenous Slopes Under Constant Water Table with and Without Surcharge Load

Figure 6a and b compares two-layered and multi-layered slopes under varying surcharge pressure of 0, 50 and 100 kPa at a water table of 1.5 m. With increase in surcharge load (0 to 100 kPa), FOS of the two and multi-layered slope decreased from 1.803 to 1.306 and 1.902 to 1.279, respectively. Due to the weak stratum beneath the toe, all these cases reported base failure involving rotational slides in which the failure surface slipped below the toe.

Table 5 presents the slice analysis results of mid slice 13. When surcharge load increases from 0 to 100 kPa, base shear force increases from 30.7 to 41.9 kN and 26.3 to 36.54 kN for two-layered and multi-layered slopes. The increased base shear force values of multi-layered slopes are due to saturated clayey layers near the toe. The surface loading/surcharge on the slope crest influenced normal force components, as evident from the slice analysis results (Table 5). Normal force increased from 185.7 to 219.8 kN and for multi-layered, normal force increased from 190.2 to 207 kN. It

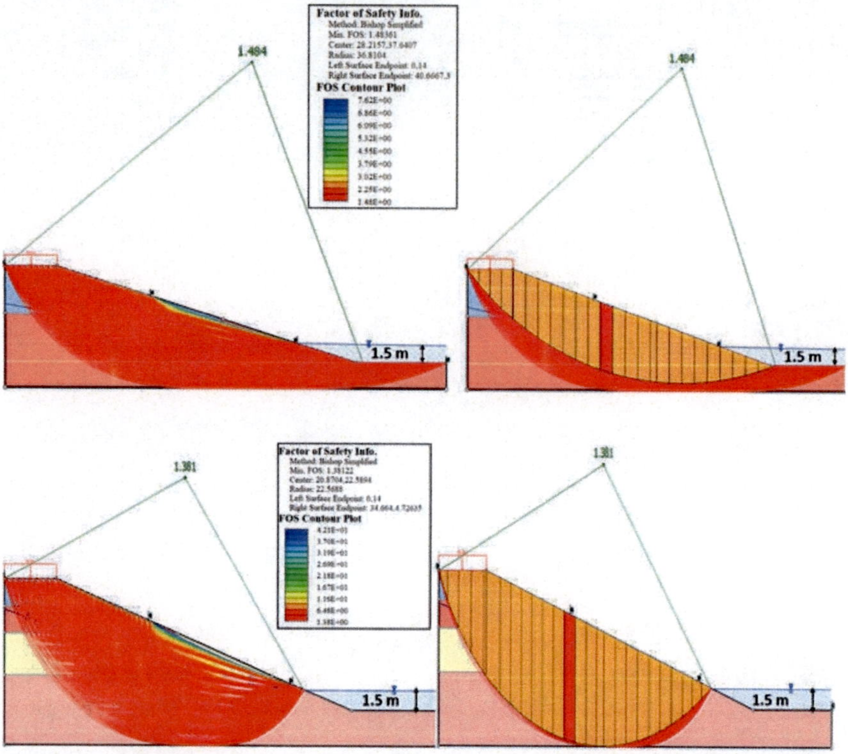

Fig. 5 Variation of the factor of safety of two-layered and multi-layered slopes at a water depth of 1.5 m

Table 4 Two-layered and multi-layered slopes for slice 13 for surcharge load of 50 kPa

Type of slope	Depth of water table (m)	Surcharge load (kPa)	Factor of safety (FOS)	Pore pressure (kPa)	Base shear force (kN)	Base normal force (kN)
Two-layered	2.5	50	1.488	49.96	43.89	194.77
Multi-layered	2.5	50	1.475	84.29	34.71	188.56
Two-layered	1.5	50	1.484	39.20	57.90	223.7
Multi-layered	1.5	50	1.381	60.83	32.87	176.35

is observed that an increase in the surcharge load enhanced the normal component of forces, increasing the stability of slopes. As an outcome, it is established that increasing the surcharge load increased the base shear but had no influence on the normal force at the slope base.

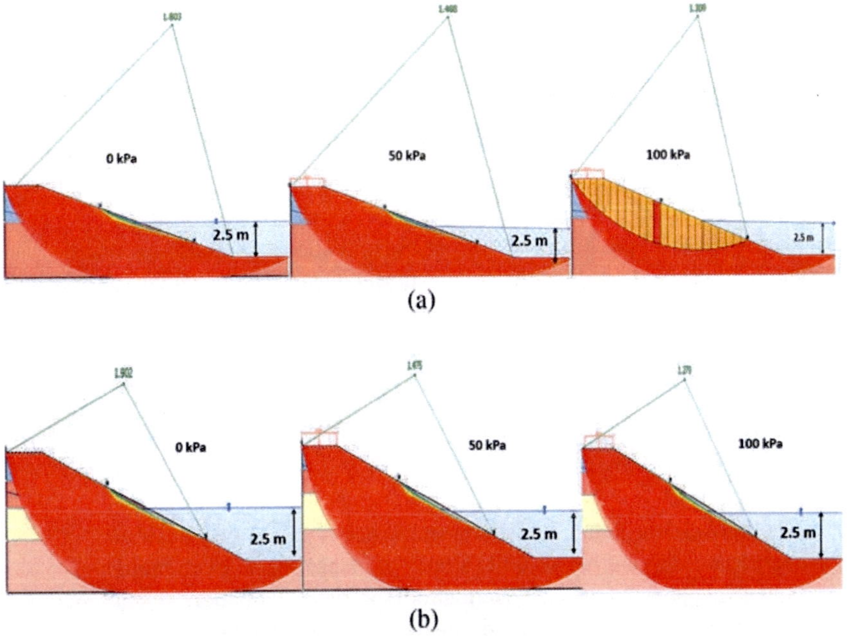

Fig. 6 a—b Variation of the factor of safety of two-layered and multi-layered slopes under varying surcharge pressure of 0, 50 and 100 kPa for a water table of 2.5 m

Table 5 Two-layered and multi-layered slopes for slice 13(mid slice) for a water table of 2.5 m

Type of slope	Surcharge load (kPa)	Factor of safety	Pore pressure (kPa)	Base shear force (kN)	Base normal force (kN)
Two-layered	0	1.803	49.96	30.70	185.7
Two-layered	50	1.488	49.96	38.89	194.3
Two-layered	100	1.309	49.96	41.90	219.8
Multi-layered	0	1.902	84.29	26.30	190.2
Multi-layered	50	1.475	84.29	34.71	198.87
Multi-layered	100	1.279	84.29	36.54	207.01

5 Conclusions

The failure of natural and artificial slopes has resulted in the death and severe destruction of humankind/nature. So civil engineers must check the safety of natural and artificial slopes of excavation. The present study determined and compared factors of safety, forces developed along the potential rupture surface and shear strength of the soil. The conclusions arrived from the present study are listed as:

- The prevalence of the water table and weak soil at the base levels drastically reduced the slope strength of multi-layered slopes.
- The pore water pressure increased for multi-layered slopes due to impermeable soil layers at the toe region.
- The effect of surcharge accounted through distributed line load over the crest contributed to the gravitational force component, which reduced the factor of safety of finite slopes.

These studies can assist in the determination of the probability of landslides on natural or artificial slopes and analyze the failure mechanisms. The research can guide in redesigning of slopes and formulate preventative measures for deteriorating slopes and failed embankments.

6 Scope for Future Studies

The present studies assessed the influence of parameters such as surcharge load, water table, infill properties on the stability of finite slopes in planar space. Even though HYCRAN is a powerful geotechnical-engineering tool for linear and nonlinear analysis of geotechnical problems, substantial experimental research is needed to examine the variability of rupture surface parameters. Thus, full-scale experimental studies and the comparison of analytical modelling results could be the scope of future work.

References

1. Griffiths DV, Lu N (2005) Unsaturated slope stability analysis with steady infiltration or evaporation using elastoplastic finite elements. Int J Numer Anal Methods Geomech 29:249–267. https://doi.org/10.1002/nag.413
2. Huat BBK, Ali FHJ, Low TH (2006) Water infiltration characteristics of unsaturated soil slope and its effect on suction and stability. Geotech Geol Eng 24:1293–1306. https://doi.org/10.1007/s10706-005-1881-8
3. Zhu D (2008) Investigations on the accuracy of the simplified Bishop method. In: Landslides and engineered slopes. From past to future, pp 1055–1057. https://doi.org/10.1201/9780203885284-c138
4. Gavin K, Xue J (2008) A simple method to analyze infiltration into unsaturated soil slopes. Comput Geotech 35:223–230. https://doi.org/10.1016/j.compgeo.2007.04.002
5. Muntohar AS, Liao HJ (2010) Rainfall infiltration: finite slope model for landslides triggering by rainstorm. Nat Hazards 54:967–984. https://doi.org/10.1007/s11069-010-9518-5
6. Ji-Cheng W, Xiao-Nan G, Shi-guo M (2014) Effects of pore-water pressure distribution on slope stability under rainfall infiltration. Electron J Geotech Eng 19 H, pp 1677–1685
7. Johansson J (2014) Impact of water-level variations on slope stability
8. Shadabfar M, Huang H, Kordestani H, Muho EV (2020) Reliability analysis of slope stability considering uncertainty in water table level. ASCE-ASME J Risk Uncertain Eng Syst Part A Civ Eng 6:04020025. https://doi.org/10.1061/ajrua6.0001072

9. Sazzad MM, Haque MFm (2014) Effect of surcharge on the stability of slope in a homogeneous soil by FEM. In: 2nd international conference on advances in civil engineering 2014, pp 314–318
10. Manna B, Rawat S, Zodinpuii R, Sharma KG (2014) Effect of surcharge load on stability of slopes—testing and analysis. Electron J Geotech Eng 19 N, pp 3397–3410

Effect of Distance on Bearing Capacity of Strip Footing on Slope

Baishali Nandi, Debjit Bhowmik, and Monowar Hussain

1 Introduction

The problem of limited bearing capacity of strip footing on slope attracts the investigation of footing on slope in detail manner to improve its bearing capacity by changing various parameters. The various parameters which can influence the bearing capacity of strip footing on slope are soil strength, foundation width, foundation depth, slope angle, slope height, inclusion of reinforcement, etc. The soil materials involved in this study are *c*-ϕ soil which was collected from NIT Silchar site and the soft soil which was collected from the lake bed of NIT Silchar, during the month of June-July 2019.

The study includes investigation of stress–strain behaviour of soils obtained from unconfined compressive strength (UCS) tests to evaluate Young's modulus (E), elastic and plastic strain values which are required for numerical simulation in ABAQUS software. In this paper, the bearing capacity of strip footing on a *c*-ϕ soil slope overlying soft clay bed has been attempted to study.

A Variety of approaches has been made in last decade, Shaiau et al. [1] using finite-element Limit analysis of upper and lower bound theorems obtained undrained bearing capacity of strip footing on cohesive slopes, their study reveals that the critical value of strength ratio ($C_u/\gamma B$) can separate bearing capacity failure and slope failure. In their study C_u indicates undrained shear strength of soil, γ indicates unit weight of soil and B indicates width of the strip footing. Zhang et al. [2] studied on reliability analysis of slope surviving a certain groundwater condition (at max water level what is the reliability). They used past survival information and past failure information in their study to find reliability. And it is been seen that considering past survival information increases reliability of slope and makes it economical, they used direct and indirect methods (where Bayes theorem and updated variables like c, ϕ,

B. Nandi (✉) · D. Bhowmik · M. Hussain
National Institute of Technology, Silchar, Assam, India
e-mail: baishali_rs@civil.nits.ac.in

A. Sil et al. (eds.), *Recent Trends in Civil Engineering*, Lecture Notes in Civil Engineering 274, https://doi.org/10.1007/978-981-19-4055-2_37

etc. are used). Kazi et al. [3] conducted model plate load tests. D_f was varied for both unreinforced and woven geotextile-reinforced sand beds. The ultimate bearing capacity of the footing was increased when woven geotextile has wrap-around ends. Kazi et al. [4] performed a series of laboratory model tests to examine the effect of submergence on settlement and ultimate bearing capacity of a surface strip footing, resting on the reinforced sand bed. A significant reduction in settlement value was observed when the water table is at the ground level for both unreinforced and reinforced cases. Palmeira and Tatto [5] worked on nonwoven geotextile and granular filters in armoured slopes subjected to the impact of waves. The bank failure can be minimized by using the filter. Filter reduces the pore pressure. It is been seen the thickness of geotextile reduces the lower the max pore pressure. After a few hours of testing, it's seen that the turbidity values obtained in the tests with geotextile filters were similar to those obtained for the granular filter. Influence of strength and stiffness of reinforcement on the load-settlement response of geocell-reinforced sand bases was investigated by Kargar and Hosseini [6]. Shear failure mechanism was observed when geocell stiffness is inadequate and also has a small height. On the other hand, failure in bending occurred when geocell stiffness is high and has higher height. Nadaf and Mandal [7] performed a small-scale test on a model footing load test. Main purpose was to observe the footing response on fly ash slope reinforced with steel grids, jutes and cellular reinforcement. The load carrying capacity of footings situated on such slopes is significantly enhanced by placing a layer of geocell reinforcement at appropriate positions among the sloped fill. It's been seen from the test results the settlement reduces with cellular reinforcement. Mansouri and Abbeche [8] found out experimentally the bearing capacity of an eccentrically loaded square footing near a slope. Their study shows that as relative density increases (30, 50 and 70%) the ultimate bearing capacity also increases. By seeing the load-settlement curve, it can be concluded that the punching, local and general shear failure occur for 30%, 50% and 70% relative densities, respectively. Their study shows that as the eccentricity of loading increases the ultimate bearing capacity reduces. Their study also shows that the bearing capacity increases when the distance of the foundation diverges from the slope.

2 Test Materials

Soft soil and c-ϕ soil samples used in the study were collected as disturbed and undisturbed forms from the lake bed and embankment of NIT Silchar, in the Northeastern part of India, respectively.

Table 1 presents the properties of the in-situ (field) c-ϕ soil. The soft soil used in this study is presented in Table 2.

Table 1 Properties of the c-ϕ soil

Property	Values
Specific gravity, G	2.57
Liquid limit	33.5%
Plastic limit	17.7%
Field density of soil	19.2 kN/m^3
Field moisture content	22.7%
Dry density of soil	15.7 kN/m^3
cohesion value (c) and friction angle (ϕ) from unconsolidated undrained triaxial test (UU)	$c = 18.4$ kPa, $\phi = 8.8°$
Young's modulus of soil (E)	888.8 kPa

Table 2 Properties of the soft soil

Property	Values
Specific gravity, G	2.3
Liquid limit	55%
Plastic limit	34%
Bulk density of soil,	16.7 kN/m^3
Moisture content	38.4%
cohesion value (c) and Friction angle (ϕ) from unconsolidated undrained triaxial test (UU)	$c = 15.5$ kPa, $\phi = 1.4°$
Soil Young's modulus (E)	224.3 kPa

3 Laboratory Tests

Experimental work was modelled to study the effect of position of footing with respect to crest on ultimate bearing capacity of the strip footing. For all these soils the standard protector test and triaxial tests were conducted to determine maximum dry density, optimum moisture content, cohesion and angle of internal friction (ϕ). The test results of different soils tested for these properties are shown in Tables 1 and 2.

3.1 Design of Experimental Setup

The schematic diagram of the experimental setup is shown in Fig. 1.

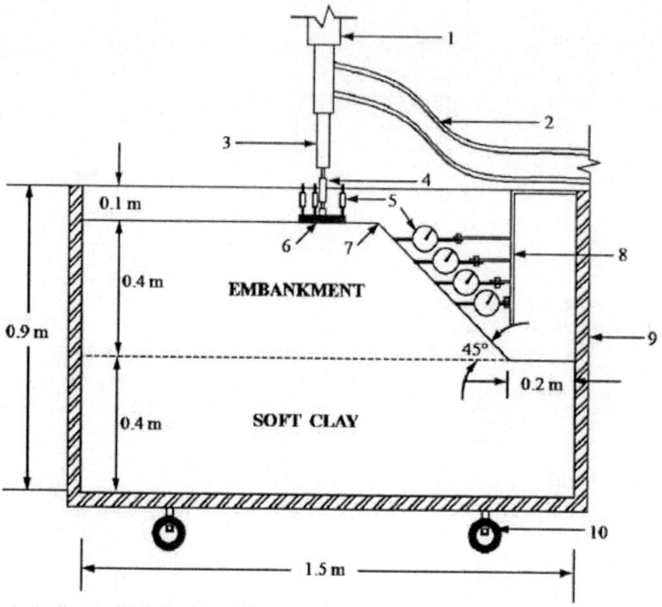

1. Hydraulic Jack, 2. Oil supply pipes, 3. Extension rod, 4. Proving ring, 5. Dial gauges, 6. Footing, 7. Crest of slope, 8. Supporting structure for dial gauges, 9. Test tank, 10. Wheels

Fig. 1 Experimental setup

3.2 Proposed Test Procedure

The primary components of the model tests are a tank, a rigid MS plate, loading system, sand and measurement devices. The internal dimensions of the tank are 1.5 m (Length, l) × 0.6 m (Width, w) × 0.9 m (Height, h). The footing is placed on the slope. The bottom, back and two sides of the tank are made of thick steel plate. The front side of the tank is made of 6 mm thick transparent Perspex sheets to observe failure mechanisms. Strip footing case has a close similarity with the plane strain condition. Plane strain conditions can be maintained by two ways: (i) there should not be any lateral movement along the width of the tank and (ii) friction between soil and wall should be kept as minimum as possible. Two steel frames are connected together by means of bolts on the side of the tank to make it sufficiently rigid. The thickness of Perspex sheets is also chosen more so that it cannot bend or deformed under the loading. These two things ensure the satisfaction of the first criteria. To reduce friction between wall and sand particles, side walls of the tank are polished very well as well as Perspex sheets are also attached.

A rigid MS plate is used as model strip footing. The dimension of the footings is 0.58 m (Length, l_f) × 0.08 m (Width, b_f) × 0.02 (Height, h_f). The length of the plate is kept almost nearer to the width of the tank to maintain plane strain condition.

To minimize end friction effects two ends of the footing is polished smooth. On one side of the plate sand particles with glues are attached to form rough base conditions.

In the experimental study, the test bed was prepared by compacting locally available c-ϕ soil at its field density for a depth of 0.4 m over a soft clay bed of height of 0.4 m. The c-ϕ soil was prepared in laboratory at its field density in the month of June-July 2019, when it was subjected to 3 days of continuous rainfall, to study its adverse effect on slope. And the soft clay bed was prepared such that it satisfies soft soil conditions.

4 Numerical Analysis

The FEM-based software used in this study is ABAQUS. In ABAQUS, the 3D the model is designed. Soil properties used in this study are from locally available soil at its field density, the properties of it mentioned in Tables 1 and 2. The model is discretized with 24,152 no. of elements. The soil slope is discretized with a total 20,064 no. of linear hexahedral elements of type C3D8R, the soft clay bed was discretized with 23,040 no. of linear hexahedral elements of type C3D8R and the footing is discretized with a total 72 no. of linear hexahedral elements of type C3D8R. The number of elements for slope model is chosen after mesh convergence study and the suitable no of elements for slope model was found to be 43,176.

4.1 Modelling Properties

The numerical analysis used for ABAQUS software includes the test bed of 0.4 m height which was modelled using the locally available soft clay collected from NIT Silchar campus overlying a c-ϕ soil slope at its field density. The required properties for this model are mentioned in Tables 1 and 2. The initial geometry entered in the software is shown in Fig. 4.

5 Results and Discussion

The numerical study using ABAQUS software was done by varying the positions of the footing from its crest position. Figure 5 shows the bearing capacity-settlement curve for various positions of footing on the slope.

The experimental model study has been performed for various footing locations from crest of the slope. It is been seen that the bulging phenomenon of soil slope predominates with the distant position of the footing. Figures 2 and 3 show the picture of experimental results obtained after failure occurred for the foundation location at 1*B* and 4*B* distance from crest, respectively. The experimental result also confirms the bearing capacity of the soil slope model to be sufficiently increased at 4*B* distance of footing from crest as compared to 1*B* distance of footing from crest. Figure 6 represents the bearing pressure versus settlement curve obtained from experimental study for footing position at 1*B* and 4*B* distance from crest, respectively. Figure 7 shows a validation curve of laboratory and experimental evaluation, which shows a close agreement between experimental and numerical results. From the load-settlement behaviour, it can be compared that with respect to footing location at crest to footing location at 4*B* distance the bearing capacity can be improved up to 76.67%.

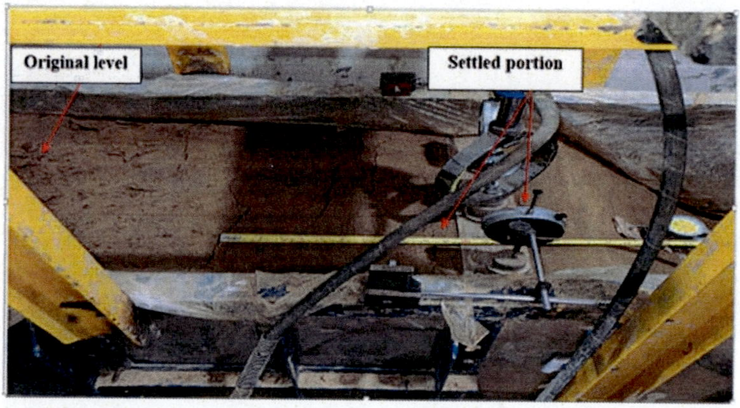

Fig. 2 Picture of experimental setup for foundation placed at 1*B*

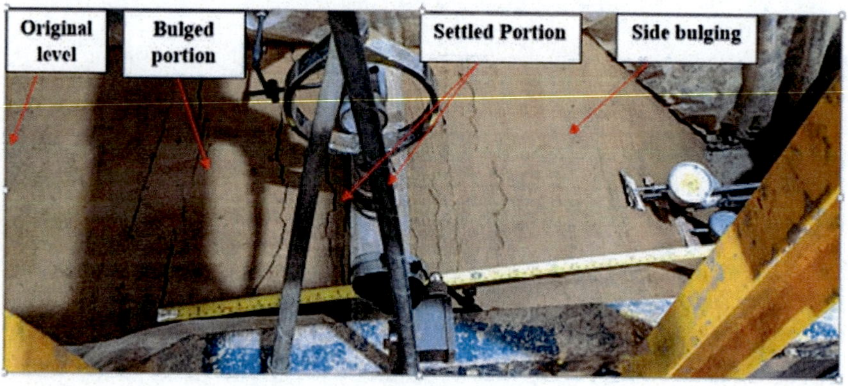

Fig. 3 Picture of experimental setup for foundation placed at 4*B*

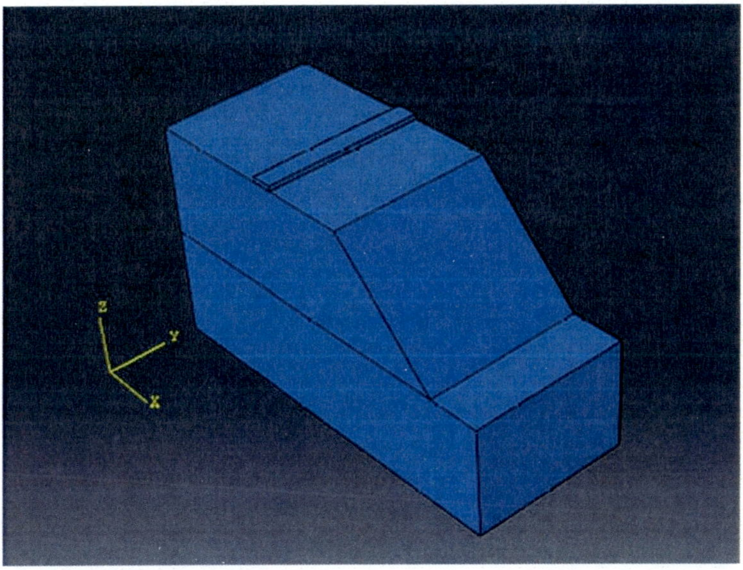

Fig. 4 Slope geometry entered in ABAQUS

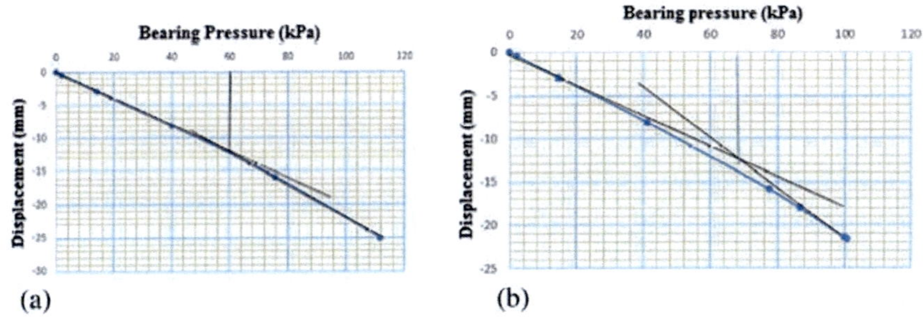

Fig. 5 Bearing pressure–displacement curve obtain from numerical analysis for footing placed **a** at crest, at **b** at 1*B*, **c** at 2*B*, **d** at 3*B*, **e** at 4*B* **f** at 5*B* distance; form crest

6 Conclusion

In this study, the bearing capacities of strip footing on soil slope overlying soft soil are calculated based on numerical analysis and experiments. Based on the results of the present study, it can be concluded that by varying footing location from the crest the ultimate bearing capacity of footing on slope increases. When the location of footing reaches 3*B* distance from the crest, it shows maximum effect; beyond that any further change in location makes negligible effect in bearing capacity. It is also observed that for footing location at 3*B* distance from the crest, the bearing capacity improved by 76.67% as compared to the footing position at crest.

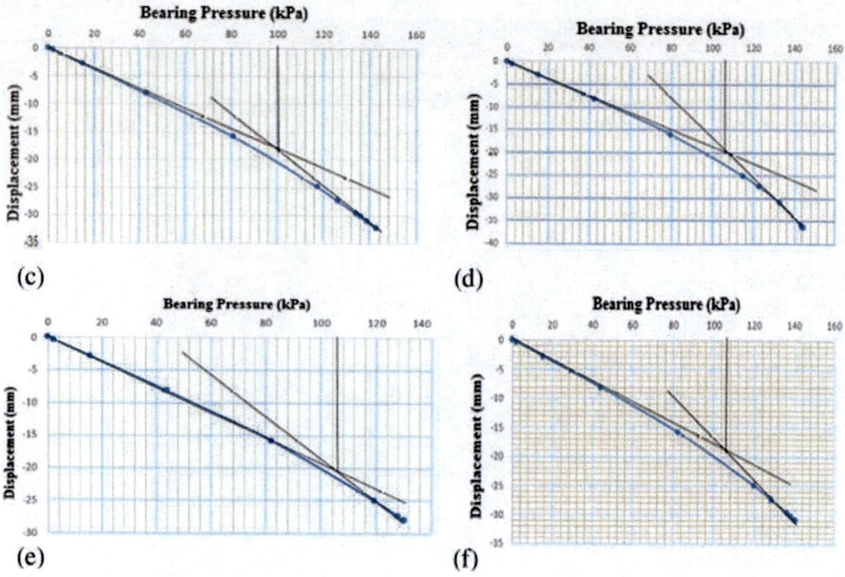

Fig. 5 (continued)

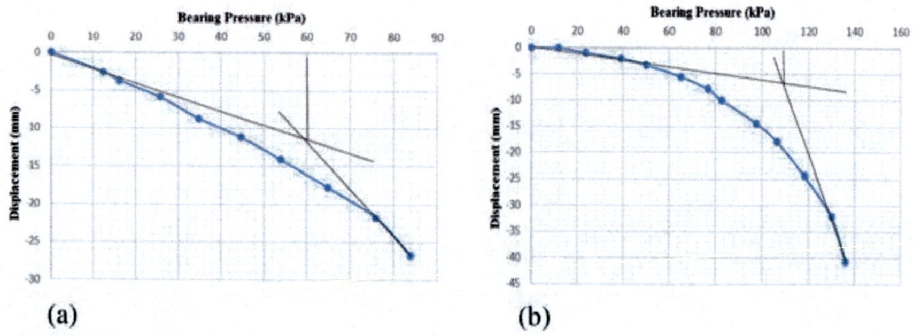

Fig. 6 Bearing pressure–displacement curve obtain from experiment for footing placed **a** at 1*B*, **b** at 4*B* distance, form crest

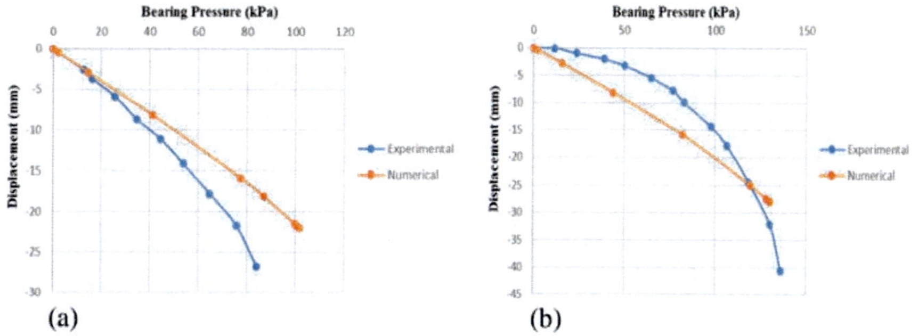

Fig. 7 Validation curve of experimental and numerical evaluation at **a** 1*B*, **b** 4*B* location of footing from crest

References

1. Shiau JS, Merifield RS, Lyamin AV, Sloan SW (2011) Undrained stability of footing on slope. Int J Geomech 11(5):381–390
2. Zhang J, Zhang LM, Tang WH (2011) Slope reliability analysis considering site-specific performance information. J Geotech Geoenviron Eng 137(3):227–238
3. Kazi M, Shukla SK, Habibi D (2015) Behavior of embedded strip footing on sand bed reinforced with multilayer geotextile with wraparound ends. Int J Geotech Eng 9(5):437–452
4. Kazi M, Shukla SK, Habibi D (2015) Effect of submergence on settlement and bearing capacity of surface strip footing on geotextile-reinforced sand bed. Int J Geosynthetics Ground Eng 1(1):4
5. Palmeira EM, Tatto J (2015) Behaviour of geotextile filters in armoured slopes subjected to the action of waves. Geotext Geomembr 43(1):46–55
6. Kargar M, Mir Mohammad Hosseini SM (2018) Influence of reinforcement stiffness and strength on load-settlement response of geocell-reinforced sand bases. European J Environ Civil Eng 22(5):596–613
7. Nadaf MB, Mandal JN (2017) Behavior of reinforced fly ash slopes with cellular mattress and strips under strip loading. J Hazard Toxic Radioactive Waste 21(4):04017019
8. Mansouri T, Abbeche K (2019) Experimental bearing capacity of eccentrically loaded foundation near a slope. Studia Geotechnica et Mechanica 41(1):33–41

Evaluation of Engineering Properties of Laterite Blocks for Construction Purposes from Selected Zones of Coastal Karnataka Region—A Case Study

Bhagyashree, H. N. Udayashankara, and Purushotham G. Sarvade

1 Introduction

Laterite blocks are of marked importance for construction of buildings compared to the several rock forms that are found in the sub-tropical and tropic region. In coastal regions, Laterite rocks are an important constituent in building construction due to its abundance. Laterites are formed when materials undergo physical and chemical weathering which is complex. These soils are enormously weathered and consist of large amounts of alumina and iron oxides. They also contain quartz and other minerals. They are found in tropics and sub-tropics in large amounts and are usually found just below the surface in areas with high rainfall. The colour varies from red to yellowish-brown, black to violet, brown, and it depends frequently on the amount of oxides of iron present. The process of weathering is composite and hence the procedure takes ages to complete. The laterite growth is affected by geology, physiography, climate affect and other such major agents during the weathering process. Further, the lateritic soil or rocks that are found within a few yards surrounding one another can have varying properties and adds up to the difficulties. However, these laterites can be very alike to other laterites which are situated hundreds of kilometres away. The chemical, mineralogical and physical properties differentiate the laterites from other laterites from different places.

Laterites are classified into two kinds: Laterites of aluminous origin and Laterites of ferruginous origin. Laterites having more iron content is classified as Ferruginous laterites and laterites having more alumina content is classified as aluminous laterites. Ferruginous laterites are darker than aluminous, higher in density and is being used as an eco-friendly construction material (Table 1).

Bhagyashree (✉) · H. N. Udayashankara · P. G. Sarvade
Manipal Institute of Technology, MAHE, Udupi, Manipal, India
e-mail: bhagya.shree@springernature.com

A. Sil et al. (eds.), *Recent Trends in Civil Engineering*, Lecture Notes in Civil Engineering 274, https://doi.org/10.1007/978-981-19-4055-2_38

Table 1 Laterite properties

Characteristics	Aluminous laterites	Ferruginous laterites
Colour	Whitish rose to red	Rust to dark brown
Density	Low	High

1.1 Geology of the Area

Dakshina Kannada/Udupi district is situated in a peninsular region. It is one of the coastal districts of Karnataka and is located along the west coast of India. The district is between Western Ghats on the east and the Arabian sea on the west. The peninsula is composed of geologically ancient rocks of diverse origin and most of them have undergone metamorphism. It represents a stable land of the earth's crust. Laterite formations cover large area along the coast and little to the interior of the district and are of considerable thickness.

1.2 Literature Review

Laterites are residual ferruginous rock masses, normally originating in the tropics and has nearby inherited relationship with bauxites. It is profoundly endured soil mass, high in optional oxides of aluminium, iron. It is either stiff or equipped for solidifying on exposure to dampness and drying. Laterites are found in various regions of India, broadly utilized as structural commodity in areas of Kerala, Karnataka, Andhra Pradesh and Goa [1].

Tests have been performed for deciding the uniaxial compressive strength (UCS), changing sample size and extent, moisture content and cyclic wetting and drying so as to evaluate possible variations in their final strength. Acquired UCS are between 1.5 and 5 MPa [2]. To the extent the size impacts are concerned, it is interesting to see that larger blocks result in lower uniaxial compressive strength. This observation is of high significance since infers that the genuine size of laterite dimension stone (LDS) utilized for brickwork applications definitively affects the available rock quality. In this manner, for the definition of standardization codes, it is important to illuminate test conditions and sample measurements further to UCS alone. In view of the result acquired from the different tests completed on the laterite specimens, low water absorption and high compressive strength is found in specimens situated in the top part of laterite profiles. Hardness of laterites is observed to be in the scope of 1–4 as indicated by Mohr's scale and diminished with depth for all laterites [3].

The on-site attributes of laterites like shading, surface, structure and induration were found by field study done previously. Freshly extracted laterite building blocks from various portions of the quarry as provided on a commercial scale, was assessed for their engineering qualities. The mineralogical and geotechnical characteristics of this soil was observed to be subject to site profile features of the stone, viz., shading, surface and structure. The procedure of choosing laterite for building purposes dependent on field remarks recorded by specialists previously. Samples were gathered as standard size blocks according to Indian standard specifications or laterite stone brick work for deciding engineering property. The bedding surface of every example was marked at the time of accumulation. Scratch tests were done on the samples to find the hardness of the blocks dependent on crystal arrangement denoted by the Mohs scale of hardness [2, 4].

Laterite soil is commonly found in the tropical region through weathering process which supports the development of oxides of iron, aluminium, titanium and manganese. Hence, all of these process separate silicate minerals into earth minerals, for example, kaolinite and first-class iron materials. Iron and aluminium oxides are found in laterite soils, and alongside the regular variation of ground water table, these oxides result in the rosy dark coloured shading that is seen in lateritic stones.

It is seen in vast quantity in the regions of high temperature and humidity, generally located just underneath areas of vast plains and backwoods in zones with sufficient rainfall. The colour can differ from yellowish-darker to red, dark coloured, violet to dark, dependent generally on the concentration oxides of iron [3]. This weathering procedure is complex in light of the fact that it takes huge number of years and the procedure to operate, for example, physiography, topography and atmosphere influence the laterite development. The trouble is additionally aggravated by the way that laterites, which are found inside a couple of feet of one another, can have different properties; in any case, every one of these laterites can be like different laterites found a large number of miles away [4]. The similarities and changes between all laterites are commonly estimated as far as their concoction, geological and physical characteristics. Greater part of the lateritic soil is arranged into at least one framework relying on their properties [5].

The exceptional characteristic of laterites to indurate when exposed to atmospheric conditions because of dehydration has been widely discussed [6]. The induration degree differs that ranges from more or less loose coherent mass that is easily hand broken, to the most toughened blocks that are difficult to break with even a chisel or hammer [7]. The hardening of laterites is because of the growth of integral sesquioxides (Fe2O3 and Al2O3), hastened, condensed and thickened as an outcome of dehydration [8]. The induration process of laterite exposed to the atmosphere by the development of two centimetre hard crust on an unprotected wall material due to dampening and drying has been reported by Alexander and Cady. It is decided that the toughening of laterite is mostly due to the water content and drying out of iron oxides and prearrangement of constituents [9]. The depth of hardening of laterite profiles differs due to fluctuation in the water table [10].

2 Methodology

2.1 *Laterite Test Samples*

Laterites are the most common building stones of Dakshina Kannada and Udupi district, and are quarried mostly in monsoon months. Until recent times, manual quarrying with digging and cutting with pick-axes was adopted. However, machine quarrying with cutting-saws is becoming more and more popular. At the quarry, the laterites when fresh before exposure to atmospheric agents are very soft and can be easily cut and dressed due to the presence of quarry sap. But when these quarried blocks are exposed to weathering agents due to the irreversible chemical reactions taking place, the blocks become hard and can withstand structural loads. The ability to withstand different types of loads depends upon many parameters, and this investigation aims at determining some of the important ones, for the different quarries spread, over Dakshina Kannada and Udupi district. The quarries were visited, inspected, and sampling were done. Sufficient number of block samples were cut from the parent laterite outcrops. The samples were labelled with numbers according to site location assigned and marked well in the records. These samples were transported and stored in the laboratory yard. In all, a total of 90 stones were collected, i.e. six blocks from 15 quarries.

The samples were collected from 15 quarries in Dakshina Kannada and Udupi District. Six samples from each quarry were procured. The 15 quarries being Hiriyadka, Shirva, Niddodi, Yellur, Mudarangadi, Jantra, KG road Kalyanpura, Vamanjoor, Mijar, Moodshedde, Yedapadavu, Polali, Kaikamba, Bajpe, and Gurupura. Basic properties were studied to check their feasibility as a building material. Proper selection is necessary for building construction so that the structure remains in good serviceable condition.

Compressive Strength Test: This test is done as per IS: 1121 (part I)—1974 for compressive strength of natural building stones. Any unevenness observed in the bed faces will be grounded to offer two flat, similar faces. The specimen is then submerged in water at normal ambient temperature for twenty-four hours. It is drained from water and any excess moisture was drained at room temperature. The sample is to be positioned with smooth faces horizontal between 2 to 3 thin plywood pieces, individually having 3 mm thickness, prudently arranged at the centre between plates of the universal testing machine. The load will then be applied axially at a constant rate of 140 kg/cm^2 per minute till failure occurs and extreme load at failure will be noted.

Water Absorption Test: Test is accomplished in accordance to IS: 1124—1974. Any specimen will be dehydrated in a ventilated oven at a temperature of 105 to 115 °C until it achieves uniform weight. The sample is then ventilated to ambient temperature and weight (W1) is then noted. The dehydrated specimen is then entirely immersed in pure water at a temperature of 27 $\pm$ 2 °C for twenty-four hours. The sample is taken out and any drops of water is to be rubbed off with the moist cloth.

The sample is then be weighed (W2) within three minutes after the specimen is removed from water.

Specific Gravity: The test is performed according to IS: 1124—1974 for laterites. The sample is placed in a 1000 ml cylinder and is then filled with clean water till it reaches 1000 ml mark. The quantity of water added is noted down. The water in the large cylinder is then drained and the sample is taken out and dried in the oven 100 to 110 °C for over 24 h. It is then cooled and weight is noted down.

Hardness: The hardness of the laterite block stones will be found using Mohs hardness scale. Minerals with known hardness will be used to find the hardness of laterite blocks.

A mineral with a given hardness rating will scratch other minerals of the same hardness and all samples with lower hardness ratings.

3 Results and Conclusion

The test samples collected from sampling sites were tested for specific gravity, water absorption, compressive strength and hardness. Results are tabulated in Table 1.

According to Mohs scale of hardness the value of hardness for the laterite stone samples ranges from 2 to 3. Colour varied from Reddish brown to Brownish yellow.

Specific Gravity: It can be seen from the test data presented in Table 2 that the specific gravity of laterites ranges from 2.41 to 2.74 with an average of 2.58 for fifteen quarries of the district. The variation could be due to variation in mineral content or voids.

Water Absorption: The water absorption property has been investigated for laterites and varies from 5.45 to 12.77% for 24 h of immersion in water. This wide range accounts for varying chemical composition and porous nature. Basically, lower the water absorption capacity stronger the laterite. Maximum water absorption for laterite blocks should be below 15%.

Compressive Strength: Compressive strength is one of the basic properties of the laterites which guides the selection for their use. Therefore, laterite samples were tested for compressive strength for both dry and soaked condition. It can be seen that the compressive strength presented in Table 2 for soaked condition varies from 0.83 to 3.36 N/mm^2 and varies from 1.71 to 5.73 N/mm^2 for dry conditions. Highest value being that of Niddodi quarry.

3.1 Mechanical Properties

Hardness: According to Mohs scale of hardness, the hardness of the laterite samples ranged from 2 to 3.

Effect of soaking on Compressive Strength: From Table 2, the compressive strength for both soaked and dry can be seen and it can be observed that compressive

Table 2 Experimental Results of the laterite stones

No.	Location	Specific gravity	Water absorption %	Compressive dry	Strength (N/mm^2) soaked
1	Hiryadka	2.42	7.70	3.9	2.55
2	Shirva	2.46	7.41	2.7	1.65
3	Yellur	2.57	8.81	2.06	1.12
4	Moodushedde	2.64	5.45	3.64	2.22
5	Jantra	2.59	6.59	3.26	2.01
6	KG Road Kalyanpura	2.54	10.64	2.91	1.07
7	Mudarangadi	2.47	6.56	2.73	1.86
8	Vamanjoor	2.51	5.80	5.31	2.74
9	Mijar	2.63	7.35	2.73	1.88
10	Bajpe	2.48	12.48	2.82	0.83
11	Yedapadavu	2.41	8.90	3.32	2.36
12	Niddodi	2.62	8.04	5.73	3.08
13	Polali	2.66	10.26	2.48	1.82
14	Kaikamba	2.60	12.77	1.71	1.24
15	Gurupura	2.74	6.49	4.23	3.36

strength gets reduced when laterites are in saturated condition. The reduction is about 50 to 60% for 24 h.

This investigation is a part of comprehensive study of behavioural characteristics of Coastal Karnataka laterites, i.e. Dakshina Kannada and Udupi district. This work is limited to the study of some of the important physical and engineering properties of laterites of 15 quarries of both the districts. After investigation, the following conclusions are arrived.

1. According to Mohs scale of hardness the value of hardness for the laterite stone samples ranges from 2 to 3. Colour varied from reddish brown to brownish yellow.
2. It can be seen from the test data presented in Table 2 that the specific gravity of laterites ranges from 2.41 to 2.74 with an average of 2.58 for fifteen quarries of the district. The variation could be due to variation in mineral content or voids.
3. There is wide variation in the physical and engineering properties of laterites from the different quarries of Coastal Karnataka mostly due to varying structure, chemical composition, etc. Thus, the term 'Laterite' as used in Coastal Karnataka covers wide spectrum of properties making the selection difficult without proper investigation.
4. There is significant reduction in the compressive strength when the stones are in saturated condition.
5. Properties of laterites have to be investigated before using them for construction as their strength can vary from place to place and with depth. Laterites with lower

strength can be used for partition walls if they do not meet the requirements for other structural works.

The behavioural characteristics data presented in this thesis for laterites of different quarries located in various Taluks of Dakshina Kannada and Udupi can be effectively made use of, for the selection of laterite quarries for different structural works like foundation, masonry to meet required specifications.

Limitations of this study includes selecting few number of quarries, whereas an extensive study can be carried out if more quarries were considered, also the number of tests carried out may not be enough to accurately describe the characteristics of laterites.

Advantages and disadvantages of this study:

Advantages—It helps us know which quarry is most suitable for the selection of the laterites for any construction purpose and it helps us find the nearby quarries.

Disadvantages—Since only few quarries are considered it may not give accurate knowledge about the quarries. There can be some other quarries which may have good quality laterite blocks which were not considered for this study.

References

1. Maklur N, Parag N (2018) Study of laterite stone as building material. Int J Eng Res 7:223–226. https://doi.org/10.5958/2319-6890.2018.00063.6
2. Errol Dsouza U, Kumar D (2017) A study on laterite stones as building material. SAYADHRI Int J Res 3:14–17
3. Kasturba AK et al (2005) Investigation of lateritic stones for building purpose from Malabar region, Kerala state, SW India—Part 1: field studies and profile characterisation. J Constr Build Mater 21:73–82. https://doi.org/10.1016/j.conbuildmat.2005.07.006
4. Gidigasu MD (1976) Laterite soil engineering—pedogenesis and engineering principles. Elsevier Scientific Publishing Company, Amsterdam
5. Varghese T, Bygu G (1993) Laterite soils—their distribution, characteristics and management. Technical monograph No. 1. State committee on science technology and environment, Government of Kerala, Thiruvananthapuram, India
6. Raychaudhuri SP (1981) The occurrence, distribution, classification and management of laterite and laterite soils. Cahiers ORSTOM. Serie Pedologie (France) 18
7. Lawane A et al (2011) Characterisation of laterite stone as building material in Burkino Faso: state of the art of the on-going research and its perspectives. Journees Scientifiques du 2iE
8. Benzy A et al (2016) Strength study on lateritic blocks. Int J Sci Technol Eng 2:338–341
9. Kasthurba AK et al (2006) Weathering forms and properties of laterite building stones used in historic monuments of western India. Struct Anal Hist Constructions, New Delhi
10. Nasheed M et al (2018) Case study on comparison between laterite stone and concrete block. Int J Latest Eng Manage Res (IJLEMR) 3:81–88

A Study on Aquaculture Waste Leachate Transport Through Soil

T. V. Nagaraju, B. M. Sunil, and Babloo Chaudhary

1 Introduction

Globally aquaculture practice gaining more potential and growing rapidly. In India, many parts of east and west coast aquaculture were expanding day by day [1]. Despite of great potential in terms of production and empowerment to local fellows, intensive farming posses many problems related to environmental concern and social consequences [1, 2]. Moreover, still many parts of the country not having proper guidelines about sustainable practice. Intensive farming and rapid expansion, characterized by large tanks with high density of shrimp, high rate of feed usage, usage of antibiotics, disinfectants, chemicals, and minerals, has led to formation of algae bloom, hypoxia, metal traces, greenhouse gas emission, and salinity [3–5].

The main challenge for farmers is to operate best ways to divert or treat aquaculture effluent, increase the rate of aerobic decomposition of aquaculture contaminate sludge mixed with soil, and to counteract eutrophication after crop period [6]. Many studies have suggested that aquaculture was playing vital role in the global market and developments in aquaculture industry. However, unfortunately, very limited studies were focused on environmental issues due to intensive aquaculture practices.

Recent years, 110 emerging contaminants (including antibiotics, pharmaceuticals, and industrial compounds) were witnessed in Tinan city, Tiwan. These contaminants lead to affect both human health and the ecosystem. So, there is a need of continuous monitoring of contaminated water metrics [7]. Moreover, intensive aquaculture practice results in out flow of large quantities of phosphorus (P) into the nearby irrigation and freshwater canals which stimulate algae bloom. So, this allows undesirable alterations in the freshwater bodies and possesses threat to human and animal health [8].

T. V. Nagaraju (✉) · B. M. Sunil · B. Chaudhary
Department of Civil Engineering, National Institute of Technology Karnataka, Mangaluru 575025, India
e-mail: varshith.varma@gmail.com

A. Sil et al. (eds.), *Recent Trends in Civil Engineering*, Lecture Notes in Civil Engineering 274, https://doi.org/10.1007/978-981-19-4055-2_39

Other hand, aquaculture intensity also significantly increased in terms of density of seed, usage of feed, minerals, and chemicals. In aquaculture ponds, feeding rate of shrimp depends on the previous feeding animal response. In general, the feeding rate was ranging from 30 to 80 kg/day/acre, due to excess or over feeding leads to formation of ammonia (NH_3). Ammonia accumulation increases in the aquaculture pond leads to toxic substance and eventually shrimp may die. Further, accumulation of ammonia could also pollute ground water [9]. Ammonium is a common substance discharge from the landfill leachate and other industrial wastewater [10, 11]. Many studies have been focused, particularly effect of ammonia from landfill on ground water.

Ammonia nitrogen (TAN) is common constituent found in most of the old landfill leachates. Most cases TAN ranges from 500 to 2000 mg/L and TAN exceeds 1000 mg/L leads to hinder the microorganism activity. Even ammonia presence in leachate in a long term is the most concern if it does not show any declined trend. Moreover, ammonia has remained as major source of ground water and subsoil contamination in old landfills [12, 13]. In aquaculture sludge, nitrogen compounds such as ammonia (NH_4^+), nitrite (NO_2^-), and nitrate (NO_3^-) are identified high concentration due to intensive farming. Further, which cause negative impact on human health and environmental deterioration [14]. Nevertheless, these nitrogenous compounds mainly ammonia (dominant fraction in aquaculture sludge) may influence the subsoil by ion exchange. Further affect the clay texture and hydraulic conductivity of the clay [15, 16].

This paper presents the physico-chemical properties of aquaculture leachate and the soil-leachate interaction. This paper also focuses on the effect of aquaculture leachate on environment.

2 Experimental Investigation

2.1 Materials

This study used expansive clay collected from Bhimavaram, India at a depth of 0.5 m below ground surface. The collected expansive clay was classified as high compressibility clay (CH) having fines 98%, liquid limit as 86%, and plastic limit as 23%. Aquaculture sludge samples were collected from bottom of the aquaculture ponds in various locations in delta region of Andhra Pradesh, India. Selection of samples was based on the intensity of farming and age of pond (1, 3, and 7 years). Leachate was collected from the aquaculture sludge using magnetic stirrer, characteristics of the leachate was tabulated in the Table 1.

P1, P2, and P3 are aquaculture ponds of age 1 year, 3 years, and 7 years, respectively.

Table 1 Physico-chemical characteristics of leachate used in this study

Property	Aquaculture ponds		
	P1	P2	P3
pH	8.2	7.8	9.1
Dissolved oxygen, DO (ppm)	4.0	3.5	4.6
Electrical conductivity, EC (μs/cm)	1045	8980	4520
Total dissolved solids, TDS (ppm)	560	2450	3020
Chemical oxygen demand, COD (ppm)	58	45	84
Biological oxygen demand, BOD (5 days) (ppm)	12.6	11.2	13.6
Ammonium, NH_4 (ppm)	2.42	1.5	3.55
Nitrate, NO_3 (ppm)	6.6	4.16	8.84
Nitrite, NO_2 (ppm)	0.48	0.33	0.17
Sulfate, SO_4 (ppm)	742	815	110
Sodium, Na (ppm)	164	160	214
Sulfur, S (kg/acre)	111	162	252
Potassium, K (kg/acre)	152	536	835
Phosphate, P (kg/acre)	420	460	484

2.2 *Percolation Experiment*

To understand the behavior of leachate-clay interaction, percolation experiments were conducted. The column leachate setup was fabricated in accordance with Jullien and Lecomte [17]. The expansive clay was pulverized and passed from 4.75 mm sieve was compacted in fabricated column. The compaction was done in three layers from the bottom of the column at their respective optimum moisture content (OMC) and maximum dry density (MDD). The column having dimensions of length 70 cm and diameter 10 cm. Filter papers with the support of porous sintered steel plates were kept both sides of the compacted clay to prevent the particles pore distribution during swelling.

Aquaculture waste sludge leachate collected after magnetic stirring process was continuously injected into the compacted soil column with constant flow rate (0.2 ml/sec). Hydraulic behavior of the clay and leachate clay was determined based on the Darcy's law ($k = Ql/Ah$, where Q is the discharge, A is the cross-sectional area of the column, h is the head, and l is the length of the compacted clay) [18]. The schematic representation of fabricated column setup is portrayed in Fig. 1. Moreover, leachate characteristics and hydraulic behavior was monitored and determined for 7, 30, 60, and 90-days.

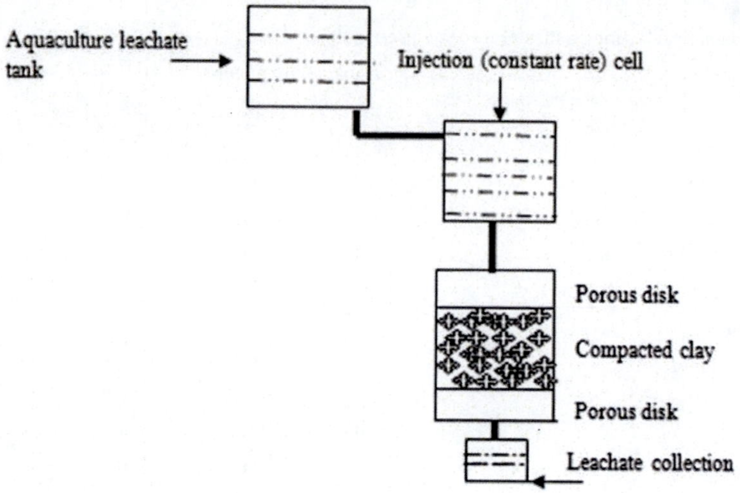

Fig. 1 Schematic of the column injected cell [17]

3 Discussion of Test Results

3.1 *Effect of Leachate on Index Properties of Clay*

Table 2 shows the test results of effect of leachate on clay properties. In this investigation, virgin expansive clay (C1), clay exposed with leachate for 7, 30, 60, and 90-days were C2, C3, C4, and C5, respectively, are considered. From test results, it is clearly evident that before and after exposure to aquaculture waste leachate, clay properties were drastically changed.

Index properties of clay such as liquid limit and plastic limit improved with increased aquaculture leachate exposure. Particle size distribution of the leachate exposed clay with curing shows improvement in particle size. This is due to the chemicals presence in the leachate reacts with the clay particles.

Aquaculture leachate exposure duration decreased the cation exchange capacity (CEC), this indicates the ion exchange takes place between the clay particles and leachate (NH_4^+ and K^+). Further, decreases the plasticity characteristics or binding property of the clay surfaces. According to few previous research reports, the sequence of the ion exchange is Al^{+3} > Ca^{+2} > Mg^{2+} > NH^{4+} > K^+ > Na^+ > Li^+ [18, 19]. However, in this study, when clay exposed with aquaculture leachate, only ammonium and potassium show significant variation compared with Ca^{+2} and Mg^{+2}.

Table 2 Effect of clay properties before and after exposure with aquaculture leachate

Property	Clay exposure with leachate				
	C1	C2	C3	C4	C5
Sand (%)	0.4	0.5	0.5	0.5	0.6
Silt (%)	14.4	11.2	12.4	22.4	27.4
Clay (%)	85.2	88.6	87.8	77.1	72.2
Liquid limit (%)	84	86	81	78	74
Plastic limit (%)	22	23	22	23	24
Plasticity Index (%)	62	63	59	55	50
Cation exchange capacity (CEC), (mEq/100 g)	50.4	47.6	40.4	36.7	28.6
Ammonium, NH_4 (mEq/100 g)	0.18	6.45	29.24	58.46	64.32
Sodium, Na (mEq/100 g)	0.78	15.8	27.63	46.42	51.12
Calcium, Ca (mEq/100 g)	54.21	55.5	47.56	48.22	49.12
Potassium, K (mEq/100 g)	3.16	17.6	22.45	28.43	35.14
Magnesium, Mg (mEq/100 g)	1.37	1.12	1.42	3.18	3.16

3.2 *Effect of Leachate on Hydraulic Behavior of Clay*

The percolation experimental data shows the effect of leachate on hydraulic behavior of the clays. Test results were summarized in the Table 3.

Leachate characteristics of P1, P2, and P3 were, P3 leachate was collected from the 7 years age pond with intensive farming. The hydraulic conductivity for the natural clay value of 0.42×10^{-7} cm/s. Leachate exposure increased significantly increased the hydraulic behavior of the clay exposed with aquaculture leachate. The hydraulic conductivity (k) of the clay sample exposed to leachate (1 year age tank) in Table 3, the increase is from 1.41×10^{-7} m/sec to 1.44×10^{-7} cm/sec for 3 to 7-days. This can be attributed due to the ion exchange between the clay particles and chemicals presence in the leachate. Further, due to ion exchange, interface layer improves, and surface forces were decreased. This result is in accordance with the landfill leachate clay interaction, NH^{4+} and K^+ cations had significant role in cation exchange process [20]. So, presence of chemical compounds (ammonia, potassium,

Table 3 Effect of leachate on hydraulic behavior of clays

Property	Leachate (based on age of tank)	Hydraulic conductivity of soil exposed with leachate (days)	
		3-days	7-days
Hydraulic conductivity, k (cm/s)	P1	1.41×10^{-6}	1.44×10^{-6}
	P2	1.55×10^{-6}	1.59×10^{-6}
	P3	1.51×10^{-6}	1.62×10^{-6}

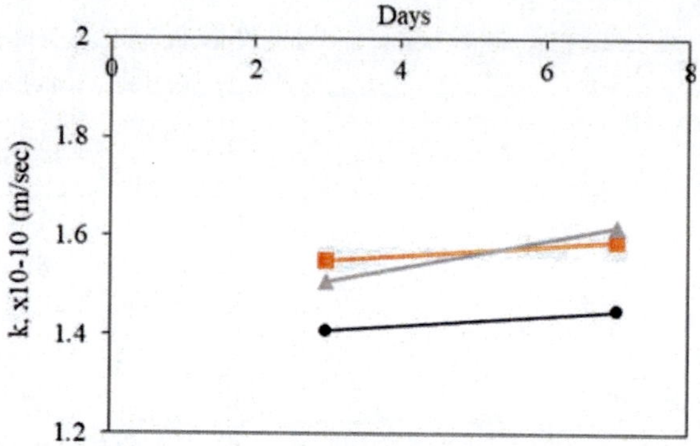

Fig. 2 Effect of leachate exposure duration on hydraulic conductivity (k)

and calcium) favors the percolation of leachate. However, due to improvement in hydraulic conductivity, groundwater pollution takes place (Fig. 2).

4 Conclusions

Aquaculture contains a wide range of pollutants: aquaculture solid waste, sludge in the bottom of the tank, and toxic substance (heavy metals, organic compounds, and chemicals). Where aquaculture contaminants enter the groundwater, significant changes in the subsoil, and water parameters have been demonstrated from the above study.

Physico-chemical characteristics of the aquaculture sludge leachate show the high concentration of ammonia and biological oxygen demand (BOD). pH of the long-term stagnated leachate exhibits high alkalinity (pH > 9).

Aquaculture leachate exposure leads to decrease in plasticity characteristics of clays and cation exchange capacity. Further, decreases the binding property and surface forces between the clays.

Hydraulic behavior of the expansive clay and clay interacted with aquaculture leachate shows significant negative impact on the permeability due to the ionic exchange. The clay particles reacted with chemicals or ammonia presence in the leachate shows higher permeability. Even, a long-term clay-leachate interaction exhibits hazardous environmental consequences.

References

1. Jayanthi M, Thirumurthy S, Muralidhar M, Ravichandran P (2018) Impact of shrimp aquaculture development on important ecosystems in India. Glob Environ Chang 52:10–21
2. Islam M, Yasmin R (2017) Impact of aquaculture and contemporary environmental issues in Bangladesh. Int J Fish Aquat Stud 5(4):100–107

3. Datta S (2012) Management of water quality in intensive aquaculture. Respiration 6:602
4. Cao L, Wang W, Yang Y, Yang C, Yuan Z, Xiong S, Diana J (2007) Environmental impact of aquaculture and countermeasures to aquaculture pollution in China. Environ Sci Pollut Res-Int 14(7):452–462
5. Bunting SW (2004) Wastewater aquaculture: perpetuating vulnerability or opportunity to enhance poor livelihoods. Aquat Resour Cult Dev 1(1):51–75
6. Gowen RJ (1994) Managing eutrophication associated with aquaculture development. J Appl Ichthyol 10(4):242–257
7. Lai WWP, Lin YC, Wang YH, Guo YL, Lin AYC (2018) Occurrence of emerging contaminants in aquaculture waters: cross-contamination between aquaculture systems and surrounding waters. Water Air Soil Pollut 229(8):1–12
8. Sugiura SH (2018) Phosphorus, aquaculture, and the environment. Rev Fish Sci Aquaculture 26(4):515–521
9. Wu D, Cheng M, Zhao S, Peng N, Hu R, Hu J, Liang Y (2020) Algal growth enhances light-mediated limitation of bacterial nitrification in an aquaculture system. Water Air Soil Pollut 231(2):1–9
10. Brennan RB, Clifford E, Devroedt C, Morrison L, Healy MG (2017) Treatment of landfill leachate in municipal wastewater treatment plants and impacts on effluent ammonium concentrations. J Environ Manage 188:64–72
11. Dos Santos HAP, de Castilhos Júnior AB, Nadaleti WC, Lourenço VA (2020) Ammonia recovery from air stripping process applied to landfill leachate treatment. Environ Sci Pollut Res 27(36):45108–45120
12. Mao X, Xiong L, Hu X, Yan Z, Wang L, Xu G (2018) Remediation of ammonia-contaminated groundwater in landfill sites with electrochemical reactive barriers: a bench scale study. Waste Manage 78:69–78
13. Cano V, Vich DV, Rousseau DP, Lens PN, Nolasco MA (2019) Influence of recirculation over COD and N-NH4 removals from landfill leachate by horizontal flow constructed treatment wetland. Int J Phytorem 21(10):998–1004
14. John EM, Krishnapriya K, Sankar TV (2020) Treatment of ammonia and nitrite in aquaculture wastewater by an assembled bacterial consortium. Aquaculture 526:735390
15. Gautier M, Muller F, Le Forestier L (2017) Interactions of ammonium-smectite with volatile organic compounds from leachates. Clay Miner 52(1):143–158
16. Sun WJ, Xu G, Wei G, Zhang WJ, Sun DA (2021) Effects of ammonium ion and bentonite content on permeability of bentonite-clay mixture. Environ Earth Sci 80(4):1–10
17. Jullien A, Proust C, Le Forestier L, Baillif P (2002) Hydro-chemio-mechanical coupling effects on permeability and swelling behaviour of a Ca smectite soaked by Cu solutions. Appl Clay Sci 21(3–4):143–153
18. Domenico PA, Schwartz FW (1990) Physical and chemical hydrogeology. Wiley, New York, United States of America
19. McBride MB (1994) Environmental chemistry of soils. Oxford University Press, New York, United States of America
20. Gautier M, Muller F, Le Forestier L, Beny JM, Guégan R (2010) NH4-smectite: characterization, hydration properties and hydro mechanical behaviour. Appl Clay Sci 49(3):247–254

Performance Analysis of Different Geosynthetic Reinforcements in Unpaved Roads of Soft Subgrade

Gautam, Sudeep Biswas, Monowar Hussain, and Debjit Bhowmik

1 Introduction

The load from unpaved roads is transferred to a base course layer, distributing the load evenly on the subgrade. The subgrade accepts the oncoming weight, whilst the aggregate layer distributes it. If the subgrade is of weak soil, there are chances of permanent deformation in the pavement in the long run due to incoming heavy traffic loads and other environmental factors. When rural unpaved roads are laid on soft or weak soil subgrade, the regular maintenance of such roads is not a very feasible option due to economic considerations, and it may disrupt the movement of traffic frequently. The conventional road construction process comprises replacing weak subgrades with a thick aggregate base course to facilitate sufficient support to the pavement surface. But the application of geosynthetics can considerably reduce the amount of aggregate required in the base layer. Barksdale et al. [1] observed that geosynthetic supports the load coming on the pavement by acting as a tension membrane. However, to function the effect, sufficient rutting depths are expected to develop in the pavement. The performance of reinforced roads is affected by geosynthetic material, deployment site, subgrade bearing strength, aggregate grading, etc. [2–4]. Perkins [5] found that when the vehicular load acts on the aggregate layer, the aggregates try to move out laterally. To prevent that, the aggregates have to be restrained by subgrade material or reinforcement. The combined action between the base layer and the geosynthetic provides a medium for transferring load through shearing action from the base layer to the geosynthetic. Geotextile effectively reduces the intrusion of soil mass into the void space present on the other side of the geotextile. Geotextile acts as an excellent separator if used at the interface of two different soil types [6]. Giroud et al. [7] found that flexible pavements without reinforcement have higher stress values when

Gautam (✉) · S. Biswas · M. Hussain · D. Bhowmik
Department of Civil Engineering, National Institute of Technology Silchar, Silchar, India
e-mail: gautam_rs@civil.nits.ac.in

A. Sil et al. (eds.), *Recent Trends in Civil Engineering*, Lecture Notes in Civil Engineering 274, https://doi.org/10.1007/978-981-19-4055-2_40

compared to pavements reinforced with geosynthetics. Geosynthetic is typically used between the subbase and the subgrade layer or flexible pavement's base layer.

The increment in pavement strength because of geosynthetic reinforcement is credited to mechanisms: lateral restraint, increase in bearing capacity, and tension membrane effect. Bergado et al. [8] performed modified CBR tests to account for the impact of geotextile on the California bearing ratio value. The use of geotextile was found to increase the field CBR value by 67–73%. Comparable findings were also concluded by Gautam et al. [9]. Kameel et al. [10] performed static and cyclic tests to acknowledge the advantage of geogrid reinforcement in the pavement. The outcomes emphasized that the CBR and the modulus of elasticity value increase with reinforcement. The maximum CBR value was achieved when reinforcement was provided at a depth of 72–76% from the top of the specimen. Latha and Nair [11] carried out model studies and field studies on unreinforced and reinforced unpaved road sections constructed on the weak subgrade. They placed geotextile, geogrid, and geocell between subgrade and base course. The load-bearing potential of the reinforced section increased, and there was a subsequent decrease in rut depth. Geosynthetics were found effective, and geocell imparted cushioning effect, resulting in reduced rutting and surface heave. Murad et al. [12, 13] studied the performance of geogrid and geotextile stabilized paved and unpaved roads upon the application of cyclic loads. The reinforced road bearing capacity and elastic modulus increase when geosynthetics are grouped as per the aperture shape and manufacturing polymer material. The results showed a decrease in rut depth in pavements. It was also found that geosynthetics installed at the interface of the base and subgrade performed better than those installed in the base-aggregate layer. The initial cost of buying and placing geosynthetic in pavements can be more, but geosynthetic reinforcement's overall impact has various economic advantages. Moreover, geosynthetic reinforcement greatly decreases the thickness of the base layer. Pameria et al. [14] performed extensive scale tests and concluded that geosynthetic reinforcement could be very cost-effective as there is a significant reduction in the maintenance of roads. Various researchers have tried fibres and naturally available textiles, such as coir, jute, and bamboo, and used them for reinforcement. Bera et al. [15] used jute textiles and fly ash to enhance the soil strength. The compressive strength of soil improved by 525% when reinforced with four layers of jute geotextiles after treating with fly ash.

In the present work, the performance of different combinations of geosynthetic reinforcement viz. woven and nonwoven geotextiles and geogrid in soft soil-aggregate system has been observed through model plate load tests. In addition, the change in load resisting capacity and cross-section profile of soil-aggregate systems corresponding to different geosynthetic combinations has been studied.

2 Materials and Test Methods Applied

The soil and aggregates locally available in Silchar, northeastern India, are considered. ASTM D 4318 [16] is used to evaluate soil index parameters. The soil's liquid

Table 1 Specifications of geosynthetics

Particulars	Nonwoven geotextile	Woven geotextile	Geogrid
Mass per unit area (gsm)	250	240	190
Thickness (mm)	1.5	1	1.5
Tensile strength (kN/mm)	15	45	27.2
Elongation (%)	50	25	16
Apparent opening size (micron)	80	75	–
Mesh size (mm × mm)	–	–	10 × 10

and plastic limits are found to be 55% and 29%, respectively. According to USCS, it is considered to be clayey soil with low compressibility (CL). The test soil's optimal moisture content and maximum dry unit weight are calculated to be 23% and 14.61 kN/m^3, respectively. In the laboratory, the impact, crushing, and abrasion values of aggregates were determined to be 28.4%, 27.3%, and 28.3%, respectively. High-density polythene biaxial geogrid, nonwoven, and woven geotextile are used for the reinforcement. Table 1 presents the specifications of the geosynthetic considered for the present work.

The northeastern part of India is one of the wettest parts of India as it receives heavy rainfall throughout the year. Rainwater seeps into the subgrade of the unpaved roads, making it very soft and reducing the subgrade's load-bearing strength. A CBR test of the subgrade used in rural roads is required under IRC-SP 20 [17], and before construction begins, it is crucial to assess the critical water content to which the pavement will be subjected. Therefore, during the rainy season at the NIT Silchar campus, regularly, a set of soil samples was collected. The average water content (w %) of the soil during this period was found to be around 35%. As a result, the soil sample for the CBR test was prepared corresponding to this moisture content. The soil with 35% moisture content had a CBR value of 1.9%, suggesting it to be a highly moist and soft subgrade. A model plate load test is conducted in a steel tank measuring 1 m × 1 m × 1 m. The soil and the granular base layer were filled up to 600 mm and 210 mm in height of the tank, respectively. Before adding to the tank, the clayey soil was sun-dried, and initial moisture content was noted. The additional amount of water required to bring the moisture content of the soil to 35% was added, and the soil was rigorously mixed. The mixed soil was then wrapped and packed with plastic sheets for seven days to attain moisture equilibrium. To attain the appropriate density, a calculated amount of the soil was deposited in five layers and compacted with a rammer in the tank. Figure 1 shows the schematic arrangement of the experiment setup. Table 2 presents the different soil-aggregate systems evaluated during the test.

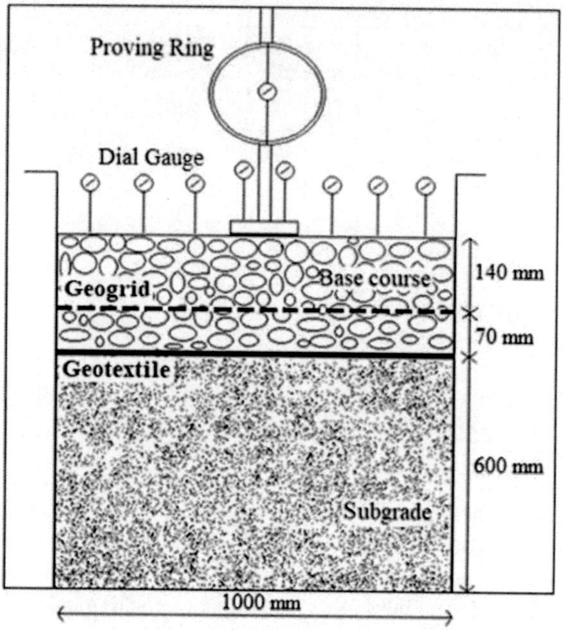

Fig. 1 Schematic diagram of the test arrangement

Table 2 Different test combinations

Nomenclature	Test combinations
S'	Soil ($w = 23\%$, i.e. optimum moisture content)
S	Soil ($w = 35\%$)
SA	Unreinforced soil-aggregate system
SANWG	Nonwoven geotextile-reinforced soil-aggregate system
SAWG	Woven geotextile-reinforced soil-aggregate system
SAG	Geogrid-reinforced soil-aggregate system
SAGNWG	Geogrid and nonwoven geotextile-reinforced soil-aggregate system
SAGWG	Geogrid and woven geotextile-reinforced soil-aggregate system

Over the subgrade layer, the aggregate layer was placed and compacted to the relative density of 80% in three layers with water content at 4.8%. The base layer was prepared using aggregates of different sizes applying the Fuller and Thompson Equation (1907) to achieve a good packing density. The basic equation is

$$P = 100\left(\frac{d}{D}\right)^n$$

Here,

Table 3 Grading of aggregates for the base course

Aggregate size (mm)	Weight passing the sieve (%)
25–20	10.56
20–16	9.44
16–13.2	7.34
13.2–10	9.42
10–6	14.26
6–4.75	5.4
4.75–2.36	12.86
2.36–1.18	8.99
1.18–0.425	8.69
0.425–0.075	7.56
0.075–pan	5.47

P % aggregate finer than the sieve
D maximum aggregate size to be used
d aggregate size being considered
n parameter for maximum aggregate density, n = 0.5

Table 3 shows the gradation of aggregates used in the base layer as per the Fuller and Thompson equation.

During reinforcing, geosynthetic was kept at the subgrade-base layer interface. However, when the arrangement of geogrid along with geotextile was adopted together for reinforcement, the geotextile was kept at the interface, whereas the geogrid was installed 70 mm above the interface in the aggregate layer (i.e. about one-third of the thickness of the base layer). The reason to not use geogrid immediately with geotextile was to make provision for interlocking of aggregate and avoid slipping of geogrid over the geotextile [18]. The load on the prepared system was exerted with the help of a footing having 150 mm diameter (D) and 20 mm thickness by a hydraulic jack. The magnitude of the applied load was assessed using a proving ring, and the reaction of prepared arrangements corresponding to the applied load was monitored in eight dial gauges of 0.01 accuracy, four located on either side of the footing at distances 0.5D, D, 2D, and 3D as shown in Fig. 2. During loading, footing settlements (s) of about 30% of D were measured using dial gauges, and loads on the proving ring were recorded at equal intervals.

3 Result and Discussion

In this section, the observations drawn from the different experimental arrangements are presented. Load-settlement graphs are plotted to evaluate the impact of geosynthetic and variation in load resisting capacity of the different test arrangements upon the reinforcement with geosynthetic. The profile of the systems corresponding to

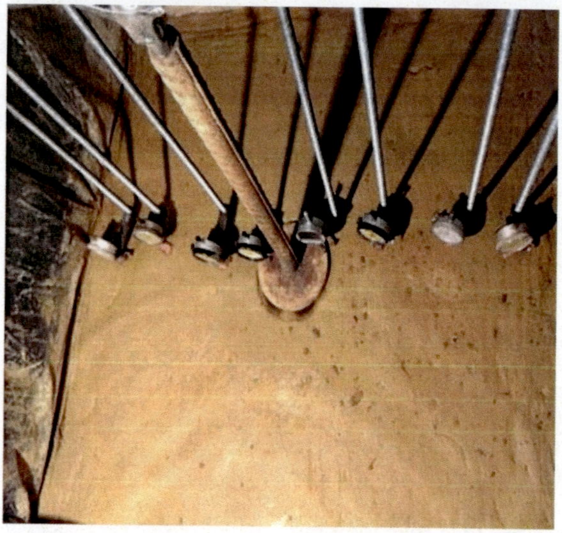

Fig. 2 Placement of dial gauges in tank

different applied loads is plotted to understand the effect of settlement and heaving of the subgrade and base layer under the influence of the coming load. For unreinforced and differently reinforced systems, the observed values of the proving ring and eight dial gauges are plotted into a bearing pressure and footing settlement graph. With the use of aggregate over the soil subgrade, the bearing capacity increased effectively, which can be attributed to the mechanism of uniform load distribution over the subgrade through the aggregate layer. Figures 3 and 4 show the settlement of the footing in the subgrade and base layer, respectively. The results obtained depict that the use of geosynthetics enhances the load-bearing potential of the pavement. The subgrade and base layer can take more load after reinforcement corresponding to a particular settlement value, as shown in Fig. 5. The geocomposite comprising of geogrid and woven geotextile demonstrated the greatest increase in load-carrying potential. Nonwoven geotextile reinforcing had little impact on the test system's strength and is not recommended for use in unpaved roads, but the previous research works have shown that nonwoven geotextile plays an important role in drainage criteria. Hence, nonwoven geotextile can be used to restrict soil erosion in waterlogged areas. As the intermixing of aggregates is eliminated, geotextile proved to be an excellent separator. The interlocking of aggregate occurred with the use of geogrid, which ultimately increased the strength of the test system.

Despite the aggregates interlocking in the geogrid, there was some diffusion of aggregates into the soil layer and vice versa. Aggregates smaller than the mesh size of geogrid were found to penetrate in the subgrade. Geocomposite exhibited both the functions as geogrid provided reinforcement and geotextile acted as a separator. The reinforcement effect became more pronounced as settlement depth increased. The tensile effect of geosynthetic provided an upward component of force, increasing resistance to the approaching load. The improvement factor was obtained by dividing

Fig. 3 Settlement of footing in the soil bed

Fig. 4 Settlement of footing in the base layer

the load-bearing capacity of the reinforced system by that of the unreinforced soil-aggregate system for a particular settlement value. Figure 6 shows the improvement factor (IF) for the plate load tests calculated corresponding to different settlement-footing diameter ratios (s/D (%)) to account for the improvement in the strength using different reinforcing materials over the unreinforced soil-aggregate system. The maximum increment in strength is seen in geocomposite consisting of geogrid and woven geotextile with IF value equals 2.27 at s/D value of 30%.

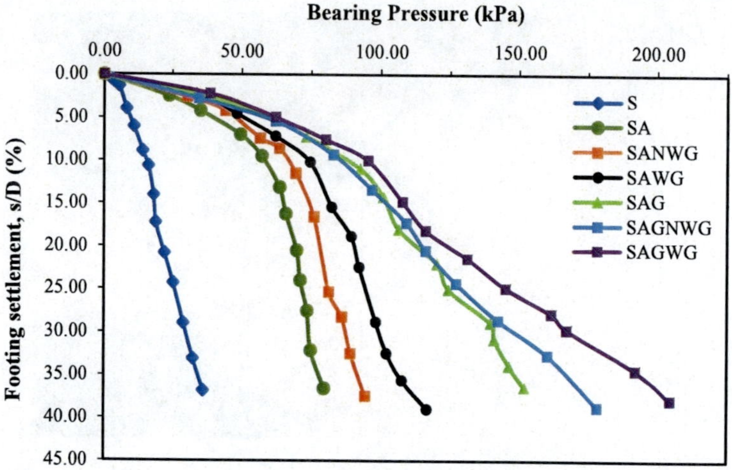

Fig. 5 Bearing pressure-footing settlement response of unreinforced and reinforced systems

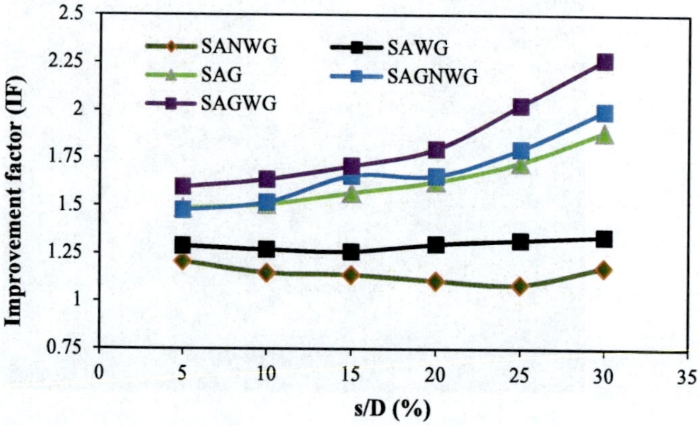

Fig. 6 Improvement factor for different reinforcement systems at different levels of settlement

4 Evaluation of Cross-Section Profile

The cross-section profiles for all the different systems were plotted for the data obtained from the eight dial gauges installed during the test. The geosynthetic distributes the load uniformly to the larger area as the heaving effect is observed (refer to Fig. 7, 8, 9, 10, 11, 12 and 13). Large settlement occurred immediately below the footing, and there is very nominal heaving for the unreinforced systems. The heaving effect is observed more in geogrid and geocomposites so was the load-carrying capacity for a particular range of settlement. All the reinforced systems could take higher pressure for a particular settlement than the unreinforced system. The

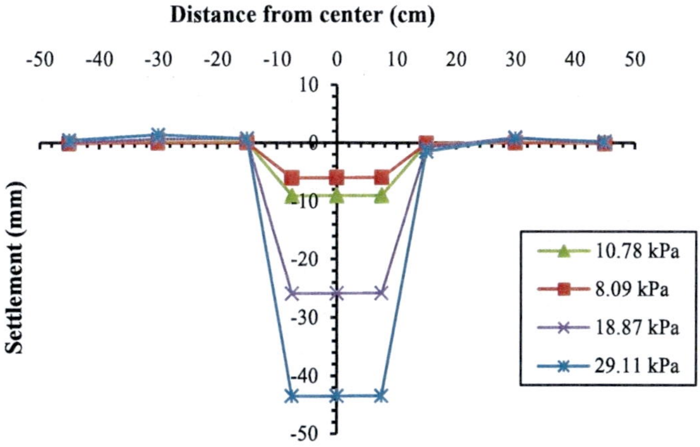

Fig. 7 Cross-section profile for unreinforced subgrade soil (S)

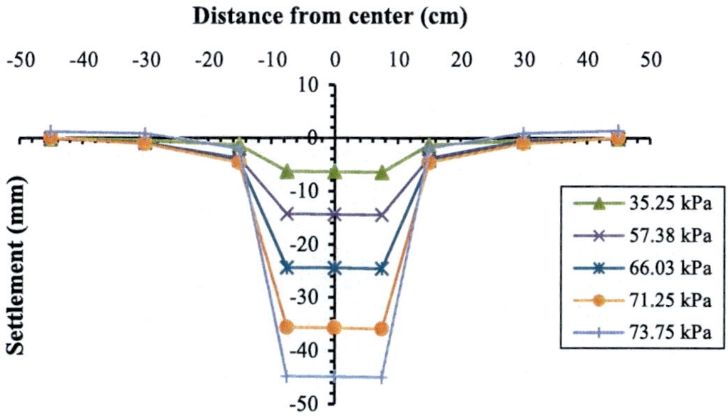

Fig. 8 Cross-section profile of SA system

load-carrying capacity improved with an increase in the tensile strength of geosynthetics. At larger deformation of the geosynthetic under the applied load, the tensile effect became more profound. It is observed that the geosynthetics provide proper support to the base layer and distributions of the load evenly. The mode of failure of the soil-aggregate system changed from punching to general with the inclusion of reinforcing materials.

5 Conclusions

From the outcomes of the present work, the conclusions derived are:

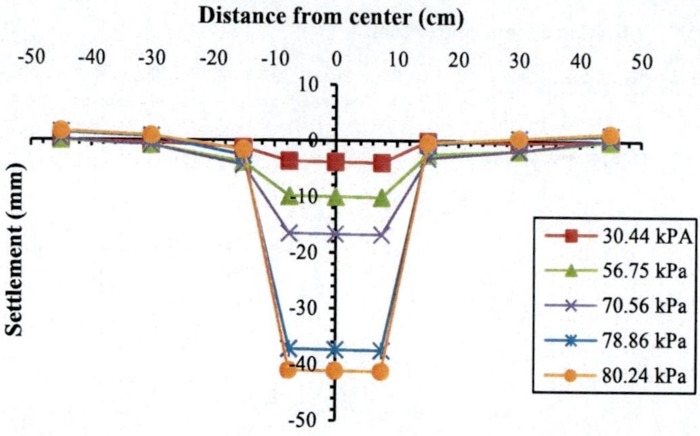

Fig. 9 Cross-section profile of SANWG system

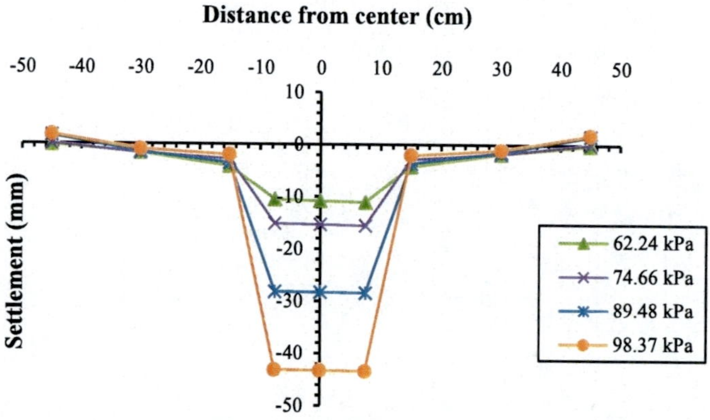

Fig. 10 Cross-section profile of SAWG system

- With the introduction of geosynthetic reinforcement, the performance of the soil-aggregate system was enhanced.
- The geotextile isolates the subgrade from the base layer, preventing the movement of aggregates. However, the application of nonwoven geotextile did not improve the system's strength significantly.
- The interlocking of aggregates in the geogrid aperture boosted the strength and allowed it to withstand higher pressure. Though geogrid restricted the movement of larger aggregates but some extent of intermixing of aggregates smaller than its mesh occurred size between the base layer and the subgrade.
- In the model plate load test, the geocomposite comprising of geogrid and woven geotextile exhibited the highest increase in strength of the test system, with an improvement factor (IF) of 2.27 at s/D value of 30%.

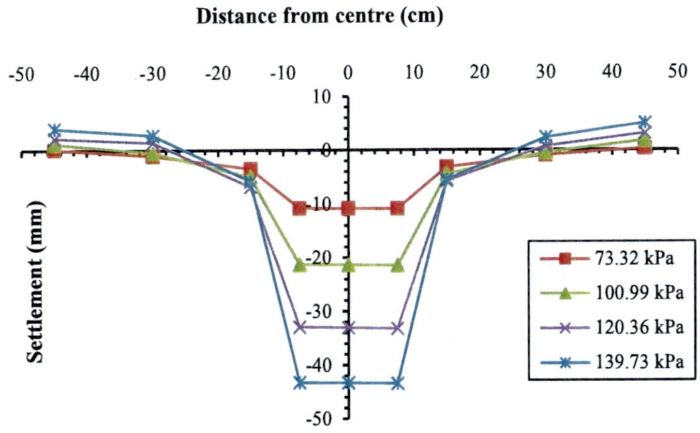

Fig. 11 Cross-section profile of SAG system

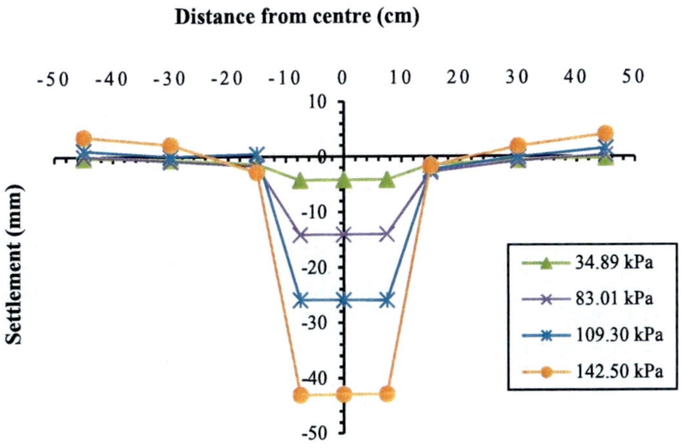

Fig. 12 Cross-section profile of SAGNWG system

- The progressive increase in the improvement factor with the increase in footing settlement can be attributed to the introduction of the tensile effect of geosynthetic that becomes profound with higher deformation.

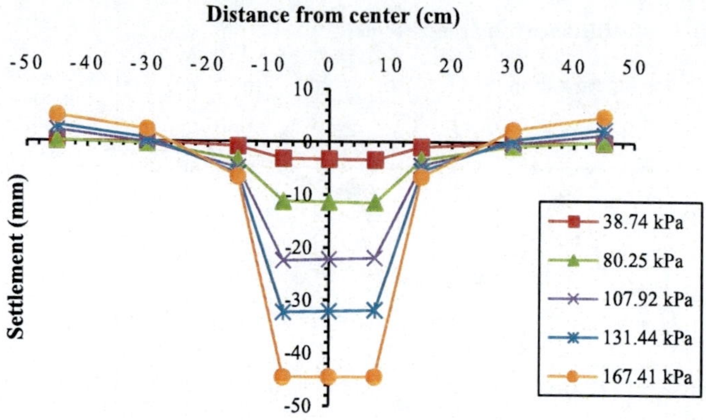

Fig. 13 Cross-section profile of SAGWG system

References

1. Barksdale RD, Baschus RC (1983) Design and construction of stone columns, Report No. FWHA/RD-83/026 federal highway administration office of engineering and highway operations research and development Washington DC
2. Ismail I, Raymond GP (1995) Geosynthetic reinforcement of granular layered soils. Geosynthetics '95, Nashville, TN, USA, IFAI, St. Paul, MN, USA 1:317–330
3. Broms BB (1977) Triaxial tests with fabric-reinforced soil. CR Coll Int Soils Textiles, pp 129–133
4. Kinney TC, Abbott J, Schuler J (1998) Benefits of using geogrids for base reinforcement with regard to rutting. Transp Res Rec 1611(1):86–96
5. Perkins S (1999) Geosynthetic reinforcement of flexible pavements: laboratory based pavement test sections. Report No. FHWA/MT-99-001/8138, US Department of Transportation, Federal Highway Administration, Washington DC, USA
6. Rowe RK (2005) Long-term performance of contaminant barrier systems. Geotechnique 55(9):631–678
7. Giroud JP, Noiray L (1981) Geotextile reinforced unpaved road design. J Geotech Eng ASCE 107(9):1223–1254
8. Bergado DT, Youwai S, Hai CN, Voottipruex P (2001) Interaction of nonwoven needle-punched geotextile under axisymmetric loading conditions. Geotext Geomembr 19(5):299–328
9. Hussain M, Bhowmik D (2022) Effect of geosynthetic reinforcement on CBR strength of soft soil-aggregate system. In: Ground improvement and reinforced soil structures, Springer, Singapore, pp 685–696
10. Kameel MA, Chandra S, Kumar P Behaviour of subgrade soil reinforced with geogrid. Int J Pavement Eng 5(4):201–209
11. Latha GM, Nair AM (2014) Geosynthetics in unpaved roads. Indian J Geosynthetics Ground Improv 3(2)
12. Abu-Farsakh M, Hanandeh S, Mohammad L, Chen Q (2016) Performance of geosynthetic reinforced/stabilised paved roads built over soft soil under cyclic plate loads. Geotext Geomembr 44(6):845–853
13. Abu-Farsakh MY, Gu J, Voyiadjis GZ, Chen Q (2014) Mechanistic–empirical analysis of the results of finite element analysis on flexible pavement with geogrid base reinforcement. Int J Pavement Eng 15(9):786–798
14. Palmeira EM, Antunes LGS (2010) Large scale tests on geosynthetic reinforced unpaved roads subjected to surface maintenance. Geotext Geomembr 28:547–558

15. Bera AK, Chandra SN, Ghosh A, Ghosh A (2009) Unconfined compressive strength of fly ash reinforced with jute geotextiles. Geotext Geomembr 27(5):391–398
16. ASTM D4318 (2005) Standard test methods for liquid limit, plastic limit, and plasticity index of soils. ASTM International, West Conshohocken, PA
17. IRC-SP-20 (2002) Rural roads manual. Indian Road Congress
18. Hufenus R, Rueegger R, Banjac R, Mayor P, Springman SM, Bronnimann R (2006) Full-scale field tests on geosynthetic reinforced unpaved roads on soft subgrade. Geotext Geomembr 24:21–37

The June 2018 Karinchola Landslide, Kerala State, South India

Nirmala Vasudevan, Kaushik Ramanathan, and Anunandh Gopakumar

1 Introduction

Landslides are downslope movements of masses of earth, debris, or rock. They often have several causative factors, such as intense and prolonged rainfall, earthquakes, volcanic activity, changes in groundwater levels, deforestation, and indiscriminate construction [1, 2]. Landslides can have catastrophic results, including loss of life, livestock, infrastructure, vegetation, and habitat. During the period 1998–2017, around 4.8 million people were affected by landslides with over 56,000 deaths [3, 4].

In India, landslides occur in the Himalayas, the northeastern mountains, the Western Ghats, and the Eastern Ghats [5–7]. Many of these slides are triggered by rainfall but are caused, at least in part, by ill-planned and poorly executed human activity. In fact, recent studies report that human activities are increasingly responsible for landslide occurrences the world over [4, 8]. In this paper, we examine possible causes—both natural and human-induced—of the June 2018 Karinchola landslide in the Western Ghats mountain range and suggest mitigation measures.

N. Vasudevan (✉)
Amrita Center for Wireless Networks & Applications, Amrita School of Engineering, Amrita Vishwa Vidyapeetham, Amritapuri, India
e-mail: nirmalav@am.amrita.edu

N. Vasudevan · A. Gopakumar
Department of Physics, Amrita School of Arts & Sciences, Amrita Vishwa Vidyapeetham, Amritapuri, India

K. Ramanathan
Department of Civil Engineering, Amrita School of Engineering, Amrita Vishwa Vidyapeetham, Coimbatore, India
e-mail: rkaushik@am.amrita.edu

A. Sil et al. (eds.), *Recent Trends in Civil Engineering*, Lecture Notes in Civil Engineering 274, https://doi.org/10.1007/978-981-19-4055-2_41

2 Study Area

Karinchola locality lies in Kattippara Panchayat[1], Kozhikode District, Kerala State, South India (Fig. 1). The landslide occurred on a south-facing slope of a 2-km long east–west trending ridge called Karincholamala (Fig. 2). The slide thus has been variously called the Kattippara slide, the Karinchola slide, or the Karincholamala slide [10]. The landslide (11°28'14.88" N, 75°55'16.68" E) initiated at an elevation of about 320 m, approximately 140 m below the peak of the slope (Fig. 2).

Fig. 1 **a** India, highlighting Kerala State, **b** Kerala district map showing Kozhikode and other districts, **c** Kattippara Panchayat and Karinchola landslide in Kozhikode, **d** Western Ghats mountain range along the western coast of India

[1] The word Kattippara means "hard rock" in the local language, Malayalam; the region comprises many hills and hard rock is seen all around.

A Panchayat is an administrative unit consisting of a few villages.

Fig. 2 Karinchola slide on the east–west trending Karincholamala ridge [9]

Kattippara is characterized by denudation hills[2] and valley flats[3] [11, 12]. The predominant rock type in the region is hornblende-biotite gneiss of an Archean migmatite complex [12, 13].

Vegetation is sparse in the vicinity of the slide and rock faces are visible in some places. Coconut plantations began to be cultivated on the lower reaches of the slope nearly 40 years ago; rubber is the other predominant plantation crop in and around Karinchola.

Kozhikode District receives approximately 3,500 mm of annual rainfall, of which around 2,500 mm falls during the main monsoon season from June to September. During this season, moisture-laden winds from the southwest are intercepted by the Western Ghats, bringing rain to the region. The southwest monsoon is followed almost immediately by the northeast monsoon from October to December. Approximately 450 mm of rain falls during the northeast monsoon, very little falls during January and February, and pre-monsoon showers occur during the summer months from March to May [14–17].

The mean monthly temperature in Kozhikode District varies from about 20 °C in December to about 30 °C in April [17].

3 The June 2018 Extreme Rainfall Event

In 2018, Kerala experienced exceptionally heavy rainfall. Both the 2018 and the mean monthly rainfall data for Kozhikode District are provided in Table 1, while the 2018 and the mean weekly rainfall data during a few select weeks for a few select districts (Idukki, Palakkad, Malappuram, Kozhikode, Wayanad, and Kannur, refer Fig. 1b) are provided in Table 2.

[2] Hills formed due to erosion/weathering.

[3] The low, flat land bordering the channel of a stream (Merriam-Webster Dictionary).

Table 1 Monthly rainfall (mm) for Kozhikode District [14]

	Jan	Feb	Mar	Apr	May	Jun	Jul	Aug	Sep	Oct	Nov	Dec
2018	7.7	0.9	36.9	79.0	563.0	1081.8	1037.4	836.0	29.3	267.8	57.5	35.0
Mean	2.3	4.1	15.1	80.6	257.1	894.0	951.7	500.6	244.2	262.5	130.7	29.9

Table 2 Weekly rainfall (mm) for a few select districts [18]

		Pre-monsoon[a]		Southwest monsoon[a]	
		10–16 May	24–30 May	7–13 June	14–20 June
Idukki	2018	79.8	148.5	466.1	134.9
	Mean	45.9	65.2	147.2	151.0
Palakkad	2018	89.6	115.8	295.3	196.2
	Mean	32.3	40.4	97.8	116.6
Malappuram	2018	60.6	183.4	288.6	301.3
	Mean	52.8	58.6	147.5	165.7
Kozhikode	2018	70.7	282.1	355.7	311.2
	Mean	55.8	89.1	206.5	219.7
Wayanad	2018	53.2	121.0	397.6	236.7
	Mean	37.1	52.6	138.3	161.8
Kannur	2018	60.6	208.3	332.1	255.5
	Mean	47.5	83.8	176.4	216.3

[a] The 2018 southwest monsoon season commenced on 29 May [19]

At least 10 landslides occurred in Idukki District from 9 to 11 June [20], 4 in Palakkad from 12 to 13 June [21], 9 in Malappuram from 14 to 15 June [21], 27 in Kozhikode from 12 to 14 June [22], 8 in Wayanad on 14 June [22], and 1 in Kannur on 12 June [22]. Table 2 shows that there was heavy rainfall in the weeks leading up to the slides. Landslides were also reported in neighboring districts in the states of Karnataka and Tamil Nadu [23, 24].

4 The June 2018 Karinchola Landslide

A rain gauge at Thamarassery (a town 6 km south of Karinchola) recorded a rainfall of 242 mm in the 18-hour period from 11:00 am, 13 June, to 4:00 am, 14 June [25]; whereas, a gauge at Vythiri (a town some 16 km northeast of Karinchola) recorded a rainfall of 152 mm in the 24-hour period from 13 to 14 June (personal communication).

Several landslides occurred in and around Kattippara on 14 June 2018. At around 3:00 am, a landslide occurred at Vettiozhinja Thottam, nearly 700 m south of the Karinchola slide. Shortly after this event, another one occurred at Kalvari, around

1 km east of the Karinchola slide. Six goats died in the Vettiozhinja Thottam slide, but there was no loss of human life in either landslide [22].

The Karinchola slide occurred at around 5:00–6:00 am. Investigations carried out shortly after the event (29 June 2018) revealed that the landslide initiated as a translational debris slide at the interface between the overburden and the underlying rock before transforming into a debris flow with multiple runout paths [22]. Fourteen people died and a few others sustained injuries. Five houses were washed away and three houses and coconut plantations were damaged. A road located at the base of the slope, a random rubble (RR) masonry retaining wall, culverts, electric poles, electric transmission lines, and two electric transformers located along the road were all washed away [22]. Some details of the landslide are provided in Table 3.

After the slide, residents were relocated, work in the plantations was stopped, and most construction activity was prohibited. However, people continue to visit their plantations to collect coconuts and other produce. Therefore, the slope material was partially removed and the construction of a 6-m wide concrete road began in 2019.

Table 3 Landslide parameters

Parameter	Details
Slide name, location	Karinchola landslide, Poonoor–Kattippara road
District, state	Kozhikode, Kerala
Toposheet number	49 M/15
Latitude, longitude, elevation	11°28′14.88″ N, 75°55′16.68″ E, 320 m
Date and time of occurrence	14 June 2018, 5:00–6:00 am
Failure mechanism	Rainfall-triggered planar failure at overburden-rock interface
Type of landslide	Translational debris slide–debris flow
Rate of movement	Very rapid to extremely rapid
Activity	Active
Depth	Shallow ($\leq$ 5 m) [22]
Height, length	~ 240 m, ~ 650 m (estimated using Google Earth [9])
Area	~ 10,000 m^2 (estimated using Google Earth)
Average slope	39° (estimated using Google Earth)
Geotechnical properties	Bulk density, moisture content, particle-size distribution (sieve analysis), liquid and plastic limits, and direct shear test results provided in [10]

5 Discussion

5.1 *Landslide Causes*

Rainfall was the trigger as well as one of the main causes of the landslide. There was excessive rainfall in the weeks leading up to the slide, resulting in an increase in the soil pore water pressure and a consequent decrease in the cohesive strength of the soil.

In addition to rainfall, construction activities also were responsible for destabilizing the slope. The slope was cut vertically for the construction of rural roads. Some months before the landslide, a JCB and other small equipment were used to excavate the slope and construct a water storage tank on the upper reaches of the hill [26]. While the excavation of the slope would have destabilized it to some extent, the storage dam also contributed to the landslide—during the slide, a dislodged boulder rolled into the tank and displaced water onto an already saturated slope.

The large-scale clearing of natural vegetation for coconut and rubber plantations also would have disturbed the soil, making it loose and reducing its shear strength. However, it is possible that the natural balance has been at least partially restored over time.

5.2 *Possible Mitigation Measures*

The Karinchola landslide left behind a considerable amount of unconsolidated slope material in its wake, which may be mobilized when heavy rains strike the region. The material must be cleared to minimize landslide risk. Endemic vegetation should be planted to strengthen the soil cover. A thorough study of the drainage must be carried out to ensure that no natural waterways are obstructed leading to increase in pore pressure and reduction in shear strength.

Most importantly, the locals should be educated about the impact of human activity on landslide susceptibility. An RR retaining wall had been in place before the 2018 landslide, suggesting that there were signs of instability prior to the event. There had also been landslides in the vicinity of Karinchola. Even so, people were engaging in activity that was clearly detrimental to slope stability. Now after the landslide, a concrete road is being constructed and slope material has been removed for this purpose. We need to ensure that the recent activities do not increase landslide risk.

6 Conclusion

Rainfall and indiscriminate construction were perhaps the major causes of the 2018 Karinchola landslide. The landslide left unconsolidated material on the slope, and there is a possibility of this material failing in future. We are hopeful that once the situation caused by the COVID-19 pandemic eases, we will be able to visit the site, carry out more detailed investigations, and suggest further remedial measures if needed.

Acknowledgements We gratefully acknowledge the Chancellor of our University, Sri Mata Amritanandamayi Devi, for Her help and support, Mr. Kevin (Rudran) Degnan for reading our manuscript with great care and providing thoughtful suggestions, Mr. Ranjith N. Sasidharan for preparing the images used in Fig. 1, and the Kattippara-Karincholamala Rehabilitation Committee for sharing a photograph taken soon after the occurrence of the landslide (shown in Fig. 2).

References

1. Cruden DM, Varnes DJ (1996) Landslide types and processes. Special Report, Transp Res Board, Natl Acad Sci 247:36–75
2. Ramkrishnan R, Karthik V, Unnithan MS, Balaji RK, Vinu AM, Venugopalan A (2017) Stabilization of seepage induced soil mass movements using sand drains. Geotech Eng J SEAGS & AGSSEA 48(4):129–137
3. Center for Research on the Epidemiology of Disasters (CRED) (2018) Economic losses, poverty & disasters 1998–2017. United Nations Office for Disaster Risk Reduction
4. Froude MJ, Petley DN (2018) Global fatal landslide occurrence from 2004 to 2016. Nat Hazards Earth Syst Sci 18(8):2161–2181
5. National Disaster Management Authority, Government of India (2019) National Disaster Management Plan, 2019. New Delhi
6. Vasudevan N, Ramanathan K (2015) Geotechnical characterization of a few landslide-prone sites in India. In: The sixth international geotechnical symposium on disaster mitigation in special geoenvironmental conditions, 509–512. Indian Geotechnical Society, Chennai Chapter
7. Vasudevan N, Ramanathan K (2016) Geological factors contributing to landslides: case studies of a few landslides in different regions of India. In: IOP conference series: earth and environmental science 30(1): 012011. IOP Publishing
8. Ramkrishnan R, Sharma A, Viswanathan K, Ravichandran D (2020) Determination of critical phreatic level for landslide assessment using a scaled down laboratory model. Intl J Geotech Eng 23:1–7
9. Image of Karincholamala ridge. Google Earth, earth.google.com/web/
10. Aswathi P, Rangaswamy K (2020) Stability analysis of Kattipara and Meppady regions of hill soil slope. In: Local site effects and ground failures: select proceedings of the seventh ICRAGEE, 201–216. Springer Nature, Singapore
11. Geological Survey of India (GSI) (2002) Geomorphological map of India
12. Department of Mining and Geology, Government of Kerala (2016) District Survey Report of Minor Minerals (except river sand), Kozhikode District, 8–10. Government of Kerala
13. Geological Survey of India (GSI) (2005) District resource map of Kozhikode
14. Hydromet Division, India Meteorological Department (IMD), Customized Rainfall Information System (CRIS). Ministry of Earth Sciences, Government of India. http://hydro.imd.gov.in/hydrometweb/(S(zstahg451ibwwiu1izxoi555))/DistrictRaifall.aspx. Last accessed 20 May 2021
15. India Meteorological Department (IMD). https://mausam.imd.gov.in/thiruvananthapuram/other_services.php/. Last accessed 20 May 2021
16. Ramesh MV, Vasudevan N (2012) The deployment of deep-earth sensor probes for landslide detection. Landslides 9(4):457–474

17. Warrier UC, Babu PM, Sudheesh M, Deshpande RD (2016) Studies on stable isotopic composition of daily rainfall from Kozhikode, Kerala, India. ISOT Environ Health Stud 52(3):219–230
18. India Meteorological Department (IMD). https://mausam.imd.gov.in/Thiruvananthapuram/mcdata/wwr2.pdf. Last accessed 20 May 2021
19. India Meteorological Department (IMD) (2018) End of season report for the 2018 south-west monsoon. https://reliefweb.int/sites/reliefweb.int/files/resources/20181102_pr_351.pdf. Last accessed 20 May 2021
20. Sulal NL, Archana KG (2019) Note on post disaster studies for landslides occurred in June 2018 at Idukki District, Kerala. Geological Survey of India (GSI), Thiruvananthapuram
21. Sachin R, Vishnu CS (2019) Note on post disaster landslide studies in Malappuram and Palakkad districts, Kerala during June 2018. Geological Survey of India (GSI), Thiruvananthapuram
22. Praveen KR, Kumar TN, Vishnu CS (2019) Note on post disaster studies for landslides occurred in June 2018 at Kozhikode, Wayanad and Kannur districts, Kerala. Geological Survey of India (GSI), Thiruvananthapuram
23. Maruthi KV, Bhat AA, Srivastava AK (2018) Report on post disaster investigation of landslide occurrences along Madikeri-Makutta and Madikeri-Koyanad-Sulya road sections, Kodagu District, Karnataka. Geological Survey of India (GSI), Bengaluru
24. Press Trust of India (PTI) (2018) Major landslide on CBE-Siruvani dam road; No Casualties
25. D'Souza C, IMD hourly precipitation data. https://craigdsouza.in/pages/datasets/. Last accessed 20 May 2021
26. Abraham J (2018) Man proposes, nature disposes. The New Indian Express, 17 June 2018, https://www.newindianexpress.com/states/kerala/2018/jun/17/man-proposes-nature-disposes-1829223.html Last accessed 20 May 2021.

Prediction of Seismic Wave Velocity Using Artificial Intelligence

Abir Sarkar and Debjit Bhowmik

1 Introduction

Northeast India is one of the most seismically active regions in India. As per IS 1893:2002 on a seismic hazard map Assam is classified as the most hazardous region (seismic zone V) in terms of occurrence of earthquakes. Thus, the study of seismic wave characteristics is very important for researchers as well as industrialists in this region. The primary objective of a geotechnical engineer is to evaluate a project site in terms of suitability of the foundation soil for a particular structure. It is mostly achieved by conducting destructive tests such as drilling or boring along with a suitable geophysical test which is non-destructive by nature. Since geophysical testing equipments are very sensitive and can only be conducted when certain specific site conditions are maintained, the common practice is to predict the wave velocities from well-established correlations. The seismic refraction test being very rapid and economical is the easiest test to be conducted compared to other geophysical tests.

With recent advancements in the field of computation, artificial neural networks (ANNs) have shown better performance and greater efficiency across various fields due to its ability to capture and represent complex relationships between a set of input and output data. The application of ANN in the field of geophysical investigation has been mostly done for the selection of first breakpoints [1, 2] in the arrival travel time plots. Moreover, ANN has been successfully applied for generating soil profiles [3, 4]. Dumke and Berndt [5] recently suggested that machine learning shows a promising potential for predicting primary wave velocity (V_p) profiles. Thus in the present study, an attempt has been made to generate a velocity profile in combination with seismic refraction surveys. The present study aims to establish the correlation

A. Sarkar (✉)
Department of Civil Engineering, Swami Vivekananda University, Barrakpore, West Bengal, India
e-mail: abirsarkar2007@gmail.com

D. Bhowmik
Department of Civil Engineering, National Institute of Technology Silchar, Cachar, Assam, India

A. Sil et al. (eds.), *Recent Trends in Civil Engineering*, Lecture Notes in Civil Engineering 274, https://doi.org/10.1007/978-981-19-4055-2_42

of V_p with basic geotechnical properties. The study also aims toward developing an ANN for quick prediction of V_p from bore log data. Seismic refraction surveys were conducted for the determination of V_p of different soil layers, and boreholes were drilled to collect disturbed and undisturbed samples for determining basic soil properties. Many previous researches [6–12] suggest that the correlation between shear wave velocity (V_s) and N to be exponential in nature. However, Lee [13] has conducted a detailed study in order to find out the best statistical model and suggested the nature of the fit could be linear or quadratic in nature. Castagna et al. [14] suggested V_p and V_s be linearly correlated. Thus, establishing a relationship between V_p and N would further be helpful in determining V_s from N values if site-specific linear model parameters are established for V_p and V_s. The contradictions in the previous studies are mostly due to the fact that these parameters are site-specific, and most of the authors have suggested the particular correlations should be used only for the location under study. Thus, developing separate numerical models of seismic wave velocity and N values for different locations is necessary. Therefore in the present study, a separate zone-specific parameter has been incorporated for the neural network model with the intention of obtaining an ANN that can be used to predict the velocity profile at any zone within the site under consideration. Four sites were chosen inside the campus of NIT Silchar for the development of the empirical relationship between V_p and other basic soil properties. For the development of ANN model depth factor, site class, N value, and bulk density were considered as inputs and velocity as output.

2 Geology of Study Area

The campus of NIT Silchar roughly covers 650 acres of land at the southernmost part of the state Assam in India. Geologically the campus can be divided into two major groups' unconsolidated deposits and semi-consolidated deposits. The unconsolidated deposits mostly consist of alluvial deposits comprising mainly of sand, silt, and clay with gravel and occasional organic contents, whereas the semi-consolidated rocks are mostly exposed in the form of hillocks comprising shale, sandstone, pebble, and boulder beds [15].

The selected sites for the study are shown in Fig. 1, and for simplicity, the sites will be referred to as sites A, B, C, and D throughout. Site A is near the auditorium, Site B is near the playground, Site C is near Gate no. 3, and Site D is near the guest house.

3 Geophysical Investigations

Traditionally SPT was mostly popular in geotechnical engineering for indirectly determining different engineering properties of the soil correlating with SPT value.

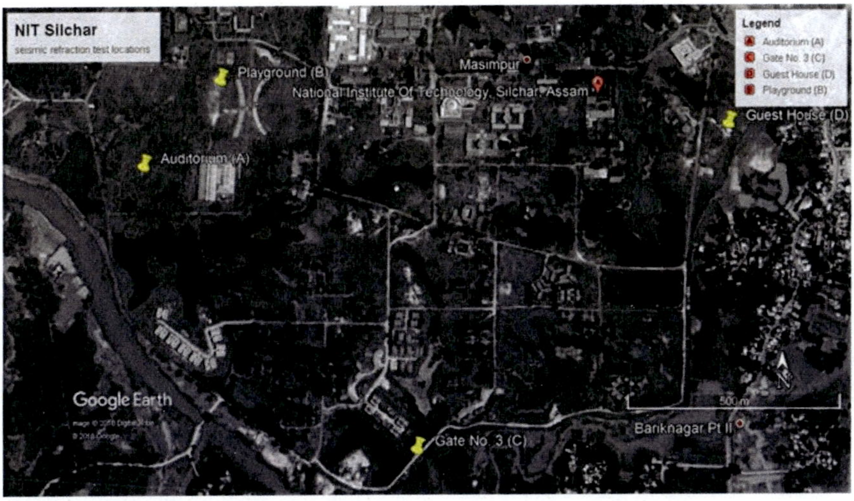

Fig. 1 Location map for all sites under study: NIT Silchar (Google Earth Pro, 2019)

However, recent advancements in geophysical techniques allow in situ direct measurements of different properties of soil. Various geophysical testings include the determination of elastic wave velocities, resistivity values, discontinuities or voids below ground surface, etc. Popular geophysical techniques used nowadays are seismic refraction survey, Multi-channel Analysis of Surface Wave (MASW), cross-hole test, seismic downhole or uphole test, electrical resistivity test (ERT), ground-penetrating radar (GPR), etc. However, the execution of such advanced techniques in urban areas can be rather difficult since each of these tests suffers from certain limitations such as ERT demands the test site to be protected since electric currents pass through the electrodes. Hellman et al. [16] while discussing ambiguities with ERT, suggested seismic profiling where ERT fails to detect a low contrast resistivity boundary between a layer of shale of clay. MASW test is frequency-dependent, and skilled professionals are required for interpreting raw data since it involves dispersion followed by inversion of the waves passing through different soil layers [17]. GPR is also frequency-dependent, and proper interpreting skills are required. High-frequency equipment is costlier, and if the wavelength recorded during the investigation is not enough small, then the minute details of the site might not possible to be captured. Seismic downhole or uphole tests require one borehole [18, 19] and can provide an accurate velocity profile at greater depth, but gives slight error near the surface. The cross-hole test gives the most reliable velocity profile since the calculations involve direct waves traveling from the source to the receiver in a straight line but requires two boreholes, and hence, it is comparatively costlier than other geophysical tests. Seismic refraction surveys, on the other hand, do not require any skilled labors for collection of data and processing of the data can be done with medium-skilled professionals (interpretations from travel—time data). Moreover, seismic refraction surveys do not require any specific field preparation such as boreholes and site protection. It is also independent of frequency since frequency does not affect the P-wave velocity [20], thus gaining an advantage over other geophysical tests. However, the seismic refraction survey has the limitation of detecting a low-velocity layer [21] below a higher velocity layer as the wave is required to be totally refracted back from between two layers according to Snell's law.

4 Artificial Neural Network

ANN is a complicated computational tool which unlike most mathematical models tries to simulate the architecture and internal operational features of the human nervous system. Any typical ANN architecture consists of an input layer, an output layer and a number of hidden layers that depend upon the complexity of the model [22, 23]. The number of neurons in each layer might differ depending upon the problem in hand and is commonly known as processing elements (PE). In this study, an interconnected feed-forward multilayered perceptron has been considered. Typical neural network architecture is shown in Fig. 2a [24]. The backpropagation (BP) algorithm used in the present model is the most commonly used training algorithm for a multilayered feed-forward network in which the activations are propagated from input to the output layer in the first phase followed by a backward phase in which the error between the observed and desired output is backpropagated in order to adjust the weights and bias in each PE [25, 26].

For any jth PE, the process can be summarized by Fig. 2b and Eqs. 1 and 2. The result obtained from Eq. 1 is processed through a transfer function as shown in Eq. 2. A simple logistic sigmoid function was used in this study as a transfer function as shown in Fig. 3 and can be represented by Eq. 3.

$$I_j = \theta_j + \sum_{i=1}^{n} w_{ji} x_i \tag{1}$$

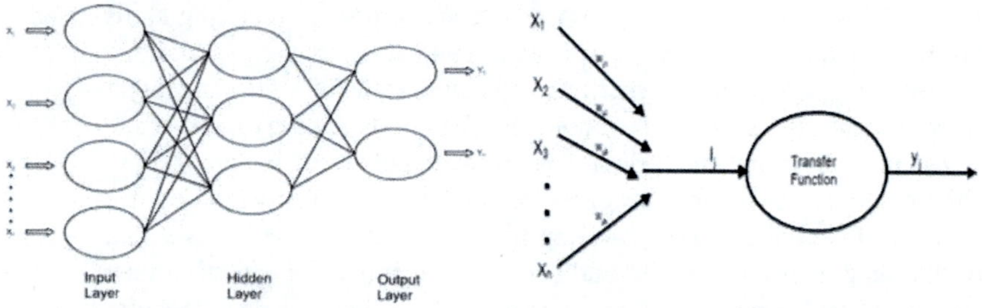

(a) basic network structure (b) operation of a typical PE

Fig. 2 Typical ANN architecture

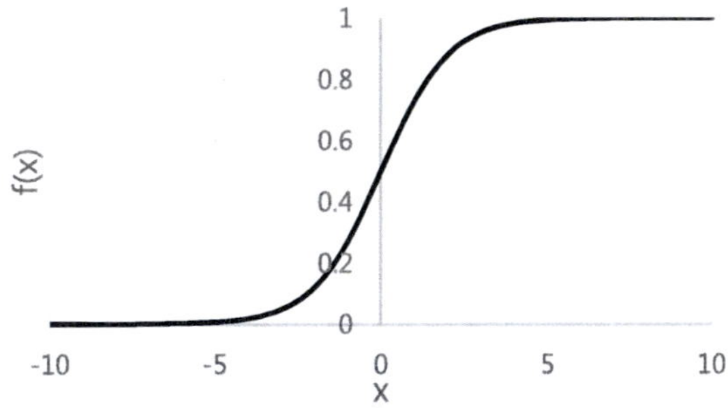

Fig. 3 Logistic sigmoid curve

$$y_j = f(I_j) \tag{2}$$

$$f(x) = \frac{1}{1 + e^{-x}} = \frac{e^x}{1 + e^x} \tag{3}$$

where

I_j = activation level of jth PE;

w_{ji} = weight of the connection between jth and ith PEs;

x_i = input from ith PE where $i = 0, 1, 2, \ldots, n$;

θ_j = bias for jth node;

y_j = output from jth node;

$f(.)$ = transfer function;

This process is repeated until a threshold error is reached or the maximum number of iterations also known as epochs is reached. This combined process of feed-forward and backpropagation results in the training of the network by adjusting the weights of the PE, and completion of one such adjustment is often referred to as one training cycle [27].

5 Seismic Refraction Survey for Soil Profiling

In order to conduct an active seismic refraction survey, the selection of proper energy input to the subsurface is important so that the energy is able to penetrate to the desired depth as required for the project. For the current study, a sledgehammer impact on a metallic plate was used since it was found to be capable of easily reach up to depths of 30 m, for further depths explosives (usually used in industries for deep exploration) can also be used as an active energy source. The generated waves then travel to the subsurface at a particular velocity depending on the elastic properties of the material through which they are traveling. From the theory of elasticity, Eqs. 4 and 5 show how wave velocity depends upon the density of the medium. Thus, the density of the layer was considered as an additional parameter for the development of the velocity models.

$$V_p = \sqrt{\frac{K + (4/3)\mu}{\rho}} = \sqrt{\frac{\lambda + 2\mu}{\rho}} = \sqrt{\frac{M}{\rho}} \tag{4}$$

$$V_s = \sqrt{\frac{G}{\rho}} \tag{5}$$

where

V_p = Primary wave velocity

V_s = Shear wave velocity

K = bulk modulus = 1/compressibility

M = P-wave modulus

λ = Lame's constant

G = shear modulus

ρ = density.

When the waves encounter an interface where the density or velocity changes significantly, a part of the energy is reflected back to the surface and the remaining is transferred to the lower layer [28].

When the velocity of the lower layer (V2) is higher than that of the upper layer (V1), a part of the energy is also critically refracted along the particular interface as shown in Fig. 4. This critically refracted wave travels along the interface at the velocity of the lower layer and continually refracts energy back to the surface which is known as head waves. The geophones spread out in the linear array on the surface, record the incoming refracted and reflected waves. All typical seismic refraction tests analyze the travel time of the first energy to arrive at the geophones. The first arrivals are either from the direct waves or critically refracted waves. A typical example of the selection of the first breakpoint is shown in Fig. 5 where the position of each geophone (in meters) is shown along the X-axis and the time (in milliseconds) along the Y-axis for a single shot closer to the first geophone. This first arrival time information is the only information received from the P-waves that propagate to the geophones [29]. The geophones closer to the source pickup the direct waves first and the geophones further from the source beyond the critical distance pickup the critically refracted waves first since they travel at a faster velocity in the lower layer and hence reach the geophones before the direct waves which travel at a constant velocity at ground level. The first arrival time for geophone closer to the shot point or the source is less than those farther away from the shot point simply shows that the wave travels along a straight path thus taking a long time to reach the final geophone. The tangent at

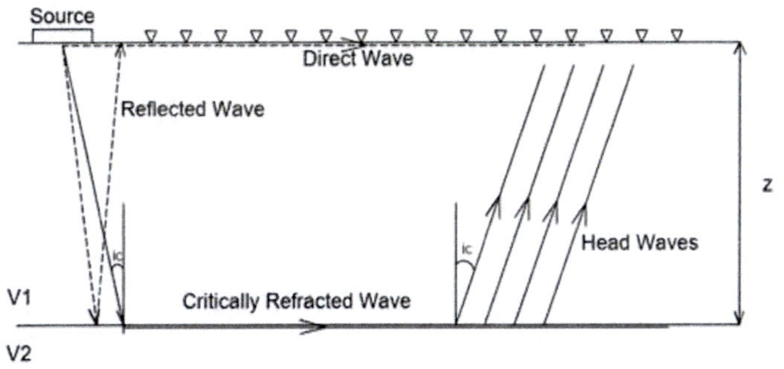

Fig. 4 Ray path diagram

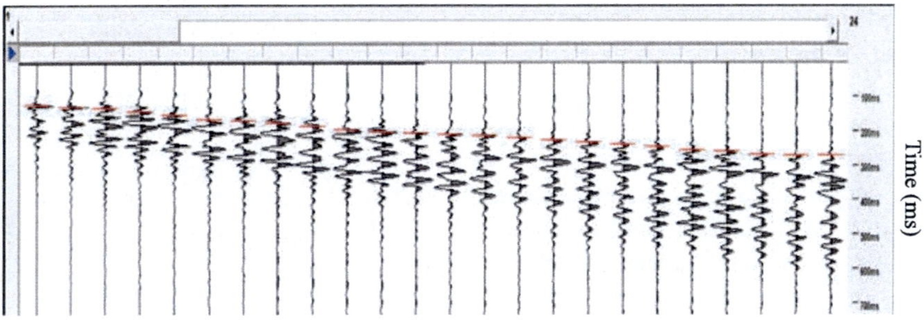

Fig. 5 Selection of first break points

any point on the distance v_s time plot simply produces the velocity and is the most convenient method of conducting a seismic refraction analysis [2].

Conventional methods such as delay—time or generalized reciprocal method does not take into account certain factors which affect the velocity structure such as lateral discontinuity, heterogeneity. Research over past years [30, 31] has shown seismic refraction tomography (SRT) to deliver more accurate results. It does not demand the model to be split into continuous layers of constant velocity. An SRT model is typically made up of a large number of small velocity grid cells or nodes [32]. Software packages Rayfract, GeoCT—II, SeisImager are few of the most widely used packages with their own commercial refraction tomography codes.

5.1 *Field Survey*

Seismic refraction surveys were carried out at sites A, B, C, and D as shown in Fig. 1. A total of 24 geophones having a natural frequency of 4.5 Hz was used for recording the arrival times of an impact caused by a sledgehammer of 6 kg weight

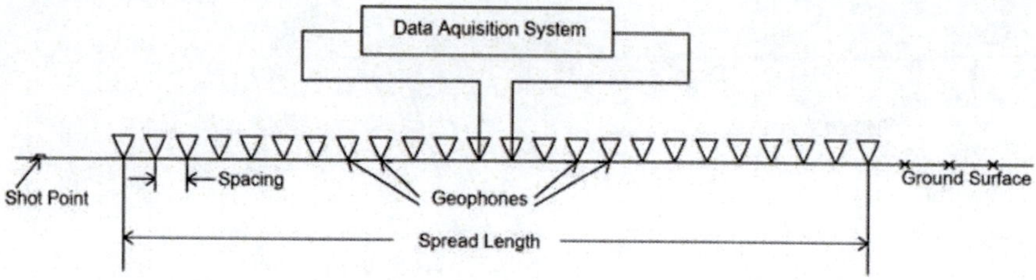

Fig. 6 Typical field test setup for seismic refraction survey

on a strike plate positioned at different stations. Figure 6 shows a typical test setup and terminologies used for the seismic refraction survey. The shot point is the station point where the impact load is applied with the help of a hammer. The term spacing is the distance between each geophone and spread length is referred to as the overall length covered for the survey.

At sites A, B, and D, total of 7 shots were taken, whereas at Site C 25 shots were taken [33]. Since Site C was located near a water body, multiple shots had to be taken due to poor signal-to-noise ratio. The reason for the poor signal-to-noise ratio lies in the geological location of Site C. The site is closer to a water body, and from previous studies on dispersion characteristics of P-waves [34], it has been understood that the velocity through pore water is greater than that through soil skeleton. Thus in order to obtain a better signal-to-noise ratio, 25 shots were used in Site C.

The raw data from the geophones were then processed in Rayfract which uses Wavepath Eikonel Tomography (WET) inversion algorithm [35] to generate the layers responsible for refraction. Further to get a better signal, shots were repeated at the same station [36–38] and the average of the response was estimated. A total of 3 stacks were taken at each site for every shot point and travel times from the geophones spaced at 2 m were recorded in the data acquisition system. Field investigation at Site C is shown in Fig. 7.

5.2 *P-Wave Velocity Profiles*

The P-wave velocity profiles for all the sites are shown in Fig. 8 after processing of raw field data.

For the refraction survey at sites A, B, and D, four shot points were selected beside geophone numbers 1, 6, 18, and 24, two shot points at an offset distance of 12 m from either end of the spread and one central shot point between geophone number 12 and 13. The geophones were spaced at 2 m from each other. At Site C, all 25 shots were taken beside each geophone and a central shot between geophone numbers 12 and 13 was taken.

(a)

(b)

Fig. 7 Investigation at local site **a** geotechnical investigation **b** geophysical investigation

Fig. 8 One-dimensional V_p (m/s) profiles using seismic refraction test

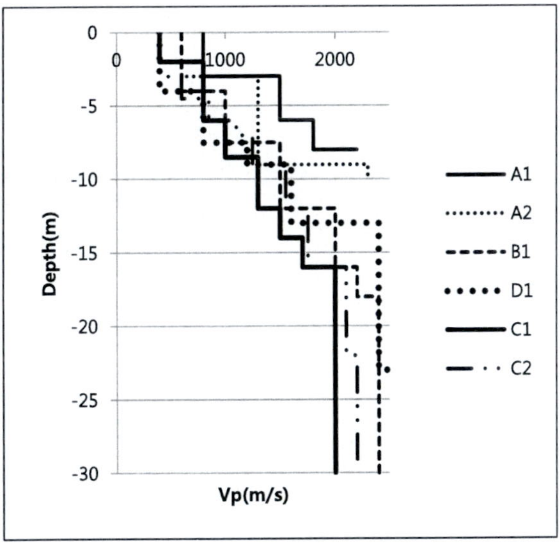

5.3 Geotechnical Investigations

Two boreholes were drilled at each sites A and C, whereas one borehole was drilled at each sites B and D. The boreholes at Site A were 10 m deep, and the remaining boreholes were 30 m deep each. SPT N values were gathered through standard procedures as per IS 2131(1983). Routine laboratory tests on disturbed samples collected at 1.5 m interval from all boreholes and on-site investigation during drilling suggest the top layer to be consisting of silty sands with poorly graded gravels (SM, SM-GP), followed by inorganic clay layer of medium plasticity (CL) followed by inorganic clay layer of high plasticity (CH). Organic silty clay (OL) of low plasticity was observed at Site C, and a second borehole also confirms its existence.

Previous studies Maheswari et al. [9] found out both corrected and uncorrected *N* values predict wave velocities with the same accuracy. However, Thaker and Rao [10] suggested uncorrected *N* value gives better correlation than corrected *N* values. In the present study, all correlations are predicted based on uncorrected *N* values. The variation of the uncorrected SPT *N* values with depth for the various sites is shown in Fig. 9. The water table at sites A, B, and D were observed at 4.5 m, and at Site C, the water table was observed at 3.5 m. At Site C, organic layer was encountered at depths 9–12 m at BH-1 and at depths 7–9 m at BH-2. This low-velocity organic layer was not detected in the refraction survey, and layer change was only observed when a higher velocity layer was encountered.

Thus, the velocity obtained for the organic layer was not accurate. Therefore, the obtained P-wave velocity for the organic layer at Site C was not used for the prediction of correlations between *N* value and P-wave velocity. The variation of bulk density is shown in Fig. 10, and again a sudden decrease in the bulk density from 1.66 to 1.26 g/cc was observed at C1, where the organic layer was present.

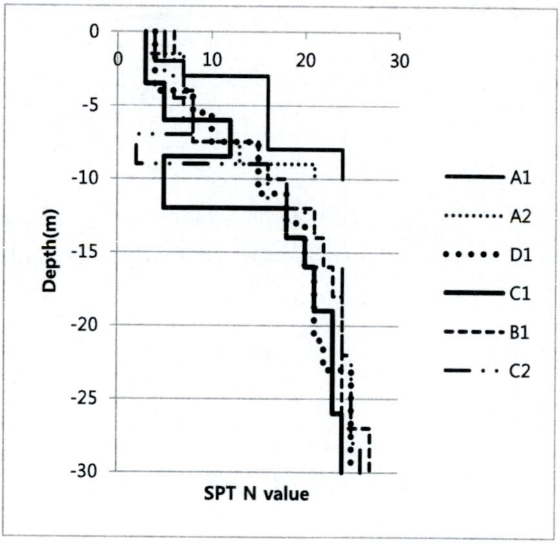

Fig. 9 Variation of SPT *N* with depth

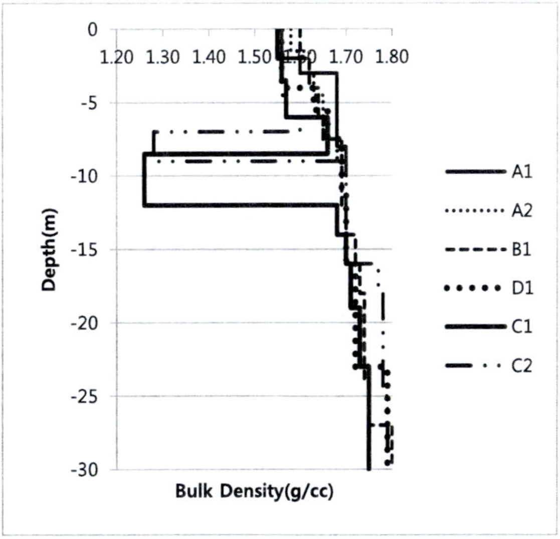

Fig. 10 Variation of bulk density with depth

6 Development of ANN Model

The artificial intelligence-based approach was applied to the seismic refraction data in combination with the borehole data to determine the soil profile at NIT Silchar campus. The number of PEs in the input and output layer is based on the geometry of the problem, and those of the hidden layers are usually determined by trial and error [39–41]. It is well known that the seismic wave velocities are dependent on the parameters depth, density, *N* value. Moreover, a zone-specific parameter termed as 'site class' is introduced in the present study to take care of the variation of soil properties within a site.

Therefore, our input layer consists of four Pes, namely site class, *N* value, depth factor, and density, whereas the output layer has only one neuron that is V_p and through multiple trial and error the best performance of the network was obtained with a single hidden layer with 10 PEs. The regression analysis was conducted between the network output V_p (predicted) and the target V_p (observed at the field). The development of the neural network can be briefly broken down into three phases, namely training, testing, and validation phase. Among the total dataset 70% for training, 15% for testing, and 15% for validation. A total of 70 datasets were obtained from the geotechnical and geophysical investigations. Details of the statistical parameters of the dataset are given in Table 1. Out of all 70 datasets, 13 datasets obtained from borehole 2 at Site C (C2) were not used for ANN training but for the validation of the ANN model. Further, the performance of the ANN was also compared with simple statistical models.

To test the efficiency of the network at other sites and under completely different environmental conditions, it was trained with a different dataset given by Anbazghan et al. [6] for Lucknow City. The entire location was divided into three site classes

Table 1 Statistical parameters of all data used for ANN and statistical models

Statistical index	Depth (m)	SPT *N*	Density (g/cc)	V_p (m/s)
Mean	11.836	15.785	1.667	1595.714
Standard deviation	9.035	8.138	0.111	675.436
Maximum	30	27	1.8	2600
Minimum	0	2	1.26	400
Range	30	25	0.54	2200

based upon both geophysical and geotechnical investigation. The site classes 1 and 2 were defined for the zones where the low-velocity region was observed at depths of 7 m and 12 m, respectively. Site Class 3 was defined for the zones where no low-velocity region was observed. For training of the ANN, only the depths were considered at which N, bulk density, and shear wave velocity were available, whereas all the other data were ignored and were not used for ANN training. Figure 11 shows the dataset for Site Class 1 and consists of 7 boreholes (B1, B6, B7, B13, B18, B20, and B21), Fig. 12 shows the dataset for Site Class 2 and consists of 5 boreholes (B2, B8, B10, B11, and B22), and finally Fig. 13 represents the dataset for Site Class 3 consisting of 5 boreholes (B3, B4, B5, B12, and B16). A total of 109 data pairs were used for training the ANN and the data division (in percentage) for training, testing, and validation phase was kept the same as that during the development of the ANN at NIT Silchar campus. The performance was similarly tracked as earlier during all three phases by analyzing the predicted and observed V_p. Moreover, the dataset from B17, B19, and B23 boreholes was not used for training the proposed ANN and was compared with the network predicted velocity profiles at the particular borehole locations. The entire development of the ANN was done in MATLAB 2016a environment with neural network toolbox.

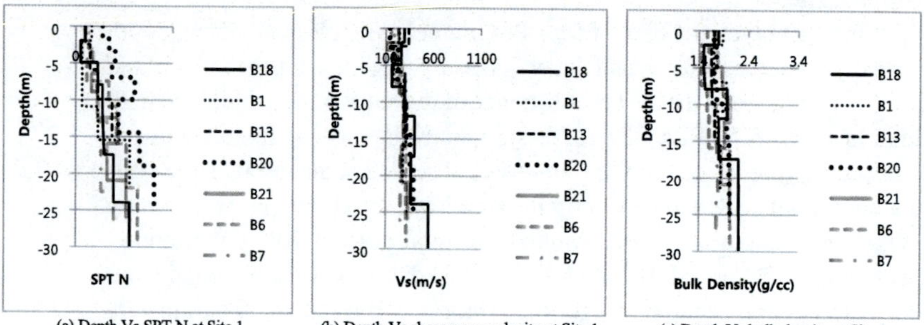

(a) Depth Vs SPT N at Site 1 (b) Depth Vs shear wave velocity at Site 1 (c) Depth Vs bulk density at Site 1

Fig. 11 Dataset at Lucknow city for site class 1 [6]

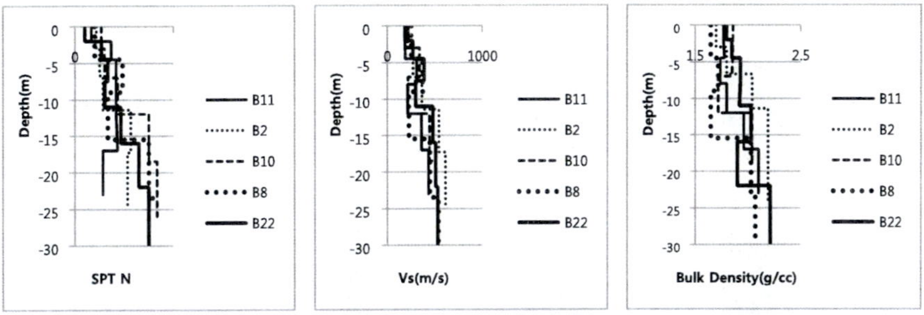

(a) Depth Vs SPT N at Site 2 (b) Depth Vs shear wave velocity at Site 2 (c) Depth Vs bulk density at Site 2

Fig. 12 Dataset at Lucknow city for site class 2 [6]

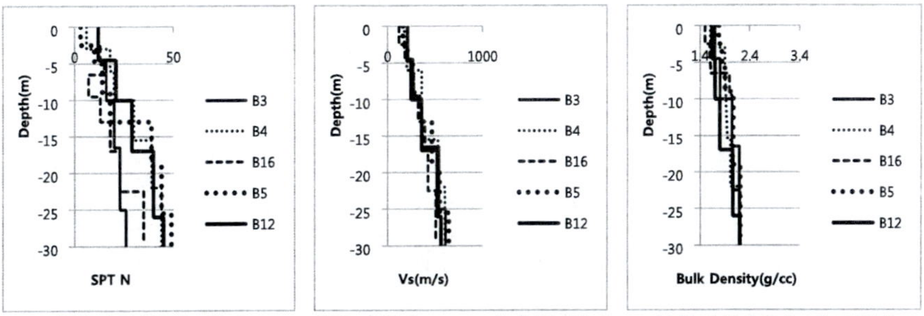

(a) Depth Vs SPT N at Site 3 (b) Depth Vs shear wave velocity N at Site 3 (c) Depth Vs bulk density at Site 3

Fig. 13 Dataset at Lucknow city for site class 3 [6]

7 Development of the Statistical Model

The inter-correlation between all the parameters is shown in Table 2, and it can be seen from the Pearson correlation test that all of them are well interconnected [42]. Thus, individual relationships were determined with V_p in order to understand their nature of the relationship. It can be seen that N shows the highest correlation coefficient with V_p followed by density and the least correlation coefficient was observed for depth factor.

For two given sets of data, X and Y having means $\overline{x}$ and $\overline{y}$ forming n pairs of (x_i, y_i), Pearson's correlation r_{xy} is given by Eq. 6.

Table 2 Pearson correlation test

	Depth	SPT N	Density	V_p
Depth	1	0.869	0.647	0.846
SPT N	0.869	1	0.816	0.928
Density	0.647	0.816	1	0.653
V_p	0.846	0.928	0.653	1

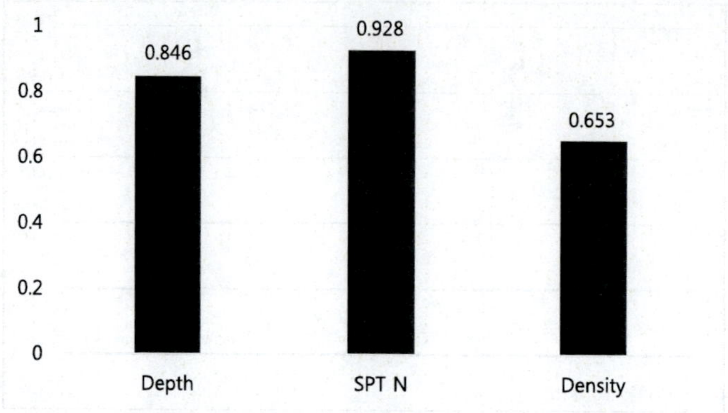

Fig. 14 Relative importance of each input on the output

$$X = \{x_1, x_2, x_3, \ldots, x_n\}$$
$$Y = \{y_1, y_2, y_3, \ldots, y_n\}$$

$$r_{xy} = \frac{\sum_{i=1}^{n}(x_i - \overline{x})(y_i - \overline{y})}{\sqrt{\sum_{i=1}^{n}(x_i - \overline{x})^2}\sqrt{\sum_{i=1}^{n} y - \overline{y})^2}} \tag{6}$$

The relative importance of each input parameter on the primary wave velocity is shown in Fig. 14. It can be seen that N shows the highest correlation coefficient with V_p followed by density, and the least correlation coefficient was observed for depth factor.

An attempt has been made to fit a curve between V_p and the rest of the variables. The statistical models linear, quadratic, cubic, exponential, and power were considered for curve fitting and can be represented by Eqs. 7, 8, 9, 10, and 11, respectively. The R-square coefficient (R^2) was used for performance evaluation which tells us how well the independent variables account for dependent variable measurements [43]. R^2 value close to unity indicates good similarity between predicted and actual values.

$$y = b_1.x + c \tag{7}$$

$$y = b_1.x^2 + b_2.x + c \tag{8}$$

$$y = b_1.x^3 + b_2.x^2 + b_3.x + c \tag{9}$$

$$y = c.{b_1}^x \tag{10}$$

$$y = c.x^{b1} \tag{11}$$

where

$y =$ independent variable,

$x =$ dependent variable,

$b_1, b_2, b_3, c =$ constants.

Figure 15 shows the different relationships between V_p and depth factor.

The correlation coefficient for the various models is shown in Table 3. The highest correlation coefficient was observed for the quadratic model and can be represented by Eq. 12. Polkowski and Grad [44] confirmed that V_p and z follow second- or third-order polynomial.

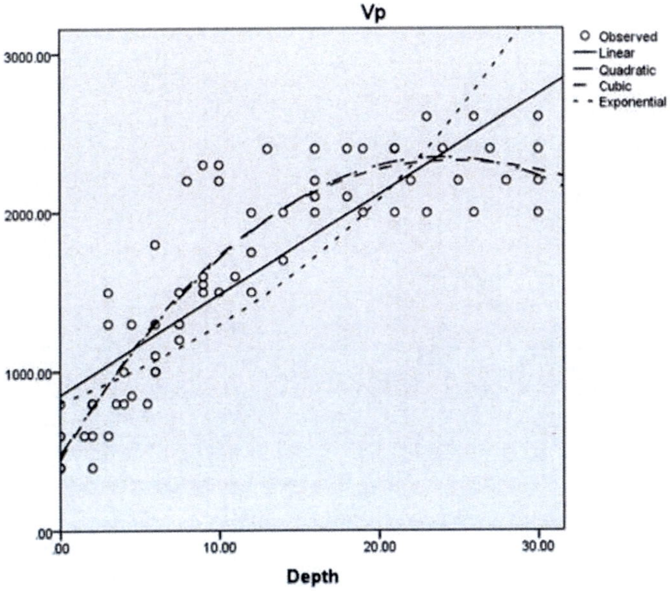

Fig. 15 Curve fitting between V_p and depth

Table 3 Model summary between V_p and z

Equation	Model summary	Parameter estimates			
	R^2	c	b_1	b_2	b_3
Linear	0.713	852.669	63.086		
Quadratic	0.841	477.562	156.043	−3.256	
Cubic	0.83	449.266	170.601	−4.585	0.031
Exponential	0.639	805.647	0.047		

$$V_p = 156.043z^2 - 3.256z + 477.562 \quad (12)$$

where z is the depth factor.

Similarly, Fig. 16 shows the various curve fit of V_p with N. The linear model fits best, with a coefficient of regression of 0.902 as shown in Table 4 and can be represented by Eq. 13.

$$V_p = 82.464N + 264.429 \quad (13)$$

where N is the SPT N value.

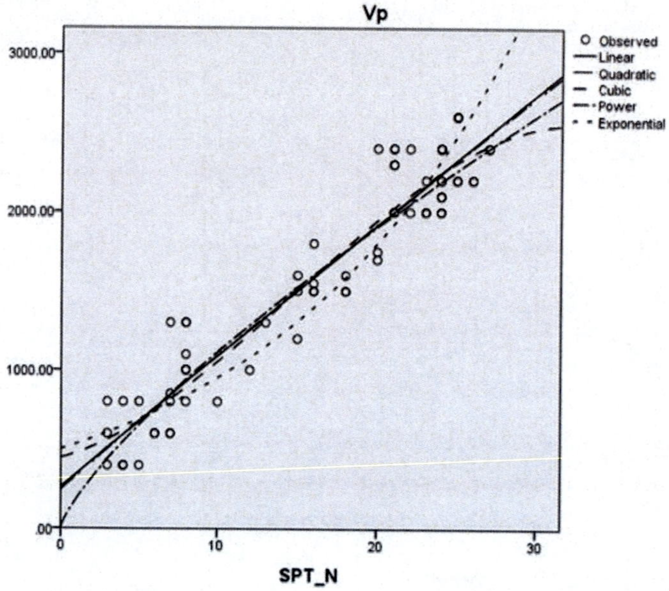

Fig. 16 Curve fitting of V_p with N

Table 4 Model summary between N and V_p

Equation	Model summary	Parameter estimates		
	R^2	c	b_1	b_2
Linear	0.902	264.429	82.464	
Quadratic	0.901	248.689	85.425	−0.099
Power	0.871	188.415	0.772	
Exponential	0.856	502.226	0.064	

The curve fit between bulk density and V_p is shown in Fig. 17.

The relationship between V_p and bulk density was found to be quadratic in nature with an R^2 of 0.821 as shown in Table 5. Since bulk density shows the least correlation coefficient, it is the least sensitive of all the parameters and can be represented by Eq. 14.

Since the relationship between V_p and N is linear, whereas with the other parameters it is quadratic, thus a linear correlation can only be possible if all the terms relating to nonlinearity are considered separately [45].

$$V_p = 34279.989\rho^2 - 7698.131\rho - 34258.26 \tag{14}$$

where ρ is the bulk density.

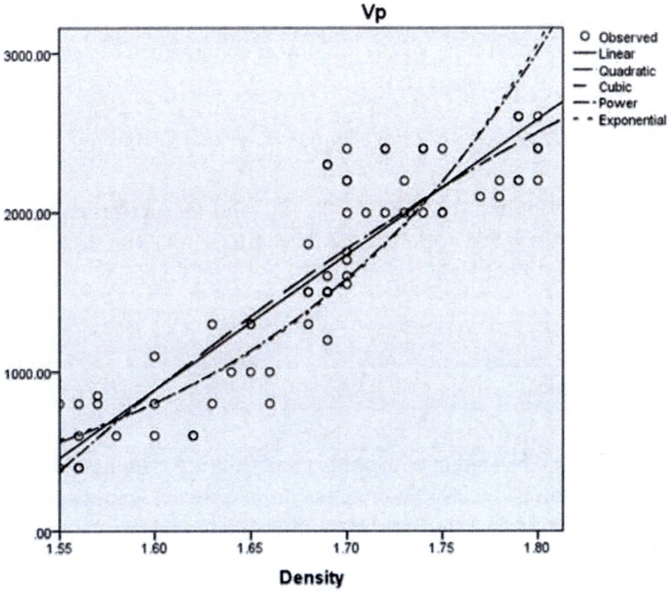

Fig. 17 Curve fitting of V_p with bulk density

Table 5 Model summary between V_p and bulk density

Equation	Model summary	Parameter estimates		
	R^2	c	b_1	b_2
Linear	0.817	−12,704.270	8492.832	
Quadratic	0.821	−34,258.260	34,279.989	−7698.131
Power	0.810	4.086	11.233	
Exponential	0.802	0.018	6.686	

Table 6 Coefficients for linear regression

Model	c
(Constant)	1768.886
Depth	38.254
SPT N	67.618
Density	−1914.899
ρ^2	606.926
z^2	−0.973

In order to convert this nonlinear problem to linear, two additional parameters z^2 (square of depth) and ρ^2 (square of density) were taken into account for linear correlation with V_p. The linear correlation coefficients for all the parameters are shown in Table 6.

Finally, V_p can be represented by Eq. 15 which gives a correlation coefficient of 0.906 (whereas without consideration of the nonlinear parameters ρ^2 and z^2 a coefficient of regression of 0.89 was observed).

$$V_p = 38.254z + 67.618N + 606.926\rho^2 - 1914.899\rho - 0.973z^2 + 1768.886 \quad (15)$$

The statistical model was applied to calculate the $V_{p,}$ and the performance of the model was checked by finding the correlation between the model predicted V_p (output V_p) and the actual V_p (target V_p).

8 Results and Discussion

The overall performance of the ANN along with performance during training, testing, and validation phase is shown in Fig. 19. The regression of network-generated V_p (output V_p) and actual V_p (target V_p) during the training phase and can be represented by Eq. 16. Similarly, Eqs. 17, 18, and 19 represent the regression between the predicted and target V_p during validation, testing, and overall progress phase, respectively (Fig. 18).

$$(\text{Output } V_p)_{\text{training}} \sim= 0.98 * (\text{Target } V_p)_{\text{training}} + 27 \quad (16)$$

$$(\text{Output } V_p)_{\text{validation}} \sim= 0.97 * (\text{Target } V_p)_{\text{validation}} + 13 \quad (17)$$

$$(\text{Output } V_p)_{\text{testing}} \sim= 0.93 * (\text{Target } V_p)_{\text{testing}} + 30 \quad (18)$$

$$(\text{Output } V_p)_{\text{overall}} \sim= 0.97 * (\text{Target } V_p)_{\text{overall}} + 30 \quad (19)$$

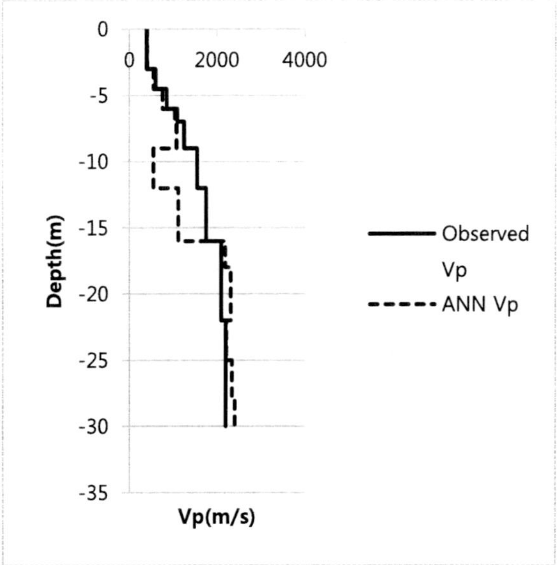

Fig. 18 Desired V_s predicted primary wave velocity at Site C

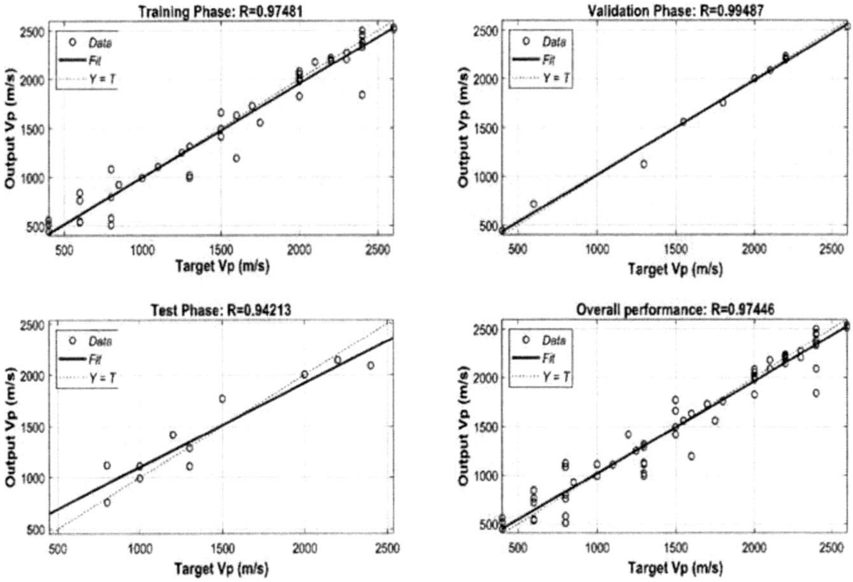

Fig. 19 Performance of ANN on local dataset

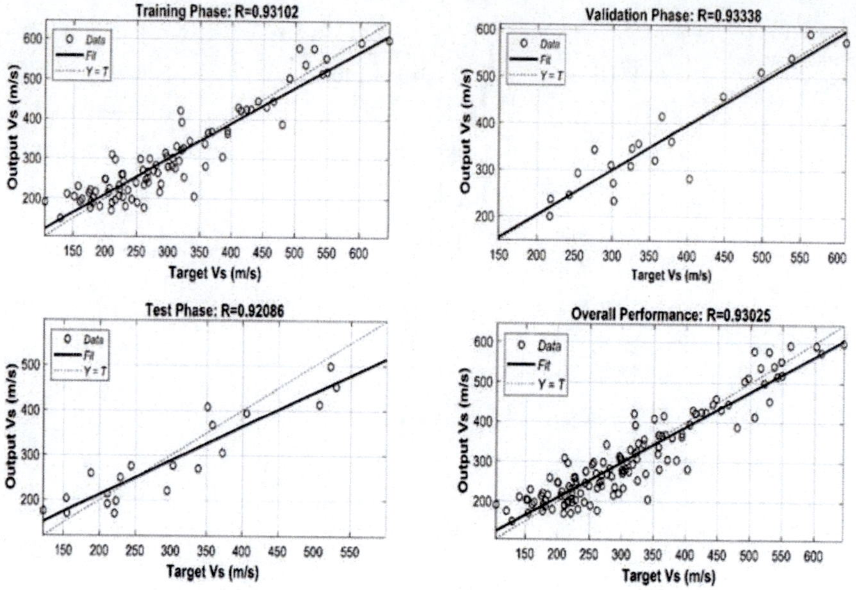

Fig. 20 Performance of the ANN on Lucknow city dataset

If the slope is assumed to be one, then from Eq. 21, the overall network prediction differs from the observed velocity profile by 30 m/s.

The prediction of V_p from the proposed ANN and statistical model was compared to that of the actual V_p at Site C and is shown in Fig. 21. It is very interesting to note that the prediction of the ANN varies significantly at a depth of 7 m, and this is due to the failure of the seismic refraction test to determine the velocity of the organic layer. The proposed network gives us a V_p of 551 m/s for the layer containing peat soil which shows good agreement with earlier studies where the V_p of peat soil was reported to be in the range of 211–534 m/s [46] (Figs. 19 and 20).

The statistical model, on the other hand, was unable to predict the velocity of the organic layer and like the results obtained from the seismic refraction test, it shows a consistent increase in the velocity profile. The performance of the model is shown in Fig. 21 and can be represented by Eq. 20 with an R^2 value of 0.906.

$$(\text{Output } V_p) = 0.9985 * (\text{Target } V_p) + 41.084 \tag{20}$$

If the slope is assumed to be unity, then the statistical model predicted values roughly differ from the actual observed V_p by 41.01 m/s.

The performance of the same network on the dataset from Anbazghan [6] is shown in Fig. 20 the prediction of V_s at B23, B17, and B19 are shown in Fig. 22. The network performed well and gave an overall correlation coefficient of 0.93 and the velocity profile generated for site classes 1, 2, and 3 agrees well with the observed velocity profile from MASW test. Regression analysis between network predicted and actual

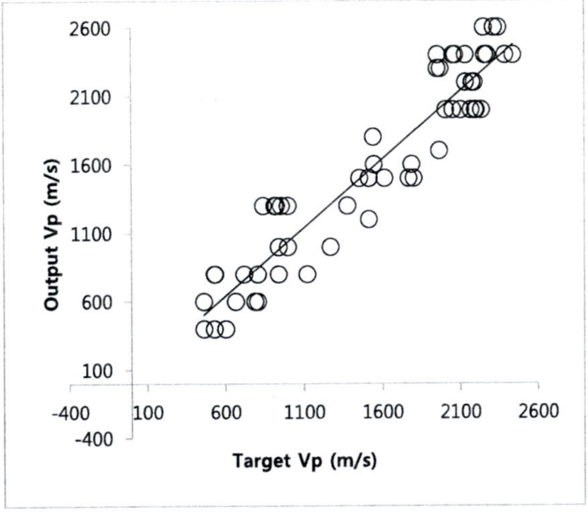

Fig. 21 Desired V_s predicted primary wave velocity from the statistical model

V_s during training, validation, testing phase, and overall training is shown in Eqs. 21, 22, 23, and 24, respectively.

$$(\text{Output } V_s)_{\text{training}} \sim= 0.9 * (\text{Target } V_s)_{\text{training}} + 39 \tag{21}$$

$$(\text{Output } V_s)_{\text{validation}} \sim= 0.82 * (\text{Target } V_s)_{\text{validation}} + 37 \tag{22}$$

$$(\text{Output } V_s)_{\text{testing}} \sim= 0.79 * (\text{Target } V_s)_{\text{testing}} + 57 \tag{23}$$

$$(\text{Output } V_s)_{\text{overall}} \sim= 0.88 * (\text{Target } V_s)_{\text{overall}} + 38 \tag{24}$$

If the slope in Eq. 24 is assumed to be one, then the network predicted shear wave velocity differs from actual shear wave velocity by 38 m/s.

9 Conclusions and Future Scope

Statistical and ANN models were successfully developed for seismic velocity profiling of typical soil at NIT Silchar campus. Seismic refraction tests were conducted for measuring the P-wave velocity, SPT N values were obtained from fresh boreholes, and bulk density was calculated at various depths.

Regression analysis was performed keeping V_p as the dependent variable and others as an independent variable for developing a statistical model, and ANN model was developed with the output layer having V_p as the only PE, whereas the input

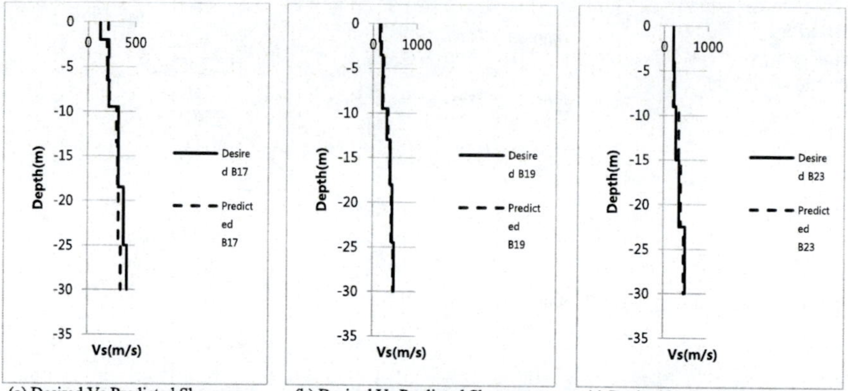

(a) Desired Vs Predicted Shear wave velocity at Site 1

(b) Desired Vs Predicted Shear wave velocity at Site 2

(c) Desired Vs Predicted Shear wave velocity at Site 3

Fig. 22 Prediction of the velocity profile at Lucknow city from the proposed ANN

layer containing the other variables, thus consisting of 4 PEs and the hidden layer consisting of 10 PEs. The key findings from the study are listed below:

1. The seismic refraction test suffers from a drawback being incapable of determining or detecting a low-velocity layer, whereas in combination with ANN training, it was able to detect as well as predict the velocity of the organic layer with decent accuracy.
2. Most statistical models are site-specific but the ANN architecture can be trained to generate velocity profile (either V_p or V_s profiles) at any site by following the steps given below:

    (a) Collecting a reasonable amount of dataset containing velocity profile, SPT *N* value and density such that all variations in the geotechnical, as well as geophysical exploration, are covered for the location under study.
    (b) Based on the velocity profiles or *N* value variation, the site needs to be classified into categories that represent a similar trend with respect to depth and then train in the proposed ANN model.
    (c) Based on the geotechnical investigation, the site for which velocity profile is to be generated must be identified under suitable site class as defined in the previous step.
    (d) Simulating the network with the geotechnical parameters as well as site class would now generate the velocity profile.

3. The ANN model outperforms the statistical model ($R^2 = 0.906$) with an R^2 value of 0.97 when compared with directly measured wave velocities.
4. Since statistical correlations are easy to use, Eq. 17 can be used to instantly generate the velocity profile within the NIT Silchar campus from simple geotechnical exploration.

The application of ANN across other geophysical investigation methods can be explored since most methods are based upon certain assumptions and often include some erroneous results. The ability of the NN to study the pattern of the data points might be helpful to eradicate some of the errors if not all. Other machine learning techniques can also be explored for better prediction.

Acknowledgements The authors are grateful for the funding received from TEQIP-III, NIT Silchar for conducting the experiments on the site.

Conflict of Interest The authors declare that they have no conflict of interest.

References

1. Sabbione JI, Velis D (2010) Automatic first-breaks picking: new strategies and algorithms, p 75
2. Akram J, Eaton DW (2014) A review and appraisal of arrival-time picking methods for downhole microseismic data. Geophysics 81:67–87. https://doi.org/10.1190/GEO2014-0500.1
3. Caglar N, Arman H (2007) The applicability of neural networks in the determination of soil profiles. Bull Eng Geol Environ 66:295–301. https://doi.org/10.1007/s10064-006-0075-9
4. Choobbasti AJ, Farrokhzad F, Mashaie SR, Azar PH (2015) Mapping of soil layers using artificial neural network (case study of Babol, Northern Iran). J South African Inst Civ Eng 57:59–66. https://doi.org/10.17159/2309-8775/2015/v57n1a6
5. Dumke I, Berndt C (2019) Prediction of seismic P-wave velocity using machine learning. Solid Earth 10:1989–2000. https://doi.org/10.5194/se-10-1989-2019
6. Anbazhagan P, Uday A, Moustafa SSR, Al-Arifi NSN (2016) Correlation of densities with shear wave velocities and SPT N values. J Geophys Eng 13:320–341. https://doi.org/10.1088/1742-2132/13/3/320
7. Ohta Y, Goto N (1978) Empirical shear wave velocity equations in terms of characteristic soil indexes. Earthq Eng Struct Dyn 6:167–187. https://doi.org/10.1002/eqe.4290060205
8. Ulugergerli E, Uyanik O (2007) Statistical correlations between seismic wave velocities and SPT blow counts and the relative density of soils. J Test Eval 35:100159. https://doi.org/10.1520/JTE100159
9. Uma Maheswari R, Boominathan A, Dodagoudar GR (2010) Use of surface waves in statistical correlations of shear wave velocity and penetration resistance of Chennai soils. Geotech Geol Eng 28:119–137. https://doi.org/10.1007/s10706-009-9285-9
10. Thaker TP, Rao KS (2011) Development of statistical correlations between shear wave velocity and penetration resistance using MASW technique. In: 2011 Pan-Am CGS geotechnical conference: Toronto, Ontario, Canada
11. Tan CG, Majid TA, Ariffin KS, Mohamad N (2012) Site-specific empirical correlation between shear wave velocity and standard penetration resistance using MASW method. CHUSER 2012 - 2012 IEEE Colloq Humanit Sci Eng Res 94–97 (2012). https://doi.org/10.1109/CHUSER.2012.6504288
12. Anbazhagan P, Kumar A, Sitharam TG (2013) Seismic site classification and correlation between standard penetration test N value and shear wave velocity for Lucknow city in Indo-Gangetic basin. Pure Appl Geophys 170:299–318. https://doi.org/10.1007/s00024-012-0525-1
13. Lee SHH (1990) Regression models of shear wave velocities in Taipei basin. J Chinese Inst Eng Trans Chinese Inst Eng A/Chung-kuo K. Ch'eng Hsuch K'an 13:519–532. https://doi.org/10.1080/02533839.1990.9677284

14. Castagna JP, Batzle ML, Eastwood RL (1984) Relationship between compressional and shear-wave velocities in classic silicate rocks. SEG Tech Progr Expand Abstr 1984(50):582–584. https://doi.org/10.1190/1.1894108
15. Ministry of Water Resources Guwahati (2013) "Ground water information booklet, Cachar District, Assam." Guwahati
16. Hellman K, Ronczka M, Günther T, Wennermark M, Rücker C, Dahlin T (2017) Structurally coupled inversion of ERT and refraction seismic data combined with cluster-based model integration. J Appl Geophys 143:169–181. https://doi.org/10.1016/j.jappgeo.2017.06.008
17. Ivanov J, Tsoflias G, Miller RD, Peterie S, Morton S (2016) Impact of density information on Rayleigh surface wave inversion results. J Appl Geophys 135:43–54. https://doi.org/10.1016/j.jappgeo.2016.09.011
18. Ghazi A, Moghadas NH, Sadeghi H, Ghafoori M, Lashkaripur GR (2015) Empirical relationships of shear wave velocity, SPT-N value and vertical effective stress for different soils in Mashhad, Iran. Ann Geophys 58. https://doi.org/10.4401/ag-6635
19. Balia R, Pirinu A (2018) Geophysical surveying of the ancient walls of the town of Cagliari, Italy, by means of refraction and up-hole seismic tomography techniques. Archaeol Prospect 25:147–153. https://doi.org/10.1002/arp.1596
20. Cha M, Cho GC (2007) Compression wave velocity of cylindrical rock specimens: engineering modulus interpretation. Japanese J Appl Phys Part 1 Regul Pap Short Notes Rev Pap 46:4497–4499. https://doi.org/10.1143/JJAP.46.4497
21. Chen J, Hubbard SS, Gaines D, Korneev V, Baker G, Watson D (2010) Stochastic estimation of aquifer geometry using seismic refraction data with borehole depth constraints. Water Resour Res 46. https://doi.org/10.1029/2009WR008715
22. Caglar N, Elmas M, Yaman ZD, Saribiyik M (2008) Neural networks in 3-dimensional dynamic analysis of reinforced concrete buildings. Constr Build Mater 22:788–800. https://doi.org/10.1016/j.conbuildmat.2007.01.029
23. Erzin Y, Cetin T (2014) The prediction of the critical factor of safety of homogeneous finite slopes subjected to earthquake forces using neural networks and multiple regressions. Geomech Eng 6:1–15. https://doi.org/10.12989/gae.2014.6.1.001
24. Chu Y, Liu S, Cai G, Bian H (2019) Artificial neural network prediction models of heavy metal polluted soil resistivity. Eur J Environ Civ Eng 0:1–21.https://doi.org/10.1080/19648189.2019.1585962
25. Shahin MA, Jaksa MB, Maier HR (2008) State of the art of artificial neural networks in engineering. Electron J Geotech Eng 13:1–25
26. Lai J, Qiu J, Feng Z, Chen J, Fan H (2016) Prediction of soil deformation in tunnelling using artificial neural networks. Comput Intell Neurosci. https://doi.org/10.1155/2016/6708183
27. Debnath P, Dey AK (2017) Prediction of laboratory peak shear stress along the cohesive soil-geosynthetic interface using artificial neural network. Geotech Geol Eng 35:445–461. https://doi.org/10.1007/s10706-016-0119-2
28. Nordiana MM, Saad R, Saidin M, Shyeh SK, Ali N, Kiu YC, Jinmin M, Rao RR (2013) 2-D resistivity and seismic refraction surveys for subsurface studies at Bukit Bunuh, Perak. Electron J Geotech Eng 18F:1109–1119
29. Sompotan AF, Pasasa LA, Sule R (2011) Comparing models GRM, refraction tomography and neural network to analyze shallow landslide. ITB J Eng Sci 43:161–172. https://doi.org/10.5614/itbj.eng.sci.2011.43.3.1
30. Sheehan JR, Doll WE, Mandell WA (2006) An evaluation of methods and available software for seismic refraction tomography analysis. J Environ Eng Geophys 10:21–34. https://doi.org/10.2113/jeeg10.1.21
31. Jacob T, Samyn K, Bitri A, Quesnel F, Dewez T, Pannet P, Meire B (2018) Mapping sand and clay-filled depressions on a coastal chalk clifftop using gravity and seismic tomography refraction for landslide hazard assessment, in Normandy, France. Eng Geol 246:262–276. https://doi.org/10.1016/j.enggeo.2018.10.007
32. Imhof AL, Sánchez M, Calvo C, Martín A (2011) Application of seismic refraction tomography for tunnel design in Santa Clara Mountain, San Juan, Argentina. Earth Sci Res J 15:81–88

33. Desper DB, Link CA, Nelson PN (2015) Accurate water-table depth estimation using seismic refraction in areas of rapidly varying subsurface conditions. Near Surf Geophys 13:455–465. https://doi.org/10.3997/1873-0604.2015039
34. Wu S, Chen L (1991) Dispersion characteristics of elastic waves in saturated soils. In: International conferences on recent advances in geotechnical earthquake engineering and soil dynamics, p 3
35. Schuster GT (1993) Wavepath eikonal traveltime inversion: theory. Geophysics 58:1314. https://doi.org/10.1190/1.1443514
36. Behm M, Brückl E, Chwatal W, Thybo H (2007) Application of stacking and inversion techniques to three-dimensional wide-angle reflection and refraction seismic data of the Eastern Alps. Geophys J Int 170:275–298. https://doi.org/10.1111/j.1365-246X.2007.03393.x
37. Uhlemann S, Hagedorn S, Dashwood B, Maurer H, Gunn D, Dijkstra T, Chambers J (2016) Landslide characterization using P- and S-wave seismic refraction tomography—the importance of elastic moduli. J Appl Geophys 134:64–76. https://doi.org/10.1016/j.jappgeo.2016.08.014
38. Sharafeldin SM, Essa KS, Youssef MAS, Karsli H, DIab ZE, Sayil N (2019) Shallow geophysical techniques to investigate the groundwater table at the Great Pyramids of Giza, Egypt. Geosci Instrumentation Methods Data Syst 8:29–43. https://doi.org/10.5194/gi-8-29-2019
39. Lam HF, Ng CT (2008) The selection of pattern features for structural damage detection using an extended Bayesian ANN algorithm. Eng Struct 30:2762–2770. https://doi.org/10.1016/j.engstruct.2008.03.012
40. Shahin MA, Jaksa MB, Maier HR (2009) Recent advances and future challenges for artificial neural systems in geotechnical engineering applications. Adv Artif Neural Syst 308239. https://doi.org/10.1155/2009/308239
41. Javadi AA, Rezania M (2009) Applications of artificial intelligence and data mining techniques in soil modeling. Geomech Eng 1:53–74. https://doi.org/10.12989/gae.2009.1.1.053
42. Deb S, Ahmed MA (2019) Quality assessment of city bus service based on subjective and objective service quality dimensions: case study in Guwahati, India. Benchmarking 26:567–589. https://doi.org/10.1108/BIJ-11-2017-0309
43. Cheng MY, Cao MT (2016) Estimating strength of rubberized concrete using evolutionary multivariate adaptive regression splines. J Civ Eng Manag 22:711–720. https://doi.org/10.3846/13923730.2014.897989
44. Polkowski M, Grad M (2015) Seismic wave velocities in deep sediments in Poland: borehole and refraction data compilation. Acta Geophys 63:698–714. https://doi.org/10.1515/acgeo-2015-0019
45. Banerjee S, Guha B, Bandyopadhyay G (2016) A post factor analysis of financial ratios of selected IPOs and its impact on grading: an empirical inquest. J Bus Stud Q 8:23–34
46. Mad Said MJ, Zainorabidin A, Madun A (2015) Soil velocity profile on soft soil using seismic refraction. Appl Mech Mater 773–774:1549–1554. https://doi.org/10.4028/www.scientific.net/AMM.773-774.1549

Evaluation of Stability of Embankment Dams

V. Divya, M. Balaji, N. Chandan Gowda, C. Chandana, and G. M. Jamuna

1 Introduction

Slope stability is the ability of a naturally occurring or engineered soil slope, like embankments, cuttings, open-pit mining, excavations and landfills, to withstand ground movement. A previously stable slope may become unstable, due to the reduction in shear stresses and increasing strains, leading to a mass movement. The natural or man-made disasters such as cyclones, floods, swelling or leaching caused by seeping water, erosion of slope surfaces, earthquakes and human activities, also contribute to the failure of slopes leading to loss of life and property. Hence stability analysis bears a significant phase in the safe design of artificial and natural slopes.

The increasing demand for engineered cut and fill slopes in construction projects has led to the development of investigative tools and stabilization techniques that helped to solve the potential stability problems efficiently. Liu et al. [1] used Limit Equilibrium Method (LEM) and two Finite Element Methods (FEM) for analyzing and comparing the slope stability in two directions. They observed that the factor of safety (FOS) obtained from the two finite element methods is higher than those obtained from LEM. Shole and Belayneh [2] studied the effect of core shape and side slopes in the stability analysis of embankment dam using modelling software. They concluded that the factor of safety increases as the side slope changes from steep to that of gentle in all loading conditions. The determination of slope stability for nonlinear failure criterion of soil was investigated by [3]. Their results indicated that the discretization technique proposed can be successfully extended to assess the stability of layered slopes and also, various complex stability problems of slopes.

V. Divya (✉)
Department of Civil Engineering (VTU RRC), CMR Institute of Technology, Bengaluru 560037, India
e-mail: divya.v@cmrit.ac.in

M. Balaji · N. Chandan Gowda · C. Chandana · G. M. Jamuna
Department of Civil Engineering, CMR Institute of Technology, Bengaluru 560037, India

A. Sil et al. (eds.), *Recent Trends in Civil Engineering*, Lecture Notes in Civil Engineering 274, https://doi.org/10.1007/978-981-19-4055-2_43

Many researchers have assessed the construction of road embankments using waste products such as recycled construction and demolition waste [4, 5] fly ash [6], etc. and have confirmed their potential to replace the natural materials in the construction. Experimental and on-site studies were conducted by [7] to investigate the potential application of recycled gypsum in clayey embankment construction. Based on their results they reported improvement in strength of clayey soil with addition of the recycled gypsum and was also found to have minimum environmental impact. The stabilization of weak embankments using sustainable methods such as Ground Granulated Blast Furnace Slag, stone dust, lime and bio-enzymes has been investigated and compared by [8]. They concluded that lime and bio-enzyme are the most potential solutions, based on their comparative studies.

In the current study, it is proposed to analyze the stability and seepage of embankment dam, modelled under different construction material characteristics and loading conditions, using the numerical modelling tools of GeoStudio software. The construction materials used in the embankment dam are such that the shear strength of the natural soil is enhanced, providing a solution for the safe disposal of waste materials as well as the economy is achieved. Further, comparative studies have been performed to suggest the most feasible condition that enhances the stability of the modelled embankment dam.

2 Methodology

A. The soil sample was collected from the field, pulverized and subjected to laboratory tests for the determination of index and engineering properties.
B. The waste materials viz construction demolition waste and tyre shreds were obtained and characterised for its properties. The proportions of the waste materials were fixed based on literature review.
C. Stability analysis was performed using SLOPE/W and SEEP/W tools of GeoStudio software and factor of safety (FOS) was determined for embankment dams modelled with and without berms and horizontal drainage blankets.
D. Comparative study of FOS for the different conditions proposed and identification of the most feasible condition of stability.

3 Characterization of Materials

The natural soil (NS) sample used in the present study was collected at a depth of 2.0 m from the ground surface. It possessed a specific gravity of 2.67 and was dark red in colour. The soil sample was classified as CL as per [9] and attained a maximum dry density of 17.5 kN/m^3 at an optimum moisture content of 15.6%. The shear parameters of soil are very crucial with respect to the safety and stability of any embankment. The present sample was estimated to possess a cohesion of 10 kPa and angle of shearing resistance of 35°.

The construction demolition waste (CDW), consisting of mainly crushed bricks and aggregates, was obtained from a nearby site and pulverized prior to the characterization. The gradation of CDW material was determined by mechanical dry sieving method and was obtained as uniformly graded. The material had specific gravity of 2.65 and angle of internal friction of 43°. The proportions of CDW material to be added to the natural soil were fixed as 5% and 10%, respectively based on [10].

The waste tyres collected for the present study were shredded to a size of 3 mm × 6 mm with a specific gravity of 1.12. The tyre shreds (TS) were used in volume proportions of 2.5 and 5% of the natural soil, based on [11].

4 Numerical Simulation

The numerical analysis was performed using SLOPE/W and SEEP/W tools of GeoStudio software. The Limit equilibrium theory has been used by the SLOPE/W to determine the factor of safety against failure. The analysis has been carried out using the Entry and Exit method in SLOPE/W. SEEP/W is used for the estimation of quantity of water seeping through the embankment dam and foundation. SEEP/W works based on finite element method and helps in the identification of phreatic lines and the velocity vectors. The overall height of the dam including the foundation was taken as 22 m with water level on the upstream at 20 m. The side slopes adopted at upstream and downstream sides were 2 Horizontal: 1 Vertical. The shear characteristics of the different proportions of embankment materials are given in Table 1. In the present study the embankment dam was modelled for the following stability conditions:

1. Provided with a berm of width 4 m on downstream slope.
2. Provided with a horizontal drainage blanket 10 m wide at the toe.
3. Provided with both berm at downstream and drainage blanket at the toe.

Table 1 Shear characteristics for various material proportions

Material type	Dry density (kN/m^3)	Cohesion (kPa)	Shearing resistance (degrees)
Natural soil (NS)	17.50	10	35
NS + 2.5% TS	18.00	7.0	38.50
NS + 5% TS	18.34	6.5	39.60
NS + 5% CDW	18.64	9.1	36.56
NS + 10% CDW	19.46	8.5	38

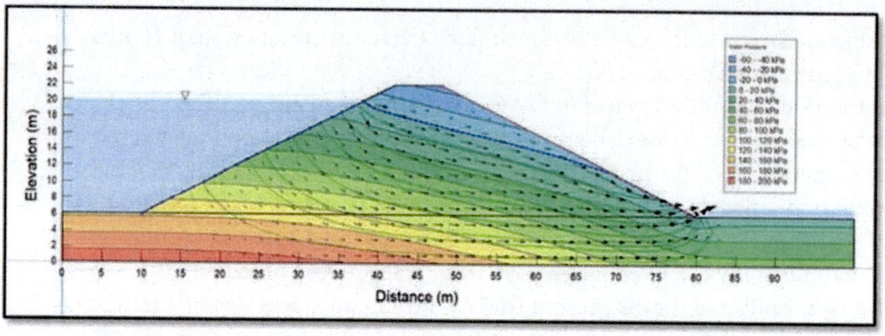

Fig. 1 Pore water pressure contours with flow path without berm and drainage blanket

5 Results and Discussions

5.1 Seepage Analysis Using SEEP/W

The SEEP/W tool of GeoStudio software has been used for estimating the pore water pressure through the dam. Figure 1 represents the pore water pressure contours developed under steady seepage conditions. The stability of the downstream slope is critical under steady seepage conditions. It can be observed that the phreatic line crosses the downstream toe and hence makes the downstream end more vulnerable to failure. The results also indicate that the rate of average pore water seepage into the foundation, at normal water levels, is very high.

A berm of width 4 m was introduced in the downstream slope to control the erosion and runoff. Berms help in controlling erosion and increase the stability of embankment dams, according to [12]. To avoid piping failure at the dam base, a horizontal drainage blanket/filter of width 10 m was provided in the downstream side. Figure 2 shows the pattern of pore pressure contours representing this condition. The results indicate a zero-pore pressure development at downstream as a result of providing berm and drainage filter.

5.2 Stability Analysis Using SLOPE/W

The stability of the upstream and downstream slopes was checked using SLOPE/W tool of GeoStudio software. The critical slip surface was obtained using Morgenstern Price method of analysis. The models were simulated for natural soil (NS) and different proportions of natural soil blended with tyre shreds (TS) and construction demolition waste (CDW). The various volumetric proportions of waste materials were found to increase the shear strength of the natural soil of the embankment (Table 1). The stability analysis was performed for steady seepage and rapid drawdown conditions which were crucial at downstream and upstream slopes, respectively.

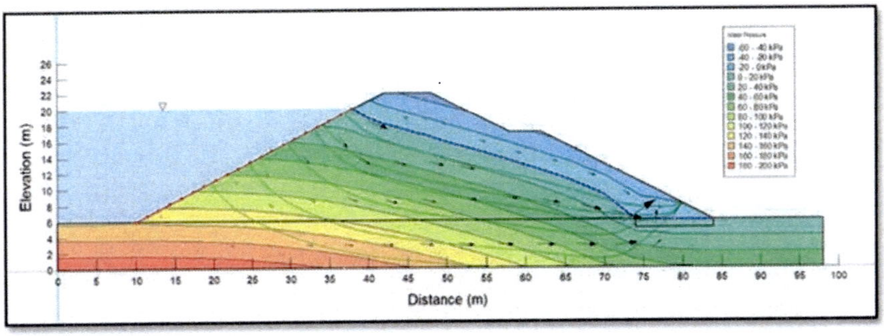

Fig. 2 Pore water pressure contours with flow path with both berm and drainage blanket

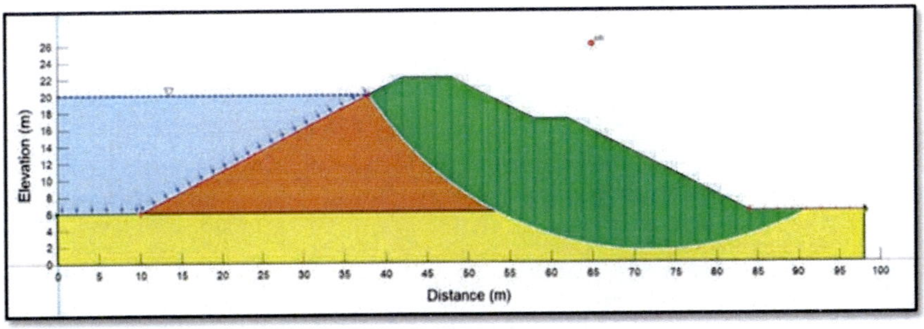

Fig. 3 Critical slip surface obtained under steady seepage condition

Figure 3 shows the critical slip surface obtained under steady seepage when the downstream was provided with berm and horizontal drainage filter.

In rapid drawdown conditions, it is assumed that continuous seepage occurs in the dam embankment and thereby the upstream side is emptied rapidly. Hence, this condition is critical only at upstream slope. The results of factor of safety (FOS) obtained from the stability analysis of downstream slopes for different soil-waste material proportions are presented in Fig. 4. The FOS results for rapid drawdown are shown in Fig. 5. It can be observed that maximum stability is achieved when the embankment soil is treated with 10% construction demolition waste and both berm and filter are provided at downstream. It could be because addition of CDW reduces the permeability of the embankment and thereby helps in reducing the pore pressure increment within the dam.

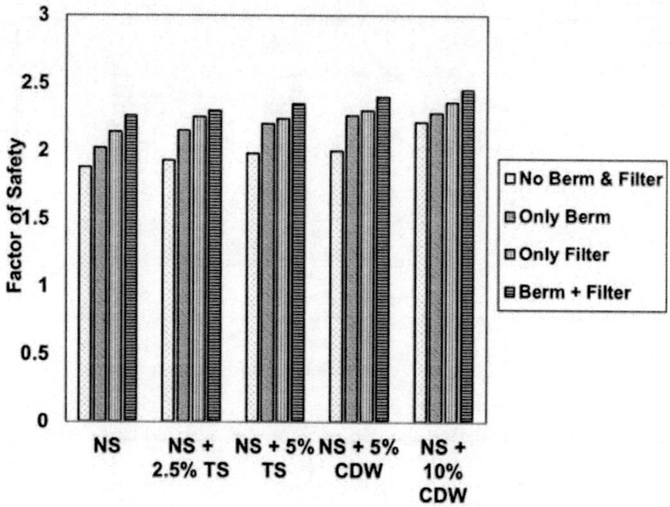

Fig. 4 FOS for different soil-waste material proportions under steady seepage conditions

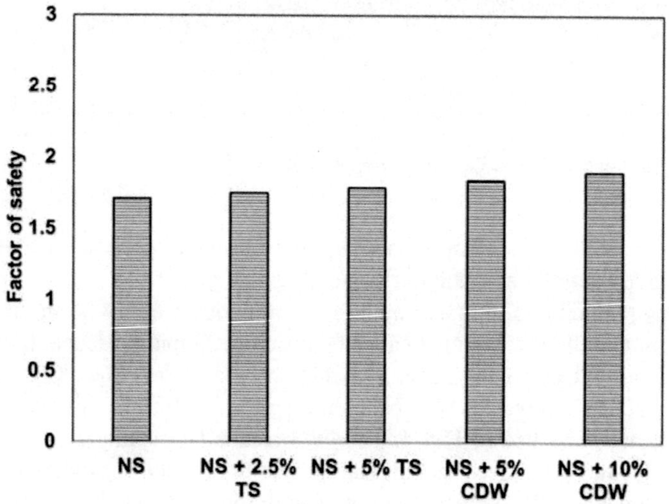

Fig. 5 FOS for different soil-waste material proportions under rapid drawdown conditions

6 Conclusions

The following could be concluded based on the current study:

1. The waste materials that are difficult/cannot be recycled can be used as potential construction materials in embankment dams and hence can contribute to safe disposal of waste and economy in the construction.

2. A minimal value of cohesion is required to increase the stability of embankment slopes.
3. Factor of safety was found to be maximum (2.45) at a combination of natural soil treated with 10% of construction demolition waste.
4. The use of berm in the downstream slope increases the stability of embankment slopes.
5. The use of drainage filter controls the pore pressure development at dam base and downstream toe.
6. The stability at rapid drawdown conditions was found to be lesser than steady seepage conditions.

References

1. Liu SY, Shao LT, Li HJ (2015) Slope stability analysis using the limit equilibrium method and two finite element methods. Comput Geotech 63:291–298. https://doi.org/10.1016/j.compgeo.2014.10.008
2. Shole DG, Belayneh MZ (2019) The effect of side slope and clay core shape on the stability of embankment dam: Southern Ethiopia. Int J Environ Sci Technol 16:5871–5880. https://doi.org/10.1007/s13762-019-02228-3
3. Li YX, Yang XL (2019) Soil-slope stability considering effect of soil-strength nonlinearity. Int J Geomech 19:04018201. https://doi.org/10.1061/(asce)gm.1943-5622.0001355
4. Cristelo N, Vieira CS, De Lurdes LM (2016) Geotechnical and geoenvironmental assessment of recycled construction and demolition waste for road embankments. Proc Eng 143:51–58. https://doi.org/10.1016/j.proeng.2016.06.007
5. Sangiorgi C, Lantieri C, Dondi G (2015) Construction and demolition waste recycling: an application for road construction. Int J Pavement Eng 16:530–537. https://doi.org/10.1080/10298436.2014.943134
6. Šešlija M, Radović N, Vatin N (2015) Construction of road embankment with waste materials. Appl Mech Mater 725–726:596–601. https://doi.org/10.4028/www.scientific.net/amm.725-726.596
7. Ahmed A, Ugai K, Kamei T (2011) Laboratory and field evaluations of recycled gypsum as a stabilizer agent in embankment construction. Soils Found 51:975–990. https://doi.org/10.3208/sandf.51.975
8. Divya V, Asha MN (2020) Evaluation of sustainable methods for embankment stabilization. IOP Conf Ser Earth Environ Sci 573.https://doi.org/10.1088/1755-1315/573/1/012008
9. IS:1498 (1970) Classification and identification of soils for general engineering purposes. Indian Standard
10. Sharma RK, Hymavathi J (2016) Effect of fly ash, construction demolition waste and lime on geotechnical characteristics of a clayey soil: a comparative study. Environ Earth Sci 75:1–11. https://doi.org/10.1007/s12665-015-4796-6
11. Cabalar AF, Karabash Z, Mustafa WS (2014) Stabilising a clay using tyre buffings and lime. Road Mater Pavement Des 15:872–891. https://doi.org/10.1080/14680629.2014.939697
12. Calamak M, Kilic Y, Yanmaz AM (2020) On the stability of moderate height berm-type earthen dams: the Hancagiz dam example. Geotech Geol Eng 38:4169–4179. https://doi.org/10.1007/s10706-020-01286-6

Oxidative Degradation of Rhodamine B Dye in Wastewater Using Microwave-Assisted Fentons Reaction

Indrashis Saha and Richa Pandey

1 Introduction

Treatment of agricultural and industrial wastewater is a problem for conventional treatment plants worldwide [1]. Effluents from textile industries often contain non-biodegradable dyes. Many physico-chemical methods have already been applied to eradicate dyes from wastewater such as coagulation, solvent extraction, adsorption, floatation, and hyperfiltration [2]. However, these methods solely transfer the pollutant and the organic dyes from one media to another. Toxic pesticides are present in agricultural wastewater. So, for the effective removal/degradation of such organic pollutants from wastewater, advanced oxidation processes (AOPs) are heavily used. Degradation of toxic organic compounds and non-biodegradable pollutants are caused by the highly reactive ·OH radicals which is known to be generated from the AOPs [3]. Fenton process is one such advanced oxidation process. The chemistry involves the reaction between hydrogen peroxide (H_2O_2) and ferrous ions (Fe^{2+}) (Eq. 1) to produce highly reactive hydroxyl radicals (·OH) under acidic conditions. The produced hydroxyl radicals oxidize the organic pollutants (Eq. 2). Among all the advanced oxidation processes, Fenton's process diminishes toxicity and results in enhanced color removal performance and improved biodegradability [4].

$$Fe^{2+} + H_2O_2 \rightarrow Fe^{3+} + \cdot OH + OH^- \tag{1}$$

$$\text{Organics} + \bullet OH \rightarrow \text{Products} \tag{2}$$

I. Saha
University of Calcutta, Kolkata, West Bengal 700 009, India

R. Pandey (✉)
Birla Institute of Technology, Mesra, Ranchi, Jharkhand 835215, India
e-mail: richapandey@bitmesra.ac.in

A. Sil et al. (eds.), *Recent Trends in Civil Engineering*, Lecture Notes in Civil Engineering 274, https://doi.org/10.1007/978-981-19-4055-2_44

In comparison with stand-alone oxidation processes, a combination of different AOPs results in enhanced removal efficiency and is more efficient in wastewater treatment. This is because of the generation of a significant amount of free radicals to initiate radical chain mechanism and due to the higher degree of the energy performance of combined AOPs [5]. External energies like light, ultrasound, or microwave enhance the rate of Fenton's reaction.

The photo-Fenton reaction assisted by UV or sunlight gives enhanced reaction rates and faster mineralization of organic pollutants over ordinary Fenton's process. For the case of the Fenton process assisted under sunlight, H_2O_2 which is the primary oxidant reacts with Fe^{2+} to generate Fe^{3+} and one hydroxyl radical is produced. The resulting Fe^{3+} absorbs the UV irradiation, while it further reacts with H_2O_2 to reproduce the initial Fe^{2+} and another hydroxyl radical is produced as shown in the following reactions.

$$Fe^{2+} + H_2O_2 \rightarrow Fe^{3+} + OH^{\bullet} + OH^{-} \quad (3)$$

$$Fe^{3+} + H_2O + h\nu \rightarrow Fe^{2+} + OH^{\bullet} + H^{+} \quad (4)$$

The use of ultrasound in AOPs is a relatively new technique that involves the employment of low to medium frequency (20–1000 kHz) and high energy ultrasonic waves for improved degradation with high removal performance. There are several theories including the hotspot theory which explains the mechanisms of sonication-induced chemical reactions [6].

The following reactions show the free radical mechanism proposed for the production of OH by the Fenton process.

$$OH^{\bullet} + H_2O_2 \rightarrow HO_2^{\bullet} + H_2O \quad (5)$$

$$Fe^{3+} + HO_2{}^{\bullet} \rightarrow Fe^{2+} + H + O_2 \quad (6)$$

$$Fe^{2+} + HO_2{}^{\bullet} \rightarrow Fe^{3+} + HO_2{}^{-} \quad (7)$$

$$Fe^{2+} + OH^{\bullet} \rightarrow Fe^{3+} + OH^{-} \quad (8)$$

The generation of free hydroxyl radicals follows a chain cyclic mechanism with step (3) serving as the chain initiation, steps (5) and (6) as termination, and the cycle (3)–(5)–(6) forms the cyclic chain which is the site for the evolution of O_2. From the reactions above, we can see that the Fenton reaction is based on the free radical mechanism. Therefore, the Fenton reagent involves several different other cations of metals (Mn^{+}), the overall reaction can be characterized as follows.

$$M^{(n+)} + H_2O_2 \rightarrow M^{(n+1)} + OH^{-} + OH^{\bullet} \quad (9)$$

The advantages of microwave-assisted Fenton's process are reduction of reaction time, increase in the selectivity of the reaction, low activation energy, reduction of equipment size, ease of control, and increase in the yield and removal % [7]. Granular activated carbon (GAC) helps in increasing the temperature of the reaction and also catalyzes the rate of reaction [8]. Microwave provides selective heating of the material but studies have shown that under certain reaction conditions it does not provide selective heating which may be due to the combination of different factors like lower target compound concentration which decreases the heating efficiency and may be due to similar dielectric properties between the target compound and the matrix.

A plethora of research work is available on the degradation of Rhodamine B dye. Rhodamine B dye is toxic to living beings and is known to be carcinogenic. So, removal of Rhodamine B is necessary from a health and safety point of view. Horikoshi et al. [9] examined the characteristic of a device for UV-Vis irradiation of the cationic dye (Rhodamine B). The results show a significant decomposition of the dye by a simultaneous process of photocatalysis and photosensitization. Wang et al. [10] investigated the integrated process of MW and CFA catalyzed Fenton-like system on Rhodamine B dye which was taken as the model dye. The CFA used was characterized to analyze the physicochemical property and the influence of radical scavengers on the removal efficacy in the resulting solution. Additionally, a comparison was made between the efficiency of different Fenton like process. The experimental data were optimized to get the final results. Finally, analytical treatments like mass spectrometry and liquid chromatography were performed to analyze the experimental results. Hong et al. [11] in their study investigated the removal efficacy of both Rhodamine B and Methylene Blue dye by MW-H_2O_2 system at alkaline pH and studied the effects of pH on the degradation performance. The authors also investigated the MW thermal effect which was performed to understand the mechanism of degradation. Gayathri et al. [12] studied the removal performance of Rhodamine B separately by microwave-assisted and UV-irradiated process under various conditions and the optimal dosage of the operational parameters were found out. Later both the MW and UV irradiated processes were combined, and the experiments were undertaken, and the results were compared with the results of the individual systems. The catalyst which was used was ZnO and a considerable amount of synergy is observed in the results. It was observed that MW enhanced photocatalytic degradation even in a system with a low amount of dissolved oxygen. This was because of the transfer of dissolved oxygen to the catalyst surface from the bulk.

Many research works have been published by different combinations of Fenton reactions for the degradation of different industrial dyes. Ozdemir et al. [13] have studied the oxidative degradation of C.I. reactive yellow 145(RY 145) by employing both Fenton and Sono-Fenton processes treated from synthetic textile wastewater. The optimum conditions as reported by the authors are pH = 3, $[Fe^{2+}] = 20$ mg/L, for both Fenton and sono-Fenton process, whereas the only difference was in dosage of H_2O_2, $[H_2O_2] = 20$ mg/L for stand-alone Fenton and $[H_2O_2] = 15$ mg/L in sonolytic Fenton process. About 91% color removal was obtained via Fenton process in comparison with 95% via sono-Fenton process, whereas 47% COD removal was achieved in stand-alone Fenton process and 51% in sonolytic Fenton process. The later was executed at 80 W power and 35 kHz ultrasonic frequency. For stand-alone Fenton and sonolytic Fenton processes the optimum conditions the molar ratios of $[Fe^{2+}]/[H_2O_2]$ were 0.60/1 and 0.80/1, respectively, which indicates that integrated process of Fenton with ultrasound improved the degradation performance even at low concentration of H_2O_2 in comparison with stand-alone Fenton. Lan et al. [14] in their work have used the ultrasonic Fenton process to investigate the degradation of naproxen (NPX). The pH was set at 3 and ultrasonic power was set at 90% and 100% removal efficiency was achieved for only 10 min reaction time in the combined sono Fenton process. Bremner et al. [15] have studied the removal efficiency of aqueous phenol solutions with the aid of sono Fenton process at ultrasonic frequencies ranging from 20 to 1142 kHz. Both TOC degradation and removal of aromatics increased on increasing the frequency of ultrasonic vibration. The TOC degradation was 29% after 6 h sonication. Siddique et al. [16] investigated the synergistic effect of US-Fenton (US intensity:8 Wcm^{-2}) on the removal efficiency of Reactive Blue 19. With UV-Fenton process, 78% degradation was reported for 30 min of reaction time, which is 50% for Fenton reaction alone. The optimum pH was observed to be 3.5. Nam et al. [17] studied the mechanism behind the degradation of non-volatile organic compounds with the aid of sonication. By employing spin-trapping method that tracks OH•, they also identified the major reaction sites which involves OH• for the degradation process. Patil and Gogate [18] used orifice as the cavitation purpose to couple the Fenton oxidation process with hydrodynamic cavitation to investigate the degradation of methyl parathion. The dosage of H_2O_2 was fixed at 100 mg/L, and the $FeSO_4$ concentration was varied. The H_2O_2:$FeSO_4$ ratio was varied from 1:0.5 to 1:4. They observed that the removal efficacy enhanced with the increase in the ratio of H_2O_2:$FeSO_4$. Neppolian et al. [19] studied the sonolytic degradation of MTBE and they reported that degradation rate increases with the increase in power density of the ultrasound system. The removal efficiency of 20% of MTBE was reported at the power density of 22 W/L and with the increase in power density to 48 and 76 W/L the removal efficiency increased to 74% and 84%, respectively. The effect of the addition of H_2O_2 as the operational parameter, on the removal process of 2,4-dinitrophenol (DNP) by the aid of ultrasonic waves, was studied by Gou et al. [20]. They reported the optimum concentration by varying H_2O_2 dosage from 100 to 800 mg/L and also observed that enhanced dissipation of power favors the degradation process. Ranjit et al. [21] in their study worked on determining the removal efficiency of DCP by employing the sono Fenton process for wastewater treatment where they reported that there is a reduction in the required concentration of Fe^{2+} (50%) and H_2O_2 (31%). They have also established that in comparison with the classical Fenton process which is most efficient at pH 3.5, sono Fenton process has higher removal performance even at higher pH of 5. Ma et al. [22] used standalone ultrasound and combined ultrasonic Fenton process for efficient degradation of carbofuran. The initial dosage of carbofuran was set at 20 mg/L at 250 °C and pH of 3. They reported that the ultrasonic process oxidizes more than 40% of carbofuran in a duration of 120 min. Compared to stand alone ultrasonic process, the removal efficiency was aggrandized to even more than 99% with 40% mineralization for the combined ultrasonic Fenton process within 30 min of reaction time.

Zhang et al. [23] reported a comparative study of the capabilities of three advanced oxidation processes (AOP) for the reduction of petroleum hydrocarbons. The three AOPs employed are stand alone conventional Fenton reaction process, stand alone ultrasonic process and combined ultrasonic Fenton's reaction process. They have found that under experimental conditions, the stand alone ultrasonic process and conventional Fenton's process resulted in a reduction of 22.6 and 13.8% but for the combined ultrasonic Fenton's process, a TPC reduction of 43.1% was achieved. Liu et al. [24] employed microwave-assisted Fenton process to degrade methylene blue dye. The influence of various operational parameters like initial dosage of the organic dye pollutant, H_2O_2 dosage, and Fe^{2+} dosage was studied to determine the optimal conditions. Later, the degradation results were analyzed by UV-vis spectroscopy.

Various studies have been conducted on the removal of pollutants from wastewater. However, studies on the microwave-assisted Fenton method for decolorization of Rhodamine B dye are scant. In the present study, degradation of Rhodamine B (RhB) dye has been investigated by treating it with a microwave-assisted Fenton process. The influence of various operational parameters on dye degradation has been reported. Color removal was studied by varying process parameters like initial pH, $FeSO_4$ dosage, and H_2O_2 dosage at fixed microwave power (450 W) to investigate the optimum process parameters. The experiment for degradation of RhB was done by keeping the concentration of RhB fixed at 200 mg/L, $FeSO_4$ at 3.60E−04 mol/L, and H_2O_2 at 2.74E−02 at pH 3.0. The efficacy of microwave-assisted Fenton process over classical Fenton process is hence studied.

2 Experimental Details

2.1 Materials

In this study, Rhodamine B (Rhodamine 610, C.I Pigment Violet 1, C.I. No. 45170) was selected as the model dye for the treatment. It was manufactured by Loba-Chemie, Mumbai, India with molecular formula $C_{28}H_{31}ClN_2O_3$, and molecular weight 479.02 g/mol was procured, and it's wavelength of maximum absorbance is 554 nm. Hydrogen peroxide (H_2O_2, 30% w/v) which is a colorless liquid, of molecular weight 34.0, in stable form, was obtained from Merck Specialties Private Limited, India was used to prepare an aqueous stock solution. Crystalline Ferrous sulfate heptahydrate ($FeSO_4.7H_2O$) was collected from Sisco Research laboratories Pvt Ltd, Mumbai-400099, India. It is of molecular weight 278.02 and 0.5% (w/v) of $FeSO_4$ solution has been used for this experiment. The pH of the solution was adjusted by sodium hydroxide (NaOH), and sulfuric acid (H_2SO_4), of the analytical scale was purchased from Merck (Germany) and was used uptill pH 4.5. In order to quench the Fenton's reaction 1.04 gm of anhydrous sodium bisulfite ($NaHSO_3$),

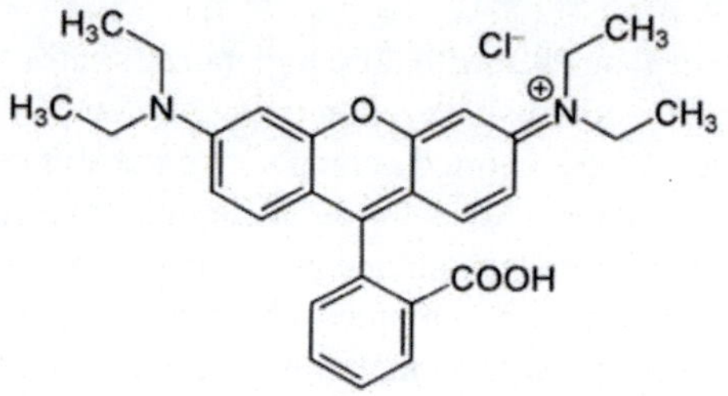

Fig. 1 Chemical structure of Rhodamine B

from LobaChemie was added to the samples. Rhodamine B is a frequently used tracer dye with water for the determination of the rate and flow direction which utilizes the fluorescence property of the dye for easy detection. The dye is suspected to be carcinogenic (Fig. 1).

2.2 *Experimental Procedure*

Microwave-assisted Fenton treatment of pollutants in wastewater

About 200 ppm concentration of pollutant solution (Rhodamine B dye solution) has been taken in a 500 ml glass beaker and the pH of the reaction mixture has been adjusted by using 0.2 (N) H_2SO_4 or NaOH. The sample containing the beaker has been kept in an ice bath. Proportional amounts of 0.5% (w/v) ferrous sulfate ($FeSO_4$) solution and 30% (w/v) hydrogen peroxide (H_2O_2) have been added to it and the mixture has been exposed to microwave (MW) in a microwave oven (Model no: MW73AD-B/XTL). The microwave power was set at 450 W. Samples obtained from Fenton's reaction have been withdrawn from time to time, the reaction has been quenched with 1.04 gm in 2 ml sodium bisulfite ($NaHSO_3$) solution. The residual dye content was analyzed by UV-vis spectrophotometer against a calibration curve. The process parameters like initial pH of the solution, $FeSO_4$ dosing, and H_2O_2 dosing were later optimized to investigate the maximum percent of degradation of RhB dye (Fig. 2).

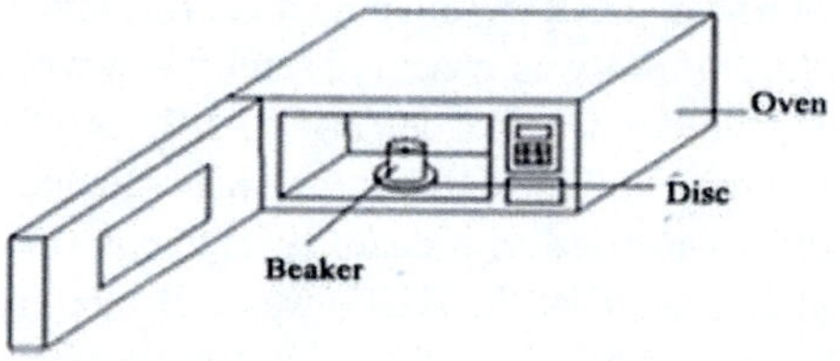

Fig. 2 Schematic diagram of experimental set up for microwave-assisted Fenton process

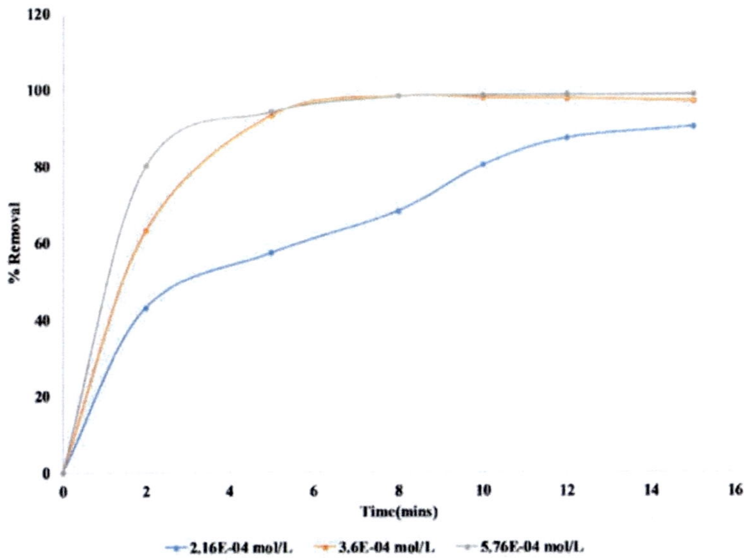

Fig. 3 The effect of $FeSO_4$ dosage on the decolorization of RhB. [RhB] = 200 mgL^{-1}, $[H_2O_2]$ = 3.92E−02 mol/L, MW power = 450 W

3 Result and Discussions

3.1 Effect of $FeSO_4$ Dosage

The $FeSO_4$ dosage is one of the major operational parameters that determines the treatment yield and removal performance. We can also predict the economy of the microwave-assisted Fenton process. In the present study, the effect of Fe^{2+} dosing on the degradation efficiency was studied at a fixed pH of 3 and at a fixed H_2O_2 dosage of 3.92E−02 mol/L and RhB concentration of 200 mg/L. From Fig. 3, it is evident that decolorization of RhB increased with increasing Fe^{2+} concentration in solution. It can be observed that there is an insignificant increase in the percent decolorization of RhB after 3.60E−04 mol/L dosing of Fe^{2+}. So, 3.60E−04 mol/L of Fe^{2+} dosage has been taken as optimum at which percent decolorization of RhB obtained was 99.49% after 15 min of reaction.

In the present study, it was observed that there has been a gradual increase in decolorization with increasing in Fe^{2+} dosage in the experimental range. The reason behind this may be because of the low concentration of Fe^{2+} compared to the concentration of the dye.

3.2 *Effect of H_2O_2 Dosage*

In Fenton chemistry for the determination of the overall efficiency of the degradation process, the initial dosage of H_2O_2 is a crucial operational parameter. With the initiation of the Fenton reaction assisted by MW irradiation, it is known that H_2O_2 dissociates into hydroxyl radicals which is the main primary oxidant for the degradation of pollutants.

$$H_2O_2 + \text{MW irradiation} \rightarrow OH^{\bullet} + OH^{\bullet} \tag{10}$$

The role of hydrogen peroxide dosage on RhB decolorization has been studied by keeping the initial dye concentration fixed at 200 mg/L, pH at 3 and $FeSO_4$ concentration at 3.60E−04 mol/L as shown in Fig. 4. The addition of H_2O_2 dosage into RhB aqueous solution has positive effect on degradation of RhB, but it was observed that there was an optimum H_2O_2 concentration at 2.74E−02 mol/L. About 99.16% RhB dye was decolorized at optimum H_2O_2 concentration. Further increase in H_2O_2 concentration after 2.74E−02 mol/L did not significantly enhance the decolorization. Radical scavenging effect of H_2O_2 is the reason for this behavior.

$$H_2O_2 + OH^{\bullet} \rightarrow H_2O + {\bullet}OH_2 \tag{11}$$

3.3 *Effect of pH*

The solution pH is an important operational parameter that has a strong influence on the efficiency of microwave-assisted Fenton reaction. The dosage of the catalyst, ferrous ion (Fe^{2+}), and the generation of hydroxyl radicals (•OH) is greatly influenced by the initial pH of the solution. In the present study, the effect of initial pH on RhB degradation was investigated with pH varied in the scale from 2.5 to 4.5 at a fixed $FeSO_4$ dosage of 3.60E−04 mol/L, H_2O_2 dosage of 2.74E−02 mol/L, and RhB concentration of 200 mg/L. It can be observed from Fig. 5 that maximum

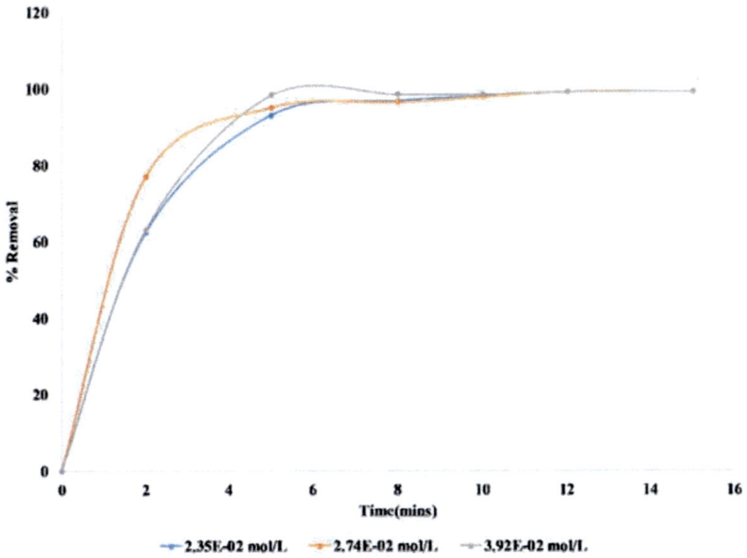

Fig. 4 The effect of H_2O_2 dosage on the decolorization of RhB. [RhB] = 200 mgL^{-1}, [$FeSO_4$] = 3.60E-04 mol/L, MW power = 450 W

decolorization of RhB was found at pH 3. About 99.16% RhB dye has been decolorized at optimum pH. Figure 5 clearly explains that maximum decolorization of the Rhodamine B was achieved at pH 3. As shown in Fig. 5, with the decrease in initial pH from 3 to 2.5 caused a decrease in the degradation efficiency. The lower decolorization at pH < 3 is due to inhibition of OH• formation. In the present study, at pH greater than 4 coagulation of the ferric hydroxyl complexes decreases the removal performance. At pH > 4, there is a larger amount of Fe(II) complexes that are formed resulting in the suppressed generation of free radicals. At higher pH, ferric oxyhydroxides precipitates which inhibits the regeneration of ferrous ion resulting in decreased rate of degradation of pollutant.

3.4 Suitability of Microwave-Assisted Fenton Process Over Normal Fenton Process

The experiment for degradation of RhB was done by keeping the concentration of RhB fixed at 200 mg/L, $FeSO_4$ at 3.60E−04 mol/L, and H_2O_2 at 2.74E−02 at pH 3.0. From the Fig. 6, it is clear that the rate of degradation of RhB is much higher in microwave-assisted Fenton process over the normal Fenton's process because of the increased generation of •OH radicals which is the primary oxidant for the efficient removal of organic pollutants. About 99.16% decolorization of RhB dye has been

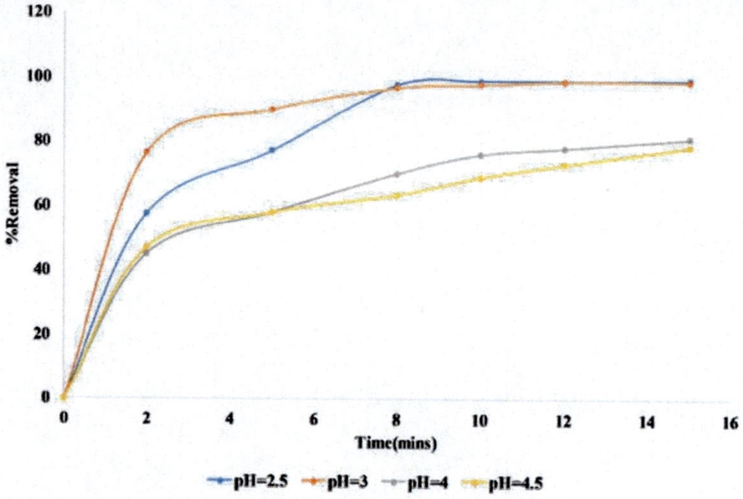

Fig. 5 The effect of initial pH on the decolorization of RhB. [RhB] = 200 mgL^{-1}, [H_2O_2] = 2.74E−02 mol/L, [$FeSO_4$] = 3.60E−04 mol/L, MW power = 450 W

achieved by MW-Fenton process, while 95.19% of RhB dye has been decolorized by normal Fenton's process after 15 min of reaction.

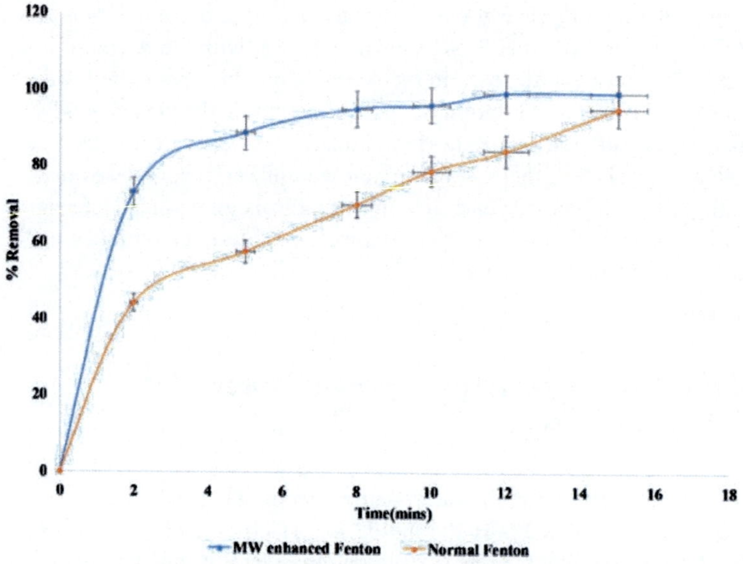

Fig. 6 Removal % of MW-assisted Fenton's process with Fenton's process on the decolorization of RhB under identical condition. [RhB] = 200 mgL^{-1}, [H_2O_2] = 2.74E−02 mol/L, [$FeSO_4$] = 3.60E−04 mol/L

For the microwave-assisted Fenton process, hydrogen peroxide which is the main oxidizing agent reacts with ferrous ions to produce highly reactive hydroxyl radicals, followed by the generation of intermediate complex [$Fe(OOH)^{2+}$] on the oxidation of Fe^{3+}. The complex on further reaction dissociates into Fe^{2+} and •OOH under microwave irradiation. This isolated free-floating Fe^{2+} further reacts with H_2O_2 to produce OH and thus it follows a cyclic mechanism. The action of Fenton reagent assisted by MW irradiation led to enhanced production of •OH radicals from the recovered Fe^{2+} ions. The primary reason behind the increased rate and improved removal efficiency is that MW could accelerate the oxidizing action of H_2O_2 to produce more hydroxyl radicals and other reactive oxygen intermediates.

The main drawback of the normal Fenton process is that as the reaction progresses and if the reaction time is large enough, the catalyst Fe^{2+} under the action of H_2O_2 rapidly gets converted to Fe^{3+} which then further reacts with H_2O_2 but much more slowly leading to a decreased production rate of Fe^{2+} in the oxidation reaction. As a result of the reduction in the amount of Fe^{2+} the Fenton reaction slows down.

Table 1 shows the comparisons between various literatures and the current study.

4 Conclusion

From the present study, it has been observed that acidic pH (pH = 3) is favorable for the degradation of RhB dye. Increase in H_2O_2 dosage and $FeSO_4$ dosage gave positive result on the degradation of RhB. In this study, it has been revealed that MW-assisted Fenton process is a more effective process in comparison with normal Fenton's process under same experimental conditions. The optimum dose of the process parameters was fixed at pH = 3.0, [$FeSO_4$] = 3.60E−04 mol/L, [H_2O_2] = 2.74E−02 mol/L at 15 min of treatment of dye of concentration 200 mg/L. However, the parametric optimization needs to executed for exploring the study further. Apart from microwave assisted, the normal fenton can be combined with ultrasonic degradation or combined effect of ultrasonic and microwave for improved efficiency. It is important to further analyze the techno-economic feasibility of the developed system in a pilot scale. From a chemical engineer's perspective, the microwave-assisted Fenton process needs to be commercialized for better results, but detailed kinetics and thermodynamics of the method needs to be studied. Additionally, the scale-up properties needs to be explored with a detailed LCA study for environmental sustainability. Optimizing microwave power and extraction of MW heat are the two critical parameters for the application of MW in real-life wastewater treatment.

Table 1 Comparison of the removal performance of various literatures

Process	Optimum pH	Time taken	Catalyst and dosage	Dye and initial dye concentration	Removal efficiency (%)	References
Microwave-enhanced Fenton's process	3	65 min	Fe^{2+}, 20 mgL^{-1}	Methylene Blue dye, 50 mg L^{-1}	98	[23]
Photo Fenton's process	6.12	120 min	SiO_2/Fe_3O_4 DMSHSs, 0.2 g L^{-1}	Rhodamine B dye, 20 mg L^{-1}	95.2	[25]
Microwave assisted Fenton reaction	–	1 min	Copper ferrite (CF) mesoporous nanocubes/reduced graphene oxide (rGO) aerogel, 15 mg	Rhodamine B dye, 30 mg L^{-1}	95.7	[26]
			rGO aerogel alone		44.6	
			CF nanocubes alone		91.1	
			Mixture of rGO + CF nanocubes		50.0	
			CuO/rGO aerogel		44.3	
			Fe_2O_3/rGO aerogel		85.8	
			Mixture of CuO/rGO + Fe_2O_3/rGO		79.3	
Microwave irradiation	2.4	7 min	Nano Fe_3O_4, 0.4 g L^{-1}	Rhodamine B dye, 300 mg L^{-1}	97.55	[27]
Microwave-enhanced Fenton like system	4	5 min	Nano Fe_3O_4, 1.25 g L^{-1}	Rhodamine B dye, 100 mg L^{-1}	99	[28]
Microwave assisted Fenton's process	3	15 min	$FeSO_4$, Fe^{2+}, 0.054686 g L^{-1}	Rhodamine B dye, 200 mg L^{-1}	99.16	Current study

Acknowledgements The author is thankful to Department of Mechanical Engineering, Birla Institute of Technology, Mesra and Dr. Richa Pandey to support and carry out the study.

References

1. Garu P (1992) Textile industry wastewater treatment. Water Sci Technol 97–103
2. Thinakaran N, Baskaralingam P, Thiruvendga Ravi KV, Panneerselvam P, Sinaveasan S (2008) Adsorptive removal of acid blue 15: equilibrium and kinetic study. Soil Air Water 36:798–804
3. Kang SF, Liao CH, Chen MC (2002) Pre oxidation and coagulation of textile wastewater by the Fenton process. Chemosphere 46:923–928
4. Ilha CEG, Santos AJMG, Desouza JR (2009) Degradation of Monoazo pigments Red 53:1 and Red 48:1 by Fenton, photo Fenton and UV/ peroxide reactions. Clean: Soil, Air, Water 37:799–805
5. Remya N, Lin J (2011) Current status of microwave application in wastewater treatment—a review. Chem Eng J 166:797–813
6. Merouani S, Hamdaoui O, Saoudi F, Chiha M (2010) Sonochemical degradation of Rhodamine B in aqueous phase: effects of additives. Chem Eng J 158:550
7. Salvador R, Casal B, Yates M, Martin-Luengo MA, Ruiz-Hitzky E (2002) Microwave decomposition of a chlorinated pesticide (Lindane) supported on modified sepiolites. Appl Clay Sci 22:103–113
8. Bo LL, Zhang YB, Quan X, Zhao B (2008) Microwave assisted catalytic oxidation of p-nitrophenol in aqueous solution using carbon supported copper catalyst. J Hazard Mater 153:1201–1206
9. Horikoshi S, Hidaka H, Serpone N (2002) Environmental remediation by an integrated microwave/UV–illumination method: characteristics of a novel UV-Vis microwave integrated irradiation device in photo degradation process. J Photochem Photobiol Ar 185–189
10. Wang N, Hu Q, Du X, Xu H, Hao L (2019) Study on decolorization of Rhodamine B by raw coal fly ash catalyzed Fenton-like process under microwave irradiation. Adv Powder Technol 30(10):2369–2378
11. Hong J, Yuan N, Wang Y, Qi S (2012) Efficient degradation of Rhodamine B in microwave-H_2O_2 system at alkaline pH. Chem Eng J 191:364–368
12. Gayathri PV, Yesodharan S, Yesodharan EP (2019) Microwave/Persulphate assisted ZnO mediated photocatalysis (MW/PS/UV/ZnO) as an efficient advanced oxidation process for the removal of RhB dye pollutant from water. J Environ Chem Eng 7(4):103122
13. Ozdemir C, Oden MK, Sahinkaya S, Kalipci E (2011) Color removal from synthetic textile wastewater by Sono-Fenton process. Clean-Soil Air Water 39(1):60–67
14. Lan RJ, Li JT, Sun HW, Su WB (2012) Degradation of naproxen by combination of Fenton reagent and ultrasound irradiation: optimization using response surface methodology. Water Sci Technol 66:2695–2701
15. Bremner DH, Molina R, Martínez F, Melero JA, Segura Y (2009) Degradation of phenolic aqueous solutions by high frequency Sono-Fenton systems (US-Fe_2O_3/SBA-15-H_2O_2). Appl Catalysis B Environ 90:380–388
16. Siddique M, Farooq R, Price GJ (2014) Synergistic effects of combining ultrasound with the Fenton process in the degradation of reactive blue 19. Ultrason Sonochem 21:1206
17. Nam SN, Han SK, Kang JW, Choi H (2003) Kinetics and mechanisms of the sonolytic destructionof non-volatile organic compounds: investigation of the sonochemical reaction zone using several OH• monitoring techniques. Ultrason Sonochem 10:139–147
18. Patil PN, Gogate PR (2012) Degradation of methyl parathion using hydrodynamic cavitation: effect of operating parameters and intensification using additives. Sep Purif Tech 95:172–179

19. Neppolian B, Jung H, Choi H, Lee JH, Kang JW (2002) Sonolytic degradation of methyl turt butyl ether: the role of coupled Fenton process and persulphate ion. Water Res 36:4699–4708
20. Gou Z, Zheng Z, Zheng S, Hu W, Feng R (2005) Effect of various sono-oxidation parameters on the removal of aqueous 2,4-dinitrophenol. Ultrason Sonochem 12:461–465
21. Ranjit PJD, Palanivelu K, Lee CS (2008) Degradation of 2,4-dichlorophenol in aqueous solution by Sono-Fenton method. Korean J Chem Eng 25:112–117
22. Ma YS, Sung CF, Lin JG (2010) Degradation of carbofuran in aqueous solution by ultrasound and Fenton processes: effect of system parameters and kinetic study. J Hazard Mater 178:320–325
23. Zhang H, Zhang Y, Zhang D (2007) Decolorization and minerisation of CI reactive black 8 by Fenton and ultrasound/Fenton methods. Color Technol 123:101
24. Liu ST, Huang J, Ye Y, Zhang AB, Pan L, Chen XG (2013) Microwave enhanced Fenton process for the removal of methylene blue from aqueous solution. Chem Eng J 215:586–590
25. Wu X, Nan Z (2019) Degradation of rhodamine B by a novel Fe_3O_4/SiO_2 double-mesoporous-shelled hollow spheres through photo-Fenton process. Mater Chem Phys 227:302–312
26. Yao T, Qi Y, Mei Y, Yang Y, Aleisa R, Tong X, Wu J (2019) One-step preparation of reduced graphene oxide aerogel loaded with mesoporous copper ferrite nanocubes: a highly efficient catalyst in microwave-assisted Fenton reaction. J Hazard Mater 378:120712
27. Vieira Y, Silvestri S, Leichtweis J, Jahn SL, de Moraes Flores ÉM, Dotto GL, Foletto EL (2020) New insights into the mechanism of heterogeneous activation of nano–magnetite by microwave irradiation for use as Fenton catalyst. J Environ Chem Eng 8(3):103787
28. Yan P, Gao LB, Li WT (2014) Microwave-enhanced Fenton-like system, Fe_3O_4/H_2O_2 for rhodamine B wastewater degradation. In: Applied mechanics and materials. Trans Tech Publications Ltd, vol 448, pp 834–837

Water Resource Engineering and GIS Remote Sensing

Estimating the Impact of Urbanization on Nitrogen Dioxide Emission Over New Delhi Region Using Satellite Imaginary Model—Pre and During COVID Era

C. Ramprasad, M. K. Deekshitha, R. Gayathri, and M. Preethika

1 Introduction

In the developing country like India, urbanization plays an important role in development and economy of the country. World Health Organization (WHO) declared that India has highest death rate from chronic respiratory disease and asthma [1, 2]. Urban populations interact with environment by changing their food consumption, energy, water and land. Delhi is the second most populated city with high pollution in India, as a step towards sustainable development needs to consider both pros and cons of urbanization. Rapid population increase and unplanned growth create an urban sprawl with negative economic, social and environmental consequence (Fig. 1). Because of urbanization, anthropogenic activities have increased which play an important role in the concentration of atmospheric parameters such as Carbon Monoxide (CO), Nitrogen Dioxide (NO_2), and Particulates Matters (PMs). During the Coronavirus disease 2019 (COVID-19) lockdown, vehicular movements, anthropogenic industrial emissions and discharges and other commercial energy-consuming activities were restricted [3–5].

There are many studies reported that the atmospheric pollutants levels of various regions over the world have significantly decreased during the lockdown period. The particulate matters (PM_{10} and $PM_{2.5}$), Carbon Monoxide (CO), Nitrogen Dioxide (NO_2) and Sulphur Dioxide (SO_2), concentrations over the Indian sub-continent showed a significant variation before and during lockdown phase [4, 6–8]. Few other studies showed that vehicular pollution, industrial emissions and power plant emissions have contributed lesser towards the atmospheric gaseous pollutants [9–12]. Huang et al. [13] have determined that the secondary pollution concentrations during the lockdown phase were improved as the result of inequities in the primary pollution

C. Ramprasad (✉) · M. K. Deekshitha · R. Gayathri · M. Preethika
School of Civil Engineering, Centre for Advanced Research in Environment (CARE), SASTRA Deemed to Be University, Thanjavur 613401, India
e-mail: ramprasad@civil.sastra.edu

A. Sil et al. (eds.), *Recent Trends in Civil Engineering*, Lecture Notes in Civil Engineering 274, https://doi.org/10.1007/978-981-19-4055-2_45

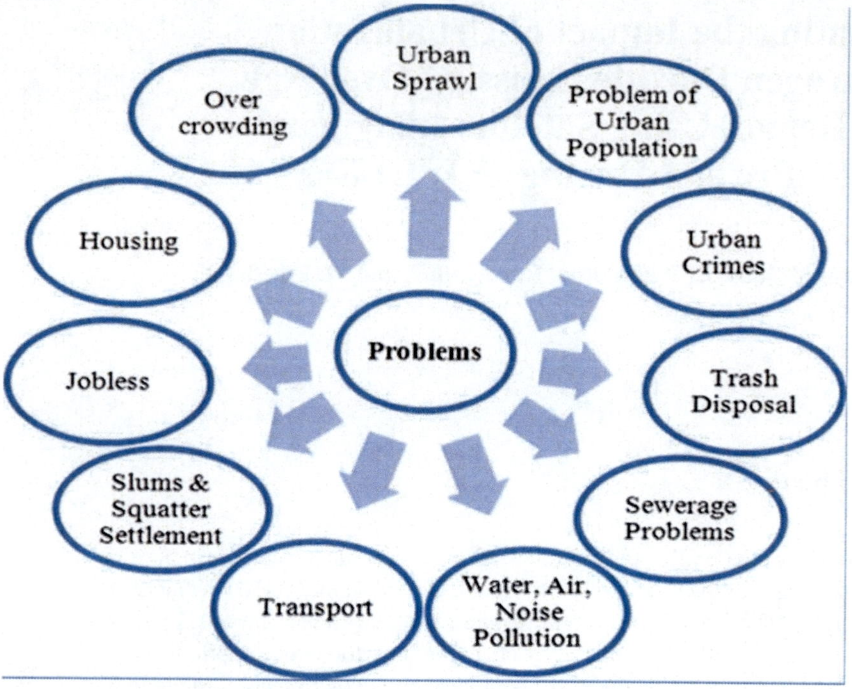

Fig. 1 Major consequences of urbanization

production and concentration in tropospheric region. The hourly measurements based taken on thermo-scientific samplers and analysers were performed over 1500 stations and the concentration of NO_2, O_2, CO, SO_2, $PM_{2.5}$ and PM_{10} were estimated. It was found that the overall reduction of NO_x about 60–70% in eastern China, 70–80% of which related to road traffic and 20–25% from industry and power plants. Emission reduction has estimated using the bottom up inventory mode of multi-resolution emission inventory of China. Response of air quality and regional coupled dynamical found that the reduction in NO_x and other atmospheric pollution, and its emissions during China's COVID-19 led to increase in O_3 and enhance the secondary pollution.

Nanda et al. [14] had mainly focused on the trace gas emissions during the pre and post-Diwali days using the satellite derived spatio-temporal variation data's during 2008–2017. The base station real-time measurement showed that the concentrations of NO_2, SO_2, CO and O_3 was 1.5–7 times higher during than the NAAQS standard safety limits during Diwali days over various major cities in India. The most sensitive and affected part was the Indo-Gangetic plain region, with an abundance of pollutants concentrated over these regions followed by the central and other parts of western India. In the present study, change in the impact of urbanization over atmospheric Nitrous Dioxide concentration over New Delhi before and during COVID-19 pandemic periods was analysed by means of user-friendly common platform GIOVANNI.

2 Materials and Methods

2.1 *Study Area*

The present study focused on one of the metropolitan cities in Indian region, the New Delhi. The GPS coordinates of the region is 77.168° E, 28.6084° N and 77.4316° E, 28.653° N. It has huge deviation in the temperature and rainfall readings between summer and winter. The hot season starts from mid of April through October and Heavy rainfall runs especially between June and September that have an average measure of 137.2 mm. The average temperature ranges between 22 °C on coldest days and 39 °C on the hottest days. In the cold season, the minimum average rainfall of 12.7 mm over November to March and mild warm temperature ranging between 7 and 28 °C on warmest days. The average blowing of wind is 10.1 km/h. Humidity is relatively high during hot season, an average value of 34.9 and 28.2% during cold season. New Delhi city which undergoes lots of pollution which causes reduction in air quality, i.e. Air Quality Index values has been recorded among the cities with the worst air quality in the world, and Delhi has always ranked first in having the worst air quality. Over 70% of the New Delhi population lives in urban areas, where the major exceedances of NO_2 pollution occur. According to studies done by the WHO over the years, a large number of deaths in the region occur due to chronic respiratory diseases and asthma. The region also suffered the highest per-capita economic loss amounting to INR 2.71 lakh crores due to air pollution in 2019.

2.2 *The NO_2 Satellite and Measurement Data*

A two-month period (April and May) during the pre-COVID period (2019) and during COVID period (2020) was selected. The tropospheric vertical column density (VCDs) of daily averaged NO_2 measured by the NASA Aura satellite, Ozone Monitoring Instrument (OMI) sensor at 0.25° spatial resolution over New Delhi were obtained from an open-access data portal (GIOVANNI). The data were processed by masking for the New Delhi administrative boundary, and the concentration for the lockdown weeks (April and May, 2020) and before lockdown phase (April and May, 2019) was computed.

3 Results and Discussion

3.1 Tropospheric NO_2 Distribution Over New Delhi

Nitrogen Dioxide is a globe's worst pollutant contributing to a large-scale air pollution. The major source of NO_2 is the fossil fuels, vehicle operating with petroleum, diesel and gasoline engines contribute to increase in NO_2 in the atmosphere. According to the Greenpeace report 2018, the coal run power plants in India are one of the largest sources of NO_2 and Nitrogen Oxide air pollutants. India's capital New Delhi is specifically a crucial location for NO_2 from the power plants, manufacturing and transport. The NO_2 gas from nearby power plants gets spread into cities like Ghaziabad, Noida, Gurugram, etc. Excessive intake of NO_2 into human body leads to chronic lung disease, lung damage and respiratory disease. The minor sources of NO_2 are from residences, agricultural waste burning, etc. Long-term exposure to NO_2 increases the mortality rate worldwide.

The tropospheric Nitrogen Dioxide concentrations over the New Delhi region was measured from the NASA Aura satellite OMI sensor, and the values during and pre-COVID time are tabulated in Table 1. The average tropospheric NO_2 levels before the COVID time was during April 2019 1.214E16 molecules per sq. cm and was 0.953E16 molecules per sq. cm during May 2019. The levels of tropospheric NO_2 decreased during the COVID lockdown and was found to be 0.591E16 molecules per sq. cm during April 2020 and 0.709E16 molecules per sq. cm during May 2020 (Fig. 2a, b). The concentrations of tropospheric NO_2 are in good agreement with Ul-Haq et al. [15]. The authors have reported a concentration of 100E13 molecules per sq. cm over the South Asia, including Indian sub-continent. Additionally, using satellite data over India, an increasing annual trend of tropospheric NO_2 concentrations was also reported by Ramachandran et al. [16].

The overall average tropospheric NO_2 concentrations during lockdown April and May 2020 reveal a 38.4% reduction. The results indicate that lockdown and subsequent restrictions on major anthropogenic activities resulted in the reduction of NO_2 levels. However, during the year 2019, the tropospheric NO_2 concentrations between the April and May saw a decrease of 21.6%. However, during the year 2020 between the months April and May, the tropospheric NO_2 concentrations increased around 20%. It may be due to few relaxations on transportation activities, and industrial operations might have caused a slight increase in the levels. Similar trend was seen in many of the earlier studies [6, 17]. The reduction in vehicular emissions, no thermal power plant operations, shorter lifetime of NO_2 and high dispersion rate are the vital cause for the decline in pollutant concentration [18, 19].

In the recent studies by researchers, the primary contributor for the NO_2 emissions in India is the emissions from the fossil fuels extraction, biomass burning and scorches, including crop residue burning [20–22]. The tropospheric chemical budget, atmospheric reactive chemistry and depletion of ozone layer largely depend on the NO_x emissions and formation of secondary pollutants [23, 24]. The data collected from the OMI satellite images shows that the concentrations of NO_2 decreased

Table 1 Concentrations of Nitrogen Dioxide over New Delhi pre and during COVID-19

Apr-19	NO_2 (E16) 1/cm^2	Apr-20	NO_2 (E16) 1/cm^2	May-19	NO_2 (E16) 1/cm^2	May-20	NO_2 (E16) 1/cm^2
01	1.4	01	0.36	01	1.19	01	1.05
02	1.4	02	0.39	02	0.93	02	0.33
03	1.4	03	0.35	03	0.93	03	0.35
04	1.55	04	0.31	04	1.2	04	0.86
05	1.59	05	0.39	05	1.2	05	0.83
06	1.17	06	0.35	06	0.87	06	0.85
07	1.13	07	0.36	07	0.87	07	0.84
08	1.19	08	0.23	08	0.87	08	0.55
09	1.33	09	0.22	09	0.95	09	0.33
10	0.91	10	0.63	10	0.85	10	0.52
11	1.05	11	0.7	11	0.85	11	0.45
12	1.08	12	0.63	12	0.9	12	0.5
13	1.05	13	0.39	13	1.15	13	0.45
14	1.15	14	0.72	14	1.15	14	0.81
15	1.17	15	0.8	15	0.78	15	0.84
16	1.17	16	0.75	16	0.78	16	0.86
17	1.17	17	0.69	17	0.75	17	0.84
18	1.19	18	0.79	18	0.85	18	0.95
19	1.22	19	0.78	19	0.85	19	0.79
20	1.22	20	0.78	20	0.85	20	0.85
21	1.25	21	0.77	21	0.85	21	0.86
22	1.55	22	0.74	22	0.9	22	0.94
23	1.62	23	0.79	23	0.95	23	0.92
24	1.05	24	0.65	24	0.95	24	0.93
25	1.05	25	0.63	25	0.85	25	0.65
26	1.15	26	0.63	26	0.81	26	0.89
27	0.88	27	0.75	27	0.81	27	1.04
28	0.95	28	0.81	28	1.15	28	0.43
29	1.19	29	0.75	29	1.25	29	0.4
30	1.19	30	0.75	30	1.29	30	0.38

during the lockdown phase (Figs. 3 and 4) and provides an even opener to apprehend the contribution of various sources in the lack of primary anthropogenic emission sources. In India, the livestock activities are predominant, and there is a prospect to explore the ammonia neutralization by nitrogen dioxide [25]. During the lockdown phase, the concentration of tropospheric NO_2 decreased by an average of 38.25%. The results prove that the NO_2 levels were decreased drastically during COVID

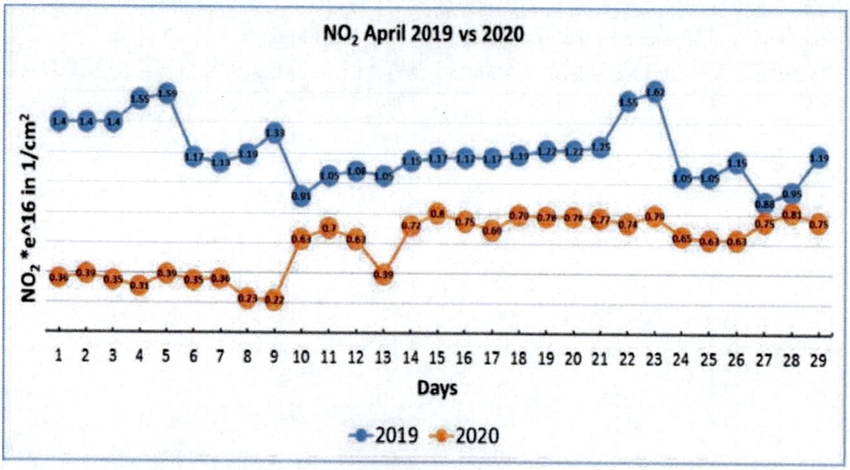

(a). Tropospheric Nitrogen dioxide concentration over New Delhi during April 2019 and 2020

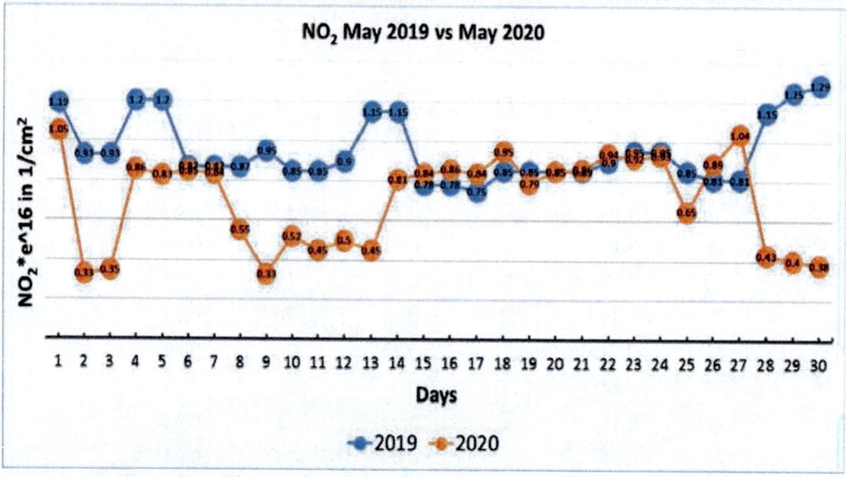

(b). Tropospheric Nitrogen dioxide concentration over New Delhi during May 2019 and 2020

Fig. 2 **a** Tropospheric Nitrogen Dioxide concentration over New Delhi during April 2019 and 2020. **b** Tropospheric Nitrogen Dioxide concentration over New Delhi during May 2019 and 2020

lockdown phase. From the study, it is evident that NO_2 levels were under the World Health Organizations permissible levels (80 μg per cu. m) of NO_2 in the atmosphere during lockdown phase. Conversely, the crop residue burning, natural (forest fire) and household solid biomass uses were common during the same period, which attributed to slightly increased NO_2 levels.

The National Clean Air Programme 2016 have provided the goal of reducing the atmospheric pollution by 20–30% over India. The lockdown has substantiated the possibility of achieving the goals by proper policy planning in restricting the

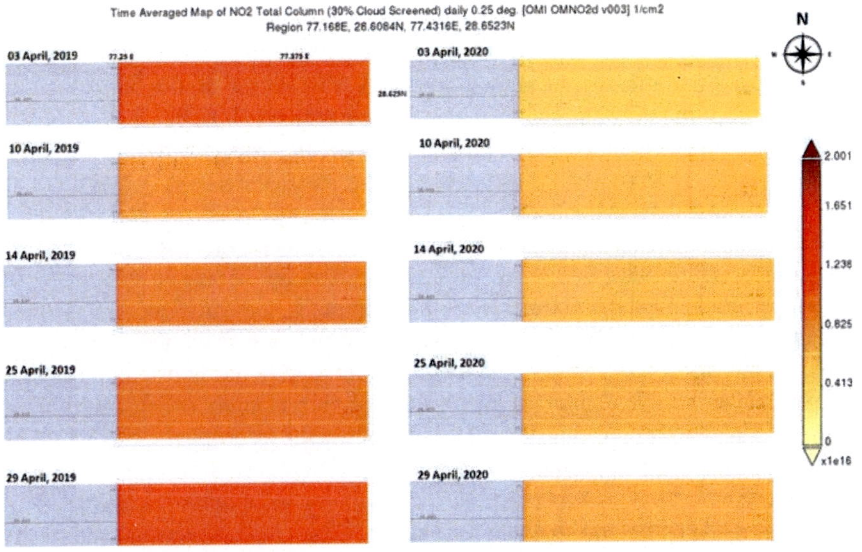

Fig. 3 Time averaged tropospheric Nitrogen Dioxide map for New Delhi during April

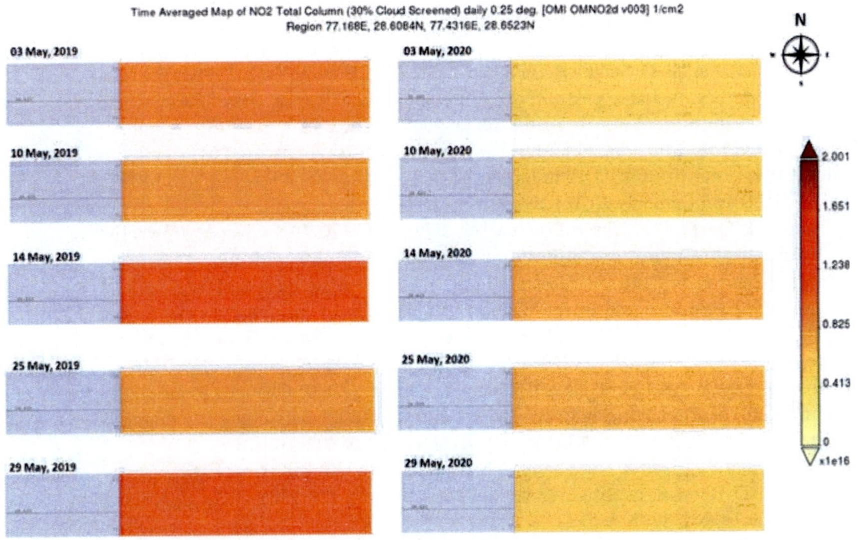

Fig. 4 Time averaged tropospheric Nitrogen Dioxide map for New Delhi during May

precursors of secondary aerosols, shorter lockdowns, reduce stubble burning and fuel switching techniques must be practised in transport and industrial sectors.

4 Conclusion

The study investigated the effect of lockdown due to COVID on the atmospheric parameters like the tropospheric Nitrogen Dioxide (NO_2) over the New Delhi, India. The obtained results show that before the imposed lockdown, the tropospheric NO_2 concentrations were 1.083E16 molecules per sq. cm. There was a decrease of 51.2% and 25.2% during the April and May before (2019) and during the lockdown phase (2020), respectively. The average tropospheric NO_2 after the lockdown showed a 38.25% reduction over the New Delhi. As the absence of major source of NO_2 emission activities like vehicular movements, biomass burning, industrial process and products manufacturing revealed the significant reduction in emission. Despite the economic and industrial deterioration, the air and water quality has boosted up in a progressive manner, in this temporary lockdown. The current study concludes that restricting the major anthropogenic emissions will reduce the tropospheric NO_2 concentrations and thereby reduce the ozone depletion. Additionally, the study results can help the planners and policy makers to formulate a pollution reduction strategies and improve the air quality for a health environment.

References

1. Gordon T, Kalpana B, Sagnik D, Sanjay R, Jonathan T, George T, Anurag A (2018) Air pollution health research priorities for India: perspectives of the Indo-US Communities of Researchers. Environ Int 119:100. https://doi.org/10.1016%2Fj.envint.2018.06.013
2. Sipra S, Muhammad MA, Muhammad I, Ehtesham H, Hafiz MHS (2020) Can $PM_{2.5}$ pollution worsen the death rate due to COVID-19 in India and Pakistan? Sci Total Environ 742:140557. https://doi.org/10.1016/j.scitotenv.2020.140557
3. Chen H, Juntao H, Qingyan F, Yusen D, Hang X, Jianmin C (2020) Impact of quarantine measures on chemical compositions of $PM_{2.5}$ during the COVID-19 epidemic in Shanghai, China. Sci Total Environ 743:140758. https://doi.org/10.1016/j.scitotenv.2020.140758
4. Jain S, Tanya S (2020) Social and travel lockdown impact considering coronavirus disease (COVID-19) on air quality in megacities of India: present benefits, future challenges and way forward. Aerosol Air Qual Res 20(6):1222–1236. https://doi.org/10.4209/aaqr.2020.04.0171
5. Wang Y, Yuan Y, Qiyuan W, ChenGuang L, Qiang Z, Junji C (2020) Changes in air quality related to the control of coronavirus in China: implications for traffic and industrial emissions. Sci Total Environ 731:139133. https://doi.org/10.1016/j.scitotenv.2020.139133
6. Sharma S, Mengyuan Z, Jingsi G, Hongliang Z, SriHarsha K (2020) Effect of restricted emissions during COVID-19 on air quality in India. Sci Total Environ 728:138878. https://doi.org/10.1016/j.scitotenv.2020.138878
7. Mahato S, Swades P, Krishna GG (2020) Effect of lockdown amid COVID-19 pandemic on air quality of the megacity Delhi, India. Sci Total Environ 730:139086. https://doi.org/10.1016/j.scitotenv.2020.139086
8. Chauhan A, Ramesh PS (2020) Decline in $PM_{2.5}$ concentrations over major cities around the world associated with COVID-19. Environ Res 187:109634. https://doi.org/10.1016/j.envres.2020.109634
9. Fan C, Ying L, Jie G, Zhengqiang L, Abdelrazek E, Mona A, de Leeuw G (2020) The impact of the control measures during the COVID-19 outbreak on air pollution in China. Remote Sens 12(10):1613. https://doi.org/10.3390/rs12101613

10. Zhang R, Yuzhong Z, Haipeng L, Xu F, Tzung-May F, Yuhang W (2020) NO_x emission reduction and recovery during COVID-19 in East China. Atmosphere 11(4):433. https://doi.org/10.3390/atmos11040433
11. Sun W, Yue Z, Jingxiang L, Jianzhong W (2019) Assessment of multi-air emissions: case of particulate matter (dust), SO_2, NO_x and CO_2 from iron and steel industry of China. J Clean Prod 232:350–358. https://doi.org/10.1016/j.jclepro.2019.05.400
12. Singh V, Shweta S, Akash B, Amit PK, Suman M, Khaiwal R (2020) Diurnal and temporal changes in air pollution during COVID-19 strict lockdown over different regions of India. Environ Pollut 266:115368. https://doi.org/10.1016/j.envpol.2020.115368
13. Huang X, Aijun D, Jian G, Bo Z, Derong Z, Ximeng Q, Rong T (2021) Enhanced secondary pollution offset reduction of primary emissions during COVID-19 lockdown in China. Natl Sci Rev 8(2):nwaa137. https://doi.org/10.1093/nsr/nwaa137
14. Nanda C, Yogesh K, Amitesh G, Mitra D (2018) Spatio-temporal distribution of pollutant trace gases during Diwali over India. ISPRS J Photogramm Remote Sens Spatial Inform Sci 4(5):339–350. https://doi.org/10.5194/isprs-annals-IV-5-339-2018
15. Ul-Haq Z, Salman T, Asim DR, Muhammad A, Khalid M, Parvez S (2015) Satellite remote sensing of total ozone column (TOC) over Pakistan and neighbouring regions. Int J Remote Sens 36(4):1038–1054. https://doi.org/10.1080/01431161.2015.1007255
16. Ramachandran A, Nayan KJ, Shashikant AS, Jayaprasad P (2013) Recent trends in tropospheric NO_2 over India observed by SCIAMACHY: identification of hot spots. Atmos Pollut Res 4(4):354–361. https://doi.org/10.5094/APR.2013.040
17. Biswal A, Tanbir S, Vikas S, Khaiwal R, Suman M (2020) COVID-19 lockdown and its impact on tropospheric NO_2 concentrations over India using satellite-based data. Heliyon 6(9):e04764. https://doi.org/10.1016/j.heliyon.2020.e04764
18. Sidhu MK, Khaiwal R, Suman M, Siby J (2017) Household air pollution from various types of rural kitchens and its exposure assessment. Sci Total Environ 586:419–429. https://doi.org/10.1016/j.scitotenv.2017.01.051
19. Chowdhury S, Sagnik D, Kirk RS (2018) Ambient $PM_{2.5}$ exposure and expected premature mortality to 2100 in India under climate change scenarios. Nat Commun 9(1):1–10. https://doi.org/10.1038/s41467-017-02755-y
20. Gurjar BR, Khaiwal R, Ajay SN (2016) Air pollution trends over Indian megacities and their local-to-global implications. Atmos Environ 142:475–495. https://doi.org/10.1016/j.atmosenv.2016.06.030
21. Rana A, Shiguo J, Sayantan S (2019) Black carbon aerosol in India: a comprehensive review of current status and future prospects. Atmos Res 218:207–230. https://doi.org/10.1016/j.atmosres.2018.12.002
22. Sembhi H, Wooster M, Zhang T, Sharma S, Nimish S, Agarwal S, Boesch H (2020) Post-monsoon air quality degradation across Northern India: assessing the impact of policy-related shifts in timing and amount of crop residue burnt. Environ Res Lett 15(10):104067. https://doi.org/10.1088/1748-9326/aba714
23. Van Der A, Eskes RJHJ, Boersma KF, Van Noije TPC, Van Roozendael M, De Smedt I, Peters DHMU, Meijer EW (2008) Trends, seasonal variability and dominant NO_x source derived from a ten year record of NO_2 measured from space. J Geophys Res Atmos 113(D4):1–12. https://doi.org/10.1029/2007JD009021
24. Li F, Paul N, Steven P, Judith P (2018) Effects of greenhouse gas increase and stratospheric ozone depletion on stratospheric mean age of air in 1960–2010. J Geophys Res Atmos 123(4):2098–2110. https://doi.org/10.1002/2017JD027562
25. Aneja VP, William HS, Jan WE, Sailesh NB, Mukesh S, William B (2012) Reactive nitrogen emissions from crop and livestock farming in India. Atmos Environ 47:92–103. https://doi.org/10.1016/j.atmosenv.2011.11.026

Detection of Oil Spills Using Sentinel-1 Data

Koushik Das, P. Janardhan, and H. Narayana

1 Introduction

The oil spill is leaking of a liquid petroleum hydrocarbon into the environment, especially from marine vehicles. An oil spill occurs mainly due to the release of unrefined petroleum products during the oil production process. It is a form of pollution by human activities which harms the marine ecosystem and also it can be a disastrous consequence for society; economically, environmentally and socially. So, it is necessary to reduce the oil spill disasters risk to protect the environment and to diminish financial losses. Advanced remote sensing technologies can identify the parties accountable for pollution and also identify the minor spills before they cause extensive damage. Manual detection of the oil spill is not feasible as it is a prolonged process and also it is very difficult to collect the marine data manually. Therefore, the task can be fulfilled by satellite remote sensing techniques. The backscatter of SAR signal over the ocean is mainly a result of the sea roughness and it is determined by small surface waves usually called gravity capillary waves. Oil fields decrease the sea surface roughness and enhance the backscatter. It causes spills to appear darker than the spill-free area. However, the contrast between polluted and non-polluted areas depends on different parameters. In the present paper, the study area is near Mauritius, where a Japanese ship named M. V. Wakashio, struck a coral reef which causes an oil spill of over 1000 tonnes into the Indian Ocean. The ship was transporting approximately 4000 tonnes of oil. The accident took place offshore of Pointe

K. Das (✉) · P. Janardhan
Department of Civil Engineering, National Institute of Technology Silchar, Silchar, Assam 788010, India
e-mail: koushik_rs@civil.nits.ac.in

H. Narayana
Department of Civil Engineering, National Institute of Technology Goa, Farmagudi, Goa 403401, India

A. Sil et al. (eds.), *Recent Trends in Civil Engineering*, Lecture Notes in Civil Engineering 274, https://doi.org/10.1007/978-981-19-4055-2_46

Fig. 1 Location of the study area over the World Ocean base map (*Sources* ESRI and Copernicus Open Access Hub)

d'Esny, south of Mauritius (Fig. 1). This paper aims at identifying and mapping this oil spill using Sentinel 1 data.

Remote sensors detect the properties of the sea surface, they send signals to the water surface and then observe the changes in the reflected signals. Mainly two types of sensors are used in remote sensing viz. active and passive. The advantages of active sensors over passive sensors are that they can obtain the measurements without depending on time and weather. Mainly used remote sensors for detecting oil spills are laser fluorosensors, microwave sensors and optical remote sensors. Microwave sensors are prevalent sensors to detect an oil spill, specifically active sensors such as radars [1]. The most used active microwave sensor is Synthetic Aperture Radar (SAR). Airplanes and satellites equipped with SAR are the main systems to monitor marine oil pollution. The key satellites used to monitor oil spills are SEASAT, ALMAZ, ERS-1 &2, RADARSAT-1 &2, ENVISAT, TerraSAR-X, etc. [2]. The important three steps of oil spill mapping are the detection of spots having the probability of being oil spill, extraction of statistical parameters of these spots and classification of these spots for discriminating the lookalikes from oil spills [3]. The age of oil slick can also be predicted from the geometrical shape of the slick and the information of wind history [4]. For detecting dark spots there are several methods based on various segmentation techniques viz. thresholding, region growing, level set segmentation and so on [5–7]. The use of neural networks for dark spot detection using edge detection, adaptive thresholding method and classification of the dark spots using another artificial neural network is an important study for oil spill detection [8]. Passive microwave sensors are also used for oil spill remote sensing. The passive microwave sensors measure the reflection of the water surface after sending the radiation from satellites and then measuring the reflection they can detect the presence of oil. The apparent emissivity factor of water is 0.4, which for oil it is 0.8 [9]. By detecting this emissivity difference, a passive microwave sensor can detect oil on the water surface. Using the SSM/I satellite data the presence of oil can be detected [10]. The thermal radiation was measured by the SSM/I from the earth's surface and atmosphere in both vertical and horizontal polarizations at 19.35–85.5 GHz [11]. Landsat imagery from thermal infrared sensors can also be used to identify the source of oil leakage in the deep sea and thick oil accumulations [12]. Different materials have different reflectivity and absorbance at different wavelengths so the emissivity is also different. The solar radiation reflected from the target is used to make the image of the earth's surface by using visible, short wave infrared and infrared sensors by optical remote sensing. But due to various limitations like the cloud cover, lack of sunlight, weather conditions, etc. the use of optical sensors is limited [1].

2 Dataset Used

Sentinel-1 Level 1 GRD products of VV polarization have been used in this study. The Sentinel 1 level-1 ground range detected (GRD) products consist of focused SAR data that was detected, multi-looked and projected to ground range using an earth ellipsoid model. The ground range detected Sentinel 1 datasets can be in one of the three resolutions: full resolution (FR), high resolution (HR) and medium resolution (MR). For interferometric wide swath (IW) and extra-wide swath (EW) modes of acquisition, the GRD products are available in medium and high resolution but for wave (WV) mode of acquisition the datasets are available in medium resolution and medium, high and full resolutions are available for strip map (SM) mode. In this study, level-1 GRD products of EW (Fig. 2) and IW (Fig. 3) modes and VV polarizations have been used for 10th August 2020 and 16th August 2020, respectively. The data used for 10th August was taken on the day and for 16th August it was taken on the night.

3 Methodology

The Sentinel 1 data of the sea area near offshore of Pointe d'Esny, South of Mauritius is procured and analyzed. The Sentinel 1 data was collected for July and August 2020. The collected datasets are processed in SNAP software to detect dark spots over the study area. The Sentinel Application Platform (SNAP) is an ESA open-source software that is used for the scientific utilization of earth observation missions in

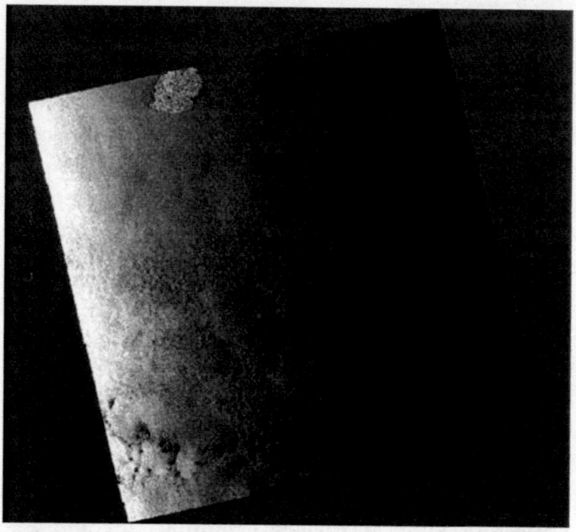

Fig. 2 Sentinel-1 GRD data for 10th August 2020 (*Sources* Copernicus Open Access Hub)

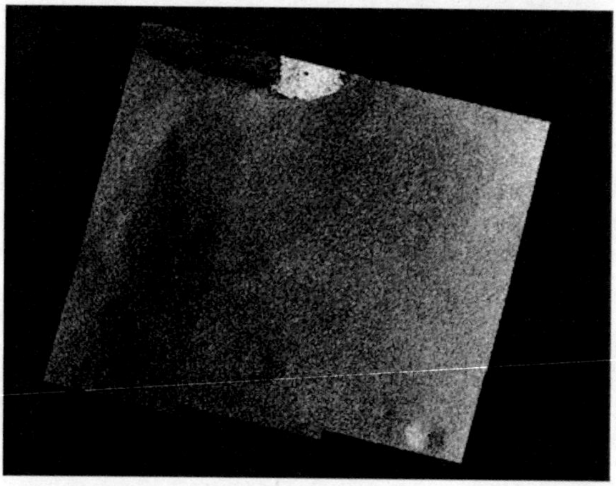

Fig. 3 Sentinel-1 data for 16th August 2020 (*Source* Copernicus Open Access Hub)

particular for the use of data from ESA missions and ESA-third party missions and from the Sentinel missions. SNAP is available to download from the Science Toolbox Exploitation Platform (STEP). The specific functions for processing optical and radar images are available in this software. The processing of the image, i.e. subsetting, calibration, speckle filtering, terrain correction, etc. and image export can be easily done in this software by using the basic processing functions. The ellipsoid correction is applied over the data for correcting its location. As the specific region of the selected dataset is the area of interest so image subsetting is done. To subset the image there are two ways: by selecting pixel coordinates and by selecting geo coordinates; pixel coordinates have been used in this particular study to subset the image. After subsetting the image, it was important to improve the contrast of the

Fig. 4 The Sentinel 1 data of 10th August 2020 after subsetting (*Sources* Copernicus Open Access Hub and SNAP)

area that is affected by the oil spill and the area that is free of pollution and for this purpose, speckle filtering is done for this data. After these processing steps, the oil spill detection is done for the selected area and the oil spill map is produced for the study area for respective dates (Figs. 4 and 5).

4 Results and Discussion

The evolution of the pixel values in dB scales of the backscatter all over the line is shown by the profile plot (Figs. 6b and 7b). In this case, one line is selected (Figs. 6a and 7a) over the study area which covers the oiled and oil-free area. From the profile plot, the differences between the backscatter values can be seen for oiled and oil-free areas. The differences between the backscatter values are used for oil spill mapping and the oil spill map is produced for 10th (Fig. 8) and 16th August (Fig. 9). The oil spill map for the above-mentioned dates is shown by overlaying the oil spill map over the study area (Figs. 10a and 11a) and also over the world ocean map (Figs. 10b and 11b).

Fig. 5 The Sentinel 1 data of 16th August 2020 after subsetting (*Sources* Copernicus Open Access Hub and SNAP)

5 Conclusions

Satellite-based SAR systems are the most useful sensors for oil spill detection, for both day and night time observation. This paper presents a procedure for oil slicks detection on sea surface which is achieved by analyzing the profile plots of the dark spots and the oil spill mapping is done. The software application for identifying dark spots and mapping oil spills is described in this study and it shows that Sentinel-1 data can be easily used for this purpose.

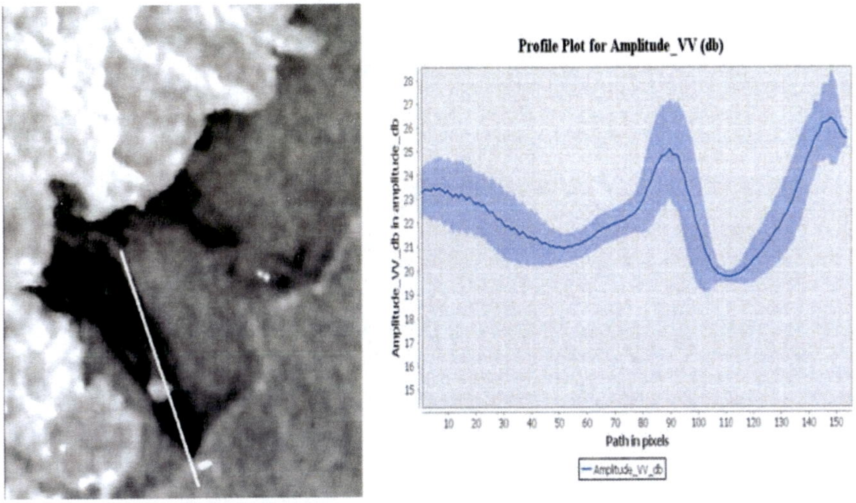

(a) (b)

Fig. 6 **a** The line for profile plot of 10th August 2020 (*Sources* Copernicus open access hub and SNAP) **b** Profile plot for 10th August (*Source* SNAP)

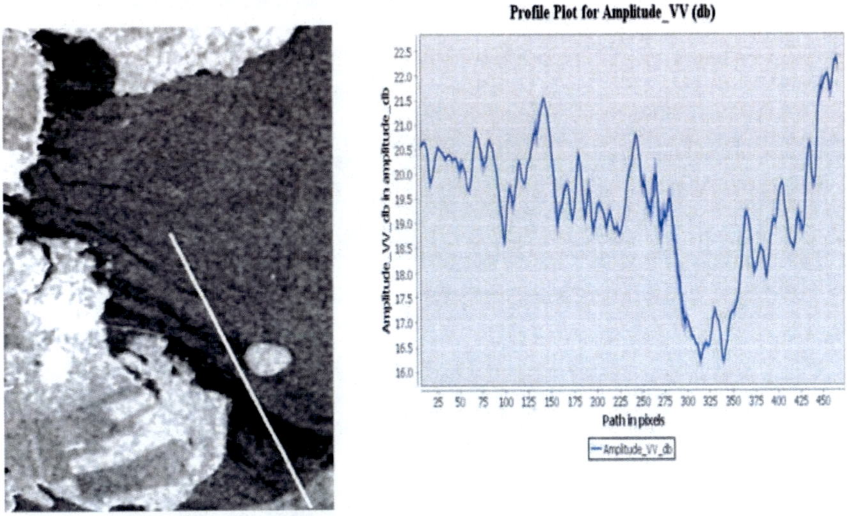

(a) (b)

Fig. 7 **a** The line for profile plot of 16th August 2020 (*Source* Copernicus Open access hub). **b** Profile plot 16th August (*Source* SNAP)

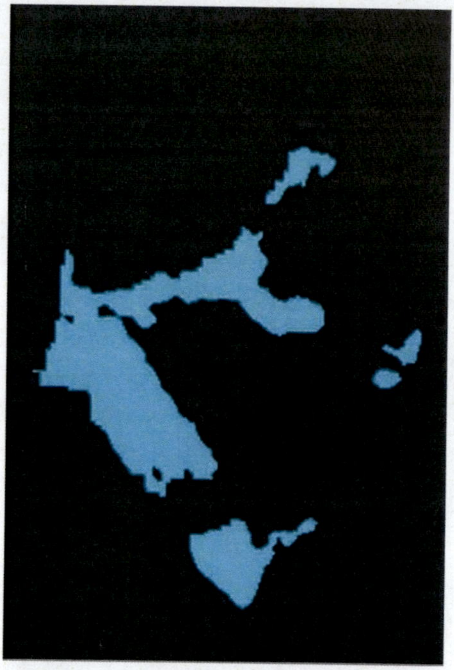

Fig. 8 Oil spill map for 10th August (*Source* SNAP)

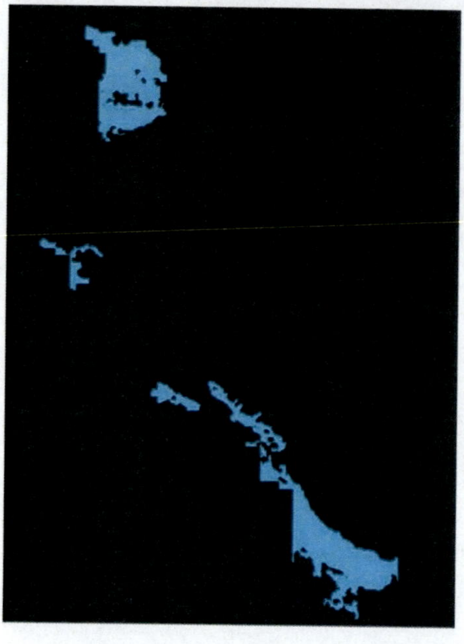

Fig. 9 Oil spill map for 16th August (*Source* SNAP)

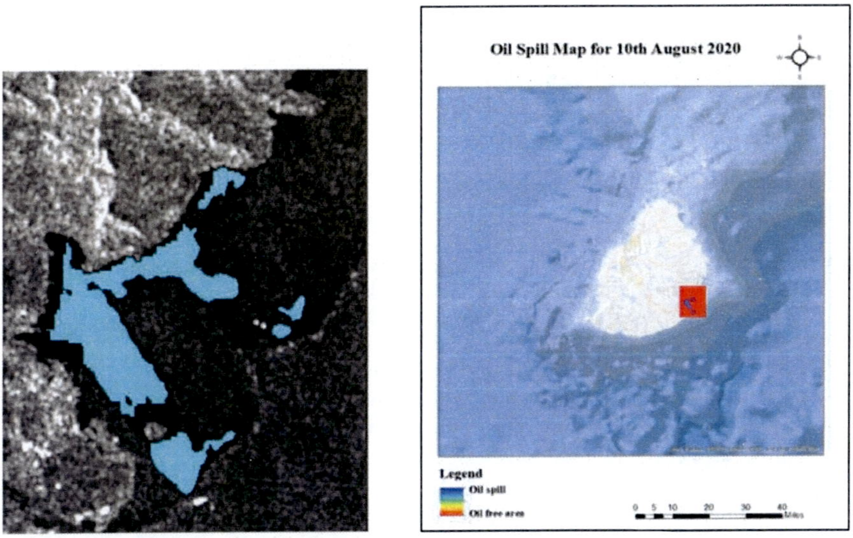

(a) (b)

Fig. 10 **a** Oil spill map over the study area for 10th August (*Sources* Copernicus Open Access hub and SNAP). **b** Oil spill map over the World Ocean Base map for 10th August (*Sources* Copernicus Open Access hub, SNAP and ESRI)

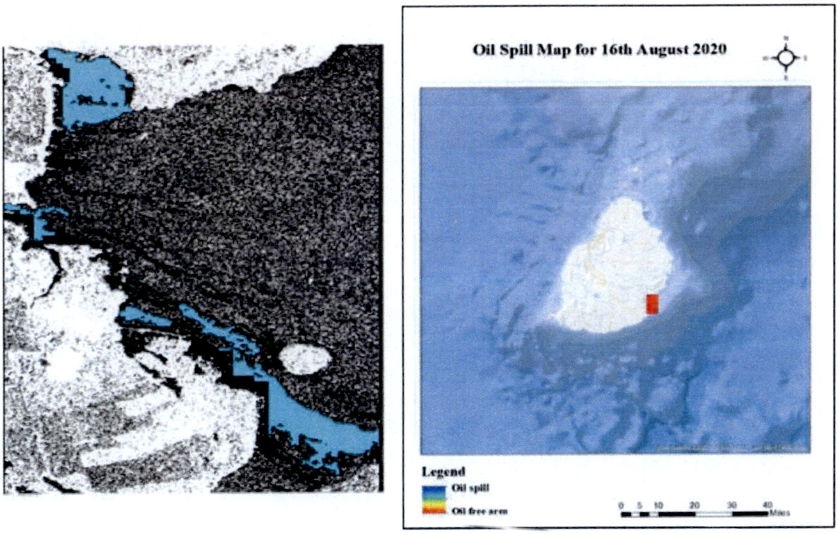

(a) (b)

Fig. 11 **a** Oil spill map over the study area for 16th August (*Sources* Copernicus Open access hub and SNAP). **b** Oil spill map over the World Ocean Base map for 16th August (*Sources* Copernicus Open access hub, SNAP and ESRI)

References

1. Fingas M, Brown C (2014) Marine Poll Bull 83:9–23
2. Topouzelis KN (2008) Sensors 8:6642–6659
3. Schistad Solberg AH, Storvik G, Solberg R, Volden E (1999) IEEE Trans Geosci Remote Sens 37:1916–1924
4. Espedal HA, Wahl T (1999) Int J Remote Sens 20:49–65
5. Huang B, Li H, Huang X (2005) Int J Remote Sens 26:1145–1156
6. Karathanassi V, Topouzelis K, Pavlakis P, Rokos D (2006) Int J Remote Sens 27:5235–5251
7. Solberg AHS, Brekke C, Husoy CO (2007) IEEE Trans Geosci Remote Sens 45:746–754
8. Singha S, Bellerby T, Trieschmann O (2012) IGARSS 5630-33
9. Ulaby FT, Richard KM, Adrian KF (1981) Microwave remote sensing-active and passive 1 and 3. Artech House, Kansas
10. Calla OPN, Dadhich HK, Singhal S (2013) Indian J Radio Space Phys 42:52–59
11. Sun N, Weng F (2008) IEEE Trans Geosci Remote Sens 46:1006–1016
12. Xing Q, Li L, Lou M, Bing L, Zhao R, Li Z (2015) Aquat Proc 3:151–156

Analysing the Changing Variations of Indian Ocean Dipole (IOD)—Indian Summer Monsoon Rainfall (ISMR) Relationship Across Northeast India

Partha Pratim Sarkar, Prashanth Janardhan, and Parthajit Roy

1 Introduction

The meteorological conditions over the North Indian Ocean and surrounding landmass are predominantly influenced by monsoons. It is due to the monsoons that the sea surface temperatures (SSTs) on eastern side of the Indian Ocean become warmer as compared to its western counterpart. The Northeast part of India is responsible for greater freshwater input into the eastern part of the Ocean in the form of precipitation, surface run off and numerous river systems. This also results in lower basin salinity on the eastern end. The thermocline, which acts as a boundary between warm water of the oceanic mixed layer and the cold water underneath, is much deeper on the eastern side of the Indian Ocean. The unevenness in the SST leads to increased convection in the east than in the west.

In the equatorial region of the Indian Ocean, the wind flow behaviour is different than the flow in the Bay of Bengal and the Arabian Sea. The winds across the equatorial Indian Ocean are weaker during the monsoons but relatively strong westerly winds develop during the spring (April-May) and during the fall (October-November) [1]. These strong winds force eastward currents along the equatorial region. The warmer layers of the ocean are therefore transported towards the east, and as a result, the thermocline is deeper in the east than in the west as it accumulates near the boundaries. The thermocline develops and becomes steeper during the spring and the fall. Therefore, these strong winds play an important role in defining the thermocline level in the equatorial Indian Ocean. This validates the strong convection process in the east. The anomaly in the above-described neutral phase during certain

P. P. Sarkar (✉) · P. Janardhan · P. Roy
Department of Civil Engineering, National Institute of Technology Silchar, Silchar, Assam 788010, India
e-mail: Parthasarkar.nits@gmail.com

P. Janardhan
e-mail: prashanth@civil.nits.ac.in

A. Sil et al. (eds.), *Recent Trends in Civil Engineering*, Lecture Notes in Civil Engineering 274, https://doi.org/10.1007/978-981-19-4055-2_47

years represented by opposite signs of SST aberrations in the east and west is known as the Indian Ocean Dipole (IOD).

IOD has 2 phases, namely positive IOD and negative IOD. In the former phase, which generally occurs during September-October, eastern Indian Ocean across the equator becomes unusually cold and the western Indian Ocean becomes unusually warm. The colder SSTs result in reduction of atmospheric convection in the east and warmer SST enhance convection in the west. The equatorial jets become weaker which in turn results in lesser movement of warm water creating a shallower thermocline. The overall sea level drops in the eastern part and rises in the centre. The thermocline comparatively becomes much deeper in the centre and in the west. The complete reverse of this is the negative phase of IOD. Negative IOD can be regarded as a further developed phase of the neural or normal condition. This study mostly focuses on the positive phase of IOD and understanding its teleconnections with the Indian summer monsoon rainfall (ISMR).

IOD events are identified using Dipole Model Index (DMI) which can be defined as the difference between western and eastern SST anomalies along the equatorial Indian Ocean [2, 3]. Mean anomalies are obtained for the spatial grid 50° E–70° E and 10° S–10° N in the west and 10° S–Equator and 90° E–110° E in the East. For a positive IOD, the standard deviation is expected to be higher than 1 for a running period of 3–4 months for the west minus east anomalies (positive) and vice versa. It has been observed that the initiation of IOD is heavily linked to the seasonal cycle due to the thermodynamics of air-sea interaction between an atmospheric anticyclone located to the east of the island of Sumatra and the Ocean below which is dependent on the seasonal wind cycle [4, 5]. The objective of the present work is to understand the relationship between IOD and ISMR and assessing its impacts on the summer monsoon rainfall distribution across Northeast India. Influence of IOD on the rainfall distribution during the withdrawal phase of the monsoon is also investigated. The next section highlights the IOD association with ISMR. Section 3 presents an overview of the climate and weather conditions across Northeast India. Section 4 utilizes the DMI and ISMR data sets to broadly depict the teleconnections, which is followed by the results (Sect. 5). The last section presents the concluding remarks.

2 IOD and Indian Summer Monsoon Rainfall

During the early 1980s, the discovery of a strong El Nino Southern oscillation (ENSO) event [6] sparked a major interest among the researchers in understanding the seasonal variability of Indian summer monsoon rainfall (ISMR) related to such anomalous climatic conditions. Pant and Parthasarthy [7]; Rasmusson and Carpenter [8]; Singh and Pai [9] and Singh et al. [10] owed this to the inter-annual variation of the coupled ocean- atmosphere system across the Pacific. Rasmusson and Carpenter [8] observed a marked correlation between monsoon rainfall and ENSO events. For the period 1875–1979, the El Nino events resulted in lesser rainfall in 21 out of 25 such cases. Nine out of eleven such seasons produced significant aberrations [11].

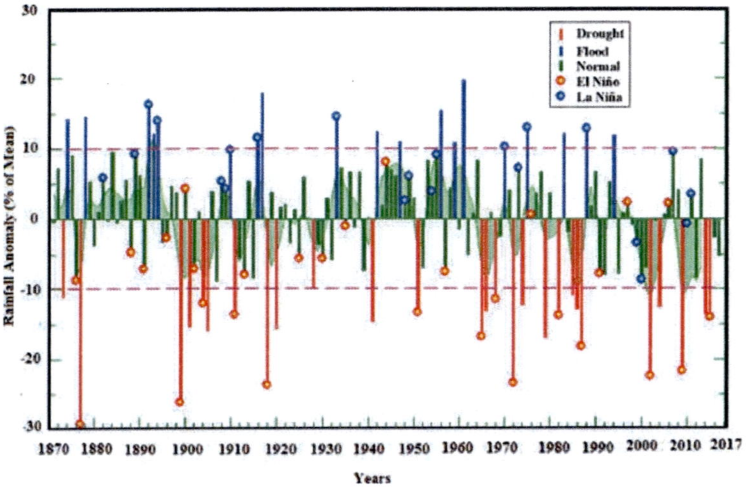

Fig. 1 Time series evolution of ISMR anomalies as percentage of mean across India for the period 1870–2017

The drought years in the country mostly coincided with the El Nino (warm) events. However, the droughts of 1974 and 1985 coincided with La Nina (cold) events. The excess rainfall seasons were observed to be influenced by La Nina events with exception like in 1997, which was a strong El Nino year. Figure 1 shows the time series Indian summer monsoon rainfall anomalies expressed as percentage departures from its long-term mean for the period 1870–2017. Figure shows the droughts, floods, El Nino and La Nina events.

The relationship between DMI and ISMR is considerably weaker. For strong positive IOD events (greater than 1 standard deviation), it was observed to have associated with positive aberrations of ISMR. For the period of 1958–1997, eight such positive IOD events (1961, 1963, 1967, 1977, 1983, 1994, 1993 and 1997) coincided with positive ISMR anomalies [12]. During the same period, two negative IOD events coincided with negative ISMR anomalies. It was also observed that the correlation behaviour of ISMR-ENSO and ISMR-IOD is different from each other, i.e. when ISMR-ENSO correlation is strong (negative), ISMR-IOD (positive) correlation is weak and vice versa. The DMI correlation with ISMR anomalies for the period 1960–2016 is depicted in Fig. 2. Both the DMI and ISMR anomalies are normalized by their respective standard deviations. These climatic anomalies (ENSO and IOD) certainly influence ISMR, but we cannot draw a direct relationship between them as they cannot explain all the droughts and excess monsoon seasons. Coupled climate models such as AGMCs and OAGMCs have also suggested that positive IODs enhance ISMR [13, 14].

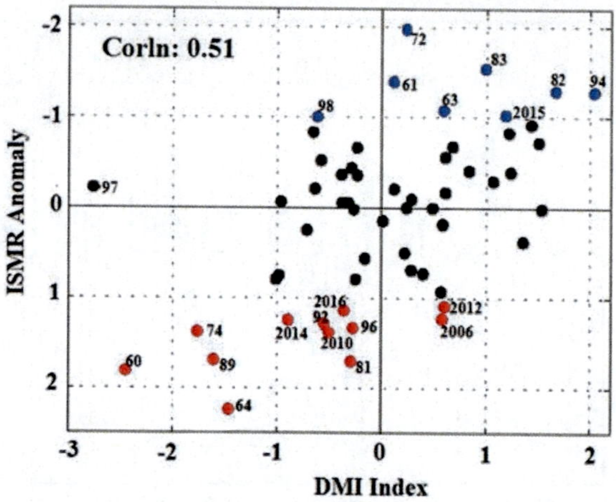

Fig. 2 Scatter diagram depicting normalized ISMR anomalies versus DMI anomalies for the period 1960–2016 (*ISMR data source* http://tropmet.res.in and *DMI source* http://psl.noaa.gov)

3 Climate and Weather Across Northeast India

Northeast (NE) India (located between 22° N and 29° N) comprising 6 states, namely, Assam, Arunachal Pradesh, Manipur, Mizoram, Nagaland and Tripura, has a subtropical climate and receives the highest average rainfall as compared to other regions of the country. Northeast part of India along with the western part of Western Ghats are the only regions that receive very heavy rainfall during the retreating monsoon. During the retreating monsoon, in the months of October and November, the southwest monsoon winds become weaker and start to withdraw from the North Indian region. In our present study, emphasis is on the entire NE region due to its uniqueness of topography and for the fact that this region is the only region across the whole country that receives very heavy rainfall after the withdrawal of southwest monsoon. The NE region along with some important weather stations covering the area is shown in Fig. 3 It is the rainiest region and Cherrapunji, a place in Meghalaya receives an average annual rainfall of 11,420 mm, which is the highest in the world. Barring a few places, most of the region receives an average rainfall of around 2000 mm yearly. The number of rainy days in this part of the country is comparatively much higher, varying from 90 to 120 days. There are two main river systems in this region, Brahmaputra and Barak. The monsoons occur in the summer, which coincides with the rainy season starting from mid-April until mid-October. The rainiest months are June and July.

Most of the region receives 90% of its rainfall during the summer months from the southeast monsoon. The monsoon burst for the regions downstream of Brahmaputra is in the month of June. Moving eastwards and upstream of Brahmaputra, the burst shifts to July. Unopposed southwest monsoon jets incepting from the Bay of Bengal move swiftly towards north of Bangladesh and enters the country in Meghalaya which are then obstructed by the plateau in that region and divides it into two section of wind jets. It is very interesting to note that Cherrapunji and Shillong, 50 km apart from

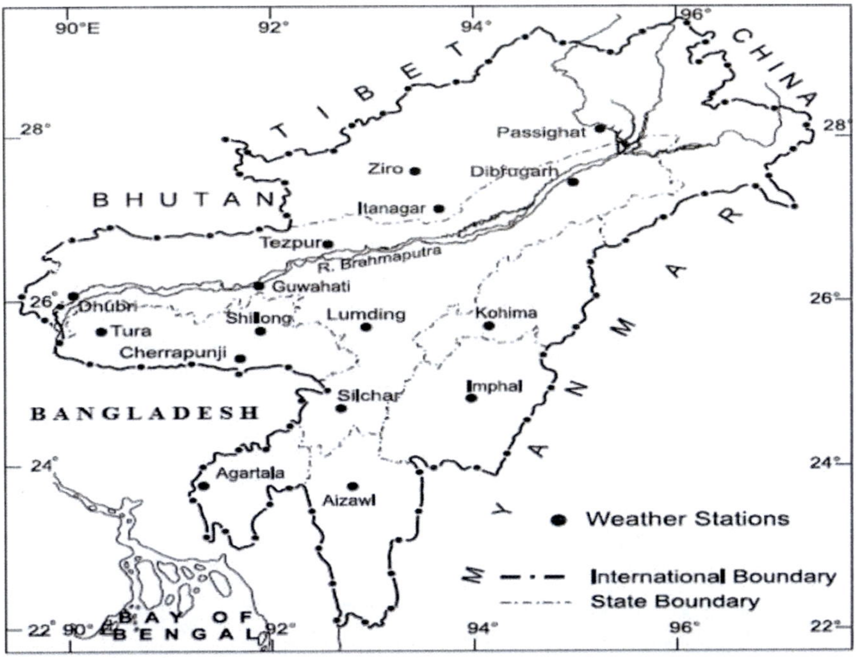

Fig. 3 Map showing the locations of a handful of weather stations across Northeast India

each other, have drastic difference in average annual rainfall. Cherrapunji receives 11,420 mm mean annual rainfall while Shillong receives only 2420 mm average annual rainfall. The rain shadow effect due to the plateau of that region plays a very important role. It also restricts rainfall in Guwahati region and has the lowest amount of rainfall (1717 mm) in the Brahmaputra valley. The rainfall declines as one move towards the east along the Brahmaputra valley. The central region receives the lowest rainfall, which again rises in the Northeast ward direction and reaches a maximum at Passighat in Arunachal Pradesh. At Passighat, the Brahmaputra emerges from the terrains and moves into the plains in Assam. Figure 4 shows the isohyets (in mm) in the region. The effect of rain shadow axis on the rainfall in the region can be visualized through Fig. 4. The variations in rainfall are considerable as we move towards the central region from either West or East.

This region has a prolonged winter lasting from early November until mid-march. The temperatures in this region are also very contrasting in nature. Northeast India has many mountainous ranges and valleys. The temperatures in the Brahmaputra and Barak valley are somewhat similar with average temperatures around 28–31 °C during the summer months while the mean temperatures in winter are around 16–17 °C. Hilly regions of similar altitudes however show divergent temperatures like Ziro in Arunachal and Shillong in Meghalaya. Ziro has a height of 1595 m ASL and Shillong is at 1476 m ASL. Shillong has 14.7 and 5.7 °C as the average Max. and Min. temperatures in the month of January, and Ziro has 12.9 °C and −0.2 °C in the

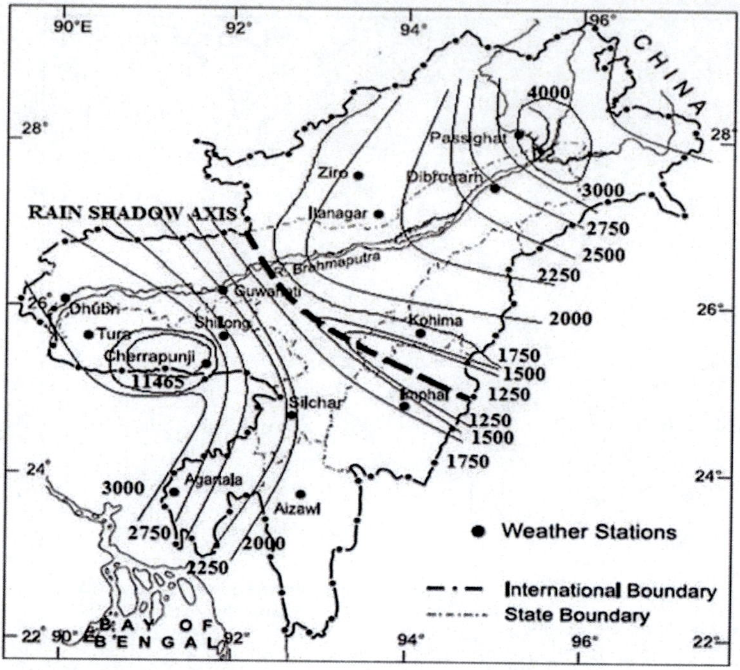

Fig. 4 Isohyets map showing the variations in annual rainfall (Isohyets in mm) in Northeast India

month of January. Shillong is nearly 50 km from Bangladesh, and Ziro is in the hilly terrains deep in Arunachal Pradesh. The hilly regions of Meghalaya as mentioned before are influenced by sea and sea winds filled with humidity.

There are three distinct rainfall zones in the Brahmaputra valley with the variation in rainfall as we move from east to west. A rainfall zone located to the west of Guwahati receiving over 2000 mm of rainfall with exception of Dhubri, which receives nearly 2900 mm rainfall annually. A lesser rainfall zone of the central region as we move from Tezpur and Nagaon regions to Lumding. Lumding region receives the least rainfall of 1200 mm annually. The last zone located to the east of Jorhat up to Passighat in the state of Aruncahal Pradesh. There is a steep rise in rainfall in this zone, 2000 mm west of Jorhat to nearly 2600 mm at Dibrugarh annually, and over 4000 mm eastwards of Passighat. The average rainfall and number of rainfall days of few weather stations across the Northeast region during the monsoon period 1960–2019 is shown in Figs. 5 and 6, respectively.

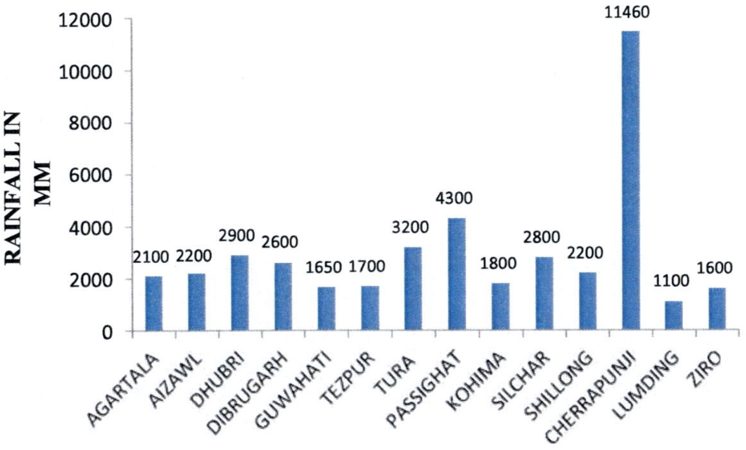

Fig. 5 Average rainfall (in mm) for some weather stations for the monsoon period 1960–2019 from Agartala to Ziro

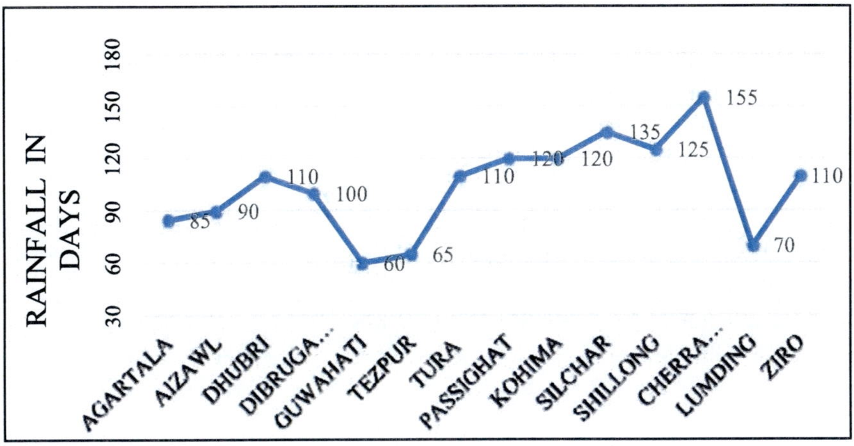

Fig. 6 Average number of rainfall days for the weather stations for the monsoon period during 1960–2019 from Agartala to Ziro

4 Data and Methodology

The summer monsoon rainfall data for the Northeast region for the period 1960–2019 has been utilized from the Indian institute of Tropical Meteorology (www.tropmet.res.in). The DMI index for the same period is employed from the data archival section of National Oceanic and Atmospheric Administration (NOAA).

The DMI index is maximum during the months of October and November (during spring). The amplitude of DMI index is minimum during March-April, which increases from the month of May and peaks during the autumn. Future time leads correlation between DMI index and western and eastern poles' SST with monthly and seasonal monsoon rainfall for the Northeast region have been computed. To define the teleconnections between DMI and onset of monsoon over Northeast India the two future lead time ahead correlations related to April and May rainfall have been computed. Similarly, correlations between the SSTs in western pole during April-May and onset date of monsoon over Northeast India have also been enumerated.

In this work, Pearson, Kendall and Spearman's rank correlation approaches are employed to study the strength of association between the above-mentioned variables.

Pearson correlation (μ) deliberates the strength of linear association between the variables (i.e. the DMI index and the ISMR anomalies) and is calculated as shown below:

$$\mu = \frac{\sum \left(V1\frac{y}{m} - V2\frac{y}{m}\right)}{\sqrt{(V1\frac{y}{m} - \overline{V}1_m)^2 \frac{N}{y=1} (V2\frac{y}{m} - \overline{V}2_m)^2}} \quad (1)$$

where $V1\frac{y}{m}$, $V2\frac{y}{m}$ are variables which denote DMI and the ISMR anomalies for 'm' month of 'y'th year $\overline{V}1_m$, $\overline{V}2_m$ are monthly average of the variables, and N denotes the total number of years.

Kendall correlation (r) investigates the dependencies of the involved variables based on the feature ranking and is calibrated as shown below:

$$r = \frac{N_{\text{corvergaent}} - N_{\text{divergent}}}{\frac{1}{2}n(n-1)} \quad (2)$$

where N is the total number of variable sets. For both the variable sets $V1$ and $V2$, a sample pair of DMI and ISMR anomalies may be called as convergent if the ranks of both the elements in $V1$ and $V2$ are same and can be termed as divergent if the ranks differ.

Spearman's correlation is a nonparametric analysis that does not carry any assumptions about the underlying distribution and can be considered a better judge of identifying monotonic behaviour between the variables. It is an appropriate correlation analysis considering the variables in our study are on a least ordinal scale. The data sets in our consideration were first ranked before proceeding with the Spearman's correlation evaluation. The ranking of the variables helps to compare whether on increasing one variable the other follows a monotonic relation with respect to it or not. The correlation can be better understood using the following equation given below:

$$\rho = 1 - \frac{6\sum d^2 i}{n(n^2-1)} \quad (3)$$

where

ρ = Spearman's correlation,

$d^2{}_i$ = The difference between the ranks of corresponding variables,

n = Number of observations.

1st of June has been considered as the normal date of onset of monsoon in the region, say for example, if onset date is 27th May, then onset anomaly date is considered as −5 and if the onset date is 6th June, then it is considered as 5.

5 Results

5.1 *Correlation Between DMI Index and Summer Monsoon Onset Over Northeast India*

The correlations of summer monsoon onset date across the Northeast region with DMI and western pole SST anomalies were observed as negative, which indicates that positive IOD during April-May could play a hand in delaying monsoon over the region. The same is supported by the observational correlations with SST anomalies in the western pole. Therefore, warmer SST anomalies in the West were found to restrict the onset of summer monsoon. SSTs in the month of May significantly correlated with the onset of monsoon with the coefficient of correlation of +0.45, significant at the 99% level as shown in Fig. 7. From the figure, it can be observed that correlations between the DMI index and western pole SST anomalies (positive phase of IOD) are at 95% and 99% significant levels for the month of April and May, respectively. With modest increase in DMI index for the month of May, SST anomalies for the month of May increased significantly. It was also observed that for 70%, such scenarios SST anomalies in the western pole intensified during May.

5.2 *Seasonal Rainfall Correlations*

Observations from Fig. 8 show that the SST anomalies in the western pole are negatively correlated to ISMR over Northeast India. Most of the lead-time correlations are generally negative. This behaviour of positive phase of IOD is similar across the entire region. Most of Northeast India is land locked except for parts of Tripura and Meghalaya, which are more influenced by sea winds and therefore are more negatively correlated. The correlations between western pole of IOD and monsoon rainfall are significant at the 95% level for 70% of the Northeast India. The best obtained one, two and three month lag-lead correlations between DMI index and seasonal rainfall for the six different states are given in Table 1. −0.7 coefficient of

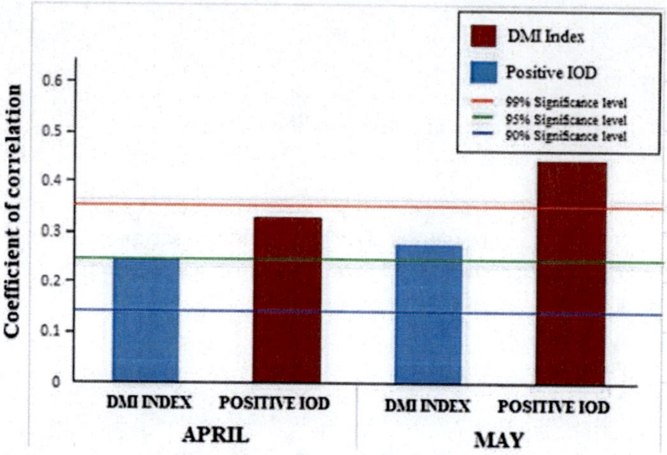

Fig. 7 Correlations between DMI index and western pole SST anomaly (p IOD) for the months April and May for the monsoon onset over Northeast India

correlation (CC) for Meghalaya and central Assam is significant at the 99% level for one month lag achieved using the nonlinear Spearman's rank correlation. The CCs for the same regions are −0.6 for two and three months significant at the 99% level. From the table below, we can observe that the lag CCs are considerably higher than the lead CCs. A possible explanation for this occurrence can be the drop in SSTs observed during the withdrawal phases of summer monsoon during the recent time period horizon (1991–2019) for both positive and negative IOD events.

Table 1 One, two and three month lag-lead CCs for the different regions in NE India

States	One month lag CC	Two month lag CC	Three month lag CC	One month lead CC	Two month lead CC	Three month lead CC
Eastern Assam	−0.75	−0.70	−0.61	−0.38	−0.27	−0.12
Central Assam	−0.72	−0.68	−0.62	−0.35	−0.25	−0.12
Western Assam	−0.65	−0.61	−0.58	−0.34	−0.23	−0.09
Eastern Arunachal Pradesh	−0.58	−0.52	−0.46	−0.28	−0.19	−0.07
Western ArunachalPradesh	−0.61	−0.56	−0.47	−0.32	−0.21	−0.09
Meghalaya	−0.78	−0.71	−0.67	−0.41	−0.31	−0.19
Manipur	−0.59	−0.52	−0.47	−0.32	−0.26	−0.07
Mizoram	−0.56	−0.50	−0.45	−0.30	−0.23	−0.08
Nagaland	−0.58	−0.51	−0.43	−0.28	−0.19	−0.08
Tripura	−0.61	−0.55	−0.48	−0.31	−0.21	−0.10

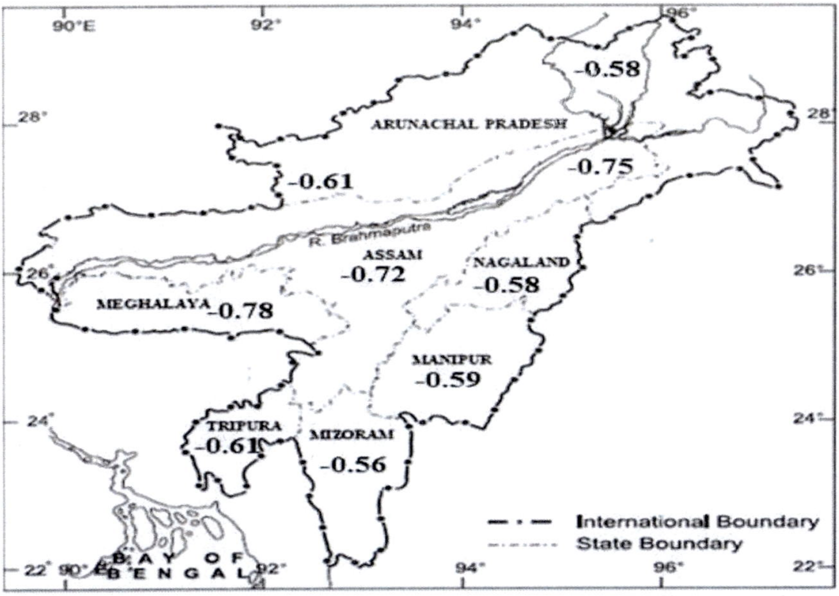

Fig. 8 Lag correlations between western pole (positive IOD) and seasonal rainfall for the six different states in Northeast India

5.3 *Monthly Rainfall Correlations*

Monthly correlations between western pole of IOD and DMI index for the months June–August are shown in Fig. 9 for the period 1960–2019. The coefficient of correlation was observed to be negatively correlated with during the positive phase of IOD events for June-August rainfall months. All the three lead-time (June-August) correlations are consistent for the region. Two months lag correlations are more prominent for August rainfall. SSTs in the eastern pole also yield a similar pattern of relationship with rainfall over eastern and central part of Assam, Meghalaya, Tripura and eastern Mizoram. From Fig. 10, it can be observed that SSTs in the eastern pole are positively correlated to seasonal rainfall over Northeast region with observed low correlations for certain regions in Arunachal Pradesh and Nagaland. However, the observed correlation of negative IOD phase and June-August monthly rainfall was found to be much weaker than its positive IOD counterpart. For strong phases of p IOD events, the onset date of monsoon over the region is observed to be delayed. Cyclones originating in the Bay of Bengal region also play an important role in this regard.

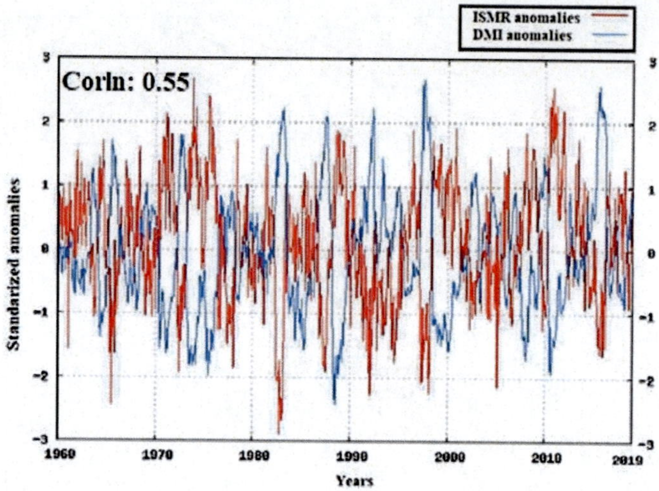

Fig. 9 Correlations between standardized ISMR anomalies and DMI anomalies for the months June-August during the period 1960–2019

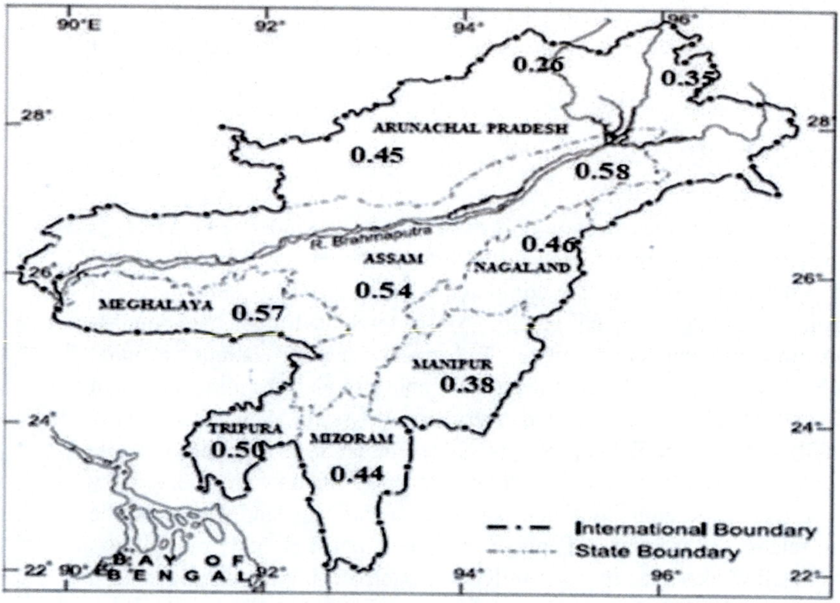

Fig. 10 Lag correlations between eastern pole (negative IOD) and monthly rainfall (June-August) for the six different states in Northeast India

6 Conclusions

Research in the last decade exhibits IOD as an important factor that affects the inter-annual and seasonal climate across the Indian sub-continent [15]. IOD plays significant role in the onset as well as in the variation of magnitude of ISMR. IOD relationship with ENSO affecting the ISMR is debatable as IOD anomaly events are generated by the India Ocean without being externally forced by coupled Ocean Atmospheric processes in the Pacific [16–19]. Research has shown that IOD can be generated either by ENSO or by sea-air interaction processes in the Indian Ocean. However, the mechanisms that generate IOD without the presence of ENSO events in still unclear. In this work, the teleconnections between IOD and ISMR across the Northeastern region of India has been established using a correlational approach. The study has brought out the following results:

(1) Positive/negative phase of Indian Ocean Dipole and warmer/colder SSTs in the western pole during April-May caused delayed/early onset of monsoon over Northeast India.
(2) The IOD climatic anomaly seems to influence the summer monsoon activity over the central and western Assam, Meghalaya and Mizoram more than over other parts of the region. Positive IOD during April-May, i.e. stronger western pole is associated with decline in rainfall over Northeast India. The correlations are significant at the 99% level for more than 50% of the region. Influence of western pole is greater on the summer monsoon rainfall in this region.
(3) This study also becomes significant as IOD events will certainly affect the retreating monsoon which commences with the beginning of the withdrawal of the southwest monsoon [mid-September-November] and lasts until early January. It is a 3-month long process where it starts from the peninsula in October and from the extreme southeastern tip by December. Retreating monsoons are different from the usual summer monsoons and results in heavy rainfall in the western part of Western Ghats and specially Northeast India. Hence, one of the key findings of this work, i.e. the DMI index of preceding one/two/three months, i.e. June, July and August can provide a good indication of summer rainfall activity over the entire region during the withdrawal of monsoon, i.e. in the month of September, it becomes even more significant.
(4) The SSTs of the western pole during March-April are associated with the detrimental monsoon rainfall over Northeast India during the onset of monsoon, i.e. in the month of June. The SST correlations are significant at the 99% level for western and central Assam, Meghalaya and Mizoram and at the 95% level for the other states in the region.

References

1. Wyrtki K (1973) An equatorial jet in the Indian Ocean. Science 181:262–264
2. Saji NH, Goswami BN, Vinayachandran PN, Yamagata T (1999) A dipole mode in the tropical Indian Ocean. Nature 401:360–363
3. Murtugudde R, McCreary JP, Busalacchi AJ (2000) Oceanic processes associated with anomalous events in Indian Ocean with relevance to 1997–98. J Geophys Res 105:3295–3306
4. Annamalai H, Murtugudde R, Potemra J, Xie SP, Liu P, Wang B (2003) Coupled dynamics over the Indian Ocean: spring initiation of the zonal mode. Deep Sea Res Part II(50):2305–2330
5. Li T, Wang B, Chang CP, Zhang Y (2003) A theory for the Indian Ocean dipole-zonal mode. J Atmos Sci 60:2119–2135
6. Bjerkenes J (1969) Atmospheric teleconnections from the equatorial Pacific. Mon Wea Rev 97:163–172
7. Pant GB, Parthasarathy B (1981) Some aspects of an association between the southern oscillation and Indian summer monsoon. Arch Meteorol Geophys Biokl 29:245–251
8. Rasmusson EM, Carpenter TH (1983) The relationship between eastern equatorial Pacific sea surface temperatures and rainfall over India and Sri Lanka. Mon Weather Rev 111:517–528
9. Singh OP, Pai DS (1996) An oceanic model for the prediction of SW monsoon rainfall over India. Mausam 47:91–98
10. Singh OP, Ali Khan TM, Rahman S (2000) Changes in the frequency of tropical cyclones over north Indian Ocean. Meterol Atmos Phys 75:11–20
11. Kumar KK, Rajagopalan B, Cane MA (1999) On the weakening relationship between the Indian monsoon and ENSO. Science 284:2156–2159
12. Ashok K, Gaun Z, Yamagata T (2001) Impact of the Indian ocean dipole on the relationship between the Indian monsoon rainfall and ENSO. Geophys Res Lett 28:4459–4502
13. Guan Z, Yamagata T (2003) The unusual summer of 1994 in east Asia: IOD teleconnections. Geophys Res Lett 30:L01540
14. Ashok K, Guan Z, Saji NH, Yamagata T (2004) Individual and combined influences of ENSO and the Indian Ocean dipole on the Indian summer monsoon. J Climate 17:3141–3155
15. Sartimbul A, Erfan R, Ali MF, Khaidar Z (2018) Anomaly technique for expression the Indian Ocean Dipole in South Java Sea. Int J Eng Technol (7):28–30
16. Abram NJ, Gagan MK, Cole JE, Hantoro WS, Mudelsee M (2008) Recent intensification of tropical climate variability in the Indian Ocean. Nat Geosci 1:849–853
17. Horii T, Hase H, Ueki I, Masumoto Y (2008) Oceanic precondition and evolution of the 2006 Indian Ocean dipole. Geophys Res Lett 35:L03607
18. Hrudya PH, Varikoden H, Vishnu R (2020) A review on the Indian summer monsoon rainfall, variability and its association with ENSO and IOD. Meteorol Atmosp Phys
19. Behera SK, Ratnam JV (2008) Quasi-asymmetric response of the Indian summer monsoon rainfall to opposite phases of the IOD. Sci Rep 8:1–8

Structural Audit of Mutha River Bridge Using Institutionally Developed Displacement Sensors

Husain Anandwala and Hanamant Magarpatil

1 Introduction

Bridges are structures built over rivers, roads and other obstacles that carry roads, railways, paths, etc. In India there are many old bridges which were constructed during the British Rule and are even used today. Loads on bridges act very differently than any other structure. They are used every day and throughout the day. They are exposed to the atmospheric conditions and continuous traffic which can cause the deterioration of the bridge structure. Bridges are exposed to fatigue loading as the loads acting on the structure varies continuously throughout the day. It can be very high during the day time and very low during night which can cause deterioration of the structure.

Structural Auditing of a structure is normally done in a span of three to five years, which depends upon the guideline established by the authorities. Bridge auditing has gained importance during recent years after few accidents on old bridges took place. For the safety of the people, it is very important to regularly audit a bridge structure. The method of auditing for building and a bridge are not totally same even if they are of RCC. As the components in a bridge and a building are not same. The first and the most important step in audit of any structure is visual inspection.

After visual inspection is completed, other tests are performed as needed based on the results obtained. Rebound Hammer test and Ultrasonic Pulse Velocity test are carried out on the bridge members considered as critical during the visual inspection. It was also found out during the inspection that most of the decks showed exposed reinforcement and corrosion was seen in them so sensor test was considered to check the displacement of the decks. The bridge undertaken for the study is Mutha

H. Anandwala (✉) · H. Magarpatil
School of Civil Engineering, Dr. Vishwanath Karad MIT World Peace University, Pune, India
e-mail: anandwalahusain@gmail.com

H. Magarpatil
e-mail: hanamant.magarpatil@mitwpu.edu.in

A. Sil et al. (eds.), *Recent Trends in Civil Engineering*, Lecture Notes in Civil Engineering 274, https://doi.org/10.1007/978-981-19-4055-2_48

River Bridge which is situated on the Mutha river on Mumbai–Bangalore National Highway near Warje, Pune. It is situated on the lane from Bangalore toward Mumbai. The bridge has a span of 174.15 m. it has total of 9 decks, 8 piers and 2 abutments. Length of each deck is 19.35 m. Width of the bridge is 12 m. Length of each pier is 9.65 m and the width is 1.85 m.

Permission required for the audit work was taken from the National Highways Authority of India (NHAI) department. And according to the NHAI officials, plans for bridge structure where not available. The sensors can be used to gets the displacement of the decks under different loading conditions. In this study the sensors were used to check the displacement of the individual decks under the constant loading condition. Normally this analysis is done using software's such as ETABS where a model is prepared and then the loads are applied but, in this study, an onsite method is used to check the condition of the decks.

2 Methodology

2.1 *Visual Inspection*

The first and the most important step to be considered in auditing of a structure is visual inspection. By using this method, we can check the condition of the structure. Drawings of the structure can be studied if they are not available can be prepared. Safety of the structure is needed to be checked if not safe the critical sections are to be marked for further testing. Changes or alterations in the structure are also checked by analyzing the drawings.

Because the bridge is located on the national highway, due to heavy traffic, it was necessary to pay attention for personal safety when visually inspecting the bridge during peak hours. It is easy to enter under the bridge from both ends. There is a straight road at one end and from other end it was required to walk down from a sloping ground. Since all bridge components are easily accessible, it is easier to carry all test instruments on site and perform all tests on selected components.

While conducting visual inspection, it was observed that there are many horizontal and vertical cracks present in the piers of the bridge. Most of the deck girders had exposed reinforcement and the reinforcements had started to corrode. Following critical sections were selected for further testing.

Pier Numbers. 1, 2, 3, 4, 6, 7, 8, 9.
Abutment Numbers. 1, 2.
Deck Numbers. 1, 2, 3, 4, 5, 6, 7, 8, 9.

Expansion joints were filled with bitumen material to close the gaps in the joints. The improper work done to close the expansion joints is cause water seepage through it below on the piers. As seen from Figs. 1 and 2, same type of bitumen filling is observed in all the expansion joints. Due to the seepage of water on the decks and piers algae growth was observed.

Horizontal and Vertical Cracks were observed on the piers surfaces as seen in Fig. 3. Concrete from the surface of the piers has started to deteriorate in many places. It was even observed that the quality of concrete was not satisfactory according to the present specifications. Random mixture of sand was seen in the concrete used in the piers and which can be easily scraped of using a spatula. Concrete railing on the bridge for the side walk was observed to be damaged in many places (Fig. 4).

Cracks were observed in every deck girder and the concrete cover had started to come off. As see in Figs. 5 and 6, concrete cover of the girder has come off and

Fig. 1 Expansion joints filled with Bitumen material

Fig. 2 Water seepage from expansion joints

Fig. 3 Horizontal and vertical cracks in the piers

Fig. 4 Surface damage on the piers

reinforcements are totally exposed. Corrosion was also observed in the reinforcement. Repairing work was also observed to be done on the piers where plaster was applied to fill up the horizontal and vertical cracks. Pier number 5 being surrounded by water was observed from some distance, horizontal and vertical cracks were seen and concrete deterioration was also observed.

After observing damage to the deck girder, large amount of vibrations was also felt when standing on the bridge deck. So, deflection test on the decks using sensors was decided to be conducted. During the visual inspection, all factors need to be considered when selecting key sections for further testing, and what tests are required.

Fig. 5 Damage to girder on deck 4

Fig. 6 Damage to girder on deck 6

2.2 *Deck Displacement Test*

The test chosen to find out the condition of the bridge decks is Deck displacement test. There is total three components in the whole setup of the test:

I. Displacement Sensor.
II. Recording Instrument.
III. Laptop for getting the results.

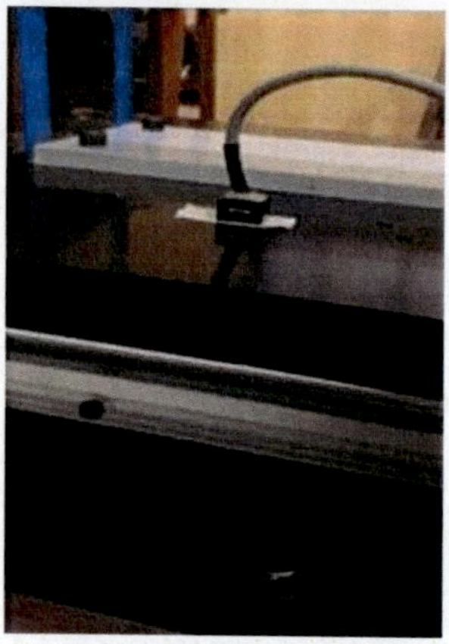

Fig. 7 Displacement sensors position

The experimental equipment was developed in the institutional laboratory by Prof. Dr. H. R. MagarPatil. These sensors can measure displacement occurring on site and can give real time results. The sensor measures displacement in millimeters, and its sensitivity is two decimal places.

As shown in Fig. 7 the displacement sensors are small black components. In each sensor, the recorder will record in one direction, which will be its main direction. The direction is engraved on them. Sensors record readings in all three axes, i.e., X, Y and Y and it can measure acceleration, displacement and velocity. The sensors are to be placed at the center of the bridge decks and displacement reading will be shown in the Y direction when a two-axel truck having weight of 18.5 ton is moved over the bridge deck.

The instrument shown in Fig. 8 is the recording instrument. The whole apparatus works on external power supply. It has four connection points, from where you can use wires to connect the sensor. Laptop is to be connected to the device using the connecting cord in the main port which is in front of the device. The recording device records the values obtained from the sensors and sends them to the laptop so that the user can evaluate readings such as acceleration, velocity and displacement. The graphs of displacement, acceleration and velocity readings can be viewed on laptop screen where they are shown in real time. Figure 9 shows the result screen on the laptop when the instrument is setup. Only the displacement data was used for study and was shown on the Y direction.

All the decks were considered for the sensor test as all deck girders showed cracks and concrete deterioration. The sensor was placed in the center of the bridge deck, and the apparatus was installed together. For reference, a zero-load reading was taken. As expected, zero reading of acceleration, velocity and displacement were shown on the laptop as seen in Fig. 10. When the setup was installed and connected, a two-axel truck was passed over the deck to measure the real time deflection readings. The channel in which the sensor was connected showed the displacement reading other showed zero readings as no sensors were connected. The software used on the laptop to get the readings from the recording device is "KAMPANA". The sensors

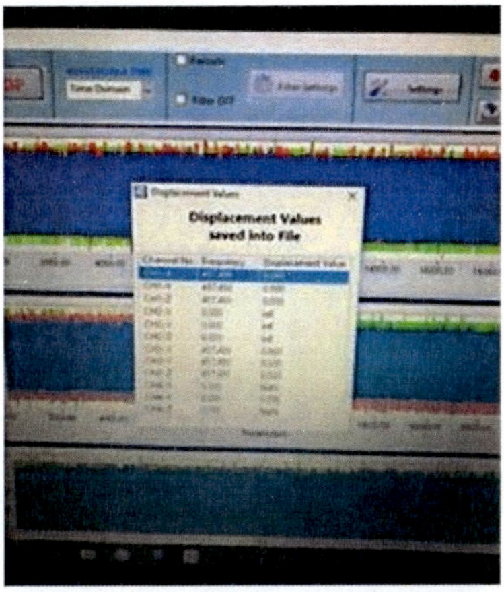

Fig. 8 Zero displacement when no load is passed

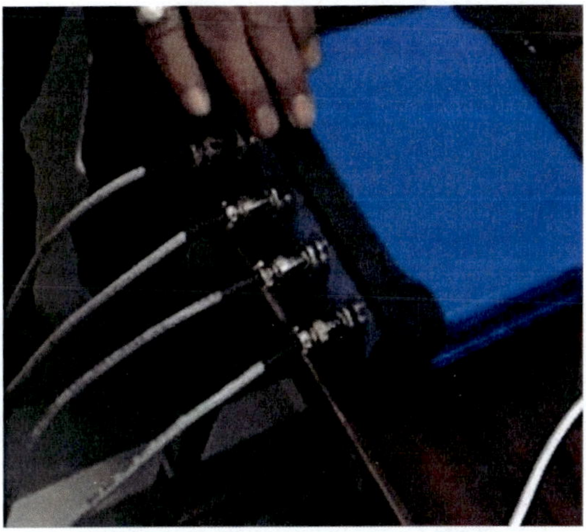

Fig. 9 Recording device connected to the sensors using wires

were placed on the ground and held in position by using the double-sided tape. The sensor was connected to the recording device using a wire, which was connected to the channel of the device. A USB wire was connected to the laptop from the device to access the readings. A UPS was used on the bridge to power the whole setup. Once the equipment was in place and running the “KAMPANA” software on the laptop was opened. Vehicle was signaled to move once the application was running. The option “record log” was clicked to record the readings of displacement which can be analyzed later. It was required to be careful that no other vehicle passes over the bridge at the same time. By not allowing any other vehicle we could make sure that the displacement was not occurring because of any other additional load. The readings of deck displacement obtained are mentioned in the Result chapter.

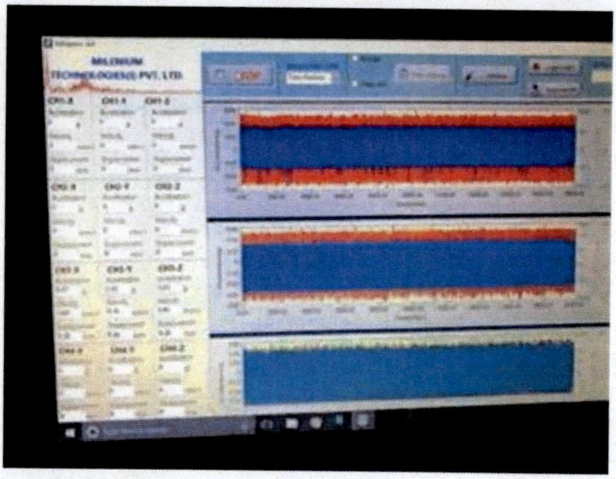

Fig. 10 Laptop screen showing the readings

3 Analysis and Design

3.1 Rebound Hammer Test

Rebound hammer test was conducted on Pier number 1, 2, 3, 4, 6, 7 and 8 and two abutments. Pier number 5 was not accessible due to river water around so it was not considered for the test. The rebound hammer test reading was obtained at a height of one meter over the ground level. For this test, total 21 readings were taken at the same height. By taking the average of these readings, and by using the compressive strength versus rebound number graph the corresponding compressive strength was calculated. A grinder stone was used to level the surface before taking the readings.

A Total of 21 readings were noted down at each location (as per IS 13322 Part 2). Readings all around the piers and on whole length of abutment was taken so that all the sides are duly tested. Test procedure was followed as per the IS code. During the visual inspection of the bridge, a visual inspection was conducted under the guidance of Prof. Dr. H. R. MagarPatil, and the critical sections for the test were selected. The Rebound Hammer number obtained are shown in Table 1.

After recording all the rebound numbers obtained from the test, the graph of the relationship between rebound numbers and compressive strength was used to obtain the compressive strength of the component. Value from all faces is averaged as shown in Table 1 and then compressive strength is obtained which is shown in the Results chapter.

Table 1 Rebound hammer test readings

S. No.	Abutment 1	Pier 1	Pier 2	Pier 3	Pier 4	Pier 6	Pier 7	Pier 8	Abutment 2
1	45	34	32	26	24	25	27	42	41
2	44	38	30	25	32	24	26	45	50
3	40	30	28	28	34	26	27	46	26
4	45	44	36	34	28	27	31	40	28
5	38	34	32	25	30	27	26	41	43
6	46	33	32	26	30	20	26	30	38
7	44	30	35	30	30	25	38	40	42
8	40	32	32	32	28	23	28	34	28
9	44	28	28	32	30	21	37	28	30
10	36	26	34	33	33	32	26	30	29
11	30	26	28	30	40	30	24	38	31
12	32	32	42	34	34	26	28	28	36
13	45	32	33	30	30	26	27	32	32
14	30	28	28	34	32	28	29	31	27
15	38	26	26	35	37	27	26	34	37
16	36	38	25	33	29	32	24	28	33
17	45	34	34	34	37	28	38	25	37
18	40	30	30	38	29	24	32	28	35
19	36	32	32	27	33	25	24	27	33
20	46	26	35	30	31	27	26	28	35
21	43	34	36	27	30	26	33	34	32
Average	40.142857	31.761905	31.809524	30.619048	31.476190	26.142857	28.714286	33.761905	34.428571

3.2 *Ultrasonic Pulse Velocity Test*

The second test carried out is Ultrasonic Pulse Velocity Test. The components selected for the test are pier number 1, 2, 3, 4, 6, 7, 8 and abutment 1 and 2. As pier 5 was surrounded by river water it was not considered for the test. Similarly, as in the rebound hammer test, a total of 21 readings were recorded around the pier and the total length of the abutment at a height of 1 m and they were averaged. Then the average velocity in km/sec from the table was checked against the table given in IS 13322 Part 1. According to the velocity obtained the table gives difference between the quality of concrete. Same procedure used for the ultrasonic pulse velocity test was obtained here. A grinder stone was used to smooth the surface of pier and abutment. Petroleum jelly was used to cover the transducers to remove the air gap between it and the concrete surface which can cause fluctuations in the readings. The table below lists the readings obtained from the test. Readings at a height of one meter around the entire pier were taken, and attention was given to avoid the area covered by steel bars for testing. Table 2 shows average velocity readings obtained from the test from pier 1, 2, 3, 4, 6, 7, 8 and abutment 1 and 2.

The rebound hammer test and the ultrasonic pulse velocity test are combined to compare the rebound number and the ultrasonic velocity to obtain the quality of concrete. If the rebound number shows good compressive strength and the quality of concrete is good then the concrete can be considered as safe. When the concrete quality is seen as medium then proper care must be taken to compare the readings but if the concrete quality seen as doubtful or poor then no comparison must be made and the concrete must be considered to be repaired using proper repair techniques. IS 13322 Part 1 and IS 133322 Part 2 were studied to perform Rebound Hammer and Ultrasonic Pulse Velocity Test.

The quality of the concrete compared to the velocity obtained is compared from the quality table in IS 13322 Part 1 and is shown in the Results chapter.

Table 2 UPV test velocity of piers and columns

S. No.	Element	Time (in micro sec.)		Velocity (km/s)	
		$L = 350$ mm	$L = 750$ mm	$L = 350$ mm	$L = 750$ mm
1	Abutment 1	165	324	2.121212	2.314815
2	Pier 1	147	340	2.380952	2.205882
3	Pier 2	137	310	2.554745	2.419355
4	Pier 3	134	229	2.611940	3.275109
5	Pier 4	95	218	3.684211	3.440367
6	Pier 6	145	264	2.413793	2.840909
7	Pier 7	145	250	2.413793	3.000000
8	Pier 8	130	255	2.692308	2.941176
9	Abutment 2	127	245	2.755906	3.061224

4 Results and Discussion

The maximum displacement of the deck as referred from AASTHO-IFRD section on deck displacement is given by [8]

$$\Delta = L/800$$

For Bridge without sidewalk and,

$$\Delta = L/1000$$

For Bridge with sidewalk.

where $L =$ Total span of the bridge.

It was observed during visual inspection that the bridge does have a side walk thus $L/1000$ formula is to be considered. Also, during the visual inspection, when standing on the bridge, heavy vehicles will feel a lot of vibration when they cross the bridge. As most of the bridge deck girders were having some cracks with concrete deterioration and reinforcement where also exposed. Some amounts of Vibration were felt on every deck so test was to be conducted on all the decks. Vibration on the deck is structurally unsafe because it will cause the concrete in the girders to further deteriorate and expose more steel bars. Table 3 shows the displacement readings of the deck 1, 2, 3, 4, 5, 6, 7, 8 and 9 as all were selected for testing.

$$\Delta = L/1000$$

where $L =$ Total span of the bridge.

$L = 174.15$ m

$$\Delta = 174.15/1000$$

Table 3 Displacement readings

S. No.	Deck No.	Deck displacement (mm)
1	1	41.94
2	2	52.64
3	3	73.37
4	4	108.36
5	5	98.76
6	6	125.21
7	7	107.31
8	8	96.86
9	9	82.11

$$\Delta = 0.17415 \text{ m}$$
$$\Delta = 0.5713 \text{ ft}$$

Thus, the displacement of the decks obtained from the test should be not more than 0.17415 m or 174.15 mm or 0.5713 ft.

From Table 3 we get the maximum value of displacement as 125.21 or 0.12521 m. Since this value and other displacement value that we get are less than the value obtained from AASTHO-IFRD formula, thus we can conclude that the decks are safe. The deflection readings show that severe damage to the bridge has not yet occurred. But due to further vibrations in the decks and removal of the concrete cover from the girders which is exposing and corroding the reinforcement can increase the deck deflections in the future and make the bridge unsafe. Bearings were seen to be damaged below the bridge decks, so it could further increase the vibrations. Proper decks treatment is required, bearings should be installed and the girder are to repaired as early as possible before further damage to the decks. There is already seen some amount of gap in the expansion joints and the vibrations may lead to further increase in the gap, causing unbalanced load distribution to the piers. Moreover, the penetration of water into internal components will cause corrosion, which may cause serious problems in the future, which will be more financially difficult and long-term process as compared with the current situation.

After getting proper results from this test, we can say that this test can be used on site to get real time deflection of the bridge decks. Therefore, this test can be used for structural review of bridge decks or any other components that require displacement testing. The compressive strength of the piers and abutments are mentioned in Table 4. The average of all 21 readings was taken. From the graph of rebound hammer versus compressive strength, the compressive strength of all elements is found and displayed in the table (Fig. 11).

Abutments show better compressive strength than the piers. As abutment 1 has compressive strength of 34.60 MPa, which is highest among all. Pier 6 shows the lowest at 15.59 MPa. This can be due to inconsistency of the materials used during the construction, as it was concluded during the visual inspection that the concrete used

Table 4 Rebound hammer test results

S. No.	Member	Average	Strength (N/mm^2)
1	Abutment 1	40.14286	34.60
2	Pier 1	31.76191	22.92
3	Pier 2	31.80952	23.00
4	Pier 3	30.61905	21.20
5	Pier 4	31.47619	22.42
6	Pier 6	26.14286	15.59
7	Pier 7	28.71429	18.35
8	Pier 8	33.76191	25.26
9	Abutment 2	34.42857	26.34

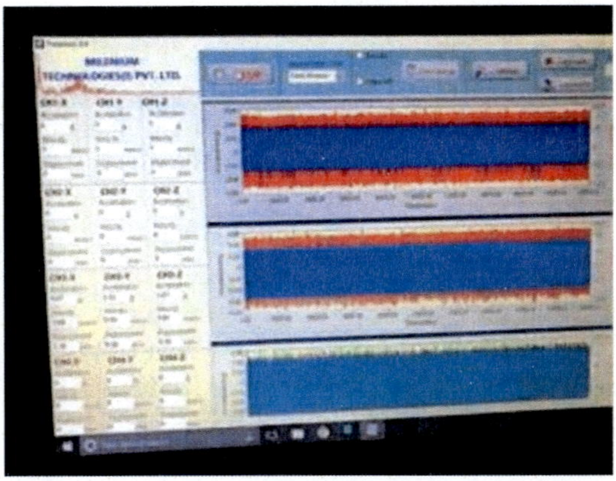

Fig. 11 Displacement reading of deck 6

for construction is not up to the specifications. Random sand materials were seen in the concrete. The surface concrete in many places was seen to been deteriorating which is to be looked after immediately. The velocity values from the ultrasonic pulse velocity test were compared to the quality table in IS 13322 Part 1 and the quality of each abutment and pier is given in Table 5.

Piers and abutments showing very low velocity means that there is no longer a homogeneity of materials, and there are many internal cracks or discontinuities in the members. As seen during the visual inspection that most of the piers and abutments were observed to have many horizontal and vertical cracks which may be showing very low velocity in the above test. Readings were taken on all faces of the piers and abutments. It was observed that when the quality of concrete was shown as doubtful the value of rebound hammer for the same was also low so a correlation can be made between both the test to analyze the results. Thus, from Tables 4 and 5, we can conclude that abutment 1 and 2 show good compressive strength and the velocity obtained from the pulse velocity test is also satisfactory. So, it can be concluded the

Table 5 UPV test concrete quality

S. No.	Element	Time (in micro sec.)		Remark	
		$L = 350$ mm	$L = 750$ mm	$L = 350$ mm	$L = 750$ mm
1	Abutment 1	165	324	Doubtful	Doubtful
2	Pier 1	147	340	Doubtful	Doubtful
3	Pier 2	137	310	Doubtful	Doubtful
4	Pier 3	134	229	Doubtful	Medium
5	Pier 4	95	218	Good	Medium
6	Pier 6	145	264	Doubtful	Doubtful
7	Pier 7	145	250	Doubtful	Medium
8	Pier 8	130	255	Doubtful	Doubtful
9	Abutment 2	127	245	Doubtful	Medium

condition of the abutments is good only surface treatment is required. Where as in pier numbers 3 and 4 show low compressive strength of the concrete but the velocity test shows results from Medium to Good. So, it can be said that the piers have lost its surface hardness but the homogeneity of the element is still maintained.

Further pier numbers 1, 2, 6, 7 and 8 show very low compressive strength and the quality of concrete found from the pulse velocity test is also from doubtful to medium. This confirms that the surface hardness and the strength of the elements is reduced. The inside of the columns is damaged, cracked and lost homogeneity. Many of these piers have shown some chunks of concrete to have been removed from its surface. Proper maintenance repair of these elements is required to be done. Thus, by using this combined test we can recognize the condition of the piers and abutments and also by conducting the tests it can be verified.

5 Conclusion

From the results obtained from the tests conducted above we can understand that the bridge is in Okay condition. The results obtained from the tests were not totally satisfactory. Many critical areas mentioned above needed to be looked as early as possible. And proper repair techniques are to be used so the safety of the bridge and its life span can be increased. The bridge is located on a national highway, so very high traffic load flows through the bridge every day so before any further deterioration it important to start the repair work for the safety of the people going over it. The main purpose of this study was to use the institutionally developed displacement sensor for the audit of the bridge structure. After getting satisfactory results from the tests, it can be concluded that the test can be used on site to get real time displacement reading of the bridge elements. Whole setup of the test is very concise and can be fitted in a small suitcase. As a UPS was used as a power supply for the equipment if continuous power supply is managed on site the equipment can be permanently fitted on site. So, by installing it full-time real time data of whole day can be obtained. Getting real time data is very important that it can be done using this sensor displacement method. This method can be added to the list of methods for structural health monitoring of bridges and decks.

Normally audit work is done every three to five years on every bridge, by using this method the results of acceleration, velocity and displacement can be obtained in a very short period of time. This setup can be installed on those critical bridges whose life span is about to end or is in critical condition. By doing this the critical bridges can be kept under observation. Being under observation serious damage can be caught and disasters can be avoided. As the main aim of doing structural audit is to avoid disasters and save human life. This method can be further developed by using the acceleration and velocity data obtained using the sensors. The sensor was placed at the mid span of the decks in this study. If the sensors are placed under the deck surface more accurate and precise readings can be obtained. And further development can be made to this method to get real time data directly by using a wireless less device into the office. So, by doing this it won't be necessary to go on the site every time and thus time and efforts to go on site can be saved.

References

1. Mahadik B, Jaiswal MH (2014) Structural audit of buildings 5(4):411–416
2. Indian Railway Institute of Civil Engineering (2014, October) Non destructive testing of bridges
3. Lasisi A, Sadiq O (2018, December) Combined use of ultrasonic pulse velocity and rebound hammer for structural health monitoring of reinforced concrete structures. https://doi.org/10.13140/RG.2.2.10379.75040
4. Raikwad M, MagarPatil HR (2020, July) Structural audit of Sadhu Vaswani Pul using institutionally developed displacement sensors 9(9):387–391
5. Bureau of Indian Standard (1992) Non-destructive testing of concrete-methods of test, part 2-rebound hammer IS 13311 (Part: 2)
6. Bureau of Indian Standard (1992) Non-destructive testing of concrete-methods of test, part 1-ultrasonic pulse velocity IS 13311 (Part: 1)
7. Gromicko N, Shepard K (2006) Visual inspection of concrete. Int Assoc Certif Home Insp 4(3):125–127. Available http://www.nachi.org/visual-inspection-concrete.html
8. LRFD bridge design manual, October

An Environmental Impact Assessment on Geomorphic Erosion Risk Analysis Using GIS at Jeevna Site, Jeypore, Koraput District, Odisha

Tanmoy Chatterjee, Surajit Munshi, and Duryadhan Behera

1 Introduction

The western part of the Eastern Ghats Mountains is a peripherally fragile region of the Jeevna site (92.41 km^2) which is a place in the Jeypore subdivision of Koraput district in the state of Odisha. The region is affected by the tropical monsoon climate; consequently, it has rainy season from June to July and dry winter during December-January. One of the major issues of the area is soil erosion which has a devastating effect on the environment and also on human development globally [1–4]. The main factor of soil erosion in that region is human interference with those feeble landscapes [5–7]. The human interfering and the illegal stone mining and sand mining both have created a significant problem in the region. In the area of mining zone, an environmental imbalance is created due to rapid change in the areas of land use and land cover [8, 9]. The constant change of river path depends on the hydrodynamic and morphodynamic processes which in turn depends on the slope gradient, terrain characteristics, structural parameters, variation of climatic condition, vegetation coverage, and time variation [10]. The soil in the experimental zone is generally sandy, especially sandy loam because of which the amount of normal vegetation here is quite low and biomass materials are very few. Soil erosion levels are exorbitant when rainwater flows downhill and reaches close to the ground [11–13], and uneven rainfall takes a significant part in causing the erosion [14–17]. Level of gully erosion in the river basin areas is quite high because it has gradually turned from sheet to gully erosion over many generations. The amount of normal vegetation in the studied area has decreased considerably in the last 10 years. The natural plants are the connection among soil, atmosphere, and water which serves as a criterion of global change

T. Chatterjee · D. Behera
Department of Earth Sciences, Sambalpur University, Burla, Odisha, India

S. Munshi (✉)
Department of Civil Engineering, CMR Institute of Technology, Bengaluru, Karnataka, India
e-mail: surajitm964@gmail.com

A. Sil et al. (eds.), *Recent Trends in Civil Engineering*, Lecture Notes in Civil Engineering 274, https://doi.org/10.1007/978-981-19-4055-2_49

study [18, 19]. The wide impact of GIS can be observed at different levels through environmental consultants [20]. It is used to get a complete result of soil erosion very accurately within a very short period of time [21]. The most important component for soil erosion modeling is digital elevation model (DEM) [22]. Soil erosion potentialities can be predicted on a cell-by-cell basis information with the help of the RUSLE model [23, 24]. The GIS-based risk zone mapping has been accomplished on watershed and regional scale [25–28]. GIS has become a progressive technology in research such as soil science, forestry, hydrology, and agriculture [29–33]. The main aim of this paper is to (i) recognize the areas of high-erosion risk, (ii) identify the site area with an urgent need of conservations, (iii) identify the forest and afforestation zone in the bare lands, and (iv) identification of the lowland sedimentation through soil erosion. This study focuses on summarizing the challenges of the geomorphic erosion, priority to protection, conservation, and management of soil.

2 Study Area and Data

The Jeevna site is located in an area of Eastern Ghats plateau (Odisha), the geographical location of the study area is 18° 51′ 27.27″ N to 18° 57′ 32.82″ N and 82° 26′ 46.28″ E to 82° 33′ 46.46″ E, and the land area is 91.96 km^2. This site belongs to South Odisha, and it is a part of East Central Part of India. The highest altitude of this region is 568 m (Fig. 1), and the lowest altitude is 479 m, which means it is the undulating topography covered area. The topography slope of the site mostly consisted of undulating landforms and steep hill slopes [34]. 1527 mm is the total rainfall. The basement rock in the Jeevna site is an ingredient of Eastern Ghat supergroup which is of the Precambrian age. The Genesis is the prelude rock formations of the site. Sandy clay loam is the main soil of the site. The survey area is under the tropical monsoon climate.

3 Methodology

One of the major recent issues in the world now is soil erosion. This problem is intensified by the increasingly unscientific and unrealistic human activities for which the environment has become heavily polluted. The recent advent of GIS, alongside remote sensing technology has enabled to accurately predict some RUSLE factors [35–37]. It connects very nicely with GIS, and it can provide better analysis. The RUSLE model through GIS, and remote sensing technology has helped in assessing soil erosion risk areas of the Jeevna site. The methodology specifically explains the mainstream of the RUSLE model, and this model can predict soil erosion and environmental risks. The parameters of this significant model depend on the rainfall, digital elevation model (DEM), soil types, land cover and slope of the area. The

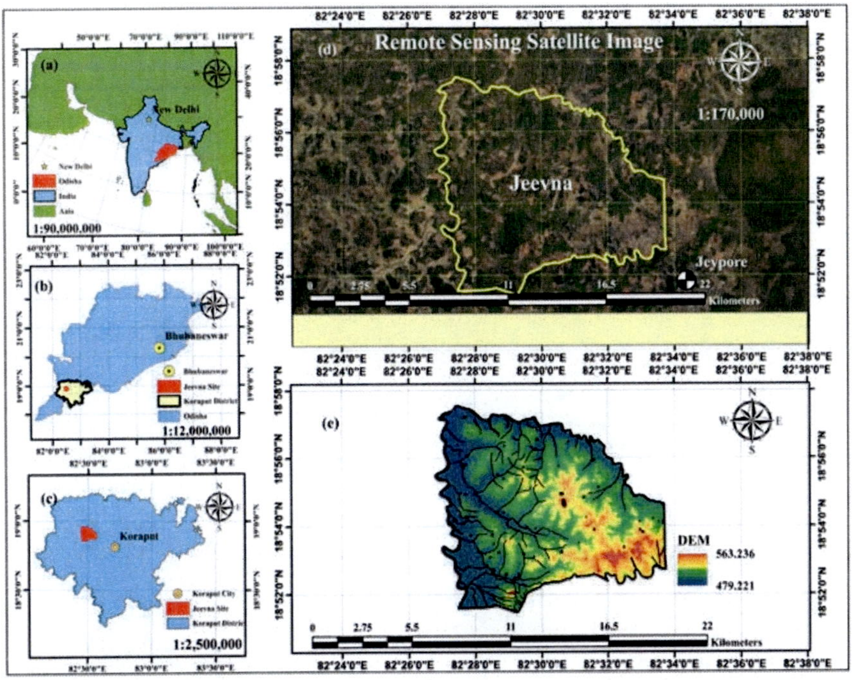

Fig. 1 **a–c** Study area in Jeevna site, India, Odisha, Koraput, **d** study area remote sensing images, **e** and DEM topography

flowchart characterized in Fig. 2 reviews all essential data used in the study. The sources of the data used for the RUSLE model are shown in Table 1.

3.1 Parameter Estimation of RUSLE Model

RUSLE model is one of the most widely used long-term and predictable methods of soil erosion worldwide. The Jeevna site has been studied especially with the help of survey of India Toposheet (65 J/5 and 65 J /9) which has been presented on a scale of 1:50,000 and identified by ArcGIS 10.2 software. The RUSEL model may designate the control over large-scale deforestation, urbanization, industrialization, and irregular human rummaging, as well as the conservation of lands into arable lands [37–41]. The base map has been created with the help of Indian remote sensing satellite and DEM data. The RUSLE model has been used to control the sediments flow expansion and sediments erosion. The model has been used to determine the annual soil loss of the particular area. The RUSLE is shown in the following equation [42].

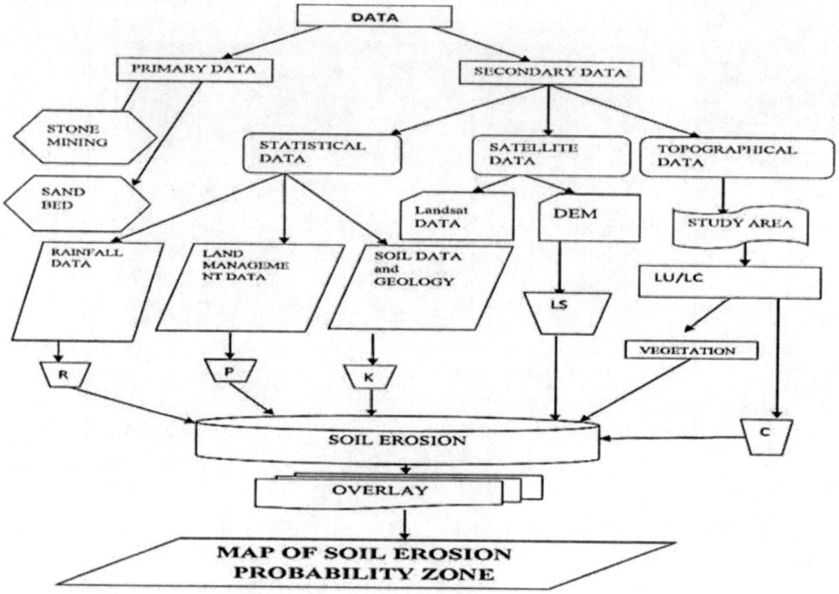

Fig. 2 Flowchart of methodology

Table 1 Description of the data

S. No.	Data types	Sources	Description
1	DEM	USGS	SRTM DEM
2	Satellite imagery	World imagery	High-resolution aerial imagery
3	Soil data	The state of	Soil Map 206–2019 5 categories
4	Rainfall	Soil survey and land	Rainfall data (2005–2020) 15 years
5	Remote sensing image	Landsat-7 and Landsat-8	30-m resolution

$$A = R \times K \times LS \times C \times P$$

where A calculates the special average soil loose and temporal average soil loss per unit area. A is articulated per ton, per hectares, per year (t ha^1/year). R [MJ mm (hm^2 ha)$^{-1}$] refers to the rainfall erosivity factor in deliberation of the erosion snowmelt runoff. K [thm^2 h (hm^2 MJmm)$^{-1}$] refers to the soil erodibility factor, which is the soil loss rate of unambiguous soil, and rainfall erosivity per unit is considered in a regular plot. L and S refer to the slope aspect factor. C refers to the coverage factor of the vegetation. P refers to the conservation measure factor. The parameters of RUSLE model can be divided into three main parts, namely erosivity, erodibility, and management factors. All parameters can be determined especially by geomorphology and rainfall properties. Raster model features cell-based demonstration, which delivers analytical capabilities for continuous data and permits for faster processing [43, 44].

3.2 *Rainfall Erosivity Factor* (**R**)

The rainfall erosivity factor reflects the consequence of rainfall intensity on soil erosion as well as necessitates permanent rainfall data for its comprehensive calculation [45]. Long-term annual rainfall from the 6 block meteorological stations and the rainfall R factor for each station were found distinctly [46]. The rainfall erosivity factor has been created from the monthly rainfall data of last 15 years (2006–2020) which is obtained from beginning to end in a systematic method from Odisha rainfall monitoring system. The work has been done with the aid of Wischmeier and Smith, 1978 [45], rainfall erosivity factor equation.

$$\sum_{1}^{12} 1.735 \times 10^{\left(1.5\log_{10}\left(\frac{P_i}{P}\right)-0.0818\right)}$$

where R is the rainfall erosivity factor (MJ mm ha^{-1} h^{-1} per year), P_i is demoted on monthly rainfall (mm); P is shown as an annual rainfall (mm). The rainfall erosivity model is created by processing rainfall data for 15 years through GIS interpolation method. Figure 3 demonstrates that the rainfall erosivity map prepared by the Odisha rainfall monitoring data of the experimental vicinity. The highest rainfall is located in the western side of that region, and lowest rainfall is situated in the north-eastern part of the region.

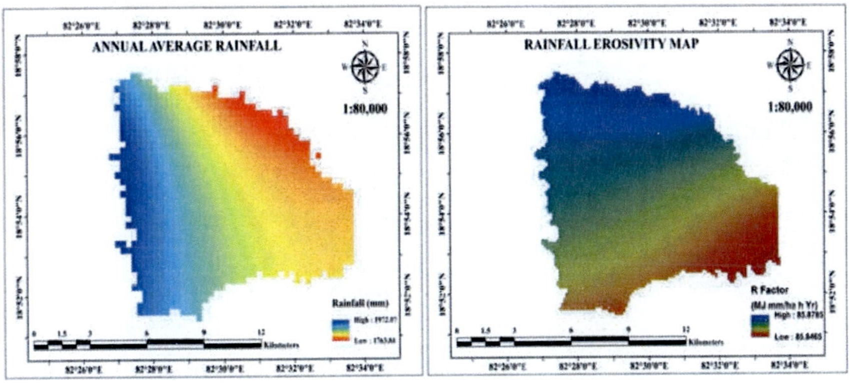

Fig. 3 Rainfall map and rainfall erosivity map (R)

3.3 Soil Erodibility Factor (K)

The prime criteria that reflect the volume of soil penetration, soil sensitivity, and erosion by runoff are the soil erodibility factor. This internal cause greatly accelerates soil erosion. The dimension of K factor is primarily influenced by the soil structure of soil organic material. The monograph is used to determine the soil K factor, which is specially controlled by the % of sand, % slit, %clay, % organic materials. The soil edibility factor equation is shown in the below equation.

$$K = f_{\text{sand}} \times f_{\text{cl-si}} \times f_{\text{org}} \times f_{\text{hisand}}$$

$$f_{\text{sand}} = (0.2 + 0.3.\exp[-0.256.m_{\text{s}}.(1 - m_{\text{slit}}/100)])$$

$$f_{\text{cl-si}} = (m_{\text{silt}}/m_{\text{c}} + m_{\text{silt}})^{0.3}$$

$$f_{\text{orgc}} = \left(1 - 0.0256.\text{orgc}/\text{orgc} + \exp\left[3.72 - 2.95.\text{orgc}\right]\right)$$

$$f_{\text{hisand}} = (1 - 0.7(1 - m_{\text{s}}/100)/(1 - m_{\text{s}/100}) + \exp[-5.51 + 22.9.(1 - m_{\text{s}}/100)])$$

The underlying soil properties related to a soil profile parameter are a function of which precise samples of permeability, slit, sand, and soil structure are analyzed [47, 48]. Where m_{s} is the present sand content (the diameter of the particles is 0.05–2.00 mm), m_{silt} is the present silt content (the particles diameter is 0.002–0.05 mm), m_{c} is the percentage of clay content (<0.002 mm diameter of particles), and 'orgc' is the present organic carbon content of the layer (%). In the general case, the lowest values observed in clay, sandy rocks, and the highest is usually observed in silt rock. Soils with high sand are highly corrosive, as they are characterized as the least resistant slit and fine sand [48, 49]. Only two soil types can be observed in the Jeevna site. Soil erodibility numerical value extends between 0 and 1. The soil is less prone to erosion if the values is closer to 0. Figure 4 shows the soil erodibility factor map prepared by the world soil type. Maximum soil erodibility is located in the southwest portion of the study area. K factor is measure by the t-h-ha/MJ/mm.

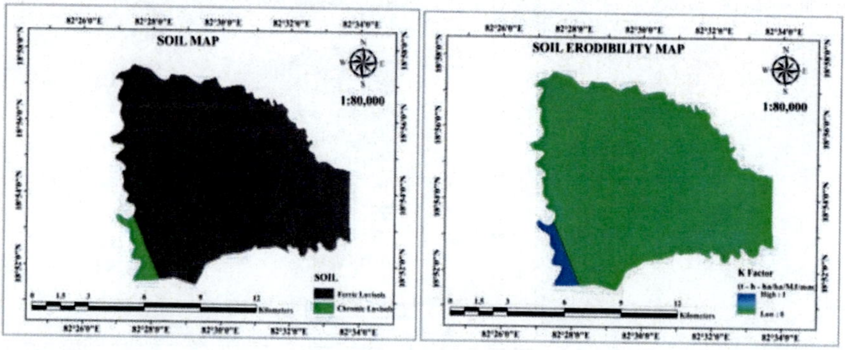

Fig. 4 Soil map and soil erodibility map

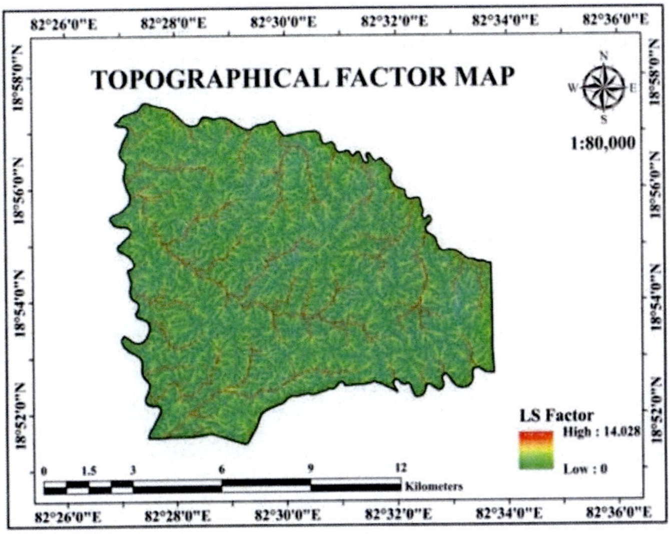

Fig. 5 Topographical factor map

3.4 *Topographic Factor (LS)*

The slope of the landforms has a propound effect on soil erosion. Soil loss is mainly associated with increase in slope. The following equation showed the subject in a well-planned way.

$$LS = (\text{slope length } 22.13/)^{0.4} \times (0.01745.\sin\theta/0.0896)^{1.4} \times 1.4$$

where θ refers to the slope. The map is based on the DEM data. Figure 5 revealed the topographical factor map of the study area. In the case of RUSEL model, the topographic factor is considered to have a topographic effect on the soil erosion [24, 50].

3.5 *Crop Management Factor* (C)

The *C* factor mainly depends on land use as well as land cover. There are 17 categories to the land use and the land cover at the Jeevna site, which are built-up area, waterbodies, agriculture land, wasteland, forest, mining, industries, and so on. The land cover map and the land use have been recategorize according to *C* factor value. The value of *C* for reservoirs and wetlands as 0 [49]. Figure 6 shows the map of the investigation area of crop management factor.

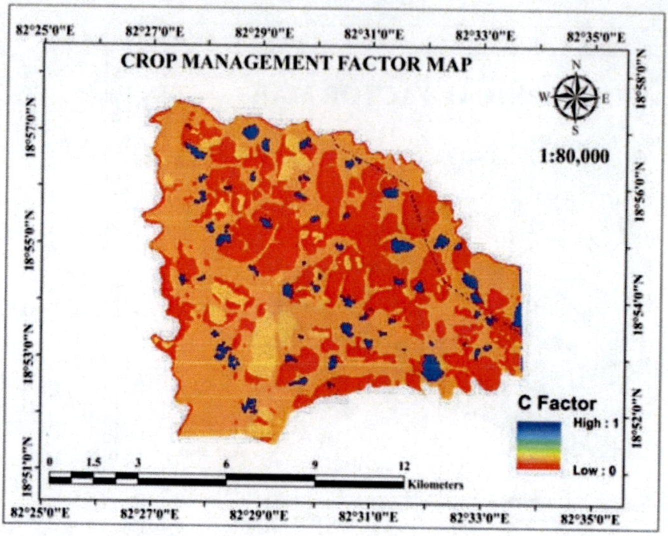

Fig. 6 Crop management factor map

3.6 *Conservation Practice Factor* (P)

The percentage of soil loss in the slope disseminating after enchanting water and soil conservation measures is called soil and water conservation constituent. The acquired *P* value is always between zero and one. The region is unaffected by soil erosion if the value is 0 and if the value is 1, then it is not possible to measure the region in terms of water conservation and soil conservation. Figure 7 demonstrated the conservation support factor map of investigated area. Most of the regions belong to the agriculture categories because the value of *P* is very low. *P* factor value is lower when soil conservation is high, and soil erosion is low for effective soil conservation [51, 52].

4 Results and Discussion

The considerable amount of plentiful rainfall due to the direct influence of monsoon winds is a good indicator of local variation in soil erosion rate [24]. To occur soil erosion, stagnant water plays a significant role in catchment area which is more sensitive than rainfall [24, 53, 54]. Due to the tropical monsoon climate, which can be observed in the summer warm humid climate, the period between July to September is the time where maximum amount of rainfall occurs [55, 56]. Raster uninterrupted grid is mainly geostatistical interpolation approach using annual inverse (IDW) distance weight [57–59]. Through the similar raster analysis, the value of *R* is range from

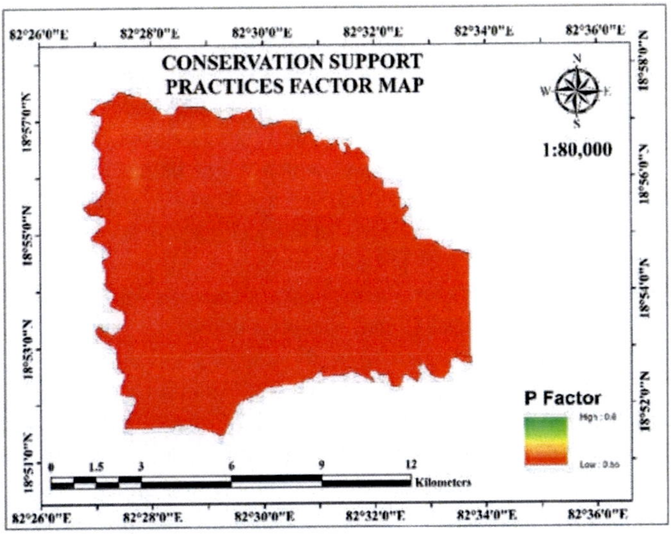

Fig. 7 Conservation support practices factor map

85.8465 to 85.8785 (MJ mm ha^{-1} h^{-1} per year). In the surveyed region, rainfall is moderate to high from this prominent *R* value. Higher rainfall value causes higher erosivity [60]. Exactly the same thing occurred here, especially in the east portion of the study area with high rainfall and high erosivity. Almost the same types of soil can be observed here (92%). A soil map has been created with the help of world soil type data, and soil erodibility map has been created by assigning *K* value. The values of *K* are located in between the range of 0.09–0.10 t-h-ha/MJ/mm. The value of *K* is very low, the permeability is also very low, and the low-moisture content is located in the area. The two most notable soil properties that control soil erosion are infiltration capacity and soil structure [50, 61]. According to the soil properties data and the following k factor equation, the *K* factor for the Jeevna site soils was calculated as 0.014 $Mgha^{-1}$ MJ^{-1} mm^{-1}. These results were alike to the similar value in the *K*-nomograph [62], which was 0.018 Mgha h ha^{-1} MJ^{-1} mm^{-1}. The amount of *K* factor usually increase during the cultivation and cultural practices and at a certain season in tropical soil [39, 63]. Cultivation and cultural practices are more prevalent in this region during the rainy season.

In particular, the topographic factors traditionally characterize and directly influence on both slope length as well as slope steepness on the erosion progression. To calculate the LS factor, DEM data is required. The flow accumulation and flow direction are processed to calculate the LS factor correctly. This value, 20.63, was calculated as the highest LS value for reservoir [38]. Similarly, the LS value of the river channel calculated as 14.0. In general, increase in the slope length (*L*) and the slope steepness (*S*), the total soil erosion per unit increases the velocity and erosivity of the overland flow in the runoff in the downslope course [25, 42, 48, 64, 65]. Human intervention also accelerates the spread of agriculture, deforestation, and the use of others land. Such information becomes tremendously appreciated to express an efficient conservation and land management plan [66–69]. Basically, the site land use is affected by the deforestation, unscientific cultivation, wasteland bushes degradation, and sand mining. *C* factor represents the impact of vegetation and soil erosion management [42]. The land use area has been classified into 17 zones. The major part is cover in the agricultural cropland and land which might be used for future cultivation. Table 2 gives the crop values used for *C* factor. These crop values have been used to create *C* factor which is very useful for the RUSLE model. The value of *C* factor in watershed is between 0 to 1 [44]. The average *C* factor value is 0.6. Lower *C* factor value is noticed in the north-central part of the area, and where the agriculture is predominant, the amount of *C* factor value is higher.

P factor is an important factor for extensive investigation of *s* loss at Jeevna site. *P* factor maps are mainly created by slope and contour. The data used in the map is collected from SRTM DEM. Although the area is made up of rocky hard rock, soil erosion has become inevitable due to inhuman and unscientific sand collection and illegal quarrying. A dimensionless ration accounts for soil erosion under *P* factor, specific management, and practice [42, 69, 70]. The general practice of farmers in such categories, such as terracing, strips cropping, and contour tillage, decreases the total volume of erosion. If the *P* factor value is low, then the better practice theory is found, and suitable control of soil erosion is located there [39]. Here, the value of

Table 2 LU/LC class and *C* factor value

Sl. No	Class name	Area (Ha.)	*C* factor
1	Builtup-builtup (urban)-transportation	25.81	0.5
2	Waterbodies-river/stream-perennial	150.29	0
3	Agricultural land-crop land-Kharif crop	1896.71	0.1
4	Agricultural land-crop land-two crop area	3173.81	0.1
5	Wastelands-gullied/ravenous land-gullied	8.02	0.05
6	Wastelands-scrub land-open scrub	2745.36	0
7	Waterbodies-lakes/ponds-perennial	38.99	0
8	Builtup-builtup (rural)-built-up area (rural)	408.33	0.5
9	Forest-forest plantation	598.54	0.13
10	wastelands-scrub land-dense scrub	47.24	0.014
11	Forest-scrub forest	5.27	0.01
12	Forest-deciduous (dry/moist/thorn)-dense/closed	73.14	0.004
13	Agricultural land-plantation-agriculture plantation	13.48	0.13
14	Forest-deciduous (dry/moist/thorn)-open	3.00	0.1
15	Waterbodies-lakes/ponds-dry	6.39	0
16	Builtup-mining/industrial area-mine/quarry	44.02	1
17	Waterbodies-canal/drain-lined	3.36	0

P factor is quite low in southwest area because of uncontrolled practices due to the short-sighted work of human beings.

Human intervention in river basin areas and sand mining needs to be stopped, otherwise the environment will be highly polluted, and there will be a hike in ecological imbalance. Finally, the soil will inevitably be eroded with far-reaching consequences.

The soil erosion high-risk zone in the surveyed area can be noticed in the south-western portion. In some places in the southwest, the average soil erosion rate is 30 tons/ha/Year, which is almost three times greater than the (10 tons/ha/Year) tolerate rate. The reason for the study area considers into high-risk soil erosion is due to the current human intervention, lack of conservation, unscientific, and improper practice of farmers that has a negative impact on the soil erosion zone [48]. Southwestern portion of the study area, the soil conservation is currently an important issue. Tree plantation in riverine areas and afforestation in bare land is the main activity to stop the soil erosion. Similar prospect given by [38] that is forest protection should be given priority, afforestation should be done on steep vacant land, maximized plant coverage, and rotational agriculture should be practiced. Silt yields are particularly threatening to water bodies, as a result the old rainwater harvesting system become a marginal area [48].

As a direct result of uncontrolled and unscientific human factors, soil erosion and hydrological processes are widely disturbed. Surface hydrology and groundwater hydrology have been significantly disrupted by a considerable number of human influences, respectively; agricultural effects, extensive deforestation, soil erosion, strong runoff and low infiltration, soil salinization, organic matter depletion, collecting excess amount of stone from the foot of plateau and collecting sand from the river side. Increase sedimentation of surface water, deposition in low laying areas, and groundwater recharge is affected by the soil erosion. Research from the Brazilian amazon has shown that the soil erosion is very low for dense forest, and the no risk of soil erosion, the healthy, and stronger groundwater storage is located [71–73]. So, the most important thing for plantation is to prevent soil erosion and to recharge the groundwater. If there are less forests, the amount of rainfall will be less, and the soil will be eroded and hydrological imbalances will be creating. In that case, it is said that the significant issues like hydrological imbalance, deforestation, soil erosion, etc., are related to climate, and the human contribution is enough to control this climate. Table 3 illustrates the distribution of land use in the surveyed area. Figure 8 shows the proposed plantation area in bare land of the surveyed area. Afforestation is the main objective at the present scenario to induce development of the studied area. The amount of agriculture land here is 5077 ha; in many places, shifting cultivation is done in the old way. Cultivation using new methods will prevent soil erosion and increase crop production. Government and non-government organization should come forward to impart complete knowledge to the farmers in this regard. The afforestation in the wasteland (2795 ha) will prevent soil erosion in the study area. Soil pH value for afforestation remains the same for a long time [74]. Therefore, plantation maintains organic fertility in the soil for a long period of time. The roots of the trees hold the soil and reduced soil erosion in the region. If grass

Table 3 Land pattern of the study area

S. No.	Land pattern	Area (ha)	Area (%)
1	Agriculture land	5077	54.9
2	Built-up area	485	5.3
3	Forest	685	7.4
4	Wastelands	2795	30.3
5	Waterbodies	199	2.1

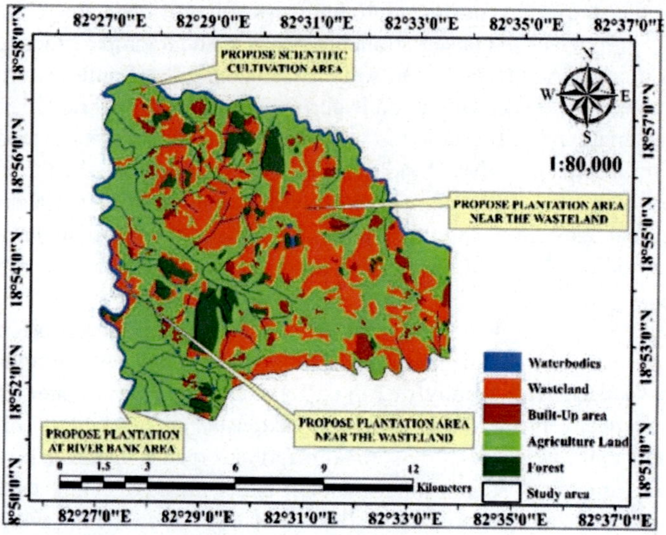

Fig. 8 Map of proposed plan for controlled the soil erosion

plantation in the wasteland area reduces soil erosion, then the legitimacy of animal husbandry needs to be formalized. The amount of forest land in the study area is very low at 7.40%. Note that it is very important to increase the forest area is this region. According to the national forest policy, the forest cover in the plains of India is 33% which is four time lesser then the study areas forest. Figure 9 demonstrated the soil loss area, sediments deposition, wasteland demarcation, and shifting cultivation area. On-spot verification of the area indicates that human intervention is sufficient for soil erosion.

4.1 *Potential Annual Soil Erosion Estimation*

With the help of GIS, the RUSLE model has been typically created with pixel-by-pixel unit, to enhance the understanding of annual soil loss. Maximum portion of soil

Fig. 9 Photos of on spot verification in the study area

loss is about 30 tons/ha/Year which is found in southwest portion of the studied area. Figure 9a demonstrates the RUSLE model in the study area. Figure 9b shows the studied area of man-made erosion. In the studied region, three distinct categories are made, i.e., low, medium, and high based on the potential soil erosion. Table 4 exposed the soil loss data of the study area. Soil erosion trend is highest in the foot hill region of the southwestern basin and southwestern plateau. Low to very low regions where the possible extent of soil erosion is very low, the whole regions can be seen excluding only the southwestern part. Soil loss is associated to land use as well as land cover. Sustainable growth of local agriculture and human interference of river bank sand mining is properly considered to be the main reasons of soil erosion. The amount of soil erosion in the studied area can be controlled by conservation technology [75–79]. The low-risk zone is accurately located across an area of 8936.68 (96.7%) hectares where the soil loss is nominal, 209.06 (2.3%) hectare area in common is a medium risk prone soil erosion area. Heavy soil erosion risks prone area is about 96.07 (1.0%) hectare.

Table 4 The soil loss data of the study area

S. No.	Group	Soil loss (Tons/ha/Year)	Area (ha)	Area (%)
1	Low	0.0–1.42	8936.68	96.7
2	Medium	1.42–6.99	209.06	2.3
3	High	6.99–30.22	96.07	1.0

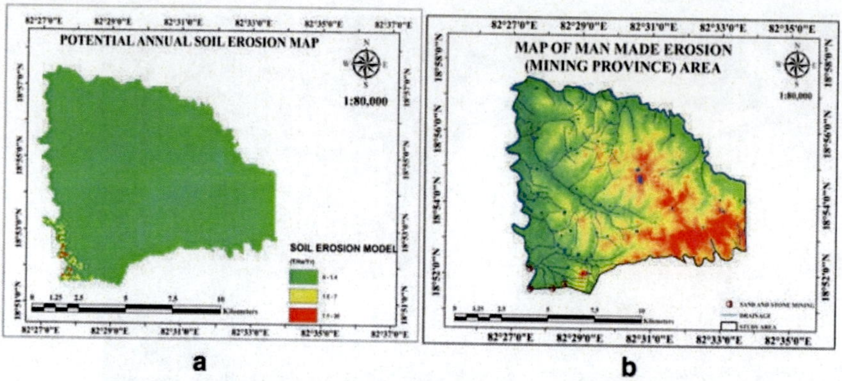

Fig. 10 **a** RUSLE map and **b** man-made erosion map

4.2 *Hydrological Impact on Soil Erosion at Jeevna Site*

As a direct result of uncontrolled and unscientific human factors, soil erosion and hydrological processes are widely disrupted. Surface hydrology and groundwater hydrology have been significantly disrupted by a considerable number of human influences, respectively; agricultural effects, extensive deforestation, soil erosion, strong runoff and low infiltration, soil salinization, organic matter depletion, collecting excess amount of stone from the foot of plateau and collecting sand from the river side. Figure 10 invariably shows the mining area of the local vicinity. Soil erosion is more prevalent in protected areas where mining is predominant and with it an inconsistent pattern of hydrology has been found there.

5 Conclusions

The RUSLE model has successfully identified specific risky areas. Soil erosion potential zone has been found through applying the GIS technology. The real-risk zone of the soil erosion has been accurately identified with the help of the Arc GIS software and as well as the field verification. Studies from this model show that the areas with high levels of erosion have higher infertility of the agriculture land. In the studied area, soil erosion is caused by pseudoscientific activities connected to humans, extensive deforestation, steep slope of local topography, heavy rainfall, shifting cultivation, deposition of sediments at river bank and by not converting the wasteland to afforestation or forest. The paramount factor that affects the soil erosion is slope and sand mining at river bank. Soil erosion will be alleviated if the higher authorities declare the sand mines illegal and stop them. Extensive afforestation along the river banks will prevent soil erosion in many cases. This comparative study will continue thoughtfully in the sustainable future, and its finding will be positively reviewed in detail to satisfactorily explain the model in a more complex way. In direct order to prevent the environmental degradation, successful efforts should be made to restore the previous conditions, especially by accurately identifying soil erosion risk zones.

Acknowledgements The authors would to acknowledge the divisional forest officer, Jeypore, Koraput, Odisha, India.

References

1. Higgitt D (1993) Soil erosion and soil problems. Prog Phys Geog 17(4):461–472
2. Martínez-Casasnovas JA, Ramos MC, Benites G (2013) Soil and water assessment tool soil loss simulation at the sub-basin scale in the alt penedÈs–anoia vineyard region (ne Spain) in the 2000s. Land Degrad Dev
3. Borrelli P, Panagos P, Märker M, Modugno S, Schütt B (2017) Assessment of the impacts of clear-cutting on soil loss by water erosion in Italian forests: first comprehensive monitoring and modelling approach. CATENA 149:770–781
4. Zeng C, Wang S, Bai X, Li Y, Tian Y, Li Y, Wu L, Luo G (2017) Soil erosion evolution and spatial correlation analysis in a typical karst geomorphology using RUSLE with GIS. Solid Earth 8:721–736
5. Bruckner H (1986) Man's impact on the evolution of the physical environment in the Mediterranean region in historical times. GeoJournal 13:7–17
6. Van Andel TH, Zangger E, Demitrack A (1990) Land use and soil erosion in prehistoric and historical Greece. J Field Archaeol 17:379–396
7. García-Ruiz JM, Nadal-Romero E, Lana-Renault N, Beguería S (2013) Erosion in mediterranean landscapes: changes and future challenges. Geomorphology 198:20–36
8. Ghose MK, Anjay K (2004) Mineral industries and their environmental aspects in Indian context. J Eng Mater Sci 11:433–437
9. Naveen Saviour M (2012) Environmental impact of soil and sand mining: a review. Int J Sci Environ Technol 1(3):125–134
10. Pan S (2013) Application of remote sensing and GIS in studying changing river course in Bankura District, West Bengal. Int J Geomatics Geosci 4(1):149–163
11. Sharma PP, Gupta SC, Foster GR (1993) Predicting soil detachment by raindrops. Soil Sci Soc Am J 57:674–680
12. Dijk AIJM, Bruijnzeel LA, Rosewell CJ (2002) Rainfall intensity-kinetic energy relationships: a critical literature appraisal. J Hydrol 261:1–23
13. Kinnell PIA (2005) Raindrop-impact-induced erosion processes and prediction: a review. Hydrol Process 19:2815–2844
14. Li CF, Geballe SGT, Burch WR (2000) Rainwater harvesting agriculture: an integrated system for water management on rainfed land in China's semiarid areas. AMBIO: J Hum Environ 29(8):477–483
15. Nearing MA (2001) Potential changes in rainfall erosivity in the US with climate change during 21st century. J Soil Water Conserv 56(3):229–232
16. Burger G (2002) Selected precipitation scenarios across Europe. J Hydrol 262:99–110
17. Endale DM, Fisher DS, Steiner JL (2006) Hydrology of a zero-order Southern Piedmont watershed through 45 years of changing agricultural land use. Part 1. Monthly and seasonal rainfall-runoff relationship. J Hydrol 3161–3112
18. Sun HY, Wang CY, Niu Z (1998) Analysis of the vegetation cover change and the relationship between NDVI and environmental factors by using NOAA time series data. J Remote Sens (in Chinese) 2(3):205–210

19. ZhongBao X, JiongXin X, Wei Z (2008) Spatiotemporal variations of vegetation cover on the Chinese Loess Plateau (1981–2006): impacts of climate changes and human activities. Sci China Ser D: Earth Sci 51(1):67–78
20. Şahin S, Kurum E (2002) Erosion risk analysis by GIS in environmental impact assessments: a case study—Seyhan Kopru Dam construction. J Environ Manage 66(3):239–247
21. Okatan A, Using GIS in the planning and conservation watersheds, Kent ve Yonetimi Insan ve Çevre Sorunları Sempozyumu'99. 17–19 Şubat 1999, cilt1, Kent ve Insan, 274–283
22. Kim JB, Saunders P, Finn JT (2005) Rapid assessment of soil erosion in the Rio Lempa Basin, Central America, using the universal soil loss equation and geographic information systems. Environ Manage 36:872–885
23. Shinde V, Tiwari KN, Singh M (2010) Prioritization of micro watersheds on the basis of soil erosion hazard using remote sensing and geographic information system. Int J Water Resour Environ Eng 2(3):130–136
24. Ganasri BP, Ramesh H (2016) Assessment of soil erosion by RUSLE model using remote sensing and GIS—a case study of Nethravathi Basin. Geosci Front 7:953–961
25. Prasannakumar V, Vijith H, Abinod S, Geetha N (2012) Estimation of soil erosion risk within a small mountainous subwatershed in Kerala, India, using Revised Universal Soil Loss Equation (RUSLE) and geo-information technology. Geosci Front 3(2):209–215
26. Chatterjee S, Krishna AP, Sharma AP (2014) Geospatial assessment of soil erosion vulnerability at watershed level in some sections of the Upper Subarnarekha river basin, Jharkhand, India. Environ Earth Sci 71:357–374
27. Reddy RS, Nalatwadmath SK, Krishnan P (2005) Soil erosion Andhra Pradesh. NBSS Publ. No. 114, NBSS&LUP, Nagpur, 76 p
28. Ramamurthy V, Nalatwadmath SK, Srinivas S, Rama Mohan Rao MS, Sivaprasad CR, Adhikari RN, Naidu LGK, Patil SL, Raizada A, Sarkar D, Singh SK, Mishra PK, Soil erosion in Karnataka. NBSS Publ. 162, NBSS&LUP, Nagpur, p 70
29. Kohl M, Gertner G (1997) Geostatistics in evaluating forest damage surveys: considerations on methods for describing spatial distributions. For Ecol Manage 95:131–140
30. Bocchi S, Castrignano A, Fornaro F, Maggiore T (2000) Application of factorial kriging for mapping soil variation at field scale. Euro J Agron 13:295–308
31. Basso B, Ritchie JT, Pierce FJ, Braga RP, Jones JW (2001) Spatial validation of crop models for precision agriculture. Agric Syst 68:973112
32. Panagopoulos T, Jesus J, Beltrão J (2002) Soil geostatistical techniques applied to sprinkle irrigation experiments. Opt Mediterranées 50:391–396
33. Rabia AH (2012) Mapping soil erosion risk using RUSLE, GIS and remote sensing techniques: the 4th International Congress of ECSSS, EUROSOIL, Bari, Italy
34. Xu Y, Tian J, Shen H (2002) Eco-environmental rehabilitation and spatial differentiation based on enlarging terrace and defarming in the loess hilly-gully region. J Nat Resour 17:430–437 (in Chinese)
35. Desmet PJ, Govers G (1996) A GIS-procedure for the automated calculation of the USLE LS-factor on topographically complex landscape units. J Soil Water Conserv 51(5):427–433
36. Nearing MA (1997) A single, continuous function for slope steepness influence on soil loss. Soil Sci Soc Am J 61(3):917–919
37. Millward AA, Mersey JE (1999) Adapting the RUSLE to model soil erosion potential in a mountainous tropical watershed. Catena 38:109–129
38. Chen T, Niu R, Li P, Zhang L, Bo D (2011) Regional soil erosion risk mapping using RUSLE, GIS, and remote sensing: a case study in Miyun Watershed, North China. Environ Earth Sci 63:533–541
39. Angima AD, Stott DE, O'Nell MK, Ong CK, Weesies GA (2003) Soil erosion prediction using RUSLE for central Kenyan Highland conditions. Agric Ecosyst Environ 97:295–308
40. Boggs G, Devonport C, Evans K, Puig P (2001) GIS-based rapid assessment of erosion risk in a small Watershed in the wet/dry tropics of Australia. Land Degrad Dev 12(5):417–434
41. Mati BM, Veihe A (2001) Application of the USLE in a Savannah environment: comparative experiences from east and West Africa. Singap J Trop Geogr 22:138–155

42. Renard KG, Foster GR, Weesies GA, McCool DK, Yoder DC, Predicting soil erosion by water: a guide to conservation planning with the Revised Universal Soil Loss Equation (RUSLE). United States department of Agriculture. Handbook no. 703. United States Government Printing Office, Washington, DC
43. Fernandez C, Wu JQ, McCool DK, Stockle CO (2003) Estimating water erosion and sediment yield with GIS, RUSLE, and SEDD. J Soil Water Conserv 58:128–136
44. Demirci A, Karaburun A (2012) Estimation of soil erosion using RUSLE in a GIS framework: a case study in the Buyukcekmece Lake watershed, northwest Turkey. Environ Earth Sci 66:903–913
45. Wischmeier WH, Smith DD (1978) Predicting rainfall erosion losses: a guide to conservation planning. Agriculture Handbook 282. USDA-ARS, USA
46. Iaquinta P, Terranova O (2004) Scenario di Rischio di Erosione Idrica. MondoGIS 43:17–20
47. El-Swaify SA, Dangler EW (1976) Erodibility of selected tropical soil in relation to structural and hydrologic parameters. In: Foster GR (ed) Soil erosion prediction and control. Soil and Water Conservation Society, Ankeny, pp 105–114
48. Farhan Y, Nawaiseh S (2015) Spatial assessment of soil erosion risk using RUSLE and GIS techniques. Environ Earth Sci
49. Prasannakumar V, Shiny R, Geetha N, Vijith H (2011) Spatial prediction of soil erosion risk by remote sensing, GIS and RUSLE approach: a case study of Siruvani River Watershed in Attapady Valley, Kerala, India. Environ Earth Sci 46:965–972
50. Mahala A (2018) Soil erosion estimation using RUSLE and GIS techniques—a study of a plateau fringe region of tropical environment. Arab J Geosci 11:335
51. Xu YQ, Peng J, Shao XM (2015) Assessment of soil erosion using RUSLE and GIS: a case study of the Maotiao River watershed, Guizhou Province, China. Environ Geol 56:1643–1652
52. Tang Q, Xu Y, Bennett JS, Li Y (2015) Assessment of soil erosion using RUSLE and GIS: a case study of the Yangou watershed in the Loess Plateau, China. Environ Earth Sci 73:1715–1724
53. Jain SK, Kumar S, Varghese J (2001) Estimation of soil erosion for a Himalayan watershed using GIS technique. Water Resour Manage 15(1):41e54
54. Dabral PP, Baithuri N, Pandey A (2008) Soil erosion assessment in a hilly catchment of North Eastern India using USLE, GIS and remote sensing. Water Resour Manage 22(12):1783e1798
55. Sen J, Sen S, Bandyopadhyay S (2004) Geomorphological investigation of badlands—a case study at Garbheta, West Medinipur District, West Bengal, India. In: Singh S, Sharma HS, De SK (eds) Geomorphology and environment. ACB publication, Kolkata, pp 204–234
56. Shit KP, Paira R, Bhunia SG, Maiti R (2015) Modeling of potential gully erosion hazard using geo-spatial technology at Garbheta block, West Bengal in India. Model Earth Syst Environ 1:2
57. Kulimushi LC, Choudhari P, Mubalama KL, Banswe TG (2021) GIS and remote sensing-based assessment of soil erosion risk using RUSLE model in South-Kivu province, eastern, Democratic Republic of Congo. Geomatics Natural Hazards Risk 12(1):961–987
58. Panditharathne DLD, Abeysingha NS, Nirmanee KGS, Mallawatantri A (2019) Application of revised universal soil loss equation (Rusle) model to assess soil erosion in "Kalu Ganga" river basin in Sri Lanka. Appl Environ Soil Sci 2019:1–15
59. Wijesundara N, Abeysingha N, Dissanayake D (2018) GIS-based soil loss estimation using RUSLE model: a case of Kirindi Oya river basin, Sri Lanka. Model Earth Syst Environ 4(1):251–262
60. Maqsoom A, Aslam B, Hassan U, Kazmi AZ, Sodangi M, Tufail FR, Farooq D (2020) Geospatial assessment of soil erosion intensity and sediment yield using the revised universal soil loss equation (RUSLE) model. ISPRS Int J Geo-Inf 9:356
61. Bready NC, Well RR, The nature and properties of soil. Pearson Prentice Hall, Singapore
62. Wischmeier WH, Smith DD, Predicting rainfall erosion losses. A Guide to conservation planning. United States Department of Agriculture, Agricultural Research Service (USDA-ARS) Handbook No. 537. United States Government Printing Office, Washington, DC
63. Roose EJ, Sarrailh JM (1989) Erodibility of some tropical soils—twenty years of records in some erosion plots under natural rainfall. Soils Fert 25:7–30

64. Onori F, Bonis PD, Grauso S (2006) Soil erosion prediction at the basin scale using the revised universal soil loss equation (RUSLE) in a catchment of Sicily (Southern Italy). Environ Geol 50:1129–1140
65. Ozsoy G, Aksoy E, Dirim M, Tumsavas Z (2012) Determination of soil erosion risk in the Mustafakemalpasa River Basin, Turkey using the revised universal soil loss equation, geographic information system, and remote sensing. Environ Manag 50:679–694
66. Bahadur KCK (2009) Mapping soil erosion susceptibility using remote sensing and GIS: a case of the upper Nam Wa watershed, Nan Province, Thailand. Environ Geol 57:695–705
67. Bahadur KCK (2012) Spatio–temporal patterns of agricultural expansion and its effect on watershed degradation: a case from the mountains of Nepal. Environ Earth Sci 65:2063–2077
68. Wu X, Wang X (2011) Spatial influence of geographical factors on soil erosion in Fuyang County, China. Proc Environ Sci 10:2128–2133
69. Farhan Y, Zreqat D, Nawaysa S (2014) Assessing the influence of physical factors on spatial soil erosion risk in northern Jordan. J Am Sci 10:29–39
70. Kushwaha NL, Yousuf A (2017) Soil erosion risk mapping of watersheds using RUSLE, remote sensing and GIS: a review. Res J Agric Sci 8(2):269–277
71. Jahun BG, Ibrahim R, Dlamini NS, Musa SM (2015) Review of soil erosion assessment using RUSLE model and GIS. J Biol Agric Healthcare 5(9) ISSN 2224-3208 (Paper) ISSN 2225-093X (Online)
72. Lu D, Li G, Valladares G, Batistella M (2004) Mapping soil erosion risk in Rondonia, Brazilian Amazonia: using RUSLE, remote sensing and GIS. Land Degrad Dev 15(5):499–512
73. Kouli M, Soupios P, Vallianatos F (2009) Soil erosion prediction using the revised universal soil loss equation (RUSLE) in a GIS framework, Chania, Northwestern Crete, Greece. Environ Geol 57(3):483–497
74. Schrijver A (2012) Tree species traits cause divergence in soil acidification during four decades of post agricultural forest development. Glob Change Biol 18:1127–1140
75. Adinarayana J (2003) Spatial decision support system for identifying priority sites for watershed management schemes. In: First interagency conference on research in the watersheds (ICRW). Benson, Arizona, pp 405–408
76. Silva RM, Santos CAG, Montenegro SMGL (2012) Integration of GIS and remote sensing for estimation of soil loss and prioritization of critical sub-catchments: a case study of Tapacura catchment. Nat Hazards 62:953–970
77. Patel DP, Gajjar CA, Srivastava PK (2013) Prioritization of Malesari mini-watersheds through morphometric analysis: a remote sensing and GIS perspective. Environ Earth Sci 69(8):2643–2656
78. Khadse GK, Vijay R, Labhasetwar PK (2015) Prioritization of catchments based on soil erosion using remote sensing and GIS. Environ Monitor Assess 187:333. https://doi.org/10.1007/s10661-015-4545-z
79. Markose VJ, Jayappa KS (2016) Soil loss estimation and prioritization of sub-watersheds of Kali River basin, Karnataka, India, using RUSLE and GIS. Environ Monit Assess 188:225

Stability Analysis of Gravity Dam Using Finite Element Method and Particle Swarm Optimization

Mrunalini A. Deshmukh and Rajendra B. Magar

1 Introduction

The assessment of structural stability of gravity dams plays a vital role in the analysis and design of gravity dams. The seismic analysis is carried out on the non-overflow section of the 33.5 m high gravity dam. The dam must be checked for different failures due to overturning, sliding, or crushing due to material failure by governing the standards given by IS code 6512-1984. Many guidelines were recommended to design and analyze gravity dams against static and dynamic loads by considering different load combinations specified in codebooks [3, 4].

The 30 m high gravity dam was numerically analyzed using the finite element with drainage gallery. The seismic structural stability analysis was executed against static and dynamic several loads and load combinations. The modal analysis was executed using four nodded plain quadrilateral elements [14]. The seismic stability analysis of a non-overflow section of gravity dam using FEM with varying earthquake intensities of 0.1–0.18 g was carried out. The obtained results concluded that it's not feasible to construct gravity dam for kh more than 0.3 g and proved that the vertical, principal, and shear stress using FEM found more conservative than the 2D gravity method [1]. The scale modeling concept was studied and explained applied in FEM. The dam structure is divided into small elements like a tri (three nodded) and quad (four nodded) element. Quad elements proved more accurate than tri elements in scale modeling. The biggest advantage of this method is that it provides changes

M. A. Deshmukh
Anjuman-I-Islam's Kalsekar Technical Campus, School of Engineering and Technology, New Panvel Affiliated to University of Mumbai, Mumbai, Maharashtra 410206, India
e-mail: mrunalini.deshmukh2000@gmail.com

R. B. Magar (✉)
Civil Engineering, Anjuman-I-Islam's Kalsekar Technical Campus, School of Engineering and Technology, New Panvel Affiliated to University of Mumbai, Mumbai, Maharashtra 410206, India
e-mail: rajendramagar69@gmail.com

A. Sil et al. (eds.), *Recent Trends in Civil Engineering*, Lecture Notes in Civil Engineering 274, https://doi.org/10.1007/978-981-19-4055-2_50

at any single node/element after completing the model [25]. The computing ability of FEM was studied to evaluate and measure stresses or moments developed at particular nodes/elements by modal analysis of 130 m concrete gravity dam. The crack commencement, crack propagation, drainage effect, and post seismic uplift pressure study were carried out. The gravity dam stability analysis validation was carried out against design loads [15]. The scale model of a 90 m high gravity dam was created using eight nodded solid elements with three translational degrees of freedom. The stress analysis under static earthquake forces was carried out using finite element-based Staad.Pro technique. The variable deflection was found very less approximately 0.002 m which is negligible [12].

The behavior for stability against seismic loads of gravity by using FEM-based Staad.Pro was reviewed. Authors found the behavior of gravity dam safe by checking stability with respect to overturning, sliding, and shear friction factors as recommended in codebooks [11]. The numerical analysis of 122 m high non-overflow monolith Pine Flat concrete gravity dam with a crest length of 560 m was studied as a case study. A concrete gravity dam's structural stability and safety against sliding on an inhomogeneous rock foundation with parallel two fault models without seepage were carried out [7]. The hydrodynamic stability analysis of concrete gravity dam against shear sliding failure situation using maximum design earthquake (MDE) was performed.

The numerical analysis of two dams with varying heights and two different dam-foundation conditions as (i) dam-foundation edge dynamic behavior and elastic behavior of fractured rock mass area and (ii) nonlinearity in dam-foundation edge the joints in fractured rock mass area both. Hence, the shear sliding failure with high stiffness rocks is more, but shear sliding displacement is reduced [2]. The "Kalhovd" concrete buttress dam was numerically analyzed with 49 buttresses in Norway. The analysis was carried out using certain key factors as interlocking between foundation and buttress. The dam was found safe in overturning and sliding failure using the finite element method. Results demonstrated an increase in load-carrying capacity with additional rock bolts, which increases stability in overturning [19]. The different tools in the finite element method for analysis of gravity dam structure were reviewed. The lateral stability is affected by the forces and moments generated due to various external loads. The safety evaluation for dam structure increases the structure's life [16].

The above studies executed by many researchers conclude that FEM permits the engineer to closely model and study the actual behavior of gravity dam against various loads and dam interaction with its foundation. FEM analysis was executed with the help of a virtual scale model. The scale modeling of a huge structure as a gravity dam was complex and tedious to develop or make any changes. The PSO method could avoid these limitations, which provides more efficient and accurate results in less time.

Particle swarm optimization is a heuristic global optimization technique of nonlinear functions introduced by Kennedy and Eberhart in 1995. It is a population-based search algorithm based on the social behavior of birds, bees, or fish schooling. It's a computational technique influenced by biological systems based on the "NO

LEADER" principle. In the PSO algorithm process, while finding an optimal solution, it follows the principle of animal society. The optimization of size and shape in structural design problems was executed by using the PSO algorithm. The authors carried out optimization on various structural elements like two bar plane truss, 10-bar plane truss, 25-bar space truss, and torque arm. Hence, the PSO algorithm proved superior to the genetic algorithm [6]. The PSO algorithm performance in structural design problems was studied. The results concluded that the PSO performs with good precision in both continuous and discrete problems [20]. The various methods of the PSO algorithm were reviewed. The basic variant PSO found maintained governing the velocity and the stable convergence. The other conditions or complex problems are resolved using modified variant PSO [13]. The shape optimization of the concrete gravity dam was carried out by considering the dam-water-foundation interface exposed to seismic loading using the proposed new hybrid method called GSA-PSO. GSA-PSO combines hybrid gravitational research algorithm (GSA) and particle swarm optimization (PSO). The optimization process is executed in three phases.

In the first phase, preliminary optimization using GSA as local search, then in the second phase, swarm initialization using the optimum result obtained by GSA, and in the final phase, PSO applied. Hence, GSA-PAO results were superior to GSA and PSO individually [13]. The single and multi-objective optimization using particle swarm optimization was reviewed. The authors explained the basic PSO algorithm and steps involved in PSO. The authors' literature review concludes that PSO performs better in the two-dimension problem than ten-dimensions [18].

The optimization of the dam section was carried out by setting its height constant and cost of the dam section with safety. A MATLAB coding is generated to optimize the dynamic programming method based on IS 1893-1984. The parameters of a dam section like the height of the dam (H), the height of reservoir water (Hu), Peak ground acceleration (Ah), and minimum width of a crest ($B2$) are optimized [3]. Authors optimized geometrical configuration, i.e., cross-sectional area using PSO, which reduced construction cost by optimizing the concrete poured during construction in a gravity dam. The PSO algorithm is proved more efficient than the conventional optimization method in solving problems where objective function and constraints are nonlinear [23]. The modified PSO, "SACPSO-FORM," was used for gravity dam stability reliability analysis on a complicated rock foundation. The SACPSO-FORM is a combination of the "First Order Reliability Method" (FORM) and "Hybrid Particle Swarm Optimization" (SACPSO). Results show that the newly proposed method is accurate, stable, and efficient in reliability analysis [24]. The authors optimized the channel cross-section, height of upstream and downstream cofferdam with a minimum cost of construction by PSO using MATLAB. The authors concluded that optimum bed width, water flow depth, and height of cofferdam are inversely proportional to side slope. In addition, the cost of construction is directly proportional to the side slope [10].

Improved PSO was applied for the dynamic changing patter of Inertia Weight "w" of a non-overflow section of the gravity dam. In the early stage of Improved PSO, the inertia weight value was high, but in the further stage, it reduces by improving the algorithm's global searching ability and meeting performance. The convergence rate of improved PSO was 31.8% higher than SPSO. Hence, improved PSO is better and more suitable for optimizing huge hydraulic structures [17]. The authors executed a back analysis of the elastic modulus of the dam body and dam foundation using improved PSO. PSO algorithms with modified parameters are used in back analysis with high accuracy [21]. The high concrete dam was analyzed using a proposed hybrid modelling method and improved PSO to predict the hydrostatic pressure deformation of the dam. The hybrid hydraulic-seasonal-time (HHST) model to simulate the effects of structural effects in operation conditions. The improved PSO algorithm is used by improving PSO with the help of inertia weight, dynamic learning factors, etc. [22]. From the literature review, it has been found that very few studies have been carried out using PSO for the stability analysis of gravity dams.

The objective of the current study is to model the dam structure and perform stability analysis with more than one fixed criterion against various static and dynamic loads acting on the dam by assuming the nonlinear deformation analysis with maximum design earthquake forces. Optimization in work is achieved by using the multi-objective algorithm in integer or discrete problems. Applying all the above conditions, structural stability analysis is executed on concrete gravity dam using 3D finite element method (FEM) and particle swarm optimization (PSO).

2 Method and Methodology

2.1 3D FEM-Based Staad.Pro Technique

The finite element method is an advantageous method for scale modeling of the dam structure. The virtual scale modelling was verified more effective and economical than a physical model [5]. Alteration in model construction and loads acting on it become easy at any point in the process. In this paper, the structural stability of the 33.50 m high gravity dam was executed on the virtual scale model in Staad.Pro technique with four nodded quad elements, i.e., plate element as shown in Fig. 1.

From Fig. 1, it can be observed that the non-overflow section of Ban Sujara dam is modelled in Staad.Pro consists of 131 nodes, and 143 four-nodded plates (elements). The scale model replicates the progression of structure, a virtual prototype with the help of beams and surfaces which they would appear in the physical world. A joint is generated anywhere two physical elements meet in the model (and at the free ends of cantilevered members, for convenience).

The basic three stages are involved in solving the physical problem using the finite element method are specified as pre-processing, processing, and post-processing stage [12]. Based on their study, the flowchart of the finite element method process for the current case study is depicted in Fig. 2.

Figure 2 shows the three stages of finite element analysis as pre-processing, processing, and post-processing. Pre-processing is a stage of creating input data for analysis. This stage defines geometry, assigning physical properties, assigning

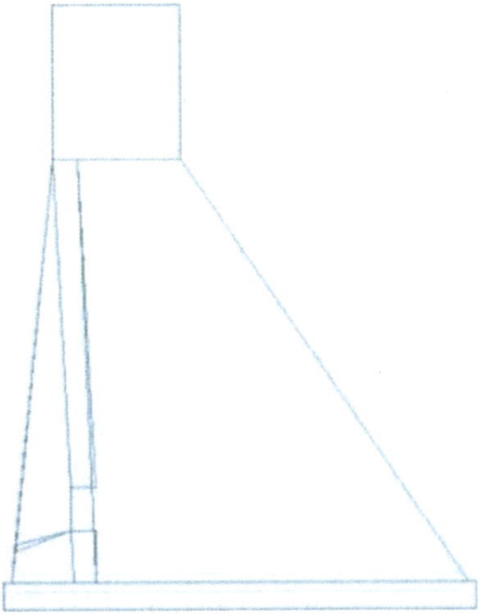

Fig. 1 Side view of along with foundation gallery-Scale model, Staad.Pro V8i

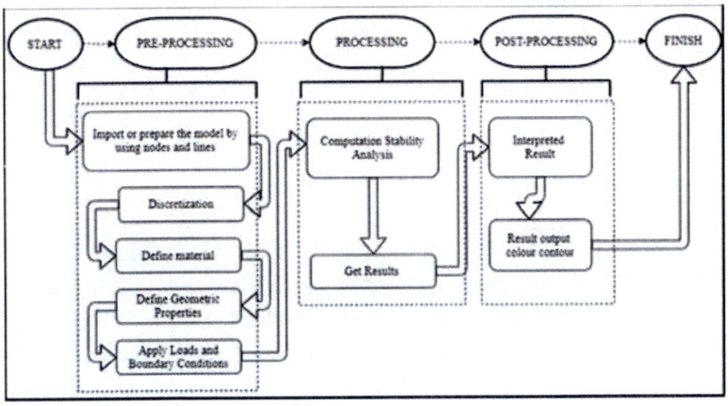

Fig. 2 Stages followed by FEM

constraint conditions, assigning load, and load combination. Processing is the core of the process of FEM. In this stage, computation of analysis is carried out by evaluating equivalent forces at all nodes. Post-processing involves the graphical presentation by contour diagrams of results obtained.

A. *Particle Swarm Optimisation (PSO).*

Particle swarm optimization helps in solving non-convex and nonlinear problems with an evolutionary algorithm. PSO is one famous higher-level design technique to find the best solution by evolving self to reach the target or goal [11].

In PSO, the authors specified the implementation process and concept of time, velocity changing particles, pbest, gbest, and lbest values, and location [3, 5, 23].

Based on the study of the above authors, the process flow diagram of PSO used in the current study is depicted as in Fig. 3.

Fig. 3 Flow diagram, PSO

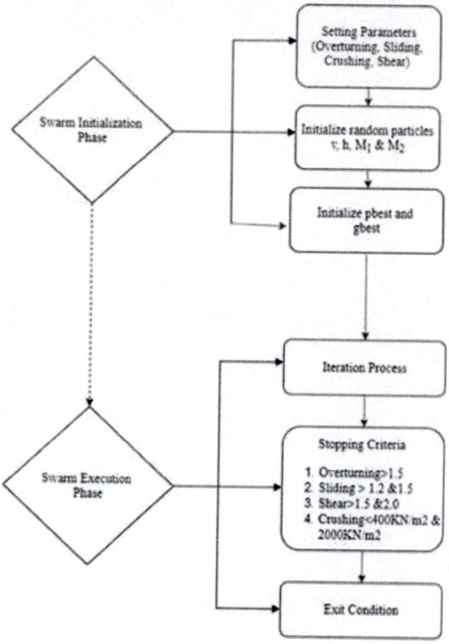

Figure 3 clearly shows the process flow diagram of the PSO algorithm executed in the current study. PSO is carried in two phases as swarm initialization and swarm execution. In the swarm initialization phase, the position and velocity are arbitrarily initialized in the search space. After that, objective values position and locations are determined. The first objective position and location is term as personal best (pbest) value and location.

The global best (gbest) value and positions are set to the objective pbest value and position respective to the gbest value and position in the entire swarm. After this, all particles move to their new position. All objective values are evaluated again. The new pbest position is updated for a new objective value better than the old pbest value; similarly, old gbest values are updated if the new gbest of a new objective value is better than the old gbest value.

Again, all particles move to their new position. The algorithm continues with changing their evaluation of pbest value, pbest position, and gbest value, gbest position till reach the specified limiting criteria or number of iterations reached.

In the current study, the random swarm initialization of vertical forces (v), horizontal forces (h), positive moment (M_1), and negative moment (M_2) are carried out in the PSO algorithm to get the pbest and gbest value till they reach the limiting criteria concerning overturning, sliding, shear, and crushing as specified by codebooks. Hence, stability analysis of concrete gravity dam is executed using PSO.

2.2 *Methodology*

The methodology applied in this work includes extensive study and analysis of FEM and PSO methods. The actual geometric data of the non-overflow section of Ban Sujara Project, Madhya Pradesh, is collected, and extensive analysis is carried out using FEM and PSO method. The methodology adopted for the analysis of the gravity dam is shown in Fig. 4.

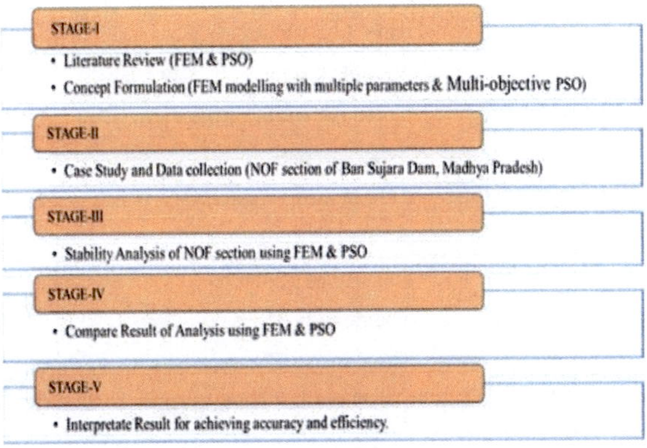

Fig. 4 Methodology

Figure 4 clearly shows the methodology used in work. Work was executed in five stages. Starting with reviewing literature by different researchers and applying it to a non-overflow section of Ban Sujara Dam, Madhya Pradesh. FEM-based Staad.Pro was used to model and analyze the non-overflow section of the gravity dam. Further, the same section is analyzed using the PSO method. The results were further compared for accuracy and efficiency.

3 Case Study and Data Collection

3.1 *Case Study: Ban Sujara Project, Madhya Pradesh*

The structural stability analysis of a non-overflow section of the dam of Ban Sujara project, Madhya Pradesh, is carried out in three stages,

Stage1: Parametric Details of Dam Non-Overflow Section.

Stage 2: Analysis of non-overflow section using FEM based Staad.Pro.

Stage 3: Analysis of non-overflow section using PSO.

To reach the dam's utmost structural stability, all 3stages look after the four major criteria as specified in codebooks. The four criteria to check structural safety for dam structure are as follows [5, 8, 9]:

1. Overturning of the dam, i.e., factor of safety against overturning (FSO)
2. Compression or crushing of base.
3. Tension in dam structure.
4. Sliding of the dam base due to shear failure, i.e., factor of safety against sliding (FSS).

The silent features and specification of a non-overflow section of the existing gravity dam of Ban Sujara Project, Madhya Pradesh as shown in Table 1.

Table 1 Silent features and specification of model of dam

Type of structure	Concrete gravity dam
Height of dam	33.5 m
The base width of the dam	27.10 m
Seismic zone	II
Safe bearing capacity	100 t/m^2
Angle of friction	35°
Density of concrete	24 KN/m^3
The density of dry silt	21 KN/m^3
The density of saturated silt	19.5 KN/m^3
Coefficient of friction (μ)	0.65–0.75 for dams

Figure 5 shows the control level of the non-overflow section of the gravity dam.

Figure 5 shows all control levels of gravity dam like Top Bank-Level (TBL) as 321.00 m, Maximum Water Level (MWL) as 316.60 m, Flood Reservoir Level (FRL) is 316.50 m, Ground Level (GL) as 290.00 m, Foundation Level (FL) as 287.50 m.

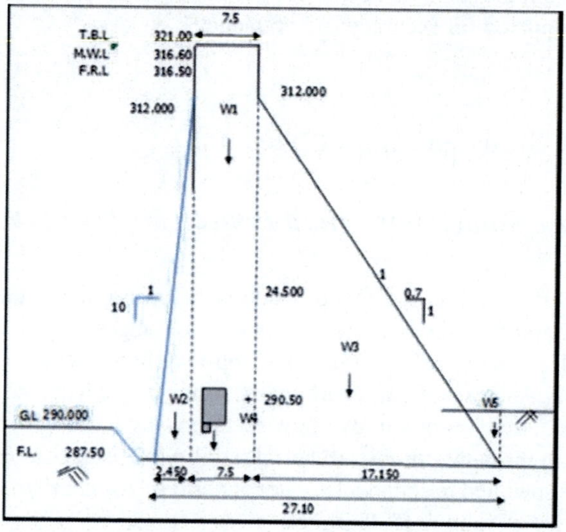

Fig. 5 Control level of gravity dam

3.2 Data Collection

The structural stability of the non-overflow section of Ban Sujara dam is executed against different static and dynamic loads like self weight of dam structure, water pressure on the upstream and downstream side of the dam, uplift pressure acting on the dam along with earthquake force in x and z-direction. As per the CWPRS Technical Report of April 2014, the peak ground acceleration values for horizontal and vertical components are 0.099 g and 0.066 g, respectively, for design basic earthquake (DBE) conditions.

To execute a detailed investigation as per seismic code, the design seismic horizontal coefficient αh is maximum at the dam's top and linearly reduces to zero at the bottom of the dam. According to codebooks, the horizontal seismic coefficient and vertical seismic coefficient are considered 0.1485 and 0.099, respectively. The horizontal acceleration due to the earthquake results in hydrodynamic pressure and inertia force into the dam's body. The horizontal seismic coefficient (α_h) is important for calculating equivalent pseudo-static inertial force in seismic analysis for dam structures [9].

According to seismic code, clause no. 3.4.2.3, IS 1893-1984/2002,

$$\text{Horizontal seismic coefficient } \alpha_h = \beta \times \mathrm{I} \times \alpha_o \tag{1}$$

where β = soil-foundation system factor = 1 for dams;

I = importance factor = 3 for dams;

α_o = basic horizontal coefficient = 0.02 for zone-II;

As per code books, the structural stability analysis of gravity dam was analyzed in all seven load combination cases as given in Table 2 [8, 9].

Table 2 Load combination cases [8, 9]

Case	I	II	III	IV	V	VI	VII
Details	Empty reservoir	FRL[a] + Zero tailwater + Normal uplift	MWL[b] + Open gates + Full tailwater at flood level + Normal uplift + Silt	Case I + Seismic forces	Case II + Seismic forces	Case III + Extreme uplift, Inoperative drains	Case V + Extreme uplift and inoperative drains

Note [a] Full Reservoir Level (FRL); [b] Maximum Water level (MWL)

4 Results and Discussion

The structural stability analysis of the non-overflow section of the Ban Sujara project concrete gravity dam is carried out using a 3D FEM-based Staad.Pro and PSO. In the finite element method, the non-overflow section of the Ban Sujara dam structure, modeled in Staad.Pro. The structure type used is "Space Frame Structure." The three-dimensional view is shown in Fig. 6.

Figure 6 shows the three-dimensional rendering view with foundation gallery shows the real view of a non-overflow section of gravity dam using plate element in model creation. The model uses 119 nodes, and four-nodded plate element 157 with a foundation gallery of 1.5 m × 2.25 m at + 290.85 m throughout the structure. The dam structure is exposed to various external loads acting on the concrete gravity dam, as shown in Fig. 7.

Figure 7 displays different external static and dynamic load distribution profiles with seven load combinations. The various external loads as water pressure, silt pressure, uplift pressure, etc., on a non-overflow section of concrete gravity dam. Water pressure acts triangular on the upstream face, maximum at the base and zero at the surface. Triangular silt pressure distribution on the upstream face in silt accumulation depth. Uplift pressure distribution on the base of the dam. The varying effect and distribution of stresses are shown in Fig. 8.

Figure 8 expresses the sample stress distribution and evaluation results for stress found in the dam structure. The effect and distribution of stresses are identified by different color contours from purple to red with ascending magnitude values. The area in which purple color contours are found is safest as having minimum stress. The area where red contours are found is the most affected area as the found magnitude of results is maximum. From the study, it is clear that red contour is found near

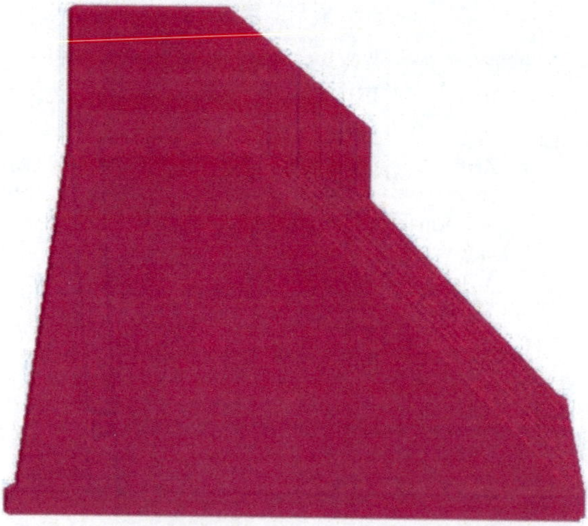

Fig. 6 Gravity dam structure, 3D view

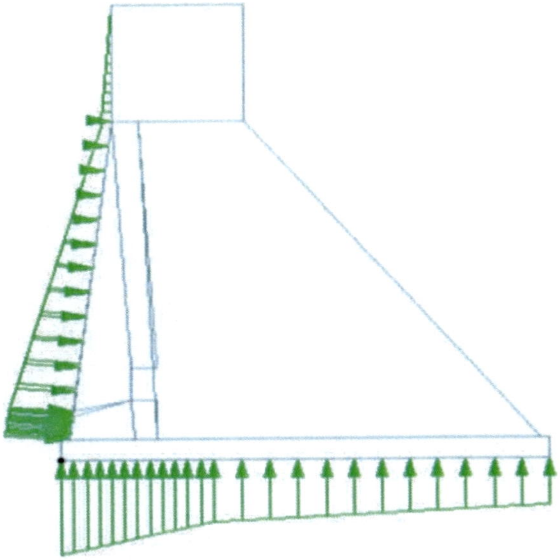

Fig. 7 Gravity dam structure, external loads

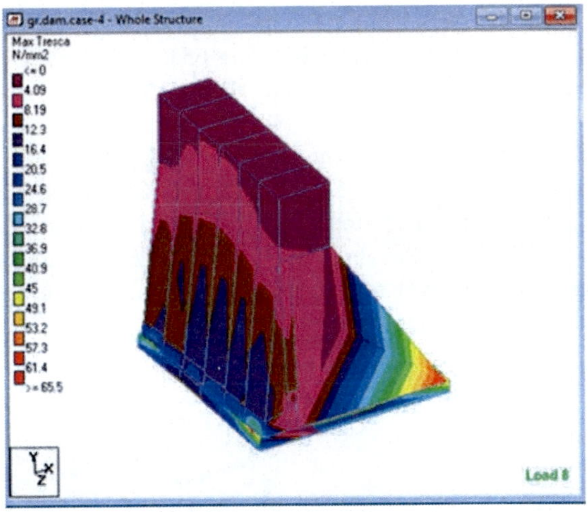

Fig. 8 Stress in structure

the toe and foundation gallery, which means the critical stress was found there. The evaluation of moments in the dam structure is as shown in Fig. 9.

Figure 9 depicts the moments caused in the x-direction in load combination V [i.e., FRL* + Zero Tailwater + Normal Uplift + EQL(+x)] in gravity dam due to different load combinations as per codebooks using FEM based Staad.Pro. The area in which purple color contour is found possesses maximum negative moment (−2176 KN/m2) and in red, areas possess the maximum positive moment (+1632 KN/m2). In the rest of the dam structure, the optimum moment is found represented by green

Fig. 9 Moment in structure

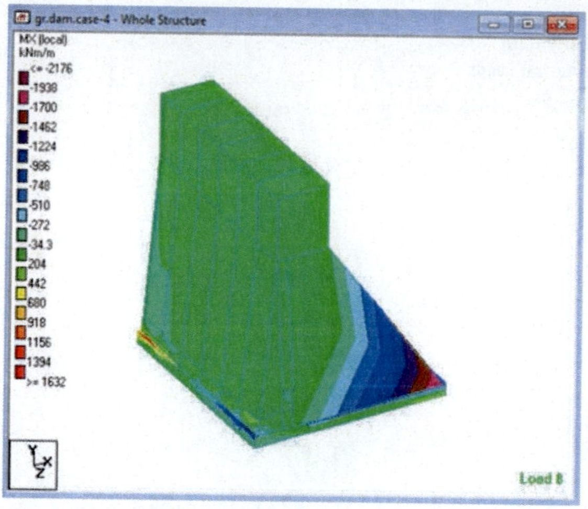

color. The maximum moment is found along with the edge at the heel and toe end of the dam.

As per codebooks, to maintain the structural stability of the gravity dam, the estimated result of check against overturning with respect to recommended limiting values is shown in Fig 10 by referring to IS 6512-1984 and IS1893-1984/2002.

Figure 10 shows safety checks results against overturning attained from Staad.Pro and PSO along with IS code limits. Overturning in a dam expresses the development of tensile stresses at the heel. The results displayed are 2.62, and 2.76 maximum overturning values attained in Staad.Pro and PSO respectively for load combination V. The minimum values found are 2.13 (load combination IV) and 1.62 (load combination II) in Staad.Pro and PSO, respectively. The evaluated overturning check results were found safe with respect to IS recommended value by codebooks. Figure 9 concluded that the dam is safe against overturning.

Fig. 10 Factor of safety against overturning

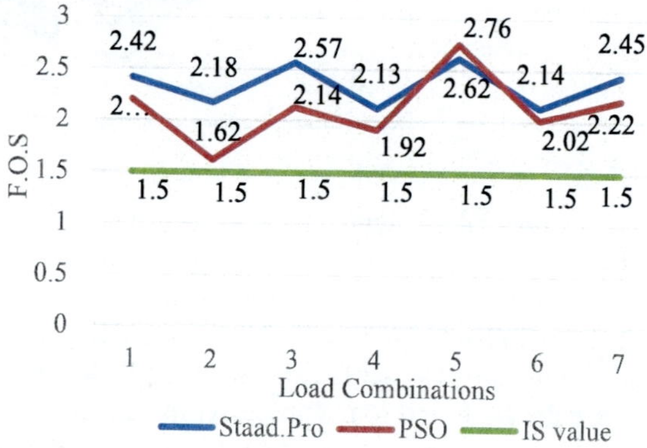

The results evaluated for a check against sliding of the dam with respect to recommended limiting values by codebooks are shown in Fig. 11.

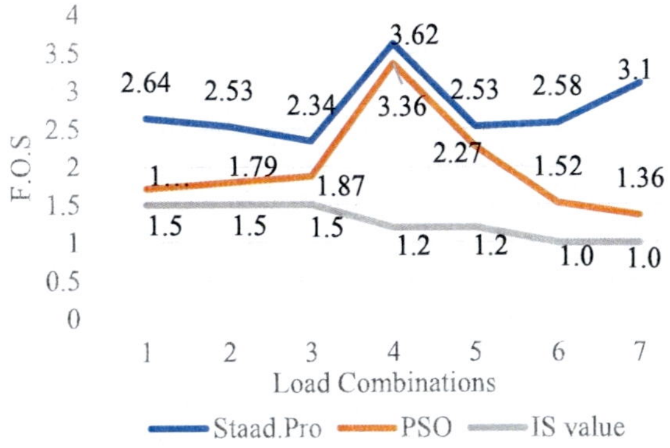

Fig. 11 Factor of Safety against Sliding

Figure 11 shows the calculated sliding check results in Staad.Pro and PSO along with IS values. Safety against sliding can potentially influence the dam due to dynamic failure along the base at the dam/foundation interface. The maximum sliding check values found are 3.62 and 3.36 in Staad.Pro and PSO in load combination IV and minimum sliding check values found are 2.34 (load combination III) and 1.36 (load combination VII) in Staad.Pro and PSO, respectively. From Fig 10 initiate that estimated sliding check values are satisfying all recommended limits in all cases. Hence, conclude that the dam structure is safe against sliding.

The shear friction factor (S.F.F) evaluated result is as shown in Fig. 12.

Figure 12 depicts the safety check results against the shear friction factor (S.F.F.). It verifies the shear condition in the dam body and at the dam-foundation interface. The estimated check maximum shear found in load combination IV with the values 6.58 and 4.72 in Staad.Pro and PSO, respectively, and minimum shear 3.85 (Load combination II) and 2.67 (load combination I) in Staad.Pro and PSO, respectively. All estimated values of shear check found in limit recommended by codebook. Hence,

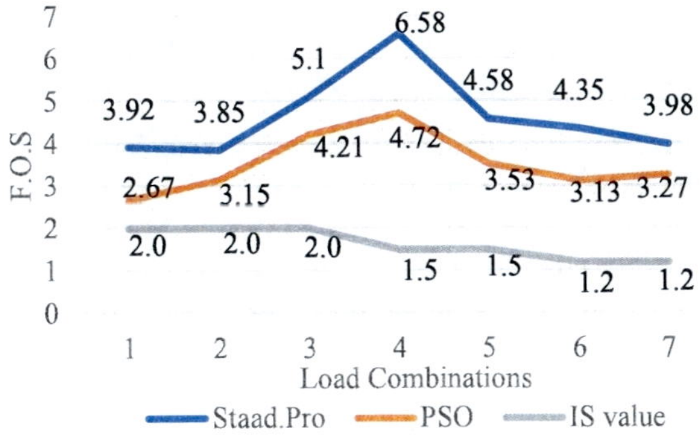

Fig. 12 Factor of Safety against Shear Friction Factor (S.F.F.)

concluded that the dam structure is safe in shear. The crushing values found in the dam structure are as shown in Fig. 13.

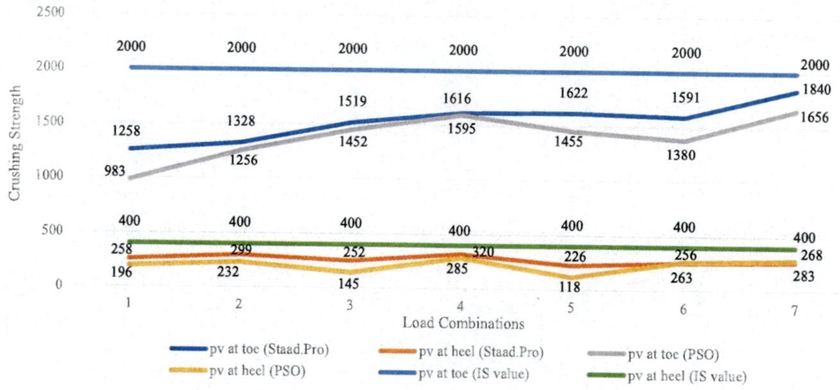

Fig. 13 Factor of Safety against Crushing Strength

Figure 13 shows calculated safety results against crushing. The crushing strength of a dam refers to the strength achieved by the material used in dam construction. The maximum crushing strength found at the toe in load combination VII is 1840 N/mm2 and 1656 N/mm2 in Staad.Pro and PSO, respectively, and minimum crushing strength found in load combination I at the toe are 1258 N/mm2 and 983 N/mm2 in Staad.Pro and PSO, respectively.

Similarly, at heel maximum crushing strength found in load combination IV are 320 N/mm2 and 285 N/mm2 in Staad.Pro and PSO, respectively, and minimum crushing strength found in load combination V are 226 N/mm2 and 118 N/mm2 in Staad.Pro and PSO, respectively. Since concluding that dam structure is safe in crushing, all estimated checks found in limit recommended by codebook.

5 Conclusion

From the modeling and analysis of concrete gravity dam in finite element method and particle swarm optimization method concluded as per following:

(1) Maximum absolute stress in the dam with foundation gallery is 51.497 N/mm^2 in FEM and 54.23 N/mm^2 in PSO.
(2) Principal stress in dam is 19.98 N/mm^2 in FEM and 20.97 N/mm^2 in PSO
(3) Shear stress in the dam is 30.945 N/mm^2 in FEM and 32.182 N/mm^2 in PSO.
(4) The critical stresses identified in the area near to foundation gallery and at the heel of the dam structure are accurate, similar, and easily calculated in PSO than FEM.

From the study of structural stability analysis of Ban Sujara project, it is found safe concerning checked recommended in codebooks. The above study observed that the result found in PSO are accurate and nullifies the probabilistic error that occurs during the assumptions in the FEM method.

Acknowledgements This research work was supported by providing the research data of the non-overflow section of Ban Sujara Project, dam, Madhya Pradesh. The author humbly acknowledges the technical support given by "Er. B. Lakshmana Rao", Former Engineer-in-Chief, Central Designs Organization, Hyderabad.

References

1. Ali H, Alam R, Haque N, Alam J, Comparison of design and analysis of concrete gravity dam. Natural Resources/2012, 3/18-28
2. Azevedo N (2018) Stability analysis of gravity dams for the maximum design earthquake. Dissertation report, FCL Faculdade of Ciencias E Technologia, Universidade Nova De Lisboa, Maio
3. Banerjee A, Paul D, Dubey R (2013) Optimization, classical and finite element analysis of concrete gravity dam. In: International Conference on Earthquake Engineering, At Skopje, Macedonia. https://doi.org/10.13140/2.1.2102.6565, Research Gate/publication/266477130, May
4. Dawlatzai K, Dominic M (2018) Structural stability and 2d finite element analysis of concrete gravity dam. Int J Eng Sci Invent (IJESI). 7(1):9–20, January. ISSN (Online): 2319–6734, ISSN (Print): 2319–6726
5. Deshpande V, Janardhana K (2019) Stability analysis of overflow dam using STAAD.Pro. Int J Eng Res Tech (IERT). Paper ID: IJERTV8IS070252, 8(7), July
6. Fourie P, Groenwold A (2002) The particle swarm optimization algorithm in size and shape optimization. Struct Multidiscip Optim 23(4):259–267
7. Ganji H, Alembagheri M (2018) Stability of monolithic gravity dam located on heterogeneous rock foundation. Arab J. Sci. Eng. 43:1777–1793
8. IS 6512-1984: Indian standard criteria for design of solid gravity dams (First Revision). First Reprint September (1998)
9. IS1893-1984/2002: Indian standard criteria for earthquake resistant design of structures (Fourth Revision). First Reprint July (1999)
10. Khudair B, Mahmood R, Sadeq N (2018) Optimal economic design of diversion structures during construction of a dam by particle swarm optimization. Int J Sci Res (IJSR), June. ISSN (Online): 2319-7064
11. Kennedy J, Eberhart R (1995) Particle swarm optimization. In: IEEE International Conference on Neural Networks 1995; Date of Conference: 27 Nov–1 Dec; Date Added to IEEE Xplore: 6 August 2002; Print ISBN:0-7803-2768-3
12. Reddy M, Rao M (2017) Design and analysis of gravity dam—a case study analysis using STAAD.PRO. Online Anveshana's Int J Res Eng Appl Sci AIJREAS 2(4), April (2017). Issn 2455–6300
13. Rini D, Shamsuddin S, Yuhaniz S (2011) Particle swarm optimization: technique, system, and challenges. Int J Comp Appl (0975–8887) 14(1), January
14. Subramani T, Ponnuvel D, Seismic and stability analysis of gravity dams using Staad PRO. Int J Eng Res Develop 1(5):44–54
15. Swetha S, Naidu G, Stability analysis of gravity dam by using STAAD Pro in time history method. Int J Magazine Eng Tech Manag Res 4(3), March. ISSN No: 2348-4845

16. Sinha D, Divya K, Singh L (2020) Analysis of a dam structure using analysis tool: a review. Int J Scient Res Civil Engineering© 2020 IJSRCE **4**(6):53–59 (2020). ISSN: 2456-6667. Publication Issue: November–December 2020
17. Si Z, Yang K, Huang L (2019) Study of the section optimization of gravity dam based on improved PSO. ACCESE 2019, IOP Conf. Series: Earth and Environmental Science 330:022007. IOP Publishing. https://doi.org/10.1088/1755-1315/330/2/022007
18. Urade H, Patel R (2011) Study and analysis of particle swarm optimization: a review. 2nd National Conference on Information and Communication Technology (NCICT). In: Proceedings published in IJCA
19. Ulfreg A (2019) Probabilistic nonlinear finite element analysis of concrete buttress dams. Civil Engineering, master's level Degree Project, 2019, Luleå University of Technology, Department of Civil, Environmental and Natural Resources Engineering
20. Venter G, Sobieski J (2003) Particle swarm optimization. Hampton, Virginia, American Institute of Aeronautics and Astronautics (AIAA) J 41(8). https://doi.org/10.2514/2.2111
21. Yang L, Su H, Wen Z (2019) Improved PLS and PSO methods-based back analysis for the elastic modulus of the dam. Elsevier
22. Yang G, Gu H, Chen X, Zhao K, Qiao D (2021) Hybrid hydraulic-seasonal-time model for predicting the deformation behavior of high concrete dams during the operational period. Wiley Online Library, J Int Assoc Struc Cont Health Monitoring, 5 January
23. Yazd H, Arabshahi S, Tavousi M, Alvani A (2015) Optimal designing of concrete gravity dam using particle swarm optimization algorithm (PSO). Indian J Sci Tech 8(12):June. https://doi.org/10.17485/ijst/2015/v8i12/70714. ISSN (Print): 0974-6846 ISSN (Online): 0974-5645
24. Wang G, Ma Z (2017) Hybrid particle swarm optimization for first order reliability method. ELSEVIER/Science Direct/Computers and Geotechnics 81:49–58
25. Zeidan B (2015) Seismic finite element analysis of dam-reservoir foundation interaction. In: International Conference on Advances in Structural and Geotechnical Engineering(ICASGE'15), Hurghada, Egypt, Conference Paper, April, pp 6–9

Hydrodynamic Analysis of a Fixed Offshore Wind Tower Using ANSYS Aqwa

Neethu Ann Paul and Anitha Joseph

1 Introduction

Wind energy is looked upon as the fastest growing clean and green renewable energy in the world. To tap its potential offshore, wind turbines erected on various types of supporting towers and platforms are currently being explored. The climate changes and global warming in the current scenario urges the need and importance of resources that are renewable, energy-efficient and eco-friendly. Different types of renewable energy include solar, geothermal, ocean and tidal energy, among which wind energy is also proved to be favourable one. Offshore wind energy has more benefits than onshore wind energy. In offshore wind energy, vast offshore areas can be used without noise and visual conflicts, with constant wind conditions. The coastline of India which includes Gujarat coast and Rameshwaram–Kanyakumari area in Tamil Nadu has good potential towards the establishment of offshore wind farms. Wind energy could play a significant role since India has a large potential for offshore wind energy resources.

2 Literature Review

This project deals with the numerical investigations on the hydrodynamic response of a fixed offshore jacket structure, namely the MJ structures. A detailed review of the literature related to the scope of the project is presented in this section.

N. A. Paul (✉) · A. Joseph
Department of Civil Engineering, Saintgits College of Engineering, Kottayam, Kerala, India
e-mail: neethuap.secm1921@saintgits.org

A. Joseph
e-mail: anitha.joseph@saintgits.org

A. Sil et al. (eds.), *Recent Trends in Civil Engineering*, Lecture Notes in Civil Engineering 274, https://doi.org/10.1007/978-981-19-4055-2_51

Raheem [2] conducted a finite element analysis of offshore jacket structure using SAP 2000. The study includes natural vibration analysis, displacement response demands, axial force and bending moment response demands of the structure. Displacement increases nonlinearly with height of platform jacket. Bending moment was concentrated at connection points between different structural systems.

Wei et al. [3] compared the feasibility of different substructures including the monopile, jacket and multipile by investigating their dynamic responses. Under deterministic and stochastic environmental conditions, Eigen value analysis and a coupled aero-hydro-servo-elastic simulation were done. Using Bladed V4.3 from Germanischer Lloyd Garrad Hassan, aero-hydro-servo-elastic analysis in the time domain was done. The mass of the monopile substructure was smaller than that of the jacket substructure. The soft-stiff design properties were shown by three substructures. Among the three substructures, jacket has got the best performance.

Di et al. [4] studied dynamic behaviour of K-type and inverted-K-type jacket structures exposed to simultaneous wave, wind and current loads using ANSYS software. It was obtained from modal analysis that the inverted-K-type jacket structure has higher natural frequencies than K-type. The K-type jacket structure showed larger displacement than that of the inverted-K-type under wave force spectrum. In the time-history responses under wind and wave-current load, the displacements and stresses of inverted-K-type jacket structure were smaller as compared to the K-type. The inverted-K-type jacket support structure has more stiffness and mechanical properties compared to K-type.

Wei et al. [5] conducted a series of time domain dynamic analyses on three models of OWT using a finite element program called USFOS. The aim of this study was to determine how dynamic amplification factors (DAF) varies with natural time period of structure and also under regular and irregular waves. The regular wave was sampled using 10th order stream function wave theory and irregular wave using JONSWAP spectrum. The DAF decreases with increasing wave height. For the regular wave analyses, the DAF in terms of base shear was always between 0.90 and 1.10, for wave heights larger than 10 m and for the irregular wave analyses, with wave heights larger than 10 m, the mean DAFs were always between 0.91 and 1.13. The effect of the natural period of the structure on the DAF was shown to be small for waves larger than 10 m because large waves have time period longer than the fundamental periods of all the models considered here.

Arvind and Desai [6] conducted static analysis of traditional jacket and three-legged modified twisted jackets with variation in twisting angles of 0°, 30°, 60° and 90°, for installing a 5-MW NREL wind turbine. Stress analyses were done to determine the maximum von Mises stresses in the models using SACS software. It was found that three-legged modified twisted jacket with more twisting under same loading and environmental conditions was able to take maximum von Mises stresses as compared to normal traditional jacket.

Cardoso et al. [7] performed aerodynamic and hydrodynamic analysis of innovative foundation using FAST and ANSYS Aqwa software during transportation phase. The foundation with self-floating capability and suction pile embedment was installed for transitional waters for a water depth of 45 m. The wind and sea state data were established to determine hydrodynamic and aerodynamic loads. The modal analysis indicates that the natural frequencies of the structure are far away from the frequency band of excitation vibrations. This prevents the structure from resonating. The response amplitude operators indicate that the structure exhibits a stable dynamic behaviour for the specified sea state. The transient structural analysis performed on the structure indicates that an increase in the wind speed and the subsequent increase on wave height and peak period of the waves, lead to an increase of the stresses in the structure. The von Mises stresses were maximum near the connections between structural members in the non-submerged part of the foundation.

Vuonga and Quana [8] analyse fatigue life of components of wind turbine support structure and predicted the expected life of offshore wind turbine. Both MATLAB code program and SACS software were used for the fatigue study. Eight loading directions were considered, spaced at 45° angle to calculate fatigue life of jacket support structure. The computation of static and dynamic transfer function of wave to support the fatigue analysis was done using Airy wave theory. Hot-spot stress in time series was computed. The results obtained were reasonable since the simultaneous interaction between wind and wind-induced wave were considered.

Taheri and Shahsavari [9] performed dynamic analysis of the offshore jacket platform using random time domain analysis. The statistical properties of the extreme sea condition were captured by simulating fifteen 1-h storm conditions of the water surface elevation. Dynamic and static responses were taken by deriving both time series of base shear and overturning moment. The modal analysis was done on twenty modes. The structural responses at each 0.5 s interval were calculated using random wave module of SACS program. DAF were generated. Using Weibull distribution function, the most probable maximum extreme value for DAF in extreme sea condition was computed. The DAF for this specific platform was 1.06 and can be stretched to other platforms with comparable dynamic characteristics.

Raheem et al. [10] performed numerical analysis of offshore jacket support structure using SACS software. The free vibration analysis, base shear, overturning moment response, horizontal and vertical displacement response were the parameters studied. First mode of vibration has the largest contribution to platform motion during environmental excitation. The maximum base shear and overturning moment occur at 140° angle of wave attack. The absolute maximum horizontal displacement responses at the mudline and the topmost deck levels in all platform legs for the three storm conditions attained for incidence angle 0°. The vertical displacements change from positive to negative displacement where maximum positive value occurs at 40° and maximum negative value occurs at 220°.

Wang et al. [11] performed an experimental and numerical modelling of jacket structure. The model was subjected to steep and breaking waves. A 1:8 scaled model of jacket structure subjected to controlled regular waves were conducted in the wave slam project. The measured wave surface elevations and water particle velocities from the experiments were validated using a fully nonlinear potential flow solver and found a fairly good agreement. The breaking waves were calculated by subdividing the calculated hydrodynamic forcing into quasi-static and an impulsive slamming force component.

Numerical investigations performed on conventional jacket structures reported are static, hydrodynamic, free vibration and fatigue analysis. The finite element software packages SESAM, USFOS, SACS and SAP 2000 are employed for the reported works. Static analysis of MJ structures using Abaqus software is reported. The MJ structures are found to have low material usage, weight of structure is low, least number of joints per each layer compared to conventional offshore jacket structure and fabrication costs are approximately 20% less expensive than conventional offshore jacket structure. Also, the ease of constructability and reduced install time is one prominent feature of MJ structures. It is more compact and can be easily transported to sites by barges, and hence, transportation cost is reduced. The modal, hydrodynamic and fatigue analysis of MJ structures are to be comprehensively investigated.

There are very limited studies that have been performed on jacket structures using ANSYS Aqwa. This software package is suitable for performing hydrodynamic analysis of offshore structures. Also, ANSYS was found to be suitable for modal and fatigue analysis of jacket structures. The main aim of the study is to identify best performing MJ structure after dynamic analysis. Modal, hydrodynamic and fatigue analysis of MJ are to be comprehensively investigated.

3 Methodology

The present study considers four different models of jacket structure of varying geometric configuration such as MJ0, MJ30, MJ60 and MJ90. For validation purpose, static analysis of two models (MJ0 and MJ60) is performed using ANSYS. The aim of study is to identify the best performing model after modal, hydrodynamic and fatigue analysis.

The typical model of MJ structure is shown in Fig. 1. The components of MJ structures are transition piece, top brace, top horizontal brace, middle brace, bottom horizontal brace and leg. The twisting of angle in MJ structures is showed in Fig. 2, where the bottom end centre and the top end centre of a leg are projected to points A and B on the X–Z plane, respectively. The centre point O is on the same plane. The angle between two vectors OA and OB is defined as the twisted angle. The variation in geometric configuration had developed four types of models such as MJ0, MJ30, MJ60 and MJ90. The model is of 70.5 m in total height and of water depth 50 m. The structures were assumed to be fixed at the bottom. The MJ structures considered in the present study were designed and modelled for supporting heavy wind turbines, such as the NREL 5-MW baseline wind turbine. The material properties of MJ structures are listed in Table 1. The sectional properties of MJ structures are shown in Table 2. Static stress analysis done using Abaqus software was used to design the dimensions of the structures so that the structures have similar level of load carrying capability [1].

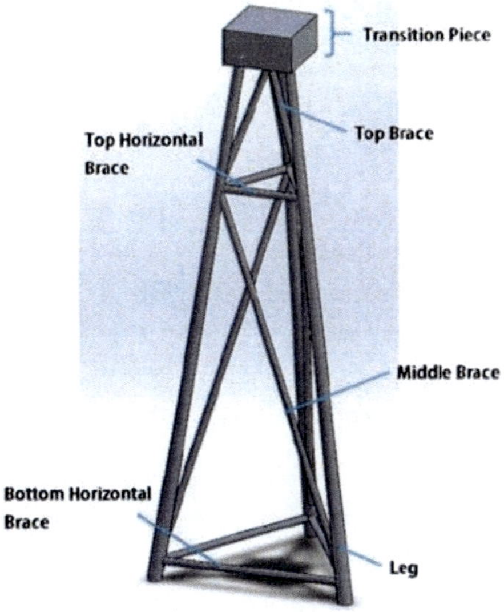

Fig. 1 Terminology of the structural members of MJ structures [1]

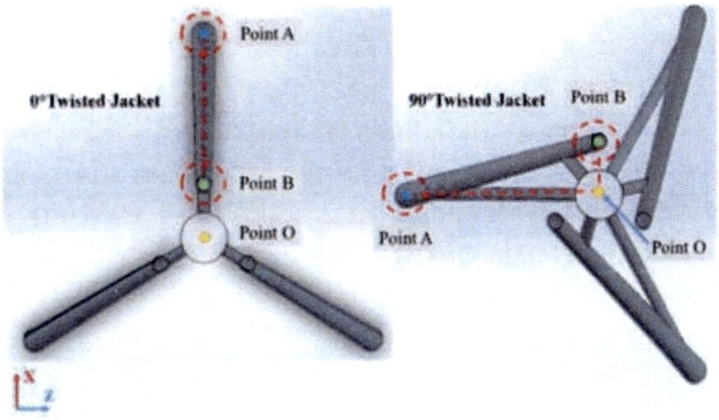

Fig. 2 Illustration for the definition of twisted angle [1]

Table 1 Material property of MJ structures [1]

Grade of steel	A36
Density of steel	7800 kg/m^3
Young's modulus	200 GPa
Poisson's ratio	0.3
Yield stress	250 MPa

4 Validation

The static structural analysis of MJ0 and MJ60 is performed in ANSYS. The MJ0 and MJ60 structures were modelled in ANSYS Static Structural as presented in Fig. 3. The simulated loads calculated using Morison equation are given in Table 3. All

Table 2 Sectional property of MJ structures [1], all dimensions are in m

	MJ0	MJ30	MJ60	MJ90
Total height	66.00	66.00	66.00	66.00
Length of leg	67.86	66.97	67.28	67.68
Length of top brace	17.23	17.15	16.93	16.63
Length of middle brace	50.85	51.10	51.12	50.89
Length of top horizontal brace	11.07	10.70	9.61	7.87
Length of bottom horizontal brace	24.41	24.38	24.32	24.16
Thickness of horizontal brace	0.03	0.03	0.03	0.03
Thickness of top brace	0.06	0.06	0.06	0.06
Thickness of middle brace	0.06	0.06	0.06	0.06
Thickness of leg	0.04	0.04	0.04	0.04
Diameter of horizontal brace	0.90	0.09	0.90	0.90
Diameter of top brace	0.90	0.09	0.90	0.90
Diameter of middle brace	1.20	1.20	1.20	1.20
Diameter of leg	1.80	1.80	1.80	1.80

wind, wave and current load in one direction (X-direction) is applied on structure as shown in Fig. 4.

The von Mises stresses obtained on the structures MJ0 and MJ60 are shown in Fig. 5. Table 4 shows comparison between von Mises stress obtained from both [1] and present study. It was found that for both MJ0 and MJ60, percentage error was only 1.24 and 2.43%, respectively.

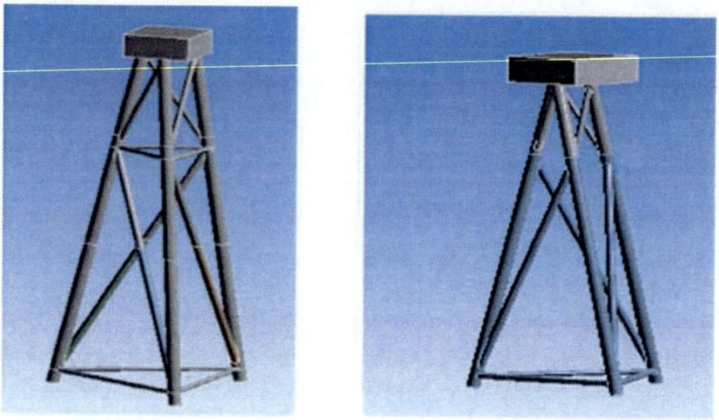

(a) (b)

Fig. 3 **a** MJ0 **b** MJ60

Table 3 Simulated loads on MJ0 and MJ60 [1]

Self-weight of baseline wind turbine machine and tower	1.859 MN
Side wind load	16.644 kN
Wind load moment	733.584 kNm
Wind load for top and horizontal brace	29.457 N/m
Wind load for leg	58.914 N/m
Wave load for top brace	31.727 N/m
Wave load for middle brace	43.0586 N/m
Wave load for leg	44.317 N/m
Current load for top and bottom brace	9.696 N/m
Current load for leg	14.544 N/m

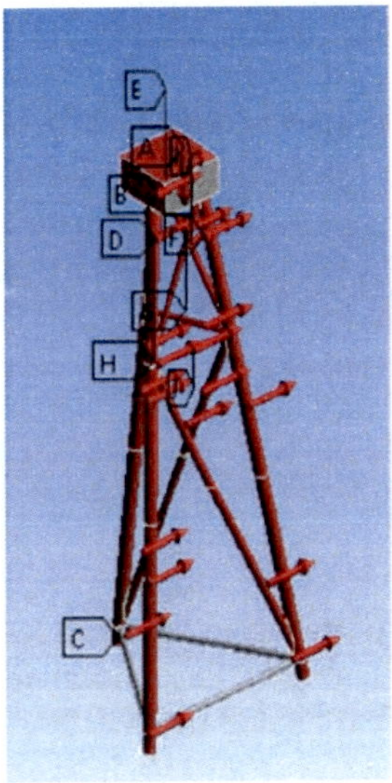

(a) MJ0

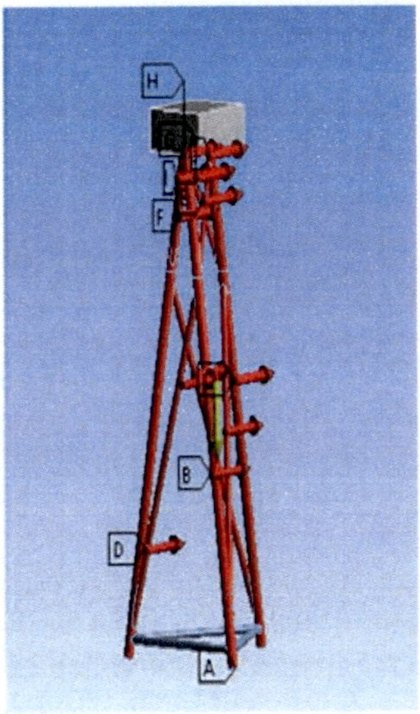

(b) MJ60

Fig. 4 Loads applied on MJ structures

5 Modal Analysis

The modal analysis can be used to determine the vibration characteristics of a structure. The natural frequencies and modes of vibration depend on stiffness, mass and distribution of mass in the structure. To evaluate the risk of resonance of structure, it is important to design the structure subjected to dynamic loads. This would cause the system to oscillate at higher amplitudes than when the force is applied at other

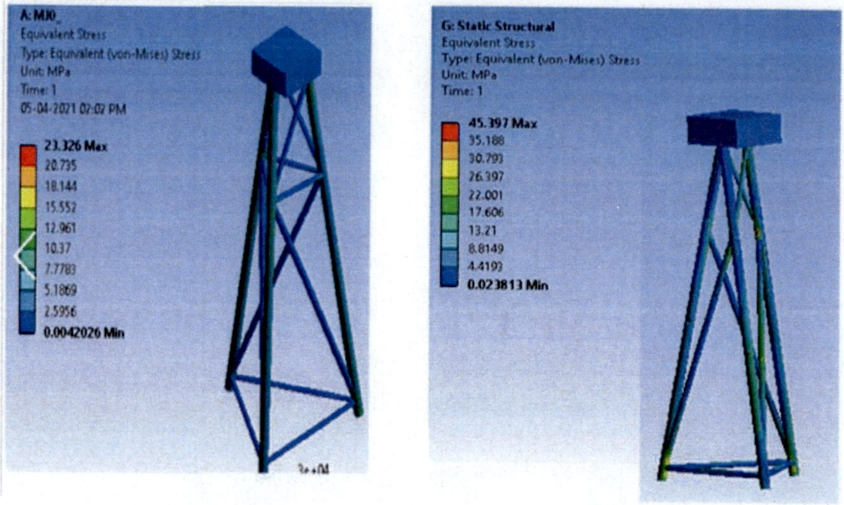

(a) MJ0 (b) MJ60

Fig. 5 von Mises stress on MJ structures

Table 4 Comparison of maximum von Mises stresses

Model	von Mises stress (MPa) [1]	von Mises stress (MPa) (present study)	% error
MJ0	23.04	23.326	1.24
MJ60	46.53	45.397	2.43

frequencies and these large oscillations can result important deformations and structural damage. It is therefore crucial to avoid the main sources of excitation transmitted to the structure causing frequencies of vibration close to their natural frequencies.

5.1 Modal Analysis of MJ0

The structure modelled in ANSYS is shown in Fig. 6 and is fixed at seabed level. The mass of structure is 3.6031×10^6 kg. The first six natural frequencies of the MJ0 structure were calculated in ANSYS and are shown in Table 5.

Fig. 6 MJ0 structure

Table 5 Natural frequencies and time period of MJ0

Mode description	Frequency (Hz)	Time period (s)
1st mode	0.14922	6.701515
2nd mode	0.14927	6.69927
3rd mode	0.30555	3.272787
4th mode	0.30847	3.241806
5th mode	0.41379	2.416685
6th mode	0.77038	1.298061

5.2 Modal Analysis of MJ30

The structure modelled in ANSYS is shown in Fig. 7 and is fixed at seabed level. The mass of structure is 3.6004×10^6 kg. The first six natural frequencies of the MJ30 structure were calculated in ANSYS and are shown in Table 6.

5.3 Modal Analysis of MJ60

The structure modelled in ANSYS is shown in Fig. 8 and is fixed at seabed level. The mass of structure is 3.5951×10^6 kg. The first six natural frequencies of the MJ60 structure were calculated in ANSYS and are shown in Table 7.

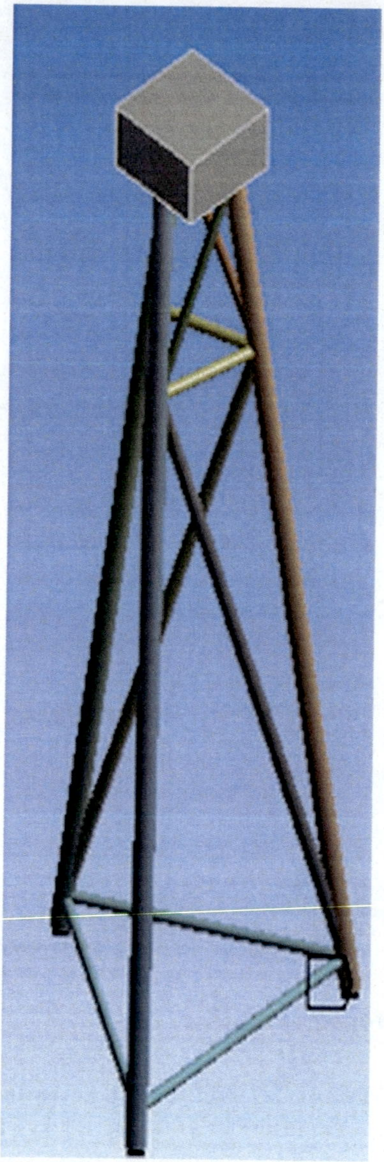

Fig. 7 MJ30 structure

5.4 *Modal Analysis of MJ90*

The structure modelled in ANSYS is shown in Fig. 9 and is fixed at seabed level. The mass of structure is 3.5919×10^6 kg. The first six natural frequencies of the MJ90 structure were calculated in ANSYS and are shown in Table 8, and corresponding mode shapes are given in Fig. 10.

Table 6 Natural frequencies and time period of MJ30

Mode description	Frequency (Hz)	Time period (s)
1st mode	0.15311	6.531252
2nd mode	0.15317	6.528694
3rd mode	0.32956	3.034349
4th mode	0.33254	3.007157
5th mode	0.41772	2.393948
6th mode	0.77432	1.291456

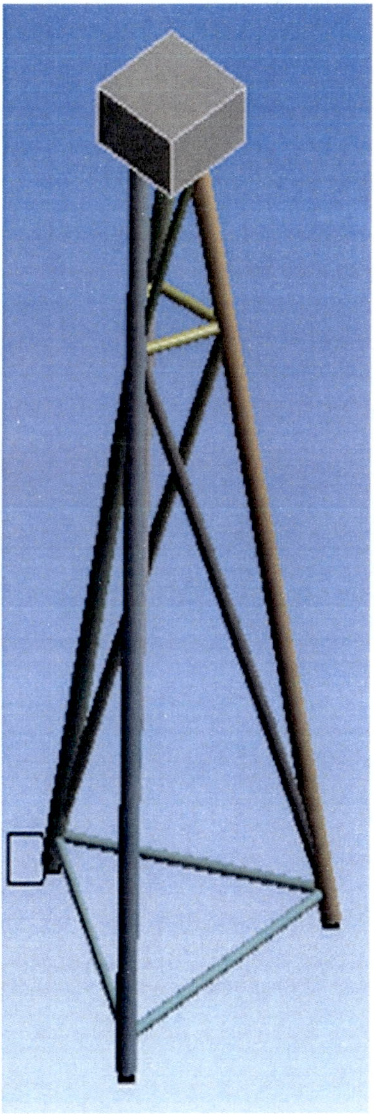

Fig. 8 MJ60 structure

Table 7 Natural frequencies and time period of MJ60

Mode description	Frequency (Hz)	Time period (s)
1st mode	0.16453	6.077919
2nd mode	0.16468	6.072383
3rd mode	0.35727	2.799004
4th mode	0.3602	2.776235
5th mode	0.42634	2.345546
6th mode	0.78079	1.280754

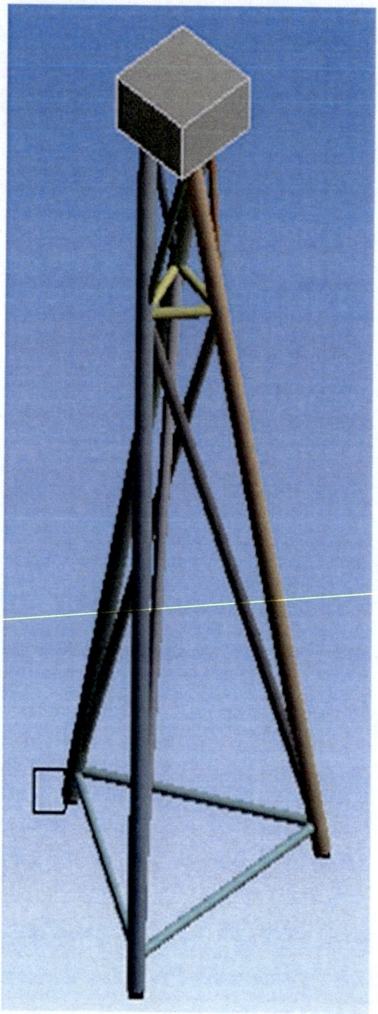

Fig. 9 MJ90 structure

Table 8 Natural frequencies and time period of MJ90

Mode description	Frequency (Hz)	Time period (s)
1st mode	0.17184	5.819367
2nd mode	0.17244	5.799119
3rd mode	0.318	3.144654
4th mode	0.32018	3.123243
5th mode	0.43307	2.309096
6th mode	0.78211	1.278593

5.5 Discussion of Results

Table 9 shows the comparison of mass, natural frequencies and time periods of MJ0, MJ30, MJ60 and MJ90. The first and second mode of vibration of structure has largest contribution to platform motion. The MJ90 structure has lowest mass with least natural time period of 5.819367 s compared to other structures. The natural time period of MJ structures lies away from wave period of the site location 11 s as shown in Table 10. Comparing the natural time period, MJ30 had a reduction of 2.54% only, MJ60 by 9.3% reduction, whereas in MJ90, 13.16% reduction is found. From modal analysis, it was found that the geometric variation increases, a reduction in mass as well as a reduction in natural time period is found (Table 9).

6 Hydrodynamic Analysis

In the present work, the dynamic response of MJ structures has been investigated using finite element software ANSYS Aqwa. The environmental site conditions of Gulf of Kutch, Gujarat, India, is chosen and is shown in Table 10. The model of sea bed created is of size $X = 287$ m, $Y = 195$ m and water depth 50 m. The mesh size for the models is 3 m coarse. In time response analysis, the structures are subjected to wind, wave and current in two cases and in total 12 load combinations are considered. Case 1 includes wind, wave and current acting unidirectional along and opposite to three legs of the structures. Case 2 includes wind, wave acting unidirectional along and opposite to leg, whereas current is made to act at an angle of 45° to the legs. The key diagrams for 12 load combinations are shown in Fig. 11.

6.1 Time Response Analysis of MJ0

The MJ0 modelled in ANSYS Aqwa is shown in Fig. 12. Tables 11 and 12 show the force and moment in the MJ0 structure under time response analysis in 12 different loading directions occurring at the centre of gravity of the structure. The MJ0 structure

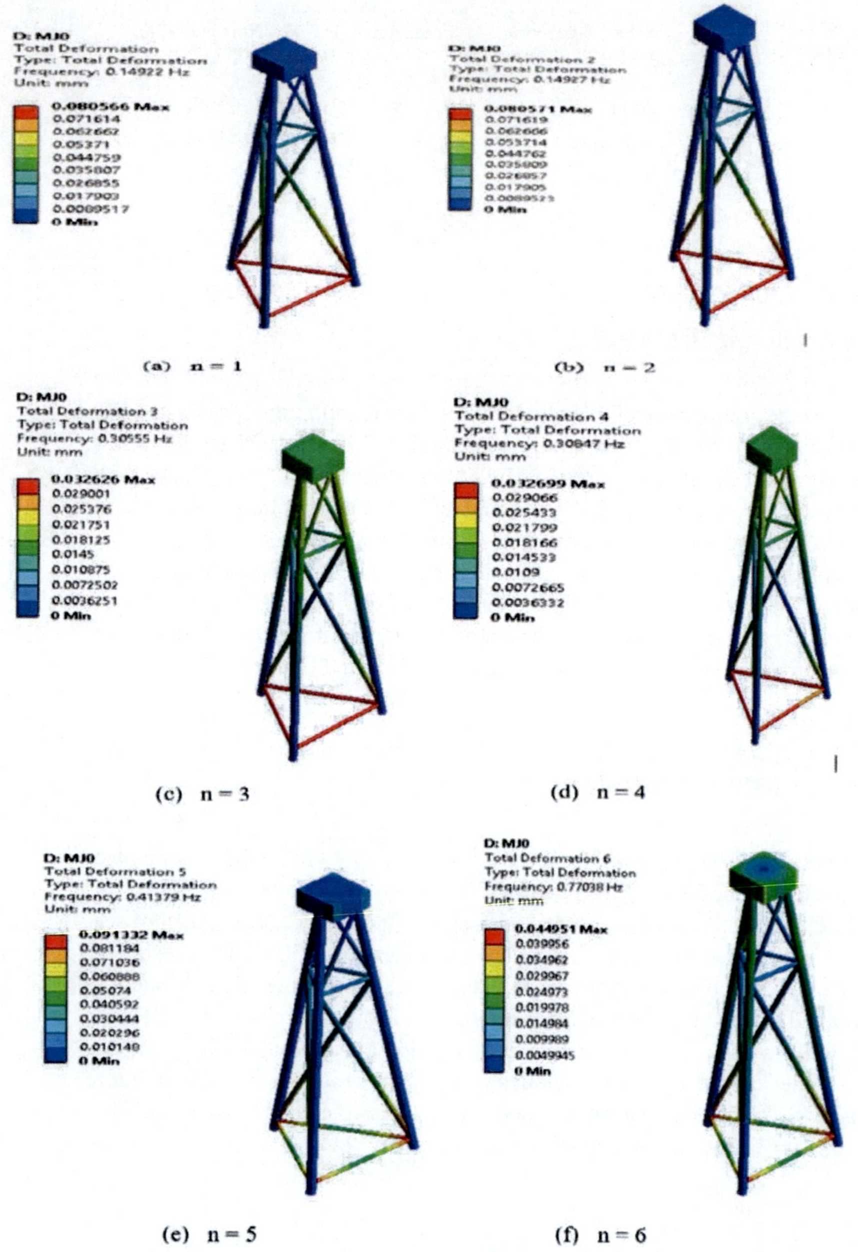

Fig. 10 First six mode shapes of MJ90

Table 9 Natural frequency of MJ structures

Type of structure	Mass × 10^6 (kg)	Natural frequency (Hz)	Time period (s)
MJ0	3.6031	0.14922	6.701515
MJ30	3.6004	0.15311	6.531252
MJ60	3.5951	0.16453	6.077919
MJ90	3.5919	0.17184	5.819367

Table 10 Environmental conditions at Gulf of Kutch, Gujarat [12]

Wind speed (m/s)	Wave height (m)	Wave period (s)	Ocean currents	
			Depth (m)	Speed (m/s)
32.778	11.58	11	0	1.25
			12.5	1.1
			25	0.92
			37.5	0.77
			50	0.45

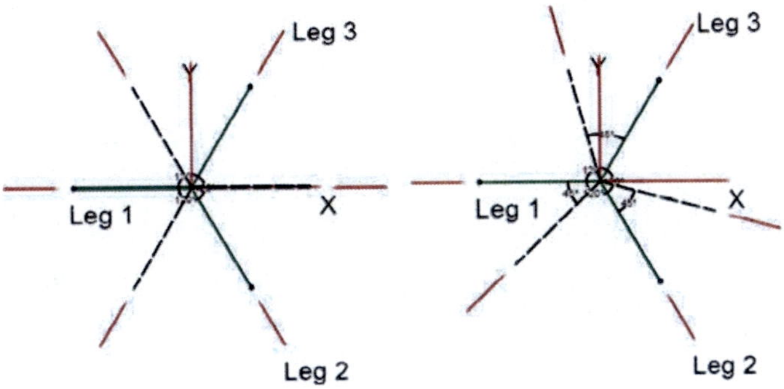

(a) Case 1 (b) Case 2

Fig. 11 Key diagram showing loading directions

is affected by maximum force of 36.656 N, when wind and wave act at 120° and current at 165° with respect to leg 1. When the wind and wave act unidirectional, along leg 2 and current at 45° to leg 2, the structure develops maximum body force and is shown in Fig. 13. The maximum moment developed on the structure is 157 Nm, when wind, wave act at 0° and the current at 45° with respect to leg 1. The structure is affected by maximum moment when the wind and wave act unidirectional, along leg 1 and current at 45° to leg 1 as shown in Fig. 15. The graphical representation of maximum force and moment is shown in Figs. 14 and 16.

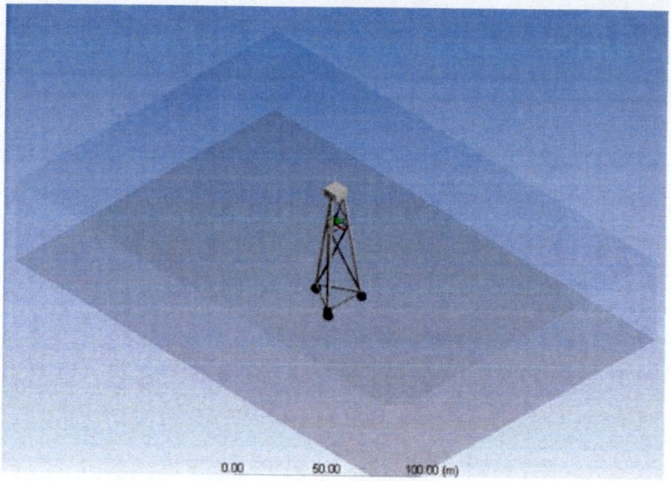

Fig. 12 MJ0 modelled in ANSYS Aqwa

Table 11 Forces and moments on MJ0 structure acting in case 1 condition

Directions	Force (N)	Moment (Nm)
Wind, wave, current @ 0°	34.898	131
Wind, wave, current @ 120°	35.016	112.75
Wind, wave, current @ 240°	32.781	128.75
Wind, wave, current @ 180°	32.625	110
Wind, wave, current @ 300°	35.031	116.844
Wind, wave, current @ 60°	31.594	112

Table 12 Forces and moments on MJ0 structure acting in case 2 conditions

Directions	Force (N)	Moment (Nm)
Wind, wave @ 0°, current @ 45°	33.25	157
Wind, wave @ 120°, current @ 165°	36.656	114
Wind, wave @ 240°, current @ 285°	33.265	113.875
Wind, wave @ 180°, current @ 225°	32.125	122
Wind, wave @ 300°, current @ 345°	35.875	114
Wind, wave @ 60°, current @ 105°	35.594	104.785

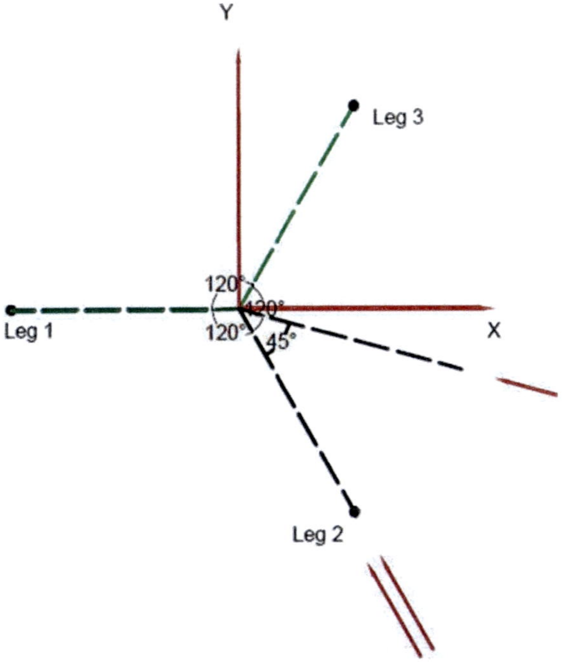

Fig. 13 Key diagram showing maximum force loading directions

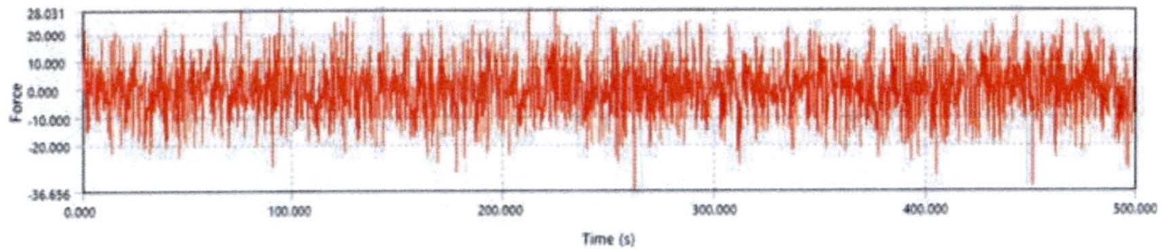

Fig. 14 Forces versus time graph of MJ0

6.2 *Time Response Analysis of MJ30*

The MJ30 is modelled in ANSYS Aqwa. Tables 13 and 14 show the force and moment in the MJ30 structure under time response analysis in 12 different loading directions occurring at the centre of gravity of the structure. The MJ30 structure is affected by maximum force of 52.625 N, when wind and wave act at 180° and current at 225° with respect to leg 1. When the wind and wave act unidirectional, along and opposite to leg 1 and current at 45° to leg 1, the structure develops maximum body force. The maximum moment developed on the structure is 132 Nm, when wind and wave act at 60° and the current at 105° with respect to leg 1. The structure is affected by

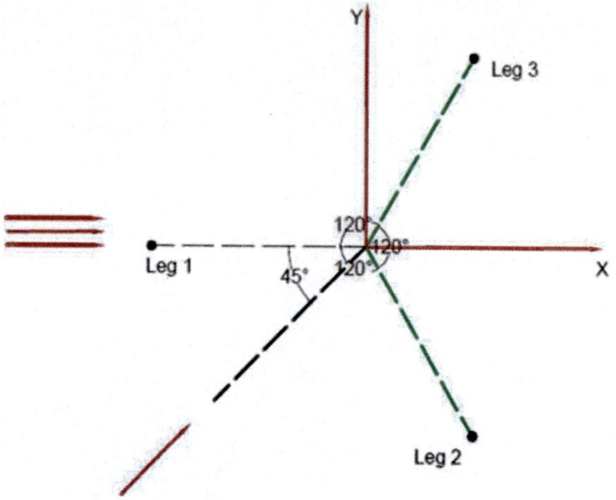

Fig. 15 Key diagram showing maximum moment loading directions

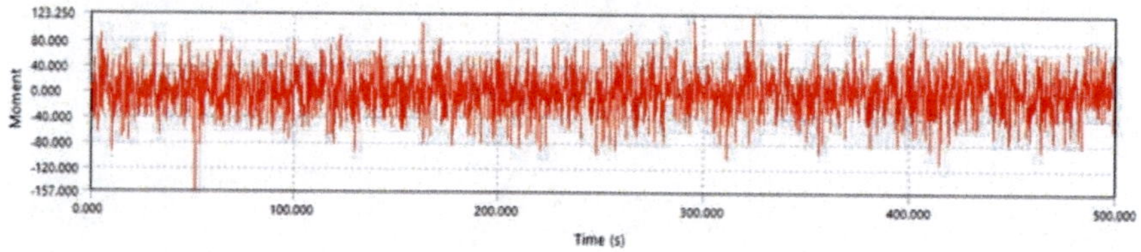

Fig. 16 Moment versus time graph of MJ0

Table 13 Forces and moments on MJ30 structure acting in case 1 condition

Directions	Force (N)	Moment (Nm)
Wind, wave, current @ 0°	47.404	103.25
Wind, wave, current @ 120°	44.875	100.5
Wind, wave, current @ 240°	50.063	104.5
Wind, wave, current @ 180°	44.824	122.375
Wind, wave, current @ 300°	48.188	108
Wind, wave, current @ 60°	45.563	120

maximum moment when the wind and wave act unidirectional, along and opposite to leg 3 and current at 45° to leg 3.

6.3 *Time Response Analysis of MJ60*

The MJ60 is modelled in ANSYS Aqwa. Tables 15 and 16 show the force and moment in the MJ60 structure under time response analysis in 12 different loading directions.

Table 14 Forces and moments on MJ30 structure acting in case 2 condition

Directions	Force (N)	Moment (Nm)
Wind, wave @ 0°, current @ 45°	49.685	102.344
Wind, wave @ 120°, current @ 165°	45.219	99.5
Wind, wave @ 240°, current @ 285°	47.563	95.5
Wind, wave @ 180°, current @ 225°	52.625	110
Wind, wave @ 300°, current @ 345°	50.469	96
Wind, wave @ 60°, current @ 105°	47.406	132

Table 15 Forces and moments on MJ60 structure acting in case 1 condition

Directions	Force (N)	Moment (Nm)
Wind, wave, current @ 0°	44.531	95
Wind, wave, current @ 120°	42	107
Wind, wave, current @ 240°	42.938	94.75
Wind, wave, current @ 180°	45.566	95.5
Wind, wave, current @ 300°	52.875	106
Wind, wave, current @ 60°	48.109	108

The MJ60 structure is affected by maximum force of 52.875 N, when wind, wave and current act at 300° with respect to leg 1. When the wind, wave and current act unidirectional, along and opposite to leg 2, the structure develops maximum body force. The maximum moment on the structure is 116 Nm, when wind and wave act at 300° and the current at 345° with respect to leg 1. The structure is affected by maximum moment when the wind and wave act unidirectional, along and opposite to leg 2 and current at 45° to leg 2 as shown in Fig. 17. The graphical representation of maximum moment is shown in Fig. 18.

6.4 *Time Response Analysis of MJ90*

The MJ90 is modelled in ANSYS Aqwa. Tables 17 and 18 show the force and moment in the MJ90 structure under time response analysis in 12 different loading directions occurring at the centre of gravity of the structure. The MJ90 structure is affected by maximum body force of 12 N in all 12 loading directions of wind, wave, and current and is shown in Fig. 19. The maximum moment developed on the structure is 138 Nm, when wind, wave and current act at 240° and also wind and wave act at

Table 16 Forces and moments on MJ60 structure acting in case 2 condition

Directions	Force (N)	Moment (Nm)
Wind, wave @ 0°, current @ 45°	42.004	95.25
Wind, wave @ 120°, current @ 165°	41.469	101
Wind, wave @ 240°, current @ 285°	50.219	95.5
Wind, wave @ 180°, current @ 225°	47.965	107.25
Wind, wave @ 300°, current @ 345°	45.938	116
Wind, wave @ 60°, current @ 105°	43.813	95

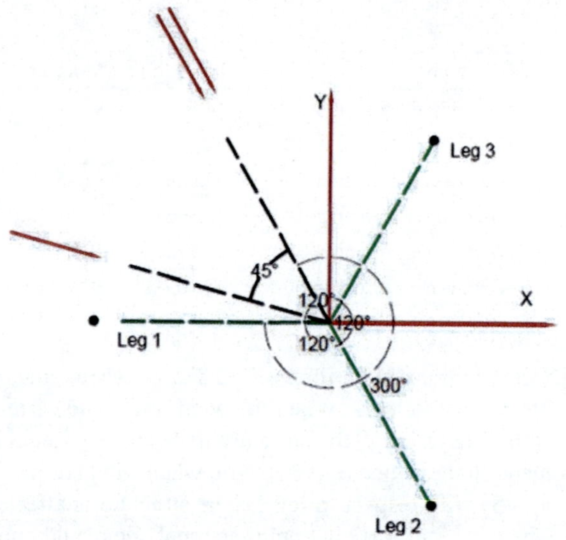

Fig. 17 Key diagram showing maximum moment loading directions

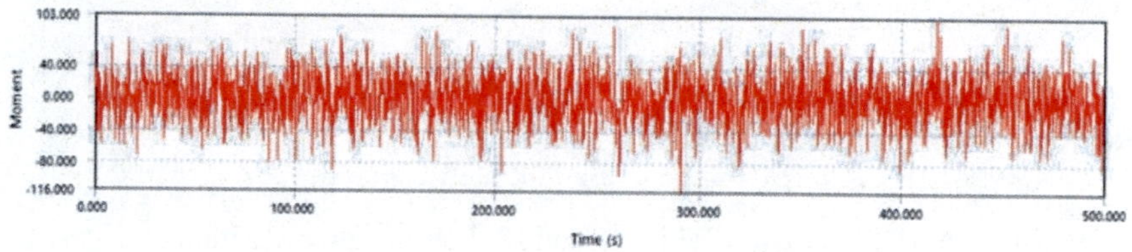

Fig. 18 Moment versus time graph of MJ60

Table 17 Forces and moments on MJ90 structure acting in case 1 condition

Directions	Force (N)	Moment (Nm)
Wind, wave, current @ 0°	12	115.75
Wind, wave, current @ 120°	12	121.5
Wind, wave, current @ 240°	12	138
Wind, wave, current @ 180°	12	121.25
Wind, wave, current @ 300°	12	124
Wind, wave, current @ 60°	12	137

Table 18 Forces and moments on MJ90 structure acting in case 2 condition

Directions	Force (N)	Moment (Nm)
Wind, wave @ 0°, current @ 45°	10	135.75
Wind, wave @ 120°, current @ 165°	12	110
Wind, wave @ 240°, current @ 285°	12	138
Wind, wave @ 180°, current @ 225°	10	117.25
Wind, wave @ 300°, current @ 345°	12	114
Wind, wave @ 60°, current @ 105°	12	125.875

240° along with current at 285° with respect to leg 1. The graphical representation of maximum force is shown in Fig. 20.

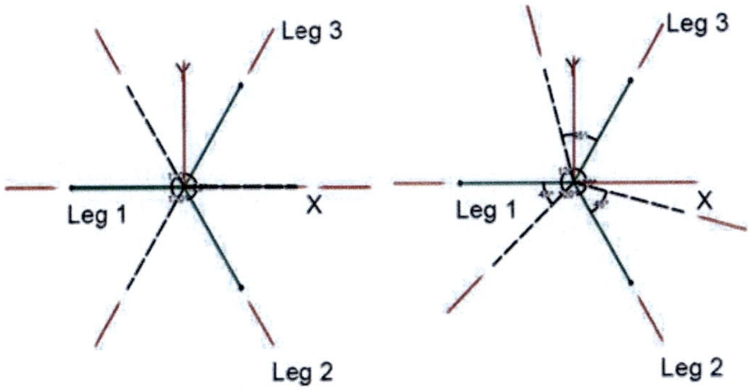

(a) Case 1 (b) Case 2

Fig. 19 Key diagram showing maximum force loading directions

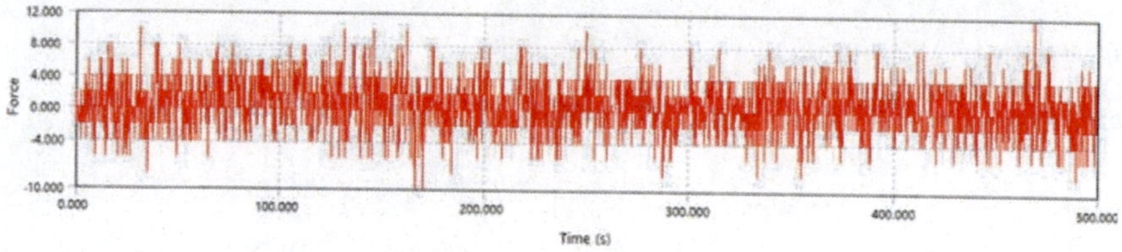

Fig. 20 Forces versus time graph of MJ90

6.5 *Discussion*

Table 19 shows the maximum forces and moment on MJ structures. All the MJ structures were subjected to the same environmental conditions of Gulf of Kutch, Gujarat. While comparing the forces, the MJ30 had an increase in forces by 43.52% whereas the MJ60 by 44.23% rise. But the MJ90 had a reduction in force by 67.266% compared to MJ0. While comparing the moment, MJ0 has developed the largest moment of 157 Nm, MJ30 had a moment reduction of 15.9%, MJ60 is reduced by 26.11% and MJ90 by 12.73% reduction. Among all the MJ structures, MJ90 develops the least force of 12 N and MJ60 develops least moment of 116 Nm. The geometric variation had significant role in development of body force and moment.

7 Fatigue Analysis

The jacket structures are affected by environmental loads such as wave, current and wind. These structures are subject to millions of stress reversals due to the action of waves during their lifetime. Fatigue failure occurs due to the cycling loading which would lead to the development of hotspot stresses and crack propagation. Hence, it is very important to predict the fatigue life of offshore structures. In the present work, ANSYS software is used to predict the design life of the OWT.

Table 19 Maximum forces and moment on MJ structures

Model	Force (N)	Moment (Nm)
MJ0	36.656	157
MJ30	52.625	132
MJ60	52.875	116
MJ90	12	138

Table 20 Maximum forces and moments on MJ0

Directions	Centre of gravity (m)	Force (N)	Moment (Nm)
X	7.6434e-6	36.656	1.75
Y	4.2547e-6	4.645	157
Z	−2.4977832	18	1.422

Fig. 21 Maximum forces and moments applied on MJ0

7.1 Fatigue Life Prediction of MJ0

The maximum force and moment obtained from hydrodynamic analysis are given in Table 20 and is applied at the centre of gravity of the model, as shown Fig. 21. The fatigue life obtained for MJ0 is 836 years.

7.2 Fatigue Life Prediction of MJ30

The maximum force and moment obtained from hydrodynamic analysis are given in Table 21 and is applied at the centre of gravity of the model. The fatigue life obtained for MJ30 is 923 years.

Table 21 Maximum forces and moments on MJ30

Directions	Centre of gravity (m)	Force (N)	Moment (Nm)
X	−6.7655e-7	23.25	132
Y	−2.4313e-5	52.625	78
Z	−2.4996278	20	11.313

Table 22 Maximum forces and moments on MJ60

Directions	Centre of gravity (m)	Force (N)	Moment (Nm)
X	7.0239e-6	22.5	116
Y	3.6791e-6	52.875	77
Z	−2.502825	18	10.625

Table 23 Maximum forces and moments on MJ90

Directions	Centre of gravity (m)	Force (N)	Moment (Nm)
X	6.6338e-6	0.188	138
Y	3.5111e-6	0.375	70.5
Z	−2.504056	12	3.5

7.3 *Fatigue Life Prediction of MJ60*

The maximum force and moment obtained from hydrodynamic analysis are given in Table 22 and is applied at the centre of gravity of the model. The fatigue life obtained for MJ60 is 1069 years.

7.4 *Fatigue Life Prediction of MJ90*

The maximum force and moment obtained from hydrodynamic analysis given in Table 23 are applied at the centre of gravity of the model for the fatigue analysis. The fatigue life obtained for MJ90 is 1277 years.

7.5 *Discussion*

Table 24 shows the design life of the four MJ structures investigated in this study. Among these, MJ0 has 836 years, the lowest predicted fatigue life. Compared to MJ0, fatigue lives of MJ30 and MJ60 are higher by 10.4% and 27.87%, respectively. MJ90 has the highest design life of 1277 years, which is about 52.75% higher than that of MJ0. It can be concluded that as the variation in geometric configuration significantly affects the design life of models.

Table 24 Design life of MJ structures

Model	Life (years)
MJ0	836
MJ30	923
MJ60	1069
MJ90	1277

8 Summary

The clean green wind energy can be efficiently used for the production of electricity. Harnessing wind energy using offshore wind turbines is a fast-developing concept. In Indian sea conditions, as of now, only the conventional jacket structure is installed. As a modification to the conventional one, MJ structures are proposed. Their structural dynamic response to wave, wind and current, pertaining to sea condition at Gulf of Kutch, Gujarat, was investigated numerically by simulations using the finite element software package ANSYS. The modal analysis performed on the MJ structures indicates that all the four models have natural time periods away from the site conditions. Hydrodynamic analysis performed on MJ structures has given the body forces and moments. The fatigue analysis performed had predicted the design life of all the MJ structures.

9 Conclusion

The dynamic behaviour of the MJ structures has been investigated numerically using the finite element software ANSYS. The hydrodynamic behaviour of MJ structures which can be used to support wind tower for a water depth of 50 m under the environmental conditions of Gulf of Kutch, Gujarat, India, was analysed using ANSYS Aqwa. For validation purpose, static analysis of two models of MJ structures was performed using ANSYS, and results are found to be in agreement with the published literature. It was found from modal analysis that the variation in geometric configurations significantly affects the natural time period of MJ structure. The time response analysis performed on the MJ structures shows that the twisting of angle significantly affects the body force and moment developed on the structure. The fatigue analysis performed shows that the MJ90 model has very high design life compared to other models.

References

1. Chen IW, Wong BL, Lin YH, Chau SW, Huang HH (2016) Design and analysis of jacket substructures for offshore wind turbines. Energies 9:264–278. https://doi.org/10.3390/en9040264
2. Raheem SEA (2014) Study on nonlinear response of steel fixed offshore platform under environmental loads. Arab J Sci Eng 39:6017–6030. https://doi.org/10.1007/s13369-014-1148-x
3. Wei S, Han J, Kim C, Lee D, Shin H, Park H (2015) Feasibility study of offshore wind turbine substructures for southwest offshore wind farm project in Korea. Renewable Energy 74:406–413. https://doi.org/doi.org/10.1016/j.renene.2014.08.039
4. Di LY, Yuan GM, Na W, Jun HL, Da C (2017) Dynamic responses of K-type and inverted-K-type jacket support structures for offshore wind-turbines. J Cent South Univ, 24:947−956. https://doi.org/10.1007/s11771-017-3497-6
5. Wei K, Myers AT, Arwade SR (2017) Dynamic effects in the response of offshore wind turbines supported by jackets under wave loading. Eng Struct 142:36–45. https://doi.org/10.1016/j.engstruct.2017.03.074
6. Arvind MK, Desai R (2018) Efficacy of twisted jacket for offshore wind turbine foundation. J Eng Res Appl 8:67–72. https://doi.org/10.9790/9622-0805046772
7. Cardoso J, Vieira M, Henriques E, Reis L (2020) Computational analysis of the transportation phase of an innovative foundation for offshore wind turbine. Ships Offshore Struct. https://doi.org/10.1080/17445302.2020.1779016
8. Vuonga NH, Quana MH (2019) Fatigue analysis of jacket support structure for offshore wind turbines. J Sci Tech Civil Eng 13:46–59. https://doi.org/10.31814/stce.nuce2019-13(1)-05
9. Taheri A, Shahsavari E (2020) Time history dynamic analysis of a new constructed offshore jacket platform in Persian Gulf due to random waves. J Mar Sci Appl. https://doi.org/10.1007/s11804-019-00098-2
10. Raheem SEA, Aal EMA, Shafy AGAA, Fahmy MFM, Omar M, Mansour MH (2020) In-place analysis for design-level assessment of the fixed offshore platform. Ships Offshore Struct. https://doi.org/10.1080/17445302.2020.1787931
11. Wang S, Larsen TJ, Bredmose H (2020) Experimental and numerical investigation of a jacket structure subject to steep and breaking regular waves. Mar Struct 72:102744–102761. https://doi.org/10.1016/j.marstruc.2020.102744
12. Oil and Natural Gas Corporation Limited. http://www.ongcindia.com

Study on Variability of Influent Characteristics Concentrations of CETP

Shanthi P. Selvam and V. M. Topkar

1 Introduction

Under the water (prevention and control of pollution) Act 1974 India, every industry has to provide adequate treatment for its effluent before disposal whether it is in stream, land, sewage system or in sea. The effluent can be treated individually by the industries, jointly by the industries themselves or under some other independent organization.

Maharashtra Industrial Development Corporation (MIDC) established an industrial estate near Thane–Belapur Road, Navi Mumbai in 1963, known as Trans Thane Creek (TTC) industrial area. This area has been divided into five blocks such as engineering zone, chemical zone, textile zone, food processing zone, electronic zone and knowledge park and into five blocks A, C, D, R and W with a total of ~3450 industries. Thane Belapur Industries Association (TBIA) initiated and promoted the formation of the TTC common effluent treatment plant (CETP) of 12 MLD capacity for two of the five blocks in the year 1997 and was later expanded as additional 15 MLD in phase II commissioned in the year 2006. Both the plants are based on extended aeration activated sludge technology (EAAST).

The focus of CETP is on the removal and treatment of settleable, and floatable solids and organic matter present in the influent to meet the stringent discharge standards set by Central Pollution Control Board (CPCB).

The performance of the CETP can be affected by the variability of the influent wastewater characteristics, inherent variability in wastewater treatment processes and variability caused by mechanical breakdown, design deficiencies and operational

S. P. Selvam (✉) · V. M. Topkar
Civil and Environmental Engineering Department, Veermata Jijabai Technological Institute, Matunga, Mumbai 400019, India
e-mail: shanthups96@gmail.com

V. M. Topkar
e-mail: vmtopkar@ci.vjti.ac.in

A. Sil et al. (eds.), *Recent Trends in Civil Engineering*, Lecture Notes in Civil Engineering 274, https://doi.org/10.1007/978-981-19-4055-2_52

failures [1]. This paper outlines variations of the influent characteristics at the plant mentioned above.

2 Previous Research

Djeddou Messaoud et al. [2] have shown the effluent variability of pollution parameter indicator of Khenchela WWTP by histogram and analysed by probability density function. The variability of the performance of the plant was analysed to determine what should be the average effluent quality to meet a certain effluent standard with a predetermined probability. The authors concluded many factors that affect WWTP performance viz. flow variability and their characteristics, the inherent variability of the behaviour of wastewater treatment processes the variability caused by failures, lack of experience of the plant operators especially in the developing countries.

Salar Niku et al. [3] analysed the performance of large number of activated sludge processes (ASP) in terms of reliability, stability and variability. Variations in the daily effluent quality of ASP were studied. A reliability model was developed to predict performance of ASP. Coefficient of reliability (COR) was developed by the authors that correlates mean constituent value to the effluent standard that must be achieved on a probability basis. The authors used an expression for the COR that can be used to compute the necessary mean value of the process (M_x) that guarantees certain reliability

$$M_x = (\text{COR}) \times X_s$$

where Xs is fixed standard mean. For example, if a 30 mg/lit standard is to be achieved 90% of the time, the process must be designed to achieve a particular mean effluent concentration.

Owusu–Ansah et al. [4] developed a reference chart to obtain COR values as a function of coefficient of variation (CV) and reliability level (50–99.9%). These charts were intended to make it easier to assess the performance of treatment plants and its corresponding reliability and compliance level. In their study, the authors calculated the compliance level of actual discharge values of parameters such as BOD, COD, total nitrogen (TN), dissolved oxygen (DO), etc., to the Ghana Environmental Protection Agency discharge standard values for two treatment plants. The authors found that some parameters meet the EPA discharge standards but its compliance level does not meet the design compliance level. The authors also calculated COR values with 95% reliability and the actual CV values for parameters such as BOD, COD, TN, DO, etc. It was found that the actual mean values were less than the mean design concentration for parameters which have achieved the design compliance level.

3 Influent Characteristics Data for the Present Study

For the present study, data was collected from TTC–CETP of 12 MLD capacity for two years 2017 and 2018. Figure 1 shows the effluent transfer system from collection to TTC–CETP for further treatment is shown below.

The influent pumped into the CETP after preliminary treatment involving removal of floatable solids and neutralization. Data collected at the inlet of CETP. Data includes the influent characteristics concentrations of quality parameters such as pH, BOD, COD and TSS. Daily data was collected. The variability in the concentrations of influent characteristics of pH, BOD, COD and TSS are observed by calculating the minimum, maximum and mean values of all months in a year. The significance of variation between the quarters is observed by Kruskal Wallis Test. H values calculated by Kruskal Wallis Test between the different quarters give the significance of the variation between them.

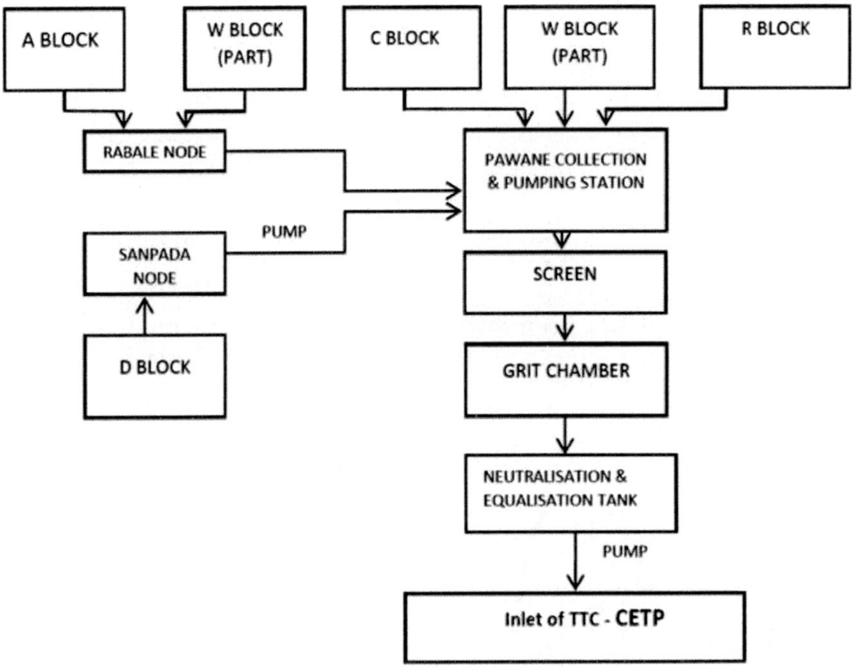

Fig. 1 Effluent transfer system

4 Results and Discussion

The monthly minimum, maximum and mean values of pH, BOD, COD and TSS at the inlet of CETP in the year 2017 are determined, and the variation plots are shown in Figs. 2, 3, 4, and 5. Each year is divided into four quarters such as Q1, Q2, Q3 and Q4. Comparison of the variations of influent characteristics is done quarterly.

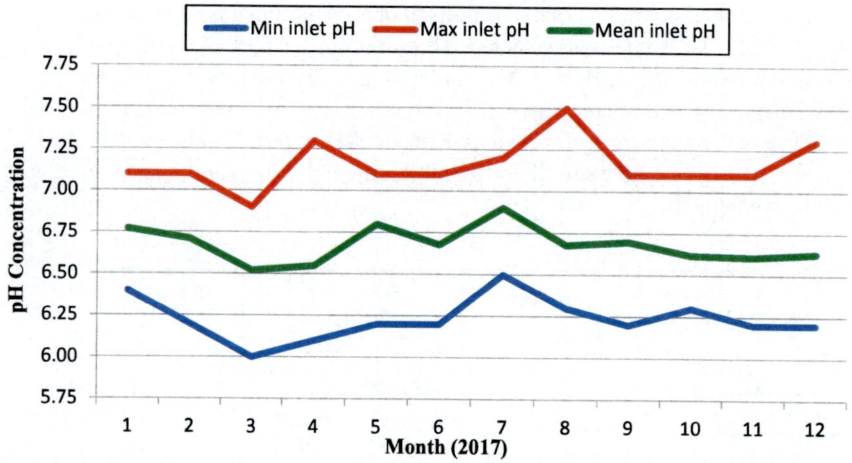

Fig. 2 Monthly variation in the minimum, maximum and mean values of pH in the year 2017

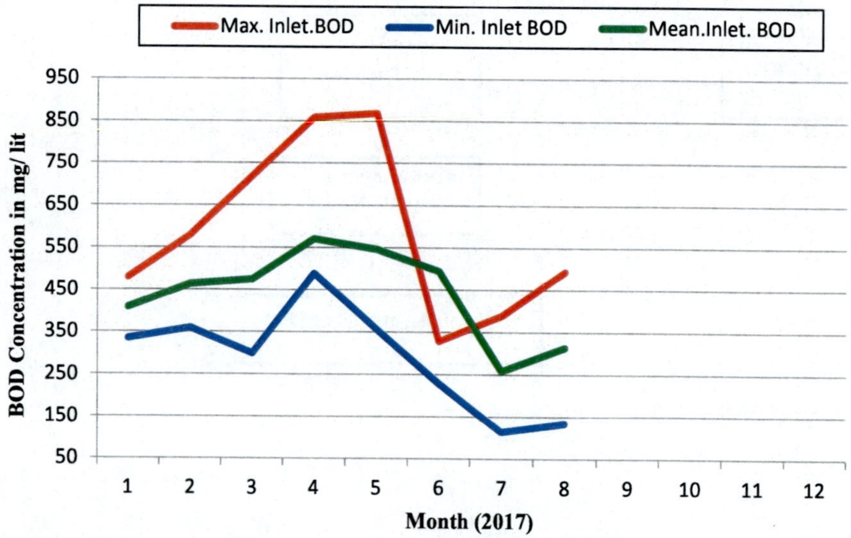

Fig. 3 Monthly variation in minimum, maximum and mean values of BOD in the year 2017. Data for September–December not shown as data is not available

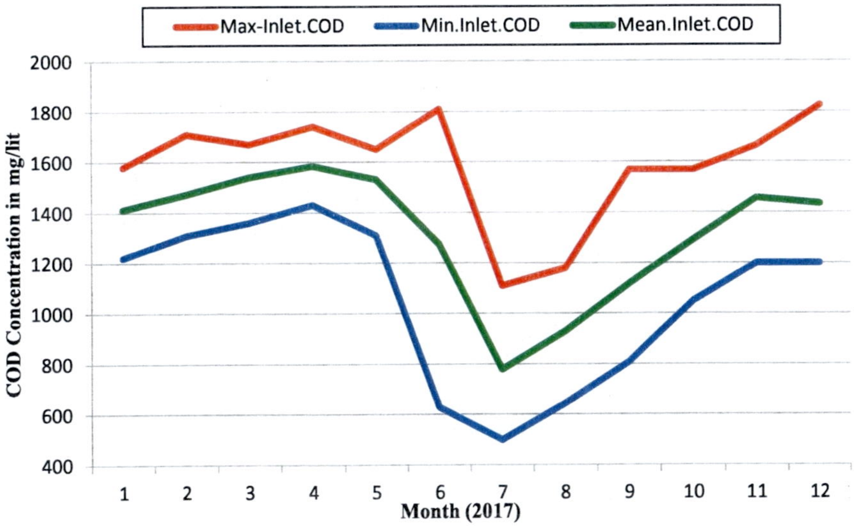

Fig. 4 Monthly variation in minimum, maximum and mean values of COD in the year 2017

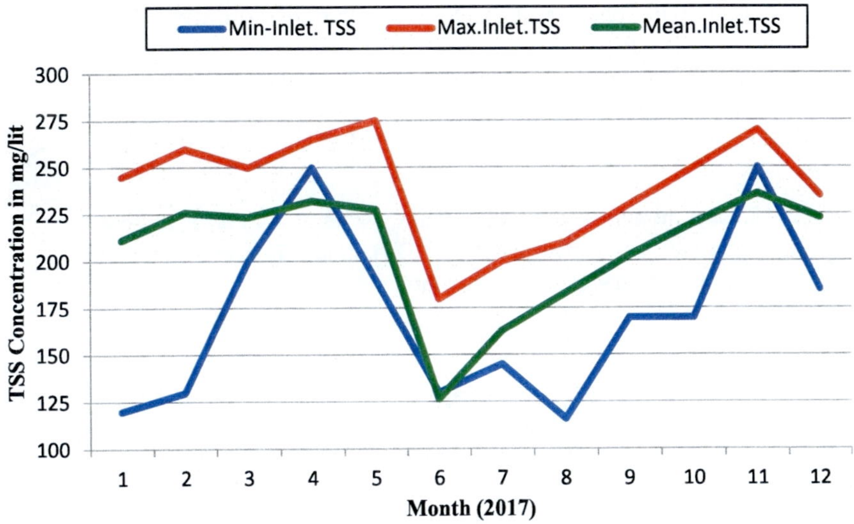

Fig. 5 Monthly variation in the minimum, maximum and mean values of TSS in the year 2017

It is observed that for the year 2017:

(a) The mean values of pH (Fig. 2) are in the range of 6.52 (Q2)—6.90 (Q3). The pH values do not change much throughout the year.
(b) The mean values of BOD (Fig. 3) are in the range 259 mg/lit (Q3—571 mg/lit (Q2). The BOD concentration for the Q3 is very less compared with the other

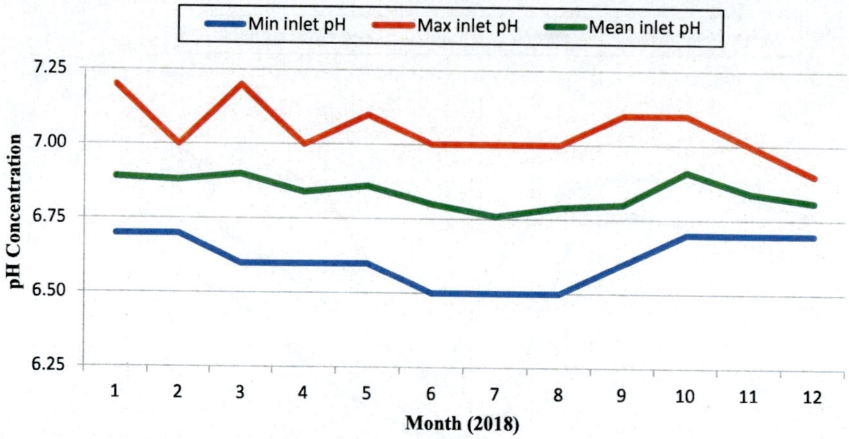

Fig. 6 Monthly variation in the minimum, maximum and mean values of pH in the year 2018

three quarters. Data for September to December is not available and hence not shown in Fig. 3.

(c) The values are in the range of 779 mg/lit (Q3)—1584 mg/lit (Q2) for COD concentration. Like BOD concentration COD concentration is also very less in Q3.

(d) The mean values of TSS (Fig. 5) are in the range of 127 mg/lit (Q2)—236 mg/lit (Q4). Like BOD and COD concentration TSS concentration also less in Q3. The higher value lies in Q4 for TSS, and in the case of BOD and COD, the higher value is in Q2.

For the year 2018, the minimum, maximum and mean values of the concentrations of influent quality parameters such as pH, BOD, COD, TSS and TDS at the inlet of TTC–CETP (12 mld) are determined, and the variation plots are shown in Figs. 6, 7, 8, and 9.

It is observed that for the year 2018:

(a) The mean values of pH (Fig. 6) are in the range of 6.76 (Q3)–6.91 (Q4) (Fig. 5). The pH values do not change much throughout the year.
(b) The mean values of BOD (Fig. 7) are in the range 286 mg/lit (Q3)–503 mg/lit (Q2).
(c) The values are in the range of 853 mg/lit (Q3)—1568 mg/lit (Q2) for COD concentration. Both BOD and COD concentrations are very less in Q3.
(d) The mean values of TSS in year 2017 (Fig. 9) are in the range of 142 mg/lit (Q3)—228 mg/lit (Q2).

For both the years, the pH concentration does not very much. All the other parameters such as BOD, COD and TSS, the minimum concentration lies in the Q3. The average influent characteristics concentrations of quality parameters such as BOD, COD and TSS at the inlet of CETP of Q3 are also lesser as compared with the other quarters. For both years, the variations in the concentrations follow the same pattern.

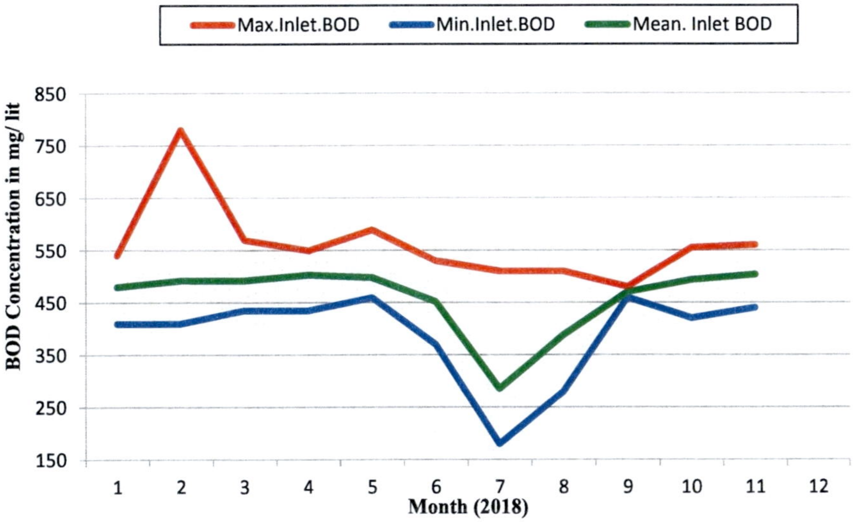

Fig. 7 Monthly variation in the minimum, maximum and mean values of BOD in the year 2018. December data is not available

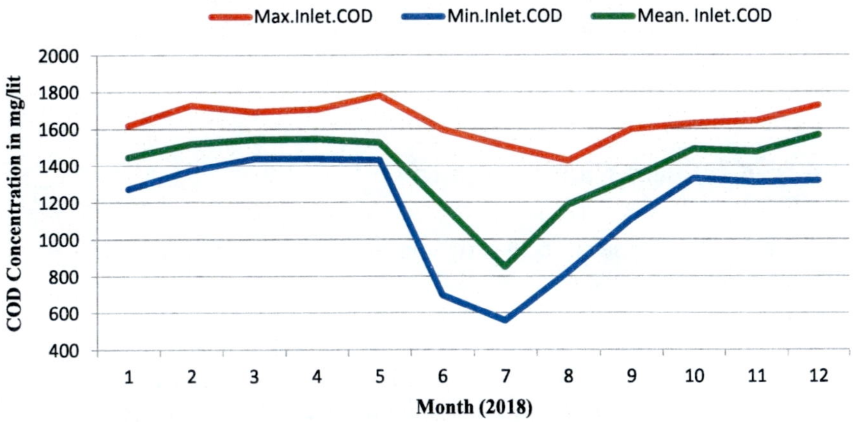

Fig. 8 Monthly variation in the minimum, maximum and mean values of COD in the year 2018

It is observed that the minimum, maximum and the mean values of July, August and September are smaller compared to the values of other months for both 2017 and 2018 of 12 MLD CETP. In the whole year, the concentrations in Q3 is very smaller. The pH values are reasonably constant due to the neutralization of the influent before it enters the CETP.

The design TTC–CETP influent characteristics concentrations is compared with the maximum value for the years 2017 and 2018 and given in Table 1.

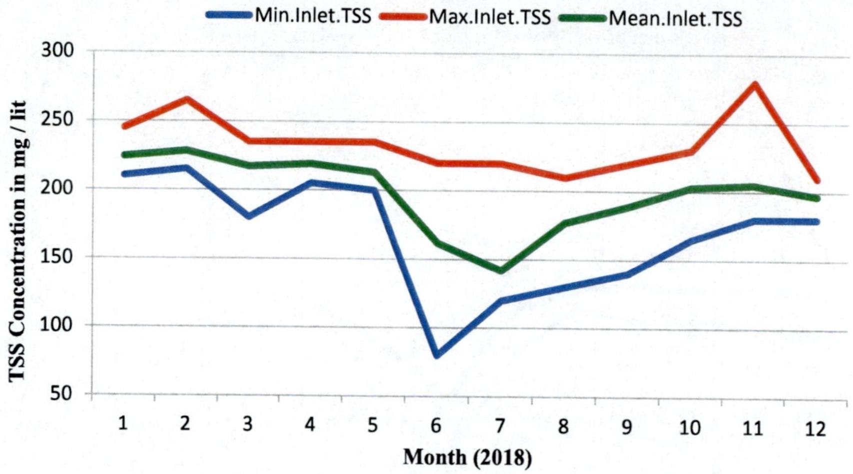

Fig. 9 Monthly variation in the minimum, maximum and mean values of TSS in the year 2018

Table 1 Design and maximum concentration of BOD, COD and TSS in 2017 and 2018

Inlet parameter of TTC–CETP	BOD in mg/lit	COD in mg/lit	TSS in mg/lit
Design concentration	1000	2600	340
Maximum concentration in year 2017	860	1810	275
Maximum concentration in year 2018	503	1568	228

5 Variability of Influent Characteristics

Two years data was analysed to see if there is significant variation of characteristic values in different quarters of a year using Kruskal Wallis Test.

Difference of the influent characteristics concentrations between two halves of the year was checked. Difference between Q1 and Q2, Q1 and Q3, Q1 and Q4, Q2 and Q3, Q2 and Q4, and Q3 and Q4 was also checked. H values are calculated for the years 2017 and 2018 and shown below. Table 2 shows the H values for the year 2017 and 2018 between two halves and different quarters.

6 Inferences of Krushkal Wallis Test

The Kruskal Wallis Test was used to check whether the difference in influent characteristic concentrations is significant over different quarters. The calculated H values are compared with the chi square values. The chi square value with 2 degrees of freedom and 3 degrees of freedom at a significance level of 0.005 are 10.597 and 12.838, respectively.

Table 2 H values of the influent characteristics concentrations of TTC–CETP for the years 2017 and 2018

	1 Half	2 Half	Q1	Q2	Q1	Q3	Q1	Q4	Q2	Q3	Q2	Q4	Q3	Q4
2018 pH	13.07		7.25		31.52		3		12.93		0.96		20.34	
2017 pH	1 Half	2 Half	Q1	Q2	Q1	Q3	Q1	Q4	Q2	Q3	Q2	Q4	Q3	Q4
	0.001		0.81		6.17		3.26		2.35		5.44		18.24	
2018 BOD	1 Half	2 Half	Q1	Q2	Q1	Q3	Q1	Q4	Q2	Q3	Q2	Q4	Q3	Q4
	37.35		0.39		74.88		1.58		83.88		0.02		70.6	
2017 BOD	1 Half	2 Half	Q1	Q2	Q1	Q3	Q1	Q4	Q2	Q3	Q2	Q4	Q3	Q4
	No value		No value		No value		No value		No value		No value		No value	
2018 COD	1 Half	2 Half	Q1	Q2	Q1	Q3	Q1	Q4	Q2	Q3	Q2	Q4	Q3	Q4
	32.24		10.6		101.5		3.11		61.07		6.87		101.9	
2017 COD	1 Half	2 Half	Q1	Q 2	Q1	Q3	Q1	Q4	Q2	Q3	Q2	Q4	Q3	Q4
	140.84		7.85		124.1		15.91		101.7		20.2		98.06	
2018 TSS	1 Half	2 Half	Q1	Q2	Q1	Q3	Q1	Q4	Q2	Q3	Q2	Q4	Q3	Q4
	120.66		42.92		123.1		84.53		48.45		10.87		56.25	
2017 TSS	1 Half	2 Half	Q1	Q2	Q1	Q3	Q1	Q4	Q2	Q3	Q2	Q4	Q3	Q4
	19.31		0.65		77.6		6.02		67.34		6.02		101.2	

Q1 and Q2, Q1 and Q4, and Q2 and Q4 have not shown significant differences since the H values for these quarterly comparison are smaller. The results of Krushkal Wallis Test H values greater than 10.597 indicate that the concentrations of influent characteristics such as COD, BOD and TSS of Q3 are significantly different from other three quarters for both the years. The Q3 is a monsoon season. Due to high runoff flows during monsoon, the effluent characteristics of the SSIs of the CETP may be varying leading to variation in the concentrations of inlet quality parameters in the Q3.

The biological treatment units at CETP are sensitive to pH value of the influent, thus it carries high importance. The H values for quarterly and half yearly comparison of pH is mostly smaller than 10.597 shows that pH influent concentrations do not vary significantly since the neutralization is done before entering into the CETP. Hence, corrosion related problems due to acidic nature of influent are avoided in biological treatment units.

After analysing two years data, it is concluded that the minimum and maximum concentrations of inlet quality parameters values in the months July, August and September are smaller compared to the values of other months. The mean values of the Q3 influent characteristics concentrations of parameters COD, BOD and TSS are also smaller compared with other quarter mean values.

Biological treatment units need consistency in influent characteristics concentrations. Changes in influent characteristics concentrations will affect the efficiency of biological treatment units and may ultimately affect the plants performance.

Acknowledgements Rajan S. Diwadkar (Senior Manager), CETP is gratefully acknowledged for providing the data for this study.

References

1. Metcalf, Eddy (2003) Wastewater engineering treatment and reuse. Tata Mcgraw Hill, New Delhi
2. Djeddou M, Bachir A, Martaud M (2014) Daily performances evaluation of municipal wastewater treatment plant using probabilistic model. In: IWA—Conference "Activated Sludge - 100 Years and Counting" Germany
3. Niku S, Schroeder E, Tchobanoglous G, Samaniego F (1981) Performance of activated sludge process: Reliability, Stability and Variability (Environmental Protection Agency, EPA Grant No. R805097-01, 1–124)
4. Owusu-Ansah E, De-Graft J, Sampson A, Amponsah SK, Abaidoo RC, Hald T (2015) Performance, compliance and reliability of waste stabilization pond: effluent discharge quality and environmental protection agency standards in Ghana. Res J Appl Sci Engg Tech 10(11):1293–1302

Modeling Glacio-Hydrologic Processes of Beas Within a Distributed, Water and Energy Balance-Based Hydrology Model

Tanmoyee Bhattacharya, Deepak Khare, and Manohar Arora

1 Introduction

In changing climate varied snow cover, glacier retreat and rainfall affect the seasonal distribution and quantity of runoff and thus influence the hydrological cycle of the mountain region. Glacier/snowmelt runoff should be taken into account to estimate reliable runoff in future climate change conditions. Due to climate variability, remoteness and limited ground station coverage Beas river basin's hydrology is challenging to model. Some research studies employ hydrology modeling to understand the relation between streamflow, melt water and Beas's climate [2, 22]. But the contribution and influence of the runoff components to the streamflow trend are still not clear for various regions of the Beas river basin. Many previous studies of the Beas river basin misrepresented quantitative assessment of snowmelt and glaciermelt contribution as snowmelt contribution using water balance, surface energy balance, temperature-index model and snowmelt runoff model [2, 13, 14, 19, 31]. Hydrological models can be considered useful for estimating river discharge, snowmelt runoff, glaciermelt runoff and baseflow, which are almost unavailable as observed data with long time series in complex basins like Beas. Very few studies estimate snowmelt and glaciermelt runoff using a conceptual hydrology model for a single discharge station of Beas [21, 22]. The non-predicted discharge in no rain season of water balance model [15], the weakly restrained boundary condition (water vapor pressure and wind speed) of surface energy balance model, low accuracy due to simplicity of the temperature-index model [16], neglection of glaciermelt contribution of snowmelt runoff model in a partly glacier-fed basin [20] and problem in parameter transfer across space and

T. Bhattacharya (✉) · D. Khare
Department of Water Resources Development and Management, Indian Institute of Technology Roorkee, Roorkee, Uttarakhand 247667, India
e-mail: bhattacharya.tanmoyee36@gmail.com

M. Arora
National Institute of Hydrology, Roorkee, Uttarakhand 247667, India

A. Sil et al. (eds.), *Recent Trends in Civil Engineering*, Lecture Notes in Civil Engineering 274, https://doi.org/10.1007/978-981-19-4055-2_53

time by the conceptual model due to no easily measurable parameters [38] make them non-feasible for mountainous Beas. The physical distributed model includes topographic indices, elevation data, soil types, glacier grid, land use, stochastic groundwater flow, precipitation, discharge, evaporation and parameterization of frozen soil in land surface for the cold region [11]. Distributed glacier energy balance models have the advantage of calculating heterogeneity of energy fluxes and melt characters for each grid cell. Variable infiltration capacity (VIC) is a physically distributed surface, macroscale hydrology model run in water and energy balance mode widely used for snowmelt/glaciermelt estimation [33, 40]. VIC considers important climate variables to calculate the sublimation and melting of ice/snow. Yet, no research has used the VIC hydrology model in Beas for snowmelt/glaciermelt studies. The non-homogeneous and sparse station data are unable to represent complete data coverage in distributed models. Gauge-based/modeled gridded datasets developed by mapping and interpolation with the predefined temporal and spatial resolution are used widely for hydrological studies of complex regions [3, 32].

In present research, an effort has been done to estimate the annual contribution of various runoff components to discharge, seasonality of discharge variation (due to substantive input from runoff components), influence of climate change on this runoff components and impact of this runoff components in managing streamflow trend of Beas using VIC-glacier hydrology model. IMD gridded rainfall product and ERA-interim gridded temperature data have been used in this study as meteorological input. The research has been carried out for lower, middle and upper altitudes of the Beas river basin at three discharge locations Pong Dam, Pandoh Dam and Manali.

2 Study Area

The Beas river basin (up to Pong Dam) is situated within Kullu, Mandi, Kangra and Hamirpur. The Beas river basin is the Indus river's tributary. The basin lies between longitudes 75°84′ E to 77°88′ E and latitudes 31°41′ N to 32°46′ N. The areas of Beas up to Pong Dam, Pandoh Dam and Manali are 12,417 km^2, 5383 km^2 and 362 km^2. The elevation of the lower altitude basin (Pong Dam), middle altitude basin (Pandoh Dam) and higher altitude basin (Manali) is 435.86 m, 857 m and 2050 m. Manali experiences a severe winter (December–March) due to severe snowfall at higher altitudes. The upper portion of Manali is snow covered and glaciated, which contributes melt water to the streamflow. About 780 km^2 catchment area of Pandoh Dam is under permanent snow and ice. Beas River is mainly used for irrigation purposes and hydropower development. Pandoh Dam was developed to divert the river to the Dehra power station of the Sutlej River. The glacier existence of Beas valley has been found toward the head of high streams and upstream of Kullu [18]. Cultivated lands and degraded forests cover a large portion of the catchment. Flash floods frequently affect the basin results loss of life and property.

3 Data and Methodology

3.1 Data Used

IMD gridded precipitation and ERA-interim gridded temperature (maximum and minimum) data interpolated to 0.01° × 0.01° resolution used as meteorological input for the VIC hydrology model. Other meteorological parameters used for the hydrology model are wind speed collected from ERA-interim, vapor pressure, shortwave and longwave radiation calculated from the algorithm developed by [26]. The pixel-to-point approach has been used to compare gridded IMD rainfall with station observation. The temperature of ERA-interim product produced by the European Centre for Medium-Range Weather Forecasts (ECMWF) [12] is the outcome of assimilated surface temperature. For the study, we have used temperature and rainfall data for the period of 1990–2009. Four major input files in ASCII format are required as VIC input database. They are soil parameter file, vegetation parameter file, vegetation library file and forcing file, respectively. Thematic layers such as LULC, DEM, soil, glacier area and ice thickness are overlaid with 0.01° × 0.01° grid of VIC hydrology model. Flow direction, slope and elevation are derived from DEM and grid fraction. We have used ASTER DEM (30 m), NASA EOSDIS LULC data for India of 100 m resolution, NBSS&LUP soil data (National Bureau of soil survey and land use planning) at 1:25,000 scale, glacier area extracted from GLIMS glacier database and initial ice thickness data collected from National Snow and Ice Data Center. Figure 1 shows DEM, soil, LULC, raingauge stations and glacier coverage of the Beas river basin.

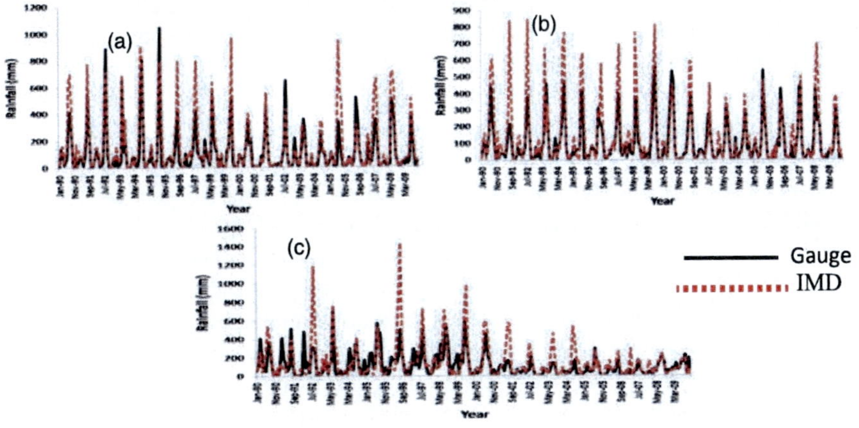

Fig. 1 Rainfall for (a) Pong Dam (b) Pandoh Dam and (c) Manali of Beas river basin from 1990–2009

3.2 Methodology

Rainfall data from IMD and raingauge are compared seasonally by statistical analysis methods such as correlation coefficient (CC), root mean square error (RMSE), relative bias (BIAS), false alarm ratio (FAR), probability of detection (POD) and accuracy (ACC). When Bias is less than/more than zero, underestimation/overestimation of the precipitation is considered. There is an acceptable correlation between precipitation products when CC is ≥ 0.40. The mean difference between observed and predicted values to provide overall model performance is given by RMSE. POD, FAR and ACC calculate the ratio of rainfall events. The temperature and rainfall data trend are computed using a modified version of the Mann–Kendall test [17, 25]. Variable infiltration capacity hydrology model is a surface water and energy balance model and can be used at spatial scale 1/8° latitude × 2° longitude. VIC hydrology model uses grid-wise daily inputs of vegetation parameter, snow/ice parameter, soil parameter, elevation and meteorological dataset [30]. The global parameter file is the most important input for the VIC hydrology model. To set the model's main options and provide the input file's path global parameter file is used. The process-based hydrology model can simulate evapotranspiration, surface runoff, baseflow, rainfall runoff, snowmelt runoff, glaciermelt runoff and other hydrology processes. The VIC hydrology model is implemented at $0.01° \times 0.01°$ resolution for the entire basin. The fractional area of LULC classes over a particular grid and rooting depth is used to derive the vegetation parameter file. The soil parameter file includes bulk density, soil moisture, saturated hydraulic conductivity, soil moisture fraction at wilting point, etc. The elevation parameter file has elevation band information (derived from DEM), elevation fraction in each grid cell, precipitation fraction, and initial ice thickness. Sub-grid variability is the unique characteristic of the VIC hydrology model. More detailed characterization of snow-related processes is possible due to sub-grid elevation bands. A routing module is used to rout the runoff of each grid cell to the outlet. The flow routing model uses the flow fraction and flow direction files. Two layers of variable thickness snowpack are modeled within the VIC hydrology model. Snowpack mechanism is simulated by mass and energy balance approach in VIC [10]. Cherkauer and Lettenmaier [10, 11] developed a method to represent frozen soil's effects on runoff in VIC. In this study, a threshold has been used for separating snowfall and rainfall. When the temperature is above threshold, precipitation is rainfall. A new class glacier is added to the mosaic scheme for the VIC-glacier model with the vegetation and bare soil classes. The elevation band needs to be arranged from high to low for the glacier model because most glaciers exist at a higher elevation. The meteorological parameter (temperature and precipitation) mean for a glaciated cell differs from a lower elevation cell. Moreover, for accurate estimation of glaciermelt, the fractional area of glaciated cell for each snowband file has been derived. The elevation band (ten bands) for each grid cell has been derived from Aster digital elevation model. File frozen soil flag for glaciated cell and glacier option is activated

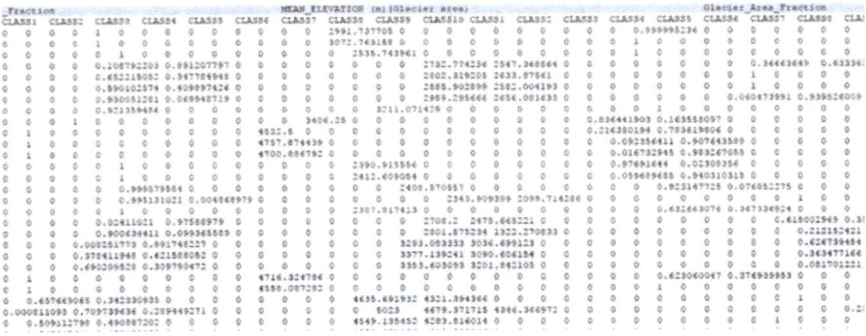

Fig. 2 Elevation band file for VIC to run glacier_module

```
##################################################################
# $Id: global.param.PENNAR,v 1.0 2011/02/09 16:45:08 vicadmin Exp $
##################################################################
# Simulation Parameters
##################################################################
NLAYER   3    #  number of soil layers
NODES    10   #  number of soil thermal nodes
TIME_STEP    12  #  model time step in hours (set to 24 if FULL_ENERGY = FALSE, set to < 24 if FULL_ENER
SNOW_STEP    12  # time step in hours for which to solve the snow model (should = TIME_STEP if TIME_STEP
STARTYEAR    1993     #  year model simulation starts
STARTMONTH   01  #  month model simulation starts
STARTDAY     01  #  day model simulation starts
STARTHOUR    00  # hour model simulation starts
ENDYEAR      2008     #  year model simulation ends
ENDMONTH     12  #  month model simulation ends
ENDDAY       31  #  day model simulation ends
FULL_ENERGY TRUE      #  TRUE = calculate full energy balance; FALSE = compute water balance only
CLOSE_ENERGY     TRUE
FROZEN_SOIL TRUE      #  TRUE = calculate frozen soils
QUICK_FLUX  FASLE     #  TRUE = use simplified ground heat flux method of Liang et al (1999); FALSE = use
NO_FLUX     TRUE  #  use no flux lower boundary for ground heat flux computation; FALSE = use constant
GLACIER     TRUE
GLACIER_OVERFLOW   TRUE
MTCLIM_SWE_CORR TRUE
IMPLICIT    FALSE
SPATIAL_FROST    TRUE 1
SPATIAL_SNOW     TRUE
SNOW_DENSITY     DENS_SNTHRM
DIST_PRCP   TRUE      #  TRUE = use distributed precipitation
PREC_EXPT   0.6 #  exponent for use in distributed precipitation eqn (only used if DIST_PRCP is TRUE)
#CORRPREC   FALSE     #  TRUE = correct precipitation for gauge undercatch
EXP_TRANS   FASLE
MIN_WIND_SPEED  0.5 #  minimum allowable wind speed (m/s)
MAX_SNOW_TEMP   0.1 #  maximum temperature (C) at which snow can fall
```

Fig. 3 Global parameter file for VIC to run glacier_module

for glacier scheme. Each grid cell's mean elevation is used to lapse grid cell's mean temperature and precipitation. The grid cell total runoff is calculated by

$$R(i) = f \times G_i + (1 - f) \times R_{\text{vic}}$$

where $R(i)$ = grid-wise total runoff, i is the number of grids; R_{vic} is VIC model calculated ice-free area runoff; f is fraction of grid cell glacier area; and G_i is runoff (mm) from glaciated area, melt water and liquid precipitation.

The equation of glaciated area melt water M_i is calculated as

$$M_i = \begin{cases} DDF \times T_i & T_i > 0 \\ 0 & T_i > 0 \end{cases}$$

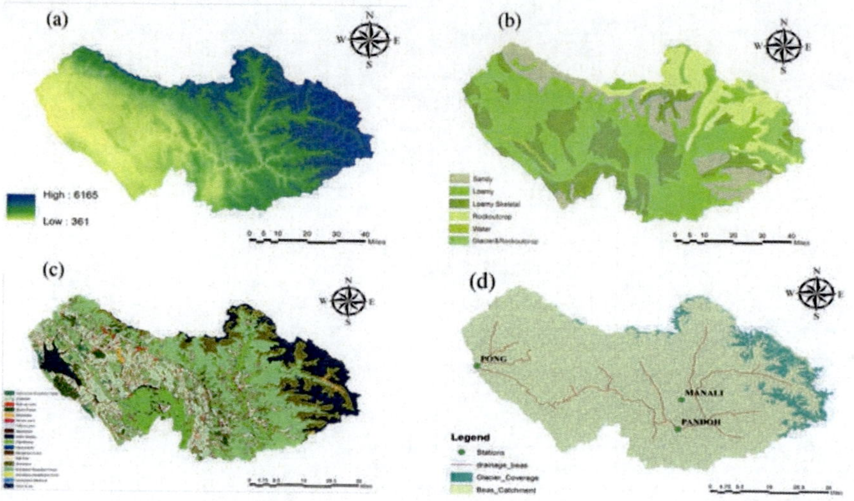

Fig. 4 **a** Dem **b** Soil **c** LULC and **d** meteorological stations and glacier map of Beas river basin

DDF = Degree-day factor (mm °C^{-1} day^{-1}). T_i = Mean daily temperature of glacier surface (°C). Snowpack on the glaciermelts first. To adjust the temperature of each grid cell glacier area 0.65 °C/100 m lapse rate of temperature is used. Regionally glacier volume is derived by volume–area scaling relation. Initially, individual basin's ice volume is calculated. The glaciated cell's initial ice volume is updated every ten years with melt water and snowfall accumulation. The glacier area at first time step is updated by reversing the relation of volume–area scaling with the updated volume [37]. The calibration and validation parameters used for the VIC hydrology model are three baseflow parameters (Ds, Dsmax and Ws), infiltration parameter (Binf) and three soil layers. The runoff increased by an increase in Binf ranges from 0.00 to 0.04. The soil depth (d_1-d_3) varied from 0.1 to 1.5 m and thicker soil depth slowed down the runoff process. The baseflow increases by increase/decrease of three baseflow parameters based on average saturated hydraulic conductivity. Our study's most sensitive calibration parameters are Ds, Ws, Binf, first- and second-layer soil depth. Dsmax and third soil layer are found less sensitive. Another three calibration parameters for snow were snow roughness (0.1), maximum temperature (0.5) and minimum temperature (−0.5). The calibration and validation parameters for Beas are shown in Fig. 5.

The statistical parameters used to evaluate the hydrology model performance are Nash Sutcliffe efficiency (NSE), root mean square error (RMSE) and percentage bias (PBIAS). NSE value indicates the perfect matching of measured and simulated values [8]. Percentage of bias (PBIAS) describes the larger or smaller value of simulated data than observed data. Trend analysis of hydrologic components has been done using a modified Mann–Kendall test. *To examine different runoff components' role in controlling the trend of total runoff percent contribution is derived by dividing*

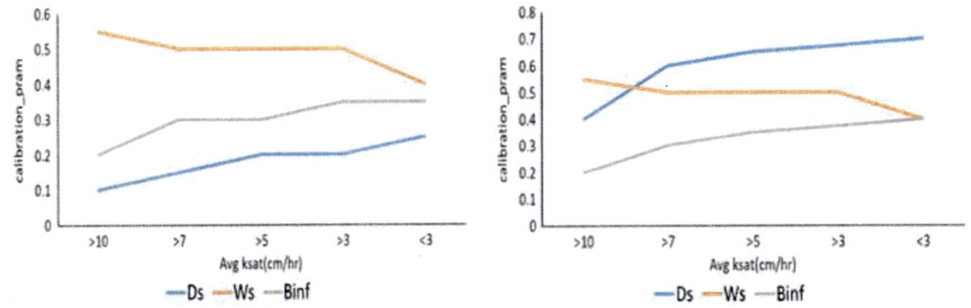

Fig. 5 Calibration parameters **a** Pandoh Dam **b** Manali

total runoff trend by trend of various runoff components. This method is used by many researchers [27, 39].

4 Result and Discussion

4.1 Characteristics and Components of Streamflow

The VIC-glacier hydrology model with gridded and remotely sensed input data is simulated. The results are compared with monthly discharge data for Beas up to Pong Dam, Pandoh Dam and Manali. IMD rainfall data have been compared with gauge rainfall data to understand the accuracy of the rainfall product (Table 1). IMD rainfall overestimates gauge rainfall except for autumn/post-monsoon season in Pandoh Dam and autumn and spring/pre-monsoon season in Manali. However, no large fluctuation is observed for bias. The gridded rainfall data correlate moderately with gauge observation in terms of correlation coefficient (CC) and RMSE. However, Pandoh Dam and Manali show a weak correlation coefficient in the winter season and a low correlation observed in the monsoon season for the Pong Dam. The reason for the weak correlation may be due to the orographic effect. The trend of IMD and gauge rainfall is quite similar. POD and ACC also show good agreement of seasonal IMD and gauge rainfall products. IMD and raingauge rainfall data are plotted in Fig. 1 for 1990–2009. The comparison proves that IMD gridded rainfall data can be used in hydrological simulations to substitute observed rainfall. Figure 6 shows the observed and simulated monthly discharge of Pandoh Dam, Pong Dam and Manali for the validation period. The model is calibrated for 1993–1998 and validated for 1999–2008 at two gauging sites Pandoh Dam and Manali. The calibration and validation period for the Pong Dam is 1990–1994 and 1995–2004. Monthly NSE, RMSE and PBIAS for Pong Dam, Pandoh Dam and Manali during calibration period are (0.24, 120.09, −34.58), (0.61, 70.39, −22.67) and (0.40, 13.87, −34.00), respectively. During validation period, the values are (0.20, 100.48, −35.00), (0.61, 72.20, −20.58) and (0.40, 15.54, −35.88), respectively. After analyzing model performance criteria, it is found that the VIC model performs satisfactorily at Pandoh Dam and Manali with underestimation as indicated by Motovilov et al. [29] and Moriasi et al. [28]. After comparing the peak flow of modeled discharge with observed data, the peak of the observed hydrograph is more than three times the peak of modeled hydrograph for Pandoh Dam and Manali. Due to reservoir regulation, impact on discharge at Pong Dam model shows unacceptable performance (in terms of all statistical indicators). The low quality rainfall dataset fails to capture the orographic rainfall pattern of the Beas river basin. Other parameters that can reduce model efficiency are sub-daily events and variation in glacier energy balance [23]. In this study, runoff variation and its components (rainfall runoff, snowmelt runoff, glaciermelt runoff and baseflow) are calculated for the lower (Beas up to Pong Dam), middle (Beas up to Pandoh Dam) and upper (Beas up to Manali) Beas river basin of North Western Himalaya during the validation period. A large fraction of runoff is contributed by rainfall at Pong Dam. The monthly variation and percent contribution of various runoff components of the Pong Dam have shown in Fig. 7. Significant rainfall contribution (49–63% of total rainfall) occurs during June–September in Pong Dam, with a maximum contribution between August and September. The contribution of snowmelt runoff to total runoff starts from March and depletes after June. Maximum 12.11–16.12% of snowmelt runoff contribution occurs to total runoff during April to June (snowmelt period). The glaciermelt contribution becomes significant after the contribution from snowmelt depletes, i.e., from July to September. The maximum glaciermelt contribution (16–17%) occurs during August and September. The baseflow contribution is substantial during the spring and winter seasons when the contribution from rain and snowmelt decreases. About 56–65% of baseflow is contributed during the winter month. The average annual contribution of Pong rainfall runoff, snowmelt, glaciermelt and baseflow to total runoff is 47.12%, 13.17%, 6.72% and 34.00%, indicating that rainfall runoff is the dominant contributor of water resources for this basin. For the middle altitude basin, (Beas up to Pandoh) monsoon discharge is sustained by a significant contribution from rainfall, snowmelt, glaciermelt and baseflow (18%, 22.00%, 17% and 36%). The baseflow is sustained from the contribution of both rain and snow/glaciermelt. For Pandoh Dam, snow/glaciermelt contribution is more than rainfall. Snowmelt contribution is dominant in April–June (53–55%), whereas the maximum rainfall contribution during the monsoon season is 17–25%. The glaciermelt contribution to discharge is significant during August and September (21–27%). Baseflow contribution is notable during the winter season, i.e., 65–73%. The monthly variation and percent contribution of various runoff components have shown in Fig. 8. The average annual contribution of Pandoh rainfall, snowmelt, glaciermelt and baseflow to total runoff is 20.89%, 26.00%, 14.58% and 37.96%. During April–June, the discharge of Manali is mainly contributed from the runoff due to snowmelt. About 60–77% of snowmelt contribution occurs during the snowmelt season (April–June). During July–September, the dominant source of runoff is snowmelt (43–48%), and rainfall runoff contribution is less (7–10%). Baseflow also has a significant contribution during winter and spring (70–77%). The glaciermelt contribution during July–September is not substantial for Manali (1.5–2.8%). The annual average contribution of rainfall runoff, snowmelt, glaciermelt and baseflow is 7.52%, 49.21%, 1.37% and

41.88%, indicating discharge of Manali mainly sustained by snowmelt and baseflow. The monthly runoff components and percent runoff contribution to runoff during the simulation period of Manali have shown in Fig. 9. The VIC hydrology model can reproduce the observed peak flow (during monsoon) and baseflow (during winter) in Beas. The seasonal characteristics of runoff components and their contribution to discharge have also been quantified in this research. The glacier ablation of Beas starts in April. The season for significant glaciermelt contribution to discharge is July–September with a high peak in July–August, agreeing with high temperature months. From October, glaciermelt contribution decrease and minimum in December, which has the same temperature trend during October–December. Snowfall and ablation of accumulated snowpack in winter are the main snowmelt source during March–June. The melt season which contributes to discharge is April–September. The monsoon rainfall plays a vital role in runoff generation for Beas as the seasonal pattern of total runoff and rainfall runoff is similar in June–September (Fig. 10).

Table 1 Statistical parameters for comparison of IMD rainfall and raingauge data

Location	Month	Statistical parameter						Trend	
		CC	RMSE	Bias	POD	FAR	ACC	IMD	Gauge
Pong Dam	MAM	0.65	28.76	56.72	0.95	0.20	0.91	−0.60	−0.51
	JJAS	0.30	160.29	44.00	0.60	0.54	0.70	−2.27	−2.13
	ON	0.94	9.30	27.73	0.95	0.24	0.80	0.52	0.53
	DJF	0.74	22.18	27.32	0.95	0.10	0.93	−2.00	−2.13
	ANL	–	–	–	–	–	–	−3.00	−2.00
Pandoh Dam	MAM	0.40	24.35	25.56	0.90	0.54	0.73	−0.76	−0.63
	JJAS	0.65	187.15	40.58	0.89	0.24	0.75	−3.01	−1.07
	ON	0.40	39.95	−2.94	0.87	0.43	0.84	0.31	0.21
	DJF	0.18	45.77	65.76	0.50	0.66	0.70	−0.71	−1.20
	ANL	–	–	–	–	–	–	−1.27	−0.10
Manali	MAM	0.45	75.76	−7.11	0.70	0.23	0.69	−5.70	−6.69
	JJAS	0.50	90.15	68.33	0.92	0.24	0.62	−5.70	−2.58
	ON	0.58	37.14	−24.93	0.70	0.60	0.89	0.54	0.40
	DJF	0.25	54.35	50.78	0.50	0.11	0.81	−2.07	−1.52
	ANL	–	–	–	–	–	–	−1.68	0.40

MAM = March–May (spring/pre-monsoon), JJAS = June–September (monsoon), ON = October–November (autumn/post-monsoon) and DJF = December–February (winter), ANL = Annual, – = Decreasing Trend

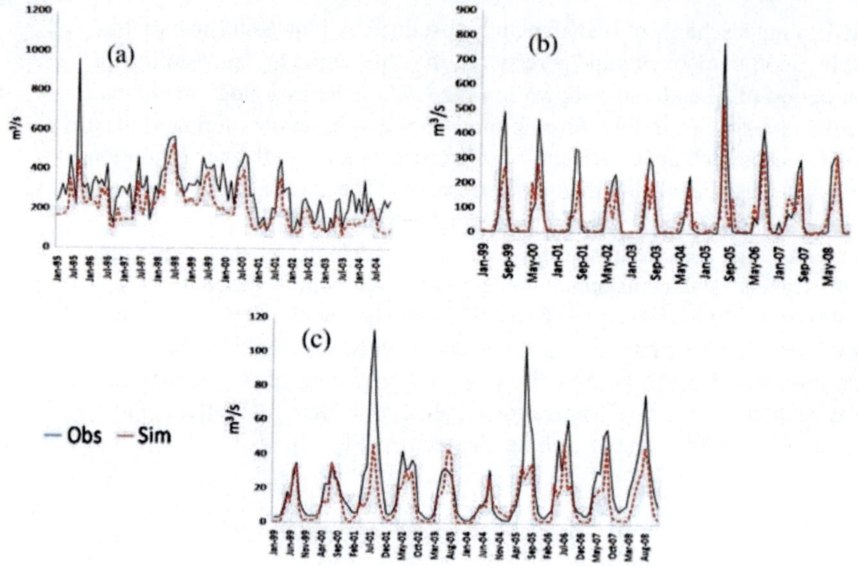

Fig. 6 Monthly observed and simulated discharge for **a** Pong Dam, **b** Pandoh Dam and **c** Manali during validation period

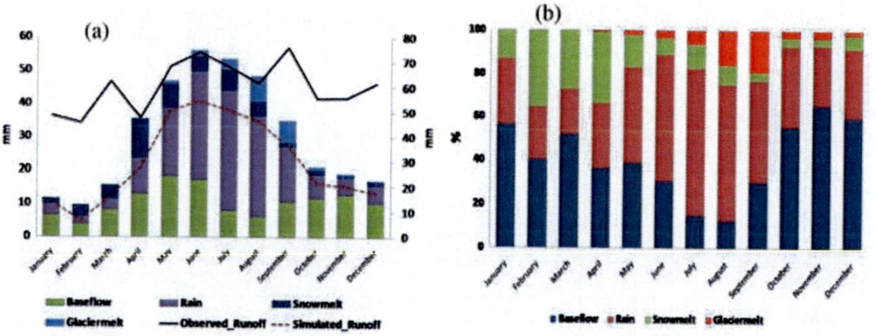

Fig. 7 Monthly **a** variation of runoff components and **b** Percent contribution of different runoff components for Pong Dam

4.2 Climate Change Impact on Runoff Components

Variation of runoff components under changing climate has been analyzed for Pandoh Dam and Manali during 1993–2008 (simulation period). The impact of climate on the runoff trend for Pong Dam has been analyzed for the whole simulation period (1990–2004). The shift in Pong rainfall is sudden rather than gradual. The year of change is 1994 and 2005. Before 1994 there was an increasing trend, and after this

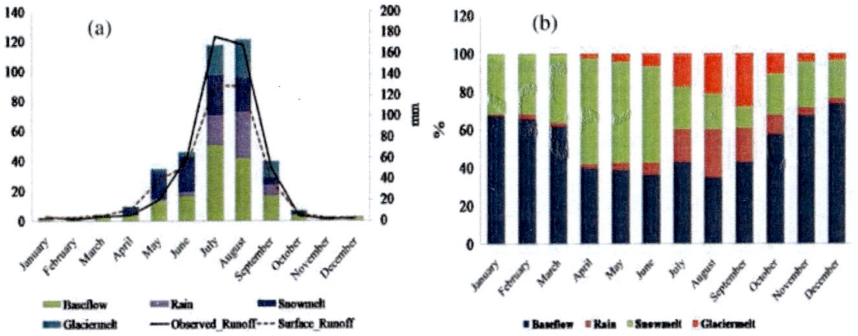

Fig. 8 Monthly **a** runoff components and **b** Percent contribution of different runoff components for Pandoh Dam

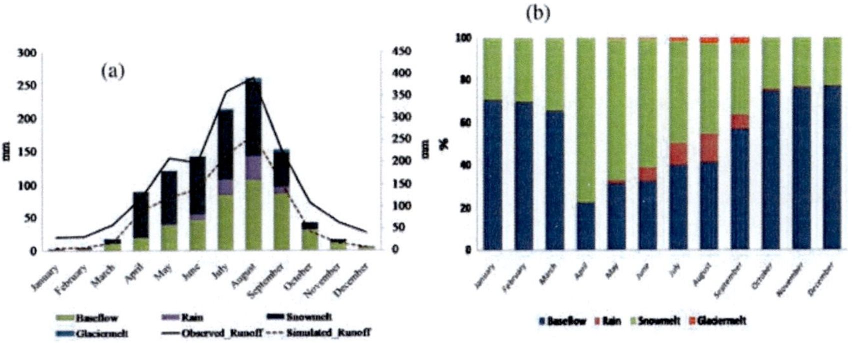

Fig. 9 Monthly **a** runoff components and **b** Percent contribution of different runoff components for Manali

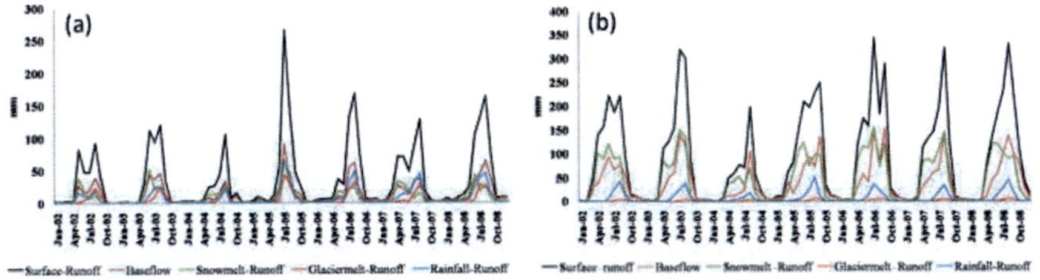

Fig. 10 Seasonal runoff components **a** Pandoh Dam **b** Manali for 2003–2008

year, decreasing trend of precipitation occurs. Again from 2005, there is a sudden increase in trend (Fig. 11f). The precipitation anomaly is not restricted to the monsoon season, but this phenomenon extends in pre-monsoon months also. The reason may be steep topography is sensitive to climate which causes extreme events and fragile ecosystems. Brar [7] found that the extreme events in North Western India are due to the fact that drier parts are getting more rainfall and wetter parts become drier.

Chandel and Brar [9] found Kangra district experienced a very high frequency of cloudburst events during 1994, 2006, 2007 and 2008. The trend analysis of Pong rainfall for 1990–2009 shows a decreasing trend of rainfall during monsoon, pre-monsoon and winter seasons, though not significant (Table 1). Pong rainfall varies 22.60–119.39 mm in pre-monsoon, 202.85–574.70 mm in monsoon, 0.41–89.08 mm in post-monsoon and 10.93–86.26 mm in winter. The knowledge of uncertain precipitation patterns is needed to properly regulate hydropower generation, agricultural management, irrigation, etc. Pong streamflow shows a declining trend (−1.63 m^3/s/year) in monsoon mainly due to decreasing trend of rainfall runoff (−1.50 mm/year). The reduced rainfall runoff is attributed to lowering trend of monsoon rainfall (Table 1). We have considered that baseflow is separated into rainfall runoff, snowmelt and glaciermelt components. Table 2 shows the trend of Pong temperature. During winter and pre-monsoon/spring season, temperature trends increase significantly. The monsoon and post-monsoon/autumn season also show a rising trend of temperature (non-significant). Glaciermelt runoff increases (0.10 mm/year) during July–September consistent with rising temperature. Temperature increase and shortage of rainfall cause less snow accumulation in winter. Moreover, due to warmer temperatures during winter, the melting of snowpack increased. This phenomenon significantly decreases the snowmelt runoff trend (−0.35 mm/year) in pre-monsoon/spring season and monsoon (−0.50 mm/year) season. Spring/pre-monsoon rainfall runoff also decreased by −0.55 mm/year due to a reduction in spring rainfall. The streamflow of the autumn season decreases by −1.20m^3/s year. Autumn rainfall runoff (Trend = −0.60 mm/year) is the dominant contributor to the decreasing tendency of autumn discharge. Autumn snowmelt runoff declines by −0.005 mm/year. The annual trend analysis shows that rainfall runoff and snowmelt runoff have decreasing trend of −2.65 mm/year and −0.21 mm/year, agreeing to a significant increasing trend of annual temperature (0.10°C/year) and downward trend of annual rainfall (−3.00 mm/year). This proves that rainfall trends mainly dominate the declining trend of annual total runoff and discharge. Rainfall contributes 64% to decrease the trend of annual total runoff. The annual decreasing trend of streamflow in Pong is −2.00 m^3/s/year. Figure 11 shows a decreasing trend of monsoon snowmelt runoff, monsoon rainfall runoff, spring snowmelt runoff and increasing glaciermelt runoff in July–September. Pong Dam is located in Kangra district, a rich cultivable area, but shows increased urbanization and land use change. The rainfall anomaly in Pong during pre-monsoon, monsoon and post-monsoon season may lead to flash floods, sudden heavy snowfall, landslides, soil erosion and drought. Protective measures such as structures and early warning systems should be taken for this region. However, the complex terrain not allows us to understand the exact influence of each raingauge station. To assess temporal and spatial variation of rainfall daily precipitation analysis and future detailed trend analysis of Pong Dam need to be done. Pandoh streamflow has a decreasing trend of −1.51 m^3/s/year during pre-monsoon/spring season (Fig. 12a), mainly due to decreasing monsoon rainfall (Fig. 12g). In contrast, glaciermelt runoff shows an increasing trend of 0.48 mm/year during July–September of 1993–2008 (Fig. 12e). The increasing/decreasing trend of glaciermelt/rainfall runoff is consistent with the increasing/decreasing trend of temperature/rainfall during this monsoon period (Fig. 12d and g). Pandoh Dam rainfall varies 16.26–98.23 mm in pre-monsoon, 200.23–478.50 mm in monsoon, 0.22–76.43 mm in autumn and 17.95–112.61 mm in winter. However, the runoff components show a decreasing trend in Pandoh. The winter precipitation is due to non-monsoonal rain, create sudden snow. Due to the decrease in Mediterranean Westerlies wind's driven non-monsoonal rain number of snowfall days' decrease. Moreover, the early increasing trend of air temperature causes instant melting of accumulated snowpack in late autumn to early spring (mid-November-mid-April), resulting in increased snowmelt runoff from the basin (Fig. 12i). Thus, there is a reduction of snowpack reserves at the beginning of snowmelt season, leading to a decrease in spring and monsoon discharge. The early shift of peak monthly runoff of Pandoh and Manali in spring also indicates the early onset of spring temperature. (Fig. 10). The significant increase of snowmelt runoff during the snow accumulation period (November–April) indicates early snowpack melt, leading to a decrease in snowmelt runoff during spring and monsoon (Fig. 12c and f). The snow cover of Beas starts to accumulate from October, increases up to February and then depletes till June. The snowmelt runoff from July to September is mainly due to fresh snowfall, which quickly melts after covering the area. Due to increased snowmelt runoff (Trend = 0.36 mm/year) in winter (December–February), discharge also rises (Trend = 0.01 mm/year) (Fig. 12j). During monsoon and post-monsoon season, temperature of Pong Dam and Pandoh Dam shows a non-significant increasing trend. In contrast, the Manali temperature has decreasing trend (Table 2). The decreasing trend of both rainfall runoff (−2.10 mm/year) and snowmelt runoff (−1.60 mm/year) in monsoon season has more impact on reduced streamflow (−2.00 m^3/s/year) as compared to the increasing trend of glaciermelt runoff (0.48 mm/year) (Fig. 12b and h). Bhutiyani et al. [6] also found decreasing trend of pre-monsoon and monsoon discharge of Beas. The temperature of Pandoh shows a significant increasing trend during the winter, pre-monsoon and post-monsoon periods. Pre-monsoon, monsoon and winter rainfall shows a decreasing trend, though not significant (Table 1, Fig. 13). Beas has less glacier coverage as compared to other glacier-dominated basins of western Himalaya. The rate of recession of smaller glaciers is more than large glaciers results in a disappearance in the near future. The VIC-Glacier model elevation band contains the information of mean elevation, snow area fraction, mean elevation for glaciated cell, precipitation fraction, band slope, and initial ice thickness (Fig. 2). For soil parameter and global parameter (Figs. 3, 4). Figure 13 presents the annual trend of various hydrologic components of Manali from 1993-2008. Figure 14a shows a strong correlation between glaciermelt runoff and temperature during the melting period (0.50). Li et al. [24] also found increased ice melting in the Himalayas with rising temperatures. The decreasing autumn streamflow (Trend = −0.32m^3/s/year) caused by declining trend of rainfall runoff (Trend = −0.10 mm/year) and snowmelt runoff (Trend = −0.36 mm/year). Increasing trend (Trend = 0.01 mm/year) of autumn glaciermelt runoff has less impact on autumn streamflow trend. The annual decreasing trend of rainfall runoff and snowmelt runoff is −1.15 mm/year and −1.02 mm/year which have a dominant contribution (rainfall runoff = 61% and snowmelt runoff = 54%) to the trend of annual total runoff. Annual streamflow for Pandoh has a reducing trend of −1.30 m^3/s/year. Annual temperature and rainfall of Pandoh show a significant increasing trend of 0.04 °C/year and decreasing trend of −1.27 mm/year. To understand the impact of glaciermelt runoff on total runoff, linear regression between the runoff components has been analyzed using a scatter plot (Fig. 14b). The negative correlation of glaciermelt contribution with total runoff is due to more contribution in dry years compared to wet years. Glaciermelt contributes 2.91% to 16.59% during dry years. Thus, glaciermelt compensates for water deficit during low monsoon years. The increasing trend of annual glaciermelt runoff (0.30 mm/year) also less influences the annual total runoff trend. The decreasing trend of spring and monsoon streamflow can create irrigation demand, disturbed ecology of the river system, drought risk, hydropower generation reduction, water supply for industry and domestic supply, reduction of stored water in reservoirs and reduction of groundwater level. The increasing trend of discharge in winter may lead to flood risk in the future. This phenomenon happens due to a significant rise in temperature (0.13 °C/year) and precipitation deficiency (Table 1) during winter and early spring. So, proper hydropower, water demand, water supply, and rivers are essential to mitigate the aforementioned hazards. The trend analysis of Manali precipitation shows a decreasing trend during all seasons except post-monsoon (Table 1). Manali rainfall varies 32.44–150.28 mm in pre-monsoon, 122.33–743.18 mm during monsoon, 0.83–108.60 mm in post-monsoon and 26.37–171.29 mm in winter. Due to warmer temperatures snowfall decreases in winter (Fig. 13f). A significant increase in mid-November to early April snowmelt (Fig. 13d) (1.13 mm/year) (due to an increase in winter (Fig. 13j) and spring temperature) results in a decrease in spring/pre-monsoon and monsoon streamflow (Fig. 13a and b). The increasing trend of glaciermelt runoff (Fig. 13e) during monsoon season (0.48 mm/year) has less influence on the trend of monsoon discharge. Figure 13c shows an increasing trend (0.10 mm/year) of winter streamflow due to the increasing trend of winter snowmelt runoff (0.12 mm/year). Bhan and Singh [4] also found decreasing trend of snowfall and rainfall in Shimla. Monsoon rainfall runoff shows a decreasing trend of −2.10 mm/year (Fig. 13h) consistent with the declining trend of monsoon precipitation (Fig. 13i). Snowmelt runoff shows a decreasing trend of −1.70 mm/year during the mid-spring to monsoon season (Fig. 13g). Autumn season streamflow increases by 0.30 m^3/s/year due to the increasing trend of snowmelt runoff (Trend = 0.05 mm/year). Abbaspour et al. 2009 also found a decrease in spring streamflow and an increase in autumn streamflow in their study. The annual streamflow of Manali decreases by −0.30m^3/s/year for 1993–2008, mainly dominated by the annual decreasing trend of snowmelt runoff (−1.93

mm/year) and rainfall runoff (−0.7 mm/year). The annual streamflow decreasing trend is accordant with a significant increasing/declining trend of annual temperature (0.10°C/year)/rainfall (−1.68 mm/year). Snowmelt runoff contributes 75% of the annual total runoff decreasing trend. Figure 10b shows the seasonal runoff components of Manali. Due to the early snowmelt at Manali, the snowpack retreats faster, resulting in future water scarcity and adversely affects the agriculture/horticulture of this area. Bhutiyani et al. [5] and Singh et al. [35] found that a rise in the long-term mean temperature causes a reduction in river discharge of North Western Himalaya (over Sutlej river basin). Singh et al. [35] suggested that shortening of the melt season, melt reduction in summer and decrease in rainfall results in decreased annual runoff during summer over North Western Himalaya. Sharma et al. [34] suggested that due to rapid snowmelt in the Jhelum river basin of North Western Himalaya streamflow increase in the pre-monsoon season. Later in the monsoon and post-monsoon season, the streamflow decrease. These findings of the studies mentioned above are very similar to our study results. However, our study is subject to uncertainty mainly due to the poor quality of precipitation data and less availability of high altitude temperature and rainfall estimates. The uncertainty of this study is also related to glacio-hydrology model parameters (degree-day factor D) and temperature lapse rate (TLR) and runoff generation parameters (d2 and Binf). The hydrology model has limitations in validating glacier mass and energy balance simulation due to a lack of reliable climate forcing data. Independent/offline glacier scheme from the hydrological model system uses a simplistic representation of the hydrologic process. Glacier scheme also cannot parameterize glacier surface albedo and blowing snow process for simulating surface mass balance [33, 40]. For sustainable water management under changing climate in Beas can be done by improving of monitoring and subsequent model efficiency using field-based data. However, uncertainty analysis is not the main consideration of this study. For proper water management to deal with present and future challenges of surface and groundwater availability uncertainties and the impact of present and future climate change on water resources needs to be quantified in Beas. Moreover, the changes in precipitation spatial pattern and land use are two critical factors that should be considered in future research.

Table 2 Trend analysis of temperature in Beas

Location	MAM	JJAS	ON	DJF	Annual
Pong Dam	0.08*	0.02	0.01	0.16*	0.10*
Pandoh Dam	0.10*	0.05	0.01	0.13*	0.04*
Manali	0.18*	−0.01	0.01	0.13*	0.10 *

Increasing/decreasing trend using significance level 0.05

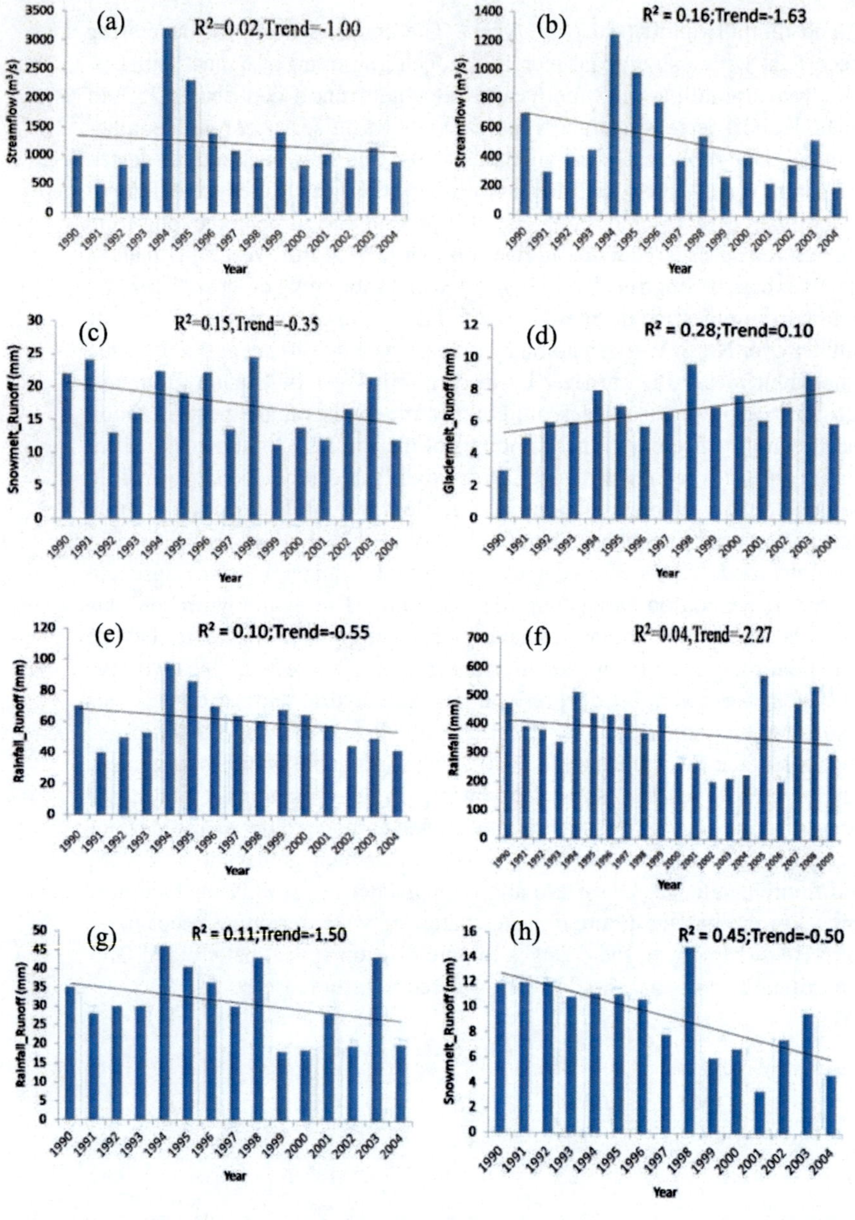

Fig. 11 Yearly trend of **a** spring streamflow **b** monsoon streamflow **c** spring snowmelt runoff **d** glaciermelt runoff **e** rainfall runoff during spring **f** rainfall during monsoon **g** monsoon rainfall runoff **h** monsoon snowmelt runoff for Beas up to Pong Dam

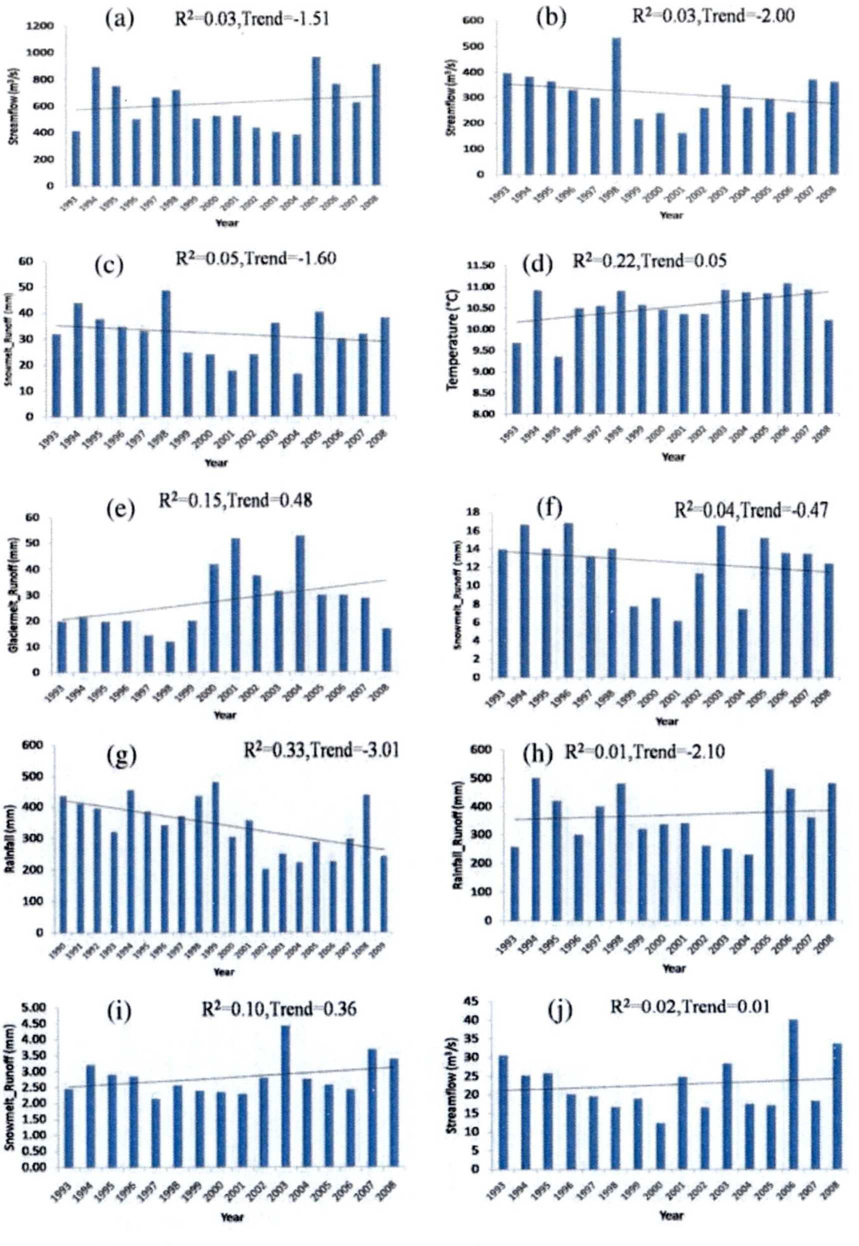

Fig. 12 Yearly trend of **a** spring streamflow **b** monsoon streamflow **c** monsoon snowmelt runoff **d** summer temperature **e** glaciermelt runoff **f** spring snowmelt runoff **g** monsoon rainfall **h** monsoon rainfall runoff **i** winter snowmelt runoff and **j** winter streamflow for Pandoh Dam

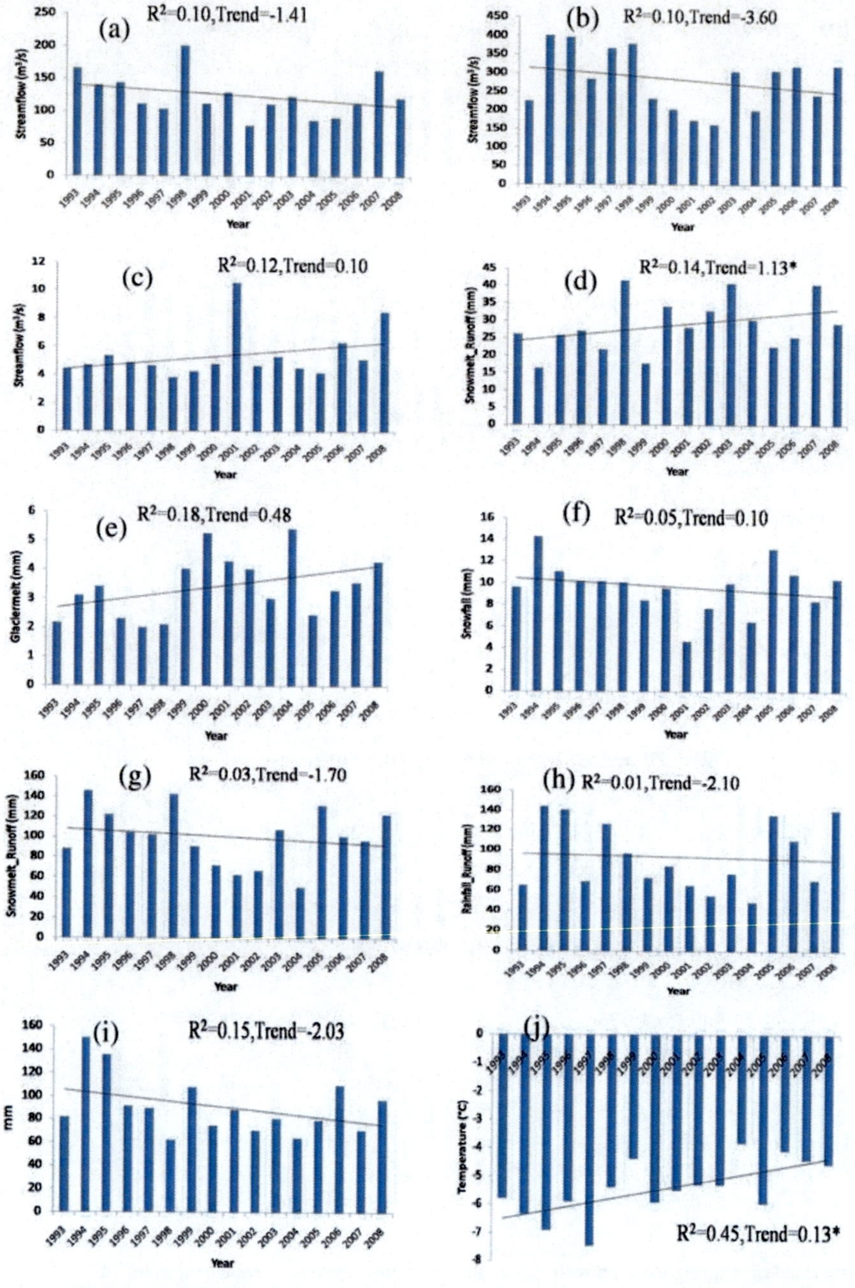

Fig. 13 Yearly trend of **a** spring streamflow **b** monsoon streamflow **c** winter streamflow **d** mid-autumn to early spring (mid-November to early April) snowmelt runoff **e** glaciermelt runoff **f** winter snowfall **g** mid-spring to monsoon (mid-April to September) snowmelt runoff **h** monsoon rainfall runoff **i** monsoon precipitation **j** winter temperature for Manali. * Trend at 0.05 significance level

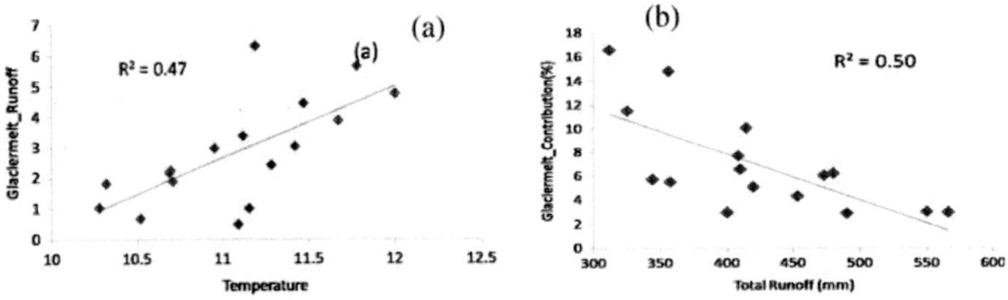

Fig. 14 Relationship between **a** glaciermelt runoff and temperature **b** glaciermelt contribution and total runoff

5 Conclusions

Hydrological processes simulation-based method by VIC-glacier model comprehensively and systematically evaluate percent contribution of runoff components and influence of rainfall/melt water in regulating discharge. Meteorological data, i.e., temperature and precipitation, have been used to simulate various river flow components, including snowmelt and glaciermelt. The findings of the study are as follows:

(i) Discharge contribution for low altitude basin (Beas up to Pong Dam) is mainly dominated by rainfall (47.12%). Snowmelt is the major contributor to total runoff (49.21%) for high altitude basin (Beas up to Manali). For the middle altitude basin (Beas up to Pandoh Dam), both rainfall (20.89%) and snowmelt (26.00%) are significant contributors to discharge. The baseflow contribution to discharge is dominant during the winter when contributions from snow/glaciermelt reduce. During winter/spring about 56–65%, 65–73% and 70–77% baseflow contribute to discharge for lower, middle and higher altitude basins. Baseflow is sustained from the contribution of both rain and snow/glaciermelt.

(ii) The increasing/decreasing trend of melt water/rainfall-induced runoff is accordant with temperature and rainfall trends. The situation of rise in temperature and rainfall deficit causes less snowfall, instant melting of accumulated snowpack in winter, reduction in melt season and pre-monsoon/monsoon rainfall trend. This climate-induced changes result in a decrease in discharge volume and timing of freshets. 0.10 °C/year increase in temperature and −1.68 mm/year decrease in rainfall result in −0.30 m^3/s/year decrease in streamflow in Manali. An annual 0.04 °C/year rise/ −1.27 mm/year decline of temperature/rainfall results in a −1.30 m^3/s/year reduction in streamflow in Pandoh. About −3.00 mm/year decrease in rainfall and 0.10 °C/year increase in temperature in Pong Dam is the main reason for the decline in annual streamflow (−2.00 m^3/s/year).

(iii) The changing climate exaggerates the snow/glaciermelt in winter that results in a decrease in pre-monsoon/monsoon melt water in Beas. Moreover, early melting of snow causes loss in the permanent snow/glacial reserve in the basin.

The warming trend with a shortage of rainfall increases winter streamflow but decreases streamflow during the pre-monsoon-post-monsoon season in different regions of Beas. The uncertainty in total runoff in summer may affect irrigation use and hydropower in the future. Reduced rainfall runoff/snowmelt runoff is the dominant factor for the declining total yearly runoff trend of Pong Dam/Manali. Total yearly runoff of Pandoh Dam falls due to decreasing trend of both rainfall runoff and snowmelt runoff.

(iv) Though glacier contributes very little to overall runoff in this watershed, our study suggests that the issue of glacier contribution to future hydrologic change should be taken into account for future water resource management.

(v) The advantage of the study is (a) possibility of simulation-based approach to analyze the trend of individual water/energy balance components and their contribution to discharge, which cannot be obtained as field data for long time series (b) the study results inform potential climate change impact on the water resource that identify knowledge gaps and associated future research priorities.

References

1. Abbaspour KC, Faramarzi M, Ghasemi SS, Yang H (2009) Assessing the impact of climate change on water resources in Iran. Water Resour Res 45(10):1–16
2. Ahluwalia RS, Rai SP, Jain SK, Dobhal DP, Kumar A (2015) Estimation of snow/glaciermelt contribution in the upper part of the Beas River basin, Himachal Pradesh, using conventional and SNOWMOD modeling approach. Hydrol Res 47(4):683–700
3. Bhattacharya T, Khare D, Arora M (2019) A case study for the assessment of the suitability of gridded reanalysis weather data for hydrological simulation in Beas river basin of North-Western Himalaya. Appl Water Sci 9(110):1–15
4. Bhan SC, Singh M (2011) Analysis of total precipitation and snowfall pattern over Shimla. J Agrometeorol 13(2):141
5. Bhutiyani MR, Kale VS, Pawar NJ (2009) Climate change and the precipitation variations in the NorthWestern Himalaya:1866–2006. Int J Climatol 30(4):535–548
6. Bhutiyani MR, Kale VS, Pawar NJ (2008) Changing streamflow patterns in the rivers of North Western Himalaya: Implications of global warming in the 20th century. Curr Sci 95(5):618–626
7. Brar KK (2000) Impact of the green revolution on the rainfall regime of Punjab. Trans Inst Indian Geogr 22:63–74
8. Brouziyne Y, Abouabdillah A, Bouabid R, Benaabidate L (2018) SWAT streamflow modelling for hydrological components' understanding within an agro-sylvo-pastoral watershed in Morocco. J Environ Sci 9(1):128–138
9. Chandel VBS, Brar KK (2010) Climate extremes and changing climate in Western Himalayas: A study of cloudburst incidences in Himachal Pradesh. Punjab Geographer. 6:29–40
10. Cherkauer KA, Lettenmaier DP (2003) Simulation of spatial variability in snow and frozen soil. J Geophys Res 108(D22):8858
11. Cherkauer KA, Lettenmaier DP (1999) (1999) Hydrologic effects of frozen soils in the upper Mississippi River basin. J Geophys Res 104(D16):19599–19610
12. Dee DP et al (2011)The ERA-interim reanalysis: Configuration and performance of the data assimilation system. Quart J Roy Meteor Soc 137:553–597
13. Dimri T, Thakur PK, Aggarwal SP (2012) Energy balance approach for snowmelt run-off estimation. In: Proceedings of the 99th Indian Science Congress, Bhubaneswar151

14. Dobhal DP, Gupta AK, Mehta M, Khandelwal DD (2013) Kedarnath disaster: facts and plausible causes. Curr Sci 105(2):171–174
15. Guo S et al (2005) A semi-distributed monthly water balance model and its application in a climate change impact study in the middle and Lower Yellow River Basin. Water Int 30(2):250–260
16. Hock R (2003) Temperature index melt modelling in mountain areas. J Hydrol 282:104–115
17. Kendall MG (1975) Rank correlation methods, 4th edn. Charles Griffin, London
18. Kumar A, Sarkar SA, Sharma N (2015) Snowmelt runoff modeling in an Indian Himalayan River Basin using WinSRM. RS & GIS Water Energy Int 58(1):65–72
19. Kumar V, Singh P, Singh V (2007) Snow and glaciermelt contribution in the Beas River at Pandoh Dam, Himachal Pradesh, India. Hydrol Sci J 52(2):376–388
20. Latif Y, Ma Y, Ma W, Muhammad S, Adnan M, Yaseen M, Fealy R (2020) Differentiating snow and Glaciermelt contribution to runoff in the Gilgit River Basin via degree-day modelling approach. Atmosphere 11:1023
21. Li L, Shen M, Hou Y, Xu CY, Lutz AF, Chen J, Jain SK, Li J, Chen H (2019) Twenty-first-century glacio-hydrological changes in the Himalayan headwater Beas River basin. Hydrol Earth Syst Sci 23(2019):1483–1503
22. Li L, Gochis DJ, Sobolowski S, Mesquita MDS (2017) Evaluating the present annual water budget of a himalayan headwater river basin using a high-resolution atmosphere-hydrology model. J Geophys Res Atmos 122(9):4786–4807
23. Li H, Beldring S, Xu CY, Huss M, Melvold K (2015) Integrating a glacier retreat model into a hydrological model—case studies on three glacierised catchments in Norway and Himalayan region. J Hydrol 527:656–667
24. Li ZX, He YQ, An WL, Song LL, Zhang W, Catto N, Wang Y, Wang SJ, Liu HC, Cao WH et al (2011) Climate and glacier change in southwestern China during the past several decades. Environ Res 6:1–24
25. Mann HB (1945) Non-parametric tests against trend. Econometrica 13:163–171
26. Maurer EP et al (2002) A long-term hydrologically based dataset of land surface fluxes and states for the conterminous United states. J Clim 15(22):3237–3251
27. Meng FC, Su FG, Li Y, Tong K (2019) Changes in Terrestrial Water Storage during 2003–2014 and Possible Causes in Tibetan Plateau. J Geophys Res Atmos 124:2909–2931
28. Moriasi D, Arnold J, Van LM, Bingner R, Harmel R, Veith T (2007) Model evaluation guidelines for systematic quantification of accuracy in watershed simulations. Trans ASABE 50(3):885–900
29. Motovilov YG, Gottschalk L, Emgeland K, Rodhe A (1999) Validation of a distributed hydrological model against spatial observation. Agric Forest Meteor 98–99:299–304
30. Nijssen B, Lettenmaier DP, Liang X, Wetzel SW, Wood EF (1997) Streamflow simulation for continental-scale river basins. Water Resour Res 33(4):711–724
31. Prasad VH, Roy PS (2005) Estimation of snowmelt Runoff in Beas Basin. India, Geocarto Int. 20(2):41–47
32. Rani S, Sreekesh S (2020) Flow regime changes under future climate and land cover scenarios in the Upper Beas basin of Himalaya using SWAT model. Int J Environ Stud. https://doi.org/10.1080/00207233.2020.1811574
33. Ren Z, Su F, Xu B, Xie Y, Xie Y, Kan B (2018) A coupled glacier-hydrology model and its application in eastern Pamir. J Geophys Res 123(24):13692–13713
34. Sharma V, Mishra VD, Joshi PK (2013) Implications of climate change on streamflow of a snowfed river system of the NorthWest Himalaya. J Mt Sci 10:574–587
35. Singh D, Gupta RD, Jain SK (2014) Study of long-term trend in river discharge of Sutlej river (N-W Himalayan region). Geography, Environment, Sustainability 7(3):87–96
36. Singh P, Bengtsson L (2004) Hydrological sensitivity of a large Himalayan basin to climate change. Hydrol Process 18(13):2363–2385
37. Su F, Zhang L, Ou T, Chen D, Yao T, Tong K, Qi Y (2013) Hydrological response to future climate changes for the major upstream river basins in the Tibetan Plateau. Global Planet Change 136:82–95

38. Sun W, Ishidaira H, Bastola S (2012) Calibration of hydrological models in ungauged basins based on satellite radar altimetry observations of river water level. Hydrol Process 26:3524–3537
39. Zhang Y, Xu CY, Hao Z, Zhang L, Ju Q, Lai X (2020) Variation of melt water and rainfall runoff and their impacts on streamflow changes during recent decades in two Tibetan Plateau Basins. Water 12:3112
40. Zhao Q, Ye B, Ding Y, Zhang S, Yi S, Wang J, Shangguan D, Zhao C, Han H (2013) Coupling a glaciermelt model to the Variable Infiltration Capacity (VIC) model for hydrological modeling in north-western China. Environ Earth Sci 68:87–101

Effect of Seasonal Changes on Water Quality of River Mutha

Chinmay Deore, Rohini More, and Sameer Shastri

1 Introduction

Mutha river originates at a village named Vegare, in the Western Ghats, about 45 km to the west of Pune and has two tributaries, Ambi and Moshi [1]. Mutha river basin after Khadakwasla dam largely consists of moderate to gently sloping plain, with massive rock- basalts domination. Soil cover is thin to moderately thick. Due to hard and massive rock terrain, the majority of the river bed is relatively shallow, rocky and wide. The Mutha river is an important source of drinking water for Pune city, it is dammed twice before it enters the city region, till the water reaches city boundaries it is clean and within acceptable limits but as it progresses forward its water quality goes on degrading due to heavy urbanisation, industrial effluent discharge and unhealthy trends of behaviour of local people residing in its vicinity [2]. On the Mutha river, there are 14 untreated sewage and natural stormwater outfalls which adds up to the river pollution load. Temperature, wind velocity, sunlight hours, precipitation and vegetation cover contribute towards the major factors of the season changes [3]. River metabolism gets affected as these seasonal parameters vary from time to time in turn affecting the water quality parameters like pH, BOD, COD, turbidity, total dissolved solids (TDS), total suspended solids (TSS) and hardness [4].

C. Deore (✉) · R. More · S. Shastri
Sinhgad College of Engineering, Pune, India
e-mail: chinmaydeore@gmail.com

S. Shastri
e-mail: ssshastri.scoe@sinhgad.edu

A. Sil et al. (eds.), *Recent Trends in Civil Engineering*, Lecture Notes in Civil Engineering 274, https://doi.org/10.1007/978-981-19-4055-2_54

2 Material and Methods

2.1 Study Area

Two kilometre stretch of Mutha river is considered in the study area; it starts from S.M. Joshi bridge at coordinates 18°30′35.2"N 73°50′25.1"E and ends at Onkareshwar temple near coordinates 18°31′15.1"N 73°51′02.3"E. Three sampling stations were considered along the two kilometre stretch, in which one sampling station collects wastewater from an open sewage pipe that outfalls in the river directly and two sampling stations collect wastewater from the flowing river. The sampling station was selected on the basis of the pollution load observed. The location of sampling stations selected are shown on the map (Fig. 1) representing red colour arrows and the numbers represent the sampling station number, i.e. S1-open sewer pipe, S2-Flowing River water near Z-bridge and S3-Flowing River water near Onkareshwar temple. Five samples were collected in the 6 months starting from December till May. The seasons which fall in that period were winter (Dec-Jan), spring (Feb-March) and summer (April–May). The seasonal factors like average temperature, average wind velocity, sunlight hours and average precipitation for the day of collections are noted down. The sample collection and testing are analysed by following standard procedures given by APHA [5] and Trivedy and Goel [6]. The parameters that are observed in the wastewater samples collected are DO, pH, COD, BOD, turbidity and hardness.

3 Results and Discussion

3.1 Weather Conditions

See Table 1.

3.2 Results for Open Sewer Pipe (S1)

See Table 2.

3.3 Results for Flowing River Water Near Z-Bridge (S2)

See Table 3.

Fig. 1 Map of study area. *Source* Google maps

Table 1 Weather conditions

Month	Avg temp (°C)	Precipitation (mm)	Sunlight (hours)	Humidity (%)
Dec 2020	20.56	0.16	9.9	58
Jan 2021	22.18	1.19	10	56
Feb 2021	21.71	0	10.5	46
Mar 2021	26.33	0.16	10.1	37
Apr 2021	28.35	0.94	10.4	38
May 2021	27.71	2.01	10.5	48

Source https://www.wunderground.com [8]

Table 2 Parameter variation at S1 station

	Dec 2020	Jan 2021	Feb 2021	Mar 2021	May 2021
DO (mg/l)	1.47	3.8	3.2	2	4
BOD (mg/l)	157.5	140	160	143	70
COD (mg/l)	320	220	246	214	140
pH	7.5	7.5	7.3	7.7	7.5
Turbidity (NTU)	5	8	5	7	12
Hardness (mg/l)	12	9	12	11	8

Table 3 Parameter variation at S2 station

	Dec 2020	Jan 2021	Feb 2021	Mar 2021	May 2021
DO (mg/l)	4.82	0.6	2	1.7	3
BOD (mg/l)	120	220	100	113	80
COD (mg/l)	194	368	160	200	170
pH	7.6	7.2	8	7.8	7
Turbidity (NTU)	8	10	14	10	21
Hardness (mg/l)	28	30	24	34	27

Table 4 Parameter variation at S3 station

	Dec 2020	Jan 2021	Feb 2021	Mar 2021	May 2021
DO (mg/l)	2.56	0.8	1.2	1.4	2.9
BOD (mg/l)	136	100	147	133	90
COD (mg/l)	220	120	220	270	160
pH	7.5	7.7	7.3	6.8	8
Turbidity (NTU)	9	14	15	10	24
Hardness (mg/l)	120	137	126	133	100

3.4 Result for Flowing River Water Near Onkareshwar Temple (S3)

See Table 4.

4 Effect on Wastewater Parameters Due to Seasonal Change

4.1 Effect on COD and BOD

The observations made in Tables 2, 3 and 4 point out towards the direction that changes in temperature, humidity, precipitation, sunlight hours and humidity make a difference and maybe the reason for the change in parameters. As in the winter season (Dec and Jan) it could be observed that at S1, S2 and S3 stations the BOD and COD values are moderately high, Highest BOD value was observed at S2, i.e. 220 mg/l and the highest COD value was also observed at S2, i.e. 368 mg/l. The sunlight exposure to the water was for 10 h, sunlight provides energy for the physiochemical processes in the wastewater. Precipitation observed was little hence it seems to not affect it. During the spring season (Feb-Mar), the temperature rises to 26 °C and the BOD and COD values are the highest in this period. Even increase can be observed at all

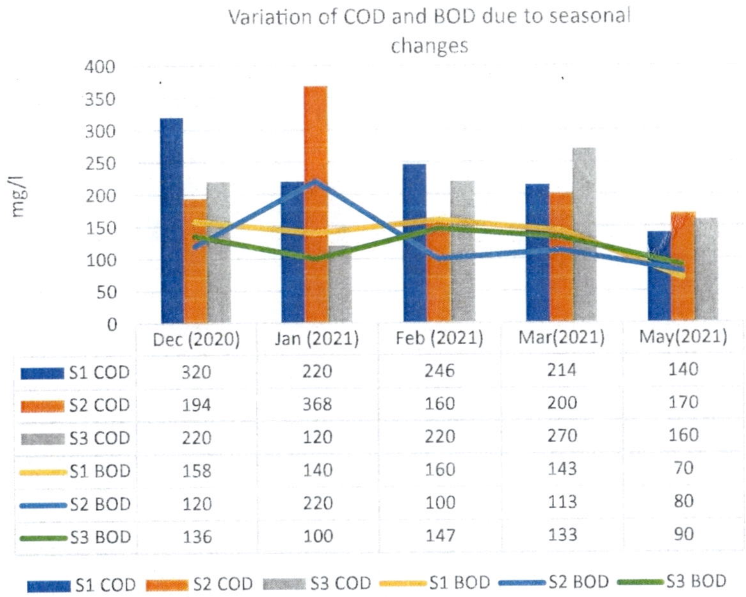

	Dec (2020)	Jan (2021)	Feb (2021)	Mar(2021)	May(2021)
S1 COD	320	220	246	214	140
S2 COD	194	368	160	200	170
S3 COD	220	120	220	270	160
S1 BOD	158	140	160	143	70
S2 BOD	120	220	100	113	80
S3 BOD	136	100	147	133	90

Fig. 2 Variations of COD and BOD at S1, S2 and S3

stations with the highest value of COD and BOD at station S3, whereas precipitation was the lowest which indicates a high concentration of wastewater. It had maximum sunlight hours and least humidity which is also considered as a factor for the increase in the rate of evaporation. In pre-monsoon (May) the COD and BOD values at all stations are the lowest, the temperature increases to the highest in the April month and then starts lowering as the monsoon season starts to approach, with the increase in the precipitation can be observed and is the highest among other months, hence can be considered as an influencing agent in the parameters. The sunlight exposure is also the highest during April and May (Fig. 2).

4.2 *Effect on DO*

Dissolved oxygen is an important parameter that can instantly give the health of a water body. The lowest value of DO can be observed in January for S2 and S3 stations and for S1 station in December, whereas the highest value of DO can be observed for all stations in May, the major difference for the month of May is the increase in the amount of precipitation and increased sunlight hours of 10.5 h, which is the highest amongst all other observed months. Hence it can be a factor of influence for the increase in DO (Fig. 3).

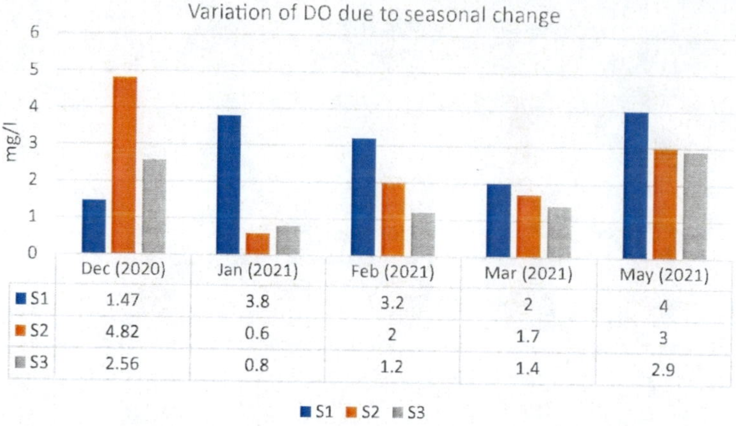

Fig. 3 Variations of DO at S1, S2 and S3

4.3 *Effect on pH*

The pH value indicates the chemical condition of the solution. The wastewater contains microbes that help in disintegrating the pollutants and react with them, these microbes are pH-sensitive as well as temperature-sensitive [7] and hence make a difference in the overall parameters of the wastewater. According to the analysis the most chemically base value of the pH is observed at S2 and S3 in February and May with a value of 8. The most acidic sample was S3 in March with a pH value of 6.8, the change in seasonal parameters might influence the variation of pH (Fig. 4).

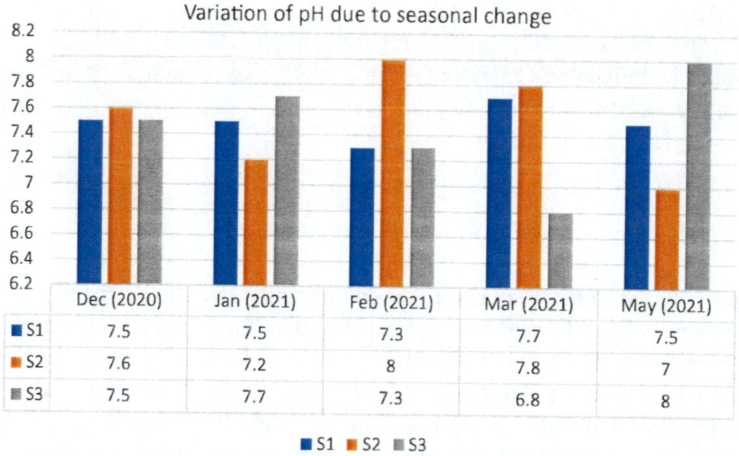

Fig. 4 Variations of pH at S1, S2 and S3

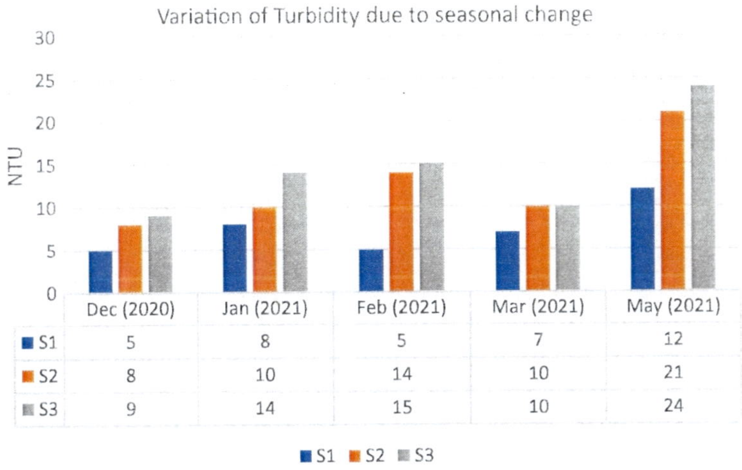

Fig. 5 Variations of turbidity at S1, S2 and S3

4.4 *Effect on Turbidity*

Turbidity is the measure of the relative clarity of a liquid. It is an optical characteristic of water and is a measurement of the amount of light that is scattered by material in the water when a light is shined through the water sample. The lowest value of turbidity can be observed in February with the observed value as 5 NTU at S1 station, when the atmospheric temperature is the highest among other months which are considered for the study, the relative humidity is also the lowest and the precipitation is considerably low. The highest value of turbidity is observed at S3 with the result of as 24 NTU in May, the highest with precipitation among other considered months (Fig. 5).

4.5 *Effect on Hardness*

Hardness is the amount of dissolved calcium and magnesium in the water. Hard water is high in dissolved minerals, largely calcium and magnesium. The lowest hardness value is observed at the S1 station throughout the study with values ranging from 8 to 12 mg/l. The highest value obtained for hardness is at the S3 station throughout the study with values ranging from 100 to 137 mg/l. It was observed that there was a Dhobi ghat situated near the S3 station which influenced the hardness values of the water. The precipitation somehow might influence the values of hardness due to the increased flow of the river water downstream due to rainwater (Fig. 6).

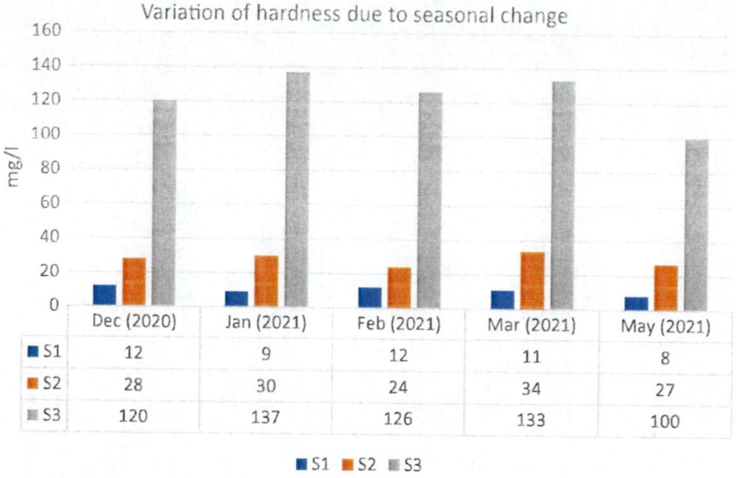

Fig. 6 Variations of hardness at S1, S2 and S3

5 Conclusion

After analysing the samples collected from the respective sampling stations and comparing them with seasonal change factors like temperature, humidity, precipitation and sunlight hours it could be observed that these changing weather conditions may be an influential factor in changing the wastewater parameters at the river Mutha, but the further detailed study is required to be sure about it.

References

1. Mutha River (2021) Wikipedia, February 18. https://en.wikipedia.org/w/index.php?title=Mutha_River&oldid=1007486623
2. Fadtare VV, Mane TT (n.d.) Studies on water pollution of Mula, Mutha and Pawana Rivers in Summer season in the Pune city region, 8
3. Rodríguez-Castillo T, Barquín J, Álvarez-Cabria M, Peñas FJ, Álvarez C (2017) Effects of sewage effluents and seasonal changes on the metabolism of three Atlantic Rivers. The Science of the Total Environment, 599–600, 1108–1118, December 1
4. Xu G, Li P, Lu K, Tantai Z, Zhang J, Ren Z, Wang X, Yu K, Shi P, Cheng Y (2019) Seasonal changes in water quality and its main influencing factors in the Dan River Basin. CATENA 173:131–140, February 1
5. APHA (1998) Standard methods for the examination of water and waste water, 20th edn. APHA, AWWA, WEF, Washington DC
6. Trivedi RK, Goel PK (1987) Chemical and biological methods for water pollution studies. Environmental Publications Karad, India

7. Comber SDW, Gardner MJ, Ellor B (2020) Seasonal variation of contaminant concentrations in wastewater treatment works effluents and river waters. Environmental Technology 41(21):2716–2730, September 18
8. Pune, India weather history | weather underground. Accessed July 10, 2021. https://www.wunderground.com/history/monthly/in/pune/VAPO/date/2020-12

Effect of Shear Connectors in the Deformability of Circular Plated Joints in Single Layered Lattice Structure

Sajith and Raghavan Ramalingam

1 Introduction

Large span single layered lattice structures like domes have been in use for decades as roofing solution for covering large open areas like stadiums. These structures are more ergonomically designed, structurally safe, and in economical point of view, a viable option for large span covering structure. Various types of domes are currently being used among which geodesic domes are popular for its simple design and construction costs. The inheritance for their rigidity from the arrangement of struts into triangular pattern makes geodesic domes a more practical solution for large span roofing. Joints of the dome or any other structural system is the critical component of the dome as they are structurally the weakest point of the dome. Among the various types of joints, circular plated gusset joints are the most common for their simplistic design. Figure 1 shows a typical circular plated joint. Shear connector have been seen to improve the overall performance of any steel or aluminum connections. The use of shear connectors in circular plated joints have recently got more importance as the use of a shear connector would suffice to improve the performance rather than increasing the geometric parameters of the circular plated joint, thereby saving weight and construction costs.

Recent studies have focused on various performance parameters of the circular plated joints and joints with shear connectors but the performance of the joints for their axial deformations are limited. Ihaddoudène et al. [2] conducted a study on effect of semi-rigid joint on steel structures. A numerical analysis on a mechanical model was conducted to incorporate the effect of semi-rigid pin joints and buckling load. Author's work signifies that a numerical analysis on the semi-rigid joint is important as the steel structures often designed are assumed to have a rigid joint. López et al.

Sajith (✉) · R. Ramalingam
Department of Civil Engineering, National Institute of Technology, Tiruchchirappalli, Tamil Nadu, India
e-mail: spurushothaman576@gmail.com

A. Sil et al. (eds.), *Recent Trends in Civil Engineering*, Lecture Notes in Civil Engineering 274, https://doi.org/10.1007/978-981-19-4055-2_55

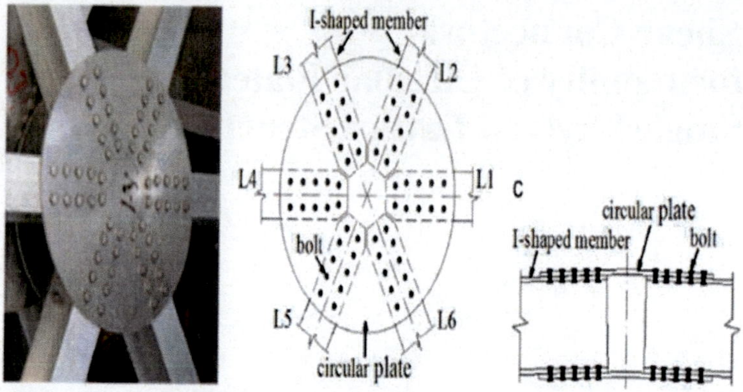

Fig. 1 Typical circular plated joint [1]

[3] studied about the geometric nonlinearities of single layered domes. He stated that the dome's stability depends on geometric parameters and joint rigidity as well as the position of the loads. The study by Tokovyy [4] on the effect of uniform pressure on diametrically opposite segment of a thin annular disk enabled to understand the mises stress pattern on a circular disk in similar load conditions in comparison to the thin annular disk. Similar to the Ihaddoundene [2], the work by Fan [5] showed the effect of axial compressive force on hollow ball socket joint's deformability. This highlighted the significance of considering the combination of real loading conditions of the joint. Ahmadizadeh [6] conducted a similar study as Fan [5] on hollow ball socket joints but with biaxial loading on the joint. Various force ratios between the loading members have caused varying slopes in load deformation curves, which can be used to apply the stiffness on global analysis of the dome while assuming specific force ratio on a particular location of the joint of the dome. Xiaonong et al. [7] have studied the effect of shear connectors on the circular plated joint on its bending stiffness and has found out that a circular plated joint with either thinner plate or large compressive force will lead to local buckling at the central area of the circular plate in the case of lateral loading on the joint but the authors did not study the deformations of the joint under axial loading conditions A paper on Component method on these joints were done by Liqiu Qiu et al. [1].

This paper focus on studying the deformability of the circular plated joints with shear connectors subjected to axial loading condition by finite element analysis using ABAQUS software. The circular plated joint is subjected to the tensile and compressive loads from the connecting members. Geometric parameters and loading conditions are changed to understand the deformability of the joint An FE model was generated based on the Circular plated joint used by Xaionong et al. [7]. Geodesic domes under application of a vertical load on the apex of the dome generates tensile and compressive members loads at the joint on the ring element of the dome. Joint at the apex have the compressive loads for all connecting members. The FE models are analyzed using loading conditions of the joint in the ring element of the dome, with tensile and compressive member forces. The comparison of the deformability of the joint with shear connector and results of the joint without shear connector from finite element analysis of the circular plated joint shows the effect of the shear connectors in the circular plated joint.

2 Finite Element Analysis of Circular Plated Joints with Shear Connector

2.1 Finite Element Analysis of Circular Plated Joint Without Shear Connector

The FE model of 6 mm based on the aluminum alloy gusset joint used in the study by Xaionong et al. [7] was validated for bending moment (Table 1). Figure 2 shows the schematic diagram of the joint. The diameter of the circular disk is 240 mm, connecting member as I member with dimensions 100 × 50 × 5 × 4 mm, bolts of size M6 were used to model the joint. The material property of the components of the joint is given in Table 2.

Table 3 shows the initial bending stiffness calculated from FEA and from the study by Xiaonong Guo et al. [7]. The relative error in the calculation of the initial bending stiffness from the numerical analysis and result obtained from the equation formulated by component method by Liqui Qui et al. [1] was 15.8% (Eq. 1). E_p denotes the elastic modulus of plate, E_b denotes the elastic modulus of the member, A_c denote the total area in contact in the joint, t_b and t_p denotes the thickness of the connecting beam and plate, respectively, R–R_c denotes the distance between the

Table 1 Designation of FE models

Plate thickness	Shear connector	Designation
6 mm	None	A
6 mm	4 mm	A1
6 mm	6 mm	A2
5 mm	None	B
5 mm	4 mm	B1
5 mm	6 mm	B2
2.7 mm	None	C
2.7 mm	4 mm	C1
2.7 mm	6 mm	C2
2.25 mm	None	D
2.25 mm	4 mm	D1
2.25 mm	6 mm	D2

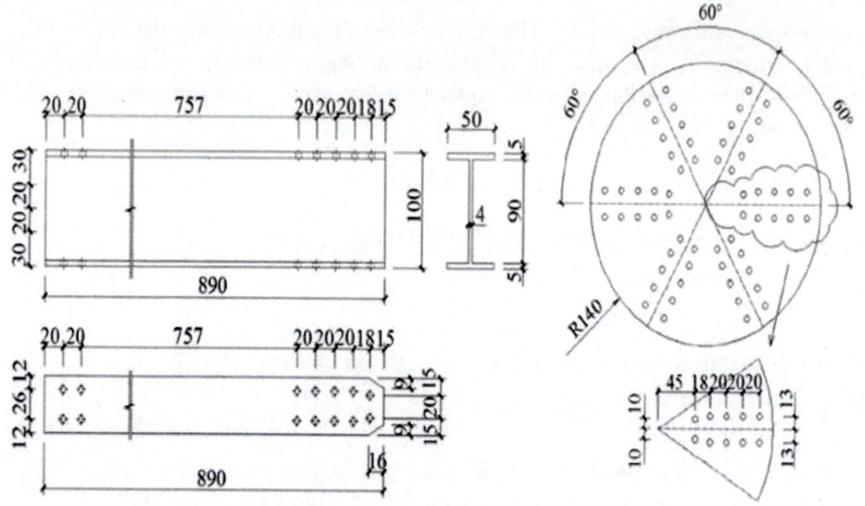

Fig. 2 Schematic diagram of the circular joint [1]

Table 2 Material data of the joint components

Component	Material	E(MPa)	$f_{0.2}$ (MPa)	f_u (MPa)
Plate and Shear Connector	Aluminum 6063-T5	69,088	184.4	213.58
Beam members	Aluminum 6063-T5	65,364	177.4	206.8
Bolts	Steel	206,000	470	725

center of the joint to the center of the bolt connection, n is the no of bolts and h is depth of the member.

The loading conditions of the joint connecting the ring members of the dome is considered for the study. The joints connecting the ring members as shown in Fig. 3 are subjected to four compression loads and two tensile loads from the members. A force ratio was defined between the loads coming on to the joint, which is the ratio of the net tensile loads to the net compressive loads, similar to the study by Ahmadizadeh [6]. Force ratios of 1, 0.75, 0.5, and 0.25 were taken. The net compressive force on the joint was taken as the minor force and the tensile forces are assumed as the major forces with the in the direction acting as shown in Fig. 4. The loading pattern and the boundary conditions of the FE model is shown in Fig. 5. The preload of all bolts was calculated in accordance to IS800:2007; 10.4.3. The FEA models are meshed with C3D8r elements with total mesh count of average of 2.65 × 10^6 elements. A

Table 3 FEA and experimental results

FEA initial bending stiffness	441 kNm
Experiment initial bending stiffness	421 kNm

parametric analysis was conducted by FEA on the circular plated joint. All contact surfaces were modeled as rough surfaces with 0.3 as friction coefficient.

$$K_f = \frac{1}{\frac{1}{0.76E_p t_p h^2} + \frac{t_b}{0.00035 E_e \mu n h^2 A_c} + \frac{(R-R_c)}{1.14 E_b I}} \tag{1}$$

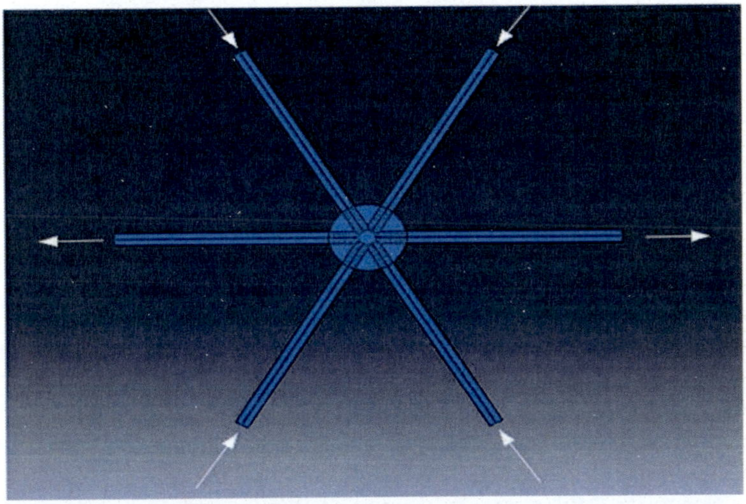

Fig. 3 Forces acting on a joint other than at the apex

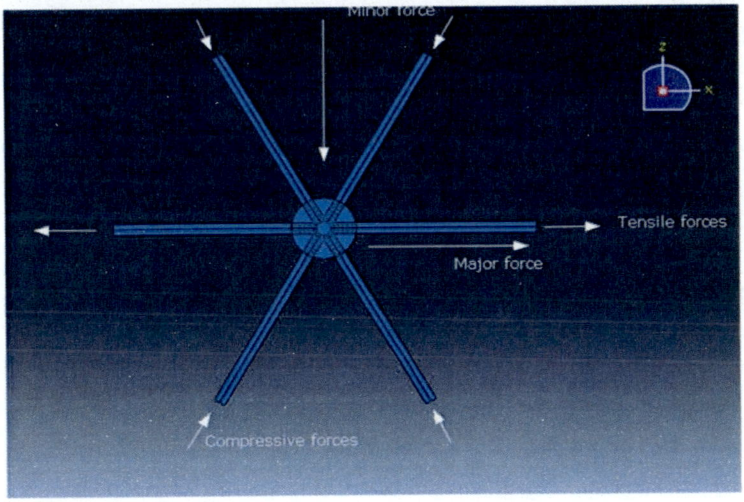

Fig. 4 Loading pattern of the joint

Fig. 5 Boundary and loading conditions of the FE model

2.2 *Finite Element of Analysis of Circular Plated Joint with Shear Connector*

The same model from the previous study, with shear connectors were analyzed using the same loading and boundary conditions. The thickness of the shear connector is varied for parametric study. The shear connector is of size 145 × 75 mm. The shear connector was introduced between the members carrying the tensile loads via joining the flanges of the members in the joint zone using two plates connected using six M6 bolts. Figure 6 shows a typical joint with shear connector. A 6 mm and 4 mm shear connector were used for the analysis. Table 1 also shows the designation of the models used along with the thickness of shear connector and the circular plate. FE models A1, A2, B1, B2, C1, and C2 follows the same force ratios and loading pattern as shown in Fig. 6.

Circular plated joint with shear connector was analyzed by varying the thickness of the circular plate and the shear connector (Table 1). Figure 7 shows the von mises stress plot of the model A2 with 6 mm circular plate and 6 mm shear connector. The stress path of the joint allows the load to be distributed to the shear connector. The

Fig. 6 Circular plated joint with shear connector [1]

Fig. 7 Von Mises stress plot of FEA model A2

stress plot also shows the higher stresses are concentrated on the bolt hole region, but the majority of the elements between the bolt holes are in elastic region. This causes the displacement of the member to be in linear with the applied load. The displacement of socket joint was higher in the direction perpendicular to the load applied in a study by [6]. Such deformations were also seen in the FEA results, thus the deformations of the models where understood.

3 Results and Discussion

On post processing the results obtained from the FE analysis, it was found out that the joint deformability is related to the thickness of the shear connector. The compressive deformation for joints A, B, C, and D on the direction of the applied load increases with increasing force ratio as the net compressive force aid in more deformation on the joint in its radial direction parallel to the tensile load. But as the thickness of the circular plate reduces the joint deforms more on the tensile force direction, making the joint stiffer to deform in the compressive load direction. Similar trend was also seen on the compressive deformability of joints with shear connector (Table 1). The thickness of the circular plate of the joint as well as the shear connector influences the performance of the joint.

As the shear connector thickness increases, the deformation in the compression load direction also reduces as the shear connector is connected to the member carrying the tensile load. This makes the joint stiffer in the tensile load direction thereby the compressive deformations are affected. For example, Figs. 8 and 9 show the force displacement curves in the compressive loading direction for force ratios 1 and 0.25 for thick plated (plate thickness > 5 mm) and thin plated joints (plate thickness

< 3 mm). The compressive deformations in the load direction had improved by maximum of 29% and 35.6% for thick and thin plated joints with shear connector, respectively, in comparison with the joint without shear connector and when the shear connector thickness increase from 4 to 6 mm the average reduction of the deformations was 6% for thick plated joints and 8.8% for thin plated joint. The effect of shear connector predominant when the circular plate thickness reduces. The shear connector makes the joint stiffer in the tensile load direction thus making it stiffer than the joint without shear connector. This trend can be seen in Fig. 9a. But as the force ratio reduces this effect diminishes. There is only a small variation in the performance ṇof the joint with 4 mm and 6 mm shear connector as the thickness of the joint varies only by 2 mm. Therefore, the deformations are close, for example Fig. 8a, A1, A2 and B1, B2 have very close values.

When the force ratio increases, the tensile deformability of the joint also increases. The deformations increase with decrease in thickness of the joint, thin plated joints C, D deforms the most as the thickness of the plate causes its reduced stiffness in the direction of the tensile load. The addition of the shear connector improves the deformability of the joint with increasing thickness of the shear connector. The deformations along the tensile load reduces maximum of 17.6% for thick plated joint and the thin plated joints deformations reduces by a maximum of 23.8%. For example, Figs. 10 and 11 shows the graphs of tensile deformation of the joint with thick and thin plated joint respectively for force ratios 1 and 0.25. The thick plated joints with shear connector have less effect in its deformability in comparison to the thin plated joints with shear connector. The thick plated joints A and B are stiffer in the direction of the tensile load in comparison with the thin plated joints C, D. Therefore, the deformations of A1, A2 and B1, B2 are very close. The experimental investigation previous done by Xiaonong et al. [7] also shows that the bending resistance of the 2.25 mm plated joint was improved with stiffener by 21.53%. Therefore, this trend in deformability can be understood. The reduction in the deformations in comparison with joints with shear connector and joints without shear connector is tabulated in Table 4.

4 Conclusions

The FE analysis of the circular plated joint with shear connector showed improvements in the deformability as compared to the joint without shear connector. The compressive deformations of the joint were improved by an average of 27% for thick plated joint (thickness of plate ≥ 5 mm) and 31% for thin plated joints (thickness of plate < 5 mm). When the plate thickness reduces the reduction in deformability was also increased. The thickness of the shear connector also played a role in reducing the deformability of the joint. The tensile deformations of the joint improved more for joints with the thin plated joints. A maximum of 18% reduction in the tensile deformation was recorded for thick plated joints (thickness of plate ≥ 5 mm) and 24.6% reduction for thin plated joint (thickness of the plate < 5 mm). In geodesic domes,

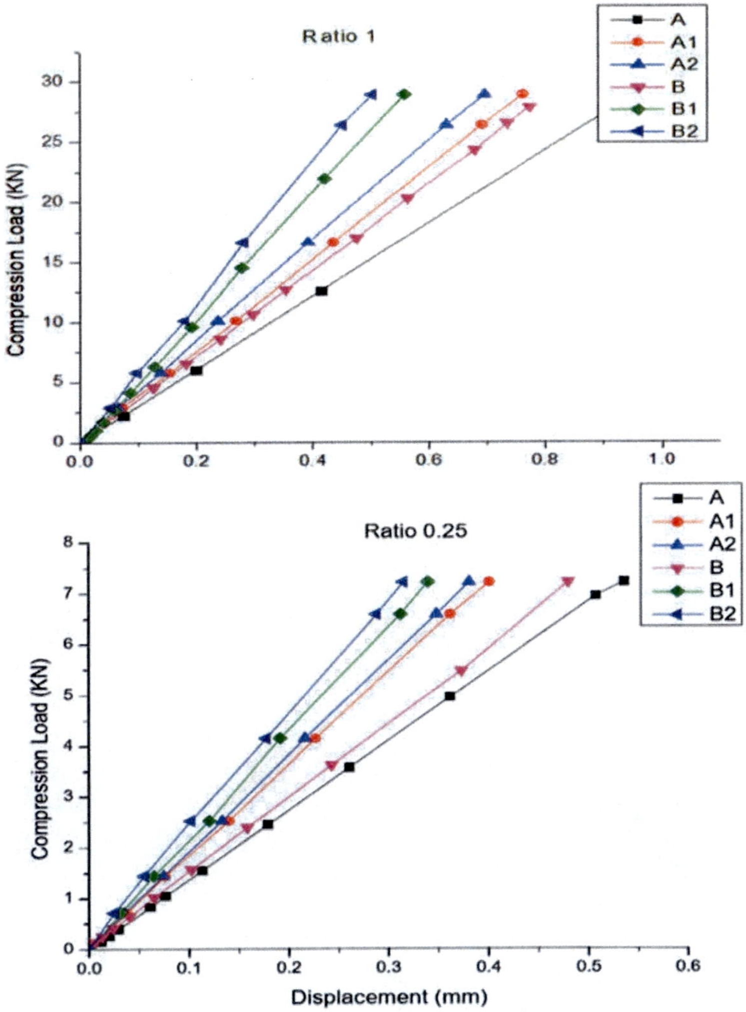

Fig. 8 Graphs showing the compressive deformation of the thick plated joints

the addition of a single shear connector between the ring members can improve the performance of the joint by a good margin by increasing the overall stiffness of the circular plated joint. These results also shows that the performance of the thin plated joints can be improved using shear connectors without increasing the thickness of the circular plate, which in turn can help in weight reduction of the dome.

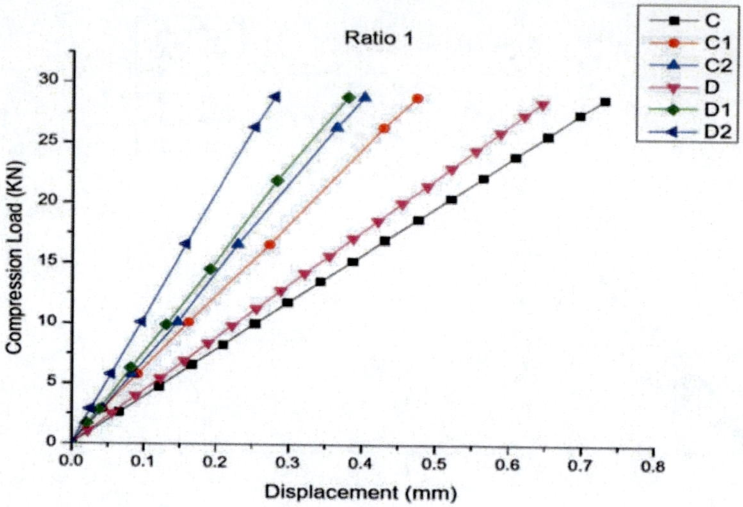

(a)

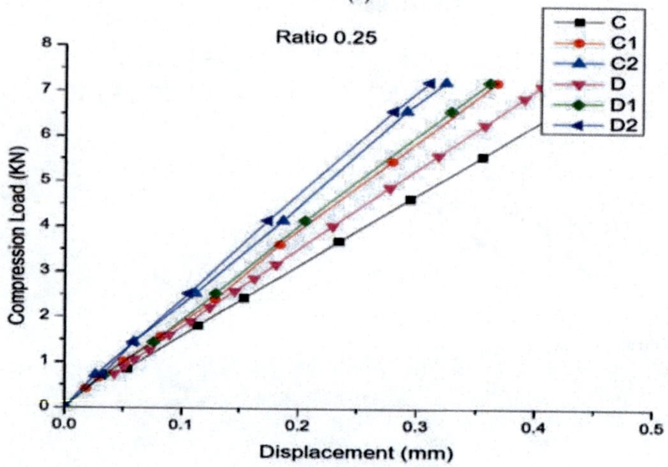

(b)

Fig. 9 Graphs showing the compressive deformation of the thin plate joints

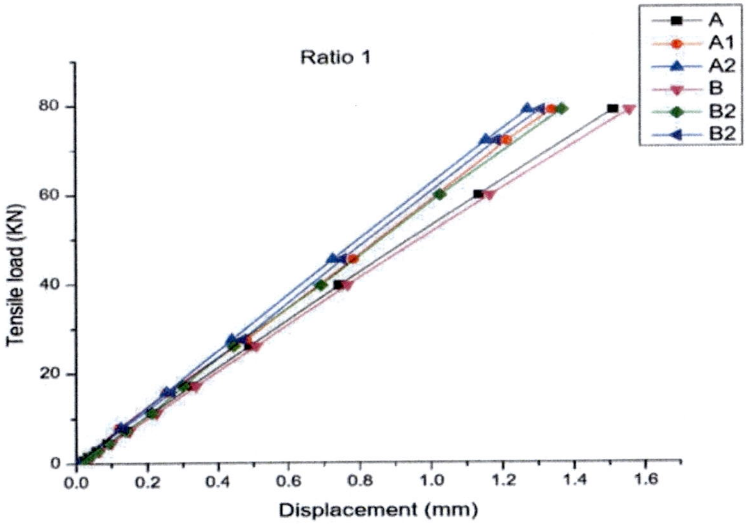

(a)

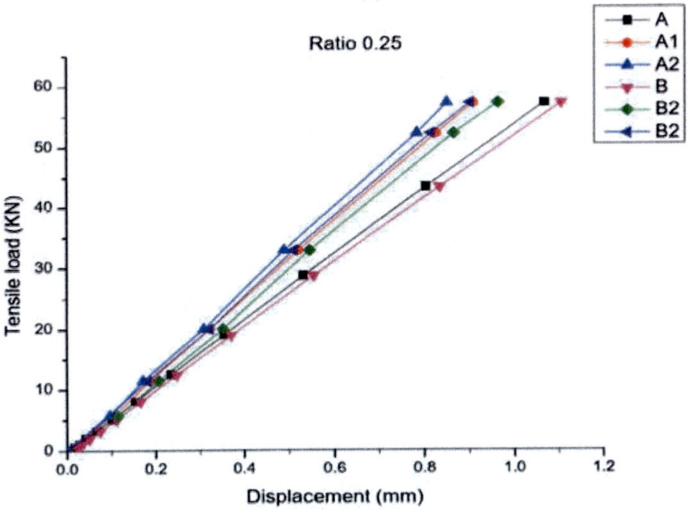

(b)

Fig. 10 Graphs showing the tensile deformations of the Thick plated joints

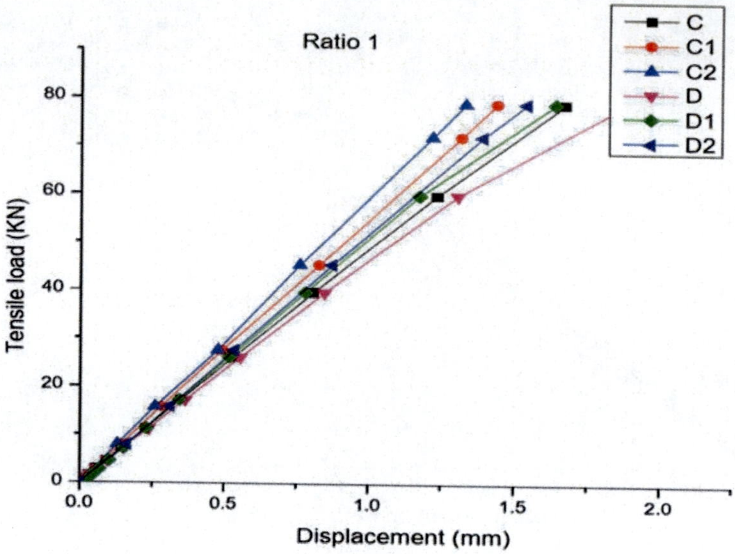

(a)

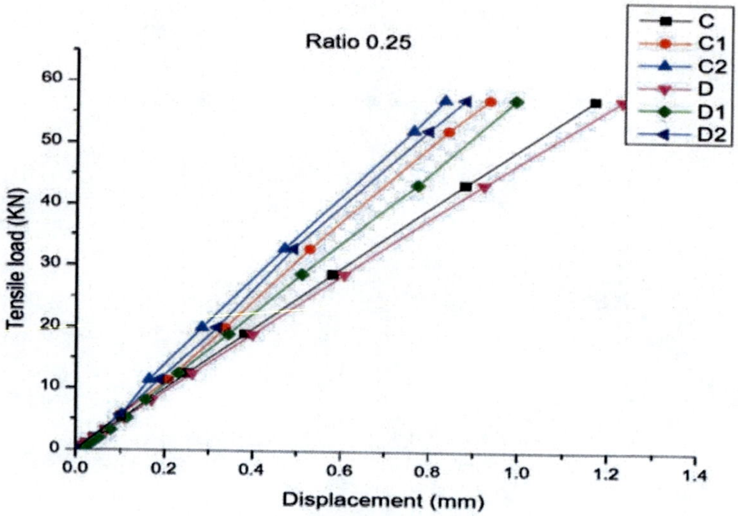

(b)

Fig. 11 Graphs showing the tensile deformations of the Thin plated joints

Table 4 Reduction in the deformation of the joints with shear connector (in percentage)

	Deformation	4 mm shear connector (%)	6 mm shear connector (%)
6 mm	Compression	22.6	27.8
	Tension	13.1	18.0
5 mm	Compression	28.5	32.1
	Tension	13.1	17.1
2.7 mm	Compression	26.6	34.7
	Tension	16.0	23.2
2.25 mm	Compression	27.2	36.9
	Tension	17.1	24.6

References

1. Guo X, Xiong Z, Luo Y, Qiu L, Huang W, Application of the component method to Aluminum alloy gusset joints. Adv Struct Eng, 18(11):1931–1946
2. Ihaddoudène ANT, Saidani M, Chemrouk M (2008) Modeling of steel frames with semi rigid Joints. Int J Appl Eng Res 3(7):955–967. ISSN 0973-4562
3. López A, Puente I, Serna MA (2007) Numerical model and experimental tests on single-layer latticed domes with semi-rigid joints. Comput Struct 85(7–8):360–374
4. Yu V, Tokovyy KM, Hung CCM (2010) Determination of stresses and displacements subjected to diameterical compression. J Math Sci 165(3)
5. Fan F, Ma H, Chen G, Shen S (2012) Experimental study of semi-rigid joint systems subjected to bending with and without axial force. J Constr Steel Res 68(1):126–137
6. Ahmadizadeh M, Maalek S (2014) An investigation of the effects of socket joint flexibility in space structures. J Constr Steel Res 102:72–81
7. Guo X, Xiongn Z, Luo Y, Qiu L, Liu J (2015) Experimental investigation on the semi-rigid behaviour of aluminium alloy gusset joints. Thin Walled Struct 87:30–40

One-Dimensional Steady Flow Hydraulic Model for Flood Inundation Mapping

Rupam Gayan and Parthajit Roy

1 Introduction

On the planet, one of the most incessant and crushing common perils is the flood. Obtaining an overall view of the flood and the extent of the affected area during a flood is a major challenge. A flood can be characterized as a high phase of a waterway because of which stream water floods the banks and immerses the minor terrains. Flood usually occurs due to severe storms and sometimes due to a combination of both storm and melting of snow in the catchment. Flood may also occur due to river course shifting or due to earthquakes which causes bank erosion. Breaching of the flood banks and blockage of the river also causes a flood. Loss of property and life, disruption of socio-economic activities, displacement of people, loss of agricultural land damage to crops, famine, epidemic diseases, etc. are some of the damages caused due to flooding. In India, flood occurs in various regions, and vast areas have been swept by flood. Regions of Brahmaputra, Godavari, Narmada, Tapti, etc. are frequently swept by flood. To plan water-powered structures, the most significant prerequisite is configuration flood size. Different measures can be embraced to control the flood to limit the harms brought about by obliterating flood.

Flood inundation maps are a set of maps that speaks to the spatial degree and profundity of flooding because of a scope of levels of water in a waterway or stream. Flood inundation maps or flood forecast information in a graphical format helps the disaster relief officials and emergency managers to be better prepared for flood problems. Hence, it is very important to provide flood-related information in graphical formats to the public, disaster relief officials, emergency managers, city planners, etc. that can be easily read. Flood inundation mapping is a process to define an area covered by water. Flood immersion mapping should be possible by the mix of geographic data framework, hydrologic models and water-powered models.

R. Gayan (✉) · P. Roy
Department of Civil Engineering, NIT Silchar, Silchar, Assam, India
e-mail: rupamgayan8@gmail.com

A. Sil et al. (eds.), *Recent Trends in Civil Engineering*, Lecture Notes in Civil Engineering 274, https://doi.org/10.1007/978-981-19-4055-2_56

The Barak basin is located amidst 89°50' E to 94°0' E longitudes and amidst 22°44' N, 25°58' N scope. The Barak basin has been divided into three sub-basins and 77 watersheds. There are many hydraulic structures in the basin especially in Manipur and Tripura which are beneficial to the people. There are no hydroelectric power projects in the selected area for study purposes. The natural river system is complex, nonlinear, time-variant and continuous. The input and output observations are stable, discrete or dynamic. Depending upon the catchment size, climate, drainage density, input–output time scale, the river system may be static (Fig. 1).

The Barak waterway reach from A.P Ghat to B.P Ghat has been selected as the study area. In between these two ghats, Jatinga, Dhaleswari and Katakhal are the tributaries of the Barak River. The Central Water Commission (CWC) has working gauging stations at Annapurna Ghat, Badarpur Ghat and Matijuri. Jatinga and Dhaleswari tributaries have no working gauging stations. So, in between the above-mentioned reach of the Barak River, the Katakhal River till Matijuri has been considered as it has a gauging station at Matijuri. The data of the above-mentioned gauging stations is available at Meghna Investigation Division, Shillong. Discharge, gauge, cross-sectional area, Manning's harshness coefficient and so on are accessible at those three destinations. The hydraulic model is prepared for the above-mentioned reaches of the Barak River and its tributary Katakhal. Upstream boundary conditions

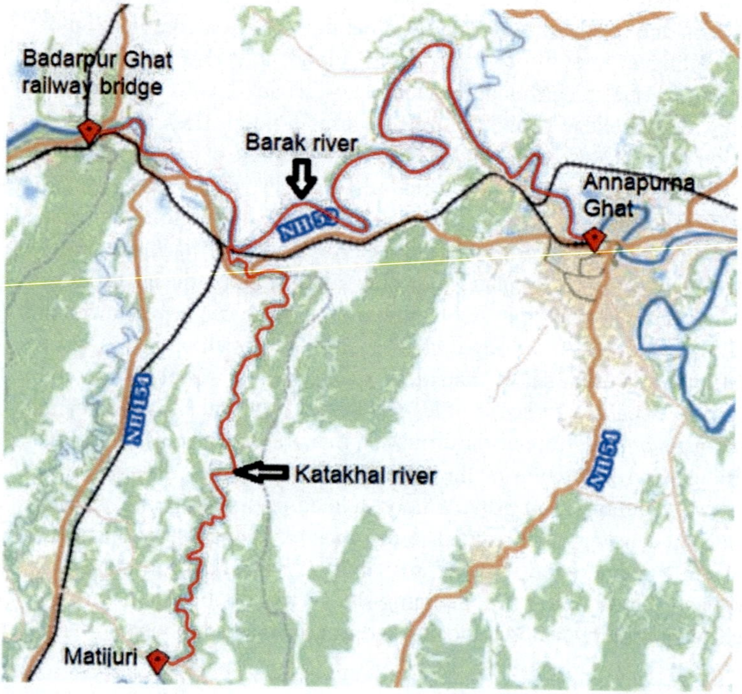

Fig. 1 Selected reach of Barak River and its tributary Katakhal. *Source* https://bhuvan.nrsc.gov.in/bhuvan_links.php

will be provided at Annapurna Ghat and Matijuri and downstream conditions will be provided at Badarpur Ghat.

2 Methodology

For this study, four Digital Elevation Models are downloaded from the USGS Earth Explorer developed by NASA. The DEMs are Shuttle Radar Topographic Mission (SRTM) V3.0 data and have a spatial resolution of 30 m. The four DEMs cover the entire Barak valley region including some portions of Bangladesh and the neighbouring areas of Barak valley. These DEMs are to be processed to obtain the DEM of the study area. The four downloaded DEMs are merged to form one single DEM in Arc-GIS by using the GIS tools. This is done because the study area covers some portions of all four DEMs. Now to obtain the DEM covering study zone, the shapefile covering the examination zone is necessary which can be used as a musk over the merged DEM to extract the required DEM. The selected river reaches or the study area lies in both Cachar and Hailakandi districts and so the shapefiles are required for preparing DEM which covers study zones. The shapefile representing the shape of country India was downloaded from the Bhuvan website developed by the Indian Space Research Organization (ISRO) and Arc-GIS was used to use this shapefile to obtain the required shapefiles. In Arc-GIS, the shapefiles of Cachar and Hailakandi districts were extracted from the shapefile of India map by using the necessary tools.

The shapefiles of Cachar and Hailakandi districts were merged and dissolved into one shapefile using Arc-GIS tools. This is the shapefile of the study area which includes the selected river reaches. This was done so that this shapefile can be utilized for obtaining the DEM of the examination region.

Using the dissolved shapefile shown in Fig. 2 as a mask over the merged DEM, the DEM regarding examination areas was extracted applying Arc-GIS tools. In Fig. 3, the examination zone's DEM has been shown along with the shapefile of the selected river reaches. This river shapefile was extracted from the Bhuvan website developed by ISRO by tracing the selected reaches and then extracting it as a shapefile. This shapefile is shown in the figure to highlight the study area over the DEM.

The extracted DEM was imported to HEC-RAS after processing in Arc-GIS. The projection was set for the model by providing the projection file of WGS 1984 UTM Zone 46 N and the DEM was converted to DTM (Fig. 4). This was done because the river geometry can be extracted from a DTM. Thereafter, the river geometry can be used for flood simulation.

The centre lines of all the reaches are drawn on the DTM in RAS MAPPER starting from upstream and ending downstream and they are connected at the junction point by providing a junction. Thereafter, the bank lines of the reaches are drawn from upstream to the downstream end. First, the left bank lines are drawn and then the right bank lines. After that, the flow path lines are drawn for the reaches by the same procedure as the bank lines from upstream to downstream. After drawing the bank lines, centre lines and flow path lines, the cross sections have been generated for all

Fig. 2 Dissolved shapefile of Cachar and Hailakandi districts

river reaches at a spacing of 100 m, and the length of the cross-sectional lines was provided as 2000–3000 m for the Barak River and Katakhal River. However, some of the cross-sectional lines generated showed error as they were crossing each other or they were intersecting more than one centre line or two bank lines or two flow paths. Those lines were deleted, and in those areas, cross-sectional lines were drawn manually keeping in view the rules (Fig. 5).

The below Fig. 6 shows the geometry data of the river reaches extracted from the DTM. Here, each of the cross sections, centre lines, flow paths and bank lines has acquired the required information from the DTM such as lateral and elevation coordinates, shape, length, etc. These data help us to create the river geometry from the DTM.

The next step is to assign harshness coefficient to the cross-segments in the geometry, thereafter provide the steady flow data, i.e. flow rates to the reaches and define the downward boundary condition. After following all the remaining steps, the simulation can be done to obtain the flood inundation maps.

For one-dimensional steady flow simulation, the discharge values for the upper reach, lower reach of the Barak waterway and the Katakhal waterway was given as 3527, 5027 and 1500 m^3/sec, respectively. The Manning's roughness coefficient for left and right banks and the main channel for all the streams were given as 0.06, 0.06 and 0.0035, respectively. The downstream limit condition regarding the Barak waterway has been set as a normal depth slope. The flood immersion maps were obtained by providing the above-mentioned information.

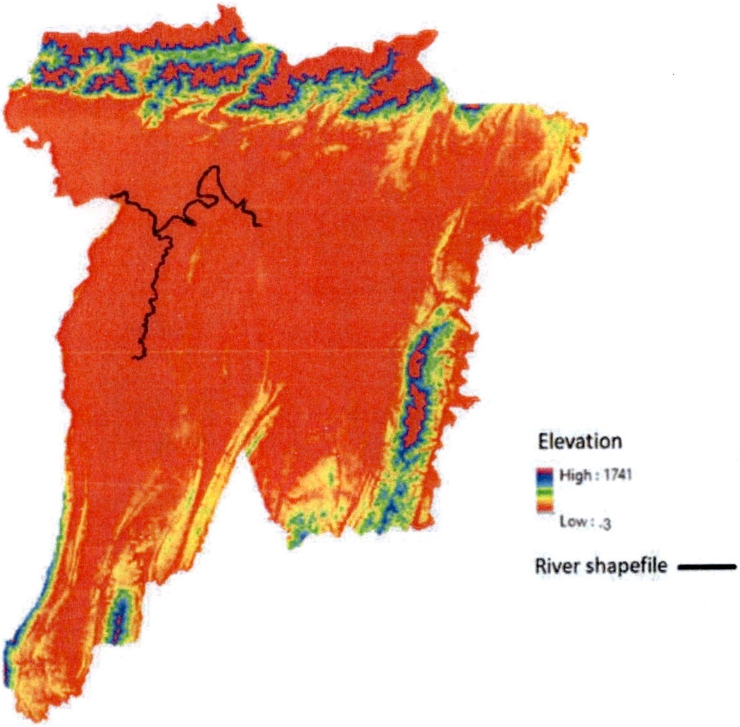

Fig. 3 DEM covering Cachar and Hailakandi districts along with the shapefile of selected river reaches

3 Results and Discussions

The simulation was carried out after providing all the required data to the extracted geometry of the study area. Flood inundation maps were observed for the provided data in the study area (Fig. 7).

In the flood inundation depth map, the variation of colour represents different values of depth of water in metres. The flood inundation depth values of greater than 0 m and a maximum of less than14 metres were found. The maximum depth occurs at the main channels, and the minimum depth is found along the left and right over banks of the river reaches (Fig. 8).

In the flood inundation velocity map, the variation of colour represents different values of velocities of water in metres per second. A maximum velocity of less than 2.35 m/s was found in the inundation map, and a minimum velocity of greater than 0.01 m/sec was found.

For the Barak River upper reach, a most extreme speed of 2.2 m/s and a base speed of 0.28 m/s were found in the primary channel. In the left overbank, a most extreme speed of 0.87 m/s and a base speed of 0.12 m/s were found. In the right overbank, a

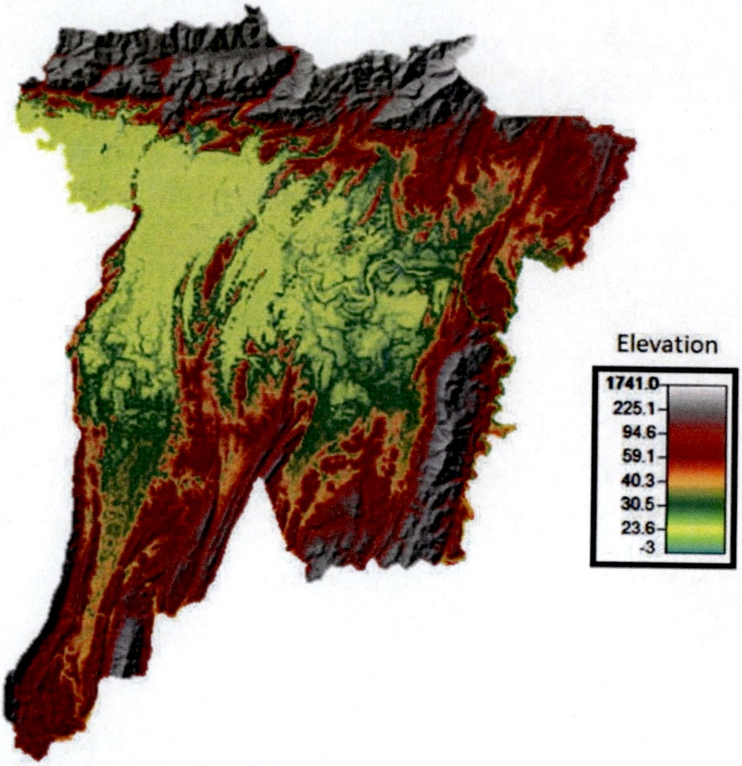

Fig. 4 Digital Terrain Model (DTM) covering Cachar and Hailakandi districts

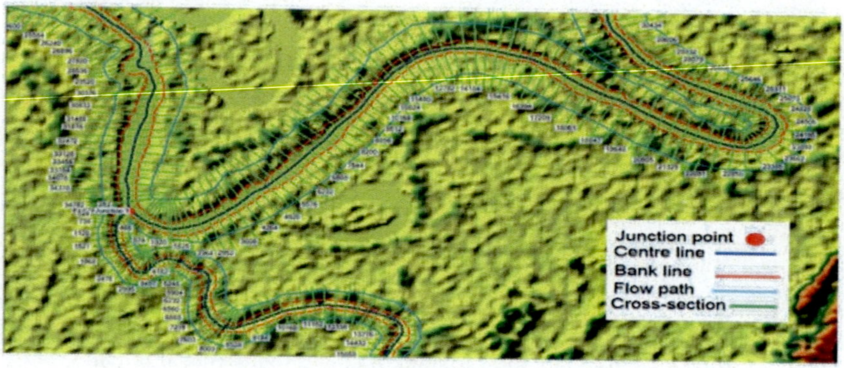

Fig. 5 Centre lines, bank and flow path lines, cross sections and junction on DTM

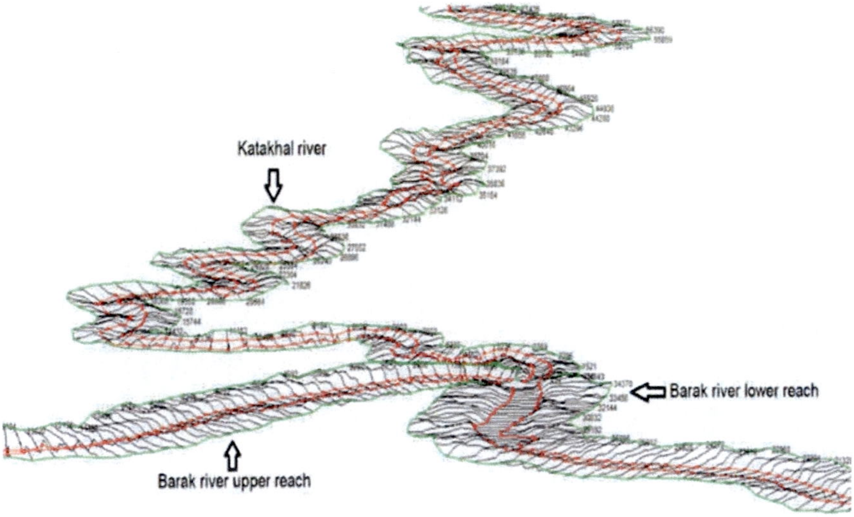

Fig. 6 Extracted geometry of the Barak River and its tributary Katakhal at the junction

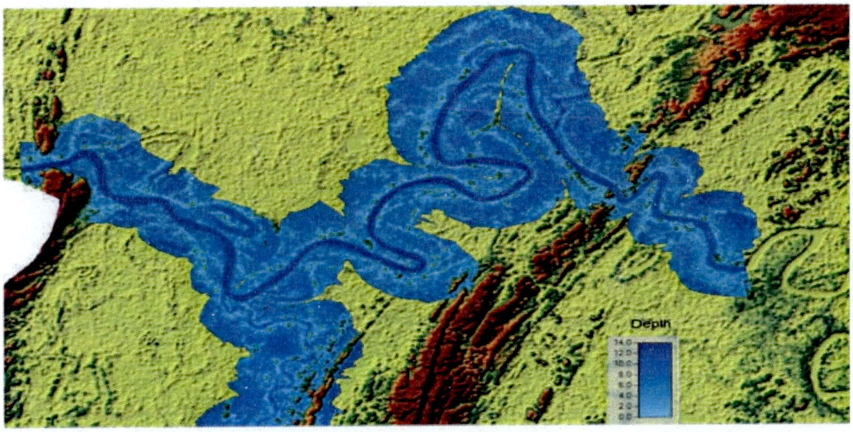

Fig. 7 Flood inundation depth map of the Barak and Katakhal rivers in metres

most extreme speed of 0.78 m/s and a base speed of 0.13 m/s have been found for the Barak River upper reach (Fig. 9).

For the Barak River lower reach, the greatest speed of 2.05 m/s and a base speed of 0.41 m/s were found in the fundamental channel. In the left overbank the greatest speed of 0.78 m/s and a base speed of 0.15 m/s have been found. In the right overbank, a most extreme speed of 0.54 m/s and a base speed of 0.17 m/s were found for the Barak waterway upper reach.

For the Katakhal stream, a most extreme speed of 1.02 m/s while a base speed of 0.10 m/s were found in the primary channel. In the left overbank a greatest speed of

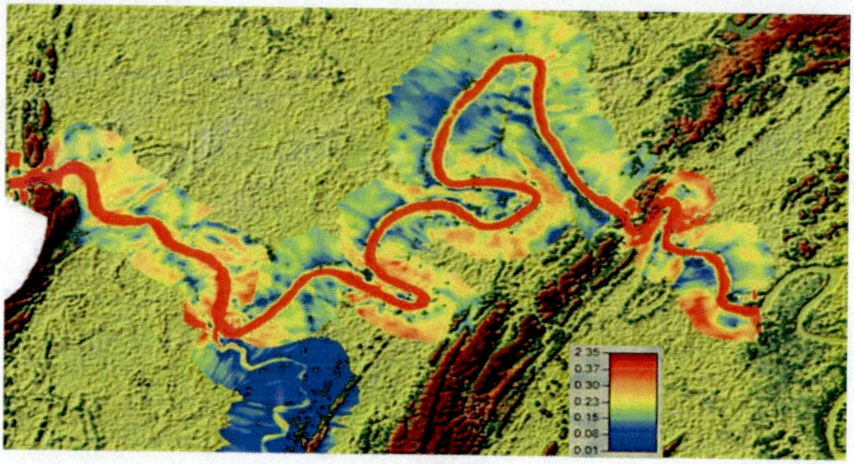

Fig. 8 Flood inundation velocity map for the Barak and Katakhal rivers in m/sec

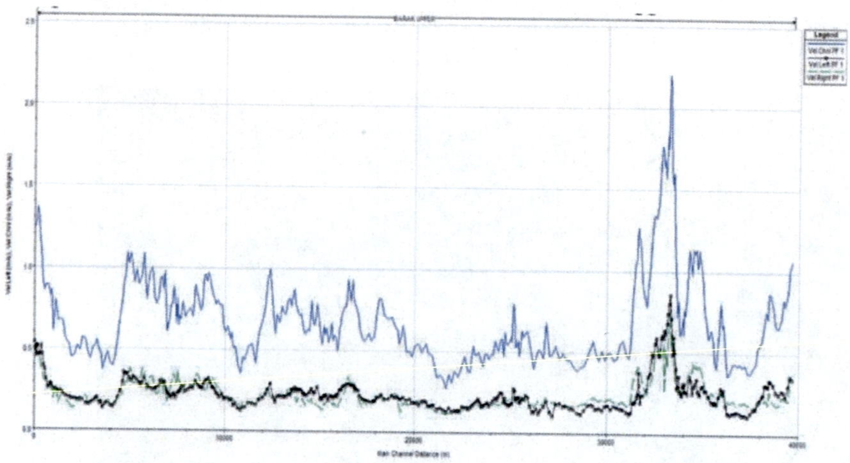

Fig. 9 Velocity profile of Barak River upper reach

0.64 m/s while a base speed of 0.06 m/s have been found. In the right overbank a most extreme speed of 0.51 m/s and a base speed of 0.05 m/s were found.

In Figs. 9, 10 and 11, the graph starts from the downstream end to the upstream end. In the graphs, the blue graphs have been used to represent the profile of average velocity in main channels, black graphs have been used to depict profiles of average velocities in the left overbank and green graphs indicate the profiles of mean velocities in the right overbank (Figs. 12 and 13).

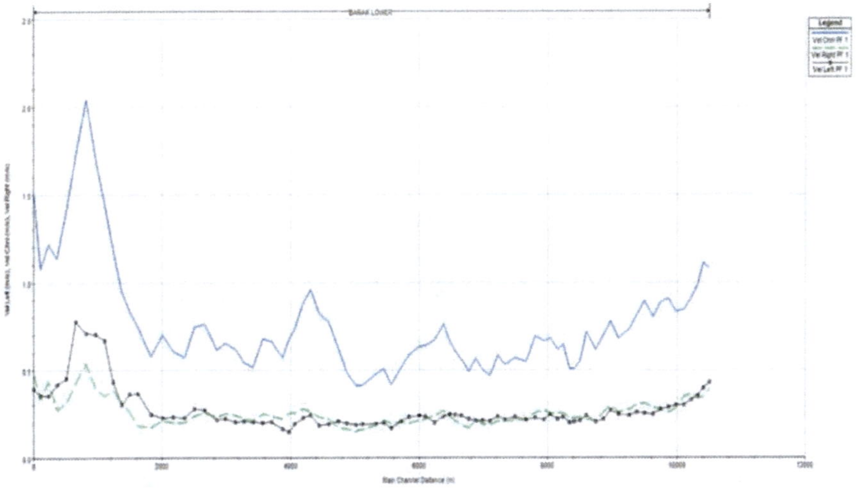

Fig. 10 Velocity profile of Barak River lower reach

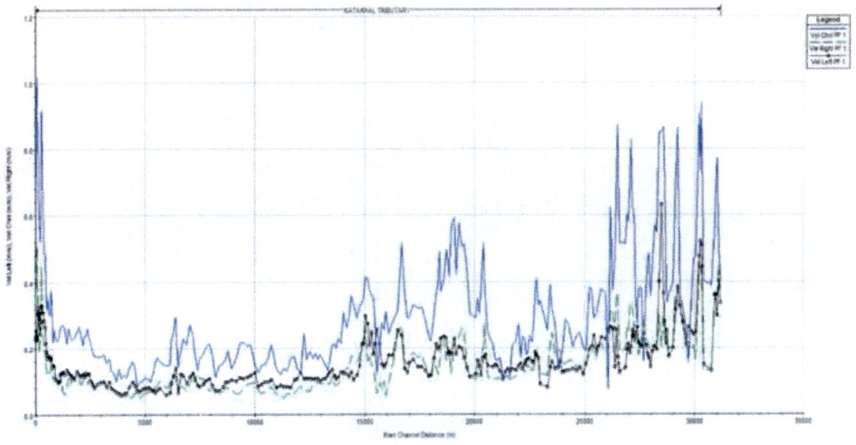

Fig. 11 Velocity profile of Katakhal River

4 Conclusion

Many advanced tools for flood modelling and prediction have been developed in the last three decades. Many of these technologies cannot be applied due to a lack of data in developing countries. The primary point of the investigation is to utilize the capability of the software HEC-RAS in developing a model for simulating the water surface profiles of the Barak waterway between A.P and B.P ghats including its tributary Katakhal. The results of the model suggest that the area under study was inundated to a great extent for the applied data. The velocity of flow was found to be

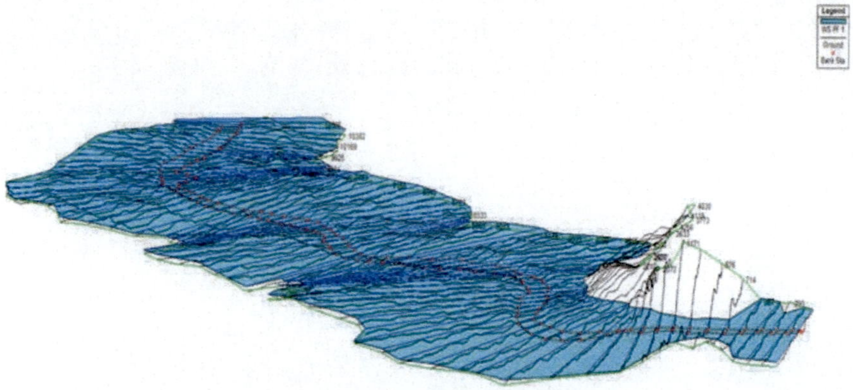

Fig. 12 Water surface profile of Barak River lower reach

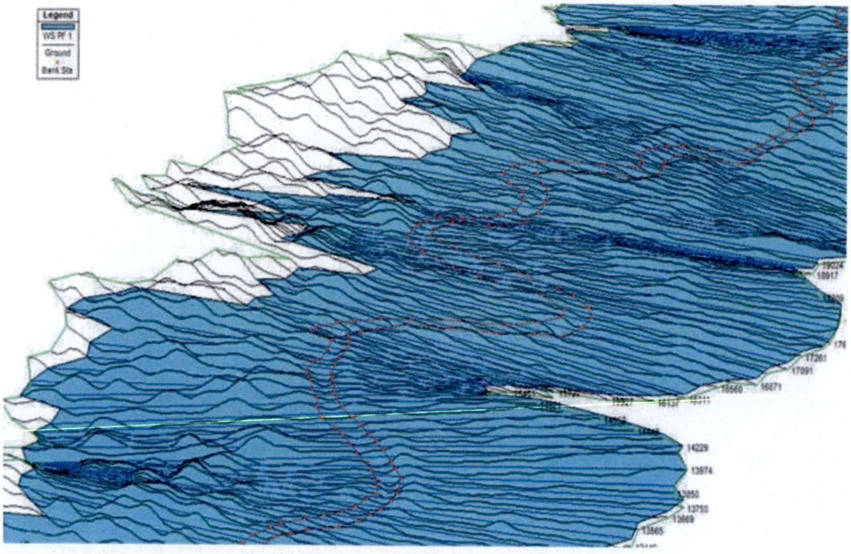

Fig. 13 Water surface profile of a reach of Katakhal River

more on the left overbank than that on the right overbank. The geometry data was extracted up to a distance of 1000–1500 m on either side of the river reaches from the centre line of the river reaches. It was seen that the maximum portion of the area covered was inundated except at a few locations. Thus, it suggests that engineering structures such as levees and dykes should be constructed along the river reaches to prevent the flooding of the crop fields, residential areas, roads, railways, etc. under such flow conditions. Incorrect topography is another problem in flood mapping that creates small errors and ultimately cause model instability and inaccuracy. So, a correct topography of the study area should be taken. On the off chance that greater complexity is added to the model, at that point various parameters will have a lesser impact on simulation and can give more accurate results. The one-dimensional flow model is not very accurate as it has certain limitations, but it has been utilized by various researchers for flood immersion mapping to great extent by providing correct topography, accurate data and engineering structures.

The one-dimensional steady flow model gives comparable results as it has certain limitations. This model can be improvised by providing the levees and bridges for better results. The stretch of the study reach can be extended beyond Annapurna Ghat and Badarpur Ghat. Providing the accurate width of the flood plains and providing the Manning's roughness coefficient according to the land use will prompt a superior model. This model can be improvised by including all the tributaries within the study area. This model can be developed into a one-dimensional unsteady flow model by providing the unsteady flow data which gives better results than the present model.

References

1. Hadi T, Feyzolahpour M, Mumipour M, Zakerhoseini F (2012) Rainfall-runoff simulation and modeling of Karun river using hec-ras and hec-hms models, Izeh district. Iran J Appl Sci 12(18):1900–1908
2. Devoir S, Narasimhan B (2013) Flood inundation mapping of Thamiraparani river basin using hec-geo ras and swat. Int J Eng Res Tech
3. İcaga Y, Tas E, Kilit M (2016) Flood inundation mapping by GIS and a hydraulic model (HEC RAS): A case study of Akarcay Bolvadin subbasin, Turkey. Acta Geobalcanica 2(2):111–118
4. Pathak P, Bhandari M, Kalra A, Ahmad S (2016) Modeling floodplain inundation for monument creek, Colorado. In: World Environmental and Water Resources Congress, pp 131–140
5. Ahmad HF, Alam A, Bhat MS, Ahmad S (2016) One dimensional steady flow analysis using HECRAS–a case of River Jhelum, Jammu and Kashmir. Eur Sci J 12:32
6. Thakur B, Parajuli R, Kalra A, Ahmad S, Gupta R (2017) Coupling HEC-RAS and HEC-HMS in precipitation runoff modelling and evaluating flood plain inundation Map. In: World Environmental and Water Resources Congress, pp 240–251
7. Bhandari R, Parajuli R, Lamichhane GR, Alyami A, Kalra A, Ahmad S, Gupta R (2018) Utilizing civil geo-HECRAS capabilities for floodplain mapping of Colorado River in Texas during Hurricane Harvey. In: World environmental and water resources congress 2018: watershed management, irrigation and drainage, and water resources planning and management. American Society of Civil Engineers, Reston, VA, pp 387–397
8. US Army Corps of Engineers, "HEC-RAS, Reference Manual," Version 5.0.
9. US Army Corps of Engineers, "HEC-RAS, User Manual," Version 5.0.

Limitation of Inter-Storey Drift Ratio in Seismic Design of Lead Rubber Bearing Isolated R. C. Building

Rajkumar Manisana and Sunil Sing Mayengbam

1 Introduction

Base isolation system using seismic isolator is one of the most recognized earthquake resistance techniques when it comes to ensure the lowering the both structural and non-structural damage and withstand throughout the strong earthquake. The decoupling phenomenal of base isolation system reduced the seismic demand like inter-storey drift, floor acceleration and base shear, eventually increasing the building period but lower the excessive strain of the building which is the main reason for damage.

Generally for non-base isolated building, design drift ratio limit is set as per the Vision 2000 [SEAOC 1995] recommended three performance limits, namely operational (max drift of 0.5%), life safety (max drift of 1.5%) and ear collapse (max drift of 2.5%). There, base isolated system particular design drift limit is not intimated. From last two decade, many researchers [1, Hamidreza et al. 2016, Jangid et al. 2014 and many more] referred UBC and FEMA to design the base isolated system which is totally a force method, to studied the behaviour of base isolated building over non isolated system. As known, the damage of building are co-related with the floor displacement, direct displacement base design (DDBD) method [2] has been introduced in which design approached are totally base on displacement rather than a forces. Recently the DDBD [3] methodology for designing the base isolated structure has been well established and adopted, which is originally drive from DDBD propose by Priestly [2].

The basic approach of DDBD is that performance level (illustrate in term of maximum displacement and inter drift ratio) of the structure should reach the target displacement set up by the designer when subjected to seismic ground motion. The methodology behind DDBD is that convert the multi-degree-of-freedom (MDOF) of

R. Manisana (✉) · S. S. Mayengbam
National Institute of Technology Manipur, Imphal, India
e-mail: mani47raj@gmail.com

A. Sil et al. (eds.), *Recent Trends in Civil Engineering*, Lecture Notes in Civil Engineering 274, https://doi.org/10.1007/978-981-19-4055-2_57

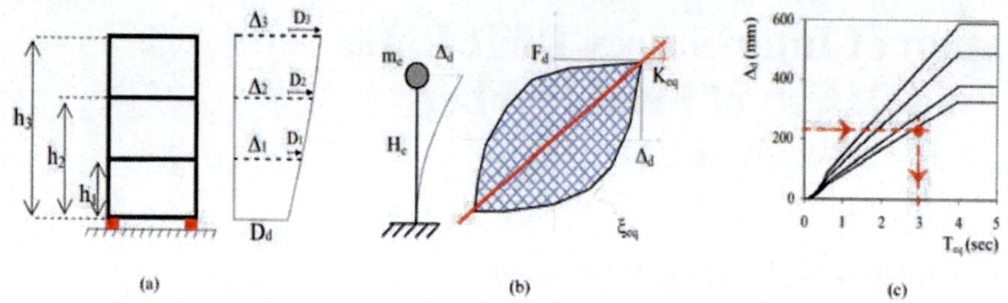

Fig. 1 Schematic representation of the DDBD method for seismically isolated buildings. Adapted with permission from reference [3]. Copyright 2010, Taylor & Francis

an original structure into a single-degree-of-freedom (SDOF) structure system. This new system involves an equivalent stiffness (k_{eq}), equivalent mass (m_e), equivalent viscous damping (ξ_{eq}), and equivalent height (H_e), respectively, as shown in below Fig. 1a and b.

In this study, especially design drift ratio of DDBD method for base isolated R.C. buildings is discussed. Base isolated structures are designed for inter drift ratio range from 0.1 to 2.5% as per the SEAOC 1995 [version 2000] to check the limitation of the storey drift, using seismic resistant design procedure so called Direct Displacement Base Design (DDBD) for three different heights 4-storey, 6-storey, and 9-storey, and each having a different building plan. Dynamic time history analysis (THA) is conducted to study the seismic responses of the respected design isolated buildings.

2 Modelling of Base Isolated R.C. Frame Building

The physical properties of base isolated R.C. frame building like sizes of beams and columns, grade of concrete (M25) and rebar (Fe415) are pre-selected as per the convenient of designer. The details and layout of BI-buildings are given in Table 1 and Fig. 2, respectively.

All the base isolated buildings are model by using DDBD [3] methodology, which design approached are totally base on displacement rather than forces. In this method, target design displacement (D_d) of 250 mm and design drift ratio of range 0.1–2.5% are taken at the beginning of the design process. In DDBD of BI-building, equivalent stiffness (k_{eq}) and damping (ξ_{eq}) of SDOF (Fig. 1) system defined in Eqs. (1) and (2) are different from that of fixed building [3].

$$k_{\text{eq}} = m_e\left(2\pi \Big/ T_{\text{eq}}^2\right) \tag{1}$$

$$\xi_{\text{eq}} = [\xi_{\text{IS}} \cdot D_d + \xi_S \cdot (\Delta_d - D_d)] \big/ \Delta_d \tag{2}$$

Table 1 Details of base isolated buildings

Model name	No. of bay			No. of story	Type of beam	Beam seizes in (m)			Type of column	Column seizes in (m)			Ratio of hc/lb (0.4–0.86)	Ratio of lb/db (5.4–12.6)
	x-axis	y-axis	z-axis			lb	db	wb		hc	dc	wc		
A1	4	4	5	4	Bm-1	4	0.36	0.28	Col-1	3.3	0.44	0.44	0.825	11.11
A2	4	4	7	6	Bm-1	4	0.36	0.28	Col-1	3.3	0.48	0.48	0.825	11.11
A3	5	4	5	4	Bm-1	4	0.38	0.28	Col-1	3.3	0.44	0.44	0.825	10.52
A4	6	6	7	6	Bm-1	4	0.38	0.28	Col-1	3.3	0.48	0.48	0.825	11.11
A5	6	6	10	9	Bm-1	4	0.40	0.30	Col-1	3.3	0.65	0.65	0.825	10
					Bm-2	4	0.36	0.26	Col-2	3.3	0.40	0.40	0.825	11.11

Bm: Beam, y_d: yield drift, **Col**: Column, **lb**: length of the beam, **db**: depth of the beam, **wb**: width of the beam, **hc**: height of the column, **dc**: depth of the column, **wc**: width of the column, D_D: *design displacement*

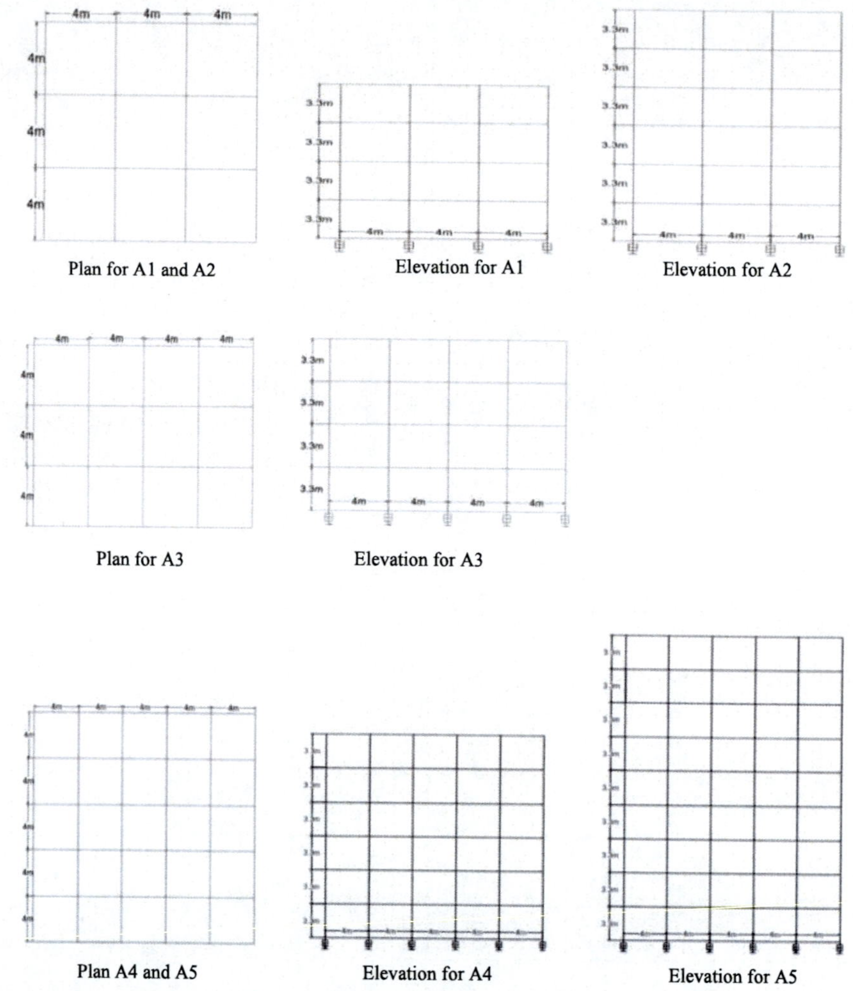

Fig. 2 Building plan and elevation for all the model, namely. Model A1, A2, A3, A4 and A5

where T_{eq} is effective time period of the BI-Building and its value is obtained from design displacement spectra (see in Fig. 1c). Δ_d and m_e are design displacement and effective mass of equivalent SDOF BI-Building which are derived based on characteristic equation (shown in Eqs. 3 and 4, respectively) of DDBD method [4].

$$\Delta_d = \sum_{i=0}^{n} \left(m_i \cdot \Delta_i^2\right) \Big/ \sum_{i=0}^{n} (m_i \cdot \Delta_i) \tag{3}$$

$$m_e = \sum_{i=0}^{n} (m_i.\Delta_i) \Big/ \Delta_d \tag{4}$$

The target displacement profile (Δ_i) of the BI-Building is calculated as sum of design displacement D_d and product of design inter-storey drift θ_d, concave deformed shape of the superstructure φ_i and c_1 (where c_1 is given by $h_1/(100^*\varphi_1)$ as shown in Eq. (5). Finally, the forced distribution F_i for Linear Static Analysis (LSA) of the BI-Building is determined by Eq. (6) [4].

$$\Delta_i = D_d + \theta_d \cdot c_1 \cdot \varphi_i \tag{5}$$

$$F_i = V_b \cdot (m_i \cdot \Delta_i) \Big/ \sum_{i=0}^{n} (m_i \cdot \Delta_i) \tag{6}$$

where, V_b is base shear of the BI-Building and it is defined as $V_b = K_{\text{eq}}{}^*\Delta_d$. Base on the design displacement the effective stiffness of the BI-Building is calculated as $k_{\text{IS}} = V_b/D_d$

3 Design of Base Isolator

All the dynamic parameters like seismic coefficients (C_a, C_v) which are again depend on soil profile type (S_A, S_B,…S_F), near source factor N_v or N_a. and other factors are taken from Table (A-2) to (A-9) of UBC-97, respectively. The basic characteristic properties of isolator like thickness (*tr*) is taken 0.22 m and diameter (*d*) of 0.55 m for all base isolated buildings. For this study, lead rubber bearing properties are calculated according to UBC-97 using the maximum vertical load as shown in Eqs. (8) to (12). Details are shown in Table 2, respectively. It was found that higher the buildings storey higher the vertical stiffness and increases the yield strength.

The design displacement (D_D) of isolator is calculated as given in Eq. (7) and the effective stiffness is calculated based on the vertical load (W) of superstructure

Table 2 Details of Lead Rubber Base Isolators (Input data for Sap2000)

BIS properties	Type of model				
	A1	A2	A3	A4	A5
For U1 effective stiffness (kN/m)	349,295	513,846	367,007	590,832	1,018,368
For U2 & U3 effective stiffness (kN/m)	349.29	513.84	367.01	590.84	1018.40
For U2 & U3 stiffness (kN/m)	321.86	473.51	338.19	544.45	938.42
For U2 & U3 effective damping	5	5	5	5	5
U2 & U3 distance from end-J (in m)	2.3×10^{-3}	2.3×10^{-3}	2.3×10^{-3}	2.3×10^{-3}	2.3×10^{-3}
U2 & U3 yield strength (kN)	0.1	0.1	0.1	0.1	0.1

acting just above the isolator (see Eq. 8), where T_D is a nature period of the isolator and its value is taken as 2.5 s as per the UBC-97

$$D_D = (g \cdot T_D \cdot S_D) / \left(4\pi^2 \cdot B_D\right) \tag{7}$$

$$k_{\text{eff}} = \left(\frac{W}{g}\right) \cdot \left(\frac{2\pi}{T_D}\right)^2 \tag{8}$$

Characteristic strength (Q) of a base isolator is given as in Eq. (9), UBC-97

$$Q = W_D / D_D \tag{9}$$

where W_D is an energy dissipated per cycle.

The post yield stiffness (k_I) is calculated based on the post to pre yield (k_2) ratio for rubber (n) which is equal to 0.1, where k_2 is given as in Eq. (10) along with the yield displacement (D_y) (distance from End–J) given in Eq. (11) as per UBC-97.

$$k_2 = k_{\text{eff}} - \left(Q / D_D\right) \tag{10}$$

$$D_y = Q / (k_1 - k_2) \tag{11}$$

4 NonLinear Time History (NLTH) Analysis

Using time history analysis, the structural response is investigated as per the dynamic equilibriums and considering short time intervals. Displacement responses were derived using nonlinear time history analysis for five spectrums compatible ground motion (scale 0.36 g) as time history loading. For this study, NLTH analyses are conducted using SAP2000, the structural responses in the time intervals are determined while ground accelerations at the structural base often at the foundation level, are taken into consideration. The below Table 3 shows the details of five seismic ground acceleration records used for the present NTHH analysis.

5 Response of NonLinear Time History Analysis (NLTH)

Base isolated buildings are analysed under five different earthquakes loading using SAP2000. The details of the earthquake ground motion are shown in Table 3, respectively. For this study, the achieved inter-storey drift ratios are calculated as per the ACI for each design isolated building by using Eq. (12).

Table 3 Details of time history compatible ground motion (PEER database)

Name	Background earthquake	Magnitude (Mw)	Duration (seconds)
TH-1	Bhuj 2001	7.0	27
TH-2	Duzce 1999	7.2	25.9
TH-3	El Centro 1940	6.9	31.9
TH-4	Gazli 1976	7.1	16.3
TH-5	N. Palm Spring 1986	5.9	20

$$\text{Inter} - \text{storey drift } = (\Delta_n - \Delta_{n-1})/h_i \tag{12}$$

where Δ_n is deflection at the floor and h_i is the floor height (shown in Fig. 1).

As per the results of the NTHA shown in Table 4, a few comparative graphs are plotted by keeping achieved IDR% in x-axis and number of storey of the BI-Building in y-axis as shown in Fig. 3a–c, respectively.

It was observed that the increase in design IDR for a BI-building, increase in achieved IDR. Summary on overall results of NLTH analyses are shown in Table 4. The minimum achieved IDR of 0.151% was found in BI-building model-A3 at design IDR of 0.18% and maximum achieved IDR of 0.450% was found in BI-building model-A5 at design IDR of 2.15%. In case of model A5, achieved IDR of 0.3% cross over the design IDR at the very low ratio of 0.15% and in model A4 achieved IDR almost reached the design IDR of 0.28% under earthquake ground motion of TH-5 respectively.

From the obtain results, comparative graph is plotted by keeping design IDR in y-axis and achieved IDR in x-axis as shown in below Fig. 4, respectively. It shows that the results of achieved IDR of BI-building having same storey height has nearly similar when compared to building having different in storeys height. It also shows that achieved IDR do not cross further value of 0.5% even though increase in design IDR upto 2.5%. The value of achieved IDR remain from 0.15% to under 0.5% even though the design IDR are increase up to maximum permitted limit of 2.5% as per the [SEAOC 1995] Vision 2000 for a non-isolated building.

6 Conclusions

The present study, the response of inert storey drift of base isolated buildings under different earthquake ground motions are taken into account. The following are the conclusion made from NLTH analysis using SAP2000 results.

- Increase in design IDR, increase in achieved IDR.

Table 4 Summary on overall achieved IDR% results of NTHA for BI-Building under different earthquake motion, for design IDR range from 0.1–2.5%

Model-A1		Model-A2		Model-A3		Model-A4		Model-A5	
Design IDR%	Achieved IDR%	Design IDR%	Achieved IDR%	Design IDR%	Achieved IDR%	Design IDR%	Achieved IDR%	Design IDR%	Achieved IDR%
0.22	0.159	0.30	0.242	0.18	0.151	0.28	0.274	0.15	0.309
0.50	0.165	0.50	0.252	0.50	0.160	0.50	0.285	0.50	0.319
0.80	0.168	0.75	0.259	0.72	0.169	0.65	0.297	0.70	0.355
1.00	0.172	1.00	0.269	1.00	0.175	1.00	0.302	1.00	0.378
1.28	0.198	1.35	0.272	1.25	0.182	1.22	0.307	1.20	0.398
1.50	0.202	1.50	0.275	1.50	0.189	1.50	0.314	1.50	0.418
1.75	0.211	1.85	0.285	1.72	0.195	1.65	0.327	1.70	0.431
2.00	0.221	2.00	0.299	2.00	0.210	2.00	0.334	2.00	0.441
2.30	0.243	2.35	0.302	2.25	0.220	2.40	0.352	2.15	0.450

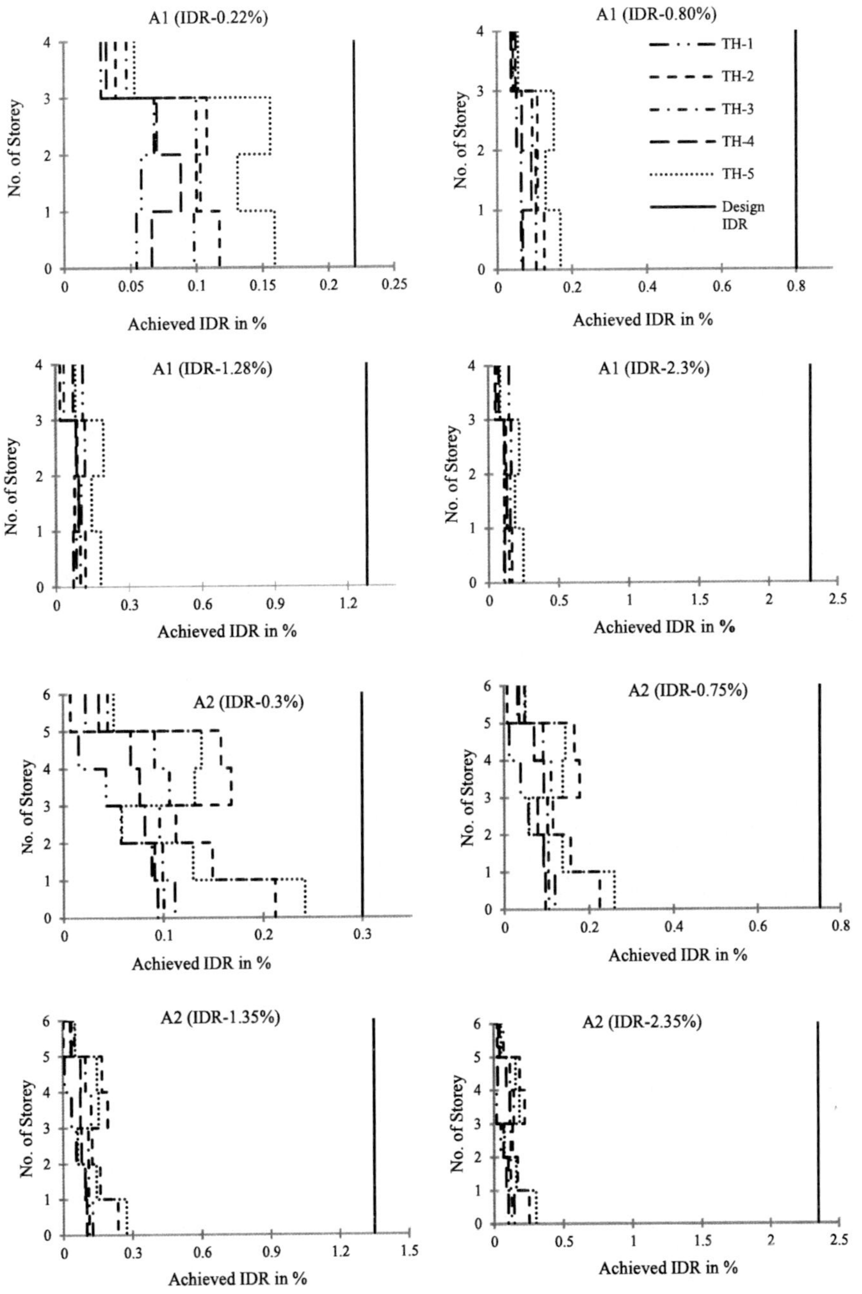

Fig. 3 **a** Inter-storey drift diagram for model A1 and A2 under different range of Design IDR 0.1–2.5% **b** Inter-storey drift diagram for model A3 and A4 under different range of Design IDR 0.1–2.5% **c** Inter-storey drift diagram for model A5 under different range of Design IDR 0.1–2.5%

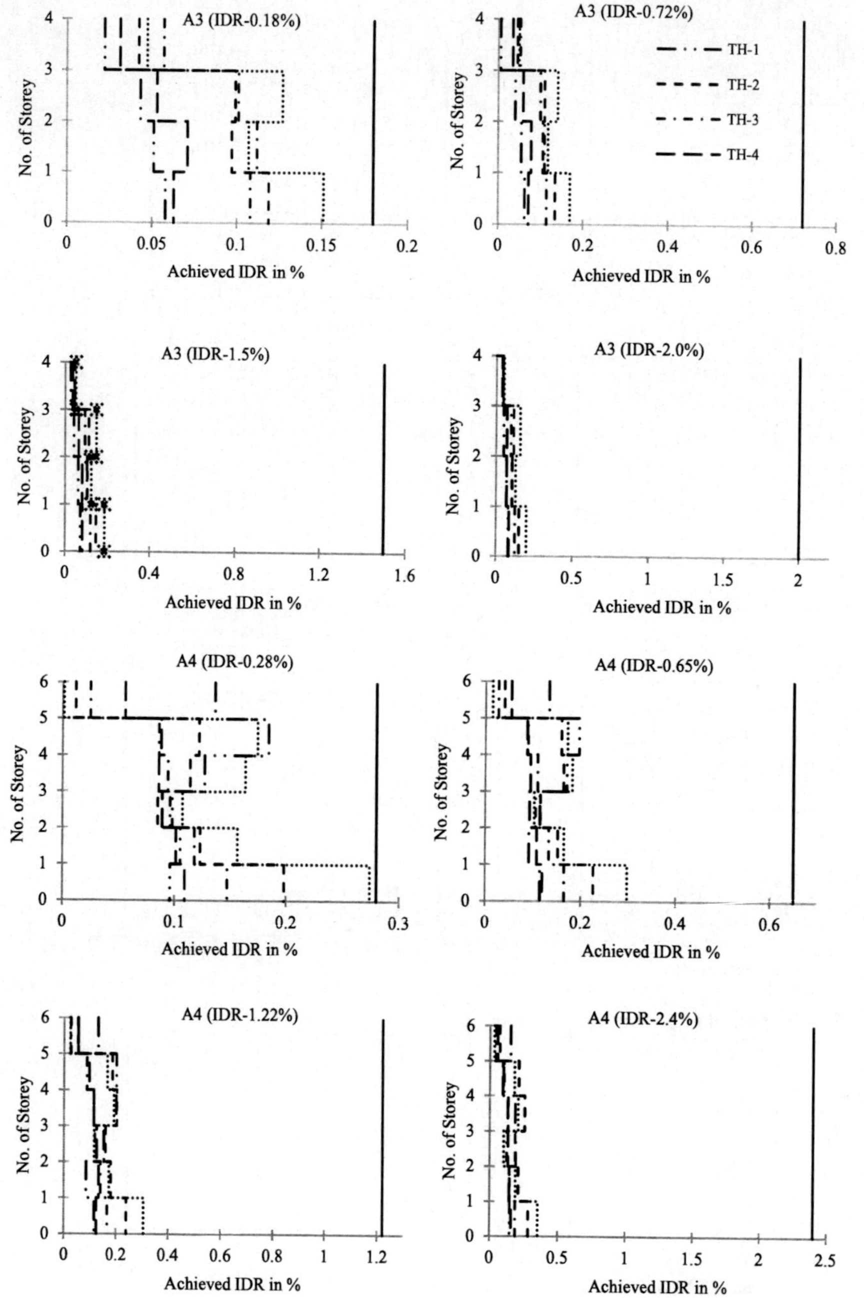

Fig. 3 (continued)

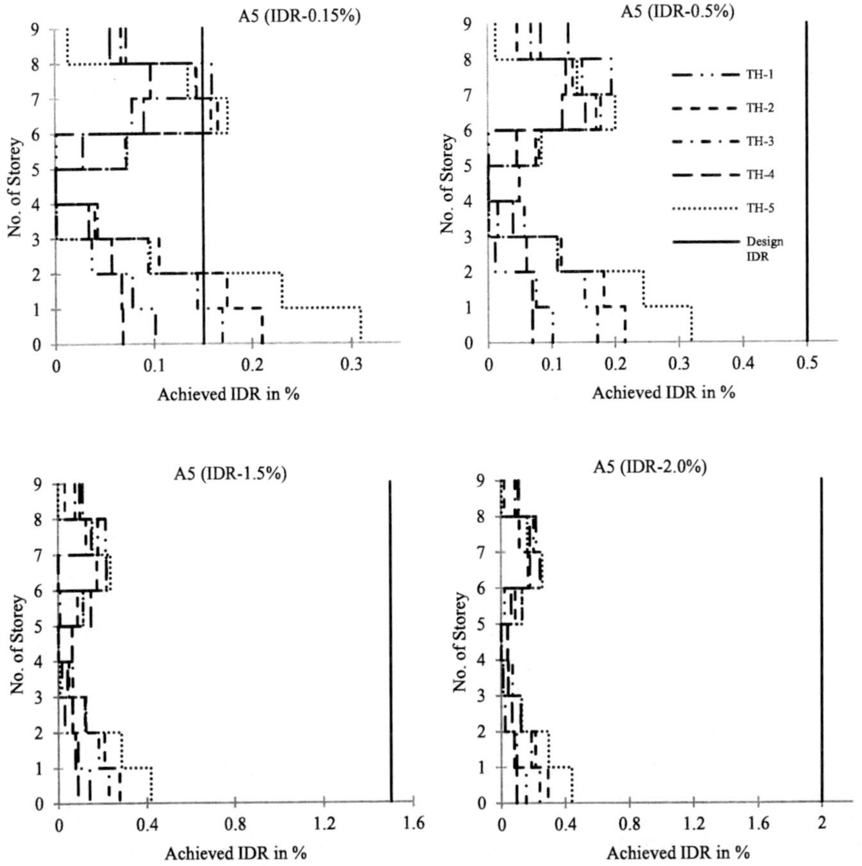

Fig. 3 (continued)

Fig. 4 Design versus achieved inter-storey drift ratio in percentage

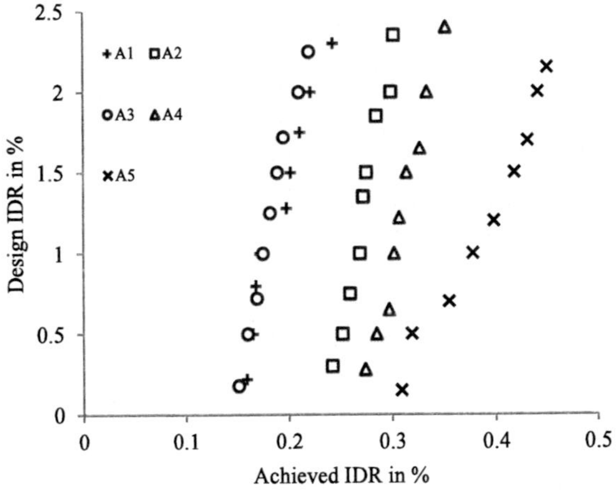

- The minimum achieved IDR of 0.151% was found in BI-building model-A3 at design IDR of 0.18% and maximum achieved IDR of 0.450% was found in BI-building model-A5 at design IDR of 2.15%.
- Average achieved IDR 0.3% is less than that of average of design IDR 1.3%.
- It was also shows that achieved IDR do not cross further value of 0.5% even though increases in design IDR up-to 2.5%.

With the current design methodology, it will not be effective if the design IDR is more than 0.5%. Therefore, from the above conclusion it was observed that the during design procedure DDBD of base isolated building, design IDR could be consider from 0.1% to 0.5%. More study is required to develop a better correlation between design and achieved IDR.

References

1. Gamit KD, Amin JA (2017) Application of DDBD and FBD methodology for 8-story RC frame using IS1893 spectra. ICRISET 1:69–79
2. Priestley MJN (2003) Myths and fallacies in earthquake engineering. IUSS Press, Pavia (Italy), Revisited
3. Cardone D, Palermo G, Dolce M (2010) Direct displacement-base design of building with different seismic isolation systems. J Earthquake Eng 14(2):163–191
4. Priestley MJN, Calvi GM, Kowalsky M (2007) Displacement-base seismic design of structures. Iuss Press
5. Tavakoli HR, Naghavi F, Goltabar AR (2014) Dynamic response of the base fixed and isolated building frames under far and near fault earthquakes. Arab J Sci Eng, Research Article,. https://doi.org/10.1007/s13369-013-0891-8
6. UBC-97, International conference of building officials [1997], Uniform building co
7. FEMA 273 and 274 (1997) NEHRP Guidelines For The Seismic Rehabilitation, Federal Emergency management agency, California, USA
8. CEN (1995) Eurocode 8: Earthquake Resistant Design of Structures—Part 1: General Rules, ENV1998–1, Parts 1–3, CEN, Brussels, Belgium
9. Matsagar VA, Jangid RS (2004) Influence of isolator characteristics on the response of base-isolated structures. Eng Struct 26:1735–1749
10. Kelly JM (1990) Base isolation: linear Theory and design. Earthq Spectra 6:223–244
11. Chopra AK (1997) Dynamics of structures: theory and application to earthquake engineering, Prentice-Hall Ltd.
12. SAP2000 Analysis Reference (2020) Structural analysis program, Computers and Structures Inn. (CSI)
13. ACI Committee 318, Building Code Requirements for Structural Concrete (ACI 318M-08) and Commentary. American Concrete Institute, Farmington Hills, MI, USA

Transportation Engineering

Laboratory Study on the Long-Term Mechanical Performance of Cold Mix Asphalt with Rice Husk Ash Filler

Pinki Deb and Kh. Lakshman Singh

1 Introduction

Recently, the issues regarding reducing energy consumption, reducing greenhouse gas emissions and managing waste materials have been highlighted worldwide. Conventional road construction technology like hot mix asphalt (HMA) production is liable for a tremendous amount of fossil fuel consumption due to heating aggregate and bitumen [1]. As an alternative to HMA, cold mix asphalt (CMA) has attracted the attention of engineers and researchers due to its benefits related to environmental protection, economic benefits and meeting the requirement related to the health and safety of the laborers [2]. CMA is a bituminous material produced by blending aggregate, premixing water and bitumen emulsion at ambient temperature [3]. The long curing time required to get full strength (generally it takes 2–24 months) [4], weak early strength and early distress due to rainfall water are the major drawbacks of using CMA [5, 6]. Incorporation of waste material either as additive or filler and proper curing of mix can be beneficial in improving the environmental aspects and possibilities to mitigate the drawbacks associated with CMA.

Yearly, India produces 120 million tons of rice paddy and 24 million tons of rice husk [7]. Rice husk ash is the by-product material produced after the combustion of rice husk [8]. Worldwide, paddy rice production is about 678 million tons (annually), leading to the production of 149.16 million tons of rice husks, and the burning rice husks leads to the formation of 37 million tons of RHA [9]. This ash leads to issues related to waste production, unnecessary landfill and overall environmental concerns. Recently, different researchers have tried to study the feasibility of using RHA in concrete and asphalt mix either as filler or additive. Fillers are the finest particles

P. Deb (✉) · Kh. L. Singh
National Institute of Technology Silchar, Silchar, India
e-mail: pinki_rs@civil.nits.ac.in

Kh. L. Singh
e-mail: lakshman@civil.nits.ac.in

A. Sil et al. (eds.), *Recent Trends in Civil Engineering*, Lecture Notes in Civil Engineering 274, https://doi.org/10.1007/978-981-19-4055-2_58

of aggregate gradation that mostly pass through 0.075 mm sieve as per MORTH [10]. The function of filler is to stabilize the mix by filling the aggregate mix's cavities and modifying the bituminous mix's consistency by attaching the aggregates together [11, 12]. The physical, chemical and morphological characteristics of filler can influence the behavior of any bituminous mix [13, 14]. In recent times, several studies have been conducted on the utilization of RHA in asphalt mix that is mostly related to bituminous mix. Sargin et al. [15] tried to utilize RHA as filler in asphalt mix at different dosages, i.e., 25%, 50%, 75% and 100% and concluded that the combination made with 50% RHA and 50% limestone filler mix exhibited better performance in terms of stability and flow parameters [15]. Another study reported no chemical interaction of using RHA in asphalt mix by Fourier transform infrared spectroscopy (FTIR) and reported improved physical properties and stable storage stability of binder after incorporating RHA [16]. In a study of the cold recycled mix, three additives, i.e., Portland cement, fly ash and RHA, were used as an additive and reported reduced permanent deformation, tensile strength and stability [17]. The incorporation of RHA shows the homogeneous dispersion of bee structure in mix that needs lower binder content and delivers paramount performance and its fine porous structure resulted in high adhesion force [18].

Curing of CMA is one of the critical design criteria which can influence the ultimate strength of CMA. Curing of CMA requires some days to months or years. The water present in the mix evaporates during the curing process. Evaporation of water leads to the bitumen emulsion attaining its original property and making it water-resistant [19, 20]. Thanaya et al. [6] reported that a properly cured CMA could attain stiffness comparable to HMA [6]. Another study conducted by Ferrotti et al. [21] on fiber-reinforced CMA and concluded an increment in mechanical properties in both the reinforced and unreinforced mixes. They further reported that the fiber-reinforced mix attains high performance at a lower curing period, whereas unreinforced mixes require a more extended curing period to attain desirable properties [21]. Two kinds of curing methods of CMA were primarily suggested by different agencies, i.e., ambient curing at 15–25 °C and accelerated curing at 35–60 °C [22]. The cheapest curing process is ambient curing, whereas mixes can attain complete curing faster in accelerated curing. Asphalt academy suggested curing for three days at 40 °C to simulate six months of field curing [23]. Curing of 14 days at 40 °C along with 20% relative humidity can simulate the curing of 1–3 years in the field [24]. Similarly, another study claimed that the curing at 40 °C for 28 days could simulate the field curing of one year [25]. The tensile strength of cured CMA at 60 °C is higher than the cured mix at 25 °C for 28 days [26]. After reviewing different literature, the CMA samples are allowed to cure for three days at 40 °C to simulate the field condition in this study.

In this study, RHA is used in CMA as a part of aggregate gradation, which passes through 0.075 mm sieve. This research aims to study the influence of RHA filler on the mechanical performance of CMA. CMA incorporating RHA has undergone additional laboratory testing such as the Marshall stability test, indirect tensile test after following a standard curing protocol. The moisture damage potential of CMA incorporating filler was studied by the index of tensile strength of mixes at conditioned and unconditioned sample mixes. All the samples were tested at different curing periods to accelerate the properties of the mix by the combined effect of the accelerated and ambient long-term curing process.

2 Materials

The coarse and fine aggregates utilized in this study are the river stone collected from the local stone quarry. The maximum and minimum nominal sizes of aggregate are 13.2 mm and 0.075 mm, respectively. Semi-dense bituminous concrete gradation is used as designed gradation for CMA that was adopted from IRC: SP: 100–2014 [27]. This continuously graded aggregate gradation is selected in this study for the good packing of aggregate particles which is described in Table 1 and physical properties are listed in Table 2.

Fillers are the finest part of aggregate gradation. The particles passing through 0.075 mm sieve is considered as filler in asphalt mix which is the part of aggregate gradation, as suggested according to MoRTH [10]. This study used two types of filler such as conventional stone dust (SD) filler and RHA filler, to the part of aggregate gradation for accelerating the mix properties. The specific gravity of SD and RHA fillers is 2.62 and 2.05, respectively.

Table 1 Gradation of aggregate [27]

Sieve sizes	Upper limit	Lower limit	Design curve
13.2	100	100	100
9.5	100	90	95
4.75	51	35	43
2.36	39	24	31.5
1.18	30	15	22.5
0.3	19	9	14
0.075	8	2	5

Table 2 Physical properties of aggregate

Physical properties	Test results
Flakiness and elongation index	16.8% and 10.5%
L.A. Abrasion value	28.65%
Impact value	23.1%
Water absorption	1.10%
Soundness	8%

Table 3 Physical properties of emulsion

Test properties of Cationic emulsion	Test results
Viscosity	55.5 s
Storage stability after 24 h	0.1%
Residue after evaporation	62.1%
Penetration on residue/100 g/5 s	8.2 mm
Ductility on residue	87 cm

Cationic emulsion collected from Bitchem Asphalt Technologies Limited, Guwahati, India, was used in this laboratory study. Different laboratory tests were carried out to assess the physical properties of the binder and are listed in Table 3.

3 Preparation and Curing of Samples

The CMA mixes were prepared using the Marshall mix design procedure with 50 blows (with Marshall hammer) on each side of the cylindrical sample. The optimum premixing water content was found to be 3% by weight of mix by visually determining the coating test of mix. In this study, all CMA samples were made with 3% premixing water content to maintain the consistency of mix. The samples were made with different dosages of emulsion varying from 7 to 10% (by weight of mix) with an increasing fraction of 0.5%. This is necessary to determine the optimum emulsion content, an essential parameter of asphalt mix design. The SD and RHA were blended with mix at 5% in mix weight to determine the optimum binder content for each type of filler mix. Sample demolding was done after 24 h of the mixes compaction to ignore the sudden breaking of sample during demolding. All the samples were cured at 40 °C in a hot air oven to simulate the field conditioning as per the mentioned works of literature and tested after 5th, 7th, 15th, 28th, 60th and 90th day of its compaction by keeping the samples at room temperature. An average of three optimum samples of CMA was fabricated for each test explained in this study.

4 Results and Discussions

4.1 Mix Design of Cold Mix Asphalt (CMA)

To find out the mix design of CMA for SD and RHA filler mixes, the Marshall samples containing each filler are subjected to Marshall stability test (ASTM D 6927–15) [28]. The samples were loaded to 50 mm/minute along with the diametrical plane. The load at which the sample fails are considered as stability of that particular mix and the corresponding deformation is the flow value. The Marshall stability versus dosage of emulsion content graph is explained in Fig. 1. The dosage of emulsion content at the highest stability is considered the optimum binder content and the composition of that particular mix is determined as the optimum mix design sample. It is found from the graph that the mix containing SD filler resulted in the best value of stability at 8% emulsion content. Whereas, the stability is highest for the mix of RHA at 9% emulsion content. RHA filler particles act as a reinforcing material to the mix and enhance the stiffness and cohesion of mix [29], improving the Marshall stability at 8.5% and 9%, respectively. The optimum stability of RHA is 20.32% higher against the optimum mix stability of SD filler. It is also observed that the RHA filler soaked more binder to get the best mix, mainly due to the lower specific gravity of RHA filler. The optimum mix design samples of SD and RHA undergo the same test to determine the long-term stability enhancement of CMA at a duration of similar curing time. The results are explained in Fig. 2.

It is concluded that the curing of samples can enhance the stability of CMA which results in the pozzolanic action of filler particles and the gradual increment of binder adhesion due to evaporation of water from the emulsion. The pozzolanic property of RHA (presence of quartz and calcite in the composition of RHA particles) is

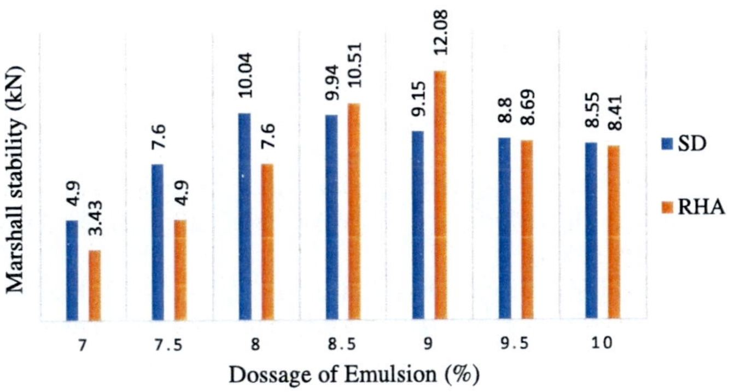

Fig. 1 Stability at different dosage of emulsion

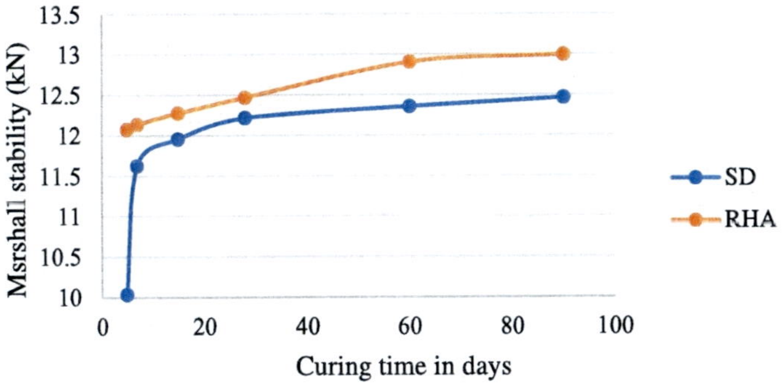

Fig. 2 Increment of stability at different curing period

responsible for the improvement of strength over time [30]. The gradual emulsion breaking during the curing time and pozzolanic action of RHA fillers [30] are involved in the improvement of stability. The stability of the optimum SD filler mix rises up to 24.20% compared to its initial strength. Besides, the increment is 7.62% less for the optimum mix of RHA filler than the SD filler mix. Nevertheless, the ultimate strength of the RHA filler mix is 4.25% higher than the ultimate strength of the SD filler mix after 90 days of curing.

4.2 *Tensile Strength Properties of CMA at Different Curing Periods*

Tensile performance of CMA is determined by indirect tensile strength test (ITS) according to ASTM D6931-12 [31]. To conduct this test, the cylindrical Marshall samples are loaded (50 mm/minute) through horizontal plate placed along the vertical plane of sample. The highest load at failure is considered as tensile strength of CMA. ASTM D6931-12 suggested formula to calculate the tensile strength:

$$\text{ITS} = \left({}^{2P}\!/_{\pi t D}\right)100$$

where P is the maximum load at failure, and t & D are the thickness and diameter of the samples, respectively. The optimum samples of two fillers are subjected to ITS test at different curing period by maintaining the curing condition explained previously. The long-term tensile performance of CMA incorporating SD and RHA filler is graphically presented in Fig. 3.

The tensile strength of the RHA filler mix is less in the initial days. While it shows an increasing pattern in the presented graph along with the curing duration. The tensile strength of optimum mix design samples (containing a similar composition of the

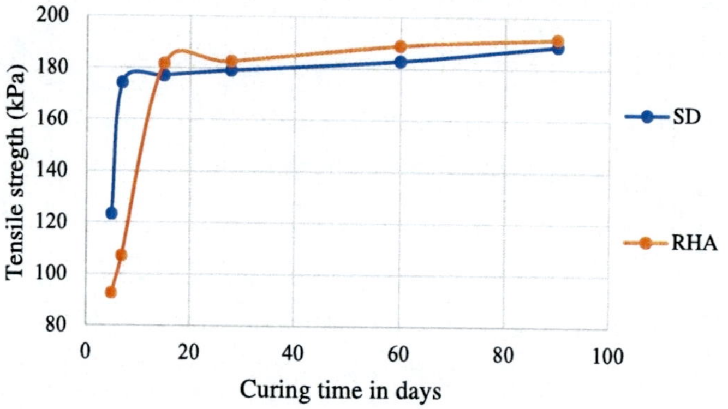

Fig. 3 Tensile strength of CMA at different curing period

material in the mix) at different curing times are explained in the graph. The gradual increment of tensile strength was explicated by the formation of integrated structure in mix that results from a uniform distribution of filler [32] and the pozzolanic reaction of filler over long time scales (long-term curing period) [30]. Results also indicated that the ultimate tensile strength (after 90 days of curing) of RHA filler mix is 1.51% higher than the ultimate tensile strength of the conventional filler mix. The enhancement of tensile strength by the effect of curing is easily observed from the graph pattern. The tensile strength of SD and RHA filler mix surges up to 53.17% and 106.87%, respectively.

4.3 *Effect of Curing on the Long-Term Durability of CMA*

The durability of CMA was measured by the index of Marshall stability ratio (MSR) and Indirect tensile strength ratio (ITSR). The retained Marshall stability test was conducted on the optimum mixes to quantify the MSR index which is the ratio of stability in dry and wet conditions. The dry and wet stability were observed after keeping the samples under water bath at 60 °C for 30 min and 60 °C for 24 h, respectively. The ratio of wet and dry stability should indicate the mix's capability to withstand the damage caused due to water stripping. The MSR value of the SD and RHA filler mix is 77.87% and 80.15%, separately, indicating that both of the mix types can restrain the water damage. Additionally, there is no significant change in the MSR index at different curing periods. The parameters of this test are tabulated in Table 4.

For determining the ITSR index, wet and dry sets of samples were prepared and subjected to the ITS test. The dry samples were tested according to the previously mentioned ITS test. The wet set of samples is cured as per the previously mentioned curing protocol. The samples are then placed at −16 °C for 16 h and subsequently soak the sample at a water bath of 60 °C for 24 h and finally placed in a water bath of 25 °C for 2 h. The samples are taken out after conditioning and tested for ITS parameter. Then tensile strength ratio of wet (ITS_{wet}) and dry conditioned samples (ITS_{dry}) are called ITSR values. The durability of CMA was observed in terms of ITSR value. The durability or moisture susceptibility of CMA is determined according to AASHTO T283 [33]. As per the code, the desirable limit of ITSR value is 80% (minimum) [33]. The ITSR value is calculated according to the formula described in AASHTO T283.

Table 4 Retained marshal stability test results (at 5th day of curing)

Properties	SD	RHA
Dry stability	10.04	12.08
Wet stability	7.82	9.68
MSR	77.87	80.15

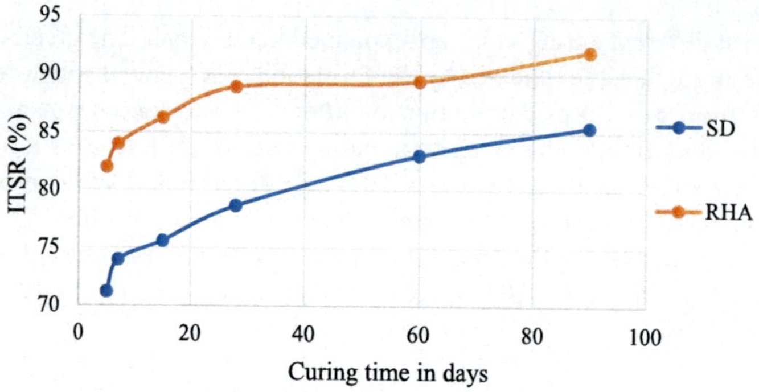

Fig. 4 ITSR or moisture susceptibility of mixes at different curing period

$$ITSR = \left(ITS_{wet}/ITS_{dry}\right)100$$

The results indicated a sharp increase in ITSR value and curing time. The result reflected that the moisture damage of SD mix is more serious and the damage potential reduced in time. At 60 days of curing, the ITSR of the SD mix is 83.04% (desirable value) which is increased by 16.596% from the initial ITSR value. The ultimate ITSR value of SD and RHA mix increased up to 20.05% and 12.31%. The mix incorporating RHA is more moisture resistant than SD filler at the initial days of curing time (Fig. 4).

5 Conclusion

This research evaluated the effect of using RHA filler on the laboratory performance CMA. Based on the outcomes, following conclusions has been made.

- RHA incorporated mixes delivered superior stability at higher binder content of 9% compared to conventional SD filler mix. However, further curing of mixes positively influences stability increment for both of the filler mixes.
- ITS result indicated that RHA might fail at lower load due to splitting. However, the tensile strength of SD filler is 33.04% higher than the RHA filler mix.
- The average increment of about 80.02% was noticed in tensile properties when the samples were subjected to the ultimate curing (90 days of curing).

The damage potential of conventional SD filler mix is lower than RHA filler mix at the initial days of curing. As the curing time increased, the moisture resistance increased. After 60 days of curing, the mix that incorporates SD filler reports the ITSR value of 83.04%. The mix made with RHA filler mixes showed water resistance even in its initial curing period (ITSR).RHA confirms to provide a positive impact in increasing the mechanical parameters of CMA. Further, this investigation can be used to design the most environment-friendly CMA with waste material and helps in managing the issues related to landfills.

References

1. Rubio MDC, Moreno F, Martinez-Echevarria MJ, Martinez G, Vazquez JM (2013) Comparative analysis of emissions from the manufacture and use of hot and half-warm mix asphalt. J Clean Prod 41:1–6
2. Dulaimi A, Al-Nageim H, Ruddock F, Seton L (2017) Laboratory studies to examine the properties of novel Cold-Asphalt concrete binder course mixture containing binary blended cementitious filler. J Mat Civil Eng ASCE, ISSN 0899-1561
3. Gomez-Meijide B, Perez I, Airey G, Thom N (2015) Stiffness of cold asphalt mixtures with recycled aggregates from construction and demolition waste. Constr Build Mater 77:168–178
4. Leech D (1994) Cold asphalt materials for use in the structural layers of roads. Project Rep. 75, Transport research laboratory, Berkshire, UK
5. Brown SF, Needham D (2000) A study of cement modified bitumen emulsion mixtures. In: Proceeding of Association of Asphalt Paving Technologists, 69, 1e22
6. Thanaya INA, Zoorob SE, Forth JP (2009) A laboratory study on cold-mix, cold-lay emulsion mixtures. Proc Inst Civ Eng Transp
7. Gidde MR, Jivani A (2007) Potential of rice husk in India a literature review. Waste to Wealth
8. Al-Hdabi A (2016) Laboratory investigation on the properties of asphalt concrete mixture with rice husk as filler. Constr Build Mater 126:544–551
9. Khan R, Jabbar A, Ahmad I, Khan W, Khan AN, Mirza J (2012) Reduction in environmental problems using rice-husk ash in concrete. Constr Build Mater 30:360–365
10. MoRTH (Ministry of Road Transport and Highways) (2013) Specifications for Road and Bridge Works (Fifth Revision). Indian Road Congress
11. Choudhary J, Kumar B, Gupta A (2018) Application of waste materials as fillers in bituminous mixes. Waste Manage 78:417–425
12. Kandal PS, Lynn CY, Parker J (1998) Characterization tests for mineral fillers related to performance of asphalt paving mixtures (NCAT Report 98-02)
13. Antunes V, Freire AC, Quaresma L, Micaelo R (2016) Effect of the chemical composition of fillers in the filler–bitumen interaction. Constr Build Mater 104:85–91
14. Choudhary J, Kumar B, Gupta A (2020) Utilization of solid waste materials as alternative fillers in asphalt mixes: A review. Constr Build Mater 234:117271. https://doi.org/10.1016/j.conbuildmat.2019.117271
15. Sargın M, Saltan N, Morova S, Serin ST (2013) Evaluation of rice husk ash as filler in hot mix asphalt concrete. Constr Build Mater 48:390–397
16. Xue Y, Wu S, Cai J, Zhou M, Zha J (2014) Effects of two biomass ashes on asphalt binder: dynamic shear rheological characteristic analysis. Constr Build Mater 56:7–15
17. Behnood A, Modiri Gharehveran M, Gozali Asl F, Ameri M (2015) Effects of copper slag and recycled concrete aggregate on the properties of CIR mixes with bitumen emulsion, rice husk ash, Portland cement and fly ash. Const Build Mat 95:172–180
18. Mistry R, Roy TK (2021) Performance evaluation of bituminous mix and mastic containing rice husk ash and fly ash as filler. Const Build Mat 268:121–187
19. Saadoon T, Garcia A, Gomez-Meijide B (2017) Dynamic of water evaporation in cold asphalt mixtures. Mater Des 134:196e206
20. Lee H, Im S, Kim Y (2012) Impacts of laboratory curing condition on indirect tensile strength of cold in-place recycling mixtures using foamed asphalt. Airfield Highway Pavement, 213e221. https://doi.org/10.1061/41005(329)19
21. Ferrotti G, Pasquini E, Canestrari F (2014) Experimental characterization of high performance fibre-reinforced cold mix asphalt mixtures. Construct Build Mater 57:117e125. https://doi.org/10.1016/j.conbuildmat.2014.01.089.

22. Doyle T et al (2010) Developing specifications for cold mix pavement materials. Submitted to Transport Research Arena Europe, Brussels
23. Kekwick SV (2005) Best practice. Bitumen emulsion and foamed bitumen materials laboratory processing. In: Proceedings of the 24th Southern African transport conference, SATC, South Africa. ISBN No. 1-920-01712-7
24. Serfass JP, Poirier JE, Henrat JP, Carbonneau X (2004) Influence of curing on cold mix mechanical performance. Mater Struct 37(5):365–368
25. National Roads Authority (2011) NRA Interim Advice Note 01/11 on Low Energy Pavements. NRA, Dublin
26. Shaowen D (2014) Effect of different fillers on performance properties of asphalt emulsion mixture. J Test Eval 42(1):126–134
27. IRC:SP:100–2014, Use of cold mix technology in construction and maintenance of roads using bitumen emulsion
28. ASTM D6927–15. Standard test method for marshall stability and flow of asphalt mixtures
29. Al-Hdabi A (2016) Laboratory investigation on the properties of asphalt concrete mixture with Rice Husk Ash as filler. Constr Build Mater 126:544–551. https://doi.org/10.1016/j.conbuildmat.2016.09.070
30. Balapour M, Ramezanianpour AA, Hajibandeh E (2017) An investigation on mechanical and durability properties of mortars containing nano and micro RHA. Constr. Build Mater. https://doi.org/10.1016/j.conbuildmat.2016.12.017
31. ASTM D6931–17. Standard test method for Indirect tensile stregth (IDT) of asphalt mixtures
32. Modarres A, Bengar PA (2017) Investigating the indirect tensile stiffness, toughness and fatigue life of hot mix asphalt containing copper slag powder. Int J Pavement Eng. https://doi.org/10.1080/10298436.2017.1373390
33. AASHTO T 283 (2014).Standard method of test for resistance of compacted asphalt mixtures to moisture induce damage

Utilization of Bamboo Fibre in the Application of Stone Matrix Asphalt

Soumyadeep Deb and Kh. Lakshman Singh

1 Introduction

In India, almost 90% of major roads are made of asphalt mixes. So, good roads are required to be constructed for superior performance and service life [1]. For heavily traffic roads, asphalt roads are needs to restrict permanent deformations. German engineers first adopted stone matrix asphalt (SMA) using a larger size of aggregates in pavements. The mechanism of SMA is stone-to-stone load transfer. SMA can handle a higher amount of load. Usually, air voids content is low in SMA, and a stabilizer is required, i.e. celluloid fibre. Modification of asphalt mix can be done in two ways: modifying the binder or modifying the aggregate-binder properties. Modification of binder can be done physically by adding solid substances or chemically by the addition of chemicals. The process of preparing asphalt mix by modifying the binder is called the wet process. Similarly, binder-aggregate adhesion properties can be improved by adding or replacing fine aggregate [2, 3]. This can be performed by the addition of natural or artificial fibres also. For high-performance roads, SMA provides better stability and deformation resistance [4]. Fibre modification of SMA is one of the most important criteria. Inclusion of fibres in asphalt mix provides better fatigue and rutting resistance [5–10]. Fibres also reduce the tendency of SMA towards moisture damage [11, 12]. In this present study, bamboo fibre has been used as a reinforcing stabilizer. Bamboo fibre is a material that is available in a huge quantity in India, especially in northeast India. The average annual production of bamboo in India is approximately about 3.23 million tonnes. The tensile strength of bamboo is very good to be used as a reinforcing agent in SMA. Water absorption and moisture damage of bamboo fibre are also less—all these qualities of bamboo

S. Deb (✉) · Kh. L. Singh
National Institute of Technology, Silchar 788010, Assam, India
e-mail: soumyadeepdeb1@gmail.com

Kh. L. Singh
e-mail: lakshman@civil.nits.ac.in

A. Sil et al. (eds.), *Recent Trends in Civil Engineering*, Lecture Notes in Civil Engineering 274, https://doi.org/10.1007/978-981-19-4055-2_59

fibre suit perfect natural fibre to be utilized in roads [13]. Recent technologies are also developing for the improvisation of bitumen to provide better performance of aggregate and binder. The addition of filler or chemical significantly modifies the binder's quality. The addition of an anti-stripping agent to bitumen provides better aggregate-binder adhesion for the long-term performance of pavement [14]. All-together, anti-stripping additive, and bamboo fibre utilization in SMA can provide a better pavement performance, which is the main objective of this experimental study. According to the literature reviews, utilization of bamboo fibre in stone matrix asphalt (SMA) has very less work in India. The impact of various sizes of fibres in SMA has not been studied yet. The performance criteria of anti-stripping modified bituminous binder added with bamboo fibre are a unique study in this research work.

2 Experimental Programme

2.1 Materials

2.1.1 Aggregate

As per the specification of MoRTH, 5th edition (2013), SMA of 19 mm binder course of aggregate gradation has been adopted in this study comprises of more amount of course aggregate and less amount of fine aggregate. Collection of aggregates was done from local areas. Physical properties of aggregates are shown in Table 1; and the particle size distribution curve is presented in Fig. 1.

Table 1 Properties of aggregate

Name of the experiment		Method	Specification as per IRC SP 79	Results
Combined flakiness and elongation test		IS: 2386 (PART-1)—1963	<30%	24%
Los Angeles abrasion test		IS: 2386 (PART-4)—1963	<25%	21%
Aggregate impact value		IS: 2386 (PART-4)—1963	<18%	13%
Water absorption		IS: 2386 (PART-3)—1963	<2%	0.63%
Aggregate crushing test		IS: 2386 (PART-4)—1963	<30%	23.5%
Specific gravity of aggregate	Coarse aggregate	IS: 2386 (PART-3)—1963		2.673
	Fine aggregate			2.642
	Filler material			2.588

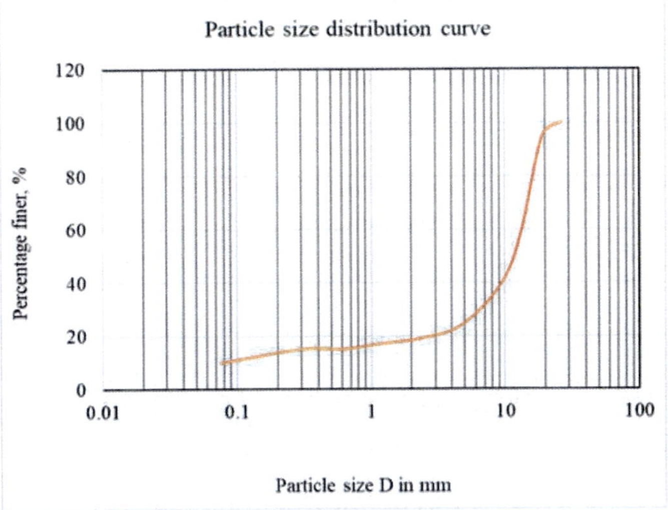

Fig. 1 Particle size distribution

Table 2 Properties of asphalt binder

Parameter	Specification as per IS 73: 2013	Test method	Results
Penetration at 25 °C, 100 g, 5 s, 0.1 mm, Min	45	IS 1203	53
Ductility at 25 °C, cm, Min	40	IS 1208	98
Softening point (R&B), °C, Min	47	IS 1205	55
Flash point (Cleveland open cup), °C, Min	220	IS 1448	255
Fire point	–	–	278
Solubility in trichloroethylene, per cent, Min	99	IS 1216	99

2.1.2 Binder

In this present study, asphalt binder of viscosity grade 30 was utilized received from Indian Oil Corporation Ltd., Assam. Four different amounts of asphalt content were selected for the preparation of SMA specimens. Properties of asphalt binder have been mentioned in Table 2.

2.1.3 Bamboo Fibre

Bamboo fibres used in this study were imported from Lakhni Fabrics Rajasthan. The fibres were segregated into two proportionate lengths by 5 mm and 10 mm

Table 3 Stone matrix asphalt requirements (MoRTH, 5th edition, 2013)

Design parameters	Requirements
Air void content, %	4.0
Bitumen content, %	5.8 minimum
Celluloid fibres	0.3% minimum (by the weight of asphalt mix)
Voids in mineral aggregates, %	17% minimum
Asphalt drain-down, %, IRC: SP: 79	0.3% maximum
Tensile strength ratio (TSR), IRC: SP: 79	85% minimum

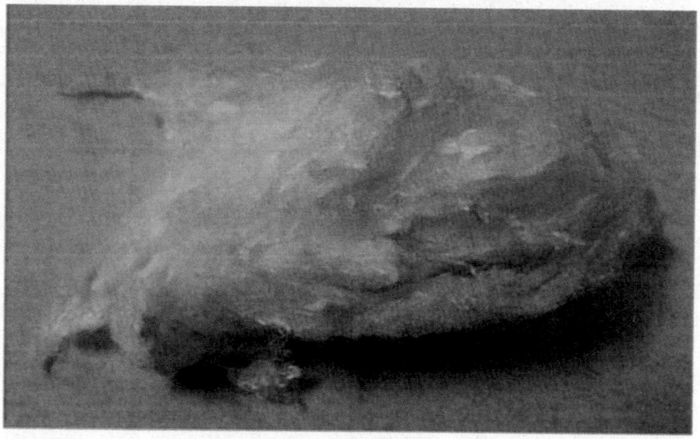

Fig. 2 Bamboo fibre

accordingly. Fibres adopted in the study were 0.3% by weight of asphalt mix [15]. The requirements of SMA for the production of asphalt mix have been mentioned in Table 3 (Fig. 2).

2.1.4 Anti-stripping Agent

An anti-stripping additive by product name of 'WETBOND-S' supplied by Petrochem Specialities, U.P., was utilized in this study. A dosage of 0.07% additive (by the instruction of the manufacturer) by the weight of binder was mixed with bitumen with the help of a closed drum mixer for 30 min in the laboratory, maintaining a temperature of 160–175 C.

2.2 Marshall Stability Test

Marshall stability test is conducted to check the compressive strength and deformation rate of asphalt specimen. It is mostly used everywhere to determine the optimum binder content. In this study, all the volumetric properties test was conducted on the Marshall sample before testing the stability test. The dry weight of sample, weight of sample in water, and surface saturated dry weight of the samples were calculated to find the volumetric results by plotting the results in graphs. Marshall stability test was conducted on the sample specimens with and without using bamboo fibre. Four asphalt mixes have been prepared for the experimental analysis. Conventional mix consists of SMA prepared with VG-30 binder without any fibre or chemical modification. The controlled mix consists of asphalt binder modified SMA with anti-stripping additive but exclusion of fibres. 5 mm FMA is the SMA mix with anti-stripping agent modified binder has 5 mm length of bamboo fibre. 10 mm FMA is the SMA mix with anti-stripping agent modified binder has 10 mm length of bamboo fibre.

2.3 Indirect Tensile Strength (ITS)

ITS is used to conducted to find out the tensile strength property of asphalt mix specimens. ITS can be conducted in two ways: dry ITS and soaked ITS. In dry ITS, samples are cured in water for 30 min at 60 °C, and in soaked ITS, samples are cured for 24 h at a temperature of 60 °C. The ratio of soaked ITS to dry ITS can be inferred as tensile strength ratio (TSR) or moisture susceptibility.

2.4 Drain-Down Test

One of the very important tests for SMA is the drain-down test by which the property of binder to drain out from aggregate can be calculated. As in SMA, most of the spaces are filled with large-sized aggregate, and there is less amount fine aggregate; bitumen covers the surface of coarse aggregate. When the pavement gets heated, bitumen starts segregating from the aggregate. So, it is necessary to know the drain-down of bitumen from the aggregate.

Table 4 Properties of asphalt binder after addition of anti-stripping agent

Parameter	Specification as per IS 73: 2013	Test method	Results
Penetration at 25 °C, 100 g, 5 s, 0.1 mm, Min	45	IS 1203	54
Ductility at 25 °C, cm, Min	40	IS 1208	96
Softening point (R&B), °C, Min	47	IS 1205	54
Flash point (Cleveland open cup), °C, Min	220	IS 1448	249
Fire point	–	–	281
Solubility in trichloroethylene, per cent, Min	99	IS 1216	99

3 Results and Discussion

3.1 *Effect of Anti-Stripping Agent on Asphalt Binder*

After adding an anti-stripping additive to the bitumen, the physical properties of the binder again tested to know the change in properties of the binder. The results obtained from the experimental tests showed in Table 4 that there was no significant change in the properties of the modified asphalt binder. As the amount of additive was mixed in bitumen was very less (0.07% by the weight of binder), the results were remains almost similar.

3.2 *Volumetric and Strength Criteria for Stone Matrix Asphalt*

Volumetric properties of conventional mix specimen and the asphalt mixes modified with fibres are presented in Fig. 3,4,5,6. The variation of volumetric properties, i.e. bulk specific gravity, air voids content, voids in mineral aggregates, voids filled by binder with respect to binder content represented in the figures. It can be observed that with an increasing amount of bitumen, the air voids decreased; no significant change was observed for conventional mix and fibre-modified specimen as SMA contains more amount of coarse aggregate and less amount of fine aggregate, so at lower binder content, air voids were more than the higher binder content. For all specimens, air voids content lies in between 4.7 and 8%, which satisfy the standard criteria. Bulk specific gravity (Gm) values observed between 2.135 for the conventional mix to a maximum value of 2.180 for 5 mm bamboo fibre-modified asphalt mix. VMA for conventional mix and fibre mixed specimens was more than 17%. VFB for conventional mix and fibre mixed specimens ranges between 60 and 78%.

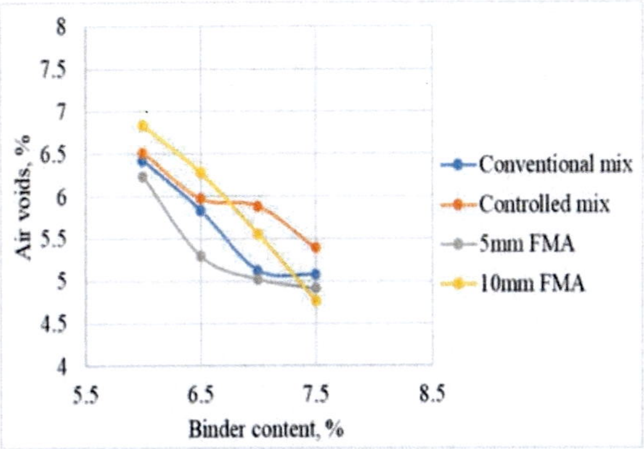

Fig. 3 Binder content vs air voids

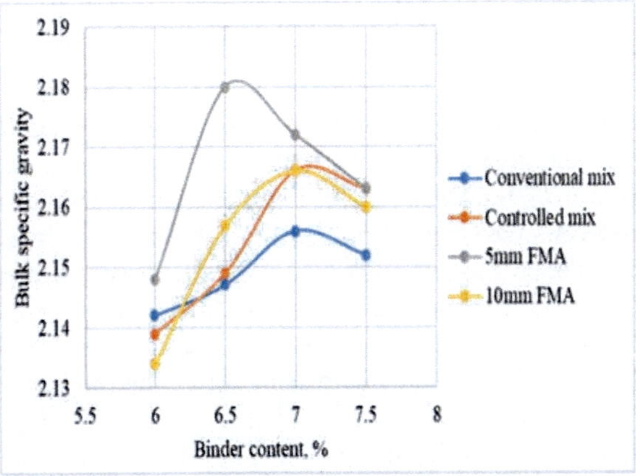

Fig. 4 Binder content vs Gm

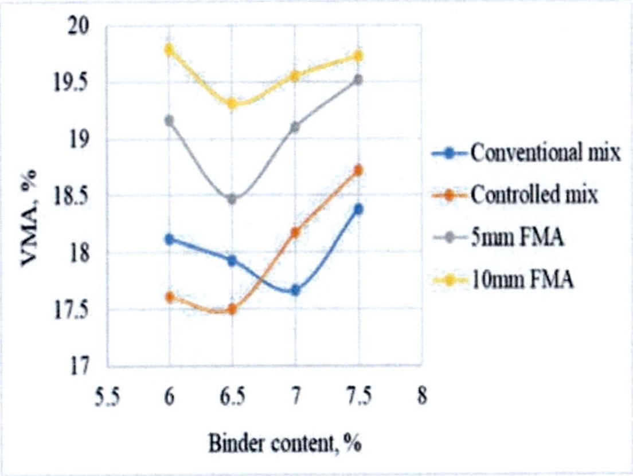

Fig. 5 Binder content vs VMA

Fig. 6 Binder content vs VFB

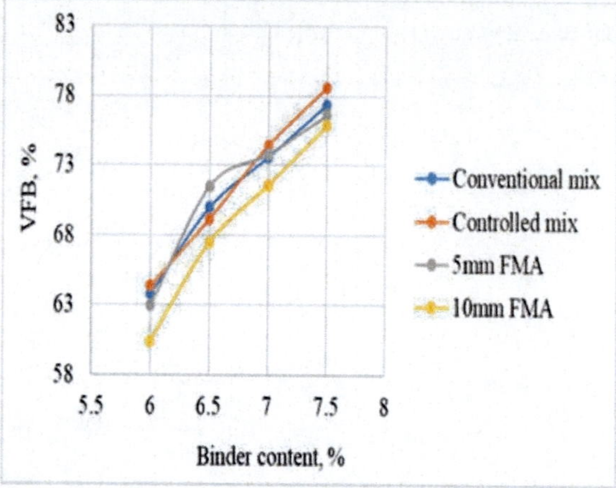

Marshall stability value and flow obtained from the experimental values are plotted in Figs. 7 and 8. From Fig. 7, it can be interpreted that; addition of bamboo fibre increased the stability value than the conventional specimen. The top peak for stability was observed at 7% binder content of 5 mm bamboo fibre, whilst lesser but no significant value was obtained from 10 mm bamboo fibre at 6.5% binder. With the increasing amount of binder content, stability values start decreasing—flow or deformation rate of asphalt specimen represented in Fig. 8. At lower bitumen content, the deformation rate was less for specimens. But with an increasing amount of bitumen content, the flow value of asphalt mix starts increasing for all the mix specimens.

Fig. 7 Binder content vs stability

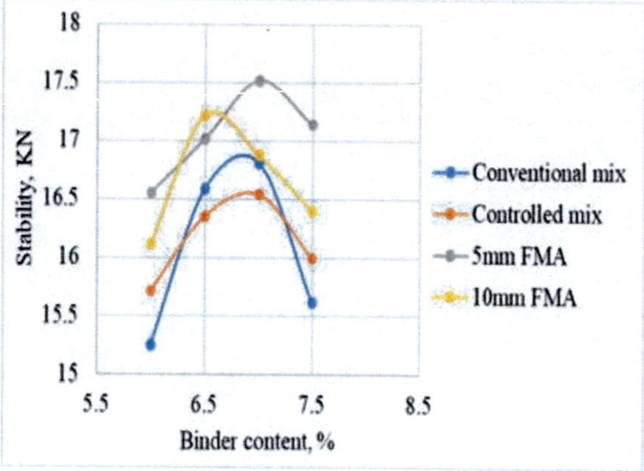

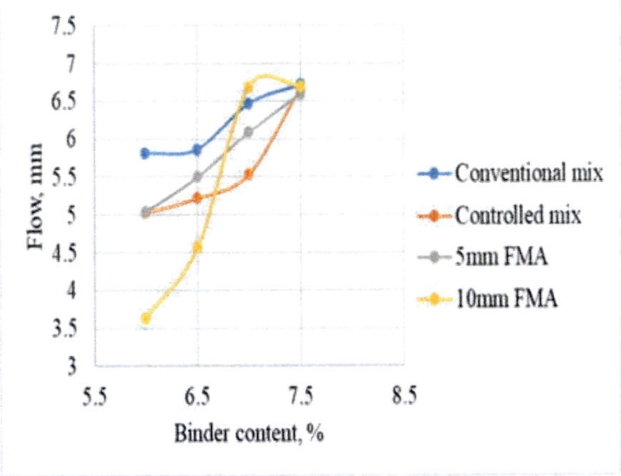

Fig. 8 Binder content vs flow

3.3 *Indirect Tensile Strength Characteristics*

Indirect tensile strength has been tested for conditioned and unconditioned specimens. Conditioned specimens kept for 24 h at 60 °C in a water bath, and for unconditioned samples, specimens were kept for 30 min at 60 °C in a water bath before the tensile strength test. In this criterion, it was observed that conditioned specimens have lesser tensile strength compared to the unconditioned ones (Fig. 9).

Fibre-modified asphalt mixes have more tensile strength compared to conventional mix. The ITS value of 10 mm bamboo fibre-modified mix has slightly high than 5 mm bamboo fibre modified. The inclusion of fibres in the asphalt mix gives better resistance to tensile cracking as the fibres act as a reinforcement in the mix to maintain stone-to-stone bonding in SMA by holding the aggregate-binder mix in place (Fig. 10).

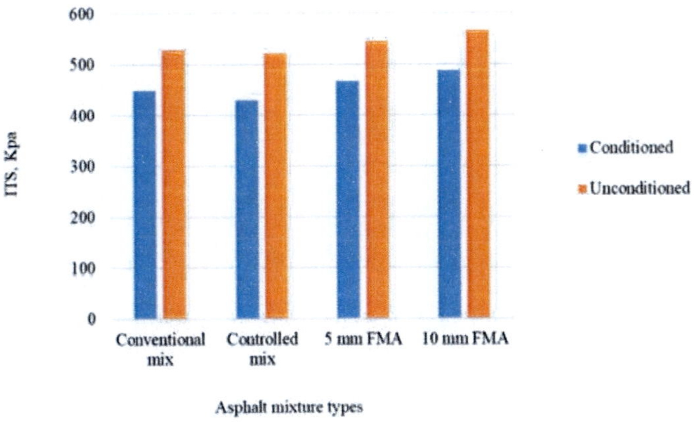

Fig. 9 Variation of ITS with different asphalt mixtures

Fig. 10 Asphalt mixture vs TSR

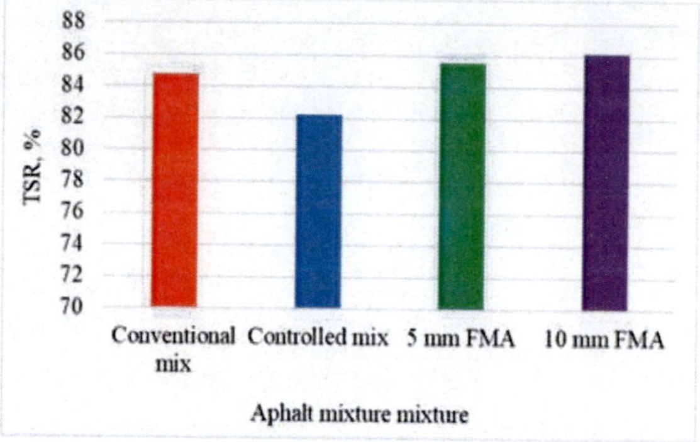

Moisture susceptibility or tensile strength ratio (TSR) of 5 mm bamboo fibre-modified asphalt mix has a higher value than the other asphalt mixes. The TSR value of conventional mix, controlled mix, 5 mm FMA, and 10 mm FMA has a value of 84.7%, 82.2%, 85.5%, and 86%, respectively. Though, all the asphalt mixes to satisfy the required criteria for tensile strength ratio.

3.4 Drain-Down Characteristics

Segregation of bitumen from aggregate, i.e. the drain-down of binder, was observed less in fibre mixed asphalt mixture compared to conventional mix specimen. It was observed from the results that fibres-modified asphalt mix required more amount of binder, which holds the bitumen in the asphalt mix and does not allow the binder to drain out. In conventional and controlled mix, missing fibre leads to the segregation of binder more from aggregate by 21% and 26%, respectively, than the fibre-modified asphalt mix specimen with 17% and 18% drain-down values. For all the cases, values satisfied the specific standard criteria (Fig. 11).

Fig. 11 Drain-down value vs asphalt mixture types

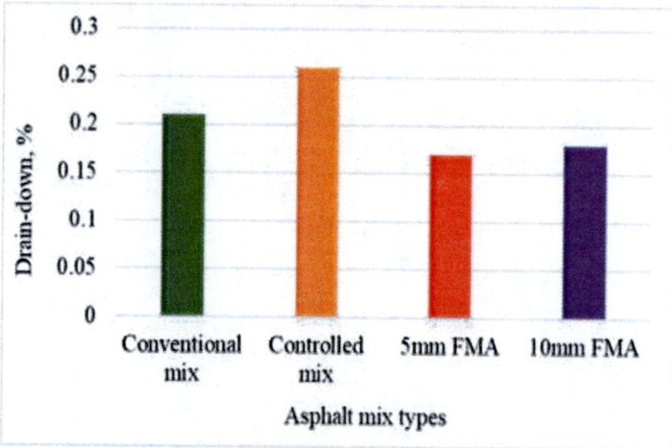

4 Conclusion

In this study, four different types of SMA specimens were prepared. Anti-stripping additive modified bitumen added with aggregates with and without the inclusion of fibres. In conventional and controlled mix, fibre was exempted for the preparation of the specimen. The other two mix designs, 5 mm and 10 mm (by length) bamboo fibres, were utilized, which was being termed as fibre-modified asphalt (FMA). The gradation of aggregates was kept constant for all specimens. Based on the strength and deformation characteristic, it can be observed that 5 mm FMA has more stability value than conventional, controlled and 10 mm FMA mix. The stability value of 5 mm FMA was approximately 4% higher than 10 mm FMA at 7% binder content but 2% less at 6.5% binder content. 10 mm FMA has less deformation at lower bitumen content, but as binder content increases, its deformation rate increased compared to 5 mm FMA. At binder content 6.5%, 10 mm FMA has 20% less deformation, but at 7% BC, it has 9% more deformation rate.

From the experimental analysis, it can be concluded that.

1. Test conducted on physical properties on bitumen almost remains unchanged after modification by the anti-stripping additive.
2. All the asphalt mix with different asphalt mix satisfied the drain-down criterion. Bitumen drain-down for FMA was less compared to conventional and controlled mix as fibres absorbed the bitumen in FMA. Drain-down of bitumen in 5 mm FMA and 10 mm FMA has no significant difference. FMA has 17% and 44% lesser than conventional and controlled mix, respectively.
3. Indirect tensile strength of 10 mm FMA was 7%, 8.4%, and 3% higher tensile strength than conventional, controlled mix, and 5 mm FMA.
4. The tensile strength ratio (TSR) of all mixes has been satisfied the criteria of a higher value of more than 80%. Amongst that, 10 mm FMA has obtained a value higher than the other mix designs.

Acknowledgements The authors like to thank the respected faculties and non-teaching staffs of the Civil Department for their technical help. Petrochem Specialities (U.P.) and Lakhani Fabrics (Rajasthan) for providing necessary materials for this study. The authors are also grateful to senior research scholars Dheeraj Deshmukh and Pinki Deb for their support and encouragement. This study and experimental analysis was carried out at National Institute of Technology, Silchar.

Disclosure Statement
Hereby, the authors declare that there is no conflict of interest between the authors.

References

1. Mallick RB, El-Korchi T (2009) Pavement engineering—principles and practice. CRC Press, Taylor & Francis Group
2. Amir M, Mortaza R, Pooyan A (2015) Effect of coal waste in hot mix asphalt compared to conventional fillers: mix mechanical properties, and environmental impact. J Clean Product 91:262–268
3. Shukla RS, Jain PK (1990) A laboratory study on the utilization of waste tyre rubber and reclaimed rubber in construction of bituminous roads. High Res Bullet, IRC 41:17–41
4. Sarang G, Lekha BM, Geetha JS, Ravi Shankar AU (2015) Laboratory performance of stone mastic asphalt mixtures with two aggregate gradation. J Mod Transport 23(2):130–136
5. Dash SB, Panda M (2016) A study on the use of natural fiber for improvement in engineering properties of dense graded bituminous mixes with coal ash. Transp Dev Econ 2:4. https://doi.org/10.1007/s40890-015-0008-z
6. Hadiwardoyo SP, Sumabrata RJ, Jayanti P (2013) Contribution of short coconut fiber to pavement skid resistance. Adv Mat Res 789:248–254
7. Huang Y, Liu Z, Liu L, Zhang Y, Xu Q (2020) Hybrid modification of stone mastic asphalt with cellulose and basalt fibre. Adv Mat Sci Eng, Article ID 5671256
8. Vishnu TB, Singh KL (2021) Experimental investigation of fatigue characteristics of asphalt mixes with different automobile waste tyres as modifiers", Iranian journal of science & technology. Trans Civil Eng. https://doi.org/10.1007/s40996-021-00588-w
9. Kiran Kumar NLN, Ravitheja A (2019) Characteristics of stone mastic asphalt by using natural fibres as additive. Materials Today Proceed 19:397–402
10. Xing X, Chen S, Li Y, Pei J, Zhang J, Wen Y, Li R (2020) Effect of different fibres on the properties of asphalt mastics. Constr Build Mater 262:120005
11. Gardiner MS (2009) Laboratory test for assessing moisture damage of asphalt concrete mixture, T.R.R.NO, 1353, p 15–23
12. Julaganti A, Choudhary R, Kumar A (2017) Moisture susceptibility of WMA mixes with modified bituminous. Pet Sci Technol 35(10):1014–1021
13. Chin SC, Tee KF, Tong FS, Ong HR, Gimbun J (2020) Thermal and mechanical properties of bamboo fiber reinforced composites. Mat Today Comm 23:100876
14. Mena IS, Hajj EY, Sebaaly PE (2015) Impact of antistrip additives on the long-term aging rheological properties of asphalt binder. J Mat Civil Eng 27(8):C 4014006
15. Ministry of Road Transport and Highways (MoRTH): Specifications for Road and Bridge Works, Indian Road Congress, New Delhi (2013)

Effect of Pocket Size of Bamboo Geocell Filled with Sand and Recycled Asphalt Pavement Over Soft Clay Bed

T. Uma Shankar, Sudeep Biswas, and Monowar Hussain

1 Introduction

Geocells are expandable three-dimensional cells used to improve the performance of various civil engineering infrastructure. The 3D nature of geocell provides confinement to the infill soil, improving the soil's overall strength. Several researchers had conducted various studies on geocells to find the advantages and mechanism in geotechnical applications [1–4]. Hegde and Sitharam [1] investigate the impact of various geocell parameters, such as height, strength and pattern, on the bearing potential of the reinforced sand bed. It was also reported that the geocell's tensile strength and surface roughness, in addition to its three-dimensional shape, have an impact on its performance. The most critical variables that govern the performance of geocell are tensile strength and surface roughness. "Short-term stability" is the most significant aspect of building a geotechnical system on soft soils. Consolidation with time results strengthens the soil. Bamboo is a biodegradable material used as reinforcement to strengthen the soil [5–8]. Therefore, locally available bamboo mats were used to construct 3D geocell and the influence of the pocket size was investigated. The photograph of the bamboo geocell is depicted in Fig. 1.

T. U. Shankar (✉) · S. Biswas · M. Hussain
Department of Civil Engineering, National Institute of Technology, Silchar 788010, India
e-mail: shankarvarma789@gmail.com

S. Biswas
e-mail: sudeep_rs@civil.nits.ac.in

M. Hussain
e-mail: monowar@civil.nits.ac.in

A. Sil et al. (eds.), *Recent Trends in Civil Engineering*, Lecture Notes in Civil Engineering 274, https://doi.org/10.1007/978-981-19-4055-2_60

Fig. 1 Bamboo geocell used in this study

2 Laboratory Model Plate Test

2.1 *Test Setup*

A steel test tank with a length, width and height of each 1 m was used in this study. A circular steel plate was used as footing of diameter 150 mm. The load was applied with a manually operated hydraulic jack through a proving ring. The subgrade bed was taken as 500 mm. Four layers of soil, and each of 125 mm was compacted for uniform compaction. The aggregates were filled up to 200 mm and compacted with a plate vibrator. To reduce sidewall friction, polyethene sheets were used to cover the tank's sides. The bamboo geocells used in the study had pocket diameters of 0.8B, 1B, 1.3B. The geocell's height and width were taken as 1B and 6.2B. The upper layer thickness was adapted as 0.33B to reduce the damage of geocell. The footing was placed at the centre of the testbed. Load on the footing and surface settlements were measured with proving ring and dial gauges. The test setup and test details are given in Fig. 2 and Table 1.

2.2 *Materials Used*

The subgrade bed was prepared using locally available soil with a specific gravity of 2.56. The soil had a liquid limit and plastic limit of 68% and 34%. According to the unified soil classification system, the soil is classified as silt with high plasticity (ML). The soils OMC and M_d were 34% and 15.4 kN/m^3. For preparing the subgrade, the undrained cohesion of the soil was considered as 15 kPa.

Fig. 2 Test setup (model plate load test)

Table 1 Details of model plate load test

Details	Type of geocell	Infill material	Variable parameters	Constant parameters
1. Test on unreinforced bed	–	Sand	–	–
2. Effect of geocell pocket size	Bamboo	Sand	d/B = 0.8,1,1.3	u/B = 0.33, h/B = 1, b/B = 6.2
3.Test on unreinforced bed	–	RAP	–	–
4. Effect of geocell pocket size	Bamboo	RAP	d/B = 0.8,1,1.3	u/B = 0.33, h/B = 1, b/B = 6.2

Note u = Depth of geocell; B = Diameter of footing; h = Height of bamboo geocell; b = Width of bamboo geocell

The infill material was taken as dry sand and recycled asphalt pavement (RAP). The sand had a specific gravity, effective particle size, coefficient of uniformity and coefficient of curvature as 2.64, 0.15 mm, 1.72 and 1.14, respectively. The internal friction angle of sand was found 34°, and the maximum and minimum void ratio was obtained as 0.81and 0.51. As per USCS the sand was listed as uniformly graded sand (SP). The specific gravity, effective particle size, coefficient of uniformity and coefficient of curvature of RAP was found as 2.43, 3.5 mm, 35, and 2.25. The internal friction angle was found as 48°. RAP was categorised as well-graded gravel with the symbol GW according to the USCS.

Table 2 Properties of bamboo geocell used

Parameters	Quantity
Mass per unit area (g/m^2)	755.24
Pocket sizes (mm)	120, 150 and 200
No. of cells	45, 35 and 25
Strip thickness (mm)	1.5
Cell depth (mm)	150
Ultimate tensile strength (kN/m)	72

The bamboo used in the study was collected from the local market of Silchar. The plane bamboo matt was cut into pieces and joined together with steel rope to create pocket size of 1.3B, 1B and 0.8B. The parameters considered for preparation of the geocell are given in Table 2.

2.3 Test Procedure

For preparing the soft subgrade bed, specific amount of water was added to the soil and kept in an airtight container for 3–4 days for uniform consistency. The parameters of the subgrade were taken as mentioned in Table 3. The soil was compacted with a steel hammer in four layers to achieve uniform compaction. The properties of subgrade bed were checked with a core cutter and found in the range of ±5%.

The aggregates were filled into the tank by measuring the weight of the aggregates. The infill material was filled up to 20 cm and compacted with the plate vibrator to achieve a relative density of 80%. The base layer was compacted for 20 ± 5 min to achieve the relative target density and found ±3%. Load and settlement were measured at regular intervals.

Table 3 Properties of soft subgrade bed

Parameters	Values
Moisture content (%)	61
Unit weight (kN/m^3)	16.48
Undrained cohesion (kPa)	15

3 Results and Discussion

3.1 *Effect of Pocket Size of Bamboo Geocell Reinforced Sand Bed Over Soft Subgrade*

Pressure-settlement response of sand base layer is presented in Fig. 3. The bearing pressure was observed as 24 kPa and 59 kPa for the unreinforced foundation subgrade and sand bed. The reinforced layer bearing capacity was found as130kPa, 191 kPa and 227 kPa for the geocell pocket size of 1.3B, 1B and 0.8B. The failure of the sand-filled bamboo geocell was observed at a footing settlement of 12%. The load-carrying capacity was found to increase with decreasing of the pocket size of bamboo geocell. Due to decreasing the pocket size, the lateral resistance of the geocell increases and reduces the infill material's movement. When pocket size increases, the bamboo geocell's confinement reduces, which leads to large settlement and surface deformation.

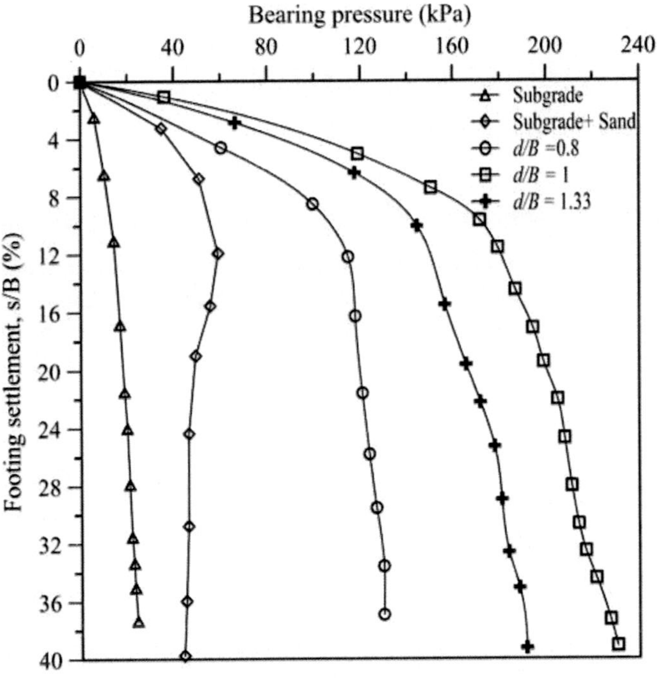

Fig. 3 Bearing pressure versus settlement for different pocket size of bamboo geocell filled with sand

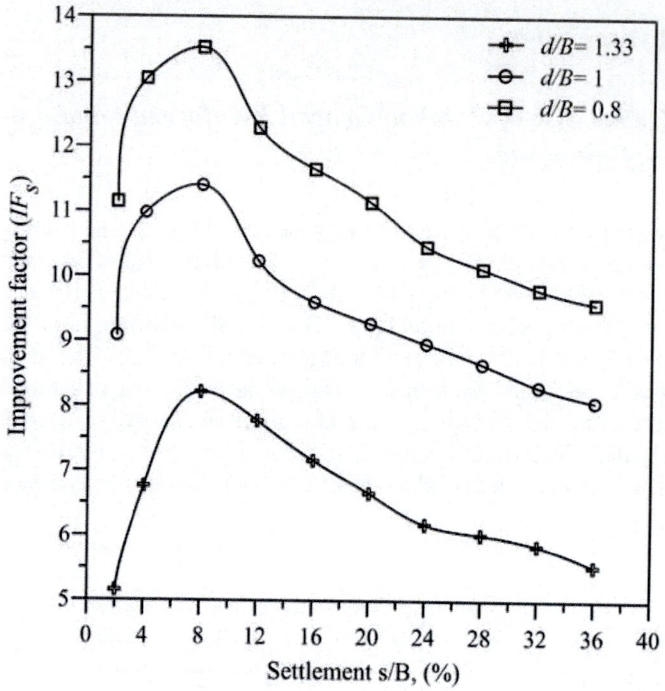

Fig. 4 Variation of improvement factors with different pocket size

Improvement factors of different pocket sizes of bamboo geocell filled with sand

The improvement factors were calculated by taking the ratio of the bearing capacity of the reinforced section to the unreinforced subgrade. Figure 4 represents the improvement factor for sand-filled bamboo geocell of different pocket sizes. The improvement factors were found as 13.6, 11.4 and 8.3 at s/B of 8% for pocket size of 0.8B, 1B and 1.3B, respectively. The smaller pocket size has more stiffness as compared to the larger pocket sizes and provides more confinement. Therefore, the improvement factor is inversely proportional to the pocket size.

3.1.1 Effect of Pocket Size of Bamboo Geocell Reinforced RAP Bed Over Soft Subgrade

Figure 5 represents the pressure-settlement response of bamboo geocell filled with RAP. The bearing capacity was observed as 185 kPa for the unreinforced RAP. The bearing capacity of the bamboo geocell filled with RAP was observed as 298 kPa, 310 kPa and 350 kPa for the pocket size of 1.3B, 1B and 0.8B, respectively. The bearing capacity of the bamboo geocell was found superior with the infill material RAP than the sand. The RAP-filled bamboo geocell's failure was found at s/B of

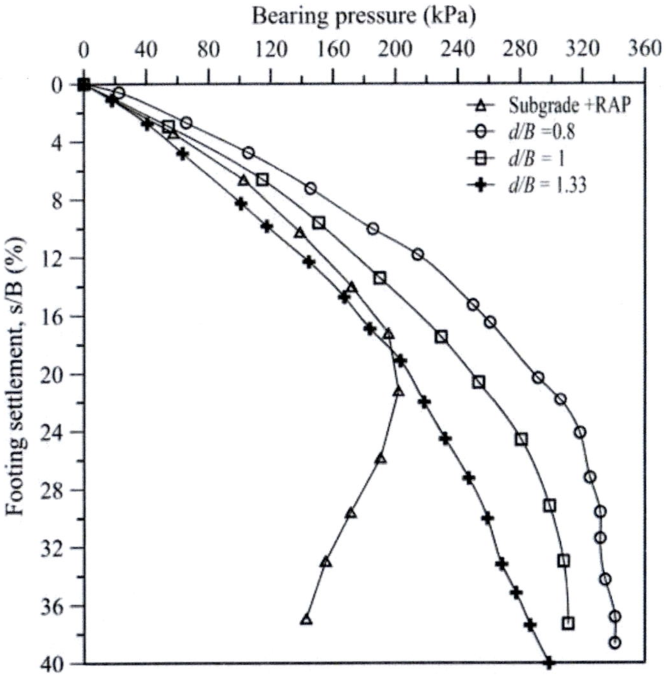

Fig. 5 Bearing pressure versus settlement for different pocket size of bamboo geocell filled with RAP

24%. The geocell performance was observed increasing with decreasing the pocket size for the RAP as similar to the sand as infill material.

Improvement factors of different pocket sizes of bamboo geocell with RAP

The improvement factor was calculated for different pocket sizes of the bamboo geocell filled with RAP as similar to the sand, which is shown in Fig. 6. From Fig. 6, it was observed that the improvement factors are 16.2, 14 and 11.2 for 0.8B,1B and 1.3B at s/B of 20%. With the decrease in pocket size, the improvement factor was observed to increase. The smaller one has more stiffness than the larger one and provides more confinement, but the improvement was marginal with decreasing the pocket size beyond 1B.

4 Conclusion

The present study consists of the variation of pocket size of bamboo geocell filled with different infill material (sand and RAP) under the circular footing. Following conclusions are extracted from the test results:

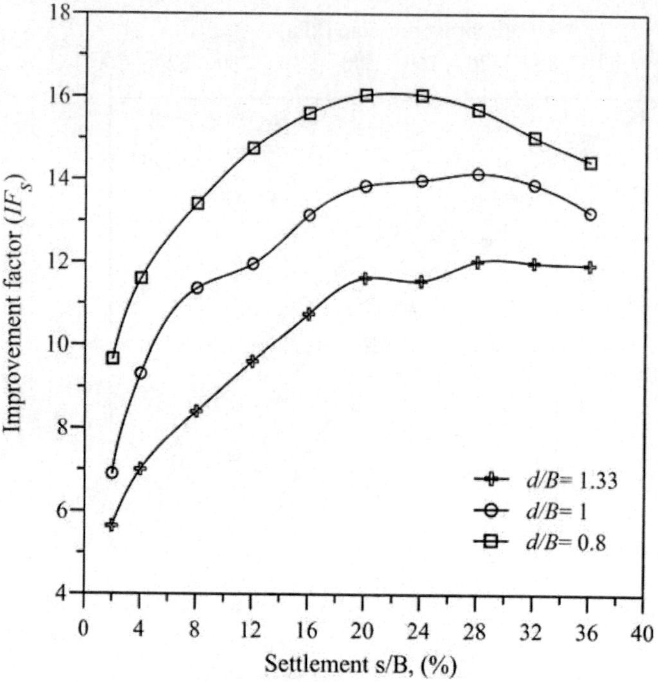

Fig. 6 Variation of improvement factors for different pocket size

- The bamboo geocell reinforced base layer showed improved bearing capacity than unreinforced sand and RAP.
- The pocket size of geocell and load-carrying capacity was found inversely proportional.
- The pocket size should be less than the footing diameter and found higher performance.
- The load-carrying capacity of bamboo geocell reinforced foundation bed was found maximum with pocket size 0.8B for both RAP and sand.
- The maximum improvement was 16.2 and 13.6 for RAP and sand-filled bamboo geocell for the pocket size of 0.8B.

References

1. Hegde A, Sitharam TG (2013) Experimental and numerical studies on footing supported on geocell reinforced sand and clay beds. Int J Geotech Eng 7(4):346–354
2. Madhavi Latha G, Nair AM, Hemalatha MS (2010) Performance of geosynthetics in unpaved roads. Int J Geotech Eng 4(3):337–349
3. Dash SK, Krishnaswamy NR, Rajagopal K (2001) Strip footing on geocell reinforced sand beds with additional planar reinforcement, 19(8), pp 529–538

4. Hegde A, Sitharam TG (2014) Effect of infill materials on the performance of geocell reinforced soft clay beds. In: Indian Geotechnical engineers conference, Chennai, pp 99–102
5. Hegde A, Sitharam TG (2015) Experimental and analytical studies on soft clay beds reinforced with bamboo cells and geocells. Int J Geosynthetics Ground Eng 1(2):1–13
6. Rajagopal K, Nithin S, Veeraragavan A (2015) The use of natural geotextiles in reinforcing the unpaved roads. 6th Int J Geotech Engineering [Google Scholar], January
7. Hegde A, Sitharam TG (2015) Use of bamboo in soft-ground engineering and its performance comparison with geosynthetics. J Mat Civil Eng 27(9)
8. Han J et al (2011) Performance of geocell reinforced RAP bases over weak subgrade under full scale moving wheel loads. J Mater Civil Engineering, 1525–1534. https://doi.org/10.1061/(ASCE)MT.1943-5533.0000286

Finite Element Analysis of Flexible Pavement with Non-linear Material Behaviour

Nabarun Dey, Barnali Debnath, and Partha Pratim Sarkar

1 Introduction

The urbanization and current progression of infrastructure requires huge amount of road networks and currently the road networks are immensely extending towards to the remote places. In India, the current "Road development plan vision 2021" has mentioned to increase the length of roads. Also, the National Highway Authority of India (NHAI) has planned to construct 22 new expressways and economic corridors by 2025. However, these roads often suffer from several distresses due to which it is required to check the performance of several roads before their application. Advances in computation tools allow the researchers to characterize the nature of pavements based in their applicability [1]. Numerical finite element (FE) packages for simulating the pavement responses under several traffic loading condition have been investigated from many viewpoints in the past few decades [2]. Finite element based structural analysis on flexible pavements has become an important mechanistic method for the analysis of flexible pavements because it can easily simulate the material behaviour [3]. However, most of these studies considered linear elastic material models, without taking the actual material behaviour. But actuality, grain-type pavement materials may behave nonlinearly, based on several factors. Some previous literature mentioned that the responses of subgrade, granular material and even the surface materials did not follow the linear pattern, rather a nonlinear stress-strain behaviour was observed [3–5]. Hence, it is very important to consider the material behaviour beyond the linear stage to enhance pavement design and to reduce the pavement distresses. For the design of flexible pavement, Indian Road Congress (IRC 37-2018) [6] has mentioned to use a software package called IITPAVE, however this software is mainly based on the elastic solutions, and it does not incorporate the nonlinear material behaviour. Hence, an attempt is adopted in the present study to

N. Dey · B. Debnath (✉) · P. P. Sarkar
National Institute of Technology Agartala, Agartala 799046, India
e-mail: brnali540@gmail.com

A. Sil et al. (eds.), *Recent Trends in Civil Engineering*, Lecture Notes in Civil Engineering 274, https://doi.org/10.1007/978-981-19-4055-2_61

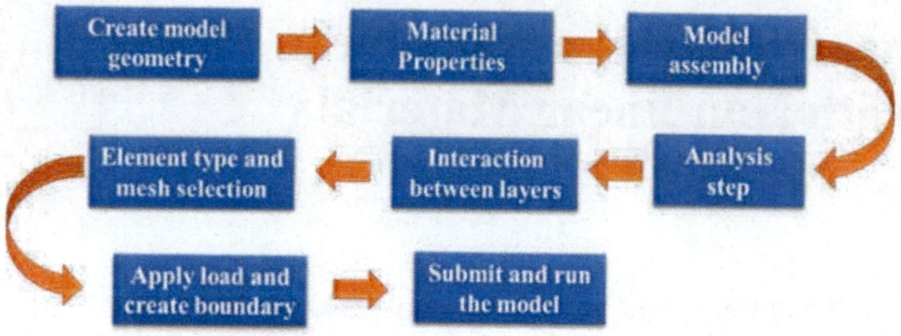

Fig. 1 Steps used for finite element modelling

incorporate the nonlinear behaviour of the pavement layers. In this research, the well-known material models are used into a 3D finite-element model developed with the software package ABAQUS. Several pavement models are prepared considering various material behaviour. Then the results obtained from nonlinear models are compared with those of the linear based program.

2 Preparation of Numerical Model

A four layer pavement is taken under consideration having bituminous surface layer, granular base and subbase layer and a fine grained subgrade layer. A pavement section is prepared with 2000 mm × 2000 mm square slabs of three different layers, and the total height is taken as 2000 mm. The thickness of bituminous surface layer is 100 mm, thickness of granular base layer is 150 mm, thickness of granular subbase layer is 300 mm and the subgrade thickness is 1450 mm. In actual condition, the subgrade extends towards infinity, and hence a much higher thickness is chosen for the subgrade. The ground contact area of the tire is taken as an equivalent loading area which is 300 mm × 150 mm and the contact pressure is kept as 0.8 MPa in static as well as dynamic loading condition. 8-nodded regular hexahedral brick element (C3D8R) is considered for all the layers. Finer mesh was applied at the vicinity of the loading area and relatively coarser mesh were provided on rest of the areas. For boundary conditions, the bottom was fixed about any kind of translations and rotations, whereas the sides are provided with symmetric boundaries. The steps used for model preparation is shown in Fig. 1. The details of the pavement model is shown in Fig. 2.

3 Model Description

In the present study, several models are prepared based on the variety of the material properties. The details of all the models are presented in the following contexts.

Fig. 2 Mesh and boundary conditions applied in the model

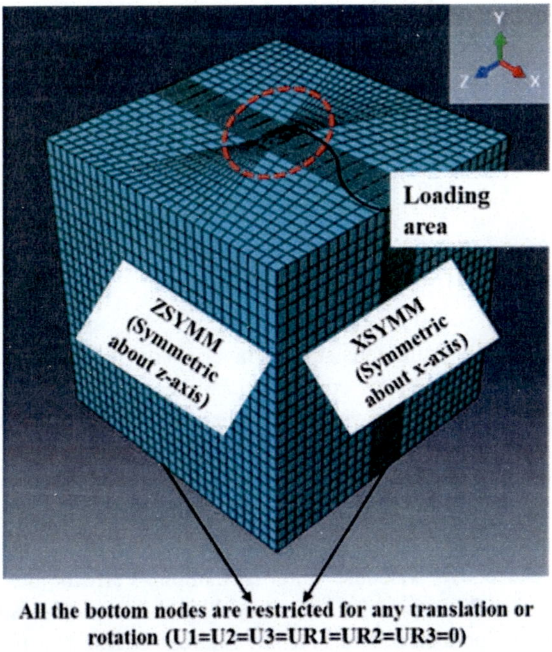

3.1 Linear Elastic Model

In linear elastic model, all the material behaviour are assumed as linear. It is a very simplified method as only elastic conditions are taken into account. The common assumptions of this model are (a) pavement layers extend infinitely in the horizontal direction, (b) the bottom layer (usually the subgrade) extends infinitely downward, and (c) materials are not stressed beyond their elastic ranges. The material properties used in this model are mentioned in Table 1.

Table 1 Input parameters for linear elastic model

Pavement layer	Elastic modulus (MPa)	Poisson's ratio
Surface	1800	0.30
Base	138	0.35
Subbase	96.6	0.35
Subgrade	55.2	0.35

Table 2 Input parameters for visco-plastic model

Pavement layer	Elastic modulus (MPa)	Poisson's ratio	*g-i*	*k-i*	tan-*i*
Surface	1800	0.3	0.43	0	33.3
Base	138	0.35			
Subbase	96.6	0.35			
Subgrade	55.2	0.35			

3.2 Visco-Elastic Model

In this model, the bituminous surface layer is considered as visco-elastic in nature, whereas the other layers are considered as linear elastic layer. Abaqus/CAE allows to directly evaluate the behaviour of viscoelastic materials by automatically creating response curves based on the input data. A viscoelastic material can be evaluated only if it is defined in the time domain and includes hyper elastic and/or elastic material data. The basic assumptions of visco-elastic models are that this model describes isotropic rate-dependent material behaviour for materials in which dissipative losses primarily caused by "viscous" (internal damping) effects must be modelled in the time domain and this model also assumes that the shear (deviatoric) and volumetric behaviour are independent in multiaxial stress states. The material data are used similar to the linear elastic model, and the Prony series data of defining viso-elasticity are taken from previous literature. The input parameters are mentioned in Table 2.

3.3 Elasto-Plastic Model

The proposed model tries to take into account the contractancy and dilatancy phenomena, the influence of the confining pressure, the influence of the initial void ratio, and the effect of partial saturation through the definition of a macroscopic cohesion. This model consists both elastic and plastic properties of the materials. The input parameters used in this model are presented in Table 3.

Table 3 Input parameters for elasto-plastic model

Pavement layer	Elastic properties		Plastic properties	
	Elastic modulus (MPa)	Poisson's ratio	Yield stress	Plastic strain
Surface	1800	0.3	–	–
Base	138	0.35	200	0
Subbase	96.6	0.35	240	0.024
Subgrade	55.2	0.35	280	0.05

Table 4 Input parameters for Drucker-Prager model

Pavement layer	Elastic modulus (MPa)	Poisson's ratio	Angle of friction	Flow stress ratio	Angle of dilation	Yield stress (MPa)	Abs. plastic strain
Surface	1800	0.3					
Base	138	0.35	20	0.795	20	0.005	0
Subbase	96.6	0.35	20	0.795	20	0.01	0
Subgrade	55.2	0.35	20	0.795	20	0.05	0

3.4 Drucker-Prager Model

This model was proposed by Drucker and Prager in 1952, and thereafter it is termed as Drucker-Prager or DP model. It can well describe pressure-sensitive materials such as rock, soil and concrete (Table 4).

4 Result and Discussion

The stress-strain relationships are obtained from all the models and presented in Figs. 3, 4, 5 and 6. It is clear from Fig. 3 that the variation of stress and strain for linear elastic model is also almost linear in nature, where the strain increases linearly with stress. For other three figures, i.e. from Figs. 4, 5 and 6, it is clear that the stress-strain relationships are nonlinear in nature. One more important aspect can also be understood from the figures that the strain value is much higher in linear elastic case as compared to visco-elastic, elasto-plastic and Druker-Prager model, whereas the stress value is much lesser in linear elastic model as compared to the other models. This indicates that the linear model overestimates the strain value and underestimates the stress value.

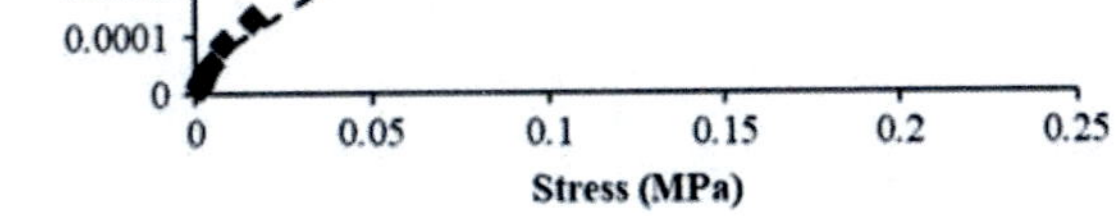

Fig. 3 Stress-strain relationship for linear elastic model

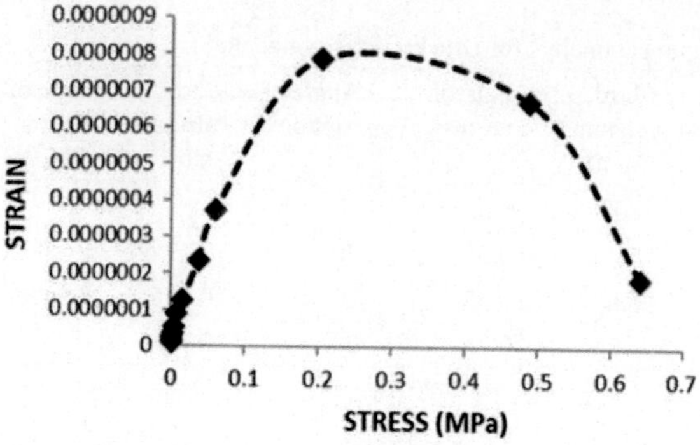

Fig. 4 Stress-strain relationship for visco-elastic model

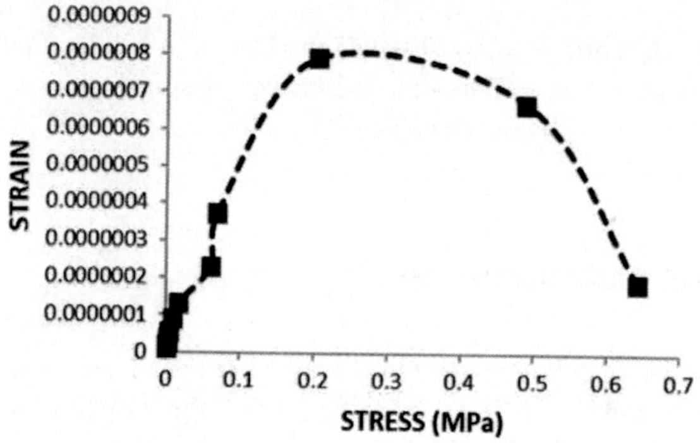

Fig. 5 Stress-strain relationship for elasto-plastic model

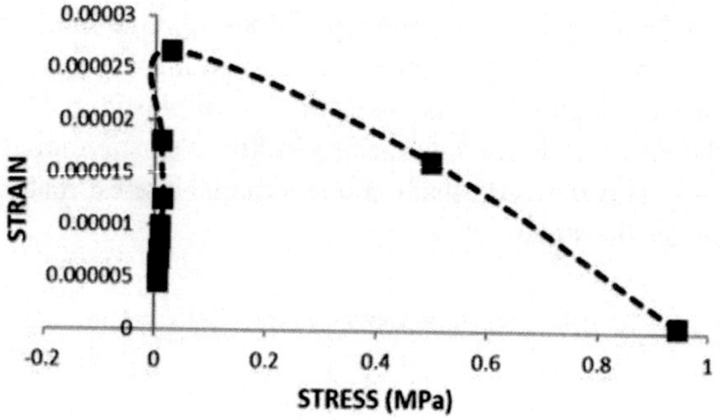

Fig. 6 Stress-strain relationship for DP model

5 Conclusion

The general conclusion of this study are:

1. Compared to conventional subgrade, vertical stress generated in the DP model is more than any other model.
2. The study clearly hints the strain decreases along as the depth progresses linearly whereas the stress variation is uniformly decreases but somewhere or in some

node point sudden spike is experienced hence, the stress variation is not uniformly decreasing along the depth.

3. The overall result points that the Elastoplastic model under static loading more accurately simulates the actual flexible pavement under normal temperature
4. From these results, it can be said that an flexible pavement is more efficiently behaving as an elasto-plastic model depicting both the elastic and plastic conditions with the top surface or bituminous surface rather than an complete elastic or simply viscoelastic model as studied in early analysis subjected to static and cyclic loading conditions.

References

1. Park DW, Martin AE, Masad E (2005) Effects of nonuniform tire contact stresses on pavement response. J Transp Eng 131(11):873–879
2. Kuo CM, Huang CW (2006) Three-dimensional pavement analysis with nonlinear subgrade materials. J Mater Civ Eng 18(4):537–544
3. Kim M, Tutumluer E, Kwon J (2009) Nonlinear pavement foundation modeling for three-dimensional finite-element analysis of flexible pavements. Int J Geomech 9(5):195–208
4. Uzan J (1985) Characterization of granular materials. In: Transportation research record. 1022, Transportation Research Board, Washington, D.C., pp 52–59
5. Brown SF, Pappin JW (1981) Analysis of pavements with granular bases. In: Transportation research record. 810, Transportation Research Board, Washington, D.C., pp 17–23
6. IRC-37 (2018) Design of flexible pavements. Indian Road Congress, New Delhi

Risk Management in Public Private Partnerships (PPP)—An Indian Case Study

K. V. Prasad, Raja Sekhar Mamillapalli, and Hanumantha Rao Pusarla

1 Introduction

The need of rapid infrastructure development is more than ever and budgetary and financing constraints have led governments to develop innovative means of financing the public projects [1]. One such option is to invite the participation of private sector in the development of public projects through Public Private Partnership (PPP). PPP has become an increasingly popular practice in many countries for the financing, construction, renovation, management or maintenance of social, transport and engineering infrastructure [2].

There have many successful PPP projects. However, a study of the unsuccessful projects lends an opportunity to identify weak areas, tighten the framework with strong foundation and help governments and private developers to manage the risks in future PPP infrastructure projects [3]. Recent studies indicate that PPP projects entail more risks than other type of projects [4]. As per the World Bank database in the developing economies, 292 PPP projects out of 8295 projects (~3.5%) failed in the last decade. Recently, the PPI database of the World Bank conducted a study to analyze the failure rates of the PPP projects. The study has brought out that the failure rate was about 2% till 2011, but now it has increased with 34 of the 1103 projects failing (i.e., about 3.1% by number of projects) or in other words project valued at $13 billion have failed out of total value of $275 billion (i.e., 4.9% by value) [5]. PPP philosophy/model of project development inherently carries many risks and requires careful consideration and planning for identification of possible

K. V. Prasad (✉) · R. S. Mamillapalli · H. R. Pusarla
National Institute of Construction Management And Research (NICMAR), Hyderabad, India
e-mail: kvprasad@nicmar.ac.in

R. S. Mamillapalli
e-mail: mrajasekhar@nicmar.ac.in

H. R. Pusarla
e-mail: hpusarla@nicmar.ac.in

A. Sil et al. (eds.), *Recent Trends in Civil Engineering*, Lecture Notes in Civil Engineering 274, https://doi.org/10.1007/978-981-19-4055-2_62

risks, sharing of risks with the party better prepared to mitigate the risk, allocation of resources rationally to reduce risk losses [6].

2 Case Study

Pimpalgaon Nashik Gonde Tollway Limited (PNGTL) is a Special Purpose Vehicle (SPV) with shareholding pattern of 48%-26%-26% by L&T by IDPL, L&T and Ashoka Buildcon, respectively. The scope of PPP contract agreement was to construct a single lane 60 km road covering the spectrum of Design, Build, Finance, Operation, Maintenance and Transfer (DBFOMT) responsibilities over a concession period of 20 years. The total investment is estimated at about Rs. 16 billion (US $270 million).

The Pimpalgaon—Nashik—Gonde road project is a six lane, 60 km route planned to connect the financial hub of India—Mumbai with states like Uttar Pradesh, Madhya Pradesh, Orissa and West Bengal among others. The road is part of the NH-3 from 380.000 to 440.000 km in the state of Maharashtra. The project comprised construction of variety of elevated structures including 6.1 km long elevated corridor, 6 Vehicular Under Passes (VUPs), 6 Pedestrian Under Passes (PUPs), 2 major bridges, 7 flyovers, and a subway. The flyover passing through Nashik city at Pathardi is India's longest integrated flyover. The project was launched on 15 January 2009. The concession agreement between NHAI and Pimpalgaon Nashik Gonde Tollway Private Limited (PNGTL) was signed on 3 July 2009. The road was expected to start operations in October 2012 but was delayed because political parties demanded for no toll-collection until the project was complete.

The partial Commercial Operation Date (COD) for the project was achieved in October 2012. However, the project SPV witnessed very weak volume of traffic due to the severe protests by the local public and subsequent diversion of the traffic to avoid the toll plaza and collection of tolls. Local users continued to pressurize the SPV to discontinue, and stop the tolling completely for the local public. When the SPV management did not heed to the demands, the protests further got intensified to political agitations. There was a tripartite support agreement signed between the State Government of Maharashtra and the SPV partners. Despite this support agreement, the State Government did not enforce any of the authority and support to the SPV. Had the State Government intervened and provided security to protect the damage to the toll plazas from the threat of protests, it would probably have aided the tolling at planned levels. When the SPV approached the Bombay High Court, the court directed the State Government and the district administration to provide full support and enforce law and order at the Toll Plaza. However, despite the directions of the High Court, the district administration failed to provide support and enforce law which forced the Project SPV to continue the project, and to survive the tolling was to be done at discounted rates. The key financials of the SPV are as in Table 1. As it can be seen, the SPV continued to incur losses from first year which kept mounting and liabilities on the project reached ~1502 crores which cannot be sustained by any developer on any mega project.

Table 1 SPV performance from inception till date

Financial year	Revenue (‘ Crs)	Profit/loss before tax (‘ Crs)	Profit after tax (‘ Crs)	Current liabilities (‘ Crs)
2009–10	–	−0.81	−0.83	2.12
2010–11	–	–	−0.04	55.35
2011–12	–	–	–	80.98
2012–13	–	−37.58	−37.58	67.92
2013–14	40.15	−32.39	−32.39	80.89
2014–15	85.89	−153.62	−153.62	95.68
2015–16	85.07	−144.67	−144.67	176.69
2016–17	176.53	−220.51	−220.51	1473.95
2017–18	–	−66.13	−66.13	1501.66

Even after the SPV achieving the partial COD in October 2012, the toll rates notified by the Authority were approximately 30% of the initial rates. This meant that the revenue collections of the SPV suffered double blow both from the traffic volume and the toll rates. Further, the toll rates were revised in 2014. However, even after this revision, the revenue collections by the SPV was significantly less than the initial estimates. This put the SPV in financial stress and the SPV had to stop paying the due share of revenue to NHAI. During this period, the SPV suffered a loss in revenue to the tune of Rs. 100 crores.

The SPV had to continue the operations at the discounted rates and substantially lesser traffic till early 2016. SPV had to finally take the inevitable route of invoking the Force Majeure Clause and seeking a termination of the concession. SPV cited lack of State support as the fundamental reason for invoking the force majeure clause. Tolling and the maintenance operations were subsequently taken over by the Authority. As per the concession agreement, if the basic reason of project failure and termination is due to any indirect political event, SPV would be eligible to receive a compensation to the extent of debts and 110% of the adjusted equity. However, if the reason of failure is proved to be due to the default of the Authority (NHAI), the SPV would be eligible to get a compensation of debt due and 150% of adjusted equity. Initially, NHAI proposed a payment of Rs. 420 crores to be paid to the concession which was rejected by the SPV and the same was referred to arbitration [7].

The sequence of events developed between the concessionaire and the authority are illustrated in Table 2.

While, the sequence of events indicated above illustrates the lack of clear policy and direction from the Authority, the turn of events also lead to disputes between the Joint Venture Partners of the SPV—L&T and Ashoka Buildcon Limited. Because of the shortfall in the foreseen revenue collection, the partners has issues in managing the cashflow for the project and the funding issues between the partners could not be resolved amicably. The dispute was referred to arbitration and Ashoka Concessions Limited was directed to pay 57.33 crores to L&T Infrastructure Development Projects Limited [8].

Table 2 Sequence of events leading to dispute and termination of the concession

S. No.	Date	Events of the project
1	16 Feb 2009	Formation of SPV by L&T IDPL 48%, L&T 26%, Ashoka Buildcon 26%
2	15 Jan 2009	Launch of the project
3	14 Oct 2012	Partial commercial operations expected to start
4	12 Mar 2014	Full commercial operations expected to start
5	25 Feb 2016	Project termination request by PNGTL to NHAI and request for payment of '1750.83 crores termination payment
7	13 Apr 2016	Project taken over by NHAI for operations and maintenance
9	21 Feb 2017	NHAI issued a termination notice retrospectively with effect from 29 March 2016 and payment of ' 423.06 crores against claim if ' 1750.83 crores
10	17 May 2017	Arbitration invocation by the SPV for adjudication of dispute
11	04 Jul 2017	Arbitral Tribunal was constituted
12	18 Jul 2017	First meeting of the Arbitral Tribunal
13	08 Sep 2017	Adjournment of the Arbitral proceedings
14	03 Nov 2018	Second meeting of the Arbitral Tribunal
15	30 Jan 2019	Honorable Delhi High Court extended the Arbitral Tribunal mandate
16	18 Feb 2019	Third meeting of the Arbitral Tribunal was held
17	04 Apr 2019	Settlement agreement between NHAI and PNGTL for Rs. 1238.06 crores
18	26 Apr 2019	Amount of Rs. 813 crores settlement by NHAI to PNGTL was paid

3 Learnings for Future Implementation

India is a country with enormous potential for infrastructure development and the needs of faster, quicker project development is growing every day. Government of India, however is constrained by the financial resources to get this done single-handedly. Even in the last five year plan, Government has sought the investment from private sector close to 53% [8]. With this being the case, it becomes essential for Government to ensure that project investments by private sector are protected by appropriate policies and risk management measures. The following sections discuss the learnings from the case study for future project implementation.

3.1 Project Lifecycle Risks

Traditionally, the Government authorities as well as private sector contractors are guided by the philosophy of evaluating the risk during the stages of construction viz., as time delays, cost escalations, design errors, etc. However, in the case of PPP projects, the Risk management has to extend to the full concession period covering all the project phases right from the stages of project conceptualization to project handover and close of concession. The Government and private sector have to be quickly develop the culture of setting up these systems in the feasibility studies itself. Many studies in the past [9–12] have covered the aspect of public objection as one of the possible risks in the development of PPP projects. However, this has not become a core element of the feasibility studies which is carried out for the screening and selection of the projects. The authorities concerned to take this as a possible risk factor during the pre-feasibility studies and sufficiently investigate before the projects are cleared through the PPP mode.

3.2 Risk Management by the Partners in the SPV

One of the essential features of a PPP project is the pooling of resources and associations of various organizations who have different interests and objectives to manage the risks on a project. In the case study, during the course of project operation, dispute developed between the SPV partners itself. The SPV organization has to be constituted into by carefully analyzing the strengths of all the parties to the contract and studies have shown that domestic and international organizations along with subcontractors having different skillsets, provide opportunities for mitigating and managing the risks on a PPP project [13].

3.3 Risk Sharing and Allocation

One of the essential success factors of PPP projects is that the structuring of PPP projects is to be done keeping in mind of the strengths and weaknesses of the public authority and the private partner. This involves the sharing of risks and rewards between the public and private sectors, arguably achieving Value for Money (VFM) by allocating risk to the party that is most capable of managing it [14–16]. The risk sharing should be borne by the party who is best equipped to manage that risk. Private sector commits a large share of the investment required for the development of a mega infrastructure project largely on the basis of stable contract with the government and the anticipated government support to mitigate the risks [17].

The PPP units and institutions set up by the Government must proactively take up project screening practices and understand the perspective of all stakeholders [18] through a detailed analysis of stakeholder associated risks [19]. Towards this, the practice of Social Impact Assessment (SIA) mandated as part of "Right to Fair Compensation and Transparency in Land Acquisition, Rehabilitation and Resettlement Act", 2013 (RFCTLARR Act) should not limit to only land acquisition, but also cover wide ranging aspects such as the ability and willingness of the users to pay for the facility as well and the agreement of the local stakeholders can be an additional requirement for the closure of Model Concession Agreement (MCA).

4 Conclusions

The present paper has taken up a case study to illustrate the developments on a mega infrastructure project and how risks transpired during the course of project development can hamper, derail and result in the failure of the project. The requirement of having robust feasibility, screening processes for the selection of a project to be taken up on PPP basis along with sound risk management practices on PPP projects cannot be undermined. The present paper has also brought out the salient learnings on the practices of Risk Management which if institutionalized in the future projects can ensure the successful delivery of PPP projects. The required finance for PPP project development is being sourced through private banks and financial institutions who lend almost 70% of the funds to PPP projects and it is very essential to ensure that their interests are protected without which there shall be a sharp decline in the lending for future PPP projects. Governments have a larger role play in ensuring fair implementation of the project as per the concession agreement and to assist the private partners in mitigating the risks associated with legal, political, social risks which shall ensure that project is implemented successfully.

References

1. Carbonara N, Costantino N, Gunnigan L, Pellegrino R (2015) Risk management in motorway PPP projects: empirical-based guidelines. Transp Rev 35(2):162–182
2. Maslova SV, Sokolov MY (2017) Risk management in public private partnership projects in health care: application of current approach and its improvement. Acad Strateg Manag J 16(4):1–17
3. Zou PXW, Wang S, Fang D (2008) A life-cycle risk management framework for PPP infrastructure projects. J Financ Manag Prop Constr 13(2):123–142
4. Rybnicek R, Placolm J, Baumgartner L (2020) Risks in public-private partnerships: a systematic literature review of risk factors, their impact and risk mitigation strategies. Public Perform Manag Rev 43(5):1174–1208
5. Xing Y, Guan Q (2017) Risk management of PPP project in the preparation stage based on fault tree analysis. IOP Conf Ser Earth Environ Sci 59(012050):1–12
6. Financial Express News article (2021) https://www.financialexpress.com/opinion/public-private-partnership-ppp-project-failures-dont-put-zombies-on-life-support/2146855/. Accessed on 10 Mar 2021
7. Kudtarkar SG (2020) Failure of operational PPP projects in India leading to private developer's apathy to participate in future projects: a case study based analysis. Indian J Fin Bank 4(2):17-27
8. The Hindu Business Line (2021) PNG Tollway may walk out of Maharashtra project. https://www.thehindubusinessline.com/news/png-tollway-may-walk-out-of-maharashtra-project/article8332228.ece. Accessed on 24 Mar 2021
9. Planning commission, Government of India, Twelfth Five Year Plan (2012–2017) Faster, more inclusive and sustainable growth, Vol 1, 2013

10. Wang Y, Wang Y, Wu X, Li J (2020) Exploring the risk factors of infrastructure PPP projects for sustainable delivery: a social network perspective. Sustainability 12:1–26
11. Ameyaw EE, Chan APC (2013) Identifying public-private partnership (PPP) risks in managing water supply projects in Ghana. J Facility Manage 11:152–182
12. Song J, Song D, Zhang X, Sun Y (2013) Risk identification for PPP waste-to-energy incineration projects in China. Energy Policy 61:953–962
13. Shen L, Platten A, Deng XP (2006) Role of public private partnerships to manage risks in public sector projects in Hong Kong. Int J Project Manage 24(7):587–594
14. Burke R, Demirag I (2019) Risk management by SPV partners in toll road public private partnerships. Public Manag Rev 21(5):711–731
15. Ng A, Loosemore M (2007) Risk allocation in the private provision of public infrastructure. Int J Proj Manage 25(1):66–76
16. Demirag I, Khadaroo I, Stapleton P, Stevenson C (2011) Risks and the financing of PPP: perspectives from the financiers. Br Account Rev 43(4):294–310
17. Ameyaw EE, Chan APC (2015) Evaluation and ranking of risk factors in public–private partnership water supply projects in developing countries using fuzzy synthetic evaluation approach. Expert Syst Appl 42:5102–5116
18. Takashima R, Yagi K, Takamori H (2010) Government guarantees and risk sharing in public–private partnerships. Rev Fin Econ 19:78–83
19. Olander S (2007) Stakeholder impact analysis in construction project management. Constr Manag Econ 25(3):277–287
20. Yang M, Chen H, Xu Y (2020) Stakeholder-associated risks and their interactions in PPP projects: social network analysis of a water purification and sewage treatment project in China. Adv Civ Eng 1–14

A Comparison of Macroscopic Traffic Stream Models Under Non-lane-Based Heterogenous Highway Traffic

Sandeep Singh, Rajesh Kumar Panda, Anisha Biswal, and S. Moses Santhakumar

1 Introduction

Many roads have insufficient capacity and often lead to congestion. Despite significant efforts, the Indian road network has not been able to keep pace with the high vehicular growth. This phenomenon can be attributed to the fact that Indian traffic is unique on its own [1]. It displays a much higher degree of heterogeneity against the traffic prevalent in other developing countries like China. Because of this considerable variation in the category of vehicles and different dynamic and static behavior, the movement of one vehicle is severely affected by the other [2]. Traffic speed is also a major controlling parameter contributing to the effective management of highways. The analysis of traffic flow on multi-lane roads is therefore vital. Traffic engineers can study and assess traffic systems on the basis of traffic flow metrics. Speed, flow, and density are the three most essential traffic parameters. Using these parameters, the highway system can be operated more efficiently.

Therefore, the researchers take advantage of traffic models to understand traffic characteristics and predict the traffic system's behavior. It is also fundamental for planning, designing road operations, and improving road serviceability. To account for the heterogeneous variability, the researchers use an equivalent unit, called passenger car unit (PCU), to interchange the traffic into a homogeneous one. In order to reduce heterogeneity variability in the traffic data, the dynamic passenger car unit (DPCU) values are calculated using the speed-area ratio method. This research has attempted to develop the relationships between speed, flow, and density using

S. Singh (✉) · A. Biswal · S. M. Santhakumar
Department of Civil Engineering, National Institute of Technology Tiruchirappalli, Tamil Nadu, Tiruchirappalli 620015, India
e-mail: sandeepsingh.nitt@gmail.com

R. K. Panda
TATA Projects Limited, Pune, Maharashtra 411015, India

A. Sil et al. (eds.), *Recent Trends in Civil Engineering*, Lecture Notes in Civil Engineering 274, https://doi.org/10.1007/978-981-19-4055-2_63

different macroscopic linear and nonlinear, single regime traffic flow models. Eventually, the basic principle of differentiation was used to find the maximum flow values for different models and based on statistical analysis, the best fit model was determined. Previous studies conducted on the same principles mainly focused on urban roads, limiting the application of macroscopic traffic steam models. The present study brings out the rational outcomes in terms of analyzing the macroscopic traffic parameters for the Indian highway traffic system.

2 Literature Review

Numerous approaches have been used in the past to estimate passenger car units (PCU) of vehicles. The basis used for most estimates is delay [3], speed [4], density [5], headway [6], and queue discharge [7]. However, these studies relate mainly to the PCU estimation of vehicles under uniform conditions of traffic. Hence, the results do not apply to the heterogeneous nature of traffic prevalent in India. Singh and Santhakumar [8] determined the DPCU and PCU for many vehicles under mixed traffic flow conditions. According to the study, the PCU values recommended by western countries' highway capacity manuals might not be applicable for capacity analysis in Asian countries.

Velmurugan et al. [9] investigated the free stream speed of various vehicles to evaluate the limit on multi-lane roadways in India. Mehar et al. [10] estimated multi-lane highway capacity through VISSIM models. They developed five different driver behaviors for various categories of vehicles based on their static and operating characteristics. Dhamaniya and Chandra [11] proposed a method to convert heterogeneous traffic volumes into homogeneous DPCUs without setting PCU factors. This study introduced a new term, the Stream Equivalency Factor (K): the ratio of traffic volume in PCU per hour to volume in vehicles per hour. A linear model proposed by Greenshields [12] was the earliest speed-density model. This model assumes speed and density to be linear, which is not valid for the entire road stretch. It can also be highly unsuitable for Indian roads due to the heterogeneity of vehicles on the roads.

Moreover, even in developing countries, there is a wide variation in traffic characteristics [13]. The vehicles have a vast difference in their physical appearance and performance abilities even though they share the same road [14]. The traffic conditions in India are therefore highly heterogeneous. For the capacity estimation, it is imperative to account for this heterogeneity under non-lane-based traffic conditions, and it is relatively tough to estimate traffic volume on the road. This discrepancy in capacity analysis instigates the need for this research work. Literature review reveals that research about the effect of heterogeneity of traffic on the performance and capacity analysis of multi-lane highways has been less and needs more studies. The major limitation of the greenshields' linear model (GLM) [12] is the linear relationship that is assumed between speed and density. Thus, in order to get an accurate speed-flow-density relationship on Indian highways, other non-linear models must be explored to arrive at the most optimal solution. Since the derived speed-density relationship is different from the actual one, the relation between speed and density was more thoroughly examined by the Greenberg logarithmic model [15], Underwood exponential model [16], Drakes's model [17], and Pipes' generalized model. Furthermore, the estimated capacity from the best-fitted macroscopic traffic stream model based on the speed-area ratio method was compared to the capacity estimated using the DPCU and PCU values suggested by the Indian Highway Capacity Manual (Indo-HCM:2017) [18] and Indian Road Congress (IRC-64:1990) [19].

3 Data Collection

Traffic data was collected using the Transportable Infrared Traffic Logger (TIRTL)-An advanced IR sensors-based device. The traffic data was collected near Kancheepuram city on Chennai-Bengaluru National Highway (NH-48). The coordinates of the site were 12.90°N (Latitude), and 79.80°E (Longitude). Figures 1 and 2 illustrate the map and aerial view of the study spot, respectively.

The NH is a four-lane divided section. It is chosen for the study so that it is free from any impact of side friction. Figure 3 illustrates the vehicle detection mechanism using the TIRTL.

Figure 3 shows the detection mechanism of the TIRTL instrument. The TIRTL instrument was set up on the lane toward Chennai, with a receiver (RX) near the kerb side and a transmitter (TX) on the median side. The TIRTL detects the vehicles that cross the IR sensor detection zone. The TX unit forms two straight and two diagonal IR beam pathways. Two beam events such as the vehicle entry and exit are recorded for any vehicle identification. This also helps in deriving vehicle speed and headway.

Further, the beam events are used to determine the inter-wheel (inter-axle) spacing and the number of vehicle axles, based on which the correct vehicle classification is done. TIRTL calibrates automatically. The real-time field recorded traffic data can be

Fig. 1 Map showing the study spot on NH-48. *Source* Google maps

Fig. 2 Aerial view of the study spot on NH-48. *Source* Google earth

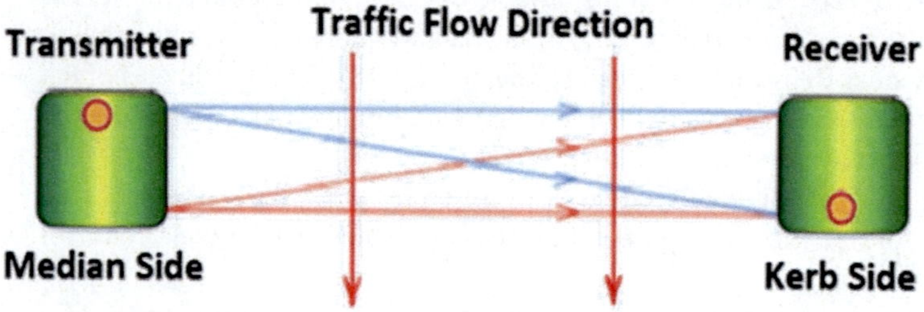

Fig. 3 Vehicle detection mechanism

viewed in the user-friendly interface, and the traffic data is stored as a.csv output file in the laptop device connected to the RX. Further, more details regarding the TIRTL instrument's working and accuracy can be inferred from the studies by Singh et al. [20, 21]. Figure 4 shows the setup of the TIRTL instrument at the study spot.

The Indian traffic is marked by a high degree of heterogeneity. The details regarding geometrics were collected from the intercity highway (NH-48) connecting

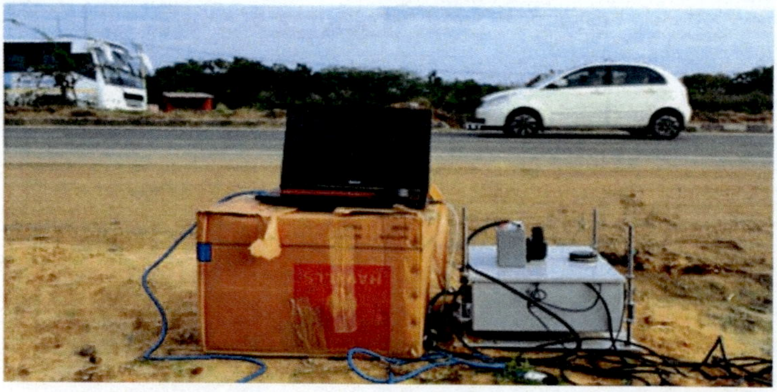

Fig. 4 Setup of the TIRTL instrument

Chennai and Bangalore. The highway section has plain terrain with a traffic lane width of 3.5 m each. There are two lanes in each direction. A paved shoulder width of 2.0 m is also provided. The median width is 5.0 m.

From Table 1, it can be observed that there were nine different categories of vehicles. A total of 42,901 vehicles were recorded for 24 h of the data collection period. Table 1 shows the traffic characteristics of the vehicles.

4 Data Analysis, Modeling, and Results

4.1 Determination of DPCU

The Indian traffic carries vehicles with different static and dynamic features. Accounting for this diversity, the DPCU factor is used for transforming all vehicles into a standard unit. The DPCU is determined using Eq. (1).

$$\mathrm{DPCU}_i = (V_c/V_i)/(A_c/A_i) \quad (1)$$

where V_c and V_i represent the speed of the Small Car (SC) and 'ith' vehicle, respectively, while, A_c and A_i represent the physical area of a SC and 'ith' vehicle, respectively.

For the evaluation of DPCU, SC was adopted as the standard vehicle category. The TIRTL records the spot speed, i.e., the time-mean speed of the vehicle; however, space mean speed calculated using the harmonic mean formula is used to represent the real-field conditions. Table 2 shows the DPCU values of each class of vehicles.

The estimated DPCU values are adopted to convert the heterogeneous traffic to homogeneous traffic and further analyze the macroscopic fundamental relationship diagrams (MFDs) using linear and nonlinear models.

4.2 Modeling Traffic Stream Parameters

Traffic flow parameters such as free-flow speed, optimum speed, jam density, and optimum density are necessary for finding the highway's capacity. The single regime and macroscopic traffic flow models were used for calculating capacity for four-lane divided road sections of the intercity highway. In this paper, the following models are used for calculating capacity,

(i) Greenshields' Linear Model,
(ii) Greenberg's Model,
(iii) Underwood's Exponential Model,
(iv) Drake's Model, and
(v) Pipes' Model.

Table 1 Traffic characteristics of the vehicles

Vehicle class	Two-wheeler	Three-wheeler	Standard/small car	Big car	Small commercial vehicle	Light commercial vehicle	Medium commercial vehicle	Heavy commercial vehicle	Multi-axle vehicle
Length (m)	1.61	1.99	4.17	4.69	3.58	5.7	8.5	9.49	12.12
Width (m)	0.59	0.98	1.35	1.52	1.24	1.47	1.84	1.92	1.96
Area (m^2)	0.95	1.95	5.63	7.13	4.44	8.38	15.64	18.22	23.75
No. of vehicles	13,677	846	18,188	2498	492	1038	4230	946	986
Share (%)	31.8	1.97	42.29	5.81	1.14	2.41	9.84	2.2	2.29
Maximum speed (kmph)	122.47	117.38	134.31	133.22	87.55	125.33	100.61	99.65	74.54
Minimum speed (kmph)	14.04	19.14	16.86	31.39	16.71	14.63	22.75	24.95	15.9
Time mean speed (kmph)	59.12	51.56	73.56	73.17	54.41	63.35	56.58	50.16	45.74

Table 2 DPCU value of the vehicles

Vehicle class	Space mean speed (kmph)	Vehicle area (m^2)	Speed ratio (V_r)	Area ratio (A_r)	DPCU (V_r/A_r)
Two-wheeler	55.47	0.95	1.27	5.93	0.21
Three-wheeler	49.03	1.95	1.44	2.89	0.50
Standard/small car	70.43	5.63	1.00	1.00	1.00
Big car	69.62	7.13	1.01	0.79	1.28
Small commercial vehicle	52.78	4.44	1.33	1.27	1.05
Light commercial vehicle	60.11	8.38	1.17	0.67	1.74
Medium commercial vehicle	54.44	15.64	1.29	0.36	3.58
Heavy commercial vehicle	48.00	18.22	1.47	0.31	4.75
Multi-axle vehicle	44.47	23.75	1.58	0.24	6.58

The traffic data was plotted using Microsoft Excel, and the essential traffic parameters were found. The maximum flow values ($q_{\max}$) using the traffic flow principle were estimated for the different models. The values found are tabulated in Table 3. Figures 5, 6, and 7 illustrate the different speed-density, speed-flow, and flow-density plots, respectively, developed using the macroscopic traffic stream models.

To compare the various models numerically, the speed-related R-squared (R^2) value along with two other measures (root mean squared error (RMSE) and the mean absolute percentage error (MAPE)) were used.

The R^2 value is a measure of the goodness of fit of the model. The higher the R^2 value, the closer the data is to the fitted regression line. But, R^2 value itself is not a reliable measure of the data fitting, and thus the researchers use other error-based measures like RMSE and MAPE.

The RMSE represents the size of a typical error since it is measured in the unit of the original data. The formula for RMSE is given in Eq. (2).

$$\text{RMSE} = \left[\frac{1}{N}\sum_{i=1}^{N}\frac{|x_{\text{est}} - x_{\text{obs}}|}{x_{\text{obs}}}\right] \times 100\sqrt{\frac{\sum_{I=1}^{N}(x_{\text{est}} - x_{\text{obs}})^2}{N}} \quad (2)$$

The MAPE is another frequently used efficacy measure that gives us a general sense of the error. The formula for MAPE is given in Eq. (3).

Table 3 Model equations and traffic parameter results

Model	Model Equation	v_f (Km/h)	v_o (Km/h)	k_j (PCU/km)	k_o (PCU/km)	$q_{\max}$ (PCU/h)
Greenshields'	$v = v_f - \left(\frac{v_f}{k_j} * k\right)$	66.75	33.37	430.0	215	7177
Greenberg's	$v = v_o * \ln \frac{k_j}{k}$	–	29.41	283.4	–	3066
Underwood's	$v = v_f * e^{\frac{-k}{k_o}}$	66.79	–	–	333	8191
Drake's	$v = v_f * e^{\frac{-1}{2}\left(\frac{k}{k_o}\right)^2}$	63.63	–	–	129	4983
Pipes' $n = 2$	$v = v_f * \left(1 - \left(\frac{k}{k_j}\right)^2\right)$	63.73	–	188.17	–	4616
$n = 3$	$v = v_f * \left(1 - \left(\frac{k}{k_j}\right)^3\right)$	62.70	–	127.86	–	3788

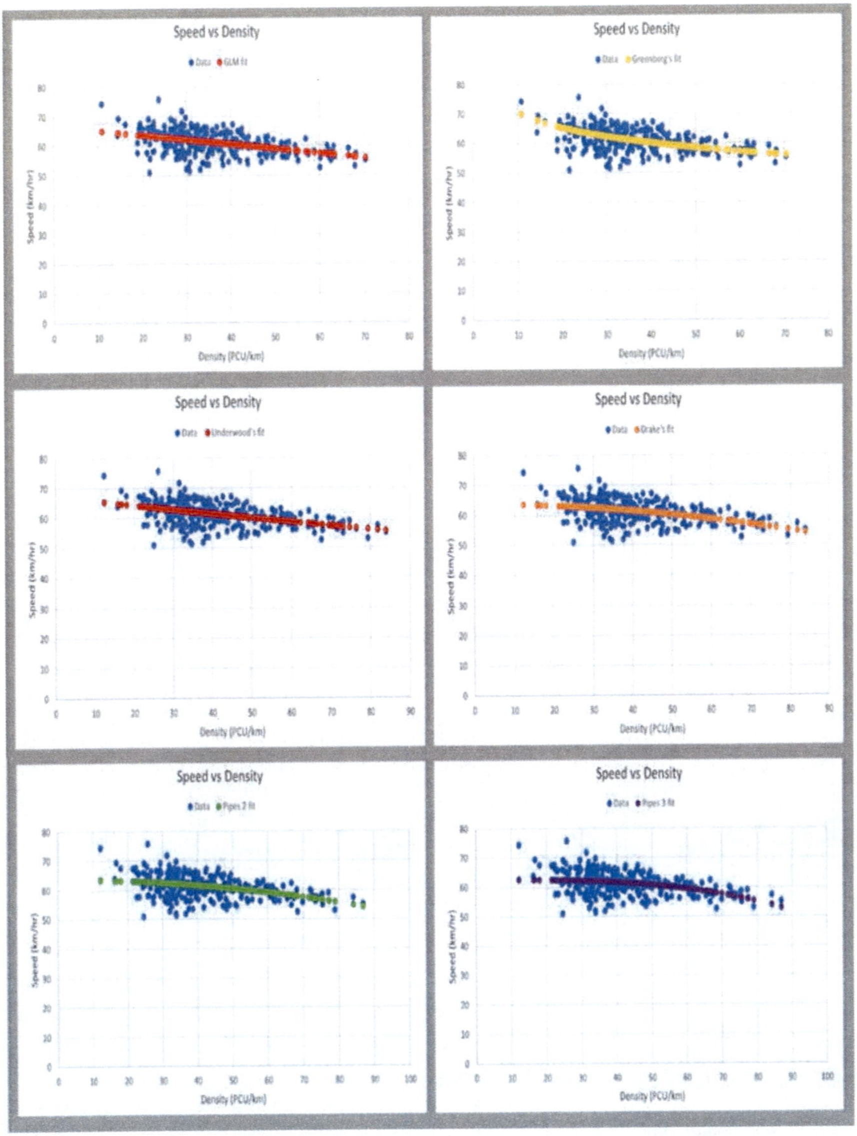

Fig. 5 Speed-density relationship plot

$$\text{MAPE} = \sqrt{\frac{\sum_{I=1}^{N}(x_{\text{est}} - x_{\text{obs}})^2}{N}} \tag{3}$$

Table 4 shows the RMSE and the MAPE values for all the models.

The GLM is a linear model that poorly maps the relationship between speed and density in real-time traffic. It is evident with the high error values of MAPE = 4.6425

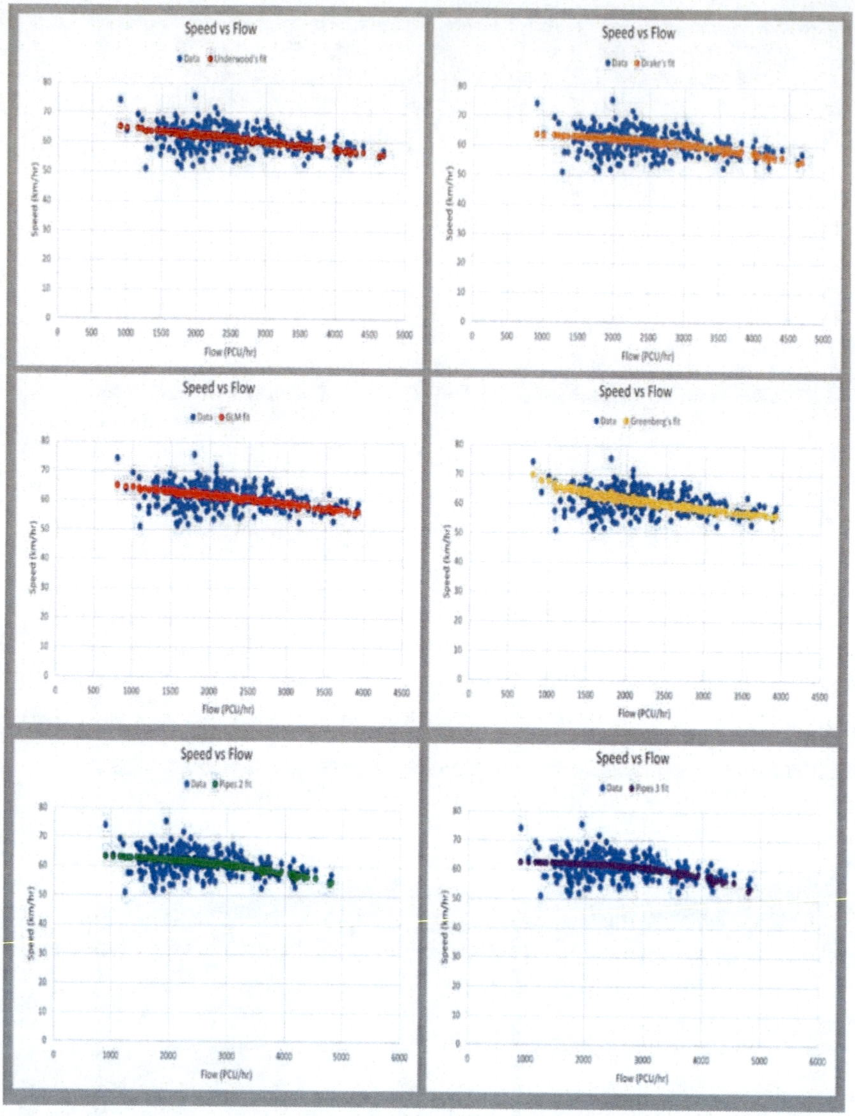

Fig. 6 Speed-flow relationship plot

and RMSE = 3.63% and low value of $R^2 = 0.39$. Several vehicle classes, intra-class variability, and weak-lane discipline make it more complicated to analyze mixed traffic assuming linearity. Greenberg's model too suffers from drawbacks due to its logarithmic nature, in which when density tends to zero, the speed tends to infinity, and vice versa. The traffic parameters, which are highly contrasting to other models, justify the high error values. This has been even stated in a paper by Khan et al. [22], which says that Greenshields' and Greenberg's model give unrealistic results for

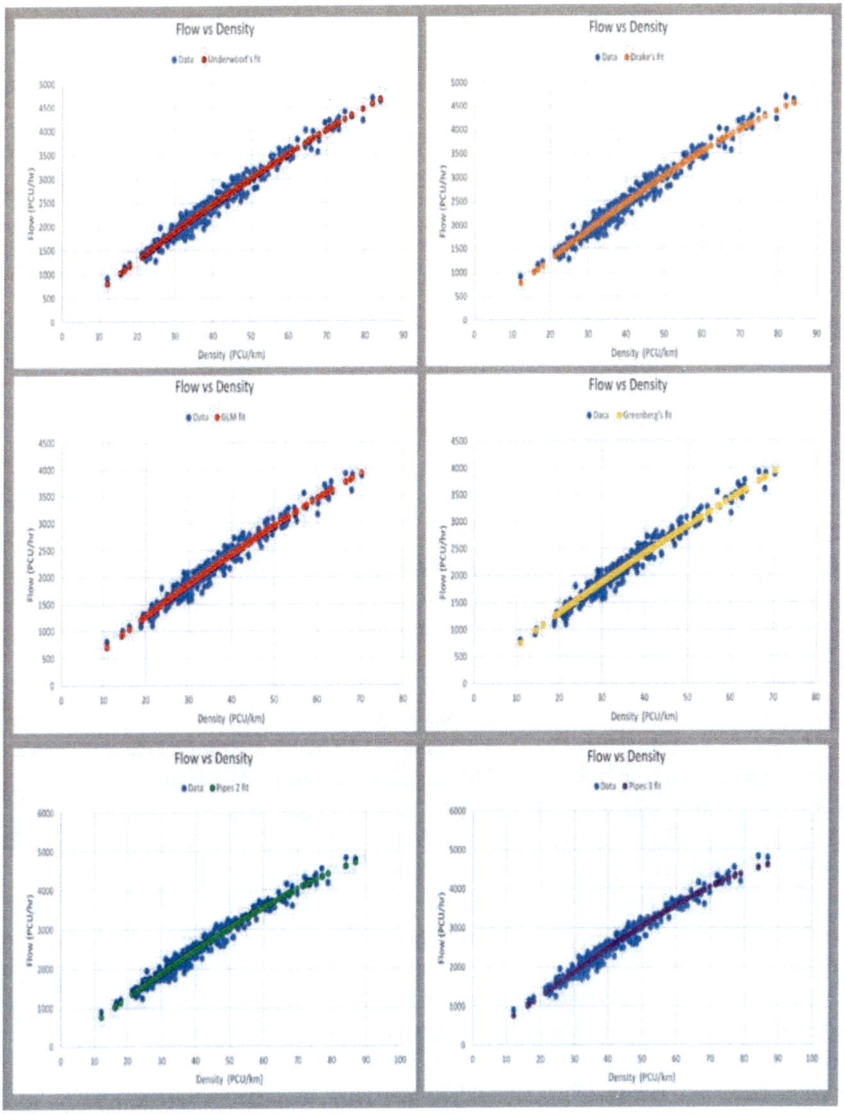

Fig. 7 Flow-density relationship plot

heterogeneous traffic. Thus, it can be inferred that those arbitrary results obtained from some of the models are due to the heterogeneous nature of the traffic under study.

The Greenshields', Greenbergs', and Underwoods' models show erratic values of optimum density and jam density. The jam density obtained for these models is above 350 PCU/km, which is impractical. However, Drake's and Pipe's generalized

Table 4 RMSE and the MAPE values for all the models

Model	R^2	MAPE	RMSE (%)
Greenshield's	0.39	4.6425	3.63
Greenberg's	0.30	10.234	8.26
Underwood's	0.44	4.6430	3.61
Drake's	0.53	3.8598	3.60
Pipes (2)	0.47	4.0038	3.61
Pipes (3)	0.56	3.5105	3.57

models show reasonable values in the range of 120 PCU/km to 200 PCU/km. Despite all these, it can be surmised that most of the characteristics of traffic data cannot be met by either of the existing functional forms. However, it can be concluded that Pipes' family models fit the traffic data the best and give the most realistic results with average free-flow speed, average jam density, and average capacity.

4.3 Comparison of Highway Capacity Using Different PCU Estimation Methods

Generally, the DPCU values are used for homogenizing the heterogeneous traffic [23]. The capacity of the highways was obtained using the best-fitted macroscopic model, i.e., the pipe's non-linear model ($n = 3$). This study estimated the capacity using the DPCU method, where indigenous DPCU values for each vehicle are used. Other PCU estimation methods which are widely used in the Indian context are the Indo-HCM (2017) and IRC:64 (1990) methods. The DPCU and PCU values for different classes of vehicles suggested by Indo-HCM (2017) and IRC:64 (1990), respectively, are shown in Table 5.

These DPCU and PCU values were used for capacity estimation, and the comparison of highway capacity was carried out. Table 6 shows the traffic parameters estimated using the DPCU values and static PCU values from Indo-HCM (2017) method and IRC:64 (1990) method, respectively, which are compared with the speed-area ratio method estimated traffic parameters.

The capacity value estimated using the DPCU values from the adopted speed-area ratio method was 3788 PCU/h. However, by adopting the Indo-HCM (2017) suggested DPCU values, the capacity value resulted in 4211 PCU/h, while by adopting the IRC:64 (1990) suggested static PCU values, the capacity value resulted in 3157 PCU/h. It can be clearly interpreted that the capacity value by the Indo-HCM (2017) DPCU values was overestimated, while the capacity value by the IRC:64 (1990) static PCU values was underestimated in comparison to the speed-area ratio method. This can be attributed to the result of the DPCU values suggested by the Indo-HCM (2017) being slightly higher than the static PCU values suggested by IRC:64 (1990). The use of DPCU and static PCU values may have amounted to

Table 5 DPCU and PCU values by Indo-HCM (2017) and IRC:64 (1990)

Vehicle class	Indo-HCM (2017) DPCU values range median		IRC:64 (1990) static PCU values
Two-wheeler	0.3–0.5	0.40	0.5
Three-wheeler	1.1–1.3	1.20	1.6
Standard/small car	–	1.00	1.0
Big car	1.4–1.5	1.45	–
Small commercial vehicle	–	–	–
Light commercial vehicle	2.7–3.3	3.10	2.8
Medium commercial vehicle	3.5–4.6	4.40	4.5
Heavy commercial vehicle	3.5–4.6	4.40	4.5
Multi-axle vehicle	6.3–7.0	6.60	6.0
Tractor	3.9–7.0	6.20	4.5
Bi-cycle	–	–	0.6

Table 6 Traffic parameters by the speed-area ratio, Indo-HCM (2017), and IRC: 64 (1990) methods

Traffic parameter	Speed-area ratio method	Indo-HCM (2017) method	IRC:64 (1990) method
Free flow speed (V_f) (Km/h)	62.70	64.57	60.12
Jam density (K_j) (PCU/km)	127.86	138.71	123.88
Capacity (Q_{max}) (PCU/h)	3788	4211	3157

overestimation and underestimation of traffic variables when they are homogenized, thus resulting in errors.

5 Conclusions

This study presents a comprehensive analysis of various methods used to model the traffic characteristics under mixed traffic conditions. This study is a case study of a national highway section in the southern part of India. For describing the traffic characteristics, five models were used. Using the different models, speed-flow-density relationships were formulated in graphs. Also, this research demonstrated the similarities and differences among the macroscopic traffic stream models. It was found that the different parameters such as R-squared values, MAPE, and RMSE were the suitable error measuring units, which helped determine the best fit model. The R-squared values were very low for all the models. This is because of the data obtained from the highway conditions, which often do not follow any trend. Further, the current study suggests using an indigenous measure for DPCU estimates to avoid overestimation and underestimation of traffic parameters. The study results substantially contributed to the view that traffic characteristics modeling could be enhanced by differentiating not only based on methods of traffic stream models but also based on the method of determination of PCU. Overall, the study results point towards the significant concerns of not carrying out traffic studies considering the difference in the various method under the non-lane-based traffic conditions. Ultimately, the study shows that traffic capacity is critical to solving traffic problems.

Considering the identified issues from past research, this study makes significant contributions (i) by developing the regime-based equilibrium traffic flow models to account for the actual traffic dynamics, (ii) by showing that the MFDs can explain the traffic mix characteristics, and (iii) by capturing the variations of MFDs on in a mixed-traffic environment. The findings of this study reinforce the point that the estimation of the traffic capacity of uninterrupted highway facilities with a higher degree of accuracy can improve the analysis of traffic characteristics, thereby enriching the traffic flow theory. The present study is conducted only for one selected section. More sections and a more extended period of traffic data can be considered for further research for the entire day and night. This study can be further extended to the development of multi-regime models.

References

1. Singh S, Santhakumar SM (2021) Empirical analysis of impact of multi-class commercial vehicles on multi-lane highway traffic characteristics under mixed traffic conditions. Int J Transp Sci Tech 11 (3):545–562. https://doi.org/10.1016/j.ijtst.2021.07.005
2. Singh S, Santhakumar SM (2022) Modeling traffic parameters accounting for platoon characteristics on multilane highways. Transp Dev Econ 8:30. https://doi.org/10.1007/s40890-022-00166-3
3. Khan A, Dhamaniya A, Arkatkar SS (2021) Effect of two wheeler proportion on passenger car units at roundabout in Indian urban scenario. Transp Lett 1–12. https://doi.org/10.1080/19427867.2021.1919455
4. Elefteriadou L, Torbic D, Webster N (1997) Development of passenger car equivalents for freeways, two-lane highways, and arterials. In: Transportation research record 1572. Transportation Research Board, National Research Council, Washington, DC, pp 51–58. https://doi.org/10.3141/1572-07
5. Webster N, Elefteriadou L (1999) A simulation study of truck passenger car equivalents (PCE) on basic freeway sections. Transp Res Part B Methodol 33(5):323–336. https://doi.org/10.1016/S0965-8564(98)00036-6
6. Krammes RA, Crowley KW (1986) Passenger car equivalents for trucks on level freeway segments. Transp Res Rec 1091:10–17
7. Al-Kaisy A, Jung AY, Rakha H (2005) Developing passenger car equivalency factors for heavy vehicles during congestion. J Transp Eng 131(7):514–523

8. Singh S, Santhakumar SM (2021) Evaluation of lane-based traffic characteristics of highways under mixed traffic conditions by different methods. J Inst Eng India Ser A 102(3):719–735. https://doi.org/10.1007/s40030-021-00549-6
9. Velmurugan S, Madhu E, Ravinder K, Sitaramanjaneyulu K, Gangopadhyay S (2010) Critical evaluation of roadway capacity of multi-lane high-speed corridors under heterogeneous traffic conditions through traditional and microscopic simulation models. J Indian Roads Congr 71(3):235–264
10. Mehar A, Chandra S, Velmurugan S (2014) Passenger car units at different levels of service for capacity analysis of multi-lane interurban highways in India. J Transp Eng ASCE 140:81–88
11. Dhamaniya A, Chandra S (2013) Concept of stream equivalency factor for heterogeneous traffic on urban arterial roads. J Transp Eng 139:1117–1123
12. Greenshields BD, Channing WS, Miller HH (1935) A study of traffic capacity. Highway Res Board Proc Nat Res Coun (USA) 4:448–477
13. Singh S, Vidya R, Shukla BK, Santhakumar SM (2021) Analysis of traffic flow characteristics based on area-occupancy concept on urban arterial roads under heterogeneous traffic scenario—a case study of Tiruchirappalli city. In: Mehta YA et al (eds) Trends and recent advances in civil engineering (TRACE-2020), advances in water resources and transportation engineering, Lecture Notes in Civil Engineering, Vol 149. Springer Nature, Singapore, pp 69–84. https://doi.org/10.1007/978-981-16-1303-6_6
14. Singh S, Barhmaiah B, Kodavanji A, Santhakumar SM (2020) Analysis of two-wheeler characteristics at signalised intersection under mixed traffic conditions: a case study of tiruchirappalli city, resilience and sustainable transportation systems. In: 13th Asia Pacific transportation development conference. American society of civil engineers (ASCE), Shanghai, China, pp 35–43. https://doi.org/10.1061/9780784482902.005
15. Greenberg H (1959) An analysis of traffic flow. Oper Res 7(4):79–85
16. Underwood RT (1961) Speed, volume, and density relationship in quality and theory of traffic flow. Yale University, Bureau of Highway Traffic, pp 141–188
17. Drake JS, Schofer JL, May AD (1967) A statistical analysis of speed-density hypotheses in vehicular traffic science. Highw Res Rec 154:112–117
18. Indian Highway Capacity Manual (Indo-HCM) Central Road Research Institute, New Delhi, India
19. Indian Roads Congress (IRC) (1994) 64 Guidelines for capacity of roads in rural areas. Indian Roads Congress, New Delhi, India
20. Singh S, Santhakumar SM (2022) Assessing the impacts of heavy vehicles on traffic characteristics of highways under mixed traffic platooning conditions. Europ Transp\Trasporti Europei 86. https://doi.org/10.48295/ET.2022.86.3
21. Singh S, Rajesh V, Santhakumar SM (2022) Effect of mixed traffic platooning by commercial vehicle types on traffic flow characteristics of highways. Periodica Polytech Transp Eng 50(4):344–356. https://doi.org/10.3311/PPtr.18200
22. Khan HZ, Imran W, Azeem S, Khattak KS, Gulliver TA, Aslam MS (2019) A macroscopic traffic model based on driver reaction and traffic stimuli. Appl Sci 9:2848. https://doi.org/10.3390/app9142848
23. Singh S, Santhakumar SM (2022) Influence of platooning of heavy transport vehicles operation on traffic flow mix of intercity highways. In: Maurya AK, Maitra B, Rastogi R, Das A (eds) Proceedings of the fifth international conference of transportation research group of India. Lecture notes in civil engineering, vol 219. Springer, Singapore, pp 447–459. https://doi.org/10.1007/978-981-16-8259-9_29

Integrated Sensor-Based Approach for Traffic Congestion Prediction Using Probe Vehicle

Varun Singh, Anurag Upadhyay, and Asit Kumar

1 Introduction

Increasing urban population and road traffic is causing the grave problem to intra-city transportation planning because of substantial increase in the vehicular traffic on road networks and increased traffic is resulting in recurrent congestion which ultimately results in unwarranted delays and accidents, especially at high-traffic density routes.

In order to mitigate congestions in urban areas, local, state and central government organizations are required to investigate and identify congestion location and take proper measures and provide facilities, information and alternate routes, etc. to the motorist. However, to accomplish this, the major challenge which any organization or agency faces is how to detect the congestion spots over the road sections effectively and efficiently.

In this paper, sensor-based methodology to collect sensor data and prediction of traffic state using probe vehicle equipped with ultrasonic sensors and GPS Sensors has been presented. The proposed methodology has been implemented for prevailing mixed traffic condition in Prayagraj City, India. On basis of the analysis of collected data, it has been concluded that integrated sensor system can suitably use for generating traffic pattern on urban roads under prevailing mixed traffic condition. Figure 1 provides the general layout of probe vehicle. The proposed methodology had been applied to the Prayagraj city, India. The sample GPS data on selected route in Prayagraj city was collected using the installed sensors in probe vehicle. The data requirement for this study includes the digitized road network and the data collected through sensors installed on probe vehicle.

V. Singh (✉) · A. Upadhyay · A. Kumar
Motilal Nehru National Institute of Technology Allahabad, Teliarganj, Prayagraj, India
e-mail: varun@mmnit.ac.in

A. Sil et al. (eds.), *Recent Trends in Civil Engineering*, Lecture Notes in Civil Engineering 274, https://doi.org/10.1007/978-981-19-4055-2_64

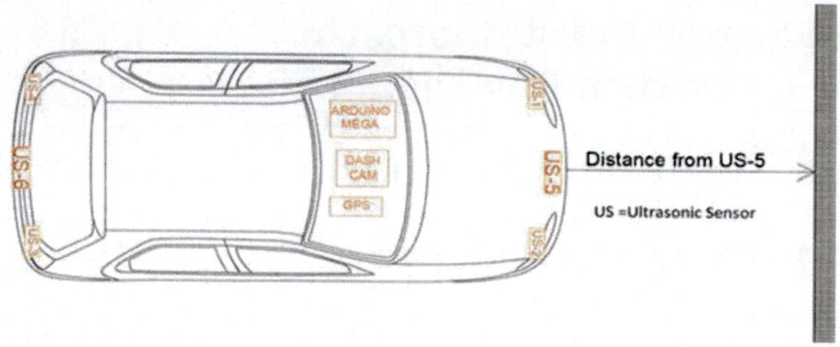

Fig. 1 Probe vehicle configuration

2 Related Works

This section reviews the previous works on the traffic state prediction for urban road networks and driver-specific predictions. Various report research works were aimed at application of probe vehicles equipped with sensors for travel time estimation and speed prediction.

Hansapalangkul et al. [5] had calculated the traffic density on the road during rush and non-rush hours by using mobile phone and Global Position System (GPS) receiver. They had developed a programme to capture the cell ID of which a cellular phone is associated using a software application based on Python. Bishop and Casabona [1] developed a system for automatic determination of vehicle traffic congestion using GPS enabled smartphone. They had developed the filtering methods to eliminate the non-matching routes. Thianniwet and Phosaard [12] proposed a technique to identify road traffic congestion level from the speed of the mobile sensors. They used the vehicle velocity to determine the congestion level. Their decision tree model achieves the accuracy of around 91.29%. Wang and Tsai [14] proposed the technique to detect the traffic congestion from the single vehicle data, with their technique achieving an accuracy of 88.94%. They had used vehicle to infrastructure (V2I) and vehicle to vehicle (V2V) communication to collect the data. Lanka and Jena [7] had developed a real time traffic management and information system by using high quality real time traffic information. Cárdenas-Benítez et al. [2] uses the connected vehicle techniques and with the usage of big data set developed a congestion system by means of series of algorithm to improve the traffic flow and reduce carbon dioxide emission in the city. Sakhare and Vanajakshi [11] investigated the use of GPS data from buses and Wi-Fi and Bluetooth data from a sample of vehicles, for accurate estimation of the travel time of all vehicles on the roadway.

Liu et al. [9] analysed Beijing cab traces for different novel applications like finding correlation amongst traffic situations in different parts of the city and navigating drivers to better routes using the trajectories of experienced cab drivers.

Upadhyay et al. [13] proposed a sensor-fusion-based approach based on integration of GPS, Ultrasonic and Accelerometer sensors for generating comparatively more precise and accurate data that could be subsequently utilized for traffic state prediction.

Most of the above studies have utilized the travel time, speed and GPS positional data for the analysis of traffic congestion. However, low speed of probe vehicle and smaller instantaneous inter-vehicle space headway are the more effective parameters for investigating the traffic congestion issues related to vehicular traffic. Proposed methodology has extended the literature whilst considering the use of speed of probe vehicle and instantaneous headway for identification of congested stretches of the road network.

3 Experimental Setup

The probe vehicle was equipped with different sensors to record the vehicle movements. The specification of the sensors are given below:

(i) **Probe Vehicle:** Hyundai Santro Hatchback car (length: 3495 mm bumper to bumper and width: 1495 mm).
(ii) **Ultrasonic Sensor**: Six ultrasonic sensors module to detect the vehicle to vehicle clearance.
(iii) **GPS Receiver**: GPS Receiver with Ublox NEO-M8 GPS chip to record vehicle movement. GPS Receiver logged the positional and speed data at the frequency of 1 Hz.
(iv) **Arduino Mega Microcontroller**: The Arduino Mega 2560 is microcontroller boards have been used for processing the data as acquired through sensors and storing them in csv file format.
(v) **Dash Cam**: Aukey Dash Cam with 170 Degree Wide-Angle Lens, and Loop Recording and Night Vision and 1080 p video recording.

In order to assess the quality of the sensor, qualitative analysis of the sensors has been carried using experimental setups and testing results along with their interpretations are presented in the Table 1.

4 Proposed Methodology

Proposed congestion detection methodology comprises different steps as summarized below:

(a) **Development of Spatial Database of Road Network**: The spatial database of road network of Prayagraj City for spatial visualization and subsequent analysis has been modelled and developed in UTM projection system (Zone 44 North).
(b) **Data Acquisition using Sensors**: The probe vehicle was traversed on specified route in City to collect the data from different sensors from starting location A

Table 1 Qualitative assessment of ultrasonic sensor HC-SR04

S. No	Type of effect	Effect on sensor	Procedure	Remark
1	Rain	Rain—no effect	A setup was made as shown in the laboratory. A series of showers were placed (to recreate rain of different intensities). Sensor placed at a known distance from the wall. The distance measured does not change with the presence of rain	Assuming no damage to sensor (as they are made waterproof). Data generated from the sensor is reliable and analysis can be done on it
2	Voltage	Measurement below the rated voltage (5 V) is not feasible	Sensor was kept at a known distance and it was verified by serial monitor value on a desktop. The signal strength was reduced in the code (Arduino), and new values were observed in serial monitor	The sensor framework, stops working immediately after the voltage falls below 5 V. So unreliable data is note captured till the discharge of the battery
3	Field of view (directivity) Horizontal Vertical	The sensor captures an elliptical shape of field of view. Accuracy is volatile	The transmitter of the ultrasonic sensor is placed, and a tape is placed in front of it, which is denoted as the transmission axis The object is moved towards the transmission axis along the tape placed on orthogonal line. The readings are observed on the serial monitor. When the actual distance is captured	FOV (full cone): horizontal ~ 9° 5′, vertical ~ 4°, Range: tested from 25 to 100 cm The smaller the FOV, more accurate is the sensor in distance measurement. But for traffic determination purposes, a larger FOV, gives more chance for obstacle detection. The FOV of this sensor is adequate for determining the clearance

to end location B as shown in Fig. 5. Total length of the route was 35.53 km and numbers of data points as captured by sensors were 2290. Table 2 provides the description of the dataset.

(c) **Pre-processing of Acquired data**: Pre-processing of acquired data from different sensors has been carried out to develop the integrated database with time stamp. Positional and other data from the sensors are re-sampled at the rate of 1 data point/sec. This database comprises the positional and speed data as acquired from GPS receiver, vehicle to vehicle spacing data from ultrasonic sensors to compute space headway and GPS Locations.

(d) **Map matching of trajectory data**: Due to GPS measurement errors and road geometric errors in digital maps, the GPS locations derived from probe vehicles may not match to road segments [3]. Thus, a map matching procedure is utilized to match these locations (as acquired through GPS sensor) to network segments. A map matching procedure based on shortest path analysis and geospatial operations was utilized to map match these locations. The map matching procedure is discussed in Sect. 5.

(e) **Point Statistics for Speed and Space Headway**: Subsequently, point statistics analysis based on the speed values at instantaneous GPS locations and average space headway as derived through averaging of spacing captured by ultrasonic sensors (labelled as U1, U2, U3, U4, U5 and U6 in Fig. 1) has been carried out using ArcGIS software. The space headway (s) is computed on bases of following expression:

$$s = L + \frac{\sum_{i=1}^{6} c_i}{6} \tag{1}$$

In Eq. (1), L is the length of probe vehicle in metres and c is the inter-vehicle clearance in metres derived from ultrasonic sensors US-1, US-2, US-3, US-4, US-5, US-6. As shown in Fig. 2, the point statistics tool performs a neighbourhood operation that generate an output raster data where the value for each output cell is a mean of the headway or speed of any input point features that fall within three numbers of cells around the point location of GPS trajectory as shown in Fig. 3. The cell size is computed by ArcGIS software on basis of division of minimum of width or height of the extent of the GPS track (UTM 44N projection system) divided by 250.

(f) **Creation of Fuzzy Membership for Raster Dataset**: Subsequently the fuzzification of output of raster layers of point statistical analysis that signifies the variation of average speed and headways has been carried out to generate raster layer for fuzzy overlay analysis. The fuzzification has been carried out using fuzzy membership tool of ArcGIS software. The Fuzzy Small transformation function with default parameters value of has been adopted since the smaller headways and speed are indicative of congestion and likely to be a member of the fuzzy set. The default midpoint and spread parameters adopted for headway are 5.25 m and 5 m, respectively, and for speed as 23.87 km/h and 5 km/h, respectively. As shown in Fig. 4, fuzzy overlay analysis of type addition has been performed between the raster layers representing the headway and speed to identify possible congested segments. It has been considered that combination of speed and headway fuzzy layers generate more conspicuous congestion zones than the individual parameters.

(g) **Reclassification**: Finally, the outer raster layer generated using fuzzy overlay analysis is reclassified on basis of average pixel value of output raster. Cells of the output raster having pixel value lesser than the average pixel value are

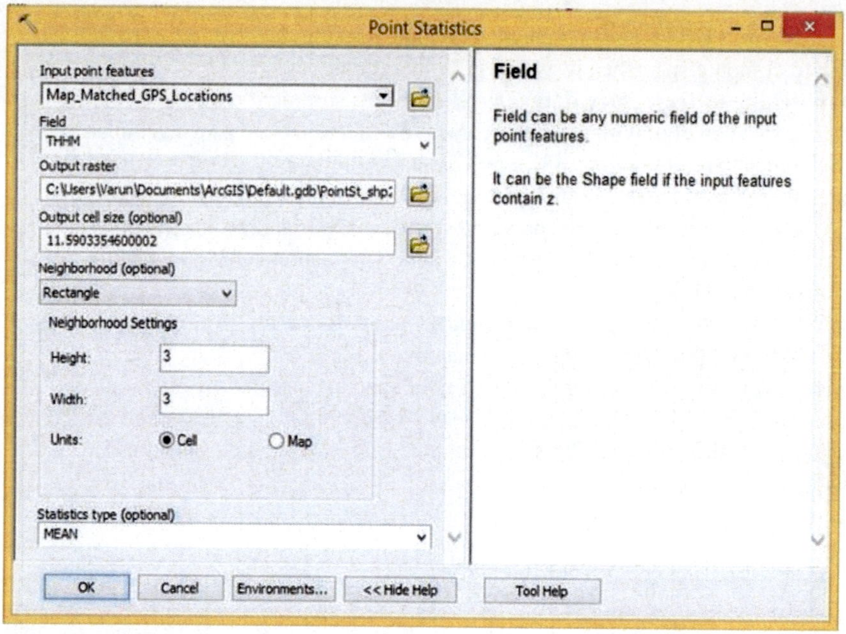

Fig. 2 Point statistics tool for statistical analysis

Fig. 3 Raster layer generated using point statistical analysis

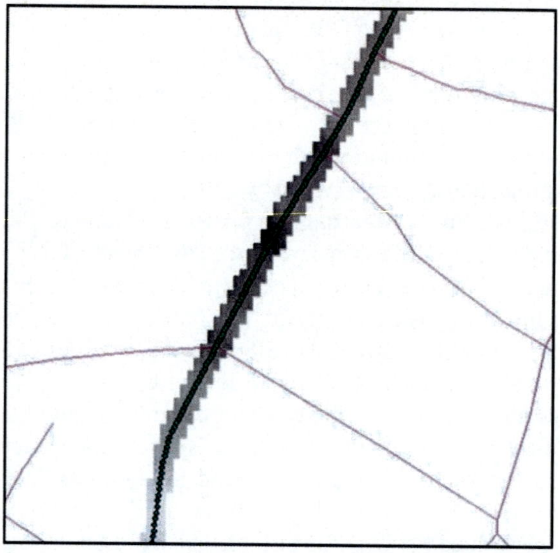

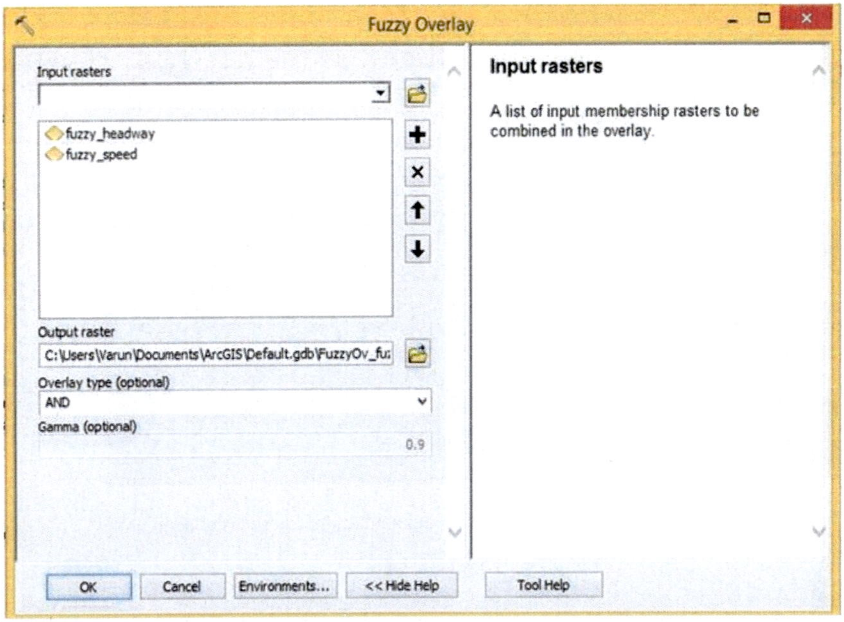

Fig. 4 Fuzzy overlay analysis in ArcGIS

assigned value of 1, otherwise 0 value is assigned to pixel value of the cells. In this way, it is assumed the cells having value of 1 in reclassified raster represents the congested section on roads.

(h) **Detection**: The congested road segments predicted in the previous section, needs verification. Hence, final detection of the results were verified using the traffic scenario captured through dash camera as shown in Fig. 5.

5 Map Matching of GPS Points

Map Matching procedure based on geospatial analysis has been developed and implemented to map match the location the GPS locations to road segments. The steps of the adopted map matching procedure are shown in Fig. 6. It is composed of different steps. GPS point dataset is map matched to the segments. GPS point dataset is converted to GPS track and centroid of each segment of the road segment is selected. Buffer of 50 m is generated around the GPS track. Spatial intersection of centroid points with buffer is carried out subsequently to select candidate road segments for map matching on basis of intersected centroids. Marker points at 100 m intervals are generated along the GPS Tracks. Shortest path is subsequently generated along the selected candidate road segments whilst connecting the marker points. In case of elevated road section, sliding-window filtering has been adopted to acquire the change in elevation values (Z value) in case of overbridges/road under bridge (RUB)

Fig. 5 Video grab of traffic scenario captured using GPS equipped dash camera installed in probe vehicle

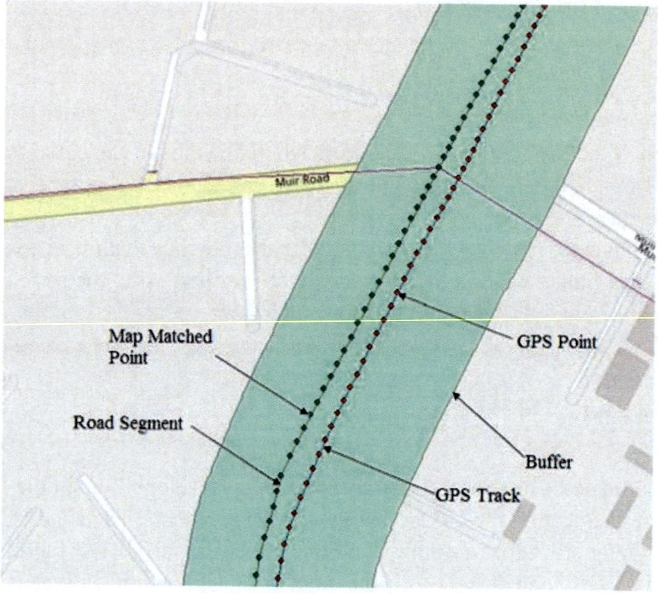

Fig. 6 Map matching of GPS trajectory data

and map match the points to corresponding road section. Using the attribute-based matching between the shortest path and the candidate road segments, relevant candidate road segments are extracted for map matching. GPS points are snapped on selected candidate road segments.

6 Results and Discussion

The output raster layer that has been generated using fuzzy overlay analysis using fuzzy raster layers representing the headway and speed is shown in Fig. 7. The output raster layer is further reclassified on basis of mean pixel value of the layer so as to derive sections having pixel values lesser than average pixel value of the output raster. Figure 8 depicts the reclassified raster and in the figure, road sections depicted in red colour are the congested sections. Figure 9 depicts the variation of speed and average headway which is continuously varying along direction of travel of probe vehicle from location A towards location B.

It can posited by analysing the variations of speed and headways that the region of the city encompassing the Central Business District are congested due to continuous movement of the traffic with the region as depicted in Fig. 8. Moreover, due to unregulated traffic at unsignalized intersections, speed of probe vehicle is lesser than its average speed (19.47 km/h). It can be observed from speed and headway variations during the course of the travel that speed and headways concurrently reduces significantly on the congested sections of traversed route. The standard deviation for speed data (8.7 km/h) is comparatively more than standard deviation of headway data (0.520 m). Therefore, the effect of speed variation on traffic congestion is more significant than headways.

Moreover, number of vehicles (in Passenger Car Unit) observed ahead of the probe vehicle at particular instance has been analysed using the dash cam to analyse the effect of mixed traffic on speed of probe vehicle. Instantaneous vehicles ahead of probe vehicle are counted based upon the analysis of video captured by dash camera and vehicle counts are converted to equivalency factor viz. Passenger Car Unit (PCU). Figure 10 depicts the (i) subplot between PCU versus elapsed time and (ii) subplot between speed of probe vehicle and elapsed time. Elapsed time in minutes in reckoned from the start of the journey. It is evident from Fig. 10 that with the increase of the PCU, the speed of the probe vehicle decreases owing to presence of other vehicle ahead of the probe vehicle and decreased instantaneous space headway. It can posited that there is a direct correlation between number of vehicles observed ahead of probe vehicle and congestion level.

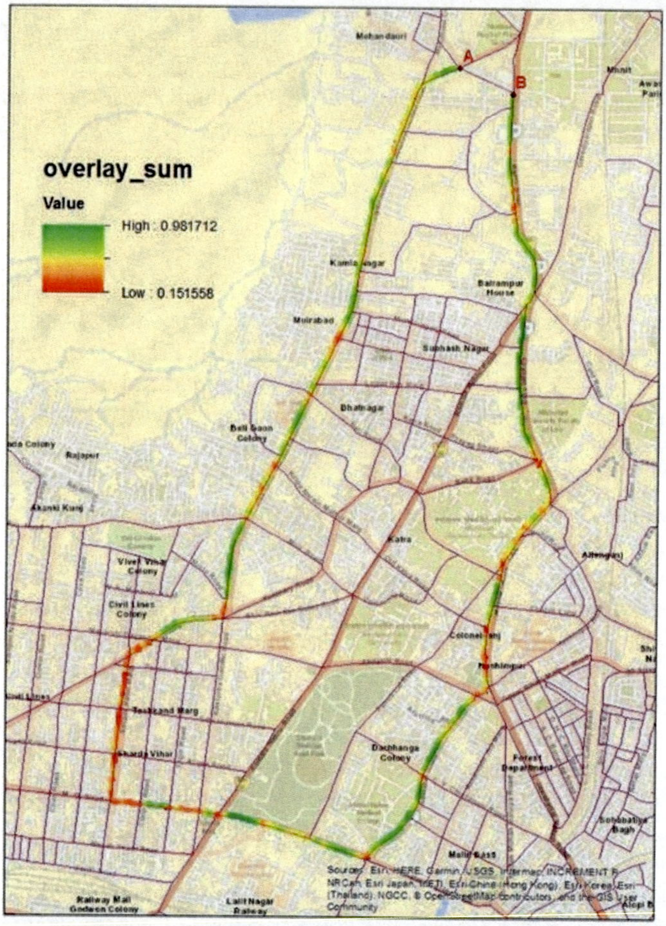

Fig. 7 Map with colour ramping depicting congested sections (as generated using fuzzy overlay analysis)

7 Conclusions

Due to increasing vehicular traffic in urban areas especially in developing countries, traffic congestion is becoming a serious problem, which is moreover resulting in reduced traffic mobility and increasing vehicular pollution.

Probe vehicles equipped with positional and other sensors have emerged viable solution to collect and analysis of the traffic data for identification of congested road sections. Mostly the speed and traffic density are important macroscopic traffic parameters that are used for the prediction of traffic congestion on road sections. Microscopic traffic parameters like space and time headways that are indirectly derived from the macroscopic traffic parameters viz. traffic density and traffic volume

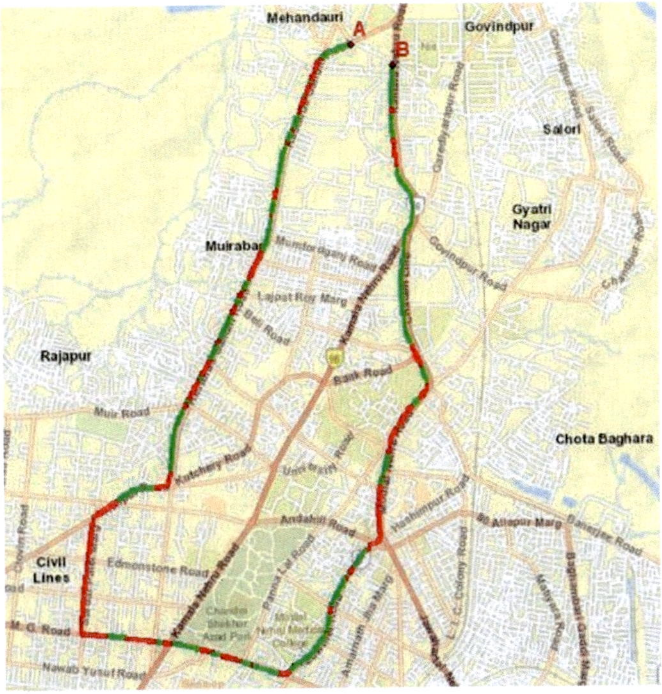

Fig. 8 Reclassified raster derived from fuzzy overlay

are utilized for subsequent analysis of traffic congestion. However, headways are difficult to measure directly from field traffic studies. Henceforth, they are not frequently analysed for congestion studies. In this paper, a methodology for identification of congestion zones using probe vehicle equipped with sensors has been proposed on basis of application of geospatial analysis, integration of macroscopic and microscopic parameters viz. speed and space headway, respectively, has been proposed for identification of congested zones.

Using the proposed methodology, congested road segments over the city can be identified and proper traffic calming measures can be adopted the city administration. The proposed methodology can be extended for the traffic state analysis and identification of congested road stretches with the utilization of advanced deep learning approaches, incorporation of image processing techniques for object-based identification of moving vehicles. These extensions are the focus of future research by the authors.

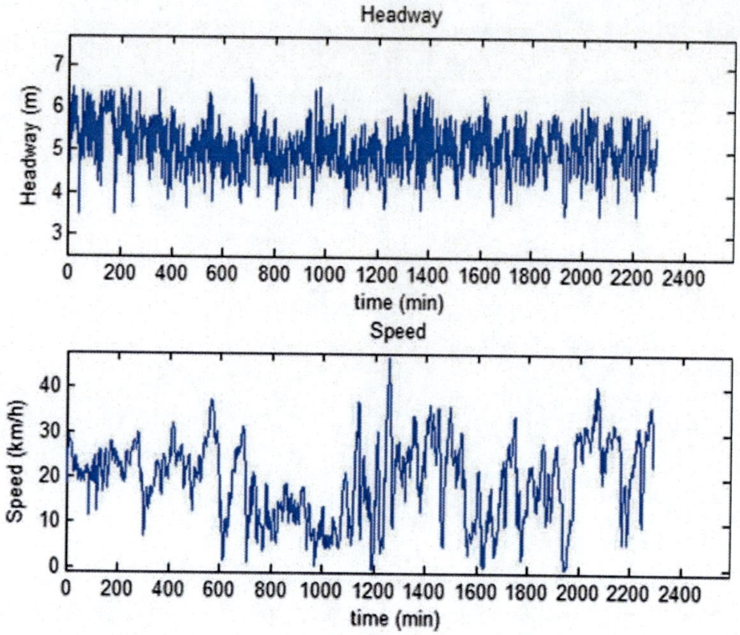

Fig. 9 Headway and speed variation in the direction of travel from A to B

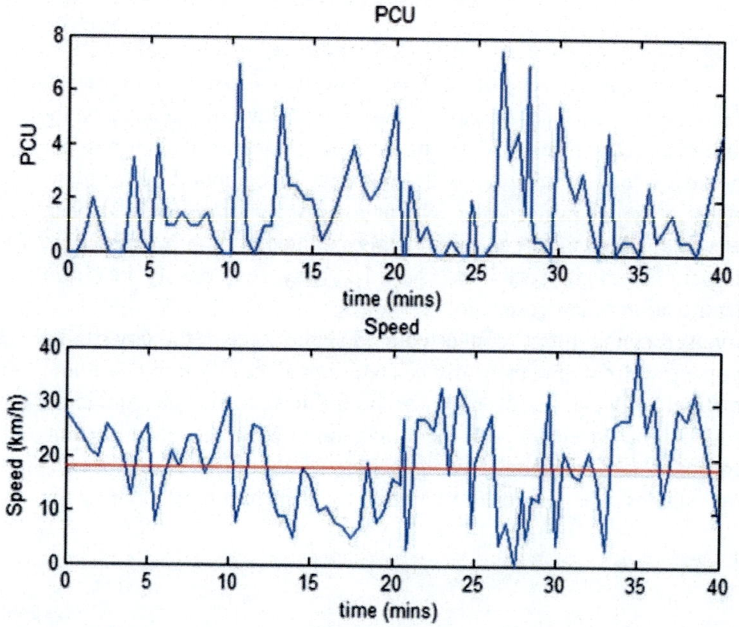

Fig. 10 Variation of (i) elapsed time versus PCU and (ii) elapsed time versus speed (red Line plot denotes the average speed (19.47 km/h) versus elapsed time plot)

References

1. Bishop B, Casabona J (2009) Automatic congestion detection and visualization using networked GPS unit data. In: Proceedings of the 47th annual southeast regional conference, pp 1–4
2. Cárdenas-Benítez N, Aquino-Santos R, Magaña-Espinoza P, Aguilar-Velazco J, Edwards-Block A, Medina Cass A (2016) Traffic congestion detection system through connected vehicles and big data. Sensors 16(5):599
3. Chen BY, Yuan H, Li Q, Lam WH, Shaw SL, Yan K (2014) Map-matching algorithm for large-scale low-frequency floating car data. Int J Geogr Inf Sci 28(1):22–38
4. ESRI (2019) Online ArcGIS documentation. https://doc.arcgis.com/en/, last accessed 21 June 2021.
5. Hansapalangkul T, Keeratiwintakorn P, Pattara-Atikom W (2007) Detection and estimation of road congestion using cellular phones. In: Proceedings of the 7th international conference on ITS telecommunications. IEEE, pp 1–4
6. Jensen AF, Larsen TV (2014) Travel-time estimation in road networks using GPS data. http://projekter.aau.dk/projekter/files/61070977/1181652577.pdf. Accessed 07 Feb 2014
7. Lanka S, Jena SK (2014) Analysis of GPS based vehicle trajectory data for road traffic congestion learning. In: Kumar KM, Mohapatra D, Konar A, Chakraborty A (eds) Advanced computing, networking and informatics, Vol 2. Springer, Cham
8. Lianxia X, Quan L, Minghua L, Zhong L (2007) Map matching algorithm and its application. In: International conference on intelligent systems and knowledge engineering (ISKE)
9. Liu W, Zheng Y, Chawla S, Yuan J, Xing X (2011) Discovering spatio-temporal causal interactions in traffic data streams. In: KDD, pp 1010–1018
10. Quiroga CA, Bullock D (1998) Travel time studies with global positioning and geographic information systems: an integrated methodology. Transp Res Part C Emerg Technol 6(1):101–127
11. Sakhare R, Vanajakshi L (2019) Reliable corridor level travel time estimation using probe vehicle data. Transp Lett 12(8):570–579
12. Thianniwet T, Phosaard S, Pattara-Atikom W (2009) Classification of road traffic congestion levels from GPS data using a decision tree algorithm and sliding windows, In: World congress on engineering, Vol 1, pp 1–3
13. Upadhyay A, Kumar A, Singh V (2020) Traffic data collection and visualization using intelligent transport systems. In: Ahmed S, Abbas S, Zia H (eds) Smart cities—opportunities and challenges. Lecture Notes in Civil Engineering, Vol 58. Springer, Singapore. https://doi.org/10.1007/978-981-15-2545-2_12
14. Wang C, Tsai HM (2013) Detecting urban traffic congestion with single vehicle. In: 2013 international conference on connected vehicles and expo (ICCVE). IEEE, pp 233–240

Structural and Economic Characteristics of Buildings with Different Fire Ratings as Per Indian Codes

Ankit Karan, Neelam Rani, and M. Abdul Akbar

1 Introduction

Fire is one of humanity's oldest inventions, and in many circumstances, it is still beyond human control. Fire may erupt in any structure, killing numerous people, and destroying million-dollar assets before eventually destroying the entire structure in a way that makes it dangerous to reuse.

Following the demolition of the Twin Towers in New York City on September 11, 2001, worldwide structural fire research surged. The nature of fire's influence on diverse materials and buildings is what researchers are most interested in. The globe is attempting to lessen the impact of fire on human lives and the economy, as well as on the environment.

In general, structural elements or systems are built to meet a certain level of fire resistance. The amount of fire resistance required varies depending on the building's importance, occupant type, number of floors, and floor space. The major goal of fire resistance rating is to provide residents enough time to flee the structure, fire fighters enough time to put out the fire, and to prevent progressive collapse.

In the United States alone, fire caused 3430 deaths, 97,800 injuries, and $14 billion in total damage in 2007 (Karter 2008) Fig. 1. Due to a variety of benefits over other materials, reinforced concrete (RC) structural systems are commonly utilised in high-rise buildings and other constructed infrastructure. The most important structural components of any RC construction are the foundation, column, beam, and slab.

Fire is one of the most severe environmental situations to which structures may be exposed over their lifespan, thus providing suitable fire safety measures for structural

A. Karan (✉) · N. Rani · M. A. Akbar
Department of Civil Engineering, Dr. B R Ambedkar National Institute of Technology Jalandhar, Punjab, India
e-mail: ankitk.sc.19@nitj.ac.in

N. Rani
e-mail: ranin@nitj.ac.in

A. Sil et al. (eds.), *Recent Trends in Civil Engineering*, Lecture Notes in Civil Engineering 274, https://doi.org/10.1007/978-981-19-4055-2_65

Fig. 1 Fire outbreak in US Karter 2008

parts are an essential aspect of building design. In contrast to beams, columns, and slabs, foundations are not sensitive to fire. The fire resistance of reinforced concrete members is generally established using prescriptive approaches, which are based on either standard fire resistance tests or empirical calculation methods.

2 Literature Review

High-strength concrete (HSC) shows 15–20% loss of compressive strength when exposed to temperature of 100–300 °C [1]. Whereas normal strength concrete (NSC) shows no such strength loss. No specimen was able to sustain the load beyond 700 °C. About one-third of these specimens failed in the temperature range of 320–360 °C whilst being heated under a constant preload [2]. HSC appears to be more prone to spalling in a fire than NSC. Spalling occurred in the thicker 200 mm flange where the cover of the steel was large (75 mm). In the 150-mm flange, the cover was 25 mm, and there was no spalling also no spalling occurred in the Web. [3] Concrete densified by means of Silica fume may explode during heating, and presence of steel fibre does not reduce the risk of explosion. Lightweight concrete is not recommended in place of normal weight concrete when spalling is concerned.

A major fire test programme was conducted on a full-scale eight-storey steel-jiramed building. The fire tests on the full-scale frame showed that the actual global and local structural behaviour of buildings are different, and typically far better; than that shown in standard small-scale fire tests. The paper discusses in detail the behaviour of the building observed during the fire tests, and preliminary conclusions are presented [4].

This paper is an attempt to summarise the behaviour of different structural components of RC buildings under elevated temperature [5]. It has been observed that concrete is more affected than reinforcing bars when structure is exposed to fire. It has also been noticed that load level and axial restraint have minor influence on the thermal response of RCC beams. [6] The NSC M25 and HSC M70 and M100 exhibit a linear response. The deflection of a slab decreases as we increase the cover of slab and grade of concrete. M100 with a cover of 50 mm shows a minimum deflection when exposed to high temperature and pressure, whereas maximum deflection is shown by M25 grade with 30 mm cover [7]. This paper deals with the effect of fire on the flexural strength of reinforced concrete M25 beam with clear cover 25 mm and M30 beam with clear cover 30 mm. It has been observed that the flexural strength for beams exposed to fire at 550 and 750 °C for 60 and 120 min was less than that for the reference beam by about 34.84% and 44.37%, respectively. But for 950 °C, there is significant decrease in flexural strength by about 61.99% and 64.24%, respectively. [8] This paper includes the fire resistance of concrete which tells about physical and chemical response to fire and how to enhance the fire resistance. By using lightweight aggregate and by using PP- fibres with reinforcement.

3 Methodology

Due to high increase in fire accidents in buildings and structure, many lives loss and economy losses have taken place in huge amount, so in keeping a view the losses and damages, a study is carried out on buildings subjected to different fire rating, i.e. (1 h, 1.5 h, 2 h, 3 h, 4 h). This research work is totally done on construction software "ETABS" in which we have modelled three types of building;

- 5-storey office building (base model)
- 10-storey building (containing shear wall).
- NIT-Jalandhar administration building (containing waffle slab).

From IS: 456 (2000), the values are taken for minimum dimension of reinforced concrete members for fire resistance from Table 16A [9]. From IS: 875 (part1), the values are taken for unit weight of material from Table 1 [10]. From IS: 875 (part2), the values are taken for imposed floor load for different occupancies, Table 1 [11].

3.1 Detail Discussion of Models

(a) **5-Storey office building base model**

In this, a five-storey office building is modelled which subjected to different fire rating. The material of concrete is taken M20, and the minimum requirement of width and clear cover of beam, columns, and thickness of slab is taken from IS: 456 (2000) from Table 16A (clauses 21.4 and 26.4.3 and Fig. 1). The load combinations are (DL + LL), 1.5(DL + LL). The model is then analysed on

Table 1 Specification and cost estimation of 5-storey office building for each fire rating

Specification	1 h	1.5 h	2 h	3 h	4 h
Grade of concrete	M20	M20	M20	M20	M20
Unit wt. of concrete (kg/m^3)	2548.538	2548.538	2548.538	2548.538	2548.538
Beam size (mm)	200 * 400	200 * 400	200 * 400	200 * 400	200 * 400
Column size	(400 * 400) I&II (300 * 300) III, IV&V	(400 * 400) I&II (300 * 300) III, IV&V	(400 * 400) I&II (300 * 300) III, IV&sV	(400 * 400) I&II (300 * 300) III, IV&V	(400 * 400) I&II (300 * 300) III, IV&V
Clear cover (mm)	40	40	40	40	40
Slab thickness (mm)	150	155	160	165	170
Slab clear cover (mm)	15	20	25	30	35
Live load (LL) (KN/m^2)	4	4	4	4	4
SDL (MARBLE) (KN/m^2)	0.534	0.534	0.534	0.534	0.534
Cost of concrete (Rs/m^3)	3400	3400	3400	3400	3400
Total cost (Rs)	1,846,411	1,889,429	1,932,471	1,975,460	2,018,476

ETABS shown in Figs. 2 and 3 showing the bending moment generated in the structure, cost calculation is discussed for each fire rating shown in Table 1, and percentage increase in cost is calculated.

Remark: As per Table 1, the graph is plotted in which x-axis represents time in hour, and y-axis represents cost of construction as shown in Fig. 4. The linear increase in graph is seen with increase in a time period.

(b) **10-Storey building (consisting shear wall)**

A 10-storey building is modelled which is situated in zone 4 seismic zone. The shear wall is constructed to take lateral loads and outer wall is subjected to partition wall load. The material of concrete is taken M30 for up to storey 3, M25 for storey (3–6), and M20 for storey (6–10). The value of seismic zone factor $Z = 0.24$ (IS Table 2), response reduction factor $R = 3$ (IS Table 7), importance factor $I = 1$ (IS Table 6), spectral acceleration coefficient $Sa/g = 1.577773$ (IS 6.4.5) of IS: 1893 (2000) [12]. The load combinations are (DL +

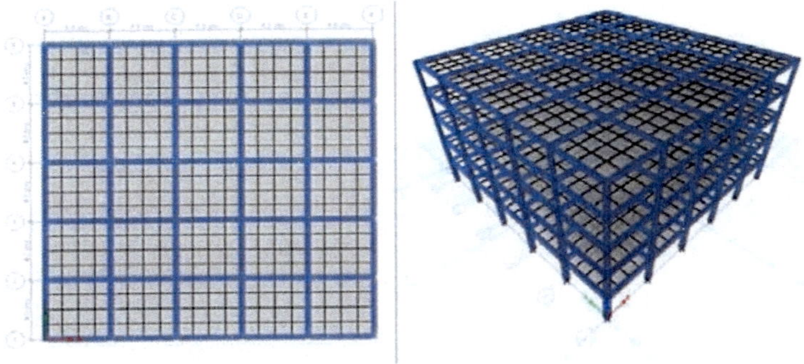

Fig. 2 Plan view and 3D view of building modelled in ETABS

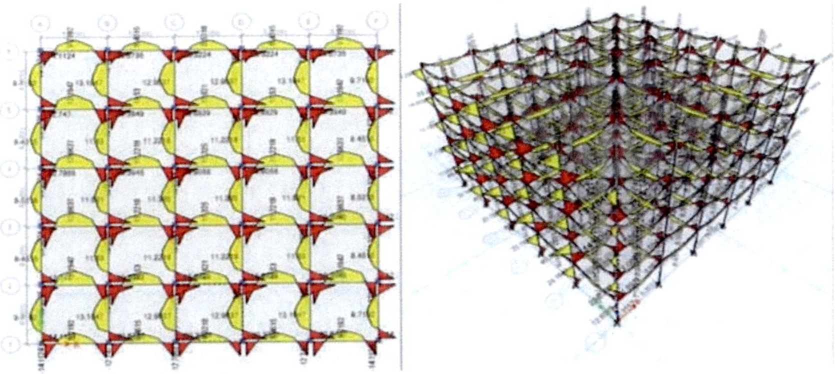

Fig. 3 Bending moment generated in building

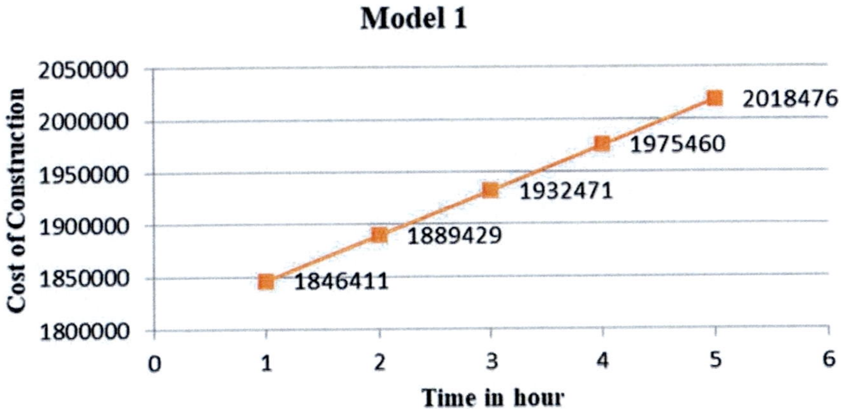

Fig. 4 Graphical interpretation of increase in the cost of construction with respect to time

LL), (DL + EQ-X), (DL + 0.8LL + 0.8EQ-X), 1.5(DL + LL), 1.2(DL + LL + EQ-X). The model is then analysed on ETABS, cost calculation is done for each fire rating as shown in Table 2, and percentage increase in cost is calculated. The following Figs. 5 and 6 show the bending moment generated in building.

Remark: As per Table 2, the graph is plotted in which x-axis represents time in hour, and y-axis represents cost of construction as shown in Fig. 7. The progressive increase in graph is seen with increase in a time period.

Table 2 Specification and cost estimation of 10-storey building for each fire rating

Specification	1 h	1.5 h	2 h	3 h	4 h
Grade of concrete	M30(0–3) M25(3–6) M20(6–10)	M30(0–3) M25(3–6) M20(6–10)	M30(0–3) M25(3–6) M20(6–10)	M30(0–3) M25(3–6) M20(6–10)	M30(0–3) M25(3–6) M20(6–10)
Unit weight of concrete (kg/m^3)	2548.538	2548.538	2548.538	2548.538	2548.538
Beam size (mm)	300 * 400	300 * 400	300 * 400	300 * 400	300 * 400
Column size M30 M25 M20	450 * 450 400 * 400 300 * 300	450 * 450 400 * 400 300 * 300	450 * 450 400 * 400 300 * 300	450 * 450 400 * 400 300 * 300	450 * 450 400 * 400 300 * 300
Clear cover (mm)	40	40	40	40	40
Slab thickness (mm)	170	170	170	170	170
Slab clear cover (mm)	35	35	35	35	35
Live load (LL)	4 KN/m^2 (0–9) Storey 1.5 KN/m^2 Roof	4 KN/m^2 (0–9) Storey 1.5 KN/m^2 Roof	4 KN/m^2 (0–9) Storey 1.5 KN/m^2 Roof	4 KN/m^2 (0–9) Storey 1.5 KN/m^2 Roof	4 KN/m^2 (0–9) Storey 1.5 KN/m^2 Roof
SDL (KN/m^2)	0.534	0.534	0.534	0.534	0.534
Shear wall (mm)	120	140	160	200	240
Wall thickness (mm)	75	90	100	170	170
PWL (KN/m)	2.1413	2.57	2.855	4.85	4.85
Cost (Rs)	7,564,899	7,880,103	8,093,865	8,681,164	8,882,285

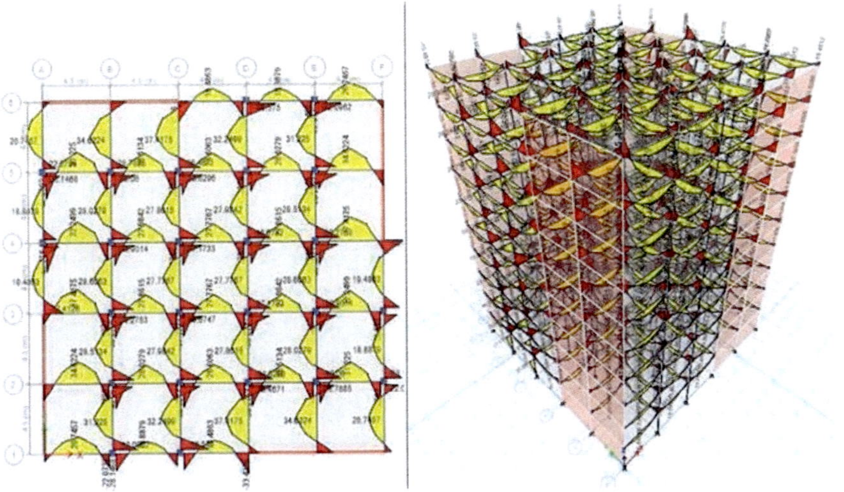

Fig. 5 Showing the bending moment generated in beams columns and slab

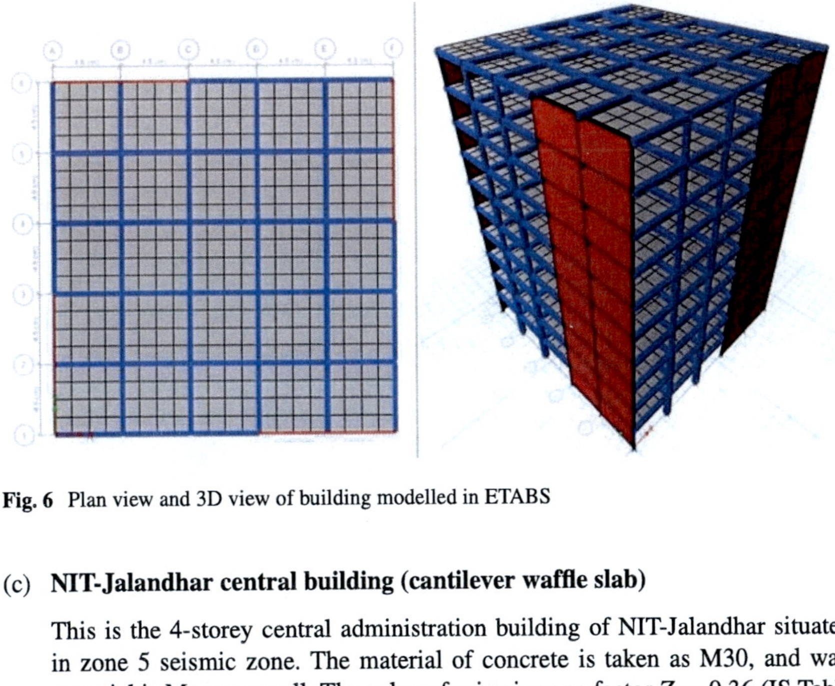

Fig. 6 Plan view and 3D view of building modelled in ETABS

(c) **NIT-Jalandhar central building (cantilever waffle slab)**

This is the 4-storey central administration building of NIT-Jalandhar situated in zone 5 seismic zone. The material of concrete is taken as M30, and wall material is Masonry wall. The value of seismic zone factor $Z = 0.36$ (IS Table 3), response reduction factor $R = 5$ (IS Table 9), importance factor $I = 1$ (IS Table 8), spectral acceleration coefficient Sa/g = 2.5 (IS 6.4.2) of IS: 1893 (2016) [13]. The load combinations are (DL + LL), (DL + EQ-X), 1.5(DL + LL), 1.5(DL + EQ-X), 1.2(DL + LL + EQ-X), (DL + 0.8LL + 0.8EQ-X).

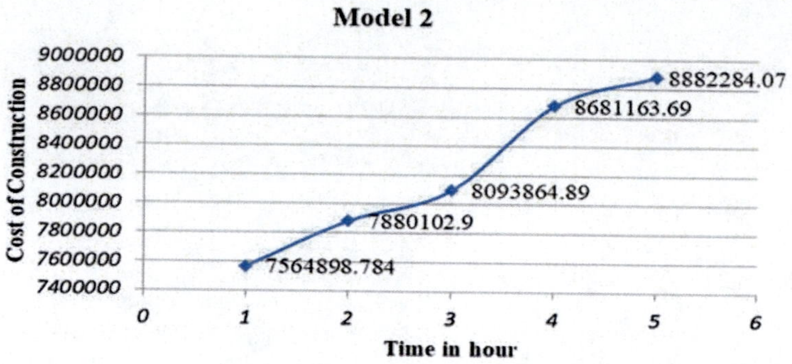

Fig. 7 Graphical interpretation of increase in the cost of construction with respect to time

The model is then analysed on ETABS, and cost calculation is done for each fire rating as shown in Table 3, and percentage increase in cost is calculated. Figures 8 and 9 show the view and deformation of the building under loading condition.

Remark: As per Table 3, the graph is plotted in which x-axis represents time in hour, and y-axis represents cost of construction as shown in Fig. 10. The progressive increase in graph is seen after 3 h of fire rating.

4 Results and Discussion

The following models are analysed in ETABS as par IS: 456 (2000), NBC (part 4) 2016, and for gravity load as per (as per IS: 875 (part 1) and IS: 875 (part 2). The percentage of cost increment is seen as we move from 1 to 4 h fire rating. The per cent of increase in cost is due to requirement of material, and reinforcement for each fire rating is different. The conclusion of result is shown below.

(a) 5-storey office building (base model) (Table 4)

(b) 10-storey building (containing sheariu wall) (Table 5)

(c) NIT-Jalandhar central building (containing cantilever waffle slab) (Table 6)

5 Conclusion

As expected, increase in a fire rating, increases the value of the structure, so as per IS: 456 (2000) and NBC (part4), the economic characteristics and effects of implementing fire rating are observed and calculate for (1 h, 1.5 h, 2 h, 3 h, and 4 h), respectively.

Table 3 Specification and cost estimation of NIT-Jalandhar administration building for each fire rating

Specification	1 h	1.5 h	2 h	3 h	4 h
Grade of concrete	M30 and Masonry wall	M30 and Masonry wall	M30 and Masonry wall	M30 and Masonry wall	M30 and Masonry wall
Unit weight of concrete (kg/m^3)	2548.538	2548.538	2548.538	2548.538	2548.538
Beam size (mm)	200 * 400	200 * 400	200 * 400	200 * 400	200 * 400 mm
Column size	400 * 400 350 * 350	400 * 400 350 * 350	400 * 400 350 * 350	400 * 400 350 * 350	400 * 400 350 * 350
Clear cover (mm)	40	40	40	40	40
Slab thickness (mm)	125	125	125	150	170
Slab clear cover (mm)	20	20	25	35	45
Live load (LL)	4 KN/m^2 (0–3) Storey 1.5 KN/m^2 Roof	4 KN/m^2 (0–3) Storey 1.5 KN/m^2 Roof	4 KN/m^2 (0–3) Storey 1.5 KN/m^2 Roof	4 KN/m^2 (0–3) Storey 1.5 KN/m^2 Roof	4 KN/m^2 (0–3) Storey 1.5 KN/m^2 Roof
SDL	0.534 (KN/m^2)	0.534 (KN/m^2)	0.534 (KN/m^2)	0.534 (KN/m^2	)0.534 (KN/m^2)
Wall thickness (mm)	75	90	100	170	170
Cost (Rs)	5,224,831	5,423,565	5,569,958	6,407,637	8,882,285

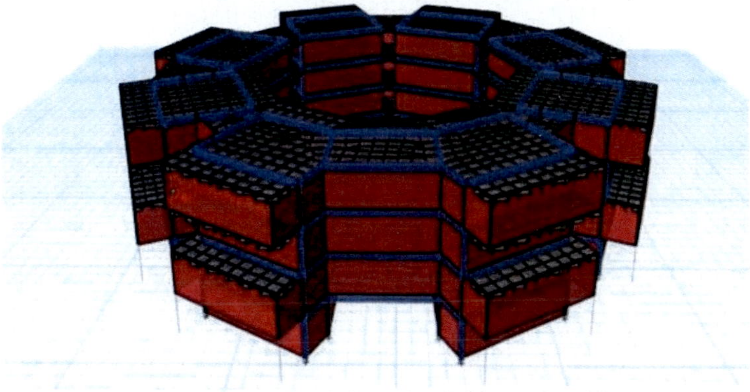

Fig. 8 Plan view and 3D view of building modelled in ETABS

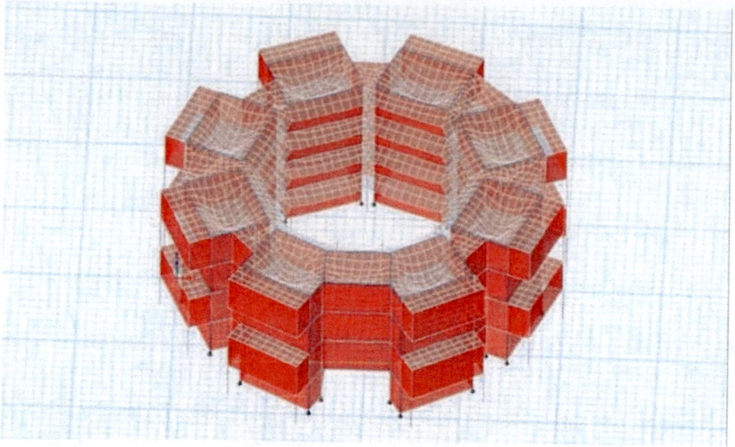

Fig. 9 Deformation of building under load

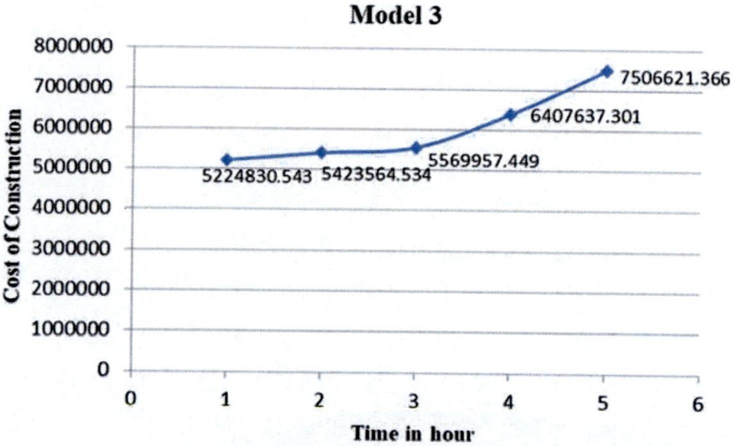

Fig. 10 Graphical interpretation of increase in the cost of construction with respect to time

Table 4 Economic difference of base model compared at different hours

Hour period (h)	Percentage difference (%)
(1–1.5)	2.32
(1–2)	4.66
(1–3)	6.98
(1–4)	9.31

Table 5 Economic difference of building compared at different hours

Hour period (h)	Percentage difference (%)
(1–1.5)	4.16
(1–2)	6.99
(1–3)	14.75
(1–4)	17.41

Table 6 Economic difference of building compared at different hours

Hour period (h)	Percentage difference (%)
(1–1.5)	3.80
(1–2)	6.60
(1–3)	22.63
(1–4)	43.67

- Minimum requirement of slab cover is seen to increase from 15 to 35 mm and thickness from 150 to 170 mm with increase in a fire rating hour for the 5-storey base model.
- The thickness of shear wall is observed to be increasing from 120 to 240 mm and partition wall thickness from 75 to 170 mm for 10-storey building.
- The maximum economic difference for first model is observed to be 9.31%, for second model 17.41%, and for third model 43.67%.
- The total volume of concrete requirement increases with increase in a hours of fire rating which let to increase the cost of overall building.

References

1. Noumowe AN, Clastres P, Debicki G, Costaz JL (1996) Transient heating effect on high strength concrete. Nucl Eng Des 166(1):99–108. https://doi.org/10.1016/0029-5493(96)01235-6
2. Sanjayan G, Stocks LJ, Spalling of high-strength silica fume concrete in fire
3. ACI Mater J 90(2):170–173, Mar 1993. https://doi.org/10.14359/4015
4. Wu Z, Huen Lo S, Hai Tan K, Leung Su K (2019) High strength concrete testsunder elevated temperature. Athens J Technol Eng 6(3):141–162. https://doi.org/10.30958/ajte.6-3-1
5. Bailey CG, Moore DB, Lennon T (1999) The structural behaviour of steel columns during a compartment fire in a multi-storey braced steel-frame. J Constr Steel Res 52(2):137–157. https://doi.org/10.1016/S0143-974X(99)00036-X
6. Ahmad S, Sadique R (2013) Unserstanding the behavior of RCC structures atelevated understanding the behavior of RCC structures at elevated temperature to mitigate fire hazard. Int Conf Chall Dis Mitig Manag 13–20
7. Reshma PS, M Prasad (2021) Effect of fire on RC slab. IJIRST Int J Innov Res Sci Technol 3. Accessed 28 May 2021. Available: www.ijirst.org
8. Chadha RP, Mundhada AR, Student FY, Structure ME (2021) Effect Of fire on flexural strength of reinforced concrete beam. Accessed 29 May 2021. [Online]. Available: www.ijert.org
9. Shams SGIA (2021) Fire resistance of concrete. Accessed 29 May 2021. [Online]. Available: www.ijetsr.com

10. B. of Indian Standards (2000) IS 456: Plain and reinforced concrete—code of practice
11. B. of Indian Standards (1987) IS 875-1: code of practice for design loads (other than earthquake) for buildings and structures, Part 1: dead loads—unit weights of building material and stored materials (incorporating IS 1911 : 1967)
12. B. of Indian Standards (1987) IS 875-2: code of practice for design loads (other than earthquake) for buildings and structures, Part 2: imposed loads
13. B. of Indian Standards (2002) IS 1893-1: criteria for earthquake resistant design of structures, Part 1: general provisions and buildings
14. Singh B (2018) IS-1893-Part-1-2016, Vol 1893, no Part 1

Paratransit Utility and Mode Choice Modelling: A Review

Waikhom Victory and Mokaddes Ali Ahmed

1 Introduction

Rapid urbanization with rising population leads to higher use on private vehicles resulting in the impact on public transportation. It needs to motivate a modal shift to public transportation towards the sustainable transport system both financially and economically [1]. The accessibility of paratransit fulfils the barrier between the public and private transportation in most of the cities. Paratransit can be described as the intermediary mode between the individually owned vehicles and the public transit that has predetermined routes and time [2]. They are operated either on shared ride or hired basis and fulfils the needs of the user. Besides, some of the paratransit modes do not have predetermined routes and time. Paratransit is basically the intermediate public transport predominantly used in urban transport mode in developing countries and provides satisfaction to users in public transportation system [3]. The literature referred paratransit as intermediate public transport, semi-formal transport, small business-run transport or more frequently as informal transport [4]. It is also known for door-to-door service, and a system consists of whole elements including transport planning, transport service transport mode that enhances economic and social benefits [5, 6]. In Thailand motorcycle-taxi, Tuk-tuk, Songtaew, and Silor-lek are available as paratransit modes. Among them, Songtaew operates as a primary public transport mode in small- and medium-sized cities as it fits around 18 commuters or more [7]. Auxiliary transport includes trucks, car chases, shutters, minibuses and microbuses [8]. Private drives and minibuses are widely used in developing countries to reduce labour costs and some inadequate transit services [9]. Paratransit offers

W. Victory (✉)
Department of Civil Engineering, National Institute of Technology Manipur, Langol Manipur, Manipur, India
e-mail: waikhomvictory@gmail.com

M. A. Ahmed
Department of Civil Engineering, National Institute of Technology Silchar, Silchar, Assam, India

A. Sil et al. (eds.), *Recent Trends in Civil Engineering*, Lecture Notes in Civil Engineering 274, https://doi.org/10.1007/978-981-19-4055-2_66

services in many countries in various forms, such as "Angkutan Kota" in Indonesia, "Jeepney" in the Philippines, "Tuk-Tuk and Songtaew" in Thailand and "Mammy Wagons" (converted trucks) and "Matatu" (converted truck) in Africa [10]. Paratransit performs a major part in Nairobi. It serves as a means of public transport in the city [11]. It also includes taxis and car chases (in developing countries), direct vehicles, and other demand-driven minivans (in developed countries) generally provide public transportation [9]. In developing countries, the mode disaggregation of this mode is limited. Around 40–50% of them out of 70–80% are motorized means of transport [10, 12]. In poor cities, the proportion of traffic is 10–15%, of which 20% is motorized [10, 12]. In most Indian cities, the percentage of private and paratransit transportation services are low [13]. In many developing countries, bridging the gap in the public transport system has become an essential and efficient means of transport [8]. In suburbs and areas without traditional public transport, paratransit performs a major task in meeting travel needs [14]. Compared to other means of public transport, Paratransit offers further advantages such as easy mobility and accessibility, optimal operating costs and low maintenance costs [15]. The service can simply be opened if necessary and closed at the same time if the number of users is low [16]. It also provides jobs for the low skilled, which benefits society [14].

2 Paratransit Services

The paratransit services are different in the developed and developing countries. In developed countries, it is mostly used for demand-driven services such as subscription buses, dial-a-ride and shared-ride taxis. But, in developing countries, due to low living standard, excessive population, low cost labour force, etc., leads to need a mode which bridges the public and private services [17]. The services provided by paratransit can be classified into three groups, i.e. feeder service, door-to-door service and fixed route shared service.

2.1 Feeder Service

This type of service is mainly associated with the transportation of passengers from their origin to the regional transit terminals such as railway stations, bus terminals and airport. These trips are generated at any time and depend on the demand of the trip makers. Autorickshaw and taxi provide the main feeder service in India.

2.2 *Door-To-Door Service*

It is a service provided by a single mode from origin to destination. Door-to-door services are often provided by taxis and seated autorickshaw. Paratransit is often a substitute for private cars for the urban middle class. This type of service is associated with work trips, shopping trips, sociocultural trips and recreational trips.

2.3 *Fixed Route Shared Service*

Shared ride fixed route service is provided by auto rickshaws in some cities depending on the demand where the city bus services are inexistent or inefficient. The fixed route service is provided by autorickshaw, Tata Sumo, Max and Tata magic.

3 Paratransit Utility

Paratransit modes will continue to provide a dominating role in the small and medium cities. The mode operates locally in larger urban areas. However, further improvement would be required to make a better choice of the commuters and to increase the utility. Besides, non-motorized paratransit cannot be ignored as it is a major source of employment generation. Considering the increasing trend of the paratransit in the developing countries, it is essential to divide non-motorized traffic lane and motorized traffic corridors. The segregation between these two types of paratransit will help in the smooth operation of a transportation network and thereby increase the utility. It can be enhanced by arranging different roads and imparting individual lanes such as beacon lanes in Indonesia or cycle-rickshaw lanes in India and Bangladesh. In addition, if non-motorized modes are restricted to serve short distance local trips, a higher speed in arterial roads can be achieved.

About 50% of trips are made by paratransit vehicles in the cities of Bandung, Java, Surabaya and Chiang Mai [17, 18]. Recently, paratransit commuters are decreased with the increased economic growth rate thereby increase in ownership of private vehicle. The percentage share of paratransit in the overall spectrum of the urban passenger transportation system is the utility of the mode. The utility of paratransit is gradually increasing and can be studied by analysing various parameters of the trips makers as well as the scope of the services provided by the paratransit vehicles. The utility of paratransit is higher in small and medium cities as compared to big cities and metropolitans. This is due to the non-availability of efficient public transport. In small cities where road networks are narrow and congested, the operations of large size vehicles such as buses are difficult.

In order to assess the utility of paratransit modes, it is necessary to study the factors which affect the utility. These factors include socio-economic characteristics of the trip makers, availability of alternative modes and behavioural parameter characteristics of the modes. In the cities of developing countries, intermediate paratransit (IPT) exhibits a major part in urban public transportation services. Paratransit acts as main transportation in few countries, while in some other countries (region), it acts a feeder to major transportation networks such as railways, metros and other public transportation systems. Paratransits are considered as an integral part of urban transportation due to its identifying features, such as low carrying capacity, low speed, low energy requirements, higher labour intensity and more dependable and small area of coverage [17]. IPT delivers crucial services in places where bus services are reluctant, improper, non-uniform or unpredictable. They are important modes which impart mobility to educational, commercial, residential, recreational and other destinations by providing door-door service. IPT vehicles often ply on the routes and neighbourhoods which are not accessible by buses. Moreover, they provide incredible services during the night shifts when bus services are stopped, and sometimes they are also the only means of getting around the cities [19].

As mentioned earlier, both non-motorized and motorized paratransit modes are operated in developing countries in various forms. Various forms of paratransit modes ranging from motorized mini buses and non-motorized are dominant in most of cities in the developing countries [17]. The urban commuter services comprised of motorized vehicles of various sizes. In Kuala Lumpur, public transport mostly comprises regular bus which constitutes 58%. In the Philippines, the transportation service is dominated by Jeepney. Minibuses in Kathmandu, Karachi and Chiang Mai constitute 50–70% of the transport network. In Dhaka and some Indian cities, cycle-rickshaw exhibits a major part. In Dhaka, almost 72% households use rickshaws for their daily trips [20]. The urban cycle-rickshaws in Bangladesh carry an average of over 30,000 passenger-miles, and nearly 100 ton-miles of goods movement is recorded yearly [21]. The motorized paratransit modes in developing countries carry 20% to more than 50% of total public transport demand. For example, the transportation demand comprised of 21% in Bangkok (Thailand), 40% in Kuala Lumpur (Malaysia), 50% in Jakarta (Indonesia) and 70% of the total public transport demand in Metro Manila (Philippines) are carried by motorized paratransit modes [17]. They also provide assistance to minute and inadequate roads conveniently, where there is lacked of additional services or inaccessible at comparatively inexpensive cost. The paratransit system is growing at an enormous rate in the developing cities. In most of the instances, the paratransit services established according to the needs and requirements of the regional circumstances. Different factors are responsible for intensified rise of paratransit. These include rapid urban population growth due to rural to urban migration and natural increase, increased in per capita incomes of the commuters, inadequate transport infrastructure and insufficient public transport system. Moreover, the convention bus services cannot fulfil the demand of numerous short trips, work trips and recreation trips like in developed nations. In addition, urban rail services are also lacked in developing countries, despite such conventional transportation needs huge funds and capital investment [17]. For example, in Mexico City, the laboriously cheap, inexpensive Metro rail services do not connect suburbs and outskirt of the city. So, a group of collective minibuses fulfilled the facility and executes the connection between the stations and suburb habitations. But today, numerous unlicensed commuter vans and autorickshaws provide immediate services with more mobility, accessibility in less time and moderate fares.

Increased in economic level of the commuter has impacted on declining of paratransit utility. With the increased in economic level, the commuters are able to afford private cars and opted out from using the non-motorized paratransit. Hence, the utility of non-motorized paratransit has been declining over the years, and simultaneously, automated paratransits came across an alarming situation to embed the space evacuated from unpowered or non-motorized paratransit. For instance, the non-motorized or unpowered paratransit has been reducing their role. Although, the requirement for single commuter paratransit raised in the cities of Delhi, Dhaka, Karachi and Colombo. However, in Bangkok, the quantity of silor is raising as a feeder service even though they have attained economic level [17]. The utility of paratransit in developing countries can be observed by considering few cases such as Kombi minibuses in South Africa, Angkots (minivans), Becaks and Ojeks in Indonesia, Dolmuses in Turkey, Fast Buses in Dakar (Senegal) and Daladalas in Dar-Es-Salaam (Tanzania). The Kombi minibus-taxi industry caters approximately 65% of daily commuters. Over the last 20 years, this market share has gradually increased, as bus and rail services have deteriorated. The minibus-taxi industry mostly caters the services to township dwellers as it offered almost a door-to-door service to commuters. In Indonesia, Angkots, Becak and Ojeks are highly preferred by the urban poor, middle-class group, shoppers, business owners, and students for the following reasons due to close connectivity to employment centres, markets, perceived to be safer and convenient than public buses, can load goods, availability at late nights, and can work with one driver on contracts [19]. In Turkey, the IPT is mainly comprised of Dolmuses which are privately owned minibuses and run on fixed routes within cities as well as in the outskirts connecting other towns and villages. They work on a fixed fee system, and cities have dedicated Dolmus stops just like buses. It can carry upto 14 passengers, but sometimes it loads more passenger (up to 25) by the operator and thus makes the trip overcrowded. In Dakar (Senegal), 75% of the daily trips are made by public transport. However, Fast Buses serve as IPT which is managed by different private owners. It consists of a fleet of 2500–3000 minibuses and provides for almost 80% of the public transport demand. The frequency of these fast buses is very high and provides cheaper service as compared with the public bus system. Ndiaga-Ndiaye (large white vans with an entry for passengers from the back) also serve as an IPT. Their main advantage is able to provide an assured seat for everyone [19]. In Dar-Es-Salaam city (Tanzania), almost 98% of users have no car and highly depend on Daladalas. Daladalas are cheap minibuses operating on fixed routes, marked with specially coloured stripes which identify their origin and destination. Daladalas carry 1.4 million passengers per day and provide the services in the suburbs [19]. IPT is cost-effective and resourceful transportation system. As it does operate in flexible routes and schedule, the operators are able to maintain maximum earning and frequent passenger turnover which ultimately ensure in increasing patronage for IPT drivers. These vehicles require less space for parking as compared to private vehicles and can carry more passengers per trip. Due to the lower investment in terms of production and operation, IPT vehicles able to provide low-cost transport service for lower and middle-income class population thus make IPT as the popular transport services in many developing countries. Further, due to their lower weight and slower speeds, these modes cause fewer damages on the roads as compared to cars and other heavy vehicles [19]. IPT vehicles are flexible in terms of demands and needs of the passengers. In response to the demands, they easily alter rates, timings, frequencies and the operation according to the market demands. Such flexibility and options make the IPT a prefer choice of the commuters over public transport.

4 Factors Affecting the Paratransit Utility

The utility of paratransit depends on certain factors, and these can be classified broadly into three categories, viz. (a) socio-economic factors, (b) availability of alternative modes and (c) behavioural parameter. The socio-economic factors are comprised of income of the individuals and travel costs. The availability of several numbers of alternative transportation mode will also affect the utility of paratransit because in such circumstances the travel demand of that particular mode will be shared with other modes. According to the requirement, people often choose their preferred and comfortable mode; for instance, a commuter might choose cycle-rickshaw over the autorickshaw depending on the location and accessibility. Nevertheless, the travel demand will raise as population increased, thus affecting efficacy of paratransit. This relationship between travel demand and the parameters is given in Fig. 1.

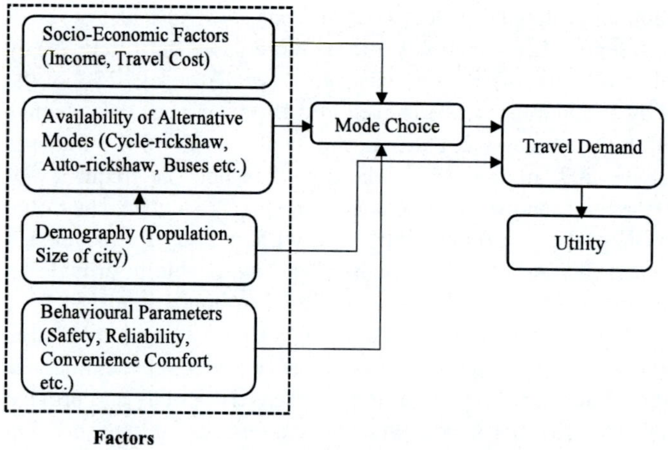

Fig. 1 Relationship between utility, travel demand and factors affecting utility

Figure 1 indicates the interrelation between the trip makers' decision-making process and their willingness to choose a particular mode. The decision of the passenger affects the demand for the particular travel mode and their utility. While making the decision or opting a mode, there are several components that affect the trip makers. Users' willingness to choose the mode is affected by socio-economic factors such as income of the trip maker and travel cost (fare). Incorporating to socio-economic factors, behavioural parameters such as safety, reliability, convenience and comfort are also important while choosing the mode. For example, if a mode is reliable, safe and comfortable as well as the fare is comparable, then the probability of choosing that particular mode will be increased. The operation of a travel mode and its demand also depends on the demographic profile of the population, the size of the city, etc.

5 Determining and Forecasting the Paratransit Utility

The forecasting in transportation was conventionally carried out based on the four-step model (FSM) that comprises trip generation, trip distribution, modal split and traffic assignment [22, 23]. The first complete implementation of FSM system was used in Chicago Area Transportation Study accompanied the model fitted by land use projection and economic estimation [24, 25]. The purpose of formulating a transport demand model is basically to estimate different types of transportation services and needs. Similarities in terms of procedures for estimating intercity freight and passenger transportation have shown. The only difference is that the former is confined to the economic-based study, while the latter depends on the population and land use pattern [26]. For travel forecasting or developing the model, it requires data on current traffic volume and travel pattern [27]. These data are necessary for analysing the existing problems and formulation of mathematical models for travel forecasting. The collection of data can be carried out through various survey techniques and methods, viz. the origin-destination (O-D) survey, traffic volume and passenger counts and stated preference survey, etc. Further, it suggests that determining the preferences of individuals in terms of likes and dislikes of the service characteristics is very much essential [27]. In order to determine the quality of available service or developing latest service and ridership forecast will require stated preference surveys. Estimating the demand for a new consumer product is very much prevalent in the marketing research, nevertheless, such approach can also apply to ridership forecasts in the field of transportation research. The basis of sampling and sample size determines the reliability of O-D survey and was studied [28]. The detailed procedures for O-D survey were postulated by Federal Highway Administration (FHWA) [29]. A survey-based method for predicting the number of passengers in the new dial-a-bus has been explained. Care must be taken when interpreting the results of the opinion survey to take into account any biases that may be reflected in the apparent results [30]. Further, presented a research analysis including non-commitment influence on common perspective of the utilization of modern transport system [31].

In the past studies, different types of analysis were used for estimating the travel demand which includes trend analysis, elasticity-based models and gravity-based models. Basically, the trend analysis is examined as a significant tool for demonstrative purpose preferably than a predictive tool [32]. Trend analysis, the projection was based on the assumptions that all the contributing factors of demand are constant over a period. The presumptions are not consistently feasible because of the difference in socio-economic attributes, degradation or improvement in the mode [33]. The aggregate values like vehicle miles travelled (VMT) or transit rides in an urban area, the trend analysis can be used appropriately provided that the abundant traffic volume data is available and the trend in terms of land development pattern and transportation network is static. Therefore, pointed out that trend analysis is not used dominantly for long-range forecasts of traffic volume [27]. Instead, gravity model is suitably used for estimating demand of freight transportation [26]. The way forward to forecast in a disaggregate form as the disaggregate trip generation is a challenge for complete forecasts of socio-economic characteristics has shown. Moreover, aggregate values of socio-economic characteristics used in trip generation models are not complicated to forecast with the current state of the art. Complication happens in disaggregating socio-economic parameters at sector level [34]. The household data are the important unit of analysis in passenger trip generation has been opinionated. Further, the forecasting process attempts to estimate different types of trips such as work, school, shopping, leisure trips, so on and so forth [26]. The needs for focusing towards a new approach of forecasting techniques which is beyond the conventional type of highway network forecasting have emphasized. So, in the recent trend, the study has been shifted from conventional highway networks to planning and management of integrated multi-modal transportation systems. This new approach is more flexible than the conventional model and tries to incorporate behavioural factors which influences the decision-making process of individuals while choosing the service. Many studies indicate that these methods are superior or comparable to conventional forecasting techniques in terms of data gathering, computational process and forecast accuracy [35].

6 Modelling in Transportation

A model is a simplified representation of a part of the real world of interest in a system that focuses on certain elements that are considered important from a particular point of view [36]. The basics of transport modelling were first used in USA during 1950s for studying Detroit and Chicago Transportation. Later, the techniques were also used in the UK in the 1960s for studying the transportation system in London city. After two decades, in the west, i.e. US and the UK, the transportation modelling shows remarkable theoretical developments [37]. As mentioned above, the transportation modelling is carried out on the line of conventional forecasting techniques or the four-step model (FSM). There are various modelling techniques which are used in transportation forecasting. However, the mode choice modelling in the form of logit model is preferably and widely used because of its simplicity and the ability to handle complex data [38]. Further, a logit model is closely related to the formula for the choice probabilities and is readily interpretable. Mode choice model has been used successfully and extensively not only in analysing economic consumer behaviour but also in transportation forecasting [38].

6.1 Logit Model

The theoretical basis for the logit function had been discussed by [39]. The attitudes affecting the work trip modal choice among three possible alternatives by using the logit model have studied [40]. The equilibrium properties of the logit model have been studied [41]. To forecast revised travel behaviour and changes in the level of service experienced by a trip maker, incremental logit model had been used [42]. The use of segmentation perspective for more projection and apprehension from user mode choice model was suggested [43]. The conclusion was made that this model could lead to better understanding of modal choice behaviour and to a reliable and stable choice model. A behavioural probabilistic logit model, appropriate for Indian commuters, has developed [44]. The logit mode choice model for non-work trips was used, and it was shown that calibration was feasible and presented no new problems over the work mode choice models [45]. The concept of minimally perceived attribute differences to logit choice model was introduced [46]. Using a binary logit model mode choice of travellers without vehicles was investigated [47]. Paired combinatorial logit model was presented [48]. Discrete choice analysis and logistic regression for the mode choice models were studied [49]. Comparative studies of the binary choice modal split models, viz. discriminant analysis, logit model and probit model were conducted [50, 51]. Further, concluded that all these three methods gave comparative results [50]. The binary choice logit model had been adopted to study the modal split between bus transit and other private modes in Bangalore city of Karnataka [52]. It is observed from the studies that logit and probit models are normally adopted in the binary choice situation. A number of studies have used these models in the multi-modal situation. A large number of variables, i.e. quantitative and qualitative with equal precision, can be handled by these two models. Probit analysis requires more time-consuming calibration procedure than the logit analysis, and it is more cumbersome. It is reported that computations in probit model will cost two to ten times as much as analogous to logit model calculations. Probit model estimation needs large specimens than logit models which leads to raise the cost of analysis [53].

6.2 Multinomial Logit Model

Luce introduced the multinomial logit model for the first time [54], and it was further developed [55]. It was suggested that the discrete choice model (logit model) developed by Luce can be derived in various ways [56]. The modelling generalized cost of travel for rural bus commuters using multinomial logit model (MNL) was studied [57]. Multinomial logit model has certain advantages [58]. Firstly, it can be interpreted as disaggregate behavioural model in relation to the probability distribution of the unobserved variables. Secondly, a number of possible disaggregate travel demand can be formulated in the MNL framework, in the form of the mean utility function depending on the application. Thirdly, MNL model has the mathematical application even in forecasting the conventional travel demand, for example, gravity travel demand model. The basic MNL model as developed by Daniel L. McFadden, 1978 [35] can be expressed as

$$\text{Utility of } i\text{th alternative} = \frac{e^{\text{mean utility of the } i\text{th alternative}}}{e^{\text{mean utility of first alternative}} + \cdots + e^{\text{mean utility of last alternative}}} \quad (1)$$

The efficacy of paratransit modes in the cities of Assam, India, by using multinomial logit model was studied and determined the advantages of the paratransit modes in Assam [59]. Multinomial logit model is operated to estimate behaviours of concert participants' in Taipei Arena using the independent variables including total travel cost, total travel time, gender, age, household income, total number of motorcycles owned by a household, trip origin, fan seniority, expected time of arrival time relative to the concert start time, single-stop trip or not and first-time visiting or not [60]. Travel mode choice decision in case of multiple alternative situations can best be modelled through multinomial logit or multinomial probit models. MNL has used to determine mode choice among three to four alternative modes [61–63]. Travel demand elasticity could be acquired for socio-economic parameters in MNL models [64]. Relative importance of perceived reliability of transportation alternatives in mode choice decisions had studied [65]. An investigation for capturing the behavioural differences in urban recreation demand has pursued [66]. An econometric model for calculating choice probabilities among nested alternatives known as nested multinomial logit models (NL) was presented [67]. It was observed from the study that choice forecasted by MNL and NL models does not vary in magnitude and sign [68]. The most acceptable form of discrete choice established on the structure of utility expansion had been provided by multinomial logit model (MNL). It gave a clear advantage to more commonly used alternative, viz. the multinomial logit model, besides it involved additional computational efforts. Three techniques for developing confidence intervals for the choice of probabilities in MNL models were described [69]. It was found that the models which are established from direct data predicted better than those are specified on prior beliefs [70]. Using logit approach, one step modal split process was developed, and it was oriented towards the requirement and attribute of small- and medium-sized cities [71]. The method for optimizing the sample used in estimating discrete choice models was tested [72]. MNL for modal split analyses for work trips in Hyderabad city in India had been used [73]. Formulation of MNL model was done to accommodate dissimilarity in responsiveness to level of service computed from observed and unobserved discrete attributes in detailed approach [74–76]. In order to regulate the part played by behavioural and choice variables, MNL model was used and explained commuter mode choice behaviour [77]. In addition, MNL model is both mathematically unambiguous and computationally controllable [78]. The mathematical probit model has been tested [79]. A comparatively special data set is required for multinomial probit where same observations were analysed at two different point of time and discussed the problems both for estimating the choice function and calculating the model attributes [72, 80, 81].

7 Importance of Logit Model

In 1958, David Cox developed logistic regression also known as logit model. This type of regression model used dependent variables which are categorical in nature. In term of economics, logistic regression would be a qualitative response or discrete choice model. The binary logistic model is used to estimate the probability of a binary response based on one or more explanatory (independent) variables. Logistic regression measures the relationship between a category-dependent variable and one or more independent variables by estimating probabilities using logistic functions. A multinomial logistic regression model is a form of regression in which the outcome variable is binary or dichotomized, and the explanatory variables are continuous variables, categorical variables or both. There is multinomial logistic regression to deal with the case of dependents with many classes. Simple linear regression is often used for regression analysis of the relationship between individual explanatory variables and individual dependent variables (responses). However, logistic multinomial regression is used conveniently for analysing model based on multiple independent variables [82]. Gradually, the logit gains momentum equally with the probit model between 1960 and 1970s. By then, the potential of the logit and its statistical application was recognized in a much wider field. The application of logistic regression is not only confined to statistics but also it has wider application in various disciplines [83]. The multinomial logit model is a much profound theory which was provided by McFadden. He used the MNL first in transportation modelling. Since the use of multinomial probit is still significantly more expensive and difficult to implement, the logit model remains the most widely used selection model. The multinomial logit (MNL) model is most often used to specify a discrete choice model [55]. In addition, the MNL model provides a convenient closed form for the underlying selection probabilities, without multivariate integration [84]. Therefore, it has been suggested that it is convenient to perform a computer analysis of the model and select many options [84]. Moreover, the mathematical simplicity and easy implementation are the key advantages of MNL [85]. Besides, the likelihood function for the specification of MNL model is generally concave and that eases the computational workload [84]. The availability of exhaustive computer software, easy and less computation burden has been made the application of logit model on a wide scale. In practical, the logit had a clear advantage over the probit in terms of computation, even with maximum likelihood [83]. Multinomial logistic regression (MLR) has various important advantages such as MLR is similar to linear regression, but its statistical diagnostic can be easily interpreted, and it can handle violations of assumptions about multidimensional normality and homoscedastic covariance matrices within a group. In addition to the benefits mentioned above, MLR continues to be popular with the following assumptions: In most cases, MLR does not assume a linear relationship between response and predictor variables. The explanatory variables do not have to be intervals. They can be used in the MLR without an unlimited number of independent variables, regardless of the general variance error term [86]. The logit model is particularly advantageous over other types of forecasting methods because it provides an important economic explanation of the causal relationship that leads to an increase in the number of car owners over time [85, 87]. However, the logit model has been criticized for "independence from irrelevant alternative properties." This property means that the probability ratio of choosing two choices does not depend on the attributes of the other choices in the selection set. Therefore, it is important to apply an "independence"-based statistical test that applies the logit specification test to multiple datasets.

8 Conclusion

Paratransit plays a vital part in cities and suburban transportation zones of developing countries. Several areas, quite fifty percent of overall conveyance demands are supported by paratransit. The fast growth in urban inhabitants, insufficient prevailing transport structures ends up in additional utilization of the paratransit vehicles. In developed countries, paratransit is usually operated as demand responsive services like shared-ride taxis, dial-a-ride and subscription buses. Lower common place of livelihood, increased population density, accessibility of low-cost workforce, etc., had anticipated an unclear arrangement of transport modes that fulfils between public and personal automobile. All types of paratransit modes are accessed in developing countries such as human-driven non-motorized mode to automated vehicles where the automated ones are commonly used. The promising characteristics of paratransit services within the metropolis are the accessibility and door-to-door assistance. The characteristics as conveyance is major breakthrough which carries two-third of conveyance passengers. The vehicles are varied as seen physically in the passenger's volume capacity, operating ranges, service patterns and regulatory frameworks. Paratransit modes are employed and controlled by freelance organizations, where majority of the vehicles are leased. This generates good employment chance and conjointly does not need vital public resources that in turn favours a significant attraction in several cities which are in the bottleneck of funds. Although paratransit system provides a valuable means of transport mode in cities, but unfortunately, there is no such valuable data and needful information in this domain. Therefore, future research and survey are highly necessary for implementations of such services. Further, governments should provide varieties of public transit with a special emphasis on paratransit system. Apart from limited research on operation and management of paratransit, there is a huge gap to be filled in other areas in the field of transportation.

From the study, it reveals that logit analysed the discrete mode choice models. Logistic regression analysis is simple, easier to implement, interpret and efficient. It not only provides the details of predictor but also interprets the direction whether positive or negative associated with it. Moreover, the results are more accurate to study. Logistic regression analysis investigated the relationship between the binary or ordinal probabilities of the response and the independent variables. In the context of maximum likelihood estimation, the probability of choice is an expression for predicting the probability that an individual will choose a particular option, and the probability function of such a model is the product of the probabilities of choice. The selection probabilities of the MNL model are relatively simple and can maximize the resulting probability function for many decisions at the same time almost immediately [88]. Besides the discrete choice model, MNL model gives more accurate results than other models with higher prediction power [89]. In the random utility model, multinomial logit model is necessary to predict the choices as it has more successful applications [90, 91].

Further, it is observed that MNL model and multinomial probit model (MNP) are generally used in the cases of multiple alternatives. However, MNL is widely used over MNP because of its simplicity over MNP which is more complex for its intractability and cost inefficiency or expensiveness where it provides little use travel demand analysis. The study is highly needed for future research and will help to the transportation planners, civil engineers and policymakers for implementing to forecast the paratransit utility.

References

1. Wongwiriya P, Nakamura F, Tanaka S, Ariyoshi R, Miura S (2020) The role of paratransit to support sustainable transportation: case study of Khon Kaen city, Thailand. Transportation Research Procedia, 48(2020): 2656–26700
2. Vuchic VR (2007) Urban transit system and technology. John Wiley & Sons, New Jersey
3. Maretić B, Abramović B (2020) Integrated passenger transport system in rural areas—a literature review. Promet—Traffic Transp 32(6):863–873
4. Boutueil V, Lesteven G, Nemett L (2020) Toward the integration of paratransit in transportation planning in African Cities. Transp Res Record 2674(9):995–1004. https://doi.org/10.1177/0361198120933270
5. Janic M (2001) Integrated transport systems in the European union: an overview of some recent developments. Transp Rev 21(4):469–497

6. Poliaková B (2013) Key success factors of integrated transport systems. In: Presentation at 13th international conference on reliability and statistics in transportation and communication 16–19 (RelStat'13), pp 83–90
7. Wongwiriya P, Nakamura F, Tanaka S, Miura S, Ariyoshi R (2016) User perception of paratransit in developing countries: a case study of Songtaew in Khon Kaen City, Thailand. In: Presentation at international conference of Asian-Pacific planning societies 25–27 August 2016
8. Cervero R (1998) The transit metropolis a global inquiry. Island Press, Washington, D.C
9. Vuchic VR (1999) Transportation for liveable cities, New Brunswick, NJ: center for Urban policy research. The State University of New Jersey, Rutgers
10. Joewono TB, Kubota H (2008) Paratransit service in Indonesia: user satisfaction and future choice. Transp Plan Technol 31(3):325–345
11. McCormick D, Mitullah W, Chitere P, Orero R, Ommeh M (2013) Paratransit business strategies: a bird's-eye view of Matatus in Nairobi. J Public Transp 16(2):135–152
12. Kaltheier RM (2002) Urban transport and poverty in developing countries. Analysis and options for transport policy and planning. (Division 44 Environmental Management, Water, Energy, Transport, Deutsche Gesellschaftfu¨r Technische Zusammenarbeit (GTZ), GmbH: Eschborn
13. Singh SK (2005) Review of Urban transportation in India. J Public Transp 8(1):79–97
14. Tarigan AKM, Susilo YO, Joewono TB (2010) Negative experiences and willingness to use paratransit in Bandung, Indonesia: an exploration with ordered probit model. In: Presentation at the 89th annual meeting of transportation research board
15. DLLAJ (Traffic and Road Transport Agency) (2001) The guide of passenger public transportation price calculation with fixed routes in Urban areas. West Java, Bandung, Indonesia
16. Cervero R, Golub A (2007) Informal transport: a global perspective. Transp Policy 14:445–457
17. Shimazaki T, Rahman M (1996) Physical characteristics of parantransit in developing countries of Asia. Journal of Advanced Transportation, Wiley 30(2):5–24
18. Soegijoko BTS (1986) The Becaks of Java. Habitat Int 10(1/2):155–164
19. Kalra K, Kumar GAN, Hoyez M (2014) Improving and upgrading IPT vehicles and services: a study. Institute of Urban Transport, Delhi
20. Islam A (1990) The role of rickshaws in the future transportation system in the Dhaka metropolitan region, Bangladesh. Asian Institute of Technology, Bangkok Thailand
21. Replogle M (1992) Bicycles and cycle-rickshaws in Asian cities: issues and strategies. Transp Res Rec 1372:76–84
22. Weiner E (2008) Urban transportation planning in the United States. In: History, policy, and practices, 3rd edn. Springer, New York
23. McNally MG (2008) The four step model in handbook of transportation modelling. In: Hensher D, Button, K (eds) 2nd end. Emerald, UK, pp 35–53
24. Weiner E (1997) Urban transportation planning in the United States: an historical overview, 5th edn. Report DOT-T-97–24, US Department of Transportation, Washington, DC
25. Hensher DA (2006) Towards a practical method to establish comparable values of travel time savings from stated choice experiments with differing design dimensions. Transp Res Part A 40:829–840
26. Meyer JR, Straszheim MR (1971) Techniques of transport planning volume one: pricing and project evaluation. The Brookings Institution, Washington, DC
27. Myer K (2004) Handbook of transportation engineering. McGraw Hill
28. Makowski GG, Chatterjee A, Sinha KC (1974) Reliability analysis of origin-destination surveys and determination of optimal sample size. Present Fift Ann Meet Transp Res Forum, IV 1:166–176
29. FHWA (1975) Urban origin-destination surveys. U.S. Department of Transportation, Federal Highway Administration.
30. Hartgen DT, Keck CA (1976) Forecasting dial-a-bus ridership in small Urban Areas in public transportation planning. Transp Res Rec 563:57
31. Chatterjee A, Wegmann FJ, McAdams MA (1983) Non-commitment bias in public opinion on transit usage. Transportation 11(4):347–360

32. Meyer MD, Miller EJ (2001) Urban transportation planning: a decision oriented approach. McGraw Hill, New York
33. Sinha KC, Labi S (2007) Transport decision making: principles of project evaluation and programming. Wiley, New Jersey
34. Chatterjee A, Khasnabis S, Slade LJ (1977) Household stratification models for travel estimation. Transp Eng J Am Soc Civ Eng NY 103
35. McFadden D (1978) The theory and practice of disaggregate demand forecasting for various modes of Urban transportation in emerging transportation planning methods. In: Brown WB, Dial RB, Gendell DS, Weiner E (eds)
36. Ortuzar JD, Willumsen LG (2011) Modelling transport, 4th edn. Wiley & Sons, New Delhi
37. Bates J (2000) History of demand modelling. In: Hensher DA, Button KJ (eds) Handbook of transport modelling. Pergamon, Oxford, pp 11–34
38. Train K (2003) Discrete choice methods with simulation. Cambridge University Press, New York
39. Sen A, Soot S, Pagitsas E (1978) The logit modal split model: some theoretical considerations. Transp Res 12:321–324
40. Lyles TE (1979) Modal split model considering carpool mode. Transp Eng J ASCE 105(2):149–164
41. Anas A (1979) Equilibrium properties of logit models. Transp Res Rec 728:8–14
42. Kumar A (1980) Use of incremental form of logit models in demand analysis. Transp Res Rec 775:21–27
43. Currim IS (1981) Using segmentation approaches for better prediction and understanding from consumer mode choice models. J Market Res xvii:301–309
44. Chari SR (1982) Modal split models for work trip travel. Indian Highw 10(1)
45. Stopher PR, Ohstrom EG, Kaltenbach KD, Clouse DL (1984) Logit mode choice models for non-work trips. Transp Res Rec 987:75–81
46. Young W, Bertram D (1985) Attribute threshold and logit mode choice models. Transp Res Rec 1037:81–85
47. Thamizh AV, Rengaraju VR, Rao KVK (1996) Trip characteristics of travellers without vehicles. J Transp Eng 122:76–81
48. Koppelman FS, Wen CH (2000) The paired combinatorial logit model: properties, estimation and application. Transp Res Part B Methodol 34B(2):75–89
49. Abdulsalam Bin Miskeen M, Alhodairi AM, Rahmat RK (2013) Modeling a multinomial logit model of intercity travel mode choice behavior for all trips in Libya. Int J Civ Archit Struct Constr Eng WASET 7(9):300–309
50. Talvitie A (1972) Comparison of probabilistic modal choice models: estimation methods and system inputs. Highw Res Rec 392
51. Stopher PR, Lavender JO (1972) Disaggregate behavioral travel demand models: empirical tests of three hypotheses. In: Transportation research forum, Oxford, Indian, U.S.A., pp 331–336
52. Veeraraghavan A, Guruvittal UK (1994) Stochastic models of Urban travel mode choice—a study in binary choice. Indian Highw 22(7):5–18
53. Albright RL, Lerman SR, Manski CF (1978) Report on the development of an estimation program for the multinomial probit model. In: Presentation at the 57th annual meeting of the transportation research board
54. Luce R (1977) Quantitative methods for analysing travel behaviour of individual: some recent developments. In: Presentation of the 3rd international conference on behavioural travel modelling, Adelaide, Australia
55. McFadden D (1973) Conditional logit analysis of qualitative choice behaviour. In: Zaremmbak P (ed) Frontiers in econometrics. Academic Press, New York
56. Ben-Akiva M, McFadden D, Abe M, Böckenholt U, Bolduc D, Gopinath D, Morikawa T, Ramaswamy V, Rao V, Revelt D, Steinberg D (1997) Modelling methods for discrete choice analysis. Market Lett Springer 8(3):273–286
57. Kumar CP, Basu D, Maitra B (2004) Modeling generalized cost of travel for rural bus users: a case study. J Public Transp 7:59–72

58. Oum T, Dodgson, Hensher DA, Morrison SA, Nash CA, Small KA, Waters II WG (2005) Transport economics. Hardwood Academic Publishers
59. Ahmed MA, Datta RN (2006) Utility of paratransit modes in the cities of Assam, India. Transp Res Rec J Transp Res Board Nat Acad 107–115
60. Chang MS, Lu PR (2013) A multinomial logit model of mode and arrival time choices for planned special events. Present Eastern Asia Soc Transp Stud 9
61. Ganek J (1976) A disaggregate model split model for work trips involving three mode choices. Transp Res Rec 610:579–612
62. Recker WW, Stevens RF (1976) Attitudinal models of mode choice: the multinomial case for selected non work trips. Transportation 5:355–375
63. Recker WW, Stevens RF (1977) An attitudinal travel demand model for non-work trips of homogeneously constrained segment of population. Transp Res 11:167–176
64. Hautzinger H (1978) Travel demand elasticity with respect to socio-economics variables: the multinomial logit case. Transp Res 12:355–356
65. Prashker JN (1979) Mode choice models with perceived reliability measures. Transp Eng J ASCE, TE 3:251–262
66. Stopher PR, Ergiin G (1979) Population segmentation in Urban recreation choice. Transp Res Re 728:59–65
67. Sheffi Y (1979) Estimating choice probabilities among nested alternatives. Transp Res Part B Methodol 13B:189–205
68. Ortuzar JDD (1983) Nested logit models for mixed mode travel in Urban Corridors. Transp Res Part A Policy Pract 17A(4):283–299
69. Horowitz J (1979) Confidence intervals for choice probabilities of the multinomial logit model. Transp Res Rec 728:23–29
70. Train KE (1979) A comparison of the predictive ability of mode choice models with various levels of complexity. Transp Res Part A Policy Pract 13A:11–16
71. Cynecki MJ, Khasnabis S, Flak MA (1982) Multi modal logit travel demand model for small and medium sized Urban Areas. Transp Res Rec 848:28–36
72. Sheffi Y, Hall R, Daganzo CF (1982) On the estimation of multinomial probit model. Transp Res Part A Policy Pract 16A:447–456
73. Chari SR, Chandrasekhar BP, Prashad CSRK (1988) Mode split analysis for Hyderabad city. In: Presentation at the national conference on transportation system studies, IIT, Bombay
74. Bhat CR (1997) Work travel mode choice and number of non-work commute stops. Transp Res Part B Methodol 31B:41–54
75. Bhat CR (1998) Accommodating variations in responsiveness to level of service measures in travel mode choice modelling. Policy Pract 32A(7):495–507
76. Bhat CR (1998b) Accommodating flexible substitution patterns in multi-dimensional choice modelling: formulation and application to travel mode choice departure time choice. Transp Res Part B Methodol 32 B (7):455–466
77. Arun KP, Ram MP, Rahman S (1999) Analysis of traveler attitudes and perceptions in explaining mode choice behavior. Transp Res Rec 1676:68–76
78. Horowitz J (1980) A utility maximizing model of demand for multi destination non work travel. Transp Res Part B Methodol 14B:369–386
79. Horowitz J (1981) Testing the multinomial logit model against the multinomial probit model without estimating the probit parameters. Transp Sci 15(2):153–163
80. Johnson L, Hensher D (1982) Application of multinomial probit to a two panel dataset. Transp Res Part A Policy Pract 16 A:457–464
81. Sparmann JM, Daganzo CF, Soheily M (1983) Linear probit models: statistical properties and improved estimation methods. Transp Res Part B Methodol 17B(1):67–86
82. Bayaga A (2010) Multinomial logistic regression: usage and application in risk analysis. J Appl Quant Meth 5(2):288–297
83. Cramer JS (2002) The origin of logistic regression. Tinbergen Institute, Netherlands, pp 1–15
84. Hausman J, McFadden D (1984) Specification tests for the multinomial logit model. Econometrica 52(5):1219–1240

85. Visser EJ (1977) Transport decisions in an age of uncertainty. In: Presentations at the third world conference on transport research, Netherlands
86. Tabachnick BG, Fidell LS (2013) Using multivariate statistics, 6th edn. Pearson, US
87. Bates JJ (1971) A hard look at car ownership modelling. Mathematical Advisory Unit Note 216 Department of the Environment
88. Kropko J (2008) Choosing between multinomial logit and multinomial probit models for analysis of unordered choice data. University of North Carolina at Chapel Hill, Master of Arts diss.
89. Zargari SA, Safari F (2020) Using clustering methods in multinomial logit model for departure time choice. Hindawi; J Adv Transp 7382569:12. https://doi.org/10.1155/2020/7382569
90. Ljubic I, Moreno E (2018) Outer approximation and submodular cuts for maximum capture facility location problems with random utilities. Europ J Oper Res 266(1):46–56
91. Haase K, Müller S (2014) A comparison of linear reformulations for multinomial logit choice probabilities in facility location models. Eur J Oper Res 232(3):689–691

A Damage Detection Technique Using Bridge Influence Surfaces for Structural Health Monitoring of Bridges

K. Lakshmi and Prathamesh Apte

1 Introduction

Any structure that undergoes loading and unloading frequently needs maintenance. The most important task becomes the detection of change of state of the structure under continuous loading over age and usage. Damage, which is defined as the undesirable change of state of the structure needs to be identified and located precisely in the structure so that the repairs and corrective measures can be implemented in the right time. This will eventually avoid the cost of repair, downtime and in the worst case the catastrophic loss to the property and the people using it. This creates a high impact on the economy and the society. Structural health monitoring (SHM) is seen as the solution to this issue and is being increasingly gaining popularity throughout the world [1, 2]. The development of data acquisition hardware and sensor technologies have given the wholesome opportunity to realize feasible SHM systems. The sensors are generally mounted on the structure of interest to measure various parameters such as strain, displacement, acceleration, rotation, etc. With the technological advances, non-contact measurements are possible to record. Such recorded signatures/signals are the responses from the structure. These responses are further analyzed to extract the meaningful information regarding the state of the structure. This is a very challenging task due to the presence of many factors which affect the features and may result in false-positive detection or false-negative detection of damage. Many features have been shown to perform well for damage detection. The popularly extracted features are the frequency, mode shapes, modal curvature,

K. Lakshmi (✉)
CSIR-SERC, CSIR Complex, Taramani, Chennai 600113, India
e-mail: lakshmik@serc.res.in

P. Apte
Indian Institute of Technology Jodhpur, Jodhpur 342037, India
e-mail: prathamesh.1@iitj.ac.in

A. Sil et al. (eds.), *Recent Trends in Civil Engineering*, Lecture Notes in Civil Engineering 274, https://doi.org/10.1007/978-981-19-4055-2_67

modal strain energy, FRF, etc. [3–5]. Apart from the modal-based features, non-modal features like principal components [6], Intrinsic mode functions [7], sources separated by blind identification [8], etc., are also been used successfully. In this paper, the bridge influence surface (BIS) for the responses of the bridge under traffic loads are used as the feature for damage detection. The influence surface is developed from the responses which arises due to the set of loads on the structure. In this work, the responses considered are the acceleration time-history measurement. The root mean square value of the acceleration time history is used as the output variable and the traffic loads are considered as the input variable to evaluate the Influence surface of the bridge.

A numerical model of a single span of a T-beam slab bridge is considered for this study. The traffic loads are simulated as point loads randomly placed on the slab bridge at the time instant of generation of the responses from the model. The responses are obtained at certain locations on the girders of the bridge, namely, the sensor locations, using combinations of random configuration of the vehicles on the carriage way. Interpolation functions are used to calculate the influence surfaces of the whole span of the bridge. A damage index is evaluated using the volumes enclosed by the influence surfaces. The results of the numerical study indicate that the proposed technique is capable of precisely identifying the location of damage in the bridge.

2 Methodology

The formulation details and the proposed methodology to detect and locate the damage on the bridge are presented in this section.

2.1 *Bridge Influence Surface (BIS)*

Bridge influence surface is defined as the response function curve of a particular point along a section of the bridge, under a passing unit concentrated load. It can be seen as the extension of 1-D influence line in two directions.

BIS is mostly used in the calibration of weigh-in-motion systems. BIS have also been applied in the performance evaluation and damage detection of bridges. It is generally observed that the methods based on the influence coefficients are insensitive to the effects of environmental variability unlike the methods based on dynamic parameters.

This enables the detection of local damage effectively. Also, influence surfaces can perform better as the surfaces contain the stiffness information of every point of the bridge, in contrast to influence lines which represent the information on stiffness of a particular section of the bridge [9]. Therefore, influence lines are useful for single dimensional damage detection of bridges [10]. BIS provides the way for assessing

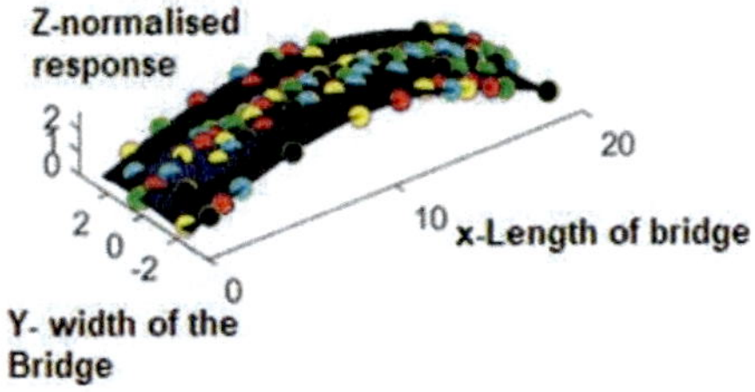

Fig. 1 Typical bridge influence surface

the performances of the connections in the lateral direction of the bridge. With the accurately identified BIS, the performance evaluation and the damage detection of bridges can be improved tremendously.

During the construction of BIS, a vehicle load is generally simulated as a point load, located in its equivalent position. Initially, the equivalent location is set at the center of vehicle load with which the initial influence coefficients of surface is generated as follows:

$$U = \frac{r}{P_{\text{equiv}}};\quad r \text{ is the response and } P_{\text{equiv}} \text{ is the equivalent load} \tag{1}$$

From the influence coefficients U, BIS is constructed. The location of the equivalent load is iteratively updated, and new influence surface is generated as discussed earlier. The above said steps are repeated till the convergence is reached. Convergence is often achieved when the percentage difference between two consecutive fitting implementations is lower than a preestablished threshold. Finally, the bridge influence surface is extracted. An example bridge surface is shown in Fig. 1.

When a group of loads are present in the bridge in the form of the traffic vehicles, an equivalent location due to the configuration of the vehicles on the bridge is calculated. When the wheel loads are given as $[P_1, P_2, P_3, \ldots, P_n]$, at coordinates $[(x_1, y_1), (x_2, y_2), + \cdots + (x_n, y_n)]$, respectively, the equivalent location is calculated as

$$X = \frac{P_1x_1 + P_2x_2 + \cdots + P_nx_n}{P_1 + P_2 + \cdots + P_n} \tag{2a}$$

$$Y = \frac{P_1y_1 + P_2y_2 + \cdots + P_ny_n}{P_1 + P_2 + \cdots + P_n} \tag{2b}$$

The responses under the equivalent location, is calculated by generating an influence surface. The vertical coordinate of the influence surface under the equivalent location is the normalized response due to the equivalent loads. The equivalent load is calculated as:

$$P_{\text{eqv}} = P_1w_1 + P_2w_2 + \cdots + P_nw_n; \tag{3a}$$

$$w_i = \frac{I(x_i, y_i)}{I(X, Y)} \tag{3b}$$

where $w_1, w_2, \ldots, w_n$ are the appropriate weights evaluated using an iterative algorithm with initial values as 1 and I is the influence surface function. The influence surface for this equivalent load is obtained using a surface-fitting algorithm. The weights are iteratively varied till the convergence is reached in terms of influence surface volumes between the recent two surfaces.

2.2 *Parameter for BIS*

Generally the influence surface is generated for the displacement response of a bridge [11, 12]. However, the measurement of the displacement responses is difficult in the real field unless costly systems like non-target vision-based methods or laser vibrometers. In view of this, the easily measurable acceleration data is employed in this research work to calculate the influence surface. Since, the influence line or surface generally can be obtained for static parameter, the root mean square (RMS) value of the acceleration time-history responses at a particular instant is used in this study. Accordingly, for 'm' sampled acceleration dataset, the RMS value is calculated as:

$$\ddot{x}_{\mathrm{RMS}} = \sqrt{\frac{1}{m}\left(\ddot{x}_1^2 + \ddot{x}_2^2 + \ldots + \ddot{x}_m^2\right)};\ \ddot{x} \text{ is the acceleration data} \tag{4}$$

The influence surfaces for RMS of acceleration data are used further to evaluate the state of the structure's health.

2.3 *Vehicle Load Evaluation Using Vision-Based Methods*

It can be seen from Eqs. (2) and (3) that the proposed method requires the weights of the vehicles, found on the bridge deck, in a particular configuration at any instant of time. This particular task is difficult in the real field as both the weights and the positions of the vehicles are required to be known to calculate the equivalent location. Therefore, the evaluation of the load and its coordinate location are proposed to be obtained using the images of the decks showing the vehicle snapshots at a particular time instant. From the images, the type of vehicle can be identified by visual observation or through any other algorithm. Once the type of vehicle is identified, its weight can be related easily. Also, the location of the vehicles on the bridge deck can be obtained by any coordinate transformation techniques. These techniques transfer the information from the local coordinate of the images to global coordinates of the bridge. However, these techniques will not be focused or discussed in this paper for the paucity of the length of the paper.

2.4 *The Step-By Step Procedure for Damage Diagnosis is Given Below*

The bridge is loaded by the group of vehicles and the acceleration responses are recorded through the accelerometers placed on the specified locations on the bridge. The positions of the combination of vehicles on the bridge at an instant are known as the Vehicle formation (VF). Such VFs are randomized using different types of vehicles, locations of vehicles, and the direction of plying of the vehicles.

2.4.1 Healthy Scenario—Construction of BIS for Every Sensor Node

In this step, initially, the healthy baseline data is created by collecting the various states of vehicle load configurations and their corresponding acceleration responses, obtained at the predefined locations on the bridge deck. Both the loading states and their responses must be time-synchronized for this proposed technique, as a BIS relates the loads and the response.

For a particular sensor, the BIS is created as follows:

To calculate the equivalent load, assume number of vehicle formations as N.

Initially, assume the weight factors, w, given in Eq. (3) as 1.

For the first vehicle formation, VF_1, P_{eqv} is evaluated as:

$$P^1_{\text{eqv}} = \sum_{i=1}^{v} P^1_i w^1_i;\ \text{v is the number of vehicles in the formation.} \tag{5}$$

The normalized response under the chosen sensor,

$$R_1 = \frac{A_1}{P^1_{\text{eqv}}}; \tag{6}$$

A_1 is RMS of the measured acceleration response at sensor node of interest and corresponding to VF_1.

Similarly, the normalized responses of all the N vehicle formations are evaluated.

The influence surface plot is generated using the N number of vehicle formation values.

From the BIS plot, the original weights w_i are updated using Eq. (3b).

Using the updated values of weights and Eqs. (5) and (6), new BIS is evaluated.

This procedure is continued till there is no change in the BIS plot of last two consecutive iterations. Finally, the converged BIS becomes the final BIS of that sensor.

Similarly, for each sensor node, BIS are created with sufficient number of VFs.

2.4.2 Test Scenario

In this scenario, the acceleration dataset obtained from a new unknown state of the structure is diagnosed. To perform the diagnosis, the synchronized loading and the responses are collected and the steps to calculate the BIS of test scenarios for every sensor, discussed in the previous healthy scenario are carried out. A damage index is proposed to identify the damage location as:

$$\mathrm{DI} = 1 - \left(\frac{\sum_{i=1}^{s} V_{\mathrm{H}}^{i} - \sum_{i=1}^{s} V_{\mathrm{c}}^{i}}{\sum_{i=1}^{s} V_{\mathrm{H}}^{i}} \right) \tag{7}$$

V_{H}^{i} is the volume enclosed by the ith BIS generated from the healthy state, V_{c}^{i} is the volume enclosed by the ith BIS generated from the current unknown state and s is the number of datasets or number of BISs.

The above damage index is evaluated for every sensor node. A keen observation of the damage index can show that value of DI lies more than 1, in the presence of damage and lies less than 1, in the healthy condition. Accordingly, the sensor node which indicates the value greater than 1, is identified as the one near the vicinity of damage.

3 Numerical Study

In order to demonstrate the proposed acceleration data-based BIS technique, a numerical example of one span T-beam slab bridge is considered. The single span T-beam slab bridge with longitudinal T-beam girders and transverse girders is shown in Fig. 2. The span of the bridge is 20 m and the width is 9 m with 7.5 m carriage way. The material of the bridge is RCC. The end conditions are simply supported. It is assumed that there are five sensors on the bottom of each girder and are numbered as L1–L15 as shown in Fig. 3.

The vehicle loads are selected in the range from 9000 to 110,000 N, such that the variety of weights from a hatchback to a heavy-truck were included in the analysis. The vehicle loads were applied as step loads on the carriage way of the bridge for various configurations. The locations in form of coordinates of the vehicles in any particular state, together constituted the vehicle formation in that state. The various combinations for vehicle formations for the states were decided taking the width of vehicles from hatchback to heavy truck into consideration. Accordingly, maximum

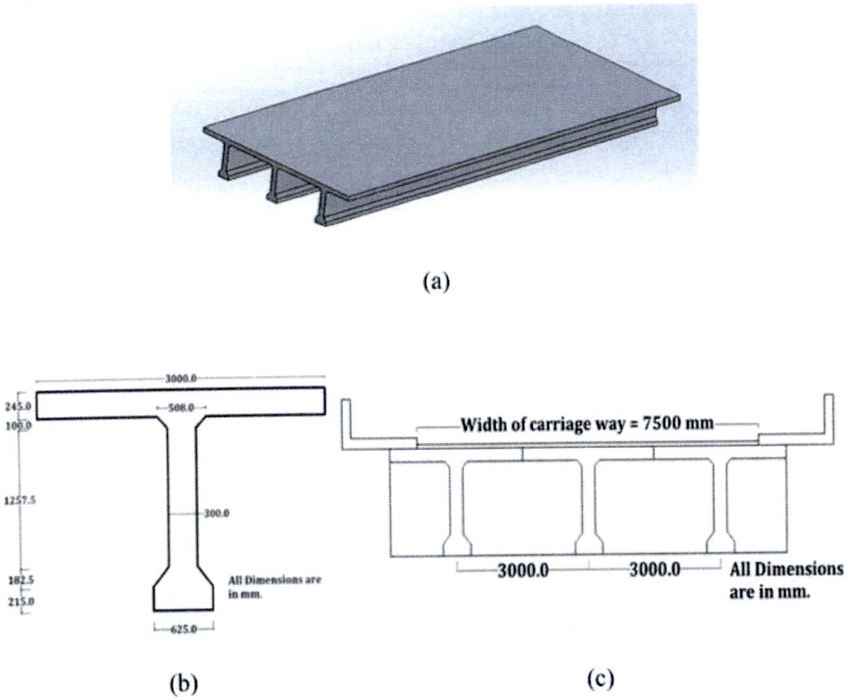

Fig. 2 T-beam slab bridge: **a** panaromic view **b** cross section of one T-grider **c** cross sectional view of the bridge

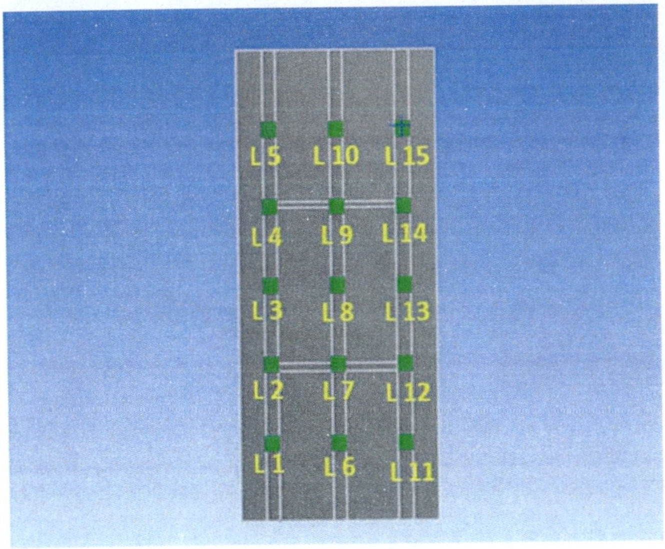

Fig. 3 Bottom view of the bridge with the sensor locations labeled as L1–L15

of six vehicles and minimum of two vehicles were considered randomly in VFs during analysis. The responses are generated for the various VF conditions.

As described in the previous section, various number of VFs were considered along the length of the bridge, from which the equivalent locations (*X*, *Y*) and the normalized responses, *R* are evaluated. The normalized responses, becomes the *Z* axis values and the equivalent locations become the *X* and *Y* axes values while plotting the BIS. In this fashion, 's' number of BISs are generated in the healthy condition.

Two unhealthy conditions of the bridge are created by inducing the damage in the form of a crack in the bridge at a location near L2 and L8 of the two girders as shown in Fig. 4. The acceleration responses were obtained in the damaged condition of the bridge. Using the healthy and damage responses, the RMS values are calculated. The BIS of healthy and damage scenarios are generated using the spline interpolation and these surfaces were further used to obtain a damage index for each sensor location. One typical BIS generated is shown in Fig. 5.

The damage indices for the sensor locations present on the girder are shown in Fig. 6 and 7 for damage scenario-1 and damage scenario-2.

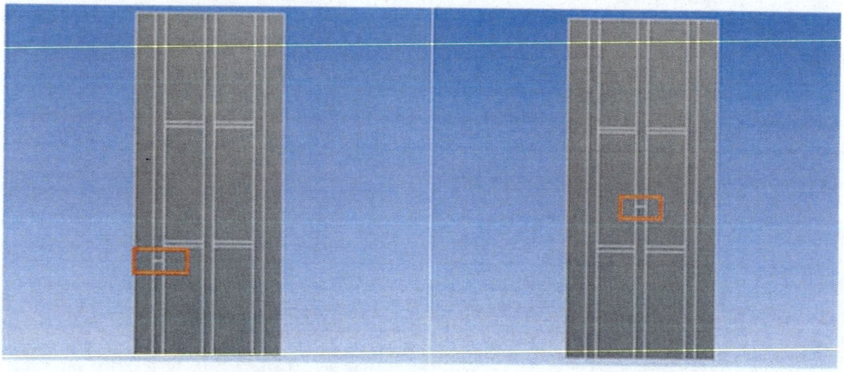

Fig. 4 Locations of the cracks induced in damage scenarios 1 and 2, respectively (orange box highlights the crack)

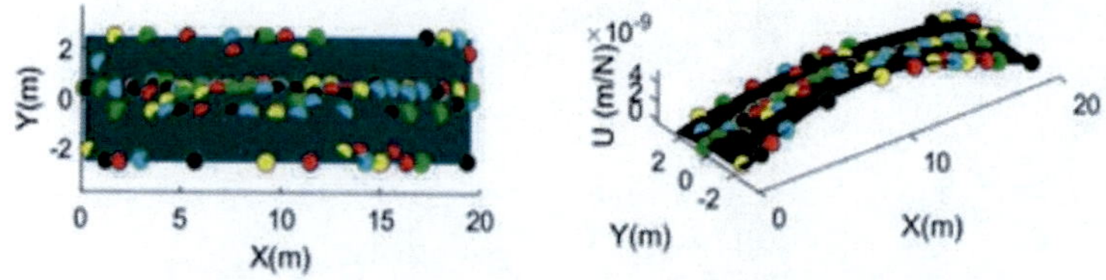

(a) Top view (b) Isometric view

Fig. 5 Bridge influence surface for normalized response at L1

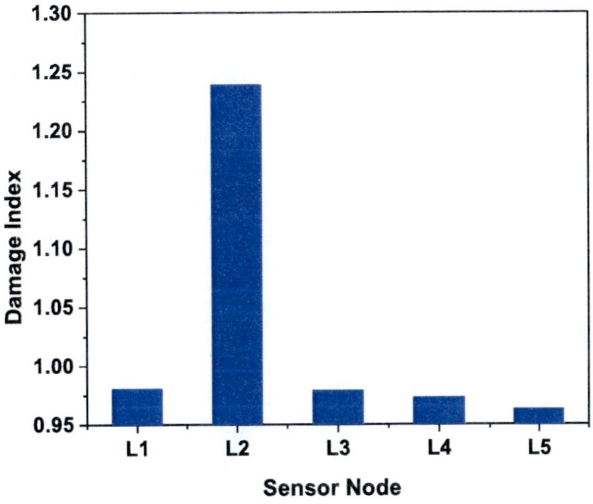

Fig. 6 Damage index of the sensor nodes L1–L5 for damage scenario-1

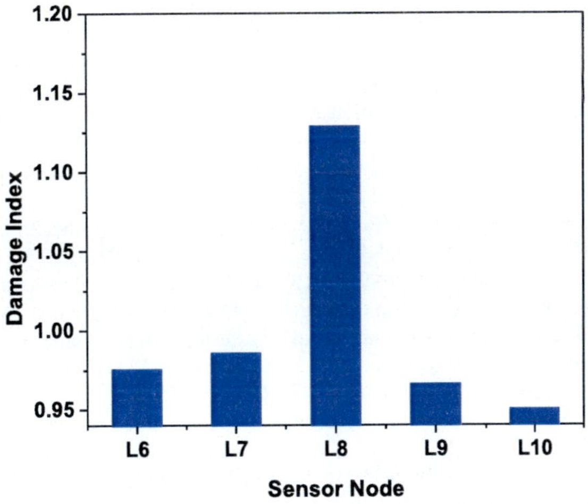

Fig. 7 Damage index of sensor node L6–L10 for damage scenario-2

From Fig. 5 and 6, it can be seen that the proposed technique identifies the location of damage precisely using the acceleration-based BIS.

4 Conclusion

In this paper, bridge influence surface for root mean squared value of the acceleration response is proposed for damage detection of bridges. The difficulty of collecting displacement data in the real scenario is eliminated in this technique by using the acceleration measurement for influence surfaces. The vehicle loading is considered as point loads and the situations of group of random vehicles plying in both the directions of the bridge carriageway are simulated. A damage index, using the change in enclosed volumes of BIS, is employed for detecting the location of damage on the bridge. A numerical study is performed using a single span, T-beam slab bridge, with two carriageways and three T-beam girders. The results of the numerical study clearly show that the proposed technique can identify the damage location successfully.

Acknowledgements This paper is being published with the permission of Director, CSIR-SERC.

References

1. PC Chang A Flatau SC Liu 2003 Health monitoring of civil infrastructure Struct Health Monit 2 3 257 267
2. MQ Feng 2009 Application of structural health monitoring in civil infrastructure Smart Struct Syst 5 4 469 482
3. S Das P Saha SK Patro 2016 Vibration-based damage detection techniques used for health monitoring of structures: a review Civ Struct Health Monit 6 477 507
4. Avci O, Abdeljaber O, Kiranyaz S, Hussein M, Gabbouj M, Inman DJ (2020) A review of vibration-based damage detection in civil structures: from traditional methods to machine learning and deep learning applications. Mech Syst Sign Proc. https://doi.org/10.1016/j.ymssp,2020,107077
5. W Fan P Qiao 2011 Vibration-based damage identification methods: a review and comparative study Struct Health Monit 10 1 83 111
6. Krishansamy L, Arumulla RMR (2018) A hybrid structural health monitoring technique for detection of subtle structural damage. Smart Struct Syst 22(5):587–609
7. Krishnasamy L, Arumulla RMR (2019) Baseline-free hybrid diagnostic technique for detection of minor incipient damage in the structure. J Perform Constr Facil 33(2):04019018
8. ARM Rao K Lakshmi 2015 Damage diagnostic technique combining POD with time-frequency analysis and dynamic quantum PSO Meccanica 50 6 1551 1578
9. MM Alamdari K Kildashti B Samali HV Goudarzi 2019 Damage diagnosis in bridge structures using rotation influence line: validation on a cable-stayed bridge Eng Struct 185 1 14
10. Chen ZW, Cai QL, Zhu S (2018) Damage quantification of beam structures using deflection influence lines. Struct Contr Health Monit 25(11):e2242
11. J Hirachan M Chajes 2005 Experimental influence lines for bridge evaluation Bridge Struct 1 4 405 412
12. Quilligan M (2003) Bridge weight-in-motion development of a 2-D multi-vehicle algorithm. PhD thesis. Royal Institute of Technology, Stockholm